Lecture Notes in Computer Science 7259

Commenced Publication in 1973
Founding and Former Series Editors:
Gerhard Goos, Juris Hartmanis, and Jan van Leeuwen

Editorial Board

David Hutchison
Lancaster University, UK

Takeo Kanade
Carnegie Mellon University, Pittsburgh, PA, USA

Josef Kittler
University of Surrey, Guildford, UK

Jon M. Kleinberg
Cornell University, Ithaca, NY, USA

Alfred Kobsa
University of California, Irvine, CA, USA

Friedemann Mattern
ETH Zurich, Switzerland

John C. Mitchell
Stanford University, CA, USA

Moni Naor
Weizmann Institute of Science, Rehovot, Israel

Oscar Nierstrasz
University of Bern, Switzerland

C. Pandu Rangan
Indian Institute of Technology, Madras, India

Bernhard Steffen
TU Dortmund University, Germany

Madhu Sudan
Microsoft Research, Cambridge, MA, USA

Demetri Terzopoulos
University of California, Los Angeles, CA, USA

Doug Tygar
University of California, Berkeley, CA, USA

Gerhard Weikum
Max Planck Institute for Informatics, Saarbruecken, Germany

Howon Kim (Ed.)

Information Security and Cryptology – ICISC 2011

14th International Conference
Seoul, Korea, November 30 – December 2, 2011
Revised Selected Papers

 Springer

Volume Editor

Howon Kim
Pusan National University
(A06) 6503 School of Computer Science
and Engineering
San-30, JangJeon-Dong, GeumJeong-Gu
Busan, 609-735, South Korea
E-mail: howonkim@pusan.ac.kr

ISSN 0302-9743 e-ISSN 1611-3349
ISBN 978-3-642-31911-2 e-ISBN 978-3-642-31912-9
DOI 10.1007/978-3-642-31912-9
Springer Heidelberg Dordrecht London New York

Library of Congress Control Number: 2012941979

CR Subject Classification (1998): E.3, K.6.5, C.2, D.4.6, G.2.1, E.4, J.1

LNCS Sublibrary: SL 4 – Security and Cryptology

© Springer-Verlag Berlin Heidelberg 2012

This work is subject to copyright. All rights are reserved, whether the whole or part of the material is
concerned, specifically the rights of translation, reprinting, re-use of illustrations, recitation, broadcasting,
reproduction on microfilms or in any other way, and storage in data banks. Duplication of this publication
or parts thereof is permitted only under the provisions of the German Copyright Law of September 9, 1965,
in its current version, and permission for use must always be obtained from Springer. Violations are liable
to prosecution under the German Copyright Law.
The use of general descriptive names, registered names, trademarks, etc. in this publication does not imply,
even in the absence of a specific statement, that such names are exempt from the relevant protective laws
and regulations and therefore free for general use.

Typesetting: Camera-ready by author, data conversion by Scientific Publishing Services, Chennai, India

Printed on acid-free paper

Springer is part of Springer Science+Business Media (www.springer.com)

Preface

ICISC 2011, the 14th International Conference on Information Security and Cryptology, was held in Seoul, Korea, during November 30 – December 2, 2011. It was organized by the Korea Institute of Information Security and Cryptology (KIISC).

The aim of this conference was to create a forum for the dissemination of the latest results in research, development, and applications in the field of information security, and cryptology. The conference received 126 submissions from 29 countries, covering all areas of information security and cryptology. The review and selection processes were carried out in two stages by the Program Committee (PC) of 52 prominent experts via the Springer OCS. First, each paper was blind reviewed by at least three PC members. Second, for resolving conflicts on each reviewer's decision, individual review reports were revealed to PC members, and detailed interactive discussion on each paper followed. Through this process, the PC finally selected 32 papers from 10 countries.

The acceptance rate was 25.4%. For the LNCS proceedings, the authors of selected papers had a few weeks to prepare for their final versions based on the comments received from the reviewers. The conference featured two invited talks delivered by Thomas Peyrin from Nanyang Technological University and Atsuko Miyaji from Japan Advanced Institute of Science and Technology.

Many people have contributed to the organization of ICISC 2011 and the preparation of this volume. We would like to thank all the authors who submitted papers to this conference. We are deeply grateful to all 52 members of the PC. It was a truly nice experience to work with such talented and hard-working researchers. We wish to thank all the external reviewers for assisting the PC in their particular areas of expertise.

Finally, we would like to thank all the participants of the conference who made this event an intellectually stimulating one through their active contribution and all Organizing Committee members who nicely managed the conference.

November 2011   Howon Kim

ICISC 2011 Organization

General Chair

Heung-Youl Youm   Soon-Chun-Hyang University, Korea

Organizing Chair

Sang-Choon Kim   Kangwon National University, Korea

Program Chair

Howon Kim   Pusan National University, Korea

Steering Committee

Man Young Rhee	Kyunghee University, Korea
Pil Joong Lee	Pohang University, Korea
Dongho Won	Sungkyunkwan University, Korea
Ju Seok Song	Yonsei University, Korea
Koji Nakao	National Institute of Information and Communications Technology, Japan

Program Committee

Joonsang Baek	KUSTAR, UAE
Alex Biryukov	University of Luxembourg, Luxembourg
Jung Hee Cheon	Seoul National University, Korea
Dooho Choi	ETRI, Korea
Yongwha Chung	Korea University, Korea
Frëdëric Cuppens	Telecom Bretagne, France
Paolo D'Arco	University of Salerno, Italy
Bart De Decker	K.U. Leuven, Belgium
David Galindo	University of Luxembourg, Luxembourg
Louis Granboulan	EADS Innovation Works, France
Matthew Green	Johns Hopkins University, USA
Johann Großschädl	University of Luxembourg, Luxembourg
JaeCheol Ha	Hoseo University, Korea
Dong-Guk Han	Kookmin University, Korea
Martin Hell	Lund University, Sweden
Seokhie Hong	Korea University, Korea
Jin Hong	Seoul National University, Korea

Jung Yeon Hwang	ETRI, Korea
David Jao	University of Waterloo, Canada
Ju-Sung Kang	Kookmin University, Korea
Ji Hye Kim	Seoul National University, Korea
Seungjoo Kim	Korea University, Korea
Taekyoung Kwon	Sejong University, Korea
Im-Yeong Lee	Soonchunyang University, Korea
Mun-Kyu Lee	Inha University, Korea
Pil Joong Lee	POSTECH, Korea
Mark Manulis	TU Darmstadt and CASED, Germany
Keith Martin	University of London, UK
Sjouke Mauw	University of Luxembourg, Luxembourg
Atsuko Miyaji	JAIST, Japan
Jose A. Montenegro	Universidad de Malaga, Spain
Kirill Morozov	Kyushu University, Japan
David Naccache	ENS DI, France
Rolf Oppliger	eSECURITY Technologies, Switzerland
Omkant Pandey	Microsoft, USA and India
Raphael C.-W. Phan	Loughborough University, UK
Bimal Roy	Indian Statistical Institute, India
Ahmad-Reza Sadeghi	Technische Universität Darmstadt, Germany
Kouichi Sakurai	Kyushu University, Japan
Palash Sarkar	Indian Statistical Institute, India
Kyung-Ah Shim	NIMS, Korea
Sang-Uk Shin	Pukyong National University, Korea
Rainer Steinwandt	Florida Atlantic University, USA
Willy Susilo	University of Wollongong, Australia
Tsuyoshi Takagi	Kyushu University, Japan
Yukiyasu Tsunoo	NEC Corp., Japan
Jorge Villar	Universitat Politecnica de Catalunya, Spain
Rijmen Vincent	Katholieke University Leuven
Jeong Hyun Yi	Soongsil University, Korea
Dae Hyun Yum	POSTECH, Korea
Jianying Zhou	Institute for Infocomm Research, Singapore
Jehong Park	ETRI, Korea

Sponsored by

National Security Research Institute (NSRI)
Electronics and Telecommunications Research Institute (ETRI)
Korea Internet & Security Agency (KISA)
Ministry of Public Administration and Security (MOPAS)

Table of Contents

Digital Signature

Side Channel Analysis II

Cryptanalysis

Efficient Implementation

Hash Function II

Cryptographic Application

Cryptographic Protocol

Improved Integral Analysis
on Tweaked Lesamnta

Yu Sasaki and Kazumaro Aoki

NTT Information Sharing Platform Laboratories, NTT Corporation
3-9-11 Midori-cho, Musashino-shi, Tokyo 180-8585 Japan
{sasaki.yu,aoki.kazumaro}@lab.ntt.co.jp

Abstract. In this paper, we show a known-key (middletext) distinguisher on the internal block cipher of tweaked Lesamnta reduced to 31
(out of 32) rounds, which is one of the hash functions submitted to the
SHA-3 competition. Moreover, we present a distinguisher for full internal
block cipher of Lesamnta with stronger assumption. Although Lesamnta
was not chosen for the second round, for its tweaked version, all previous
cryptanalysis can work no more than 24 rounds. We search for a new
integral characteristic for the internal block cipher, and discover a 19-
round integral characteristic for forward direction. We then search for an
integral characteristic for backward direction, and the characteristics can
be combined to full rounds with some assumption. The distinguisher for
the internal block cipher of Lesamnta-256 requires 2^{192} query complexity
and negligible memory. This is the best attack on Lesamnta compression
function and its internal block cipher after the tweak.

Keywords: integral attack, middletext distinguisher, known-key,
chosen-key, Lesamnta, hash, SHA-3.

1 Introduction

Hash functions are one of the most basic primitives used in many applications.
After the discovery of real collision pairs for MD5 and collision attacks for SHA-1
by Wang *et al.* [1,2], cryptographers are seeking for secure and efficient hash
functions. Based on these backgrounds, NIST started the SHA-3 competition
which determines a new hash function standard [3].

In October 2008, 51 algorithms were accepted by NIST as the first round candidates for the SHA-3 competition. In August 2009, 15 algorithms were chosen
for the second round, and in December 2010, 5 algorithms were chosen for the
third round. Lesamnta, which was proposed by Hirose *et al.* [4], is one of the
first round candidates but was not chosen for the second round.

Although it has already been out of the SHA-3 competition, Lesamnta has various interesting properties such as the efficiency in the hardware implementation
and the security. In fact, Lesamnta-LW [5], which is a successor of Lesamnta
and was proposed at ICISC 2010, has been designed and published recently.
Therefore, even if the analysis on Lesamnta does not give any impact to the

H. Kim (Ed): ICISC 2011, LNCS 7259, pp. 1–17, 2012.
© Springer-Verlag Berlin Heidelberg 2012

SHA-3 competition, Lesamnta is still an interesting and useful research target. We believe that its security evaluation will contribute to the future hash function design.

Lesamnta has a narrow-pipe Merkle-Damgård structure and its compression function consists of an internal block-cipher with the Matyas-Meyer-Oseas mode [6, Algorithm 9.41]. The internal block-cipher uses a generalized Feistel structure with 4 branches, and consists of 32 rounds. Lesamnta has a provable security from several viewpoints. Most of the case, the provable security including the security proof of MMO mode is based on the ideal behavior of the internal block-cipher. Therefore, analyzing the internal block-cipher and discovering non-ideal properties[1] is an important work to know the security margin of the tweaked version of Lesamnta.

For the original version of Lesamnta, Bouillaguet *et al.* found a distinguishing attack on the full-rounds of the internal block-cipher and a pseudo-collision attack on the full-rounds of Lesamnta [7]. After that, designers of Lesamnta proposed a tweak [8] which made these attacks invalid. Several attacks can still work for the tweaked version of Lesamnta. For example, collision and preimage attacks for 16 rounds of Lesamnta proposed by Mendel in the submission document [4] can still work. The current best attack is the one using the cancellation property proposed by Bouillaguet *et al.* [9] which finds collisions and preimages for 24 rounds of Lesamnta. Moreover, the full version of the paper proposes a key-recovery attack on the internal block-cipher reduced to 21 rounds using an integral characteristic with the cancellation property [10].

The integral attack[2], which is the main object we study in this paper, is a cryptanalytic technique for symmetric-key primitives proposed by Daemen *et al.* [11]. Then, Knudsen and Rijmen applied the integral attack for AES [12,13] in the known-key attack model [14]. They showed that integral characteristics in the forward direction and the backward direction can be combined together in the known-key model. Afterwards, Minier *et al.* proposed the middletext distinguisher to formalize a part of known-key model [15]. Their middletext distinguisher is easy to understand the known-key attack with combined integral characteristics.

The designers of Lesamnta evaluated its security against the integral attack [4, Sect. 12.6.3]. They showed a 19-round integral characteristic (in forward direction) for the internal block-cipher of Lesamnta, and demonstrated a 20-round key-recovery attack in the secret-key model with a complexity of $2^{253.7}$ decryptions. However, Bouillaguet *et al.* showed that their integral characteristic was flawed [10]. In the submission document, the designers also evaluated the resistance against known-key attacks [4, Sect. 12.6.4], and showed that 12 rounds could be distinguished in the known-key setting.

---

[1] Actually, the security proof may be fixed without the ideal property of the internal block cipher and with the discovered non-ideal property. However, we cannot determine whether or not the target is secure until a new proof is made.

[2] The integral attack is sometimes referred to as SQUARE attack, saturation attack, or multi-set analysis.

Table 1. Summary of attacks for tweaked version of Lesamnta-256

Attack	Rounds	Attack setting		Complexity/Query	Ref.
Collision	24	Comp. Func.		2^{112}	[9]
Second Pre.	24	Comp. Func.		2^{240}	[9]
Integral KR	20†	Internal BC	(secret-key)	$2^{253.7}$	[4]
Integral KR	21	Internal BC	(secret-key)	2^{192}	[10]
Integral Dist.	31	Internal BC	(middletext)	2^{192}	Ours
Integral Dist.	32(Full)	Internal BC	(middletext with 64-bit key restriction)	2^{192}	Ours

†: The impacts of the correction for the integral characteristic is not considered.
Pre, KR, Dist., Comp. Func., and BC represent Preimage, Key Recovery, Distinguisher, Compression Function, and Block-Cipher, respectively.
For all attacks, the above complexity/query is for Lesamnta-256. They can also be applied to Lesamnta-512 with additional complexity/query.

Our Contributions

In this paper, we investigate the security of the internal block-cipher for the tweaked version of Lesamnta. We first search for a longest integral characteristic for the forward and the backward directions, because the integral characteristic described in [4] is flawed and [10] only showed the good integral characteristic *with* cancellation property and there is no known good integral characteristic for the backward direction. We find a 19-round integral characteristic for the forward direction and 15-round integral characteristic for the backward direction, and find that the best combination of the forward and backward directions can reach 31 rounds. This can be turned into a middletext distinguisher for the internal block-cipher of Lesamnta with 2^{192} query complexity and negligible memory. Moreover, we point out that the above characteristic can be extended to 32 rounds by assuming an equality between two specific subkeys. We then consider the key schedule function of Lesamnta and explain how to choose the key satisfying this condition. This can be used as a full-round distinguisher with 64-bit relation of the key with 2^{192} query complexity and negligible memory. The summary of the attacks is shown in Table 1.

Lesamnta has a provable security in the ideal-cipher model. Therefore, although this attack does not threat the security of the Lesamnta hash function, these security proofs are needed to be updated.

Paper Outline

This paper is organized as follows. In Sect. 2, the specification of Lesamnta is given. In Sect. 3, several related works are explained shortly. In Sect. 4, we search for new integral characteristics with a machine experiment. In Sect. 5, our new integral characteristic for the 31-round middletext distinguisher is described. In Sect. 6, we extend the attack to a full-round chosen-key distinguisher. Finally, in Sect. 7, we conclude this paper.

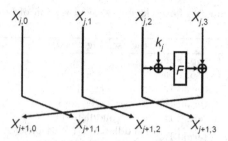

Fig. 1. Round function of Lesamnta

2 Specification of Lesamnta

Lesamnta [4] is a hash function which takes almost arbitrary length message as input and computes 224, 256, 384, and 512-bit strings as output. The main difference between Lesamnta-224/-256 and Lesamnta-384/-512 is the word-size, and Lesamnta-224 (resp. -384) and Lesamnta-256 (resp. -512) use the same internal block-cipher. Hence, in this paper, we mainly discuss only Lesamnta-256.

Lesamnta uses a Merkle-Damgård structure. For Lesamnta-256, the input message M is padded to a multiple of 256 bits, and divided into 256-bit blocks $M_0\|M_1\| \cdots \|M_{N-1}$. Then, the compression function CF : $\{0,1\}^{256} \times \{0,1\}^{256} \to \{0,1\}^{256}$ is iteratively computed as follows;

$$H_{i+1} \leftarrow \mathrm{CF}(H_i, M_i) \quad \text{for } i = 0, 1, \ldots, N-1,$$

where H_0 is the initial value defined in the specification. Finally, H_N is output as the hash value of M.

The input to the compression function CF is a 256-bit chaining variable H_i and 256-bit message block M_i. The compression function is composed of the Matyas-Meyer-Oseas mode with an internal block-cipher E_K. The output H_{i+1} is computed as $H_{i+1} \leftarrow E_{H_i}(M_i) \oplus M_i$.

The internal block-cipher E_K has a 4-branch generalized Feistel structure as shown in Fig. 1. First, a key generation function takes a 256-bit chaining variable H_i as input and computes 64-bit subkeys k_j, where $0 \le j \le 31$. By using 64-bit variables $X_{j,0}, X_{j,1}, X_{j,2}$, and $X_{j,3}$, the output of E_K is computed as follows;

$$(X_{0,0}, X_{0,1}, X_{0,2}, X_{0,3}) \leftarrow M_i,$$
$$(X_{j+1,0}, X_{j+1,1}, X_{j+1,2}, X_{j+1,3}) \leftarrow (F(X_{j,2} \oplus k_j) \oplus X_{j,3}, X_{j,0}, X_{j,1}, X_{j,2})$$
$$\text{for } 0 \le j \le 31.$$

F is a 64-bit permutation motivated by the AES-design. Because our attacks do not look inside F except for the property that F is a permutation, we omit its description.

The key generation function uses almost the same round function as the data encryption part. By using 64-bit variables $Y_{j,0}, Y_{j,1}, Y_{j,2}$, and $Y_{j,3}$, subkeys $k_0, k_1, \ldots, k_{31}$ are computed as follows;

$$(Y_{0,0}, Y_{0,1}, Y_{0,2}, Y_{0,3}) \leftarrow H_i,$$

Repeat the followings for $0 \leq j \leq 31$:

$$k_j \leftarrow Y_{j,0},$$
$$(Y_{j+1,0}, Y_{j+1,1}, Y_{j+1,2}, Y_{j+1,3}) \leftarrow (F(Y_{j,2} \oplus C_j) \oplus Y_{j,3}, Y_{j,0}, Y_{j,1}, Y_{j,2}).$$

C_j is a pre-defined constant. Note that the tweak for Lesamnta made by the designers [8] only changes the value of C_j. Because our attacks do not depend on the value of C_j, our attacks can work for both of the original and tweaked versions of Lesamnta.

3 Related Work

3.1 Integral Attack

The integral attack is a cryptanalytic technique for symmetric-key primitives, which was first proposed by Daemen *et al.* to evaluate the security of the SQUARE cipher [11]. The crucial idea of the integral attack is to collect a set of plaintexts which contains all possible values for some bytes and has a constant value for the other bytes. All plaintexts in the set are passed to the encryption oracle, and an attacker computes the XOR of all corresponding ciphertexts. If the encryption procedure does not mix the text well, several bytes of the XOR of the ciphertexts always become 0. We call this characteristic *integral characteristic*.

The application of the integral attack to AES is widely known. AES is a 16-byte block-cipher. The attacker collects 256 plaintexts where the first byte varies from 0 to 255 and the other 15 bytes are fixed to some constant, e.g. 0. After all 256 plaintexts are encrypted by 3.5-rounds, the XOR of the corresponding 256 ciphertexts always becomes 0 in all bytes.

Note that the same attack can be performed in the decryption direction.

3.2 Integral Attack in the Known-Key Setting

Known-key attack is a framework proposed by Knudsen and Rijmen [14] to evaluate the security of block-ciphers. Although the formalization of the known-key attack is still an open problem, the intuition of this model is as follows;

> *A secret key is randomly chosen and given to attackers. Attackers aim to detect a certain property of a random instance of the block cipher, where the same property cannot be observed for a random permutation.*

For the confidentiality use of block ciphers, the secret key should be kept secret, and an attacker does not know the value of the secret key in advance. However, a block cipher can be used to construct hash functions such as Davies-Meyer

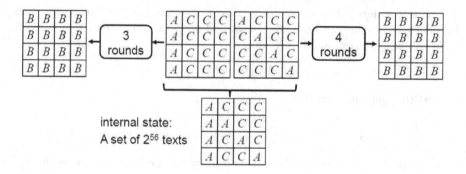

Byte A stands for "Active", which means all possible values are included and are used exactly the same number of times in the set of texts.

Byte B stands for "Balanced", which means the XOR of all texts in the set is 0.

Byte C stands for "Constant", which means the value is fixed to a constant for all texts in the set.

Fig. 2. Integral attack on AES in the known-key setting [14]

construction. In such a case, attackers can sometimes know and even specify the secret key as one of the message blocks to find unpleasant properties such as the collision. Hence, we should analyze the block cipher with the known-key setting.

Knudsen and Rijmen showed that, in the known-key model, the integral characteristic in the encryption and the decryption directions can be combined together. They proposed a known-key attack on 7-round AES with this concept. In this attack, they prepared the 3-round integral characteristic in backward and 4-round integral characteristic in forward as shown in Fig. 2. The attacker collects 2^{56} texts for the internal state and computes the XOR of the corresponding plaintexts and ciphertexts. In 7-round AES, the result always becomes 0^3.

After the publication of [14], the known-key attack combining two integral characteristics was formalized by Minier *et al.* as an n-limited non-adaptive chosen middletexts distinguisher (NA-CMA) [15, Algorithm 1]. In this framework, the oracle $\mathcal{O}$ implements either a random permutation or a random instantiation of the target block-cipher. The goal of the distinguisher is to decide which is implemented in $\mathcal{O}$. The distinguisher first determines the acceptance region $A^{(n)}$ which is the set of the plaintexts and ciphertexts determined by considering the biased property of the target cipher. For example, $A^{(n)}$ for the above known-key distinguisher on AES should be any set of 2^{56} plaintexts whose XOR is 0 and any set of 2^{56} ciphertexts whose XOR is 0. The distinguisher then chooses or computes a set of texts for some intermediate state and input them to $\mathcal{O}$ and obtain the oracle's output. Finally, the attack checks if the obtained output is in $A^{(n)}$ or not. The detailed procedure of NA-CMA by [15] is given in Appendix A.

---

[3] Knudsen and Rijmen suggest to use more strong property: confirm that the number of occurrence of the value of each byte of plaintexts and ciphertexts is the same.

The security of a pseudo-random permutation is usually defined as follows.

Definition 1. *For a pseudo-random permutation E_K parameterized by $K \in \mathcal{K}$, an advantage of distinguisher D is defined as follows.*

$$\mathrm{Adv}_D = \left| \Pr[D^{E_K, E_K^{-1}} \to 1 \mid K \in_U \mathcal{K}] - \Pr[D^{\pi, \pi^{-1}} \to 1 \mid \pi \in_U \mathcal{P}] \right|,$$

where $x \in_U X$ means that x is uniformly chosen from set X, and $\mathcal{P}$ is the set of all permutations on the text space.

Definition 2. *A pseudo-random permutation E_K ($K \in \mathcal{K}$) is secure if Adv_D is negligible for any distinguisher D.*

From the above definition, we can say that E_k is not secure even if we only show an example of a distinguisher whose advantage is not negligible.

3.3 Previous Integral Attack and Known-Key Attack on Lesamnta

The designers of Lesamnta evaluated its security against integral attack [4, Sect. 12.6.3]. They showed the 19-round integral characteristic. However, Bouillaguet *et al.* pointed out that their integral characteristic would not work, and they proposed a 20-round integral characteristic with the cancellation property, and mount a key-recovery attack on 22-round internal block-cipher of Lesamnta with the complexity of 2^{192} encryptions.

The designers of Lesamnta also evaluated the resistance against known-key attacks [4, Sect. 12.6.4]. They considered a differential cryptanalysis, and constructed a known-key distinguisher for Lesamnta-256 reduced to 12 rounds.

3.4 Chosen-Key Attack

The concept of the known-key attack can be extended to the chosen-key attack. Similar to the known-key attack, the formalization of the chosen-key attack is also an open problem. Examples of papers discussing the chosen-key attack are [16,17,18].

4 New Integral Characteristics for Lesamnta

4.1 Existing Integral Characteristics

The submission document [4] showed a 19-round integral characteristic. Bouillaguet *et al.* showed the flaw in the integral characteristic. To show the flaw, they used complicated formulae, because the 19-round integral characteristic uses multi-active-words.

Bouillaguet *et al.* also proposed a new integral characteristic. However the new integral characteristic was intended to be used with the cancellation property, and it seems difficult to be used for the middletext distinguisher, though the integral characteristic is long.

As a result, we do not know the longest integral characteristic for the forward direction, and we do not have any knowledge of the integral characteristic for the backward direction.

Table 2. New integral characteristic

Round	Inputs
0	C A A A
1	A C A A
2	A A C A
3	A A A C
4	A A A A
5	A A A A
6	A A A A
7	A A A A
8	A A A A
9	A A A A
10	A A A A
11	A A A A
12	A A A A
13	B A A A
14	B B A A
15	B B B A
16	? B B B
17	? ? B B
18	? ? ? B

4.2   Experiment and New Integral Characteristic

This section tries to find the best integral characteristic for the internal block-cipher of Lesamnta in the forward and backward directions. As shown by Bouillaguet *et al.*, a simple analysis method intended to activate one word may cause an error of the characteristic, and formulae analysis is complicated, and we may miss an optimal characteristic. We decide to use computer experiments to find good integral characteristics. This strategy has a possibility to detect wrong characteristic, but its probability is quite low.

Because computing 2^{192} texts for Lesamnta is infeasible, we consider the small experiment. First, we reduce the word size to 8 bits. Then, the block-size becomes 32 bits and computing all values for 3 words cost 2^{24} computations. Second, we replace the F function with a single S-box computation. Third, we replace the subkey in each round with the S-box output whose input is a round number. The algorithm of our experiment, which checks the integral characteristic up to R rounds is shown in Fig. 3. When the algorithm in Fig. 3 is implemented we prepare temporary variables $T_{u,0}$ for line 8, where $0 \leq u \leq R$. $T_{u,0}$ are initialized to 0. Every time $X_{u,0}$ is computed, we compute $T_{u,0} \leftarrow T_{u,0} \oplus X_{u,0}$ to update the current XOR-sum of $X_{u,0}$. If $T_{u,0} = 0$ after 2^{24} iterations, we know that the variable is balanced. Because $X_{u,1}, X_{u,2}$, and $X_{u,3}$ are just a copy of previously computed value, only updating $X_{u,0}$ is enough. Note that the XOR of 2^{24} results may happen to become 0, i.e. result in a balanced word with a probability of 2^{-8}. To avoid this event, we ran the algorithm several times with changing the S-box, sub-key, or value for the constant word, and check that balanced words are always

1. Set $X_{0,0} \leftarrow C$ for some constant C; and initialize XOR-sum;
2. FOR all possible 2^{24} values of $(X_{0,1}, X_{0,2}, X_{0,3})$ {
3.   FOR $round = 0$ to $R - 1$ {
4. $X_{(round+1),0} = X_{round,3} \oplus S(X_{round,2} \oplus S(round))$;
5. $X_{(round+1),1} = X_{round,0}$;
6. $X_{(round+1),2} = X_{round,1}$;
7. $X_{(round+1),3} = X_{round,2}$;
8.   Update the XOR-sum of $X_{(round+1),0}$;
9.   } //END of FOR
10. } //END of FOR

Fig. 3. Algorithm in our experiment

balanced for any S-box, sub-key, and constant number. Also note that to check if each word is active or not, counting the number of the occurrence of each value is necessary. Actually, we counted it in our experiment. Because investigating the active words is irrelevant to our attack and the algorithm becomes more complicated than Fig. 3, we omit its description.

As a result of the experiment, we obtain the new integral characteristic which is shown in Table 2. The most important difference from the previous characteristic is that the integral characteristic only can work up to 18 rounds. Note that this will give some impact to the 20-round secret key attack in [4]. Intuitively, the number of attacked rounds will decrease by 1 round. However, we will not discuss details because the open-key approach explained in the following sections can attack much more rounds.

5 Known-Key Attack on 31-Round Block-Cipher of Tweaked Lesamnta

As [14] showed, the integral characteristics in the forward direction and the backward direction can be combined together. Therefore, we search for the backward integral characteristic with the new approach explained in Sect. 4. Because the algorithm for the backward search is very similar to the one in Fig. 3, we omit its description.

[14] combined the most effective independent characteristics for the forward and backward directions. On the other hand, we cannot combine the most effective ones because our characteristic in Table 2 has already activated three words, and thus cannot combine the backward characteristic which activates the constant word. That is, the positions of the active words at the combining state must be identical between the forward and backward characteristics. We search for the best characteristic satisfying this condition. The result is described in Table 3.

As shown in Table 3, if we collect 2^{192} texts whose the left most word is fixed to some constant and the right three words take all possibilities, we will have one balanced word after 18-round encryption and 13-round decryption.

Table 3. 31-round characteristic for known-key attack

Round in each direction	Total round	State
13	0	B ? ? ?
12	1	A B ? ?
11	2	A A B ?
10	3	A A A B
9	4	A A A A
8	5	A A A A
7	6	A A A A
6	7	A A A A
5	8	A A A A
4	9	A A A A
3	10	A A A A
2	11	A A A A
1	12	A A A A
0	13	C A A A
1	14	A C A A
2	15	A A C A
3	16	A A A C
4	17	A A A A
5	18	A A A A
6	19	A A A A
7	20	A A A A
8	21	A A A A
9	22	A A A A
10	23	A A A A
11	24	A A A A
12	25	A A A A
13	26	B A A A
14	27	B B A A
15	28	B B B A
16	29	? B B B
17	30	? ? B B
18	31	? ? ? B

Table 4. 32-round characteristic for chosen-key attack

Round in each direction	Total round	State	Condition
14	0	B ? ? ?	
13	1	B B ? ?	
12	2	A B B ?	
11	3	A A B B	
10	4	A A A B	
9	5	A A A A	$k_5 = k_{13}$
8	6	A A A A	
7	7	A A A A	
6	8	A A A A	
5	9	A A A A	
4	10	A A A A	
3	11	A A A A	
2	12	A A A A	
1	13	A A A A	
0	14	C A A A	
1	15	A C A A	
2	16	A A C A	
3	17	A A A C	
4	18	A A A A	
5	19	A A A A	
6	20	A A A A	
7	21	A A A A	
8	22	A A A A	
9	23	A A A A	
10	24	A A A A	
11	25	A A A A	
12	26	A A A A	
13	27	B A A A	
14	28	B B A A	
15	29	B B B A	
16	30	? B B B	
17	31	? ? B B	
18	32	? ? ? B	

We can mount the 31-round middletext distinguisher with this characteristic. The property we distinguish is a partial zero-sum, namely, the distinguisher collects a set of 2^{192} plaintexts whose XOR is 0 regarding the left most word ($X_{0,0}$) and the XOR of the corresponding 2^{192} ciphertexts is also 0 regarding the right most word ($X_{31,3}$). If we rephrase it for the context of $NA\text{-}CMA$ [15], the acceptance region $A^{(n)}$ is any set of 2^{192} plaintexts whose XOR is $(0, *, *, *)$ and any set of 2^{192} ciphertexts whose XOR is $(*, *, *, 0)$, where $*$ represents an arbitrary value.

The distinguishing procedure is the same as $NA\text{-}CMA$. The 2^{192} middletexts are chosen as the leftmost word is fixed constant and other words take all values, and the belonging confirmation of the acceptable set can be realized by the examination of XOR of all plaintexts as $(*, *, *, 0)$ and XOR of all ciphertexts as $(0, *, *, *)$.

For the ideal cipher, the best way to achieve this data set is making exactly 2^{192} queries so that the XOR of the left most word becomes 0, and check that the XOR of the ciphertexts are 0 in the right most word. Because the XOR of the ciphertexts behaves truly random in the last query, the probability for satisfying $\bigoplus_{i=1}^{2^{192}} X_{31,3} = 0$ is 2^{-64}. On the other hand, for 31-round internal block-cipher of Lesamnta, by starting 2^{192} texts from the middle state, we can obtain the set of 2^{192} plaintexts and ciphertexts which achieves the property with probability 1. Hence, the distinguisher obtains a significant advantage, $1 - 2^{-64}$.

Remark 1. Even if the case for ideal cipher, we can make an element in the acceptable set with high probability if we allow one more query to the oracle for the distinguisher, since $\binom{2^{192}+1}{2^{192}} = 2^{192} + 1$ and the restriction on the acceptable set is 128 bits in total. The advantage of this case is almost 0, and it is negligible. Followed by Definition 2, we only need to show one distinguisher whose advantage is not negligible to say that a block cipher is not secure. As the above, we showed an example of a distinguisher whose advantage is significant, when we restrict to use exactly 2^{192} queries. We can conclude that 31-round block cipher of tweaked Lesamnta is not secure followed by Definition 2.

Remark 2. Aumasson *et al.* pointed out that the zero-sum property can be converted to an existential forgery attack against the prefix-MAC construction [19, Section 3.1].

6 Chosen-Key Attack on Full-Round Block-Cipher of Tweaked Lesamnta

6.1 32-Round Integral Characteristic and Analysis for the Key Schedule

The 31-round integral characteristic explained in Sect. 5 can be extended to 32-rounds by assuming an equality between two subkeys. In details, the backward characteristic is extended by one more round under the condition $k_5 = k_{13}$. We confirmed this fact by the experiment. In Step 4 of the backward version of the algorithm in Fig. 3, we replace $S(round)$ with some constant value for rounds 1 and 9. The discovered 32-round characteristic is shown in Table 4. Compare Table 4 with Table 3, the condition $k_5 = k_{13}$ makes one more B state in 11th round of the backward direction. For those who wants to verify the experiment, we show the code of the experiment written in the C-language in Appendix B.

In order to search for a key satisfying this condition, we analyze the key schedule function of Lesamnta. The computations for obtaining k_5 to k_{13} are shown in Fig. 4. Note that in the key schedule function of Lesamnta, if a 256-bit

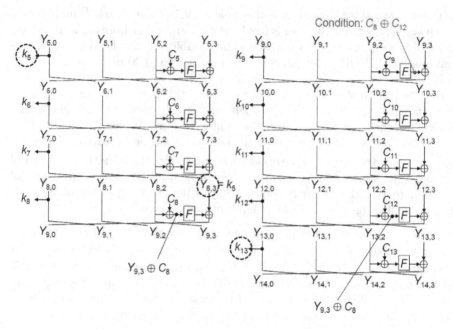

Fig. 4. Key schedule function for obtaining k_5 to k_{13}

internal state is fixed, the original key value is uniquely computed by inverting the round function. Therefore, the goal is searching for a 256-bit internal state which produces the same k_5 and k_{13}. This can be achieved by using the idea of the cancellation property [9]. Also note that, different from the known-key attack in Sect. 5, we need to use the property that the inversion of F is easily computed.

$$k_5 = Y_{8,3} = F(Y_{9,3} \oplus C_8) \oplus Y_{9,0}, \tag{1}$$
$$k_{13} = Y_{13,0} = F(Y_{12,2} \oplus C_{12}) \oplus Y_{12,3}, \tag{2}$$

Because $Y_{9,0} = Y_{12,3}$, the condition to achieve the goal ($k_5 = k_{13}$) is expressed as

$$Y_{9,3} \oplus C_8 = Y_{12,2} \oplus C_{12}. \tag{3}$$

Because $Y_{12,2} = Y_{10,0} = F(Y_{9,2} \oplus C_9) \oplus Y_{9,3}$, the condition Eq. (3) is expressed as

$$F(Y_{9,2} \oplus C_9) = C_8 \oplus C_{12}, \tag{4}$$

which is,

$$Y_{9,2} = F^{-1}(C_8 \oplus C_{12}) \oplus C_9, \tag{5}$$

In the end, to obtain the key satisfying $k_5 = k_{13}$, we first compute $Y_{9,2}$ with Eq. (5) and randomly choose values of $Y_{9,0}, Y_{9,1}$, and $Y_{9,3}$. Then we invert the round function and the resulting $(Y_{0,0}, Y_{0,1}, Y_{0,2}, Y_{0,3})$ is a desired key. Note that the complexity for choosing a key is almost one key schedule computation.

6.2   Chosen-Key Distinguisher and Its Impact

Mounting the chosen-key distinguisher with 32-round characteristic can be done by the same manner as the known-key distinguisher in Sect. 5. The property we discuss in this distinguisher is a partial zero-sum. The distinguisher computes a key and collects a set of 2^{192} plaintexts whose XOR is 0 regarding the left most word $(X_{0,0})$ and the XOR of the corresponding 2^{192} ciphertexts is also 0 regarding the right most word $(X_{32,3})$. Because the partial zero-sum property is satisfied with probability 1, we can distinguish the full-round Lesamnta block-cipher from the ideal cipher.

According to the designer's document [4, Sect. 11], the security proof of Lesamnta is based on the work by Black *et al.* in 2002 [20], which showed that the security bound of the MMO mode in terms of the preimage resistance is $2q/2^n$ under the ideal cipher model. This represents the upper-bound of distinguisher's advantage after making q queries for n-bit output. Because our distinguisher succeeds with probability 1 with $q = 2^{192}$ queries, the internal block-cipher of Lesamnta cannot achieve the security bound of $2q/2^n$.

Note that our chosen-key distinguisher can be regarded as a known-key attack for a weak-key. The attack works against any key satisfying $k_5 = k_{13}$, which is satisfied with probability of 2^{-64} for a randomly chosen key. The number of such weak-keys is $2^{256} \cdot 2^{-64} = 2^{192}$.

7   Concluding Remarks

In this paper, we revisited the integral attack on Lesamnta after the tweak. We did the experiment on the small variant of the same generalized Feistel structure, and discovered that the 19-round characteristic seems optimal for the forward direction.

We then searched for the integral characteristic suitable for the known-key attack under the framework of the middletext distinguisher. As a result, we found the 31-round characteristic that could be turned into the known-key attack on 31-round internal block-cipher of Lesamnta with 2^{192} query complexity and a negligible memory.

In addition, we discovered that the 31-round characteristic could be extended by one more round by assuming an equality between two specific subkeys. We then analyzed the key schedule function of Lesamnta, and showed that finding keys satisfying this condition was possible with complexity of one key schedule function computation. In the end, we successfully achieved the chosen-key distinguisher on the full-round internal block-cipher of Lesamnta. This attack does not threat the security of the Lesamnta hash function immediately, but invalidates some security proof made on Lesamnta. As far as we know, this is the first result that shows the non-ideal property of the full-round Lesamnta block-cipher after the tweak.

Although it has already been out of the SHA-3 competition, Lesamnta still has various interesting properties. Our results except for the full-round analysis of Lesamnta-256 can also be applied to reduced Lesamnta-LW and type-I generalized Feistel network, since we do not use any detailed property of the round function. Even if the analysis on Lesamnta does not give impact to the competition, we believe that its security evaluation will contribute to the future hash function design.

References

1. Wang, X., Yu, H.: How to Break MD5 and Other Hash Functions. In: Cramer, R. (ed.) EUROCRYPT 2005. LNCS, vol. 3494, pp. 19–35. Springer, Heidelberg (2005)
2. Wang, X., Yin, Y.L., Yu, H.: Finding Collisions in the Full SHA-1. In: Shoup, V. (ed.) CRYPTO 2005. LNCS, vol. 3621, pp. 17–36. Springer, Heidelberg (2005)
3. U.S. Department of Commerce, National Institute of Standards and Technology: Federal Register vol. 72, No. 212/Friday, November 2, 2007/Notices (2007), http://csrc.nist.gov/groups/ST/hash/documents/FR_Notice_Nov07.pdf
4. Hirose, S., Kuwakado, H., Yoshida, H.: SHA-3 proposal: Lesamnta. Lesamnta home page (2009), Document version 1.0.1, http://www.hitachi.com/rd/yrl/crypto/lesamnta/ (January 15, 2009)
5. Hirose, S., Ideguchi, K., Kuwakado, H., Owada, T., Preneel, B., Yoshida, H.: A Lightweight 256-Bit Hash Function for Hardware and Low-End Devices: Lesamnta-LW. In: Rhee, K.-H., Nyang, D. (eds.) ICISC 2010. LNCS, vol. 6829, pp. 151–168. Springer, Heidelberg (2011)
6. Menezes, A.J., van Oorschot, P.C., Vanstone, S.A.: Handbook of applied cryptography. CRC Press (1997)
7. Bouillaguet, C., Dunkelman, O., Leurent, G., Fouque, P.-A.: Another Look at Complementation Properties. In: Hong, S., Iwata, T. (eds.) FSE 2010. LNCS, vol. 6147, pp. 347–364. Springer, Heidelberg (2010)
8. Hirose, S., Kuwakado, H., Yoshida, H.: A minor change to Lesamnta — Change of round constants — Lesamnta home page (2009), http://www.hitachi.com/rd/yrl/crypto/lesamnta/ (July 18, 2009)
9. Bouillaguet, C., Dunkelman, O., Leurent, G., Fouque, P.-A.: Attacks on Hash Functions Based on Generalized Feistel: Application to Reduced-Round $Lesamnta$ and $SHAvite-3_{512}$. In: Biryukov, A., Gong, G., Stinson, D.R. (eds.) SAC 2010. LNCS, vol. 6544, pp. 18–35. Springer, Heidelberg (2011)
10. Bouillaguet, C., Dunkelman, O., Leurent, G., Fouque, P.A.: Attacks on hash functions based on generalized Feistel - application to reduced-round $Lesamnta$ and $SHAvite-3_{512}$. Cryptology ePrint Archive, Report 2009/634 (2009), http://eprint.iacr.org/2009/634 (Full version of [9])
11. Daemen, J., Knudsen, L.R., Rijmen, V.: The Block Cipher SQUARE. In: Biham, E. (ed.) FSE 1997. LNCS, vol. 1267, pp. 149–165. Springer, Heidelberg (1997)
12. Daemen, J., Rijmen, V.: The design of Rijndeal: AES – the Advanced Encryption Standard (AES). Springer (2002)
13. U.S. Department of Commerce, National Institute of Standards and Technology: Specification for the ADVANCED ENCRYPTION STANDARD (AES) (Federal Information Processing Standards Publication 197) (2001)
14. Knudsen, L.R., Rijmen, V.: Known-Key Distinguishers for Some Block Ciphers. In: Kurosawa, K. (ed.) ASIACRYPT 2007. LNCS, vol. 4833, pp. 315–324. Springer, Heidelberg (2007)

15. Minier, M., Phan, R.C.-W., Pousse, B.: Distinguishers for Ciphers and Known Key Attack against Rijndael with Large Blocks. In: Preneel, B. (ed.) AFRICACRYPT 2009. LNCS, vol. 5580, pp. 60–76. Springer, Heidelberg (2009)
16. Biryukov, A., Nikolić, I.: A new security analysis of AES-128. In: Rump session of CRYPTO 2009 (2009), http://rump2009.cr.yp.to/
17. Biryukov, A., Nikolić, I.: Automatic Search for Related-Key Differential Characteristics in Byte-Oriented Block Ciphers: Application to AES, Camellia, Khazad and Others. In: Gilbert, H. (ed.) EUROCRYPT 2010. LNCS, vol. 6110, pp. 322–344. Springer, Heidelberg (2010)
18. Nikolić, I., Pieprzyk, J., Sokołowski, P., Steinfeld, R.: Known and Chosen Key Differential Distinguishers for Block Ciphers. In: Rhee, K.-H., Nyang, D. (eds.) ICISC 2010. LNCS, vol. 6829, pp. 29–48. Springer, Heidelberg (2011)
19. Aumasson, J.-P., Käsper, E., Knudsen, L.R., Matusiewicz, K., Ødegård, R., Peyrin, T., Schläffer, M.: Distinguishers for the Compression Function and Output Transformation of Hamsi-256. In: Steinfeld, R., Hawkes, P. (eds.) ACISP 2010. LNCS, vol. 6168, pp. 87–103. Springer, Heidelberg (2010)
20. Black, J., Rogaway, P., Shrimpton, T.: Black-Box Analysis of the Block-Cipher-Based Hash-Function Constructions from PGV. In: Yung, M. (ed.) CRYPTO 2002. LNCS, vol. 2442, pp. 320–335. Springer, Heidelberg (2002)

A Formalization of Known-Key Integral Attack [15]

In this section, we cite the formalization of the known-key integral attack by [15, Algorithm. 1]. It defines an n-limited non-adaptive chosen middletexts distinguisher $(NA\text{-}CMA)$. In this framework, the oracle $\mathcal{O}$ implements either a random permutation or a random instantiation of the target block-cipher. The goal of the distinguisher is to decide which is implemented in the oracle $\mathcal{O}$. The core of the distinguisher is the acceptance region $A^{(n)}$: it defines the set of input and output values $(\mathbf{P}, \mathbf{C}) = (P_1, \cdots, P_n, C_1, \cdots, C_n)$ which lead to output 0 (i.e. it decides that the oracle implements a random permutation) or 1 (i.e. it decides that the oracle implements a random instantiation of the target block-cipher).

Algorithm 1. An n-limited generic non-adaptive chosen middletexts distinguisher $(NA\text{-}CMA)$

Parameters: a complexity n, an acceptance set $A^{(n)}$
Oracle: an oracle $\mathcal{O}$ implementing internal functions f_1 (resp. f_2) of permutation c that process input middletexts to the plaintext (resp. ciphertext)
  Compute some middletexts $\mathbf{M} = (M_1, \cdots, M_n)$
  Query $\mathbf{P} = (P_1, \cdots, P_n) = (f_1(M_1), \cdots, f_1(M_n))$ and $\mathbf{C} = (C_1, \cdots, C_n) = (f_2(M_1), \cdots, f_2(M_n))$ to $\mathcal{O}$
  if $(\mathbf{P}, \mathbf{C}) \in A^{(n)}$ **then**
       Output 1
  else
       Output 0
  end if

B   Code of the Experiment for the Chosen-Key Attack

```c
#include <stdio.h>
#include <math.h>
#include <stdlib.h>

/* S Box data values */
static int Sdata[256] = {
    0x63, 0x7c, 0x77, 0x7b, 0xf2, 0x6b, 0x6f, 0xc5,
    0x30, 0x01, 0x67, 0x2b, 0xfe, 0xd7, 0xab, 0x76,
    0xca, 0x82, 0xc9, 0x7d, 0xfa, 0x59, 0x47, 0xf0,
    0xad, 0xd4, 0xa2, 0xaf, 0x9c, 0xa4, 0x72, 0xc0,
    0xb7, 0xfd, 0x93, 0x26, 0x36, 0x3f, 0xf7, 0xcc,
    0x34, 0xa5, 0xe5, 0xf1, 0x71, 0xd8, 0x31, 0x15,
    0x04, 0xc7, 0x23, 0xc3, 0x18, 0x96, 0x05, 0x9a,
    0x07, 0x12, 0x80, 0xe2, 0xeb, 0x27, 0xb2, 0x75,
    0x09, 0x83, 0x2c, 0x1a, 0x1b, 0x6e, 0x5a, 0xa0,
    0x52, 0x3b, 0xd6, 0xb3, 0x29, 0xe3, 0x2f, 0x84,
    0x53, 0xd1, 0x00, 0xed, 0x20, 0xfc, 0xb1, 0x5b,
    0x6a, 0xcb, 0xbe, 0x39, 0x4a, 0x4c, 0x58, 0xcf,
    0xd0, 0xef, 0xaa, 0xfb, 0x43, 0x4d, 0x33, 0x85,
    0x45, 0xf9, 0x02, 0x7f, 0x50, 0x3c, 0x9f, 0xa8,
    0x51, 0xa3, 0x40, 0x8f, 0x92, 0x9d, 0x38, 0xf5,
    0xbc, 0xb6, 0xda, 0x21, 0x10, 0xff, 0xf3, 0xd2,
    0xcd, 0x0c, 0x13, 0xec, 0x5f, 0x97, 0x44, 0x17,
    0xc4, 0xa7, 0x7e, 0x3d, 0x64, 0x5d, 0x19, 0x73,
    0x60, 0x81, 0x4f, 0xdc, 0x22, 0x2a, 0x90, 0x88,
    0x46, 0xee, 0xb8, 0x14, 0xde, 0x5e, 0x0b, 0xdb,
    0xe0, 0x32, 0x3a, 0x0a, 0x49, 0x06, 0x24, 0x5c,
    0xc2, 0xd3, 0xac, 0x62, 0x91, 0x95, 0xe4, 0x79,
    0xe7, 0xc8, 0x37, 0x6d, 0x8d, 0xd5, 0x4e, 0xa9,
    0x6c, 0x56, 0xf4, 0xea, 0x65, 0x7a, 0xae, 0x08,
    0xba, 0x78, 0x25, 0x2e, 0x1c, 0xa6, 0xb4, 0xc6,
    0xe8, 0xdd, 0x74, 0x1f, 0x4b, 0xbd, 0x8b, 0x8a,
    0x70, 0x3e, 0xb5, 0x66, 0x48, 0x03, 0xf6, 0x0e,
    0x61, 0x35, 0x57, 0xb9, 0x86, 0xc1, 0x1d, 0x9e,
    0xe1, 0xf8, 0x98, 0x11, 0x69, 0xd9, 0x8e, 0x94,
    0x9b, 0x1e, 0x87, 0xe9, 0xce, 0x55, 0x28, 0xdf,
    0x8c, 0xa1, 0x89, 0x0d, 0xbf, 0xe6, 0x42, 0x68,
    0x41, 0x99, 0x2d, 0x0f, 0xb0, 0x54, 0xbb, 0x16,};

int main()
{
    unsigned int H[24];
    int round;
    int sum[24];
```

```
int n;

H[0]=0;
for(n=0;n<68;n++){
    sum[n]=0;
}


for(H[2]=0;H[2]<256;H[2]++){
    for(H[1]=0;H[1]<256;H[1]++){
        for(H[3]=0;H[3]<256;H[3]++){
            H[ 4]=H[ 0]^Sdata[H[ 3]^Sdata[ 0]]; sum[ 4]^=H[ 4];
                                                /* Set k[0]=k[8] */
            H[ 5]=H[ 1]^Sdata[H[ 4]^Sdata[ 5]]; sum[ 5]^=H[ 5];
            H[ 6]=H[ 2]^Sdata[H[ 5]^Sdata[ 6]]; sum[ 6]^=H[ 6];
            H[ 7]=H[ 3]^Sdata[H[ 6]^Sdata[ 7]]; sum[ 7]^=H[ 7];
            H[ 8]=H[ 4]^Sdata[H[ 7]^Sdata[ 8]]; sum[ 8]^=H[ 8];
            H[ 9]=H[ 5]^Sdata[H[ 8]^Sdata[ 9]]; sum[ 9]^=H[ 9];
            H[10]=H[ 6]^Sdata[H[ 9]^Sdata[10]]; sum[10]^=H[10];
            H[11]=H[ 7]^Sdata[H[10]^Sdata[11]]; sum[11]^=H[11];
            H[12]=H[ 8]^Sdata[H[11]^Sdata[ 0]]; sum[12]^=H[12];
                                                /* Set k[0]=k[8] */
            H[13]=H[ 9]^Sdata[H[12]^Sdata[13]]; sum[13]^=H[13];
            H[14]=H[10]^Sdata[H[13]^Sdata[14]]; sum[14]^=H[14];
            H[15]=H[11]^Sdata[H[14]^Sdata[15]]; sum[15]^=H[15];
            H[16]=H[12]^Sdata[H[15]^Sdata[16]]; sum[16]^=H[16];
            H[17]=H[13]^Sdata[H[16]^Sdata[17]]; sum[17]^=H[17];
            H[18]=H[14]^Sdata[H[17]^Sdata[18]]; sum[18]^=H[18];
            H[19]=H[15]^Sdata[H[18]^Sdata[19]]; sum[19]^=H[19];
            H[20]=H[16]^Sdata[H[19]^Sdata[20]]; sum[20]^=H[20];
            H[21]=H[17]^Sdata[H[20]^Sdata[21]]; sum[21]^=H[21];
            H[22]=H[18]^Sdata[H[21]^Sdata[22]]; sum[22]^=H[22];
            H[23]=H[19]^Sdata[H[22]^Sdata[23]]; sum[23]^=H[23];
            H[24]=H[20]^Sdata[H[23]^Sdata[24]]; sum[24]^=H[24];
        }
    }
}

for(round=20;round>=0;round--){
    printf("Round%02d: %02x  %02x  %02x  %02x\n",
    round,sum[round],sum[round+1],sum[round+2],sum[round+3]);
}

return(0);
}
```

Analysis of Trivium
Using Compressed Right Hand Side Equations

Thorsten Ernst Schilling and Håvard Raddum

Selmer Center, University of Bergen
{thorsten.schilling,havard.raddum}@ii.uib.no

Abstract. We study a new representation of non-linear multivariate equations for algebraic cryptanalysis. Using a combination of multiple right hand side equations and binary decision diagrams, our new representation allows a very efficient conjunction of a large number of separate equations. We apply our new technique to the stream cipher TRIVIUM and variants of TRIVIUM reduced in size. By merging all equations into one single constraint, manageable in size and processing time, we get a representation of the TRIVIUM cipher as one single equation.

Keywords: multivariate equation system, BDD, algebraic cryptanalysis, Trivium.

1 Introduction

In this paper we present a new way of representing multivariate equations over $GF(2)$ and their application in algebraic cryptanalysis of the stream cipher TRIVIUM.

In algebraic cryptanalysis one creates an equation system of the cipher being analyzed and tries to solve it. The solution will reveal the key or some other secret information. Solving the system representing a cipher in time faster than exhaustive search will be a valid attack on the cipher.

There exist several ways to represent such a system, e.g., ANF, CNF [1] or MRHS [2]. Along these representations different families of algorithms to solve equation systems have been proposed, e.g., Gröbner Basis like algorithms [3], XL [4] SAT-solving [1] and Gluing/Agreeing algorithms [5,2,6].

For the stream cipher TRIVIUM, which has an especially simple structure, one can easily construct an equation system describing its inner state constraints using some known keystream bits. Attempts at solving this system have nevertheless been unsuccessful. While reduced versions of TRIVIUM could be broken [1], there is no attack better than brute-force known for the full version.

Previous methods describe the TRIVIUM-equation system as a set of nonlinear constraints, which have to be true in conjunction. One can simplify those equation systems by joining several constraints into a single new one. Unfortunately the conjunction operation usually leads to exponentially big objects, which quickly become too big for today's computers.

H. Kim (Ed): ICISC 2011, LNCS 7259, pp. 18–32, 2012.
© Springer-Verlag Berlin Heidelberg 2012

In this paper we present a new way of representing the constraints given by a non-linear equation system. This representation allows all equations in the TRIVIUM-equation system to be merged into one single equation. The process of merging equations has asymptotically exponential complexity, but using our new technique we are nevertheless still able to complete it in practice, with an actual complexity far lower than the $O(2^{80})$-bound for TRIVIUM.

The paper is organized as follows. In Section 2 we explain the Multiple Right Hand Side equation representation and Binary Decision Diagrams as well as some operations on both constructions. The cipher TRIVIUM is also briefly described. Section 3 introduces Compressed Right Hand Side equations and shows how a solution to such equations can be found. In Section 4 we present our experimental results and explain how to reduce the TRIVIUM equation system to a single Compressed Right Hand Side equation. Section 5 concludes the paper. The appendix contains examples for several of the used constructions and algorithms.

2 Preliminaries

2.1 Multiple Right Hand Side Equation Systems

The Multiple Right Hand Side (MRHS) representation [2,5] is an efficient way to represent equations containing much inherent linearity. Equation systems coming from cryptographic primitives are well suited for MRHS representation, since cryptographic algorithms are usually built using both linear and non-linear components.

A MRHS equation is a linear system with, as the name suggests, multiple right hand sides. We write one MRHS equation as $Ax = B$, where A and B are matrices with the same number of rows, and x is a vector of variables. Any assignment of x such that Ax equals some column in B satisfies the equation.

We construct a system of MRHS equations from a cryptographic primitive as follows. First we assign variable names to the bits of cipher states at several places in the encryption process. The assignment of variables should be done such that the bits of the input and output of any non-linear component can be written as linear combinations of variables. Then we construct one MRHS equation $Ax = B$ for each non-linear component f. The rows of A are the input and output linear combinations of f. Finally, we list all possible inputs to f, with their corresponding outputs. Each input/output pair becomes a column in B. An example of this can be found in the appendix.

Following this procedure we can construct a system of MRHS equations

$$A_1 x = B_1, \ldots, A_m x = B_m$$

for any cryptographic primitive that uses relatively small non-linear components.

For a given solution to the system, there is exactly one column in each B_i corresponding to this solution. We say such a column is *correct*. If the system has a unique solution, there is only one correct right hand side in each B_i. Solving

MRHS equation systems means identifying columns in the B_i that cannot be correct, and delete them.

Several techniques for solving MRHS systems exist. One of them is called *gluing* and is used in this paper. Gluing means to merge two equations into one, making sure that only solutions that satisfy both original equations are carried over into the new (glued) equation.

Gluing two equations reduces the number of equations by one. The process of gluing can be repeated, packing all initial equations into one MRHS equation. The resulting equation is nothing more than a system of linear equations, and can easily be solved. The solution we find will necessarily satisfy all the original initial MRHS equations, so this strategy will solve the system in question.

The problem we face when applying the technique of gluing in practice, is that the number of right hand sides in glued equations tends to increase exponentially. Only when there are just a few equations remaining, with large A-matrices, will the restrictions on potential solutions be so limiting that the number of possible right hand sides rapidly decreases. As we shall see, however, the problem of exponential growth in the number of right hand sides may be circumvented using *binary decision diagrams*.

2.2   Binary Decision Diagrams

In this section we will introduce binary decision diagrams (BDDs). A BDD is a directed acyclic graph used to represent a set of binary vectors or a Boolean formula. They are mostly used in design and verification systems and were introduced by S.B. Akers [7]. Later implementations and refinements led to a broad interest in the computer science community as BDDs allow the manipulation of large propositional formulae [8,9] in compressed form. Sometimes they are used as an alternative to *guess-and-verify* solvers of propositional problems since they enable one to keep track of all satisfying assignments at once and offer polynomial time algorithms to count the number of solutions of a propositional problem given in the form of a BDD.

The use of BBDs in cryptanalysis for LFSRs was proposed by Krause [10] and successfully applied to Grain with NLFSRs by Stegemann [11].

Definition 1 (Binary Decision Diagram). *A binary decision diagram is a pair $\mathcal{D} = (G, L)$ where $G = (V, E)$ is a directed acyclic graph, and $L = (l_0, l_1, \ldots, l_{r-1}, \epsilon)$ is an ordered set of variables.*

The vertices of G are $V = \{v_0, v_1, \ldots, v_{s-1}\} \cup \{\top, \bot\}$ where all v_i denote inner vertices and contain exactly one root vertex with no incoming edges. Every inner vertex v has exactly two outgoing edges, which we call the 1-edge and the 0-edge. We call $\top$ and $\bot$ terminal vertices, they have no outgoing edges. Every vertex v is associated with a variable, denoted $L(v)$, and for all edges (u, v) we have $L(u)$ appearing before $L(v)$ in L. We always have $L(\top) = L(\bot) = \epsilon$.

We denote with $G(v)$ the subgraph of G rooted at v, i.e., the graph consisting of vertices and edges along all directed paths originating at v. For any pair of vertices u, w it holds that if $G(u) = G(w)$ then $u = w$.

There exist other definitions of BDDs which do or do not include the order L or the reducedness property of unequal subgraphs. The definition above is also known as a reduced ordered BDD and is canonical [9]. We denote the number of vertices in a binary decision diagram $\mathcal{D}$ by $\mathcal{B}(\mathcal{D}) = |G|$. The size of a BDD depends heavily on the order L. Finding the optimal ordering to minimize $\mathcal{B}(\mathcal{D})$ is an NP-hard problem [9].

In Definition 1 L induces a partial order of the vertices. We visualize a BDD by drawing it from top to bottom, with vertices of the same order on the same line, and we say that these vertices are at the same *level*. There is only one root vertex and it must necessarily associated with the first variable in L. This node associated with l_0 is drawn on top, and the nodes $\top$ and $\bot$ are drawn on the bottom. An example of a BDD can be found in the appendix.

Definition 2 (Accepted Inputs of a BDD). *In a BDD $\mathcal{D}$ every path from the root vertex to the terminal vertex $\top$ is called an accepted input of $\mathcal{D}$.*

Since every inner node is associated with a variable, we can regard every edge as a *variable assignment*. To find a variable assignment (or vector) which is accepted by the BDD, we start with an empty vector of length $|L|$. Following a path from the root vertex to $\top$ we visit at most one node at each level.

Whenever we go from v through a 1-edge, we say that $L(v)$ is assigned to 1, and $L(v) = 0$ whenever we go via a 0-edge. A path that ends up in $\top$ gives us one accepted input in terms of variable assignments. Likewise, a path from the root vertex to $\bot$ gives us a rejected input to a specific BDD. By traversing all paths to $\top$ we can build the set of all vectors which are accepted by the BDD.

If a path from the root to $\top$ *jumps* a level, i.e. the assignment to a variable l_k is undefined since the path does not contain a vertex v with $L(v) = l_k$, both assignments to this variable are accepted and we get two different variable assignments. If an accepted input jumps r levels in total we get 2^r different satisfying assignments from this path. An example of accepted inputs of a BDD can be found in the appendix.

AND-Operation on BDDs. As shown above, we can use BDDs to represent the set of vectors that satisfy a Boolean equation. By the nature of our equation systems, we need a way to merge solution sets from different equations. Below is a simple recursive algorithm which does this. A more general version of the algorithm can be found in [12].

Let $\mathcal{D}$ and $\mathcal{D}'$ be two BDDs with v_0 as the root of $\mathcal{D}$ and u_0 the root of $\mathcal{D}'$. The conjunction of $\mathcal{D}$ and $\mathcal{D}'$ into a new BDD $\mathcal{E}$ is done as follows.

First we need to define an ordering on the union of variables from $\mathcal{D}$ and $\mathcal{D}'$. Next, we set the root node of $\mathcal{E}$ at the top level, and label it $(v_0 u_0)$. Then we perform Algorithm 1, which will fill in nodes and edges in $\mathcal{E}$, from top to bottom.

The paths in the BDD that results after merging $\mathcal{D}$ and $\mathcal{D}'$ using Algorithm 1 will correspond to vectors that satisfy both Boolean equations related to $\mathcal{D}$ and $\mathcal{D}'$. One feature of the conjunction of two BDDs is that all nodes in the new

---

Algorithm 1. Merging BDDs $\mathcal{D}$ and $\mathcal{D}'$ into $\mathcal{E}$

---

while $\exists$ a node (vu) in $\mathcal{E}$ without outgoing edges **do**

    Let v^e be child of v in $\mathcal{D}$ through e-edge

    Let u^e be child of u in $\mathcal{D}'$ through e-edge

    if $L(v) = L(u)$ **then** $\triangleright$ v and u are at the same level

        Insert $(v^0 u^0)$ at level $\min\{L(v^0), L(u^0)\}$ with 0-edge from (vu).

        Insert $(v^1 u^1)$ at level $\min\{L(v^1), L(u^1)\}$ with 1-edge from (vu).

    end if

    if $L(v) < L(u)$ **then** $\triangleright$ v is higher up than u

        Insert $(v^0 u)$ at level $\min\{L(v^0), L(u)\}$ with 0-edge from (vu).

        Insert $(v^1 u)$ at level $\min\{L(v^1), L(u)\}$ with 1-edge from (vu).

    end if

    if $L(v) > L(u)$ **then** $\triangleright$ u is higher up than v

        Insert (vu^0) at level $\min\{L(v), L(u^0)\}$ with 0-edge from (vu).

        Insert (vu^1) at level $\min\{L(v), L(u^1)\}$ with 1-edge from (vu).

    end if

end while

BDD can be labelled with (vu) where v and u come from the two original BDDs. It is then not hard to see that the following upper bound holds

$$\mathcal{B}(\mathcal{E}) \leq \mathcal{B}(\mathcal{D})\mathcal{B}(\mathcal{D}'). \tag{1}$$

We will use this fact later in the paper. For a more detailed description and analysis of operations on BDDs one might consult [12,9,8]. An example of the AND-operation on BDDs can be found in the appendix.

2.3 Trivium

Trivium [13] is a synchronous stream cipher and part of the ECRYPT Stream Cipher Project portfolio for hardware stream ciphers. It consists of three connected non-linear feedback shift registers (NLFSR) of lengths 93, 84 and 111. These are all clocked once for each key stream bit produced.

Trivium has an inner state of 288 bits, which are initialized with 80 key bits, 80 bits of IV, and 128 constant bits. The cipher is clocked 1152 times before actual keystream generation starts. The generation of keystream bits and updating the registers is very simple. The pseudo-code in [13] is a good and compact description of the whole process of generating keystream as shown in Algorithm 2.

Here z_i is the key stream bit, and the registers are filled with the bits $s_1, \ldots, s_{288}$ before clocking.

For algebraic cryptanalysis purposes one can create four equations for every clock; three defining the inner state change of the registers and one relating the inner state to the key stream bit. Solving this equation system in time less than trying all 2^{80} keys is considered a valid attack on the cipher.

Small Scale Trivium. For our experiments we considered small scale versions of Trivium. While reduced versions of a cipher sometimes dismiss some structural

---

Algorithm 2. Trivium Pseudo-Code

---

for $i = 1$ to N **do**

   $t_1 \leftarrow s_{66} + s_{93}$

   $t_2 \leftarrow s_{162} + s_{177}$

   $t_3 \leftarrow s_{243} + s_{288}$

   $z_i \leftarrow t_1 + t_2 + t_3$ $\triangleright$ Keystream bit

   $t_1 \leftarrow t_1 + s_{91} \cdot s_{92} + s_{171}$

   $t_2 \leftarrow t_2 + s_{175} \cdot s_{176} + s_{264}$

   $t_3 \leftarrow t_3 + s_{286} \cdot s_{287} + s_{69}$

   $(s_1, s_2, \ldots, s_{93}) \leftarrow (t_3, s_1, \ldots, s_{93})$

   $(s_{94}, s_{95}, \ldots, s_{177}) \leftarrow (t_1, s_{94}, \ldots, s_{176})$

   $(s_{178}, s_{179}, \ldots, s_{288}) \leftarrow (t_2, s_{178}, \ldots, s_{287})$

end for

---

component of the full scale cipher, e.g. Bivium [1], we try to keep our reduced versions as close to Trivium as possible.

We scale with respect to the number of bits in the state. When we speak about Trivium-N, we are speaking about a cipher with N bits of internal state, that is, scaled down by a factor $\alpha = N/288$. The lengths of the two first registers will be 93α and 84α, rounded to the nearest integers. The length of the last register will be what remains to get N as the total number of state bits (either $\lfloor 111\alpha \rfloor$ or $\lceil 111\alpha \rceil$).

In the full Trivium, the three top positons in each register are all used as tap positions. This property is also carried over to all the scaled versions. For the tap positions appearing elsewhere in the registers, we simply scale their indices with α. For example, as 66 is used as a tap position in the full Trivium, for Trivium-N the corresponding tap position will be 66α, rounded to the nearest integer, with the following exception: Tap positions that are close to each other in the full Trivium may get the same indices in some Trivium-N if α is small enough. When this happens, we reduce the tap position of the smaller index by one, thus ensuring that all tap positions in Trivium-N are distinct. The equation systems representing Trivium-N and Trivium will then have similar structures.

3   Compressed Right Hand Side Equation Systems

With MRHS equations a clear separation between the linear and the non-linear part of an equation was introduced. Overall it yielded a much smaller representation for equations typical in algebraic cryptanalysis. Nevertheless, solving MRHS equations has been limited to relatively small-scale examples because of the problem with a big number of right hand sides.

It was shown in [7] that representing Boolean equations as BDDs is canonical with respect to the ordering of variables. This way of recording sets of assignments gives us the advantage that we may have a moderate number of nodes in a BDD, but very many paths from the root leading to $\top$. Rather than writing out all satisfying assignments, or a truth table for a Boolean equation, only a BDD is retained in memory. However, when experimenting with equations from certain ciphers, BDDs may also become too big to keep in computer memory [11].

By combining the MRHS and BDD approaches, we get a new way to handle large equation systems in algebraic cryptanalysis. We call this representation of equations *Compressed Right Hand Sides* (CRHS) equations.

Definition 3 (CRHS). *A compressed right hand side equation is written as* $Ax = \mathcal{D}$, *where* A *is a* $k \times n$-*matrix with rows* $l_0, \ldots, l_{k-1}$ *and* $\mathcal{D}$ *is a BDD with variable ordering (from top to bottom)* $l_0, \ldots, l_{k-1}$. *Any assignment to* x *such that* Ax *is a vector corresponding to an accepted input in* $\mathcal{D}$, *is a satisfying assignment.*

An easy example of a CRHS equation can be found in the appendix.

CRHS Gluing. If we are given two Boolean equations $f_1(X_1) = 0, f_2(X_2) = 0$ and we want to find vectors in variables $X_1 \cup X_2$ which satisfy both equations simultanously we can do this by investigating their individual satisfying vectors at common variables. If two vectors have the same values at common variable indices we have found a vector which satisfies both equations. This operation is part of the Gluing operation described in Section 2.1.

If we are given two CRHS equations $[C_1]x = \mathcal{D}_1, [C_2]x = \mathcal{D}_2$ and we want to compute their common solutions we use a similar technique called *CRHS Gluing*. The result of gluing both equations above is

$$\begin{bmatrix} C_1 \\ C_2 \end{bmatrix} x = \mathcal{D}_1 \wedge \mathcal{D}_2.$$

Any assignment of x such that $\begin{bmatrix} C_1 \\ C_2 \end{bmatrix} x$ is an accepted input in the conjunction $\mathcal{D}_1 \wedge \mathcal{D}_2$ gives a solution to both initial equations simultanously. Like the Gluing operation on MRHS equations the right hand side BDD contains all possible combinations of vectors from the original equations. The difference is that satisfying vectors are no longer explicit in the computer memory, but are recorded in a compressed format, namely as paths in the BDD.

It is easy to output all possible vectors from the paths in a BDD. There also exists an easy polynomial-time (in the number of nodes) algorithm to count the number of accepted inputs to a BDD. An example of CRHS-gluing can be found in the appendix.

3.1   Dependencies among Linear Combinations

The left hand side in a CRHS equation is equal to the left hand side in a MRHS equation, namely a set of linear combinations $\{l_0, \ldots, l_{k-1}\}$ in the variables of the system. If we glue several CRHS equations together, it might happen that the resulting left hand side matrix in the glued equation does not have full rank, that is, the set of linear combinations in the left hand side contains linear dependencies.

The BDD on the right hand side treats the l_i as variables, and is oblivious to the constraint that some of them should sum to zero or one. Therefore, an accepted input in the BDD may or may not satisfy the linear dependencies known to the left hand side. These paths should be taken out of the BDD in order to not produce false solutions.

The straight-forward way to remove paths that do not satisfy some linear dependency is to use the AND-operation. The number of nodes in the BDD representing a linear equation $g(l_0, \ldots, l_{k-1})$ is two times the number terms in g. It is then easy to construct the BDD for any g, and combine it with the BDD in the equation using the AND-operation. This will remove all false solutions.

4   Experimental Results

While exploring the possibilities of CRHS equations we used a software library called $Cudd$[14]. The Cudd software library implements various types of BDDs and algorithms/operations which can be performed on BDDs. The code base is optimized and usable on a personal computer even for very big BDDs.

We used Cudd together with C++ code and developed a program capable of reading different equation systems representing scaled Triviums and then gluing the equations together.

It was crucial in the experiments to find out the size of the resulting CRHS equation when gluing many of them together. This number is important to determine in order to evaluate the feasability of our method. Theoretically the size of the final CRHS equation C is upper bounded by

$$\mathcal{B}(C) \leq \mathcal{B}(c_0) \cdot \mathcal{B}(c_1) \cdot \ldots \cdot \mathcal{B}(c_{r-1})$$

when gluing CRHS equations $c_0, c_1, \ldots, c_{r-1}$ into C. This value is exponential in the number of nodes and might lead to infeasible sizes of BDDs, even for quite small versions of Trivium. However, our experiments showed that the size of the BDD for the glued CRHS equations was far smaller than the upper bound, and stayed manageable. Thus we are indeed, in contrast to MRHS equation systems, able to glue all equations in large CRHS equation systems together. For MRHS equation systems, gluing all equations together will reveal the solutions to the system. As we explain below, it is more complicated for CRHS equation systems, due to false solutions in the right hand side BDD.

In the experiments reported below, we created CRHS equation systems representing Trivium-N for various values of N. Then we glued all equations into one single big CRHS equation. We examined different aspects of the equation systems, which can tell us something about their solvability with our method. For several small scale versions we measured the following properties:

Value	Description
n	# of variables = # of initial CRHS equations
k	# of different linear combinations of variables
$\mathcal{B}$	# vertices in BDD in final equation
lc	# of linear constraints for solution
Sol.	# paths in final BDD
Mem.	Memory consumption in MB

Table 1. Experimental results

N	n	k	$\mathcal{B}$	lc	Sol.	Mem.
35	85	173	$2^{18.86}$	88	$2^{85.67}$	87
40	94	191	$2^{20.57}$	97	$2^{93.77}$	182
45	106	215	$2^{21.68}$	109	$2^{106.60}$	358
50	115	233	$2^{21.15}$	118	$2^{115.60}$	258
55	127	257	$2^{21.55}$	130	$2^{127.60}$	329
60	138	282	$2^{22.34}$	144	$2^{140.35}$	560
65	148	299	$2^{22.66}$	151	$2^{148.60}$	687
70	160	323	$2^{22.42}$	163	$2^{160.49}$	588
75	171	349	$2^{22.78}$	178	$2^{173.83}$	742

Initial equations have 4 nodes in the BDD, so we see from Table 1 that the size of the BDD after gluing all equations together is far from the theoretical upper bound. However, the growth of $\mathcal{B}$ is exponential just with a very small constant. It is worth to notice that $\mathcal{B}$ is not strictly increasing with N. We also see that the expected number of paths that satisfy all constraints given by lc is between 2^{-4} and 2^{-2}.

A point worth mentioning is that the exponential upper bound for gluing CRHS equations together is tight, in general. There *are* equations that will achieve the bound when glued together. Equation systems coming from ciphers tend to be very sparse, in the sense that each initial equation contain few variables, and each variable only appears in a few equations. This is also the case for Trivium. Two equations that do not share any variables have a linear size when glued together. As shown in (5), the gluing in this case is basically putting one BDD on top of the other. This may explain why it is particularly easy to glue together CRHS equations coming from scaled versions of Trivium.

Full Trivium. So what about $N = 288$? For full Trivium our computer ran out of memory before finishing gluing all equations together. On the other hand, we were able to glue 404 of the 666 initial equations together, producing a CRHS equation C_1 of size $2^{22.9}$. Then we glued the remaining initial equations into C_2, of size $2^{24.8}$. By using the upper bound (1) for merging two BDDs, we have then demonstrated that the single CRHS equation representing the full Trivium has a size smaller than $2^{47.7}$. The true size of the BDD for the full Trivium is probably a lot smaller than $2^{47.7}$, given that the upper bound we use has proved to be very loose for the systems we study. In any case, we know that the size of the CRHS equation representing the full Trivium is quite far from the 2^{80}-bound for a valid attack.

4.1 Solving Attempts

If a single CRHS equation gave a solution as readily as a MRHS equation, we would be done, and have an algebraic attack on Trivium with complexity much smaller than the $O(2^{80})$-bound for exhaustive search. As noted above, we can not deduce a solution straight from the CRHS equation, since we have eventually to find a path in the BDD that satisfies a number of linear constraints. For scaled Triviums, we have of course tried the straight-forward approach mentioned in Section 3.1. Gluing BDDs representing linear constraints onto the BDD of the cipher CRHS equation unfortunately makes the size grow too large very rapidly.

Another solving method we have tried works as follows. Let the set of linear constraints to be satisfied be contained in a matrix LC. We set LC at the (single) top node in the BDD, and will propagate the matrix through the whole BDD according to Algorithm 3.

Algorithm 3. Propagating linear constraints through BDD with k levels.

---

for $i = 0$ to k **do**
    for every node a at level i **do**
        if a contains matrix **then**
            Build matrix M of linear constraints present in all matrices in a
            if $l_i = 0$ is consistent with M **then**
                Send $M|_{l_i=0}$ through 0-edge
            end if
            if $l_i = 1$ is consistent with M **then**
                Send $M|_{l_i=1}$ through 1-edge
            end if
        end if
        end for
    end for

What we are bascally doing is to fix the value of l_i in LC to 0 or 1 when passing LC through a 0- or 1-edge out of a node at level i. If the linear constraints of LC would become inconsistent by sending it across an edge, the matrix is not propagated in that direction. Nodes receiving more than one LC-matrix will only keep linear constraints present in all matrices.

A node containing a matrix could be interpreted as saying "*Any path below me must satisfy the linear constraints in my matrix.*" We hope that the matrix ending up in the T-node will contain some other linear constraints than the ones we started with. If this is the case, we can repeat Algorithm 3 with increasingly large LC.

In small examples (that can be checked by hand) the method of propagating the linear constraints through the BDD works, but for Trivium-35 it did not, as there were no new linear constraints in the matrix arriving at the bottom. What we did see for Trivium-35 however, was that there is a significant amount of nodes at levels $113 - 138$ in the BDD that did not receive any matrices (due to inconsistencies). At some levels almost half of the nodes were empty. We learn from this that there is no path satisfying the linear constraints in LC that can pass through these nodes, and so they can be deleted. Hence we can use Algorithm 3 to prune the BDD, and reduce its size.

5 Conclusion and Further Work

In this paper we have introduced a new way of representing algebraic equations, and shown its advantages compared to previously known representations. With the CRHS representation it is possible to merge many more equations together, than what is possible by other approaches. Building the CRHS equation system for Trivium, we have shown that Trivium may be described by a single CRHS equation with a BDD of size $2^{47.7}$ nodes, at most.

We have not yet been able to solve big CRHS equation systems, due to the many false solutions appearing in the right hand side BDD. The problem that needs to be solved is: **How do we efficiently find a path in a BDD that satisfies a set of linear constraints?** The method of matrix propagation helps in reducing the size of the BDD, and may be an approach worth pursuing. This is a topic for further research.

Finally, we should keep in mind that the operation of merging equations in a system is a process with exponential complexity. This is also true for CRHS equations, but for systems representing versions of Trivium we can do full merging anyway, because of the structure of the system. Solving non-linear equation systems is NP-hard in general, so we cannot hope to have a solving algorithm without any exponential step in it. Gluing all equations together is an exponential step, and full gluing normally solves the system. We can then speculate that after gluing all initial equations into one, we have overcome the exponential step and that the remaining problem for finding a solution can be solved efficiently. It is not clear that the problem of finding a path in a BDD subject to a set of linear constraints must have exponential complexity in the number of nodes. Further investigation into this question is needed.

References

1. McDonald, C., Charnes, C., Pieprzyk, J.: Attacking Bivium with MiniSat. eS-TREAM, ECRYPT Stream Cipher Project, Report 2007/040 (2007), http://www.ecrypt.eu.org/stream
2. Raddum, H., Semaev, I.: Solving Multiple Right Hand Sides linear equations. Designs, Codes and Cryptography 49(1), 147–160 (2008)
3. Faugère, J.: A new efficient algorithm for computing Gröbner bases (F4). Journal of Pure and Applied Algebra 139(1-3), 61–88 (1999)

4. Courtois, N., Klimov, A., Patarin, J., Shamir, A.: Efficient Algorithms for Solving Overdefined Systems of Multivariate Polynomial Equations. In: Preneel, B. (ed.) EUROCRYPT 2000. LNCS, vol. 1807, pp. 392–407. Springer, Heidelberg (2000)
5. Raddum, H.: MRHS Equation Systems. In: Adams, C., Miri, A., Wiener, M. (eds.) SAC 2007. LNCS, vol. 4876, pp. 232–245. Springer, Heidelberg (2007)
6. Semaev, I.: Sparse algebraic equations over finite fields. SIAM Journal on Computing 39(2), 388–409 (2009)
7. Akers, S.: Binary decision diagrams. IEEE Transactions on Computers 27(6), 509–516 (1978)
8. Somenzi, F.: Binary decision diagrams. In: Calculational System Design. NATO Science Series F: Computer and Systems Sciences, vol. 173, pp. 303–366. IOS Press (1999)
9. Knuth, D.: The Art of Computer Programming. vol. 4, Fascicles 0-4, The Art of Computer Programming. Addison Wesley (PEAR) (2009)
10. Krause, M.: BDD-Based Cryptanalysis of Keystream Generators. In: Knudsen, L.R. (ed.) EUROCRYPT 2002. LNCS, vol. 2332, pp. 222–237. Springer, Heidelberg (2002)
11. Stegemann, D.: Extended BDD-Based Cryptanalysis of Keystream Generators. In: Adams, C., Miri, A., Wiener, M. (eds.) SAC 2007. LNCS, vol. 4876, pp. 17–35. Springer, Heidelberg (2007)
12. Bryant, R.E.: Graph-based algorithms for boolean function manipulation. IEEE Transactions on Computers 35, 677–691 (1986)
13. Cannière, C.D., Preneel, B.: Trivium specifications. ECRYPT Stream Cipher Project (2005)
14. Somenzi, F.: CUDD: CU Decision Diagram Package (2009), http://vlsi.colorado.edu/~fabio/CUDD/

Appendix

Example 1 (MRHS). The basic non-linear component in Trivium is the bitwise multiplication found in the function updating the registers. The new bit (x_6) coming into a register at some point is related to the old ones ($x_1, \ldots, x_5$) by

$$x_1 \cdot x_2 + x_3 + x_4 + x_5 = x_6.$$

The multiplication is the non-linear component, with inputs x_1 and x_2, and a single linear combination as output, namely $x_3 + x_4 + x_5 + x_6$. There are four different inputs to this function, hence there will be four columns in the B-matrix. The corresponding MRHS equation is

$$\begin{bmatrix} 1 & 0 & 0 & 0 & 0 & 0 \\ 0 & 1 & 0 & 0 & 0 & 0 \\ 0 & 0 & 1 & 1 & 1 & 1 \end{bmatrix} \begin{pmatrix} x_1 \\ x_2 \\ x_3 \\ x_4 \\ x_5 \\ x_6 \end{pmatrix} = \begin{bmatrix} 0 & 1 & 0 & 1 \\ 0 & 0 & 1 & 1 \\ 0 & 0 & 0 & 1 \end{bmatrix}. \tag{2}$$

Example 2 (BDD). Figure 1 shows an example BDD. The vertex v_0 is the root. Solid lines indicate 1-edges and dashed lines indicate 0-edges. In this example the order is (l_0, l_1, l_2) as indicated to the left.

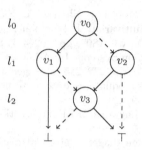

Fig. 1. Example BDD

Example 3 (Accepted Inputs). The accepted inputs for the BDD in Figure 1 are the vectors (l_0, l_1, l_2):

$$(0, 0, 0)$$
$$(0, 0, 1)$$
$$(0, 1, 1)$$
$$(1, 0, 1) .$$

One can see that $(0, 0, 0)$ and $(0, 0, 1)$ are on the same path from v_0 to $\top$. On that path no node associated with l_2 is visited, so l_2 can be assigned both values.

A Boolean equation may be characterized by its set of satisfying assignments. Building a BDD whose accepted inputs match the set of satisfying assignments, gives us another representation of the same equation. For example, the Boolean equation corresponding to the BDD in Figure 1 is $l_0 l_1 + l_0 l_2 + l_1 l_2 + l_0 + l_1 = 0$.

Example 4 (AND operation). The top half of Fig. 2 shows the BDDs of two Boolean functions. The left BDD shows $l_0 + l_1 + l_2 = 0$, the right BDD represents $l_0 l_1 + l_2 = 0$. Both BDDs share the same order of variables, and the resulting BDD of their conjunction after reduction is shown below the two original BDDs.

Example 5 (CRHS). We write equation (2) from Example 1 as a CRHS equation by converting the right hand side into a BDD.

Instead of writing out the left hand matrix of equation (2), we write down the corresponding linear combinations, and give them the names l_0, l_1, l_2.

$$\begin{bmatrix} x_1 & = l_0 \\ x_2 & = l_1 \\ x_3 + x_4 + x_5 + x_6 = l_2 \end{bmatrix} = \left\{ \begin{array}{c} \end{array} \right. \qquad (3)$$

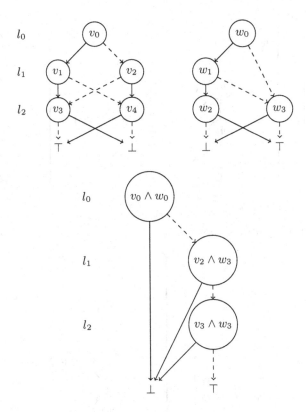

Fig. 2. AND-operation example

The right hand side of the CRHS equation is a *compressed* version of the right hand side in a MRHS equation. Every accepted input in the graph of the CRHS equation stands for one right hand side of the corresponding MRHS equation. The example above contains the edge (v_0, v_3). This edge is *jumping* over a level, i.e. every path through this edge does not contain any vertex at level l_1. That means that for a path containing the edge (v_0, v_3), the variable l_1 can take any value. The path $\langle v_0, v_3, \top \rangle$ thus contains two vectors for (l_0, l_1, l_2), namely $(0, 0, 0)$ and $(0, 1, 0)$.

Example 6 (CRHS Gluing). The following two equations are similar to equations in a Trivium equation system. In fact, the right hand sides of the following are taken from a full scale Trivium equation system. The left hand matrices have been shortened.

$$
\begin{bmatrix} x_1 & = l_0 \\ x_2 & = l_1 \\ x_3 + x_4 = l_2 \end{bmatrix} = \left\{ \vphantom{\begin{matrix}a\\a\\a\\a\\a\end{matrix}} \right.
$$

$$
\begin{bmatrix} x_4 & = l_3 \\ x_5 & = l_4 \\ x_6 + x_7 = l_5 \end{bmatrix} = \left\{ \vphantom{\begin{matrix}a\\a\\a\\a\\a\end{matrix}} \right.
$$

$$\tag{4}$$

The gluing of the equations above is

$$
\begin{bmatrix} x_1 & = l_0 \\ x_2 & = l_1 \\ x_3 + x_4 = l_2 \\ x_4 & = l_3 \\ x_5 & = l_4 \\ x_6 + x_7 = l_5 \end{bmatrix} = \left\{ \vphantom{\begin{matrix}a\\a\\a\\a\\a\\a\\a\\a\\a\end{matrix}} \right.
$$

$$\tag{5}$$

where $\bot$-paths in this last graph are omitted for better readability. Note that omitting these paths does not decrease the overall number of vertices. The resulting equation has 8 nodes where the corredsponding MRHS equation would have 16 right hand sides.

Cryptanalysis of Round-Reduced HAS-160

Florian Mendel, Tomislav Nad, and Martin Schläffer

Institute for Applied Information Processing and Communications (IAIK)
Graz University of Technology, Inffeldgasse 16a, A-8010 Graz, Austria
Tomislav.Nad@iaik.tugraz.at

Abstract. HAS-160 is an iterated cryptographic hash function that is standardized by the Korean government and widely used in Korea. In this paper, we present a semi-free-start collision for 65 (out of 80) steps of HAS-160 with practical complexity. The basic attack strategy is to construct a long differential characteristic by connecting two short ones by a complex third characteristic. The short characteristics are constructed using techniques from coding theory. To connect them, we are using an automatic search algorithm for the connecting characteristic utilizing the nonlinearity of the step function.

Keywords: differential attack, hash function, coding theory, collision.

1 Introduction

In the last years research in cryptanalysis of hash function has made significant progress. Weaknesses have been shown in many commonly used hash functions as SHA-1 [19] and MD5 [18]. These breakthrough results in the cryptanalysis of hash functions were the motivation for intensive research in this field. Especially, in the ongoing SHA-3 [12] competition several new design strategies and attack techniques have been proposed. However, it also draws the attention away from currently used hash function standards, whereas it is important to analyze these standards to achieve a better understanding of the security margin in critical applications like e-commerce and e-government systems. In this paper, we focus on the hash function HAS-160. It is standardized by the Korean government (TTAS.KO-12.0011/R1) [17] and hence widely used in Korea. It is an iterated cryptographic hash function that produces a 160-bit hash value. The design of HAS-160 is similar to SHA-1 and MD5.

In [22], Yun et al. applied the techniques invented by Wang et al. in the cryptanalysis of MD5 and SHA-1 to the HAS-160 hash function. They show that a collision can be found for HAS-160 reduced to 45 steps with a complexity of about 2^{12}. This attack was later extended by Cho et al. [3] to HAS-160 reduced to 53 steps. The attack has a complexity of about 2^{55} 53-step HAS-160 computations. Mendel and Rijmen [10] improved the attack and reduced the complexity to 2^{35} and presented an actual colliding message pair for HAS-160 reduced to 53 steps. Furthermore, they presented a theoretical attack on 59 steps. Finally, preimage attacks on 52 steps by Sasaki and Aoki [16] and on 68 steps by Hong

H. Kim (Ed): ICISC 2011, LNCS 7259, pp. 33–47, 2012.
© Springer-Verlag Berlin Heidelberg 2012

et al. [6] have been presented. Both attacks have only theoretical complexity and are only slightly faster than the generic attack which has complexity 2^{160}.

In this paper, we combine different techniques to construct a semi-free start collision for 65 (out of 80) steps of HAS-160 with practical complexity. A semi-free-start collision is a collision attack where the adversary can choose the value of the initial value (IV). The basic idea of our attack is similar to the attack on a DES based hash function by Rijmen and Preneel [15] and to the recent attack on the SHA-3 candidate Skein by Yu et al. [21]. The idea is to construct a long differential characteristic by connecting two short ones by a complex third characteristic. We show how this idea can be applied on HAS-160 resulting in a semi-free start collision. Furthermore, we present an actual colliding message pair and IV fulfilling all conditions of the differential characteristics. This is so far the best attack in terms of number of steps on HAS-160 with practical complexity.

The remainder of this paper is structured as follows. A description of the hash function is given in Section 2. In Section 3 we describe the basic attack strategy. In Section 4 the search for two short differential characteristics and the determination of a good position for the connection is explained. In Section 5 we connect the short characteristics and present the final differential path. Finally, we present a colliding message pair in Section 5.3 and conclude in Section 6.

2   Description of HAS-160

HAS-160 is an iterative hash function that processes 512-bit input message blocks, operates on 32-bit words and produces a 160-bit hash value. The design of HAS-160 is similar to the design principles of MD5 and SHA-1. In the following, we briefly describe the hash function. It basically consists of two parts: message expansion and state update transformation. A detailed description of the HAS-160 hash function is given in [17].

Message Expansion. The message expansion of HAS-160 is a permutation of 20 expanded message words W_i in each round. The 20 expanded message words W_i used in each round are constructed from the 16 input message words m_i as shown in Table 1.

For the ordering of the expanded message words W_i the permutation in Table 2 is used.

Table 1. Message expansion of HAS-160

	Round 1	Round 2	Round 3	Round 4
W_0	m_0	m_0	m_0	m_0
$\vdots$	$\vdots$	$\vdots$	$\vdots$	$\vdots$
W_{15}	m_{15}	m_{15}	m_{15}	m_{15}
W_{16}	$W_0 \oplus W_1 \oplus W_2 \oplus W_3$	$W_3 \oplus W_6 \oplus W_9 \oplus W_{12}$	$W_{12} \oplus W_5 \oplus W_{14} \oplus W_7$	$W_7 \oplus W_2 \oplus W_{13} \oplus W_8$
W_{17}	$W_4 \oplus W_5 \oplus W_6 \oplus W_7$	$W_{15} \oplus W_2 \oplus W_5 \oplus W_8$	$W_0 \oplus W_9 \oplus W_2 \oplus W_{11}$	$W_3 \oplus W_{14} \oplus W_9 \oplus W_4$
W_{18}	$W_8 \oplus W_9 \oplus W_{10} \oplus W_{11}$	$W_{11} \oplus W_{14} \oplus W_1 \oplus W_4$	$W_4 \oplus W_{13} \oplus W_6 \oplus W_{15}$	$W_{15} \oplus W_{10} \oplus W_5 \oplus W_0$
W_{19}	$W_{12} \oplus W_{13} \oplus W_{14} \oplus W_{15}$	$W_7 \oplus W_{10} \oplus W_{13} \oplus W_0$	$W_8 \oplus W_1 \oplus W_{10} \oplus W_3$	$W_{11} \oplus W_6 \oplus W_1 \oplus W_{12}$

Table 2. Permutation of the message words

step i	1	2	3	4	5	6	7	8	9	10	11	12	13	14	15	16	17	18	19	20
Round 1	18	0	1	2	3	19	4	5	6	7	16	8	9	10	11	17	12	13	14	15
Round 2	18	3	6	9	12	19	15	2	5	8	16	11	14	1	4	17	7	10	13	0
Round 3	18	12	5	14	7	19	0	9	2	11	16	4	13	6	15	17	8	1	10	3
Round 4	18	7	2	13	8	19	3	14	9	4	16	15	10	5	0	17	11	6	1	12

State Update Transformation. The state update transformation of HAS-160 starts from a (fixed) initial value IV of five 32-bit registers and updates them in 4 rounds of 20 steps each. Figure 1 shows one step of the state update transformation of the hash function.

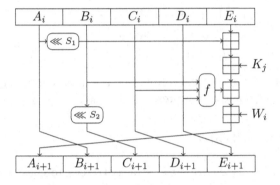

Fig. 1. The step function of HAS-160

Note that the function f is different in each round: f_0 is used in the first round, f_1 is used in round 2 and round 4, and f_2 is used in round 3.

$$f_0(x, y, z) = (x \wedge y) \oplus (\neg x \wedge z)$$
$$f_1(x, y, z) = x \oplus y \oplus z$$
$$f_2(x, y, z) = (x \vee \neg z) \oplus y$$

A step constant $K_j \in \{0, \text{5a827999}, \text{6ed9eba1}, \text{8f1bbcdc}\}$ is added in every step and is different for each round. While rotation value $s_2 \in \{10, 17, 25, 30\}$ is different in each round of the hash function, the rotation value s_1 is different in each step of a round. The rotation value s_1 for each step of a round is given in Table 3.

Table 3. Permutation of the message words

step i	1	2	3	4	5	6	7	8	9	10	11	12	13	14	15	16	17	18	19	20
s_1	5	11	7	15	6	13	8	14	7	12	9	11	8	15	6	12	9	14	5	13

After the last step of the state update transformation, the initial value and the output values of the last step are combined, resulting in the final value of one iteration known as Davies-Meyer hash construction (feed forward). The feed forward is a word-wise modular addition of the IV and the output of the state update transformation. The result is the final hash value or the initial value for the next message block.

2.1   Alternative Description of HAS-160

As one can see in the description of the step update transformation (see Figure 1) only the state variable A_i is updated in each step. The values of the other state variables are defined by A_i. Therefore, we can redefine the state update such that only one state variable is used.

$$
\begin{aligned}
A_{i+1} = & A_{i-4} \ggg s_2 + A_i \lll s_1 + \\
& f(A_{i-1}, A_{i-2} \ggg s_2, A_{i-3} \ggg s_2) + \\
& K_j + W_i
\end{aligned}
\tag{1}
$$

Note that s_2 need to be adapted accordingly if the update uses A's between two rounds. The chaining values are represented by $A_0, A_{-1}, A_{-2}, A_{-3}, A_{-4}$.

3   Basic Attack Strategy

In this section, we briefly describe the attack strategy to construct a semi-free start collision for 65 steps of HAS-160. A similar attack was done on a DES based hash function by Rijmen and Preneel [15] and recently on Skein by Yu et al. [21]. The main idea is to construct a long differential characteristic by connecting two short ones. First, proper differences in the expanded message words need to be chosen, such that they result in two short linear characteristics with low Hamming weight and hence hold with high probability. Second, we connect the two short differential characteristics by a third one. This one can have low probability, since we can use message modification to fulfill the conditions. Figure 2 illustrates the strategy.

The attack can be summarized as follows:

1. Choose an optimal position for the connection and find two differential characteristics, which hold with high probability.
2. Find a connecting differential characteristic.
3. Find inputs fulfilling the conditions and use message modification to improve the attack complexity.

To find two good characteristics and to determine an optimal position, we use a linearized model of the hash function. Finding a characteristic in a linearized hash function is not difficult. However, we aim for characteristics with high probability such that the available freedom can be used for the connection. The

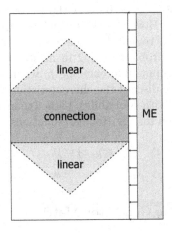

Fig. 2. Basic attack strategy. Differences occur only in the parts with background color.

probability that the linear characteristic holds in the original hash function is related to the Hamming weight of the characteristic. In general, a differential characteristic with low Hamming weight has a higher probability than one with a high Hamming weight. Finding a characteristic with high probability (low Hamming weight) is related to finding a low weight word in linear codes. Therefore, we use a probabilistic algorithm from coding theory to find good characteristics. It has been shown in the past, for instance the cryptanalysis of SHA-0 [2], SHA-1 [13], EnRUPT [7] or SIMD [8] that this technique works well for finding differential characteristics with low Hamming weight.

We are constructing different linear codes for different positions and lengths of the connecting part to determine the optimal choice. Afterwards, we use an automatic search technique to find a connecting differential characteristic. Finally, we use message modification, introduced by Wang et al. in [20], to find inputs fulfilling all conditions.

4   Finding Two Short Characteristics

As mentioned before the problem of finding characteristics for a linearized hash function which hold with high probability for the original function is related to coding theory [8,13,14]. In order to find such characteristics for HAS-160 we need to linearize the hash function.

4.1   Linearization of HAS-160

Since the message expansion is already linear, only the step update transformation has to be linearized. The nonlinear parts of this function are the modular additions and the Boolean functions f_0 and f_2 (f_1 is linear). In the attack, we replace all modular addition by XORs. For the Boolean functions we tried several different linearizations. However, the following variant turned out to be the best. The function f_0 (IF) is replaced by the 0-function, i.e. we block each input difference in f_0. This has probability $1/2$ in most cases (cf. [4]). One can see that there is exactly one input difference for f_0 where the output difference is always one. In that case we discard the characteristic. f_2 is approximated by its second input. which holds with probability higher than $1/2$. In summary we get the following approximation for the Boolean functions:

$$f_0'(x, y, z) = 0$$
$$f_2'(x, y, z) = y$$

4.2   Construction of the Generator Matrix

In this section we explain the standard approach to find collision producing characteristics for a linearized hash function. As observed by Rijmen and Oswald [14], all differential characteristics for a linearized hash function can be seen as the codewords of a linear code. Our goal is to find codewords with low Hamming weight, i.e. characteristics with high probability. Therefore, we have to include all intermediate chaining values where differences could decrease the success probability in the linear code. Based on the alternative description of HAS-160 (see Section 2.1) we include only A_i in the linear code, since the other state variables do not add any additional information to the code. This decreases the length of the code significantly and therefore also the running time of the search algorithm.

Let $\Delta A_i \in \{0, 1\}^{32}$ be the difference vector of the chaining value A_i in bit representation at step i. Then the vector

$$cw := (\Delta A_1, \cdots, \Delta A_n), \tag{2}$$

where $cw \in \{0, 1\}^{n \cdot 32}$, represents the differences in the chaining value A_i after each step of n steps of HAS-160. cw is one codeword of the linear code and therefore a differential characteristic. To construct the generator matrix for the linear code, we proceed as follows:

1. Compute cw_j with the input difference $\Delta M = e_j$, where $e_j \in \{0, 1\}^{512}$ is the j-th unit vector and ΔM the difference of the message block in bit representation.
2. Repeat the computation for $j = 1, \ldots, 512$.

The resulting generator matrix of the linear code representing linearized HAS-160 is defined in the following way:

$$G_{512 \times n \cdot 32} := \begin{pmatrix} cw_1 \\ \vdots \\ cw_{512} \end{pmatrix}. \tag{3}$$

Since we are aiming for a collision in the last step, we need to apply code shortening on the last 160 bits, i.e. ensuring that all code words are zero in the last 160 bits. This reduces the dimension and length of the code to 352 and $(n \cdot 32 - 160)$, respectively.

Using this matrix one can search for low Hamming weight codewords over all n steps. As explained in Section 3 we are looking for two short characteristics, which will be connected later. Therefore, we need to modify the linear code to include this requirement.

Modification. The easiest way to define a linear code for both characteristics simultaneously and ensuring that both use the same expanded message, is the following. Firstly, ignore t steps in the middle. Hence, we change the vector (2) to:

$$cw := (\Delta A_1, \cdots, \Delta A_l, \Delta A_{l+t+1}, \cdots, \Delta A_n). \tag{4}$$

At the beginning of the second characteristic (after step $l+t$), the state variables can have any difference, since the differences in the steps before are yet undefined. Therefore, we need to add the information to the code that after step $l + t$ all differences are possible. Hence, we add the chaining variables at step $l+t+1$ to the linear code. The construction of the generator matrix changes to:

1. Compute cw_j with the input difference $\Delta M = e_j$, where $e_j \in \{0,1\}^{512}$ is the j-th unit vector and ΔM the difference of the message block in bit representation.
2. Repeat the computation for $j = 1, \ldots, 512$.
3. Compute cw_{512+k} as follows:
   (a) Set $\Delta M = 0$ and $cw_s = e_k$, where $e_k \in \{0,1\}^{160}$ is the k-th unit vector and

   $$cw_s = (\Delta A_{l+t-3}, \Delta A_{l+t-2}, \Delta A_{l+t-1}, \Delta A_{l+t}, \Delta A_{l+t+1}).$$

   (b) Compute ΔA_i for $(l+t+1) < i \leq n$ with cw_s and ΔM as input. Hence, we get following codeword:

   $$cw_{512+k} := (\Delta A_1 = 0, \cdots, \Delta A_l = 0, cw_s, \Delta A_{l+t+2}, \cdots, \Delta A_n).$$

4. Repeat the computation for $k = 1, \ldots, 160$.

Note that $\Delta B_{l+t+1} = \Delta A_{l+t}, \Delta C_{l+t+1} = \Delta A_{l+t-1}, \Delta D_{l+t+1} = \Delta A_{l+t-2}$ and $\Delta E_{l+t+1} = \Delta A_{l+t-3}$ and therefore all possible chaining values after step $l + t$ are included in the code. The resulting generator matrix is

$$G_{672 \times (n-t+4) \cdot 32} := \begin{pmatrix} cw_1 \\ \vdots \\ cw_{672} \end{pmatrix}. \tag{5}$$

Again code shorting is applied to ensure that all codewords result in a collision after n steps.

Determining l, t and n. There exist several possible choices for the parameters l, t and n of the linear code. First of all we limit $t \leq 21$. The reason for this is simple. We have 21 words (16 message words and 5 IV words) which can be choosen freely and hence can be used for message modification to fulfill all conditions in the connecting part which is usually the most expensive part of the attack. However, we aimed for a smaller t to reduce the search space for the connecting part as well.

For the search we constructed generator matrices for $21 \leq l \leq (n - 21)$ and $t = 21$. If we have found two characteristics with high probability we reduce t.

4.3   Searching for Low Hamming Weight Codewords

We use the publicly available CodingTool Library [11] which contains all tools needed to search for codewords with low Hamming weight. It implements the probabilistic algorithm from Canteaut and Chabaud [1] to search for codewords with low Hamming weight. This iterative algorithm basically looks for small Hamming weight codewords in a smaller code. Such a codeword is considered as a good candidate for a low Hamming weight codeword for the whole code. The algorithm randomly selects σ columns of it and splits the selection in two sub-matrices of equal size. By computing all linear combination of p rows (usually 2 or 3) for each sub-matrix and storing their weight, the algorithm searches for a collision of both weights which allow to search for codewords of $2p$. Then two randomly selected columns are interchanged, followed by one Gaussian elimination step. This procedure is repeated until a sufficiently small Hamming weight is found. With this tool we can find good characteristics for different choices of l and t in few seconds on a standard PC. In Table 4 we present the best (lowest Hamming weight) characteristics we have found for different parameters. As one can see after 65 steps the Hamming weight is getting too high such that we cannot find a characteristic and conforming inputs with practical complexity.

Note that decreasing t always increases the Hamming weight, since more state variables with differences are included in the linear code. Furthermore, the Hamming weight in Table 4 includes only differences in A. To estimate the probability

Table 4. Results for the low weight search

n	l	t	Hamming weight
53	18	21	3
60	18	21	3
65	18	21	3
66	19	21	25
67	18	21	25
68	18	21	72
69	18	21	72
70	18	21	119
75	19	21	123
80	19	21	247

one has to take the differences in all state variables into account. Therefore, the probability for the linear characteristic can be roughly estimated by four times the Hamming weight of A.

Using this general approach we can cover the whole (linear) search space and allow arbitrary differences in the message words. However, it turned out that the best characteristics we have found are indeed the trivial ones which have only few differences in the message words and only a one bit difference per message word.

4.4   Short Differential Characteristics

To describe the differential characteristics we use generalized conditions which are explained in Section 5.1. We have found several different characteristics, depending on the choice of l and t. In Table 8 of Appendix A we present two short characteristics, where t is kept small. To improve readability, we used the alternative description of HAS-160 (see Section 2.1)

5   Finding Connecting Characteristics

In this section, we show how one can find a connecting differential characteristic which is the most expensive part in our attack. The main idea to find a connecting characteristic is to use the nonlinearity of the step update function. Constructing such complex characteristics is a difficult task. In [5], De Cannière and Rechberger proposed a new method to find complex characteristics for SHA-1 in an efficient way. In their concept they allow characteristics to impose arbitrary conditions on the pairs of bits (referred to as generalized conditions). Based on this they presented an efficient probabilistic search algorithm. Recently, Mendel et al.[9] extended this technique and applied it successfully on SHA-2. The basic idea of the search algorithm is to randomly pick a bit position and impose a zero-difference. Afterwards, it is calculated how this condition propagates. This is repeated until an inconsistency is found or all unrestricted bits are eliminated.

5.1   Generalized Conditions

To describe the search algorithm in more detail we first repeat the notation of generalized conditions which was introduced in [5]. Inspired by signed-bit differences, generalized conditions for differences take all 16 possible conditions on a pair of bits into account. Table 5 lists all these possible conditions and introduces notations for the various cases.

For example, all pairs of 8-bit words X and X^* that satisfy

$$\{(X, X^*) \in \{0,1\}^8 \times \{0,1\}^8 \mid X_7 \cdot X_7^* = 0, X_i = X_i^* \text{ for } 1 \le i \le 5, X_0 \ne X_0^*\},$$

can be conveniently written in the form

$$\nabla X = [7?-----x].$$

Table 5. Notation for possible generalized conditions on a pair of bits [5]

(X_i, X_i^*)	(0,0)	(1,0)	(0,1)	(1,1)	(X_i, X_i^*)	(0,0)	(1,0)	(0,1)	(1,1)
?	✓	✓	✓	✓	3	✓	✓	-	-
-	✓	-	-	✓	5	✓	-	✓	-
x	-	✓	✓	-	7	✓	✓	✓	-
0	✓	-	-	-	A	-	✓	-	✓
u	-	✓	-	-	B	✓	✓	-	✓
n	-	-	✓	-	C	-	-	✓	✓
1	-	-	-	✓	D	✓	-	✓	✓
#	-	-	-	-	E	-	✓	✓	✓

5.2 Application to HAS-160

Due to the similarities of HAS-160 to SHA-1 the adaption of the above concept can be done in a straightforward manner and can be used to find the connecting characteristic. For more details see [5,9]. We proceed as follow:

1. Pick a random unrestricted bit (?) or an unsigned difference (x).
2. Impose a zero-difference (-) or randomly a sign (u or n), respectively.
3. Check how the new condition propagates.
4. If an inconsistency occurs jump back to the point where the last sign was imposed and make a different decision.
5. Repeat this until all unrestricted bits are eliminated

Using a small number of unrestricted words reduces the search space and running time of the algorithm significantly. Therefore, we reduced this number by extending the two short linear characteristics linearly. Since there are only few differences at the end of the first linear characteristic and at the beginning of the second linear characteristic, we can extend them forward and backward respectively, without increasing the Hamming weight too much. In fact for the characteristic in Table 8 in Appendix A we extended the linear characteristics linearly forward by two and backwards by ten steps. Table 6 shows the starting

Table 6. Steps free of conditions at the beginning of the search algorithm

step	∇A	∇W
⋮	⋮	⋮
20	x-----------------x--x---------	--------------------------------
21	????????????????????????????????	--------------------------------
22	????????????????????????????????	--------------------------------
23	????????????????????????????????	--------------------------------
24	????????????????????????????????	--------------------------------
25	????????????????????????????????	x-------------------------------
26	-x-x------x---x-xxx--x-------x--	--------------------------------
⋮	⋮	⋮

point of the search algorithm using the notation of generalized conditions leaving only five words unrestricted.

Applying the above algorithm on this starting point the algorithm converges already after an hour (on a standard PC) to a complete characteristic for 65 steps. Determining the complexity of the probabilistic algorithm in general is still an open problem. Among others it depends on the hash function, search strategy, start characteristic and implementation. The complete characteristic is given in Table 8 of Appendix A. Note that with this approach we can find several different characteristics.

5.3 Finding a Message Pair

Almost all of the differences in the characteristic of Table 8 in Appendix A are within 21 steps. Since we can choose up to 21 words (16 message and 5 IV) freely we can use message modification to find efficiently inputs which fulfill all the conditions of the characteristic. The conditions for the characteristic are listed in Table 9 in Appendix A. The resulting colliding message pair and IV is given in Table 7.

Table 7. A colliding message pair and IV for HAS-160

IV	ed3c8ca6 38127dc3 bcf7b374 264eeb2b 73be1247
M	467d7948 3c433177 981f570c 6bf43c12 3dc04b7c cb85a46d 3356206e bff3ea04 9603f6ca 252c37eb 3a1d6197 479ca8d1 badbe3d9 4e23c48c c52a6189 53f1ea06
M'	467d7948 3c433177 981f570c 6bf43c12 3dc04b7c cb85a46d 3356206e bff3ea04 9603f6ca 252c37eb 3a1d6197 479ca8d1 3adbe3d9 4e23c48c 452a6189 53f1ea06
ΔM	00000000 00000000 00000000 00000000 00000000 00000000 00000000 00000000 00000000 00000000 00000000 00000000 80000000 00000000 80000000 00000000
h	4b0a28ae bc82dbb1 a4805bfd cd226435 7cb7eb52
h'	4b0a28ae bc82dbb1 a4805bfd cd226435 7cb7eb52

6 Conclusions

The progress in the cryptanalysis of hash functions in the last years shows that the security of existing standards need to be reevaluated. Therefore, we analyze in this paper the Korean hash function standard (TTAS.KO-12.0011/R1) HAS-160. The main idea of our attack is to construct two short linear differential characteristics which hold with high probability and connect them by a complex third characteristic by using the nonlinearity of the state update function. We use techniques from coding theory to search efficiently for the short characteristics and simultaneously determine an optimal position and length of the connecting characteristic. In a second step we use an automatic search algorithm to find a connecting characteristic taking the nonlinearity of the state update into account.

We present a semi-free-start collision for 65 (out of 80) steps HAS-160 with practical complexity. Extending the attack to more rounds seems to be difficult. One can always extend the size of the connecting part, but this also increases the complexity of finding the connecting characteristic, which running time is hard to estimate. If we limit the length of the connecting part to 21 steps, then the best short characteristics we can find with probability below the generic complexity of a collision attack, are for up to 65 steps.

Even though we only present a semi-free-start collision, it is a step forward in the analysis of HAS-160. This is so far the best known attack with practical complexity in terms of attacked steps for HAS-160.

Acknowledgments. The work in this paper has been supported by the European Commission under contract ICT-2007-216646 (ECRYPT II) and by the Austrian Science Fund (FWF, project P21936).

References

1. Canteaut, A., Chabaud, F.: A New Algorithm for Finding Minimum-Weight Words in a Linear Code: Application to McEliece's Cryptosystem and to Narrow-Sense BCH Codes of Length 511. IEEE Transactions on Information Theory 44(1), 367–378 (1998)
2. Chabaud, F., Joux, A.: Differential Collisions in SHA-0. In: Krawczyk, H. (ed.) CRYPTO 1998. LNCS, vol. 1462, pp. 56–71. Springer, Heidelberg (1998)
3. Cho, H.-S., Park, S., Sung, S.H., Yun, A.: Collision Search Attack for 53-Step HAS-160. In: Rhee, M.S., Lee, B. (eds.) ICISC 2006. LNCS, vol. 4296, pp. 286–295. Springer, Heidelberg (2006)
4. Daum, M.: Cryptanalysis of Hash Functions of the MD4-Family. PhD thesis, Ruhr-Universität Bochum (May 2005),
   http://www.cits.rub.de/imperia/md/content/magnus/dissmd4.pdf
5. De Cannière, C., Rechberger, C.: Finding SHA-1 Characteristics: General Results and Applications. In: Lai, X., Chen, K. (eds.) ASIACRYPT 2006. LNCS, vol. 4284, pp. 1–20. Springer, Heidelberg (2006)
6. Hong, D., Koo, B., Sasaki, Y.: Improved Preimage Attack for 68-Step HAS-160. In: Lee, D., Hong, S. (eds.) ICISC 2009. LNCS, vol. 5984, pp. 332–348. Springer, Heidelberg (2010)
7. Indesteege, S., Preneel, B.: Practical Collisions for EnRUPT. In: Dunkelman, O. (ed.) FSE 2009. LNCS, vol. 5665, pp. 246–259. Springer, Heidelberg (2009)
8. Mendel, F., Nad, T.: A Distinguisher for the Compression Function of SIMD-512. In: Roy, B., Sendrier, N. (eds.) INDOCRYPT 2009. LNCS, vol. 5922, pp. 219–232. Springer, Heidelberg (2009)
9. Mendel, F., Nad, T., Schläffer, M.: Finding SHA-2 Characteristics: Searching through a Minefield of Contradictions. In: Lee, D.H., Wang, X. (eds.) ASIACRYPT 2011. LNCS, vol. 7073, pp. 288–307. Springer, Heidelberg (2011)
10. Mendel, F., Rijmen, V.: Colliding Message Pair for 53-Step HAS-160. In: Nam, K.-H., Rhee, G. (eds.) ICISC 2007. LNCS, vol. 4817, pp. 324–334. Springer, Heidelberg (2007)
11. Nad, T.: The CodingTool Library. Workshop on Tools for Cryptanalysis 2010 (2010),
    http://www.iaik.tugraz.at/content/research/krypto/codingtool/

12. National Institute of Standards and Technology. Cryptographic Hash Algorithm Competition (November 2007), http://csrc.nist.gov/groups/ST/hash/sha-3/index.html
13. Pramstaller, N., Rechberger, C., Rijmen, V.: Exploiting Coding Theory for Collision Attacks on SHA-1. In: Smart, N.P. (ed.) Cryptography and Coding 2005. LNCS, vol. 3796, pp. 78–95. Springer, Heidelberg (2005)
14. Rijmen, V., Oswald, E.: Update on SHA-1. In: Menezes, A. (ed.) CT-RSA 2005. LNCS, vol. 3376, pp. 58–71. Springer, Heidelberg (2005)
15. Rijmen, V., Preneel, B.: Improved Characteristics for Differential Cryptanalysis of Hash Functions Based on Block Ciphers. In: Preneel, B. (ed.) FSE 1994. LNCS, vol. 1008, pp. 242–248. Springer, Heidelberg (1995)
16. Sasaki, Y., Aoki, K.: A Preimage Attack for 52-Step HAS-160. In: Lee, P.J., Cheon, J.H. (eds.) ICISC 2008. LNCS, vol. 5461, pp. 302–317. Springer, Heidelberg (2009)
17. Telecommunications Technology Association. Hash Function Standard Part 2: Hash Function Algorithm Standard (HAS-160), TTAS.KO-12.0011/R1 (2008)
18. Wang, X., Lai, X., Feng, D., Chen, H., Yu, X.: Cryptanalysis of the Hash Functions MD4 and RIPEMD. In: Cramer, R. (ed.) EUROCRYPT 2005. LNCS, vol. 3494, pp. 1–18. Springer, Heidelberg (2005)
19. Wang, X., Yin, Y.L., Yu, H.: Finding Collisions in the Full SHA-1. In: Shoup, V. (ed.) CRYPTO 2005. LNCS, vol. 3621, pp. 17–36. Springer, Heidelberg (2005)
20. Wang, X., Yu, H.: How to Break MD5 and Other Hash Functions. In: Cramer, R. (ed.) EUROCRYPT 2005. LNCS, vol. 3494, pp. 19–35. Springer, Heidelberg (2005)
21. Yu, H., Chen, J., Ketingjia, Wang, X.: Near-Collision Attack on the Step-Reduced Compression Function of Skein-256. Cryptology ePrint Archive, Report 2011/148 (2011)
22. Yun, A., Sung, S.H., Park, S., Chang, D., Hong, S., Cho, H.-S.: Finding Collision on 45-Step HAS-160. In: Won, D., Kim, S. (eds.) ICISC 2005. LNCS, vol. 3935, pp. 146–155. Springer, Heidelberg (2006)

A Characteristic

Table 8. Characteristic for 65 steps HAS-160 using generalized conditions. The rows with darkgray background represent the connecting part. The rows with lightgray background represent the two linear characteristics. All conditions can be fulfilled using message modification.

step	∇A	∇W
-4	--------------------------------	
-3	--------------------------------	
-2	--------------------------------	
-1	--------------------------------	
0	--------------------------------	
⋮	⋮	⋮
16	--------------------------------	--------------------------------
17	u-------------------------------	x-------------------------------
18	------------------u-------------	--------------------------------
19	n-----------u-------------------	x-------------------------------
20	u------------------u--u---------	--------------------------------
21	-------n-uuuuuu--u----n---u-----	--------------------------------
22	u--n---uu-nu---uu---nn--------uu	--------------------------------
23	--n-n-nnnu-n-u--nu------nu------	--------------------------------
24	uuun-nu--u-u----n-n-unnuuuuuuu-n	--------------------------------
25	--n----uu---uu-un-u-----nu-n-n--	x-------------------------------
26	-n-n------n---n-uun--u------n-	--------------------------------
27	-unu------u-n---uu---u-n-u-u---n	--------------------------------
28	--n---u---u---u--u-n---u-----u-n	--------------------------------
29	---------n---u---------n----u-n	--------------------------------
30	--u-----n-u---------u--	--------------------------------
31	--n-------n--------------n-n--	--------------------------------
32	-----------------n-----------n--	--------------------------------
33	----------n-	x-------------------------------
34	----------n----------u--	--------------------------------
35	----------------------u--	--------------------------------
36	--------------------------------	--------------------------------
37	--------------------------------	--------------------------------
38	--------------------------------	--------------------------------
39	-------------------n-----	--------------------------------
40	--------------------------------	--------------------------------
41	--------------------------------	--------------------------------
42	--------------------------------	x-------------------------------
43	--------------------------------	--------------------------------
44	--------------------------------	x-------------------------------
45	--------------------------------	--------------------------------
⋮	⋮	⋮
65	--------------------------------	--------------------------------

Table 9. Set of conditions for the semi-free-start collision for 65 steps

step	set of conditions	#
16	$A_{16,3} = 0$, $A_{16,21} = A_{15,21}$	2
17	$A_{17,3} = 1$, $A_{17,31} = 1$	2
18	$A_{18,9} = 1$, $A_{18,13} = 1$, $A_{18,8} \neq A_{17,8}$	3
19	$A_{19,18} = 1$, $A_{19,31} = 0$, $A_{19,23} = A_{17,13}$, $A_{19,27} \neq A_{18,2}$, $A_{19,9} \neq A_{18,31}$, $A_{19,24} = A_{18,31}$	6
20	$A_{20,9} = 1$, $A_{20,12} = 1$, $A_{20,31} = 1$, $A_{20,16} \neq A_{18,6}$, $A_{20,3} = A_{18,25}$, $A_{20,0} \neq A_{19,0}$, $A_{20,1} = A_{19,1}$, $A_{20,2} = A_{19,2}$, $A_{20,3} = A_{19,3}$, $A_{20,4} = A_{19,4}$, $A_{20,5} = A_{19,5}$, $A_{20,23} \neq A_{19,6}$, $A_{20,7} = A_{19,7}$, $A_{20,19} = A_{19,19}$, $A_{20,24} = A_{19,24}$, $A_{20,29} \neq A_{19,29}$	16
21	$A_{21,4} = 1$, $A_{21,9} = 0$, $A_{21,14} = 1$, $A_{21,17} = 1$, $A_{21,18} = 1$, $A_{21,19} = 1$, $A_{21,20} = 1$, $A_{21,21} = 1$, $A_{21,22} = 1$, $A_{21,24} = 0$, $A_{21,26} \neq A_{19,9}$, $A_{21,29} = A_{19,12}$, $A_{21,3} \neq A_{20,3}$, $A_{21,6} \neq A_{20,6}$, $A_{21,7} \neq A_{20,7}$, $A_{21,11} \neq A_{20,11}$, $A_{21,15} \neq A_{20,15}$, $A_{21,16} \neq A_{20,16}$, $A_{21,3} \neq A_{20,18}$, $A_{21,25} \neq A_{20,25}$, $A_{21,26} = A_{20,26}$, $A_{21,30} = A_{20,30}$	22
22	$A_{22,0} = 1$, $A_{22,1} = 1$, $A_{22,10} = 0$, $A_{22,11} = 0$, $A_{22,15} = 1$, $A_{22,16} = 1$, $A_{22,20} = 1$, $A_{22,21} = 0$, $A_{22,23} = 1$, $A_{22,24} = 1$, $A_{22,28} = 0$, $A_{22,31} = 1$, $A_{22,2} = A_{20,17}$, $A_{22,3} = A_{20,18}$, $A_{22,4} \neq A_{20,19}$, $A_{22,5} \neq A_{20,20}$, $A_{22,6} = A_{20,21}$, $A_{22,7} \neq A_{20,22}$, $A_{22,9} = A_{20,24}$, $A_{22,3} \neq A_{21,3}$, $A_{22,5} = A_{21,5}$, $A_{22,6} \neq A_{21,6}$, $A_{22,7} = A_{21,7}$, $A_{22,8} = A_{21,8}$, $A_{22,12} \neq A_{21,12}$, $A_{22,29} = A_{21,29}$, $A_{22,30} = A_{21,30}$	28
23	$A_{23,6} = 1$, $A_{23,7} = 0$, $A_{23,14} = 1$, $A_{23,15} = 0$, $A_{23,18} = 1$, $A_{23,20} = 0$, $A_{23,22} = 1$, $A_{23,23} = 0$, $A_{23,24} = 0$, $A_{23,25} = 0$, $A_{23,27} = 0$, $A_{23,29} = 0$, $A_{23,17} = A_{21,0}$, $A_{23,28} \neq A_{21,11}$, $A_{23,0} \neq A_{21,15}$, $A_{23,1} \neq A_{21,16}$, $A_{23,8} = A_{21,23}$, $A_{23,13} = A_{21,28}$, $A_{23,16} = A_{21,31}$, $A_{23,3} = A_{22,3}$, $A_{23,21} \neq A_{22,4}$, $A_{23,5} = A_{22,5}$, $A_{23,8} = A_{22,8}$, $A_{23,9} \neq A_{22,9}$, $A_{23,26} = A_{22,9}$, $A_{23,12} \neq A_{22,12}$, $A_{23,13} = A_{22,13}$, $A_{23,2} = A_{22,17}$, $A_{23,17} \neq A_{22,17}$, $A_{23,3} = A_{22,18}$, $A_{23,4} \neq A_{22,19}$, $A_{23,19} = A_{22,19}$, $A_{23,26} \neq A_{22,26}$, $A_{23,30} \neq A_{22,30}$	34
24	$A_{24,0} = 0$, $A_{24,2} = 1$, $A_{24,3} = 1$, $A_{24,4} = 1$, $A_{24,5} = 1$, $A_{24,6} = 1$, $A_{24,7} = 1$, $A_{24,8} = 1$, $A_{24,9} = 0$, $A_{24,10} = 0$, $A_{24,11} = 1$, $A_{24,13} = 0$, $A_{24,15} = 0$, $A_{24,20} = 1$, $A_{24,22} = 1$, $A_{24,25} = 1$, $A_{24,26} = 0$, $A_{24,28} = 0$, $A_{24,29} = 1$, $A_{24,30} = 1$, $A_{24,31} = 1$, $A_{24,23} = A_{22,6}$, $A_{24,24} \neq A_{22,7}$, $A_{24,12} = A_{22,27}$, $A_{24,14} = A_{22,29}$, $A_{24,17} \neq A_{23,0}$, $A_{24,1} \neq A_{23,1}$, $A_{24,18} \neq A_{23,1}$, $A_{24,27} = A_{23,10}$, $A_{24,12} \neq A_{23,12}$, $A_{24,1} = A_{23,16}$, $A_{24,17} \neq A_{23,17}$, $A_{24,19} = A_{23,19}$, $A_{24,21} = A_{23,21}$, $A_{24,16} \neq A_{23,31}$	35
25	$A_{25,2} = 0$, $A_{25,4} = 0$, $A_{25,6} = 1$, $A_{25,7} = 0$, $A_{25,13} = 1$, $A_{25,15} = 0$, $A_{25,16} = 1$, $A_{25,18} = 1$, $A_{25,19} = 1$, $A_{25,23} = 1$, $A_{25,24} = 1$, $A_{25,29} = 0$, $A_{25,17} \neq A_{23,0}$, $A_{25,20} = A_{23,3}$, $A_{25,21} = A_{23,4}$, $A_{25,22} \neq A_{23,5}$, $A_{25,25} \neq A_{23,8}$, $A_{25,26} \neq A_{23,9}$, $A_{25,27} = A_{23,10}$, $A_{25,28} = A_{23,11}$, $A_{25,30} = A_{23,13}$, $A_{25,11} = A_{23,26}$, $A_{25,17} = A_{24,17}$, $A_{25,3} = A_{24,18}$, $A_{25,8} = A_{24,23}$, $A_{25,9} = A_{24,24}$, $A_{25,12} = A_{24,27}$	27
26	$A_{26,2} = 0$, $A_{26,10} = 1$, $A_{26,13} = 0$, $A_{26,14} = 1$, $A_{26,15} = 1$, $A_{26,17} = 0$, $A_{26,21} = 0$, $A_{26,28} = 0$, $A_{26,30} = 0$, $A_{26,1} = A_{24,16}$, $A_{26,3} = A_{24,18}$, $A_{26,4} \neq A_{24,19}$, $A_{26,8} = A_{24,23}$, $A_{26,9} = A_{24,24}$, $A_{26,20} \neq A_{25,3}$, $A_{26,22} \neq A_{25,5}$, $A_{26,25} \neq A_{25,8}$, $A_{26,26} = A_{25,9}$, $A_{26,27} = A_{25,10}$, $A_{26,11} = A_{25,11}$, $A_{26,12} = A_{25,12}$, $A_{26,5} = A_{25,20}$, $A_{26,7} = A_{25,22}$, $A_{26,25} \neq A_{25,25}$, $A_{26,11} = A_{25,26}$, $A_{26,16} \neq A_{25,31}$	26
27	$A_{27,0} = 0$, $A_{27,4} = 1$, $A_{27,6} = 1$, $A_{27,8} = 0$, $A_{27,10} = 0$, $A_{27,14} = 1$, $A_{27,15} = 1$, $A_{27,19} = 0$, $A_{27,21} = 1$, $A_{27,28} = 1$, $A_{27,29} = 0$, $A_{27,30} = 1$, $A_{27,27} = A_{25,10}$, $A_{27,2} = A_{25,17}$, $A_{27,13} = A_{25,28}$, $A_{27,23} \neq A_{26,6}$, $A_{27,24} = A_{26,7}$, $A_{27,12} \neq A_{26,12}$, $A_{27,1} \neq A_{26,16}$, $A_{27,3} \neq A_{26,18}$, $A_{27,23} = A_{26,23}$, $A_{27,9} = A_{26,24}$, $A_{27,27} \neq A_{26,27}$	23
28	$A_{28,0} = 0$, $A_{28,2} = 1$, $A_{28,8} = 1$, $A_{28,12} = 0$, $A_{28,14} = 1$, $A_{28,17} = 1$, $A_{28,21} = 1$, $A_{28,25} = 1$, $A_{28,29} = 0$, $A_{28,23} = A_{26,6}$, $A_{28,4} = A_{26,19}$, $A_{28,19} \neq A_{27,2}$, $A_{28,30} \neq A_{27,13}$	13
29	$A_{29,0} = 0$, $A_{29,2} = 1$, $A_{29,10} = 0$, $A_{29,19} = 1$, $A_{29,23} = 0$, $A_{29,29} = A_{27,12}$, $A_{29,4} \neq A_{28,4}$, $A_{29,21} \neq A_{28,4}$, $A_{29,6} = A_{28,6}$, $A_{29,27} \neq A_{28,10}$, $A_{29,4} = A_{28,19}$, $A_{29,13} = A_{28,28}$, $A_{29,15} = A_{28,30}$	13
30	$A_{30,10} = 1$, $A_{30,21} = 1$, $A_{30,23} = 0$, $A_{30,29} = 1$, $A_{30,27} = A_{28,10}$, $A_{30,4} = A_{28,19}$, $A_{30,8} \neq A_{28,23}$, $A_{30,4} = A_{29,4}$, $A_{30,25} = A_{29,8}$, $A_{30,12} \neq A_{29,12}$, $A_{30,2} \neq A_{29,17}$, $A_{30,17} = A_{29,17}$, $A_{30,6} \neq A_{29,21}$, $A_{30,14} \neq A_{29,29}$	14
31	$A_{31,2} = 0$, $A_{31,4} = 0$, $A_{31,21} = 0$, $A_{31,29} = 0$, $A_{31,6} \neq A_{29,21}$, $A_{31,14} = A_{29,29}$, $A_{31,0} \neq A_{30,0}$, $A_{31,17} \neq A_{30,0}$, $A_{31,19} \neq A_{30,2}$, $A_{31,17} \neq A_{30,17}$	10
32	$A_{32,2} = 0$, $A_{32,17} = 0$, $A_{32,19} = A_{30,2}$, $A_{32,21} = A_{30,4}$, $A_{32,27} = A_{31,10}$, $A_{32,8} \neq A_{31,23}$	6
33	$A_{33,21} = 0$, $A_{33,2} \neq A_{31,17}$, $A_{33,4} \neq A_{32,4}$, $A_{33,6} = A_{32,21}$, $A_{33,14} = A_{32,29}$	5
34	$A_{34,2} = 1$, $A_{34,21} = 0$, $A_{34,6} \neq A_{32,21}$, $A_{34,19} \neq A_{33,2}$, $A_{34,17} = A_{33,17}$	5
35	$A_{35,2} = 1$, $A_{35,19} = A_{33,2}$	2
36	$A_{36,6} \neq A_{35,21}$	1
37	$A_{37,21} = 0$, $A_{37,19} \neq A_{36,2}$	2
39	$A_{39,6} = 0$	1
41	$A_{41,31} = 1$	1

An Efficient Method for Eliminating Random Delays in Power Traces of Embedded Software

Daehyun Strobel and Christof Paar

Horst Görtz Institute for IT Security
Ruhr-University Bochum, Germany
{strobel,cpaar}@crypto.rub.de

Abstract. Generating random delays in embedded software is a common countermeasure to complicate side channel attacks. The idea is to insert dummy operations with varying lengths at different moments in time. This creates a non-predictable offset of the attacking point in the time dimension. Since the success of, e.g., a correlation power analysis (CPA) attack is largely affected by the alignment of the power traces, the adversary is forced to apply additional large computations or to record a huge amount of power traces to achieve acceptable results.

In this paper, we present a new efficient method to identify random delays in power measurements. Our approach does not depend on how the random delays are generated. Plain uniform delays can be removed as well as Benoit-Tunstall [11] or improved floating mean delays [4]. The procedure can be divided into three steps. The first step is to convert the power trace into a string depending on the Hamming weights of the opcodes. After this, the patterns of the dummy operations are identified. The last step is to use a string matching algorithm to find these patterns and to align the power traces.

We have started our analysis with two microcontrollers, an Atmel AVR ATmega8 and a Microchip PIC16F54. For our practical evaluation, we have focused on the ATmega8. However, the results can be applied to many other microcontrollers with a similar architecture.

Keywords: Side channel analysis, random delays, alignment of power traces, embedded devices.

1 Introduction

Although there is a wide range of modern ciphers that allow very high levels of security, their implementations in real systems can often be broken due to their susceptibility to side channel attacks. In recent years, the side channel community spent a lot of work on practical countermeasures. In this paper, we will focus on a *hiding* countermeasure which is often adopted on embedded devices and combined with additional *masking* methods. The idea of hiding is to hinder the adversary to assign instruction operands and intermediate values to the power consumption. This is realized either by additional noise in the frequency domain or by randomized shiftings of operations in the time dimension.

H. Kim (Ed): ICISC 2011, LNCS 7259, pp. 48–60, 2012.
© Springer-Verlag Berlin Heidelberg 2012

We will concentrate on the second case, namely on the insertion of random delays during the execution of the algorithm.

An adversary who is faced with the problem of mounting a correlation power analysis (CPA) attack against a hiding protected implementation has mainly three possibilities [7]: alignment of the power traces, preprocessing of the power traces or simply retrieving a large set of measurements. While preprocessing techniques like integration or convolution of power traces and fast Fourier transformation may increase the success of an attack, acquiring a large set of measurements is not feasible in many cases. The alignment of the power traces is in fact the most promising method, since a correct alignment directly influences the effect of the countermeasure.

1.1 Alignment Techniques

There are some alignment techniques we would like to mention before we begin with our proposal. A generic approach is carried out in two steps [7]:

First, a characteristic pattern is selected which is (nearby) the attacking point and has to be aligned. This is in fact a challenging task. The attacker usually has no information about the time period that contains the relevant data for his attack. Hence, it is difficult to find the right pattern that is close enough to the attacking point such that no delays are between them. In the second step, this pattern has to be found in other traces mainly by using pattern matching algorithms. This can be, e.g., by stepwise computing the Euclidean distances or correlation coefficients between trace and pattern. The problems that may encounter with this approach are also discussed by Mangard *et al.* in [7] and include, e.g., the uniqueness of the selected pattern or intermediate results that are processed in the pattern and cause a variation in the power consumption.

Another alignment technique was proposed by Woudenberg *et al.* in [12]. In this paper, fast dynamic time warping (FastDTW) is used to find the optimal alignment between two power traces in linear time and space complexity. FastDTW, first introduced in [9], produces a warp path between time series which leads to an alignment with the minimum distance. Woudenberg *et al.* applied this algorithm to power traces to circumvent misalignment produced by random process interrupts or an unstable clock. The time complexity to align two power traces with length T is given by $O(T)$.

At CARDIS 2011, Muijrers *et al.* introduced a method called RAM (Rapid Alignment Method) [8]. It is based on algorithms that are mostly applied for object recognition in images. In a direct comparison with elastic alignment, it achieves similar results in less time (about factor 0.2).

1.2 Our Contribution

In this paper, we present an efficient method to detect and eliminate random delays in power traces. Instead of applying pattern matching algorithms on raw traces, we first apply a mapping of the power consumption of each instruction cycle to the Hamming weight of the processed instruction. This allows us to use efficient string matching techniques to identify and detect random delays. We evaluate our proposal on an AES-128 implementation with the improved floating mean countermeasure introduced at CHES 2010. After the application of our approach, we mount a CPA attack [3] and get comparable results to an attack on an unprotected implementation.

The main target device is an 8-bit microcontroller Atmel AVR ATmega8. To show that our proposal is not constrained to this specific device, we partially extend our analysis to a second microcontroller, the Microchip PIC16F54.

1.3 Structure of This Paper

We begin with a short description of the two microcontrollers with a focus on their prefetching and execution process in Sect. 2. Our approach of removing the misalignment is introduced in three steps in Sect. 3: mapping the power consumption to strings, identifying random delays and detecting them in strings. Finally, practical results on an ATmega8 are given in Sect. 4.

2 The Pipelining Concept of ATmega8 and PIC16F54

This section gives us important background information about the two devices Atmel AVR ATmega8 and Microchip PIC16F54. Both devices are 8-bit micro-controllers which use the Harvard architecture, i.e., the instruction memory and the data memory are physically separated and accessed via different buses. In contrast to the von Neumann architecture where instructions and data are shar-ing one bus system, an instruction is fetched from the flash memory while another one is executed. This basic pipelining concept is depicted in Fig. 1 and is used by both microcontrollers to maximize the performance. In Sect. 3, we will see that the prefetching mechanism leaks an essential information that we exploit for our approach.

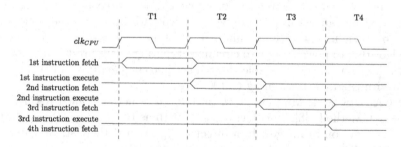

Fig. 1. Parallel instruction fetches and executions for an ATmega8 [1]

Please note that Fig. 1 can only be applied to instructions that are executed in one instruction cycle. For any other instruction, especially for program branches, the successive instruction in the execution flow need not necessarily be the next one in the program memory. An example is given by the following sequence:

```
CP R4,R10      ; compare registers R4 and R10
BREQ SUB       ; branch if equal
ADD R4,R10
...
SUB:
AND R4,R11
```

While CP is being executed, BREQ SUB is fetched into the instruction register. In the next cycle, ADD R4,R10 is fetched and, if the two registers match, the program counter is changed due to the branch instruction. Hence, the previously fetched instruction is discarded and another cycle is needed to fetch the correct instruction AND R4,R10.

The instructions are stored as 12-bit (PIC16F54) or 16-bit (ATmega8) opcodes in the program memory. If we mention the Hamming weight of an instruction, we always refer to one complete opcode including the associated registers and literals, respectively. According to the instruction set of the ATmega8 ADD, e.g., is defined by the opcode 0000 11rd dddd rrrr, with the 5-bit description of the source register r and the destination register d. Hence, the opcode for ADD R4,R10 is 0000 1100 0100 1010 resulting in the Hamming weight 5.

The length of the opcode is not the only difference between the two microcontrollers concerning their instruction sets. The main differences that have to be considered for our approach are given in Tab. 1.

Table 1. Comparison of the AVR Atmega8 and Microchip PIC16F54 instruction sets

	Atmega8	PIC16F54
Number of instructions	130	33
Opcode length	16 bits[1]	12 bits
Execution cycles (# clocks)	mostly 1, up to 4	4 or 8[2]

3   Removing Random Delays

Within this section, we give an efficient solution to remove random delays. For this purpose, we reduce the power consumption of one instruction, which may have thousands of sampling points, to only one value that is independent of the processed data. This leads to a string with the length of the number of performed instructions. The strings of several executions are then analyzed to identify the random delays. Afterwards we use the generalized Bayer-Moore-Horspool algorithm [10] to detect the random delays.

---

[1] Most of the 130 instructions have a length of 16 bits. However, there are four instructions for accessing the program memory that can also be described by a 32-bit opcode which are fetched in two clock cycles. For the rest of this paper we will only focus on 16-bit opcodes.

[2] Only if the program counter is changed by the instruction.

The final step after the detection of the random delays is to find the correct positions in the power traces and to remove the dummy operations. Because every clock cycle of the target devices is clearly distinguishable, this last step is rather trivial and is not discussed in this paper.

3.1   Conversion of Power Traces to Strings

As mentioned in Sect. 2, our target devices use a pipelining concept that prefetches every instruction on a separated bus during the execution of the predecessor. A closer look at the power traces reveals that the Hamming weight of the opcode has a characteristic impact on the power consumption. Figures 2 and 3 show the different Hamming weights in contrast to each other at a clock rate of $1MHz$. For both figures, a number of different instructions and operands were chosen randomly. All traces are single traces that have not been averaged.

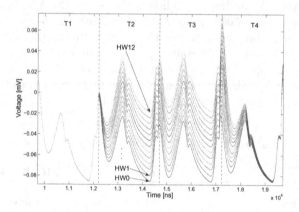

Fig. 2. Power traces of PIC16F54 opcodes with Hamming weight 0 (lowermost) to 12 (uppermost)

One can clearly recognize that the different voltage levels are related to the Hamming weights of the opcodes, especially in the second and third clock cycle (PIC16F54) and just before the rising to the second peak (ATmega8). While for the PIC16F54 a smaller Hamming weight leads to a lower power consumption, for the ATmega8 it is the other way round. These voltage levels seem to be hardly influenced by the executed instruction nor by the processed operands. Hence, these time intervals are well-suited to map the power consumption of one instruction cycle to one Hamming weight value. As a result we get a string of Hamming weights in chronological order of the prefetched opcodes. For instance, the instructions

```
INC R1    ; 1001 0100 0001 0011
ADD R4,R1 ; 0000 1100 0100 0001
CP R4,R6  ; 0001 0100 0100 0110
```

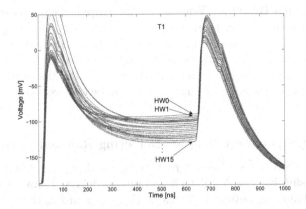

Fig. 3. Power traces of ATmega8 opcodes with Hamming weights 0 (uppermost) to 15 (lowermost). There exist no 16-bit instruction with a Hamming weight of 16.

are mapped to the string 6, 4, 5. Note that the Hamming weights do not depend on the values that are stored in R1, R4 and R6.

For the rest of the paper, the point of measurement that leaks the Hamming weight of the prefetched opcode is referred to as *index voltage*.

3.2   Identification of Random Delays

There are two ways for a random delay implementation. The first is to choose fixed points in the execution flow, where a subroutine for the generation of dummy instructions is called. The advantage of this approach is that the programmer can choose the exact points in time when the misalignment should take place. This can be, e.g., right in front of S-box substitutions. In contrast to this, the second method uses an the interrupt service routine which is triggered by a timer overflow or an external signal. In addition to the random lengths of the delays, another random parameter can be used to trigger the execution of a callback subroutine.

The identification of random delays depends on the implementation and is discussed in the following.

Fixed Calls to the Subroutine (or In-Line Implementations). Let us assume a set of strings which include random delays at fixed points in time. If the random delays of all strings are of equal length, the corresponding power traces are aligned and nothing has to be done. Different lengths can easily be recognized by comparing the strings. The expected subroutine typically consists of an initialization where the length of the random delay is set, a variable number of loops and the return instruction. A mismatch between the strings occurs at the point where the return instruction of the shortest delay collides with unfinished loops. The alignment step can then be accomplished by locating the return instructions and cutting the loops to an equal length. Note that it is not necessary to completely eliminate the random delays.

Interrupt Service Routine. When using interrupts, the overhead per delay is higher than for the first method. Some reasons are, e.g., disabling interrupts, temporarily storing intermediate values, resetting the timer, etc. A comparison between strings will lead to a mismatch directly at the first call of a subroutine. The end of the delay can be detected by a string search after the regular execution flow. In this scenario, the complete interrupt has to be removed to align the power traces.

3.3   Efficient String Matching for Detecting Random Delays

This part of the section describes different string matching algorithms and figures out the most suitable one for detecting random delays. Before we start, we introduce the following notations: Let s be the string of length n, e.g., derived from the computations proposed in Sect. 3.1, and p a pattern of length m. Each character s_i of the string s, with $i \in \{1, 2, 3, \ldots, n\}$, and p_j of the pattern p, with $j \in \{1, 2, 3, \ldots, m\}$, is taken from the alphabet Σ. In our example, s and p are strings composed of Hamming weights. Hence, $\Sigma = \{0, 1, 2, 3, \ldots, l\}$, where l is the number of bits of the opcode, e.g., $l = 16$ in case of the ATmega8.

According to Sect. 3.2, either the delay loop or the interrupt has to be detected. This is done by a string matching algorithm which tests if a pattern is included in a string and at which positions it occurs. A naive approach is to write the pattern below the string, compare the characters and shift the pattern by one position to the right, every time a mismatch occurs. As one can imagine, this is not very efficient. If the mismatch occurs at the last character p_m, the previous positions have to be checked again. For the naive approach, the worst case runtime is $\Theta((n - m + 1)m)$.

In the remaining part of this section, we describe more efficient string matching algorithms. First of all, we present the Boyer-Moore algorithm which serves as basis for the sophisticated generalized Boyer-Moore-Horspool algorithm.

Boyer-Moore Algorithm. In 1977, Boyer and Moore presented a fast string matching algorithm with the complexity of $O(n)$ (see Alg. 1).

After aligning p_1 to a character s_i, the comparison is done from right to left, i.e., we start with s_{i+m} and p_m. If a mismatch occurs at position s_{i+j} and p_j, respectively, the Boyer-Moore algorithm uses two heuristics to avoid frequent one by one shifts:

1. Bad character heuristic: Find the rightmost character, let's say p_l, in p that matches with s_{i+j}. Shift the pattern p by $j - l$ positions, but at least one position, to the right. If s_{i+j} does not occur in p, the pattern is shifted by j positions.
2. Good suffix heuristic: Move the pattern to the right, until the suffix $s_{i+j+1} \cdots s_{i+m}$ matches to a part of the pattern. If this part does not occur in p, shift the pattern by m positions.

At every mismatch, the algorithm chooses the larger shift of both heuristics (line 14). The tables Γ_1 and Γ_2 are preprocessings of the two heuristics and consist

---

Algorithm 1. Boyer-Moore algorithm based on [5]

Input : string s, length n of s, pattern p, length m of p, bad character table
Γ_1, good suffix table Γ_2
Output: vector pos containing all positions of matchings

```
1  begin
2  |   cnt ← 1
3  |   i ← 0
4  |   while i ≤ n − m do
5  |   |   j ← m
6  |   |   while j > 0 and s_{i+j} = p_j do
7  |   |   |   j ← j − 1
8  |   |   end
9  |   |   if j = 0 then
10 |   |   |   pos(cnt) ← i
11 |   |   |   cnt ← cnt + 1
12 |   |   |   i ← i + 1
13 |   |   else
14 |   |   |   i ← i + max[j − Γ_1(s_{i+j}), Γ_2(j)]
15 |   |   end
16 |   end
17 end
```

---

of the number of shifts depending on the position of the mismatch. We skip the description of the good suffix table Γ_2 and only give an algorithm for the generation of the bad character table Γ_1 (see Alg. 2). Every character of the alphabet Σ is assigned the position of the rightmost occurrence in the pattern p. If a character is not included in the pattern, the entry of the table is set to 0. For further information, especially for the computation of Γ_2, we refer to [2].

Algorithm 2. Precomputation of the bad character table Γ_1

Input : pattern p, length m of p, alphabet Σ
Output: bad character table Γ_1

```
1  begin
2  |   foreach element ϵ ∈ Σ do
3  |   |   Γ_1(ϵ) ← 0
4  |   end
5  |   for j = 1, ..., m − 1 do
6  |   |   Γ_1(p_j) ← j
7  |   end
8  end
```

Boyer-Moore-Horspool Algorithm. A simplification of the Boyer-Moore algorithm was introduced by Nigel Horspool in 1980 [6]. For patterns that have no repetitions, the good suffix heuristic does not bring any advantages. Hence, Horspool proposed the simplified Boyer-Moore algorithm, also known as Boyer-Moore-Horspool (BMH) algorithm, which only makes use of the bad character heuristic. In contrast to the original algorithm, the pattern is not aligned to the mismatched character s_{i+j}, but to the rightmost character s_{i+m}, to induce larger shifts. Hence, line 6 of Alg. 2 has to be changed to $\Gamma_1(p_j) \leftarrow m - j$ and line 14 of Alg. 1 to $i \leftarrow i + \Gamma_1(s_{i+m})$.

Generalized Boyer-Moore-Horspool Algorithm with k Mismatches.
Since we do not expect a fully correct mapping of opcodes to Hamming weights, a string matching is needed that does not only consider exact string matches, but also partial matches which include up to a predefined number k of mismatches. Tarhio et al. introduced a generalized BMH algorithm that solves the k mismatches problem in $O(nk(\frac{k}{c} + \frac{1}{m-k}))$ on average, where $c = |\Sigma|$ [10]. Additionally, a preprocessing is executed once for every pattern in $O(m + kc)$.

The first step is quite obvious. While the original algorithm stops comparing after the first mismatch, the generalized BMH algorithm continues until k mismatches are exceeded. The following shift is then determined by considering the rightmost $k + 1$ string characters $s_{i+m-k}, \ldots, s_{i+m}$. We shift the pattern to the right until at least one match occurs in the last $k + 1$ string characters.

Let us assume $k = 0$. The procedure is in fact equivalent to the exact string matching. After a mismatch only the last position s_{i+m} is focused and $\Gamma_1(s_{i+m})$ returns the number of shifts that are necessary to align s_{i+m} to a character of the pattern. If $k = 1$, we enhance our focus on the last two positions. In addition to Γ_1, another look-up table Γ_1' for the position next to last is created. $\Gamma_1'(s_{i+m-1})$ then gives us the number of shifts to align s_{i+m-1}. We can now skip all shifts smaller than $\min(\Gamma_1'(s_{i+m-1}), \Gamma_1(s_{i+m}))$, because none of the two characters will appear at the according position in the pattern. If we skip more than $\min(\Gamma_1'(s_{i+m-1}), \Gamma_1(s_{i+m}))$, probably a valid match is missed. Consequently, for considering k mismatches, $k+1$ tables have to be generated with c entries each or one $(k + 1) \times c$ table. An efficient computation of the $(k + 1) \times c$ table is given in [10].

4   Practical Results

We present our practical results by verifying an AES-128 implementation on an ATmega8 which was clocked with an $1MHz$ external quartz oscillator. The power consumption of the device was measured by means of the voltage drop over a 47Ω shunt resistor inserted on the GND line of the ATmega8.

We extended a standard AES implementation by the improved floating mean countermeasure proposed by Coron et al. at CHES 2010 [4]. The number of loop iterations of the delays is defined by random numbers that are transferred to the microcontroller and stored in the SRAM before every encryption. We chose the same parameters as proposed by Coron et al.. This includes the insertion of 10 random delays per round and three dummy rounds at the beginning and at the end of every encryption. In total, before the first S-box byte substitution occurs, 32 random delays were executed.

We computed the correlation coefficients on raw power traces for all key hypotheses using the Hamming weight power model to attack the first S-box. According to [4], around 160 000 traces should be required for a successful attack on an 8-bit AVR Atmel microprocessor. However, although we took the implementation proposed in their appendix, the correct key was already leaking out at about 35 000 traces (see Fig. 4), which is much lower than expected but still a good starting point for our approach.

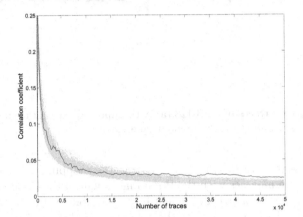

Fig. 4. CPA attack on an AES-128 implementation with improved floating mean countermeasure. The correct key guess is highlighted.

As mentioned in 3.1, the power consumption right before the rising to the second peak of the instruction is highly data independent and is therefore suitable as index when classifying the instructions to Hamming weights. This is verified by the histogram in Fig. 5 which shows the distribution of the index voltages. The intersections between the peaks were used as threshold values for the classification. In this way, we achieved a nearly perfect mapping of index voltages

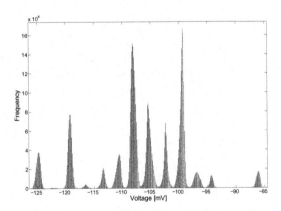

Fig. 5. Histogram of the index voltages discussed in Sect. 3.1. It visualizes a total of 1000 executions of AES with improved floating mean countermeasure on an ATmega8.

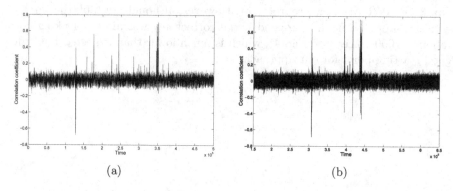

(a)   (b)

Fig. 6. CPA on 1000 traces of AES-128 (a) with improved floating mean countermeasure after alignment and (b) without countermeasure

to Hamming weights. In fact, over 99% of the the mappings were correct, i.e., the resulting value matches with the Hamming weight of the prefetched opcode.

We analyzed the set of strings to identify the variable part of the first random delay as described in Sect. 3.2. In the reference implementation of Coron *et al.* it was only a short loop of the pattern $8, 12, 5$. This is in fact too short to detect, especially in presence of faulty mappings to Hamming weights. Hence, we decided to choose an eight clock cycle pattern of the delay initialization right before the beginning of the loop. The detection process was performed by the generalized BMH algorithm with $k = 1$. Every time a match was found, the subsequent characters were sought after the loop pattern to remove them in the corresponding part of the power traces.

In total, over 98% of the delays were found and eliminated. After this preprocessing step, we mounted a CPA attack on the first S-box again. The results are given in Figs. 6(a) and 7(a). For comparison, Figs. 6(b) and 7(b) show a CPA

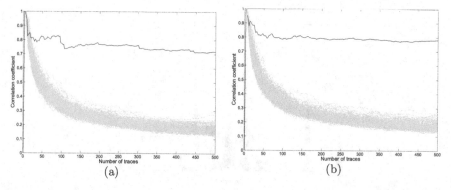

(a)   (b)

Fig. 7. Comparison between a CPA attack on AES-128 (a) with improved floating mean countermeasure after eliminating random delays and (b) without countermeasure. The traces for wrong key guesses are displayed in gray, for the correct key guess in black.

attack on an AES-128 implementation without any countermeasure encrypting the same input data.

The correlation coefficient of the correct key guess is a bit lower for the random delay implementation. However, in both scenarios the correct key is clearly distinguishable from wrong key guesses with a CPA attack using only 30 power traces.

5 Conclusion

We have shown that in future discussions about generating random delays in embedded software, the leakage of the instruction prefetch has to be taken into consideration. Many microcontrollers are using a pipelining concept that allows us to detect the Hamming weight of the prefetched instruction. With this additional information, we have presented an efficient method to identify and eliminate random delays after the acquisition of power traces by applying string matching algorithms. Our proposal has been evaluated on an 8-bit microcontroller with a reference implementation given in [4]. We have shown that the effect of the countermeasure after the application of our new method was negligible.

5.1 Discussions

Although we do not have analyzed other countermeasures like dummy rounds or random order executions, it seems to be not infeasible to detect them as well. Dummy rounds, e.g., are mostly implemented in a special routine or loop with unused registers or with registers whose values have to be stored before accessing the dummy rounds (due to register limitations). In both cases, the Hamming weight sequence may leak information about executed instructions, which includes also storing immediate values, branching to a routine or leaving a loop. The attacker could be able to exploit this leakage to distinguish between real rounds and dummy rounds. However, it highly depends on the implementation if this kind of attack is practical or not.

We did not mention that there is one instruction in the Atmega8 instruction set that does not comply with the model we introduced in Sect. 3.1. The instruction LPM Rd,Z loads one byte from the program memory to the register Rd and does not leak the Hamming weight of the opcode. Instead, the index voltage depends on intermediate values. Although this did not influence the methods proposed in this paper, it can be used to complicate the attack by frequent executions of LPM with random values during the random delay.

References

1. ATMEL. ATmega8 datasheet: 8-bit AVR with 8k bytes in-system programmable flash
2. Boyer, R.S., Moore, J.S.: A fast string searching algorithm. Commun. ACM 20, 762–772 (1977)

3. Brier, E., Clavier, C., Olivier, F.: Correlation Power Analysis with a Leakage Model. In: Joye, M., Quisquater, J.-J. (eds.) CHES 2004. LNCS, vol. 3156, pp. 16–29. Springer, Heidelberg (2004)

4. Coron, J.-S., Kizhvatov, I.: Analysis and Improvement of the Random Delay Countermeasure of CHES 2009. In: Mangard, S., Standaert, F.-X. (eds.) CHES 2010. LNCS, vol. 6225, pp. 95–109. Springer, Heidelberg (2010)

5. Duda, R.O., Hart, P.E., Stork, D.G.: Pattern Classification, 2nd edn. Wiley, New York (2001)

6. Horspool, R.N.: Practical fast searching in strings. Software Practice and Experience 10, 501–506 (1980)

7. Mangard, S., Oswald, E., Popp, T.: Power Analysis Attacks - Revealing the Secrets of Smart Cards. Springer (2007)

8. Muijrers, R.A., van Woudenberg, J.G.J., Batina, L.: RAM: Rapid Alignment Method. In: Prouff, E. (ed.) CARDIS 2011. LNCS, vol. 7079, pp. 266–282. Springer, Heidelberg (2011)

9. Salvador, S., Chan, P.: FastDTW: Toward accurate dynamic time warping in linear time and space. In: KDD Workshop on Mining Temporal and Sequential Data. ACM (2004)

10. Tarhio, J., Ukkonen, E.: Approximate Boyer-Moore string matching. SIAM J. Comput. 22, 243–260 (1993)

11. Tunstall, M., Benoit, O.: Efficient Use of Random Delays in Embedded Software. In: Sauveron, D., Markantonakis, K., Bilas, A., Quisquater, J.-J. (eds.) WISTP 2007. LNCS, vol. 4462, pp. 27–38. Springer, Heidelberg (2007)

12. van Woudenberg, J.G.J., Witteman, M.F., Bakker, B.: Improving Differential Power Analysis by Elastic Alignment. In: Kiayias, A. (ed.) CT-RSA 2011. LNCS, vol. 6558, pp. 104–119. Springer, Heidelberg (2011)

An Efficient Leakage Characterization Method for Profiled Power Analysis Attacks

Hailong Zhang[1,2], Yongbin Zhou[1], and Dengguo Feng[1]

[1] State Key Laboratory of Information Security,
Institute of Software, Chinese Academy of Sciences,
P.O. Box 8718, Beijing 100190, P.R. China
[2] Graduate University of Chinese Academy of Sciences,
19A Yuquan Lu, Beijing, 100049, P.R. China
{zhanghl,zyb,feng}@is.iscas.ac.cn

Abstract. In typical *Profiled Power Analysis Attacks*, like Template Attack (TA) and Stochastic Model based Power Analysis (SMPA), key-recovery efficiency is strongly influenced by the accuracy of characterization in profiling. In order to accurately characterize signals and noises in different times, a large number of power traces is usually needed in profiling. However, a large number of power traces is not always available. In this case, the accuracy of characterization is rapidly degraded, and so it is with the efficiency of subsequent key-recovery. In light of this, we present an efficient *Covariance Analysis based Characterization Method* (CACM for short) to deal with the problem of more accurate leakage characterization with less power traces. We perform experimental power analysis attacks against an AES software implementation on STC89C52 microcontroller, then conduct a comparative study of the effectiveness of these profiled attacks. The results firmly support the validity and efficiency of our method.

Keywords: Profiled Power Analysis Attacks, Covariance Analysis based Characterization Method, Template Attack, Stochastic Model based Power Analysis.

1 Introduction

Since Kocher first introduced Differential Power Analysis in [KJJ1999] more than a decade ago, a myriad of practical power analysis attacks have been proposed, including Template Attack (TA)[CRR2003], Correlation Power Analysis (CPA)[BCO2004], Stochastic Model based Power Analysis (SMPA)[SLP2005], Mutual Information Analysis (MIA)[GBTP2008], etc. Among these methods, TA and SMPA belong to one broad category, as they often have a profiling phase. Therefore, these attacks are referred to as *Profiled Power Analysis Attacks*. In profiled attacks, a reference device similar or identical to the targeted device is usually assumed to be available for profiling. With the help of certain reference device, an adversary characterizes the leakage of the targeted device, and then uses the result of this profiling phase for subsequent key-recovery.

H. Kim (Ed): ICISC 2011, LNCS 7259, pp. 61–73, 2012.
© Springer-Verlag Berlin Heidelberg 2012

In profiling, noises in different times are assumed to follow multivariate normal distribution, and a large number of power traces measured from reference device are needed to accurately characterize signals and noises in different times. In subsequent key-recovery, the adversary resorts to leakage characterization information produced in profiling to recover the secret key. At a high level, the adversary tries to match noises contained in sampled power traces with those of characterized ones. If the key-hypothesis is correct, the match probability will be higher.

In fact, characterization accuracy exerts a strong influence on the key-recovery efficiency. If noises are not accurately characterized, the match probability is influenced. In this case, the correct key is not easy to distinguish, and the attacker needs more power traces in key-recovery to recover the correct key. However, in the attack scenario of profiled attacks, only a few power traces are given to the attacker in key-recovery. If noises are not accurately characterized in profiling, the attacker can not successfully recover the correct key. Therefore, in order to improve the key-recovery efficiency, one feasible way for the attacker is to characterize noises accurately. For TA and SMPA, this task means a large number of power traces are needed. After looking up to references about TA and SMPA, we summarize the number of power traces needed in profiling, and results are listed in Table 1. Table 1 shows that, to reach an acceptable level of characterization accuracy, both TA and SMPA need a large number of power traces.

Table 1. Number of Power Traces Needed by TA and SMPA

Reference	Algorithm	Platform	Attack Method	Number of Power Traces
[CRR2002]	RC4	-	TA	512,000
[SLP2005]	AES	ATM163	SMPA	4,000
[GLP2006]	AES	ATM163	TA	230,000
[GLP2006]	AES	ATM163	SMPA	230,000
[OM2007]	AES	Microcontroller	TA	10,000
[LP2007]	AES	AT90S8515	SMPA	40,000

Under the assumption that one reference device is available, the attacker can operate the reference device as many times as possible and sample a large number of power traces to help accurately characterize signals and noises in different times. However, in practical scenario, it is not always the case. For example, a common countermeasure is to limit the number of operations that the attacked cryptographic device can perform in certain time interval, or that the attacked cryptographic devices can perform under one key. In these scenarios, the attacker can only record limited number of power traces. In order to make profiled attacks still powerful and practical in this scenario, one important technique route is for the attacker to find feasible approach to characterizing signals and noises more accurately with limited number of power traces. Motivated by this, we present

Covariance Analysis based Characterization Method (CACM). We have experimentally demonstrated that, compared with TA and SMPA, the attacker with CACM can use less power traces in profiling to more accurately characterize signals and noises, which helps to improve the correct key distinguish level and reduce the number of power traces needed in key-recovery to recover the correct key.

The rest of the paper is organized as follows. In Section 2, we briefly introduce two typical Profiled Power Analysis Attacks. In order to more accurately characterize signals and noises with less power traces , we give CACM in Section 3. In Section 4, we have experimentally demonstrated the advantage of CACM over SMPA and TA, in terms of the number of power traces needed in profiling and in key-recovery, the characterization accuracy for signals and noises, and the influence of correct key distinguish level. Finally, conclusions are given in Section 5.

2 Typical Profiled Power Analysis Attacks

TA and SMPA are two typical *Profiled Power Analysis Attacks*. In this section, we briefly introduce principles of TA and SMPA.

2.1 Profiling

Template Attack
In TA, the characterization of signals and noises in different times is as follows. First, for each key hypothesis k_i, input the same plaintext p, operate the reference device M times and measure the corresponding power traces $I_1, ..., I_M$. Then, for the power traces $I_1, ..., I_M$ that correspond to key hypothesis k_i, the attacker calculates their mean:$m_i = \frac{1}{M} \sum_{j=1}^{M} I_j$ to get rid of noises and obtain signals in different times. Noise at time $t_x, x\epsilon[1, l]$ can be obtained by subtracting signal from the power consumption: $R_{j,t_x} = I_{j,t_x} - m_{i,t_x}, j\epsilon[1, M]$. Finally, the attacker uses covariance matrix $C_i = Cov(R_{t_x}, R_{t_y})_{l \times l}$ to characterize the relationship of noises in different times. In this way, for each key hypothesis k_i, the attacker obtains a template which is composed of a mean vector m_i and a covariance matrix C_i.

Stochastic Model based Power Analysis
In SMPA, power consumption at time $t_x, x\epsilon[1, l]$ can be seen as $I_{t_x}(p, k) = h_{t_x}(p, k) + R_{t_x}$, where $h_{t_x}(p, k)$ is the data-dependent part which depends on p and k, and R_{t_x} denotes a random variable that is irrelevant to the targeted intermediate value. In profiling, the attacker knows the key that is used by reference device. The attacker chooses some mutually independent base functions $g_1(p, k), ..., g_u(p, k)$. $h_{t_x}(p, k)$ can be approximated by a linear combination of these mutually independent base functions: $h_{t_x}(p, k) = \sum_{j=1}^{u} a_j g_j(p, k)$. The choice of base functions relies on the targeted intermediate value. In order to

obtain the coefficient of each base function, the attacker needs to solve the following linear equation

$$
\begin{bmatrix} I_{t_x}(p_1,k) \\ I_{t_x}(p_2,k) \\ .... \\ I_{t_x}(p_{N_1},k) \end{bmatrix} = \begin{bmatrix} g_1(p_1,k) & g_2(p_1,k) & ... & g_u(p_1,k) \\ g_1(p_2,k) & g_2(p_2,k) & ... & g_u(p_2,k) \\ ... & ... & ... & ... \\ g_1(p_{N_1},k) & g_2(p_{N_1},k) & ... & g_u(p_{N_1},k) \end{bmatrix} \begin{bmatrix} a_1 \\ a_2 \\ ... \\ a_u \end{bmatrix}
$$

The attacker needs N_1 power traces in this process. Then, the attacker uses N_2 power traces to calculate the noise at time $t_x, x\epsilon[1,l]$: $R_{j,t_x} = I_{t_x}(p_j,k) - h_{t_x}(p_j,k), j\epsilon[1,N_2]$. Finally, he builds covariance matrix for noises in different times: $C = Cov(R_{t_x}, R_{t_y})_{l \times l}$. In this way, the attacker characterizes signals and noises in different times.

2.2   Key-Recovery

In key-recovery, both TA and SMPA use the Maximum Likelihood Rule(MLR) to distinguish the correct key. If the key hypothesis is correct, noises contained in the attacked power traces match those of the characterized ones best. The attacker uses N_3 power traces to distinguish the key hypothesis $k_i \epsilon [1, ..., K]$ that maximizes $\alpha(I_1, ..., I_{N_3}, k_i) = \prod_{j=1}^{N_3} Prob_{i,j}(R_j)$. For TA,

$Prob_{i,j}(R_j) = \dfrac{exp(-\frac{1}{2}(I_j - m_i)^T C_i^{-1}(I_j - m_i))}{\sqrt{(2\pi)^l \cdot det(C_i)}}$; for SMPA, $Prob_{i,j}(R_j) = \dfrac{exp(-\frac{1}{2}(I_j - h(p_j,k_i))^T C^{-1}(I_j - h(p_j,k_i)))}{\sqrt{(2\pi)^l \cdot det(C)}}$.

3   Covariance Analysis Based Characterization Method

As two typical *Profiled Power Analysis Attacks*, both TA and SMPA need a large number of power traces in profiling to accurately characterize signals and noises. For TA, in order to accurately characterize signals, the attacker needs to calculate the mean of a large number of power traces to avoid the influence of noises; for SMPA, when building the linear equation, the attacker ignores the influence of noises, which will affect the characterization accuracy. Therefore, SMPA also needs a large number of power traces to reach a relatively accurate characterization. In order to characterize signals and noises more accurately with less power traces, and to improve the key-recovery efficiency, we propose *Covariance Analysis based Characterization Method* (CACM).

3.1   Main Idea

In [Hoo2010], Hoogvorst presented Variance Power Analysis (VPA). The idea of this method is that, if an attacker can accurately characterize noise at single time, the variance of noise is small; otherwise, the variance of noise is big. The attacker can use different key hypothesis to characterize noise at single time.

The key hypothesis that makes the variance of noise the smallest is the correct key.

In profiled attacks, signals and noises in different times have to be characterized. Because noises in different times follow multivariate normal distribution, noise at single time follow gaussian distribution. If the key hypothesis is correct, noises in different times are characterized accurately, and the covariance of noises in different times is small; otherwise, noises in different times are characterized incorrectly, and the covariance of noises in different times is big. The attacker can characterize signals and noises by analyzing the covariance of noises in different times. Based on these considerations, we present CACM. In this method, an attacker first has to choose the targeted intermediate value; then, he has to choose the targeted times. The targeted times correspond to the times when the targeted intermediate value is being processed. The attacker can characterize signals and noises in these targeted times.

3.2   Characterization Procedure

The steps to implement CACM are as follows:

1) Randomly generate plaintext $p_i, i\epsilon[1, M]$ and feed these messages into the reference device. Measure the power consumption of the device when it is operating and obtain M power traces;

2) Choose the targeted intermediate value (usually, the S-Box output byte is chosen as the attacked intermediate value). The targeted times corresponding to the targeted intermediate value being processed are $t_1, ..., t_l$. When the input plaintext is p_i, the power consumption at time t_x, $x\epsilon[1, l]$ is I_{i,t_x};

3) For signal at time t_x, define some mutually independent events $e_{j,t_x}, j\epsilon[0, N]$. N is the number of events. When event e_{j,t_x} occurs, its value is 1; otherwise, its value is 0. When event e_{j,t_x} occurs, its power consumption is P_{j,t_x}. Signal at time t_x is the sum of mutually independent events' power consumption. E_{t_x} is a $M \times N$ matrix, which represents the value of N events when randomly input M plaintext;

4) Denote noise at time t_x as $J_{i,t_x}, i\epsilon[1, M]$. Covariance of noises in two times t_x and t_y, $x, y\epsilon[1, l]$ and $x \leq y$, can be computed as follows:

$$Cov(J_{t_x}, J_{t_y}) = E[J_{t_x}][J_{t_y}] - E[J_{t_x}]E[J_{t_y}] \tag{1}$$

in which

$$J_{i,t_x} = I_{i,t_x} - \sum_{j=1}^{N}(P_{j,t_x} \cdot e_{j,t_x}), i\epsilon[1, M] \tag{2}$$

(1) can also be written as:

$$Cov(J_{t_x}, J_{t_y}) = P_{t_x} X P_{t_y}^T + P_{t_x} Y_1 + P_{t_y} Y_2 + z \tag{3}$$

in which

$$X = \frac{1}{M}\sum_{i=1}^{M}(E_{t_x} \cdot E_{t_y}^T) - (\frac{1}{M}\sum_{i=1}^{M}E_{t_x})(\frac{1}{M}\sum_{i=1}^{M}E_{t_y})^T \qquad (4)$$

$$Y_1 = \frac{1}{M}\sum_{i=1}^{M}E_{t_x} \cdot \frac{1}{M}\sum_{i=1}^{M}I_{i,t_y} - \frac{1}{M}\sum_{i=1}^{M}(I_{i,t_y} \cdot E_{t_x}) \qquad (5)$$

$$Y_2 = \frac{1}{M}\sum_{i=1}^{M}E_{t_y} \cdot \frac{1}{M}\sum_{i=1}^{M}I_{i,t_x} - \frac{1}{M}\sum_{i=1}^{M}(I_{i,t_x} \cdot E_{t_y}) \qquad (6)$$

We know that the correct key makes the covariance of noises in two different times the smallest. In order to obtain the minimum of (3), we can take partial derivative of (3) with respect to vector P_{t_x} and P_{t_y}. When the partial derivative equals to zero, the covariance of noises in times t_x and t_y is the smallest.

Taking partial derivative of (3) with respect to vector P_{t_x} and P_{t_y}, and make the partial derivative equals to zero, we obtain equations (7) and (8) as follows:

$$XP_{t_y}^T + Y_1 = 0 \qquad (7)$$

$$XP_{t_x}^T + Y_2 = 0 \qquad (8)$$

Using equations (7) and (8), we obtain P_{t_x} and P_{t_y}, that is $P_{t_x}^T = -X^{-1} \cdot Y_2$, $P_{t_y}^T = -X^{-1} \cdot Y_1$. Using (4)-(6), we can obtain the value of X, Y_1 and Y_2. Therefore, the value of vector P_{t_x} and P_{t_y} can be obtained. With P_{t_x} and P_{t_y}, we can accurately characterize signals in times t_x and t_y.

For signals in different times $t_1, ..., t_l$, we repeat step 4 to accurately characterize signals in these times;

5) Using equation (2) to compute noise $J_{t_x}, x\epsilon[1, l]$. Noise in each time has to be adjusted so that they follow multivariate normal distribution. For noise at time t_x, we compute its mean m_{t_x} and its standard deviation σ_{t_x}; then, for noise $J_{i,t_x}, i\epsilon[1, M]$, we adjust its value: $J_{i,t_x} = \frac{J_{i,t_x} - m_{t_x}}{\sigma_{t_x}}$. When we have adjusted noise at each time, we can compute the covariance matrix $C = Cov(J_{t_i}, J_{t_j})_{l \times l}$ for noises in different times.

3.3   Summary

In SMPA, in order to characterize signal at single time, an attacker has to build and solve an linear equation. Under the fact that the attacker ignores the influence of noise when building linear equation, signal cannot be accurately characterized. Compared with SMPA, CACM doesn't ignore the influence of noises when characterizing signals. Therefore, CACM characterizes signals and noises more accurately than SMPA.

Compared with TA, CACM has to build only one covariance matrix for noises in different times. However, in TA, the attacker has to build a template for each key hypothesis, which means the attacker has to build many covariance matrices. Therefore, CACM has more samples to characterize noises in different times more accurately.

We will experimentally demonstrate the above mentioned advantages of CACM over SMPA and TA. Meanwhile, we note that the method used to select the targeted times is not restricted, and the attacker can use known methods to select the targeted times.

4 Experiments

In this section, we will experimentally demonstrate the advantage of CACM over SMPA and TA, in terms of the characterization accuracy for signals and noises in different times, the minimum number of power traces needed in profiling, influence of correct key distinguish level and the number of power traces needed in key-recovery.

We attack AES software implementation on an 8-bit STC89C52 microcontroller. The clock frequency of the microcontroller is set 22.1184MHZ. An Agilent DSA90404A digital oscilloscope is used to sample power traces. The sampling rate is 100Ms/s. Differential probe of digital oscilloscope is connected at two ends of a 20Ω resistor in series with the GND line of the microcontroller. We collect power traces correspond to the 1st round implementation of AES, and use these power traces to accomplish our experimental analysis. The mean of 100 power traces correspond to the same plaintext and key is calculated to reduce the influence of noise. We use CPA to find the the targeted times.

First, we show that CACM can also be an attack method, and we compare CACM with VPA in terms of the characterization accuracy for signal at single time; second, we analyze the characterization accuracy of TA, SMPA and CACM for signal at single time when there is limited number of power traces; third, for each method, we analyze the minimum number of power traces needed in profiling to accurately characterize signals and noises; finally, we analyze the correct key distinguish level and the number of power traces needed in key-recovery influenced by the characterization accuracy in profiling using different methods.

Experimental results show that, compared with TA and SMPA, CACM needs less power traces in profiling to more accurately characterize signals and noises in different times, which helps to improve the correct key distinguish level and reduce the number of power traces needed in key-recovery to recover the correct key.

4.1 Comparison with Variance Power Analysis

VPA is an attack method, with which an attacker recovers the correct key by analyzing the variance of noise at single time. The attacker can characterize signal in the process of analyzing noise. CACM is a characterization method, with which an attacker can characterize signals and noises by analyzing the covariance of noises in different times. The common factor is, in both methods attackers assume signal at single time is the sum of some mutually independent events' power consumption. Attackers can characterize signal at single time by analyzing noise.

In fact, CACM can also be used to construct practical attacks. An attacker can recover the correct key by comparing the covariance of noises computed using different key hypothesis. The key hypothesis that makes the covariance of noises the smallest is the correct key. In order to demonstrate this fact, we use 100 power traces to recover the correct key. The targeted two times correspond to the times that the 1st S-Box output byte of AES 1st round being processed. We choose an event for each bit of S-Box output byte, and each event's value equals to the corresponding bit's value. Experimental result is shown in Figure 1.

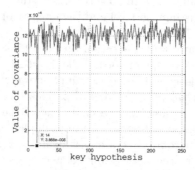

Fig. 1. Covariance of Noises in Two Different Times

Figure 1 clearly shows that the correct key 14 makes the covariance of noises in two times the smallest. We also experimentally analyzed the characterization ability of VPA and CACM for signal. The experimental result told us that with the same number of power traces and the same event choice, CACM and VPA characterize signal equally accurate. The essential difference lies in the different characterization object. VPA can only characterize signal in single time, while CACM is able to characterize signals and noises in different times.

4.2 Evaluation of Characterization Accuracy

In this subsection, we analyze the characterization accuracy of TA, SMPA and CACM for signal at single time when there is limited number of power traces. 500 power traces are used to characterize signal, and another 100 power traces are used to evaluate the characterization accuracy of each method. All traces correspond to the same key and random plaintext. We compute the correlation coefficient between the characterized signal and measured power consumption. For CACM, the same event as in 4.1 is chosen; for SMPA, we choose a base function for each bit of S-Box output described in 4.1, and the value of each base function is the corresponding bit's value. The targeted time correspond to the time that the 1st S-Box output byte of AES 1st round being processed.

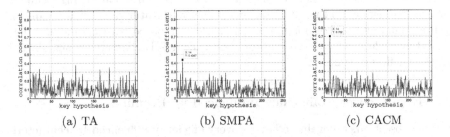

Fig. 2. Characterization Accuracy of Different Methods

Figure 2 shows that, using 500 power traces, TA cannot accurately characterize signal; using SMPA, when the key hypothesis is correct, the correlation between the measured power consumption and the characterized signal is 0.4367; finally, using CACM to characterize signal, when the key hypothesis is correct, the correlation reaches to 0.7020. Therefore, when the number of power traces is limited, CACM has the best characterization ability for signal.

4.3   The Minimum Number of Power Traces Needed in Profiling

In this subsection, we analyze the minimum number of power traces needed by each method to accurately characterize signals and noises. 3000 power traces are used to characterize signal at single time. Another 100 traces are used to evaluate the characterization accuracy. All traces correspond to the same key and random plaintext. For CACM and SMPA, we choose the same event as in subsection 4.2. The targeted time correspond to the 1st S-Box output byte of AES 1st round being processed.

Figure 3 shows that, about 2500 power traces are needed by TA to accurately characterize signal. The correlation between the measured power consumption

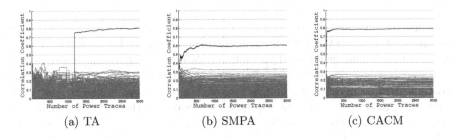

Fig. 3. Minimum Number of Power Traces Needed in Profiling

and the characterized signal is 0.8. About 600 power traces are needed by SMPA to characterize signal. The correlation between the measured power consumption and the characterized signal is 0.6. About 200 power traces are needed by CACM to make the correlation between the characterized signal and the measured power consumption reaches to 0.8. We know that power consumption in single time is composed of signal and noise, an accurate characterization for signal means an accurate characterization for noise. Therefore, compared with SMPA and TA, CACM needs less power traces in profiling to accurately characterize signals and noises in different times.

4.4   Key-Recovery Efficiency Influenced by Profiling

In this subsection, we evaluate key-recovery efficiency influenced by characterization accuracy in profiling. First, we evaluate the correct key distinguish level influenced by characterization accuracy. 300 power traces are used in profiling and 20 power traces are used in key-recovery. All power traces correspond to the same key and random plaintext. In profiling, noises at four times correspond to the 1st S-Box output byte of AES 1st round being processed are characterized. For SMPA and CACM, we choose the same event as in 4.3. In key-recovery, we use MLR to recover the correct key. To avoid the exponentiation, we can compute the inverse of the absolute value of the logarithm of the probability. From [MOP2007], we know that the correct key makes this statistical value the highest. Experimental results are shown in Figure 4.

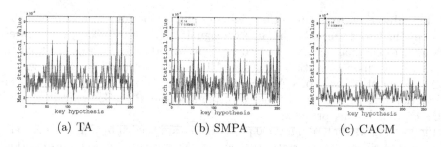

(a) TA   (b) SMPA   (c) CACM

Fig. 4. Match statistical values in Key-Recovery

We use the distinguish level which was proposed in [MMPS2009] as the metric to evaluate the correct key distinguish level influenced by the characterization accuracy of different methods. This metric is defined as follows: $D = \frac{V_{max} - V_{sec}}{V_{max}}$. $V = \{V_1, ..., V_{|K|}\}$ denotes a set of statistical value which is calculated using certain statistic tool and different key hypothesis. V_{max} denotes the largest value in the set, and V_{sec} denotes the second largest value in the set. The calculated correct key distinguish level is shown in Table 2.

Table 2 shows that, TA cannot accurately characterize noises with 300 power traces. Therefore, TA cannot recover the correct key with 20 power traces. With

Table 2. Distinguish Level of Correct key Using Different Methods

Methods	TA	SMPA	CACM
Distinguish Level	-	0.1030	0.4771

300 power traces, SMPA can characterize noises, but not much accurate. Therefore, although SMPA can recover the correct key with 20 power traces, the correct key distinguish level is merely 0.1030. Finally, CACM can accurately characterize signals and noises with 300 power traces, and the distinguish level of the correct key reaches to 0.4771, which increases about 37%. Therefore, when the number of power traces is limited, CACM characterizes noises more accurately than SMPA and TA, which induces higher distinguish level of the correct key.

Next, we evaluate the number of power traces needed in key-recovery to recover the correct key. In profiling, CACM uses 100 power traces; SMPA uses 1200 power traces; and TA uses 2500 power traces. In key-recovery, different methods use different number of power traces. We use success rate proposed in [SMY2009] as the metric to evaluate the number of power traces needed in key-recovery to successfully recover the correct key. For each number of power traces, we do 500 tests and calculate the success rate. Experimental results are shown in Figure 5.

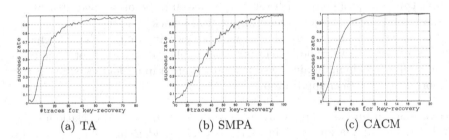

(a) TA	(b) SMPA	(c) CACM

Fig. 5. Success Rates of Distinguishing the Correct Key

We evaluate the number of power traces needed in key-recovery when the success rate exceeds 90%. We note that, because CACM characterizes signals and noises in different times more accurately, an attacker with CACM needs less power traces in key-recovery than TA and SMPA to reach a success rate 90% of recovering the correct key. The evaluation results are shown in Table 3.

Table 3. Number of Traces Needed in Profiling

Methods	TA	SMPA	CACM
Number of Traces	40	70	7

Table 3 shows that, with TA, when using 2500 power traces in profiling to characterize signals and noises, about 40 power traces are needed in key-recovery to reach a success rate 90%. With SMPA, when using 1200 power traces in profiling to characterize signals and noises, about 70 power traces are needed in key-recovery to reach a success rate 90%. Finally, with CACM, when using 100 power traces in profiling to characterize signals and noises, only about 7 power traces are needed in key-recovery to reach a success rate 90%. Experimental results demonstrated that, because CACM characterizes noises more accurately, compared with TA and SMPA, attacker with CACM can use less power traces in key-recovery to successfully recover the correct key, which makes *Profiled Power Analysis Attacks* more powerful in practical application.

5   Conclusions

In this paper, we presented *Covariance Analysis based Characterization Method* (CACM for short) to deal with the problem of more accurate characterization with less power traces. We have experimentally demonstrated the advantage of CACM over TA and SMPA in terms of the characterization accuracy for signals and noises, and the number of power traces needed in profiling and key-recovery.

We argue that CACM brings two improvements to known profiled attacks. First, CACM can characterize signals and noises more accurately, which helps improve the key-recovery efficiency of profiled attacks; second, CACM needs less power traces in profiling, which relaxes the assumption of profiled attacks and makes profiled attacks more useful in practical attack scenario.

On the other hand, evaluators can more efficiently evaluate the side channel leakage of cryptographic devices in practical application, and adopt some countermeasures to help maintain the security of cryptographic devices, which is very important in practical application.

Acknowledgements. This work is supported in part by the National Natural Science Foundation of China (No.61073178) and Beijing Natural Science Foundation (No.4112064). We acknowledge their supports. We also thank the anonymous reviewers of ICICS2011 and ICISC2011 for their insightful comments. Their comments help to improve this paper.

References

[BCO2004]   Brier, E., Clavier, C., Olivier, F.: Correlation Power Analysis with a Leakage Model. In: Joye, M., Quisquater, J.-J. (eds.) CHES 2004. LNCS, vol. 3156, pp. 16–29. Springer, Heidelberg (2004)

[CRR2002]   Chari, S., Rao, J.R., Rohatgi, P.: Template Attacks. In: Kaliski Jr., B.S., Koç, Ç.K., Paar, C. (eds.) CHES 2002. LNCS, vol. 2523, pp. 13–28. Springer, Heidelberg (2003)

[GBTP2008] Gierlichs, B., Batina, L., Tuyls, P., Preneel, B.: Mutual Information Analysis. In: Oswald, E., Rohatgi, P. (eds.) CHES 2008. LNCS, vol. 5154, pp. 426–442. Springer, Heidelberg (2008)

[GLP2006]   Gierlichs, B., Lemke-Rust, K., Paar, C.: Template vs. Stochastic Methods - A Performance Analysis for Side Chennel Cryptanalysis. In: Goubin, L., Matsui, M. (eds.) CHES 2006. LNCS, vol. 4249, pp. 15–29. Springer, Heidelberg (2006)

[Hoo2010]   Hoogvorst, P.: The Variance Power Analysis. In: COSADE (2010)

[KJJ1999]   Kocher, P., Jaffe, J., Jun, B.: Differential Power Analysis. In: Wiener, M. (ed.) CRYPTO 1999. LNCS, vol. 1666, pp. 388–397. Springer, Heidelberg (1999)

[LP2007]   Lemke-Rust, K., Paar, C.: Analyzing Side Channel Leakage of Masked Implementations with Stochastic Methods. In: Biskup, J., López, J. (eds.) ESORICS 2007. LNCS, vol. 4734, pp. 454–468. Springer, Heidelberg (2007)

[MMPS2009]   Moradi, A., Mousavi, N., Paar, C., Salmasizadeh, M.: A Comparative Study of Mutual Information Analysis under a Gaussian Assumption. In: Youm, H.Y., Yung, M. (eds.) WISA 2009. LNCS, vol. 5932, pp. 193–205. Springer, Heidelberg (2009)

[MOP2007]   Mangard, S., Oswald, E., Popp, T.: Power Analysis Attacks. Springer, Heidelberg (2007)

[OM2007]   Oswald, E., Mangard, S.: Template Attacks on Masking—Resistance Is Futile. In: Abe, M. (ed.) CT-RSA 2007. LNCS, vol. 4377, pp. 243–256. Springer, Heidelberg (2006)

[Pro2005]   Prouff, E.: DPA Attacks and S-Boxes. In: Gilbert, H., Handschuh, H. (eds.) FSE 2005. LNCS, vol. 3557, pp. 424–441. Springer, Heidelberg (2005)

[SKS2009]   Standaert, F.-X., Koeune, F., Schindler, W.: How to Compare Profiled Side-Channel Attacks? In: Abdalla, M., Pointcheval, D., Fouque, P.-A., Vergnaud, D. (eds.) ACNS 2009. LNCS, vol. 5536, pp. 485–498. Springer, Heidelberg (2009)

[SLP2005]   Schindler, W., Lemke-Rust, K., Paar, C.: A Stochastic Model for Differential Side Channel Cryptanalysis. In: Rao, J.R., Sunar, B. (eds.) CHES 2005. LNCS, vol. 3659, pp. 30–46. Springer, Heidelberg (2005)

[SMY2009]   Standaert, F.-X., Malkin, T.G., Yung, M.: A Unified Framework for the Analysis of Side-Channel Key Recovery Attacks. In: Joux, A. (ed.) EUROCRYPT 2009. LNCS, vol. 5479, pp. 443–461. Springer, Heidelberg (2009)

Correcting Errors in Private Keys Obtained from Cold Boot Attacks

Hyung Tae Lee[1], HongTae Kim[1], Yoo-Jin Baek[2], and Jung Hee Cheon[1]

[1] ISaC & Dept. of Mathematical Sciences, Seoul National University,
599 Gwanangno, Gwangak-gu, Seoul 151-747, Korea
{htsm1138,kafa46,jhcheon}@snu.ac.kr
[2] Samsung Electronics, Nongseo-Dong, Giheung-Gu, Yongin-City,
Gyeonggi-Do 446-711, Korea
yoojin.baek@samsung.com

Abstract. Based on the cold boot attack technique, this paper proposes a new algorithm to obtain the private key of the discrete logarithm (DL) based cryptosystems and the standard RSA from its erroneous value. The proposed algorithm achieves almost the square root complexity of search space size. More precisely, the private key of the DL based system with 160-bit key size can be recovered in $2^{43.24}$ exponentiations while the complexity of the exhaustive search is $2^{71.95}$ exponentiations if the error rate is given by 10%.

In case of the standard RSA with 1024-bit key size, our algorithm can recover the private key with $2^{49.08}$ exponentiations if the error rate is given by 1%. Compared with the efficiency of some algorithms [7,6] to recover the private key in RSA using Chinese Remainder Theorem, the recoverable error rate of our algorithm is quite small. However, our algorithm requires only partial information of the private key d while other algorithms require additional information such as partial information of factors of the RSA modulus N.

The proposed algorithm can also be used for breaking countermeasure of differential power analysis attack. In the standard RSA, one uses the randomized exponent $\tilde{d} = d + r \cdot \phi(N)$ instead of the decryption exponent d with the random value r. When the size of a random value r is 26-bit, it can be shown that the randomized exponent can be recovered with $2^{49.30}$ exponentiations if the error rate is 1%. Finally, we also consider the breaking countermeasure that splits the decryption exponent d into d_1 and d_2 of same size.

Keywords: Cold Boot Attack, Discrete Logarithm, RSA, Side Channel Attack.

1 Introduction

The cold boot attack [8] is a very sophisticated side channel attack and is based on the phenomenon that even though the volatile memory is cut off from its power source, the memory retains its data (some parts of which are erased) for

H. Kim (Ed): ICISC 2011, LNCS 7259, pp. 74–87, 2012.
© Springer-Verlag Berlin Heidelberg 2012

several seconds. More surprisingly, if the temperature is very low, the data of memory retains for several hours. Using this phenomenon, Halderman et al. [8] could read erroneous data from the (power-offed) memory and suggested some algorithms to recover the real data from the erroneous data in block cipher cryptosystems.

Later, based on the idea of cold boot attacks, Heninger and Shacham [7] proposed the algorithm to recover the private key from its erroneous value in RSA cryptosystems using Chinese Remainder Theorem (CRT)[1]. That is, even though the private key is known to the attacker with some bits decayed in a unidirectional way, their algorithm could recover the whole private key in a reasonable time if the error rate is assumed to be bounded above by some threshold values. Henecker et al. [6] improved the algorithm to deal with the case that the errors can occur in a bidirectional way, that is, each bit of the private key can be flipped with a certain probability. Both algorithms make use of the specific structure of RSA-CRT, thus they are not directly applied to the DL based cryptosystems or the standard RSA.

1.1   Our Results

In this paper, we firstly consider cold boot attacks on the DL based cryptosystems and the standard RSA, and propose an algorithm to recover the private key from its bidirectional erroneous value in these cryptosystems. When the order of the base group of DL problem (DLP) is q, there is a generic algorithm to solve a DLP with the square root complexity if the range of exponent is $\mathbf{Z}_q$ [11]. However, if the range of exponent in DLP is constrained to a random subset H of $\mathbf{Z}_q$ and the size of H is less than or equal to $\sqrt{q}$, the complexity of a generic algorithm for this DLP is almost same with the size of H [10]. We provide a generic algorithm to recover the private key from bidirectional erroneous key using the splitting system with additional techniques. The complexity of our algorithm achieves almost the square root of search space. For example, when the error rate is 10% in 160-bit DL based cryptosystems, the private key can be recovered in $2^{71.95}$ operations[2] by exhaustive search, however, our algorithm takes only $2^{43.24}$ operations. In case of unidirectional errors, we observed that one can recover the private key using the method in [4] and the complexity is reduced to $2^{36.38}$ operations if the error rate is less than 10%.

The proposed algorithm can be applied to the standard RSA as well as the DL based cryptosystems. While there are two algorithms [7,6] to recover the private key in RSA-CRT, there have been no research to recover the private key from the cold boot attack in the standard RSA. Our algorithm can be applied to the standard RSA and recover the decryption exponent d in time $2^{49.08}$ operations if the error rate is 1% in 1024-bit RSA. Compared with algorithms for RSA-CRT, our algorithm looks inefficient and the recoverable error rate is quite small. In the standard RSA, however, since only a faulty value of the decryption exponent

---

[1] Throughout this paper, these cryptosystems are denoted by RSA-CRT.

[2] The unit of operations for complexity is an exponentiation or a scalar multiplication.

d is given without any information of prime factors of the modulus, this result is reasonable. Moreover, it is practical scenario to obtain an erroneous value whose error rate is less than 1% [8].

Furthermore, we investigate how the previous countermeasures [9,2,3] against side channel attacks resist our algorithm. Consider the use of the blind exponent presented in [9,2], that is, one uses $\tilde{d} = d + r \cdot |G|$ instead of the private key d where r is a random integer and $|G|$ is the order of the underlying group in the cryptosystem. A larger blind factor gives better security, however, it causes the efficiency problem. We estimate the complexity for various blind factors in order to check the security of this countermeasure. For example, if 26-bit random number r is used, one can recover the private key in $2^{49.30}$ operations when the error rate is 1%. We also consider the security of countermeasure [3] that splits the private key d into d_1 and d_2 of same size. When the error rate is 1%, the private key can be recovered in $2^{79.16}$ operations in the standard RSA with 1024-bit key size.

1.2   Related Works

There are several researches [7,6] to recover the private key in RSA-CRT. However, these techniques can not be applied to the DL based cryptosystems since some additional information such as some bits of factors of the modulus are also required in these algorithms. In the standard RSA, Fouque et al. [4] proposed the algorithm to recover the private key when few bits of the private key are missing. They suggested the use of the splitting system [12], which was suggested to solve the low Hamming weight DL problem. According to their analysis, one can recover the private key when the number of missing bits in 1024-bit RSA is less than or equal to 145 (14.16%) under 2^{80} operations. We can apply their method to get the private key under the assumption that unidirectional errors are occurred in cold boot attacks.

2   Splitting System

The Hamming weight of an integer x (denoted by $\mathrm{wt}(x)$) is the number of 1's in its binary representation. Let G be a cyclic group of order q with a generator g. Then the discrete logarithm problem of Hamming weight t (DLP of weight t) is to find $\log_g y$ whenever $g, y \in G$ are given with $\mathrm{wt}(\log_g y) = t$.

Let x be an n-bit integer with $g^x = y$ and $\mathrm{wt}(x) = t$. In order to solve this DLP of weight t, Heiman and Odlyzko independently proposed an algorithm [5]. Their idea is to consider the exponent x as $x_1 + x_2$ with $\mathrm{wt}(x_1) = t_1$, $\mathrm{wt}(x_2) = t_2$ and $t = t_1 + t_2$. Thereafter, they compute g^{x_1}, yg^{-x_2} for all cases and compare the values g^{x_1} and yg^{-x_2}. When we find x_1, x_2 satisfying the equality $g^{x_1} = yg^{-x_2}$, $x_1 + x_2$ becomes the correct value. In this case, the complexity is $\binom{n}{t_1} + \binom{n}{t_2}$. Coppersmith improved Heiman and Odlyzko algorithm through the splitting system. The splitting system is defined as follows:

Definition 1. *Let n and t be even integers with $0 < t < n$. An (n, t)-splitting system is a pair (X, B) that satisfies the following properties:*
1. $|X| = n$ and B is a set of $\frac{n}{2}$-subsets of X called blocks.
2. For every $Y \subseteq X$ such that $|Y| = t$, there exists a block $B_i \in B$ such that $|B_i \cap Y| = \frac{t}{2}$.

An (n, t)-splitting system with N blocks is denoted by $(N; n, t)$-splitting system. The following lemma guarantees the existence of an $(\frac{n}{2}; n, t)$-splitting system.

Lemma 1. *([12]) For all even integers n and t with $0 < t < n$, there exists an $(\frac{n}{2}; n, t)$-splitting system for $\mathbb{Z}_n$.*

Stinson generalized Definition 1 to arbitrary integers n, t with $0 < t < n$ and extended Lemma 1 to the cases where one or both of n and t are odd [12]. We can solve the DLP of weight t in $n\binom{\frac{n}{2}}{\frac{t}{2}}$ from the lemma.

Since the DL based cryptosystems and RSA cryptosytems in this paper have the key size of 160-bit and 1024-bit, respectively, we assume the parameter n is even. To simplify the description of the splitting system, we will use the following notation. Given a and b with $0 \le a, b < \ell$ and $a \ne b$, we define

$$[a, b)_\ell = \begin{cases} \{a, a+1, \cdots, b-1\} & \text{if } a < b, \\ [a, n)_\ell \cup [0, b)_\ell & \text{if } b < a. \end{cases}$$

Additionally, we denote by $A[j]$ the value of the least significant $(j+1)$-th bit of A, and by $\lfloor a \rfloor$ and $\lceil a \rceil$ the largest integer $b \le a$ and the smallest integer $b \ge a$, respectively.

3   Correcting Errors in Private Key of DL Based Cryptosystems

The public key pk and private key sk of DL based schemes, such as ElGamal encryption scheme and Schnorr signature scheme, have the form $(\mathsf{pk}, \mathsf{sk}) = (g^x, x)$ when g is the generator of the base group G and x is an n-bit positive integer less than the group order q. We assume that through cold boot attacks an adversary can obtain the erroneous private key x', some of whose bits are same as that of the private key x.

We introduce an algorithm to recover the private key from an erroneous private key. We also extend our algorithm to the case that the private key is protected using some countermeasures against previous side channel attacks.

3.1   Basic Algorithm

We first deal with the case that the private key is stored in the memory of a device without any transformation. Assume that the erroneous private key x' is given. Then we can recover the correct private key x using the splitting system.

Bidirectional Errors. Assume each bit of the private key can be flipped independently with error rate δ [3]. First one computes the upper bound of error bits, $\lfloor n\delta \rfloor$, in the given erroneous private key x' and executes the following phase with growing the estimated number of error bits from 0 to $\lfloor n\delta \rfloor$ as in Algorithm 1.

Algorithm 1. Recovering private key from the erroneous key x'

INPUT: (g, y, x', n, δ)
OUTPUT: x such that $y = g^x$
for $t = 0$ to $\lfloor n\delta \rfloor$ **do**
  for $i = 0$ to $\lfloor n/2 \rfloor - 1$ **do**
    set $B_{1,i}$ and $B_{2,i}$ to $[i, i + n/2)_n$ and $[i + n/2, i)_n$, respectively
    set $U_{1,j}$ and $U_{2,j}$
    while possible $T_{1,\ell}$'s **do**
      set $\overline{U}_{1,\ell}$
      compute $g^{\overline{U}_{1,\ell}}$ and store $(\overline{U}_{1,\ell}, g^{\overline{U}_{1,\ell}})$ in the table **Tab**
    end while
    while possible $T_{2,m}$'s **do**
      set $\overline{U}_{2,m}$
      compute $yg^{-\overline{U}_{2,m}}$
      find $yg^{-\overline{U}_{2,m}}$ among $g^{\overline{U}_{1,m}}$'s in **Tab**
      if collision $g^{\overline{U}_{1,\ell^*}} = yg^{-\overline{U}_{2,m^*}}$ occurs **then**
        return $\overline{U}_{1,\ell^*} + \overline{U}_{2,m^*}$
      end if
    end while
    initialize the table **Tab**
  end for
end for

Fix the number of error bits in x' by t. Consider $\frac{n}{2}$ pairs of two blocks $(B_{1,i}, B_{2,i})$ such that $B_{1,i} = [i, i + n/2)_n$ and $B_{2,i} = [i + n/2, i)_n$ for $i = 0, \cdots, \frac{n}{2} - 1$. Set $U_{k,i}$ to

$$U_{k,i}[j] = \begin{cases} x'[j] & \text{if } j \in B_{k,i}, \\ 0 & \text{otherwise} \end{cases}$$

where $k = 1, 2$ and $i = 0, \cdots, \frac{n}{2} - 1$. Consider sets $T_{1,\ell}, T_{2,m} \subset B_{k,i}$ such that $|T_{1,\ell} \cap B_{1,i}| = \lceil \frac{t}{2} \rceil$ and $|T_{2,m} \cap B_{2,i}| = \lfloor \frac{t}{2} \rfloor$ for $k = 1, 2$, $i = 0, \cdots, \frac{n}{2} - 1$, and possible ℓ's and m's. Define $\overline{U}_{k,\ell}$ so that

$$\overline{U}_{k,\ell}[j] = \begin{cases} 1 - U_{k,\ell}[j] \pmod{2} & \text{if } j \in T_{k,\ell}, \\ U_{k,\ell}[j] & \text{otherwise.} \end{cases}$$

---

[3] In this paper, it is assumed that δ is the upper bound of error rate.

From $i = 0$ to $\frac{n}{2} - 1$, one does the followings: for fixed i, one computes $g^{\overline{U}_{1,\ell}}$ for all possible $T_{1,\ell}$'s and stores $(\overline{U}_{1,\ell}, g^{\overline{U}_{1,\ell}})$'s in the table Tab. Thereafter, for all possible $T_{2,m}$'s, compute $yg^{-\overline{U}_{2,m}}$ and compare it with elements $g^{\overline{U}_{1,\ell}}$'s in the table Tab. If the collision between $g^{\overline{U}_{1,\ell}}$'s and $yg^{-\overline{U}_{2,m}}$'s occurs, output $\overline{U}_{1,\ell*} + \overline{U}_{2,m*}$ for satisfying $g^{\overline{U}_{1,\ell*}} = yg^{-\overline{U}_{2,m*}}$. If collision does not occur, then delete all data in the table Tab and repeat the above process for the next i.

The following theorem gives the correctness of Algorithm 1.

Theorem 1. *When Algorithm 1 is executed with the input (g, y, x', n, δ), then it outputs the discrete logarithm x of y if the number of error bits is less than $\lfloor n\delta \rfloor$.*

Proof. From Lemma 1, there exists j such that the numbers of errors of x' in $B_{1,j}, B_{2,j}$ are $\lceil \frac{t}{2} \rceil, \lfloor \frac{t}{2} \rfloor$, respectively. Since, for all cases that the error is less than $\lfloor n\delta \rfloor$, we observe all possible $(T_{1,\ell}, T_{2,m})$ and $(B_{1,i}, B_{2,i})$, we can find $(T_{1,\ell}, T_{2,m})$ and $(B_{1,j}, B_{2,j})$ such that the set of error bits of x' is same to $T_{1,\ell} \cup T_{2,m}$. For $(\overline{U}_{1,\ell*}, \overline{U}_{2,m*})$ corresponded to such $(T_{1,\ell*}, T_{2,m*})$ and $(B_{1,j}, B_{2,j})$, the relation $g^{\overline{U}_{1,\ell*}} = yg^{-\overline{U}_{2,m*}}$ is satisfied. Hence the output of Algorithm 1 is to be the private key x. $\qquad\square$

Toy Example. Let $G = \langle 2 \rangle$ be a cyclic subgroup of Z_{2039}^* generated by 2 of order 1019. Set the public key y to 1571 and the private key x to $1110101101_{(2)} = 941$. Assume that the erroneous private key x' is given as $1010110111_{(2)}$.

With Algorithm 1, when $t = 4$, $B_{1,2} = [2, 7)_{10}$ and $B_{2,2} = [7, 2)_{10}$, $U_{1,2}$ and $U_{2,2}$ are to be $0010110000_{(2)}$ and $1000000111_{(2)}$, respectively. Then, when $T_{1,\ell}$ is $\{2, 6\}$, $\overline{U}_{1,\ell}$ is $0110100000_{(2)}$. Hence the value

$$g^{\overline{U}_{1,\ell}} = 2^{0110100000_{(2)}} \equiv 645 \pmod{2039}$$

is stored in the table Tab with $\overline{U}_{1,\ell} = 0110100000_{(2)}$.

Then, when $T_{2,m}$ is $\{7, 9\}$, $\overline{U}_{2,m}$ is $1000001101_{(2)}$. Hence one also obtains

$$y \cdot g^{-\overline{U}_{2,m}} = 1571 \cdot 2^{-1000001101_{(2)}} \equiv 1571 \cdot 1737 \equiv 645 \pmod{2039}.$$

Therefore, Algorithm 1 outputs $\overline{U}_{1,\ell} + \overline{U}_{2,m} = 1110101101_{(2)} = 941$.

Complexity. The following theorem gives the complexity of Algorithm 1.

Theorem 2. *Let n be the private key size and δ be the upper bound of error rate. Then Algorithm 1 is required less than $\sum_{t=0}^{\lfloor n\delta \rfloor} n \binom{n/2}{\lceil t/2 \rceil}$ complexity and $\binom{n/2}{\lceil (\lfloor n\delta \rfloor/2)\rceil}$ storage to recover the private key x.*

Proof. In Algorithm 1, there exist $\frac{n}{2}$ pairs $(B_{1,i}, B_{2,i})$ of blocks. For each pair, there are $\binom{n/2}{\lceil t/2 \rceil}$ candidates of $T_{1,\ell}$ and $\binom{n/2}{\lfloor t/2 \rfloor}$ candidates of $T_{2,m}$ and hence one computes $\binom{n/2}{\lceil t/2 \rceil} g^{\overline{U}_{1,\ell}}$'s and $\binom{n/2}{\lfloor t/2 \rfloor} yg^{-\overline{U}_{2,m}}$'s for each pair. Therefore, the complexity of Algorithm 1 is

$$(\text{Complexity of Algorithm 1}) = \sum_{t=0}^{\lfloor n\delta \rfloor} \frac{n}{2}\left(\binom{n/2}{\lceil t/2 \rceil} + \binom{n/2}{\lfloor t/2 \rfloor}\right)$$

$$\leq \sum_{t=0}^{\lfloor n\delta \rfloor} n\binom{n/2}{\lceil t/2 \rceil}.$$

All $g^{\overline{U}_{1,\ell}}$'s corresponded to $T_{1,\ell}$'s are stored for all $T_{1,\ell}$'s and the data stored in the table Tab are deleted if a pair $(B_{1,i}, B_{2,i})$ is changed. And the maximum number of generated $g^{\overline{U}_{1,\ell}}$'s in each pair is $\binom{n/2}{\lceil (\lfloor n\delta \rfloor/2) \rceil}$ when t is $\lfloor n\delta \rfloor$. Hence the storage for $\binom{n/2}{\lceil (\lfloor n\delta \rfloor/2) \rceil}$ elements is required.   □

Remark 1 (Unidirectional Errors). In [8], Halderman et al. observe that errors are overwhelmingly unidirectional, that is, either $0 \to 1$ or $1 \to 0$. In general, since the private key x is chosen uniformly at random, we can expect that the numbers of 0's and 1's are approximately equal. Hence we can determine the direction of errors by comparing the number of 0's and 1's in the erroneous key x'. If the number of 1's is larger than that of 0's, then the direction of errors might be $0 \to 1$ and all 0's in the erroneous key are correct in this case. Then we obtain the private key whose some bits are missing and we can recover correct private key x applying the algorithm in [4]. When n_1 bits of the private key are missing and the error rate δ is given, one recovers the real key within $\sum_{t=0}^{\lfloor n\delta \rfloor} n_1 \binom{n_1/2}{\lceil t/2 \rceil}$ complexity and $\binom{n_1/2}{\lceil (\lfloor n\delta \rfloor/2) \rceil}$ storage using the algorithm in [4].

Table 1. Complexity of exhaustive search, Algorithm 1 and unidirectional case ($n = 160$)

upper bound of error rate	complexity		
	exhaustive search	Algorithm 1	unidirection
0.03	$2^{24.69}$	$2^{19.98}$	$2^{17.21}$
0.05	$2^{43.10}$	$2^{28.99}$	$2^{24.65}$
0.10	$2^{71.95}$	$2^{43.24}$	$2^{36.38}$

In Table 1, we provide the complexity of Algorithm 1 for various upper bounds of error rate when the private key size n is 160. Also we give more precise complexity of algorithm in [4] when an unidirectional erroneous key is given with $n_1 = \frac{n}{2} + \lfloor n\delta \rfloor$ and compare them with exhaustive search which is the best algorithm before our method. In case of exhaustive search, one tries to compute for all possible cases of an erroneous key and the complexity is $\sum_{t=0}^{\lfloor n\delta \rfloor} \binom{n}{t}$ when the error rate is δ.

According to our analysis, when the error rate is 10%, the exhaustive search can recover the private key in $2^{71.59}$ operations, but our algorithm takes only $2^{43.24}$ operations. If we assume unidirectional errors, it is further reduced to $2^{36.38}$.

Within 2^{80} security, while the exhaustive search can recover the private key with at most error rate 0.118, our algorithm can recover the private key with smaller than error rate 0.343. In other words, an erroneous private key which has 54 or less error bits can be recovered within 2^{80} complexity using Algorithm 1 in contrast to that exhaustive search can recover 18 or less error bits.

3.2   Coron and Kocher's Method [9,2]

There are many methods to randomize the private key x to endure against differential power analysis. Coron [2] and Kocher [9] independently proposed the use of $\tilde{x} = x + rq$ instead of the original private key x which is less than the group order q. Here, r is a randomly chosen integer of size n_r-bit. In this case, since the relation

$$c^{\tilde{x}} = c^{x+rq} = c^x$$

is satisfied for all $c \in G$, $\tilde{x}$ is available as the private key. However one who knows $\tilde{x}$ can also decrypt the encrypted message, it is enough to recover the correct $\tilde{x}$ from erroneous private key $\tilde{x}'$.

Assume that the erroneous private key $\tilde{x}'$ is given. Then one may utilize Algorithm 1 with the input $(g, y, x', n + n_r, \delta)$ to recover the private key $\tilde{x}$ and the complexity of Algorithm 1 in this case is

$$\sum_{t=0}^{\lfloor (n+n_r)\delta \rfloor} (n + n_r)\binom{(n + n_r)/2}{\lceil t/2 \rceil}$$

with $\binom{(n+n_r)/2}{\lceil \lfloor (n+n_r)\delta \rfloor /2 \rceil}$ storage.

Table 2. Lower bound of n_r to provide 2^{80} complexity ($n = 160$)

upper bound of error rate	0.10	0.15	0.20	0.25	0.30
lower bound of n_r	155	87	45	24	10

Table 2 shows the lower bound of n_r for various upper bounds of error rate, to be required more than 2^{80} complexity for recovering the private key using our algorithm when the private key size n is 160. From Table 2, we observe that the bit size of random number r in Coron and Kocher's method has to be larger than or equal to 155 to provide 2^{80} security when the upper bound of error rate is 0.10. In this case, since the bit size of modified private key $\tilde{x}$ is to be 315, the exponentiation using $\tilde{x}$ is roughly two times slower than that using original 160-bit private key x.

3.3 Clavier and Joye's Method [3]

To randomize the exponent x, Clavier and Joye split it into two parts $(x_1, x - x_1)$ for a randomly chosen x_1 [3]. Given an element g, g^x is performed by computing $g_1 = g^{x_1}, g_2 = g^{x_2}$ and then $g_1 \cdot g_2$, where $x_2 = x - x_1$. Suppose we obtained erroneous values x_1', x_2' of x_1, x_2 with error rate δ, respectively. Now, we provide an algorithm for recovering the private keys x_1, x_2.

Algorithm. Since error rate is given, one computes the upper bound of error bits, $\lfloor n\delta \rfloor$, in the given erroneous private keys x_1', x_2'. Fix the numbers of error bits in x_1', x_2' are t_1, t_2, respectively. We have to check the all $1 \le t_1, t_2 \le \lfloor n\delta \rfloor$ to recover the keys x_1, x_2.

Algorithm 2. Recovering private key from the erroneous keys x_1', x_2'

INPUT: $(g, y, x_1', x_2', n, \delta)$
OUTPUT: x such that $y = g^x$
for $t_1 = 0$ to $\lfloor n\delta \rfloor$ **do**
   set $\overline{x_1'}$
   while possible T_1's **do**
      compute $g^{\overline{x_1'}}$ and store $(\overline{x_1'}, g^{\overline{x_1'}})$ in the table **Tab**
   end while
end for
for $t_2 = 1$ to $\lfloor n\delta \rfloor$ **do**
   set $\overline{x_2'}$
   while possible T_2's **do**
      compute $yg^{-\overline{x_2'}}$
      find $yg^{-\overline{x_2'}}$ among $g^{\overline{x_1'}}$'s in the table **Tab**
      if collision occurs **then**
         return $\overline{x_1'} + \overline{x_2'}$
      end if
   end while
end for

Let $\overline{x_1'}, \overline{x_2'}$ be the guessing key of x_1', x_2', respectively. Consider sets $T_i \subset [0, n)_n$ such that $|T_1 \cap [0, n)_n| = t_1$ and $|T_2 \cap [0, n)_n| = t_2$ where $i = 1, 2$. Define $\overline{x_i'}$ so that

$$\overline{x_i'}[j] = \begin{cases} 1 - x_i'[j] \quad (\bmod\ 2) & \text{if } j \in T_i, \\ x_i'[j] & \text{otherwise.} \end{cases}$$

First, for all possible T_1's ($1 \leq t_1 \leq \lfloor n\delta \rfloor$), compute $g^{\overline{x_i'}}$ and store $(\overline{x_i'}, g^{\overline{x_i'}})$ in the table Tab. Then, for all possible T_2's ($1 \leq t_2 \leq \lfloor n\delta \rfloor$), compute $yg^{-\overline{x_2'}}$ and compare it with elements $g^{\overline{x_i'}}$'s in the table Tab. If the collision occurs, output $\overline{x_1'} + \overline{x_2'}$.

Complexity. The complexity of Algorithm 2 is

$$\text{(complexity of Algorithm 2)} = \sum_{t_1=0}^{\lfloor n\delta \rfloor} \binom{n}{t_1} + \sum_{t_2=0}^{\lfloor n\delta \rfloor} \binom{n}{t_2}$$

$$= \sum_{t_1=0}^{\lfloor n\delta \rfloor} 2 \binom{n}{t_1}$$

with $\sum_{t_1=0}^{\lfloor n\delta \rfloor} \binom{n}{t_1}$ storage.

In Table 3, we give the complexity of Algorithm 2 to recover the private key suggested by Clavier and Joye when the private key size n is 160. Table 3 shows that when the error rate is 5%, the private key can be recovered within $2^{44.10}$ complexity using Algorithm 2. It is further reduced to $2^{37.05}$ if errors occur unidirectional.

Table 3. Complexity of Algorithm 2 and unidirectional case ($n = 160$)

upper bound of	complexity	
error rate	Algorithm 2	unidirection
0.03	$2^{25.69}$	$2^{21.95}$
0.05	$2^{44.10}$	$2^{37.05}$
0.10	$2^{72.95}$	$2^{60.51}$

Now, we compare Coron and Kocher's suggestion with Clavier and Joye's suggestion. For fair comparison, we assume that square-and-multiply algorithm is utilized for the exponentiation and the cost for one multiplication in a group G is same as that of one squaring in G. To compute one exponentiation with an n'-bit exponent using square-and-multiply algorithm, the expected numbers of multiplications and squarings are $n'/2$ and n', respectively. Clavier and Joye's suggestion requires 480 multiplications when the private key size n is 160. This cost is same when the random number size n_r is 160 in case of Coron and Kocher's suggestion. Then the recoverable error rate within 2^{80} complexity using our algorithm is 0.096 in case of Coron and Kocher's and is 0.118 in case of Clavier and Joye's. It shows that Clavier and Joye's suggestion is more tolerable than Coron and Kocher's suggestion against our proposed algorithm.

4 Applying to RSA Cryptosystem

There have been numerous algorithms to recover the RSA decryption key from its partial information. One approach is to assume that some contiguous bits of decryption keys are known [1]. The other is to deal with less restrictive information on the private key, but it is assumed that some additional partial information on prime divisors of the modulus are given. The latter approach [7,6] has been initiated mainly from the cold boot attack. These results show that if an erroneous value of private key (p, q, d, d_p, d_q) is given in RSA-CRT, one can recover the whole key with a reasonable probability. These papers assume that the decay direction of key is unidirectional [7] or bidirectional [6].

In this section, we consider the standard RSA decryption module and propose an algorithm to recover the private key of RSA cryptosystem from an erroneous decryption key d. While the attack against the standard RSA in [1] requires contiguous $\frac{n}{4}$ least significant bits of the private key d where n is the bit size of modulus N, our attack assumes random and independent errors. Our algorithm requires a lot more information than the previous, but it does not have such a restriction.

Throughout this section, N will be the RSA modulus and the public key and the private key of the standard RSA cryptosystem will be given as e and d, respectively. The bit length of the private key will be given as n and it is assumed that the bit length of the private key d is same as that of N. Let m be the message and C be the ciphertext.

Standard RSA. Assume that the erroneous private key d' is given. To recover the correct private key d using Algorithm 1, one chooses a message m chosen at random and compute $m^e \equiv C \pmod{N}$. Then he gets a pair (C, m) that satisfies the relation $C^d \equiv m \pmod{N}$. With input (C, m, d', n, δ), one executes Algorithm 1 and then obtains d as the output. In this case, the complexity of Algorithm 1 is to be $\sum_{t=0}^{\lfloor n\delta \rfloor} n \binom{n/2}{\lceil t/2 \rceil}$ with $\binom{n/2}{\lceil n\delta \rceil/2}$ storage.

Table 4. Complexity of exhaustive search, Algorithm 1 and unidirectional case in RSA ($n = 1024$)

upper bound of error rate	complexity		
	exhaustive search	Algorithm 1	unidirection
0.003	$2^{27.42}$	$2^{27.01}$	$2^{24.04}$
0.005	$2^{43.09}$	$2^{34.42}$	$2^{30.49}$
0.010	$2^{78.16}$	$2^{49.08}$	$2^{43.23}$

Table 4 provides the complexity of exhaustive search, Algorithm 1 and unidirection for upper bounds of error rate. Similarly with DL based cryptosystems, the complexity of exhaustive search is $\sum_{t=0}^{\lfloor n\delta \rfloor} \binom{n}{t}$,when the error rate is δ, which is the same with the number of possible cases of an erroneous key. When n is 1024 and the error rate is 1%, one could recover the private key d from the given erroneous key d' within $2^{49.08}$ while exhaustive search takes $2^{78.16}$. In the case of unidirectional errors, the complexity becomes $2^{43.23}$.

Coron and Kocher's Method [9,2]. In order to randomize the private key protecting against differential power analysis, Coron and Kocher proposed the use of blinding private key $\tilde{d} = d + r\phi(N)$ where r is a randomly chosen integer of size n_r-bit. In this case, as in Section 3.2, the private key can be recovered using Algorithm 1 with input $(C, m, \tilde{d}', (n+n_r), \delta)$ where $C^{\tilde{d}} = m$. And the complexity of Algorithm 1 is $\sum_{t=0}^{\lfloor (n+n_r)\delta \rfloor} (n + n_r) \binom{(n + n_r)/2}{\lceil t/2 \rceil}$ with $\binom{(n+n_r)/2}{\lceil \lfloor (n+n_r)\delta \rfloor /2 \rceil}$ storage.

Table 5. Lower bound of n_r to provide 2^{80} complexity in RSA $(n = 1024)$

upper bound of error rate	0.005	0.008	0.010	0.015	0.020
lower bound of n_r	1976	1101	699	243	26

In Table 5, we provide the value of n_r for given upper bounds of error rate to get the complexity more than 2^{80} complexity. If the upper bound of error rate is 0.008, we have to choose n_r so that $n_r > 1024$.

Clavier and Joye's Method [3]. In order to randomize the private key d, Clavier and Joye split it into two parts $(d_1, d - d_1)$ for a random number d_1 where $d_1 + d_2 = d$. In this case, we can recover the private keys d_1, d_2 using Algorithm 2 of Section 3.3, with input $(C, m, d_1', d_2', n, \delta)$. And the complexity of Algorithm 2 is $\sum_{t=0}^{\lfloor n\delta \rfloor} 2\binom{n}{t}$ with $\sum_{t=0}^{\lfloor n\delta \rfloor} \binom{n}{t}$ storage. Their complexities for various error rates are presented in Table 6.

Table 6. Complexity of exhaustive search, Algorithm 2 and unidirectional case in RSA $(n = 1024)$

upper bound of error rate	complexity		
	exhaustive search	Algorithm 2	unidirection
0.003	$2^{55.83}$	$2^{28.42}$	$2^{25.44}$
0.005	$> 2^{80}$	$2^{44.09}$	$2^{39.15}$
0.010	$> 2^{80}$	$2^{79.16}$	$2^{69.39}$

5   Conclusion

In this paper, we proposed the algorithm to recover the private key from its erroneous value using the splitting system in the DL based cryptosytems and the standard RSA. We also considered breaking countermeasures of differential power analysis. Our algorithm achieves almost the square root complexity of the size of the search space. Considering that the most efficient generic algorithm to solve DLP has the complexity of the square root of the base group order, the square root complexity can be considered optimal. It would be interesting to find the lower bound of the complexity in terms of the size of search space to recover the private key when its partial information is given.

There have been numerous algorithms to recover the RSA decryption key from its partial information. However, all the works either require some contiguous information or additional partial information on prime divisors of the modulus. Our algorithm requires a lot more information than the previous, but it does not have those restriction. It would be interesting to apply the previous lattice techniques to have more efficient algorithms in this case.

Acknowledgements. We thank the anonymous reviewers for useful comments and discussions. Hyung Tae Lee, HongTae Kim, and Jung Hee Cheon were supported by the National Research Foundation of Korea (NRF) grant funded by the Korea government (MEST) (No. 20110018345).

References

1. Boneh, D., Durfee, G., Frankel, Y.: An Attack on RSA Given a Small Fraction of the Private Key Bits. In: Ohta, K., Pei, D. (eds.) ASIACRYPT 1998. LNCS, vol. 1514, pp. 25–34. Springer, Heidelberg (1998)
2. Coron, J.-S.: Resistance against Differential Power Analysis for Elliptic Curve Cryptosystems. In: Koç, Ç.K., Paar, C. (eds.) CHES 1999. LNCS, vol. 1717, pp. 292–302. Springer, Heidelberg (1999)
3. Clavier, C., Joye, M.: Universal Exponentiation Algorithm. In: Koç, Ç.K., Naccache, D., Paar, C. (eds.) CHES 2001. LNCS, vol. 2162, pp. 300–308. Springer, Heidelberg (2001)
4. Fouque, P.-A., Kunz-Jacques, S., Martinet, G., Muller, F., Valette, F.: Power Attack on Small RSA Public Exponent. In: Goubin, L., Matsui, M. (eds.) CHES 2006. LNCS, vol. 4249, pp. 339–353. Springer, Heidelberg (2006)
5. Heiman, R.: A Note on Discrete Logarithms with Special Structure. In: Rueppel, R.A. (ed.) EUROCRYPT 1992. LNCS, vol. 658, pp. 454–457. Springer, Heidelberg (1993)
6. Henecka, W., May, A., Meurer, A.: Correcting Errors in RSA Private Keys. In: Rabin, T. (ed.) CRYPTO 2010. LNCS, vol. 6223, pp. 351–369. Springer, Heidelberg (2010)
7. Heninger, N., Shacham, H.: Reconstructing RSA Private Keys from Random Key Bits. In: Halevi, S. (ed.) CRYPTO 2009. LNCS, vol. 5677, pp. 1–17. Springer, Heidelberg (2009)

8. Halderman, J.A., Schoen, S.D., Heninger, N., Clarkson, W., Paul, W., Calandrino, J.A., Feldman, A.J., Appelbaum, J., Felten, E.W.: Lest we remember: Cold boot attacks on encryption keys. In: USENIX Security Symposium, pp. 45–60 (2008)
9. Kocher, P.C.: Timing Attacks on Implementations of Diffie-Hellman, RSA, DSS, and Other Systems. In: Koblitz, N. (ed.) CRYPTO 1996. LNCS, vol. 1109, pp. 104–113. Springer, Heidelberg (1996)
10. Schnorr, C.-P.: Small Generic Hardcore Subsets for the Discrete Logarithm: Short Secret DL-keys. Information Processing Letters 79(2), 93–98 (2001)
11. Shoup, V.: Lower Bounds for Discrete Logarithms and Related Problems. In: Fumy, W. (ed.) EUROCRYPT 1997. LNCS, vol. 1233, pp. 256–266. Springer, Heidelberg (1997)
12. Stinson, D.R.: Some Baby step Giant step algorithms for the low hamming weight discrete logarithm problem. Mathematics of Computation 71(237), 379–391 (2002)

Strong Security Notions for Timed-Release Public-Key Encryption Revisited

Ryo Kikuchi[1], Atsushi Fujioka[1], Yoshiaki Okamoto[2], and Taiichi Saito[2]

[1] NTT Information Sharing Platform Laboratories
{kikuchi.ryo,fujioka.atsushi}@lab.ntt.co.jp
[2] Tokyo Denki University
{okamoto@crypt.,taiichi@}c.dendai.ac.jp

Abstract. Timed-release public-key encryption (TRPKE) provides a mechanism that a ciphertext cannot ordinarily be decrypted, even with its secret key, before a specific time. TRPKE with pre-open capability (TRPKE-PC) provides an additional mechanism where the sender can permit a receiver to decrypt the ciphertext before that specific time if necessary. A TRPKE(-PC) scheme should be secure in following aspects: against malicious receivers, a time-server, and, only in TRPKE-PC, against malicious senders. In this paper, we mention that previous security definitions are incomplete or insufficient, and propose new ones in all aspects of the above. We also present a generic construction of a TRPKE-PC scheme. Our construction provides the first TRPKE(-PC) scheme that is provably secure with respect to the above security definitions, especially against malicious key generations of the time-server.

Keywords: timed-release encryption, malicious time-server, malicious key generation, strong decryption, generic construction, pre-open capability.

1 Introduction

Timed-release public-key encryption (TRPKE) was introduced by Chan and Blake [7] in 2005. Intuitively, TRPKE provides a mechanism whereby a ciphertext cannot be decrypted until a specific time. In addition to TRPKE, TRPKE with *pre-open capability* (TRPKE-PC), introduced by Hwang et al. in 2005 [22], has more flexibility such that a receiver can decrypt a ciphertext with the support of the sender even before a specific time. In this paper, we focus on TRPKE-PC and describe security definitions and syntax of a scheme for TRPKE-PC since it contains TRPKE.

A TRPKE-PC system consists of three entities: a *time-server*, *sender*, and *receiver*. The sender encrypts plaintext by using the receiver's and time-server's public keys and designates a *time-period* T after which the receiver is allowed to decrypt the ciphertext. At each T, the time-server periodically generates a *time signal* s_T corresponding to T and broadcasts it to all users including the receiver. The receiver can decrypt the ciphertext with his/her secret key and

H. Kim (Ed): ICISC 2011, LNCS 7259, pp. 88–108, 2012.
© Springer-Verlag Berlin Heidelberg 2012

s_T corresponding to T. This decryption is called *time-period decryption* and in TRPKE-PC, the receiver can decrypt in another way called *pre-open decryption*. This uses the receiver's secret key and a *pre-open key pok* generated together with the ciphertext. Therefore, the receiver can decrypt the ciphertext independent of the T if the sender sends *pok* to the receiver.

The TRPKE(-PC) scheme has many applications for constructing secure protocols in which information is revealed to users after a specific time. One of major applications is *timed-release commitment* in a sealed-bid auction: Bidders encrypt their bids with a closure time. To confirm bids, an auctioneer decrypts ciphertexts with a time-signal. Therefore, there is no interaction between bidders and the auctioneer.

Previous Security Definitions. Security notions for TRPKE are considered from two aspects against malicious receivers and against a time-server. The former means that no one can obtain information from the ciphertext without a corresponding time signal and the latter means no one can obtain information without the receiver's secret key. The TRPKE-PC scheme demands another security notion against senders such that no one can create a tuple of a ciphertext, time-period, and pre-open key where a plaintext decrypted with time-period decryption differs from one decrypted with pre-open decryption.

There are many security definitions for TRPKE(-PC), which can be categorized into two aspects: what an adversary issues to the decryption oracle, and if an adversary can maliciously generate its own keys. We summarize these definitions in Table 1.

Table 1. Previous Security Definitions for TRPKE(-PC)

	notion name	against	dec. query	key gen.
Cheon et al. [8]	IND-RTR-CCA2	receiver	(usk, T, c)	malicious
	IND-CCA2	time server	(T, c)	honest
Cathalo et al. [6]	IND-CTCA	receiver	(upk, T, c)	malicious
	IND-CCA	time server	(T, s_T, c)	honest
Chow et al. [10]	Type-I	receiver	(upk, T, c)	malicious
	Type-II	time server	(upk, T, c)	honest
Dent and Tang [15]	IND-TR-CPA$_{IS}$	receiver	*no dec.*	honest
	IND-TR-CCA$_{TS}$	time server	(T, c)	honest
	Binding	sender	(T, c)	honest
Chow and Yiu [11]	Type-I	receiver	(upk, T, c)	malicious
	Type-II	time server	(upk, T, c)	honest
	Binding	sender	(upk, T, c)	honest

c is a ciphertext, usk is a user's secret key, upk is a user's public key, T is a time period, s_T is the time signal at T, "malicious" means attacker can generate own keys maliciously and "honest" means attacker cannot do this.

Strong Security Notion. Several previous security notions are indistinguishability-based definitions allowing an adversary to issue a decryption query including a user's public key, then a challenger responds with a message derived from the issued ciphertext. Such a decryption oracle is called a *strong decryption* oracle and we call the security definition with such an oracle *strong security*. This might seem too strong since an adversary can issue decryption queries with an adversarial chosen public key; no one knows the corresponding secret key, even the adversary.

However, it is important to consider such a strong security definition. Many applications of TRPKE(-PC) involve multiple receivers, and some adversaries may collude with each other to maliciously generate their keys, e.g., depending on other keys. It seems natural to assume that an adversary can choose a user's key issued to the decryption oracle. Theoretically, it is interesting to consider whether a scheme that achieves strong security can exist or not.

Furthermore, in some applications of TRPKE(-PC), a strong decryption oracle makes sense. To explain this, we first discuss the security notion, *complete non-malleability*, defined by Fischlin [16]. Roughly speaking, complete non-malleability states that giving a ciphertext to an adversary does not help to produce a ciphertext of a related message under an *adversarial chosen* public key. This security notion is desired, for example, when one uses an encryption scheme as a commitment, e.g., timed-release commitment in a sealed-bid auction described in the previous section. Without complete non-malleability, the adversary may produce a related commitment and cheat, i.e., generate a ciphertext and a public key of plaintext $m + 1$ from those of plaintext m.

Recently, Barbosa and Farshim [3] clarified[1] the relation between indistinguishability with strong decryption and complete non-malleability.[2] This means that indistinguishability under a strong decryption oracle is a convenient formalization for establishing that a scheme is completely non-malleable. Therefore, strong decryption for TRPKE(-PC) is important since complete non-malleability is worth considering, and satisfying a strong security notion comprises complete non-malleability.

1.1 Our Contribution

There are many benefits for considering a strong decryption oracle as described in the previous section, so we focus on the strong security notion. We show that previous security definitions are incomplete or insufficient. We thus provide precise or strong security definitions and also propose a generic construction of a TRPKE-PC scheme that satisfies our definitions in the random oracle model.

Note that we provide new security definitions only in the context of TRPKE-PC. This is because the security definitions for TRPKE-PC are easily converted to those for TRPKE.

---

[1] They prove the equivalence for ordinal public-key encryption, and it can be applied to TRPKE-PC.

[2] Though their definition is slightly different from Fischlin's, the aim is the same.

Problems with Previous Strong Security Definitions and New Definitions. We first go with the security against malicious receivers. Several definitions were proposed, such as IND-CTCA and Type-I security, but they do not explain the behavior when an adversary issues an invalid public key or invalid ciphertext to a strong decryption oracle. They may work well for a specific scheme, but become a problem when we consider a general case, e.g., generic construction. Fujioka et al. [18] recently noticed this ambiguity and avoided it by adding a restriction in which an adversary cannot issue an invalid public key. However, this restriction is not realistic. We aim to define a strong security notion without such a restriction on an adversary.

A strong decryption oracle can be defined in roughly two ways. One is based on the principle "reply as far as possible", and the other is based on the principle "reply only if a query is valid". They are incomparable with each other since they assist the adversary with an another aspect: decrypting invalid ciphertexts under an invalid public key or checking their validity. Our definition uses the former principle since it is more suitable for a generic construction. The latter principle may lose generality since it excludes a TRPKE-PC scheme such that one cannot check the validity of both the public key and ciphertext, e.g., most RSA-based schemes.

Second, we go with the security against a malicious time-server. As described in Table 1, honest key generations are assumed with the previous definitions, But the malicious time-server, which generates its keys maliciously, may break the TRPKE-PC scheme. In a sealed-bid auction, for example, the time-server with malicious key generations can see all bids and, if it colludes with a bidder, the bidder may illegally succeed in the auction. We take Chow and Yiu's TRPKE-PC scheme [11], which can be totally broken by a such an adversary, as an example in Appendix D.

Third, we go with the security against malicious senders in a TRPKE-PC scheme. The sender knowing nothing about the time-server's secret key is assumed with the previous definitions. This implies that collusion with the time-server is beyond the definition. We thus define another strong security definition called *Strong Binding* to prevent collusion.

Generic Construction of Strongly Secure TRPKE-PC. We also propose a generic construction of a TRPKE-PC scheme that first satisfies the above new security definitions, especially against a malicious time-server with malicious key generations even in the random oracle model [4]. We use the random oracle since Fischlin [16] claimed that there is no *black-box* construction which is complete non-malleable (and is likely secure with the strong decryption oracle) in the standard model without a trusted setup. There are several schemes (e.g., [30,3,13]) that are complete non-malleable or secure with the strong decryption oracle in the standard model, but they require a trusted setup. On the contrary, our scheme does not has a trusted setup since our security allows malicious key generations of the time-server. Therefore, it seems a difficult task to construct secure schemes w.r.t. our definitions in the standard model.

Our construction consists of a public-key encryption (PKE) scheme and an identity-based encryption (IBE) scheme. It uses a simple approach, a *PKE-then-IBE* paradigm. A naive construction of this sort is insecure against a "Decrypt-then-Encrypt" attack [14,32] since a ciphertext differs from one that is decrypted partway (i.e., only IBE decryption) and re-encrypted. We prevent such an attack by encrypting with a second (IBE) scheme deterministically, which is *bounded randomness* technique [17]. To enable pre-open functionality, we use an intermediate ciphertext, encrypted with a PKE scheme, as a pre-open key. That is, we use *PKE-then-IBE*, not *IBE-then-PKE* used in Ref. [18]. In the PKE-then-IBE construction, a receiver can decrypt an intermediate ciphertext while others cannot decrypt it without the receiver's secret key; in the IBE-then-PKE construction, the receiver cannot decrypt the intermediate ciphertext.

We require the property called *extensive γ-uniformity* from both IBE and PKE, where γ is negligible. Intuitively, this means the entropy of a ciphertext for a fixed message is large *even if* a public key is invalid. If PKE(IBE) is IND-(ID-)CPA secure, we can easily convert them to satisfy this property. The original γ-uniformity defined by Fujisaki and Okamoto [19] means the same thing under uniformly and honestly chosen public keys, and γ-pk-uniformity used in [18] means the same thing under all valid public keys. We also assumes the collision resistance of the encryption, which is essential to be secure against a strong chosen ciphertext attacks. This is trivially guaranteed under a valid public key thanks to completeness.

Fujioka et al. [18] also proposed a generic construction of TRPKE with IBE-then-PKE technique and negligible γ-pk-uniformity. Their construction is, however, proven secure against only a malicious time-server with honest key generations and, against malicious receivers under the restriction of invalid key queries. It also does not provide pre-open capability.

1.2 Related Works

Certificateless encryption (CLE) [2,12] is a primitive related to TRPKE. Chow et al. [10] proposed a method for converting any *general* CLE scheme, that is, a CLE scheme with additional properties, into a TRPKE scheme. In the CLE context, security with strong decryption has been also studied. Chow et al. [10] defined Type-I and Type-II security, and Au et al. [1] pointed out that a malicious key generation center (KGC) with malicious key generations may violate the security of CLE. Hwang et al. [23] proposed a scheme secure against a time-server with malicious key generations, without strong decryption.

Timed-release public-key encryption is also closely related to multiple encryption. Zhang et al. [32] and Dodis and Katz [14], studied the security of multiple encryption and how to construct secure schemes from IND-CCA secure components. Recently, Fujioka et al. [17] studied the security of multiple encryption and constructed it from IND-CPA secure schemes.

The notion of complete non-malleability was first defined by Fischlin [16], and Ventre and Visconti [30] later formalized it in another way and proposed a complete non-malleable secure scheme with a trusted setup in the standard model. Barbosa and Farshim [3] studied and categorized the definition of a strong decryption oracle, and clarified the relation between a strong decryption oracle and complete non-malleability.

Very recently, Kawai et al. [24] showed that it is impossible for PKE to reduce the security with strong decryption oracle (called IND-SCCA) to any other weaker security notion under black-box analysis in the standard model. Furthermore, they also showed that even if the encryption system has a setup procedure, it is also impossible under setup-preserving black-box reductions. These results indicates the difficulty to construct strongly secure schemes in the standard model.

Due to space limitation, we describe traditional schemes and give a brief history of TRPKE and TRPKE-PC in Appendix A.

2   Preliminaries

We review the primitives used as components of our construction and their security notions.

Notation. Throughout this paper, λ denotes the security parameter and PPT algorithm denotes a probabilistic polynomial-time algorithm. $x \leftarrow y$ means that x is chosen from y uniformly at random if Y is a finite set; if otherwise, simply substitute y into x. For probabilistic algorithm $\mathcal{A}$, $y \leftarrow \mathcal{A}(x; r)$ means that y is the output of $\mathcal{A}$ with input x and randomness r, and if r is picked uniformly at random, r is omitted as $y \leftarrow \mathcal{A}(x)$. $x \| y$ denotes the concatenation of x and y. $\mathbb{A}$ denotes a space of a, e.g., a space of a message m is denoted as $\mathbb{M}$. $|\mathbb{M}|$ denotes the number of elements that belong to $\mathbb{M}$.

2.1   Public-Key Encryption

A public-key encryption (PKE) scheme is a tuple of algorithms as follows. A key generation algorithm PKE.KG takes 1^λ as input, and outputs a public key pk and a secret key sk. An encryption algorithm PKE.Enc takes a pk, a message m and as inputs and outputs a ciphertext $\hat{c}$. A decryption algorithm PKE.Dec takes sk and ciphertext $\hat{c}$ as inputs and outputs the plaintext m or $\bot$. These algorithms are required to satisfy $\mathsf{PKE.Dec}(sk, \mathsf{PKE.Enc}(pk, m)) = m$ for any $(pk, sk) \leftarrow \mathsf{PKE.KG}(1^\lambda)$ and any m. Throughout this paper, we use $\mathbb{M}_{\mathrm{PKE}}$, $\mathbb{C}_{\mathrm{PKE}}$, and $\mathbb{R}_{\mathrm{PKE}}$ to indicate the message space, ciphertext space, and randomness space of the encryption algorithm respectively.

We use IND-CPA secure PKE in our construction. Due to space limitation, the definition of IND-CPA security appears in Appendix B.

2.2   Identity-Based Encryption

An identity-based encryption (IBE) scheme consists of the following four algorithms. A setup algorithm IBE.Setup takes 1^λ as input and outputs a public

parameter *params* and a master secret key *msk*. An extract algorithm IBE.Ext
takes *params*, *msk*, and an arbitrary string (identity) $\text{ID} \in \{0,1\}^*$ as inputs and
outputs a decryption key d_{ID}. An encryption algorithm IBE.Enc takes *params*,
ID, and a message m as inputs and outputs a ciphertext c. A decryption algorithm IBE.Dec takes *params*, d_{ID}, and c as inputs and outputs the message m or
$\perp$. These algorithms are required to satisfy IBE.Dec($params, d_{\text{ID}}$,
IBE.Enc($params, \text{ID}, m)) = m$ for any m, any $(params, msk) \leftarrow$ IBE.Setup(1^λ),
and any $d_{\text{ID}} \leftarrow$ IBE.Ext($params, msk, \text{ID}$). Throughout this paper, we use $\mathbb{M}_{\text{IBE}}$,
$\mathbb{C}_{\text{IBE}}$, and $\mathbb{R}_{\text{IBE}}$ to indicate the message space, ciphertext space, and randomness
space of the encryption algorithm respectively.

We use IND-ID-CPA secure IBE in our construction. Due to space limitation,
the definition of IND-ID-CPA security appears in Appendix C.

2.3 γ-Uniformity and Collision Resistance of the Encryption

Fujisaki and Okamoto proposed γ-uniformity [19,20], which represents the size
of the entropy for a fixed message in PKE. We denote it as γ_{PKE}-*uniformity*.
We say a PKE scheme is γ_{PKE}-uniformity if $\Pr[\forall x \in \mathbb{M}_{PKE}; \forall y \in$
$\mathbb{C}_{PKE}; (pk, sk) \leftarrow$ PKE.KG(1^λ) $: y =$ PKE.Enc($pk, x)] \leq \gamma_{PKE}$ holds. Note that
we consider entropy under a *uniformly and honestly* chosen public key with γ_{PKE}-
uniformity, and all IND-CPA secure PKE schemes have γ_{PKE}-uniformity where
γ_{PKE} is negligible in λ. Fujioka et al. [18] extended it to γ_{PKE}-pk-uniformity, which
aims to represent the size of the entropy under *all* public keys generated by a key
generation algorithm. We also extends it to a slightly stronger property called *extensive γ_{PKE}-uniformity*. Roughly speaking, extensive γ_{PKE}-uniformity states the
size of the entropy under a public key that is *possibly invalid*. It is formally de-
scribed as $\Pr[\forall x \in \mathbb{M}_{PKE}; \forall y; \forall pk : y =$ PKE.Enc($pk, x)] \leq \gamma_{PKE}$. Our construction
requires extensive γ_{PKE}-uniformity where γ_{PKE} is negligible in λ. Although an IND-
CPA secure PKE does not generally have extensive γ_{PKE}-uniformity where γ_{PKE} is
negligible in λ, all IND-CPA secure PKEs can be converted to one that has extensive γ_{PKE}-uniformity where γ_{PKE} is negligible [20].

Yang et al. extended γ-uniformity to IBE [31] and we denote it as γ_{IBE}-
uniformity. We say IBE has γ_{IBE}-uniformity if $\Pr[\forall x \in \mathbb{M}_{\text{IBE}}; \forall y \in \mathbb{C}_{\text{IBE}}; \forall \text{ID} \in$
$\{0,1\}^*; (params, msk) \leftarrow$ IBE.Setup(1^λ) $: y =$ IBE.Enc($params, \text{ID}, x)] \leq \gamma_{IBE}$
holds. Note that all IND-ID-CPA secure IBE schemes have γ_{IBE}-uniformity where
γ_{IBE} is negligible. Similar to the PKE scheme, we use *extensive γ_{IBE}-uniformity*,
formally described as $\Pr[\forall x \in \mathbb{M}_{\text{IBE}}; \forall y; \forall \text{ID} \in \{0,1\}^*; \forall params : y =$
IBE.Enc($params, \text{ID}, x)] \leq \gamma_{IBE}$. Our construction also requires that an IBE
scheme has extensive γ_{IBE}-uniformity where γ_{IBE} is negligible in λ. Although an
IND-ID-CPA secure IBE does not generally have extensive γ_{IBE}-uniformity where
γ_{IBE} is negligible, all IND-ID-CPA secure IBEs can be converted to have negligible
extensive γ_{PKE}-uniformity, almost the same as that of PKE.

We also require collision resistance of the encryption: for all pk there is
no message pair m_0, m_1 such that PKE.Enc(pk, m_0) $=$ PKE.Enc(pk, m_1)
for PKE. Also for all $params, \text{ID}$ there is no message pair m_0, m_1 such that

IBE.Enc($params$, ID, m_0) $=$ IBE.Enc($params$, ID, m_1) for IBE.[3] This is trivially fulfilled under a valid public key because of completeness, and seems to be fulfilled by pudding more random components under an invalid key.

3   Timed-Release Encryption with Pre-open Capability

In this section, we discuss and review the definitions of TRPKE-PC and their security notions.

3.1   Syntax

A TRPKE-PC scheme is formally defined as a tuple of the below algorithms.

- Setup: A setup algorithm that takes 1^λ as input, and outputs a time-server's public key tpk and corresponding secret key tsk.
- Release: A release algorithm that takes tpk, tsk, and a time-period T as inputs and outputs a time signal s_T corresponding to T.
- KeyGen: A user key generation algorithm that takes tpk as input, and outputs a user's public key upk and corresponding secret key usk.
- Enc: An encryption algorithm that takes tpk, T, upk, and message m as inputs and outputs the ciphertext CT and corresponding pre-open key pok.
- Dec$_{\mathsf{TR}}$: A time-period decryption algorithm that takes tpk, s_T, usk, and CT as inputs and outputs $m \in \mathbb{M} \cup \{\bot\}$.
- Dec$_{\mathsf{PO}}$: A pre-open decryption algorithm that takes tpk, pok, usk, and CT as inputs and outputs $m \in \mathbb{M} \cup \{\bot\}$.

Throughout this paper, we use $\mathbb{M}$, $\mathbb{C}$, and $\mathbb{R}$ to indicate the message space, ciphertext space, and randomness space of the encryption algorithm.

These algorithms are required to satisfy $\mathsf{Dec_{TR}}(tpk, s_T, usk, CT) = \mathsf{Dec_{PO}}(tpk, pok,$
$usk, CT) = m$ for any $(tpk, tsk) \leftarrow \mathsf{Setup}(1^\lambda)$, any $(upk, usk) \leftarrow \mathsf{KeyGen}(tpk)$, any T, any m, any $(CT, pok) \leftarrow \mathsf{Enc}(tpk, T, upk, m)$, and any $s_T \leftarrow \mathsf{Release}(tsk, T)$.

3.2   New Security Notions

Security against Malicious Receivers. We define the security notion for TRPKE-PC against malicious receivers with a precise definition of strong decryption. We call this definition *IND-SCCA Security against a Type-I Adversary* (Type-I security) since it is mostly based on Chow and Yiu's definition [11].

We define the strong decryption oracle with the principle "reply as far as possible". The strong decryption oracle responds with a message m, where $(CT, \cdot) = \mathsf{Enc}(tpk, T, upk, m)$; nevertheless, both upk and CT may be invalid. We can also

---

[3] To be exact, it is sufficient for our construction that the equation is not hold with an overwhelming probability.

define it with the principle "reply only if a query is valid", which responds with $\perp$ if one of components of a query is invalid. In our security definition, we use the former principle, though it could accept an invalid query as mentioned by Barbosa and Farshim [3], by following reasons. First, these principles empower the adversary in another aspect: The adversary can receive a plaintext of even an invalid query in the former principle, and the adversary can check the validity of both the ciphertext and public key in the latter principle. Therefore, we choose one of these principles. Second, the latter principle loses generality of the construction. It requires the checkable property of both the ciphertext and public key, so it excludes TRPKE-PC schemes such that one cannot check the validity. This property is especially difficult for RSA-based construction.

Of course, one can use the latter principle when constructing a TRPKE-PC scheme with a checkable property. However, if a TRPKE-PC scheme has a checkable property, our definition is essentially equal to one with the latter principle. This is because a TRPKE-PC scheme secure w.r.t. the latter principle can be also secure w.r.t. ours by checking a validity of a public key in the encryption and checking a validity of a ciphertext in the decryption.

Fujioka et al. [18] used another approach that restricts the adversary from issuing a query including invalid components. This definition, however, implies the adversary is generates its key as honest-but-curious, *not* malicious.

We formally describe Type-I security for a TRPKE-PC scheme based on the following Type-I game between a challenger C and an adversary A.

1. C takes a security parameter 1^λ, runs the setup algorithms $(tpk, tsk) \leftarrow \mathsf{Setup}(1^\lambda)$ and give tpk to A.
2. A is permitted to issue a series of queries to some oracles. (A is allowed to make adaptive queries here – subsequent queries are made based on the answers to previous queries.)
   - Release queries: A issues a time-period T and receives a corresponding time signal s_T derived from $\mathsf{Release}(tpk, tsk, T)$.
   - Time-period decryption queries: A issues a tuple of (upk, T, CT). If a group $\{m \mid (CT, \cdot) = \mathsf{Enc}(tpk, T, upk, m)\}$ is not empty, C responds with $m \xleftarrow{\$} \{m \mid (CT, \cdot) = \mathsf{Enc}(tpk, T, upk, m)\}$. Otherwise, C responds with $\perp$.
3. A outputs a user's public key upk^*, a pair of messages m_0, m_1, and a time-period T^* that was not issued as a release query. C randomly chooses $b \in \{0,1\}$, computes a challenge ciphertext and a pre-open key $(CT^*, pok^*) \leftarrow \mathsf{Enc}(tpk, T^*, upk^*, m_b)$, and gives CT^* to A.
4. A is permitted to issue a series of queries similarly. except with the restriction that no Release query T^*, no time-period decryption query (upk^*, T^*, CT^*) are allowed.
5. At the end of this game, A outputs a guess $b' \in \{0,1\}$.

A wins a Type-I game if $b' = b$, and its advantage is defined by $\mathbf{Adv}^{\mathrm{Type\text{-}I}}_{\mathrm{TRPKE\text{-}PC}, A}(\lambda) = |\Pr[A \text{ wins Type-I game}] - \frac{1}{2}|$.

Definition 1. *A timed-release encryption scheme with pre-open capability is Type-I secure if no PPT adversary A has non-negligible advantage $\mathbf{Adv}^{Type\text{-}I}_{TRPKE\text{-}PC, A}(\lambda)$.*

Note that an adversary is not allowed to receive a pre-open key corresponding to a challenge ciphertext nor to issue pre-open decryption queries. The adversary can choose *all* user's public keys in the above game so it is logical that the adversary has the corresponding user's secret keys; hence, pre-open decryption queries are not useful for the adversary. In addition, if we allow the adversary to access a pre-open decryption oracle, TRPKE-PC schemes that can generate *pok* from a ciphertext with a sender's secret key are excluded.

Security against a Malicious Time-Server. We show a new strong security definition for TRPKE-PC against a malicious time-server with malicious key generations. The security definition is called *IND-SCCA against a Type-II⁺ adversary* (Type-II⁺ security). It is similar to the Type-II security defined in Ref. [11] except that an adversary can maliciously generate time-server's key, so Type-II⁺ security is a stronger notion.

Previous schemes that are secure with a strong decryption oracle in the standard model often embed a trapdoor in the time-server's key. Such schemes are not secure in our model because an adversary can also embed in the same way. In other words, previous schemes are secure when the adversary honestly generates own keys. This is a serious vulnerability since, if the adversary maliciously generates own keys, it can decrypt a ciphertext *without a recognition* of other participants. We take Chow and Yiu's scheme [11] as an example of the above insecurity in Appendix D.

As the same to Type-I security, one can define to check the validity in a strong decryption oracle for TRPKE-PC schemes with checkable property. We do not give it since it is easily derived from ours.

We formally describe the Type-II⁺ security for a TRPKE-PC scheme based on the following Type-II⁺ game between a challenger $\mathcal{C}$ and an adversary $\mathcal{A}$.

1. $\mathcal{C}$ takes a security parameter 1^λ as input, passes it to $\mathcal{A}$ and gets tpk from $\mathcal{A}$. $\mathcal{C}$ computes $(upk^*, usk^*) \leftarrow \mathsf{KeyGen}(tpk)$ and gives upk^* to $\mathcal{A}$.
2. $\mathcal{A}$ is permitted to issue a series of (adaptive) queries to some oracles.
   - Time-period decryption queries: $\mathcal{A}$ issues a tuple of (upk, T, CT). If a group $\{m \mid (CT, \cdot) = \mathsf{Enc}(tpk, T, upk, m)\}$ is not empty, $\mathcal{C}$ responds with $m \xleftarrow{\$} \{m \mid (CT, \cdot) = \mathsf{Enc}(tpk, T, upk, m)\}$. Otherwise, $\mathcal{C}$ responds with $\perp$.
   - Pre-open decryption queries: $\mathcal{A}$ issues a tuple of (upk, pok, CT). If a group $\{m \mid (CT, pok) = \mathsf{Enc}(tpk, \cdot, upk, m)\}$ is not empty, $\mathcal{C}$ responds with $m \xleftarrow{\$} \{m \mid (CT, pok) = \mathsf{Enc}(tpk, \cdot, upk, m)\}$. Otherwise, $\mathcal{C}$ responds with $\perp$.
3. $\mathcal{A}$ outputs a pair of messages m_0, m_1 and a time-period T^*. $\mathcal{C}$ randomly chooses $b \in \{0, 1\}$, computes a challenge ciphertext and a pre-open key $(CT^*, pok^*) \leftarrow \mathsf{Enc}(tpk, T^*, upk^*, m_b)$, and gives (CT^*, pok^*) to $\mathcal{A}$.
4. $\mathcal{A}$ is permitted to issue a series of queries similarly, except with the restriction that no time-period decryption query (upk^*, T^*, CT^*) and no pre-open decryption query (upk^*, pok^*, CT^*) is allowed.
5. At the end of this game, $\mathcal{A}$ outputs a guess $b' \in \{0, 1\}$.

$\mathcal{A}$ wins a Type-II$^+$ game if $b' = b$, and its advantage is defined by $\mathbf{Adv}_{\text{TRPKE-PC},\mathcal{A}}^{\text{Type-II}^+}(\lambda) = |\Pr[\mathcal{A} \text{ wins Type-II}^+ \text{ game}] - \frac{1}{2}|$.

Definition 2. *A timed-release encryption scheme with pre-open capability is Type-II$^+$ secure if no PPT adversary $\mathcal{A}$ has non-negligible advantage* $\mathbf{Adv}_{\text{TRPKE-PC},\mathcal{A}}^{\text{Type-II}^+}(\lambda)$.

Note that the setup phase often generate system components such as description of the group. Our construction is secure even in this situation since the construction is completely separated into a PKE and IBE scheme, and we regards hash functions as a random oracle. However, this is problematic in some situations, e.g., generating a user's key over the group generated by the setup phase. One can avoid it to restrict the adversary from generating a part of its key, i.e., system components. If so, the security is guaranteed as long as these components is determined honestly, e.g., using the group stated by ISO.

Security against Malicious Senders. In the context of TRPKE-PC, another security notion called *Binding* was defined by Dent and Tang [15]. This binding notion means that an adversary cannot make a tuple of ciphertext, time-period, and pre-open key where a plaintext that is decrypted with time-period decryption differs from one decrypted with pre-open decryption.

A previous security definition is considered for only an adversary with a user's public key. If an adversary colludes with a time-server and receives its key, the binding property is no longer guaranteed. We thus define a stronger notion called *Strong Binding*.

We formally describe the strong binding property for the TRPKE-PC scheme based on the following game between a challenger $\mathcal{C}$ and an adversary $\mathcal{A}$.

1. $\mathcal{C}$ takes a security parameter 1^λ as input, passes it to $\mathcal{A}$ and gets tpk from $\mathcal{A}$. $\mathcal{C}$ runs $(upk^*, usk^*) \leftarrow \mathsf{KeyGen}(tpk)$ and gives upk^* to $\mathcal{A}$.
2. $\mathcal{A}$ is permitted to issue a series of (adaptive) queries to some oracles.
   - Time-period decryption queries: $\mathcal{A}$ issues a tuple of (upk, T, CT). If a group $\{m \mid (CT, \cdot) = \mathsf{Enc}(tpk, T, upk, m)\}$ is not empty, $\mathcal{C}$ responds with $m \xleftarrow{\$} \{m \mid (CT, \cdot) = \mathsf{Enc}(tpk, T, upk, m)\}$. Otherwise, $\mathcal{C}$ responds with $\perp$.
   - Pre-open decryption queries: $\mathcal{A}$ issues a tuple of (upk, pok, CT). If a group $\{m \mid (CT, pok) = \mathsf{Enc}(tpk, \cdot, upk, m)\}$ is not empty, $\mathcal{C}$ responds with $m \xleftarrow{\$} \{m \mid (CT, pok) = \mathsf{Enc}(tpk, \cdot, upk, m)\}$. Otherwise, $\mathcal{C}$ responds with $\perp$.
3. At the end of this game, $\mathcal{A}$ outputs a ciphertext CT^*, a time-signal s_T^*, and a pre-open key pok^*.

$\mathcal{A}$ wins a strong binding game if $\perp \neq m_{tr} \neq m_{po} \neq \perp$, where $m_{tr} \leftarrow \mathsf{Dec_{TR}}(tpk, s_T^*, usk^*, CT^*)$, and $m_{po} \leftarrow \mathsf{Dec_{PO}}(tpk, pok^*, usk^*, CT^*)$. Its advantage is defined by $\mathbf{Adv}_{\text{TRPKE-PC},\mathcal{A}}^{\text{Strong Binding}}(\lambda) = \Pr[\mathcal{A} \text{ wins strong binding game}]$.

Definition 3. *A timed-release encryption scheme with pre-open capability has the strong binding property if no PPT adversary $\mathcal{A}$ has non-negligible advantage* $\mathbf{Adv}^{Strong\ Binding}_{TRPKE\text{-}PC,\mathcal{A}}(\lambda)$.

Note that in our definition the adversary outputs a time-signal s_T^*, not a time-period T as defined by Dent and Tang [15]. This is because if the adversary colludes with the time-server, it is natural that s_T^* may be disturbed.

4   Generic Construction of Strong Secure TRPKE-PC

In this section, we explain the general construction of a TRPKE-PC scheme. A TRPKE-PC scheme derived from the above construction is the first scheme that satisfies the above strong security definitions, especially against a malicious time-server with malicious key generations. The proposed construction consists of an IND-CPA secure PKE and an IND-ID-CPA secure IBE and uses a PKE-then-IBE technique.

Although we assume with our construction that both PKE and IBE schemes have negligible extensive γ-uniformity, which can be achieved for all IND-(ID-)CPA secure schemes. We also assumes collision resistance of the encryption.

Our proposed scheme is described as follows.

- Setup(1^λ): Run $(params, msk) \leftarrow$ IBE.Setup(1^λ) and choose hash functions H_1 mapping $\{0,1\}^*$ to $\mathbb{R}_{\text{PKE}}$ and H_2 mapping $\{0,1\}^*$ to $\mathbb{R}_{\text{IBE}}$. Then set $tpk = (params, H_1, H_2)$, $tsk = msk$ and output them.
- Release(tpk, tsk, T): See tpk and tsk as $(params, H_1, H_2)$ and msk respectively. Run $d_T \leftarrow$ IBE.Ext($params, tsk, T$), set $s_T = (T, d_T)$, and output s_T.
- KeyGen(tpk): Run $(pk, sk) \leftarrow$ PKE.KG(1^λ), set $upk = pk$, $usk = (pk, sk)$, and output (upk, usk).
- Enc(tpk, T, upk, m): See tpk and upk as $(params, H_1, H_2)$ and pk respectively. Compute $\hat{c} \leftarrow$ PKE.Enc($pk, m||r; H_1(pk, m||r, T)$) where $r \leftarrow \mathbb{R}$, compute $c =$ IBE.Enc($params, T, \hat{c}; H_2(T, \hat{c})$), set $CT = (T, c)$ and $pok = \hat{c}$, and output them.
- Dec$_{\text{TR}}$(tpk, s_T, usk, CT): See tpk, s_T, usk, and CT as $(params, H_1, H_2)$, (T, d_T), (pk, sk), and $(\tilde{T}, c)$ respectively. If $T = \tilde{T}$, compute $\hat{c} \leftarrow$ IBE.Dec($params, d_T, c$) and if $c =$ IBE.Enc($params, T, \hat{c}; H_2(T, \hat{c})$), compute $m||r \leftarrow$ PKE.Dec($sk, \hat{c}$), and if $\hat{c} =$ PKE.Enc($pk, m||r; H_1(pk, m||r, T)$), output m. Otherwise, output $\perp$.
- Dec$_{\text{PO}}$(tpk, pok, usk, CT): See tpk, usk, and CT as $(params, H_1, H_2)$, (pk, sk), and (T, c) respectively. If $c =$ IBE.Enc($params, T, pok; H_2(T, pok)$), compute $m||r \leftarrow$ PKE.Dec(sk, pok), and if $pok =$ PKE.Enc($pk, m||r; H_1(pk, m||r, T)$), output m. Otherwise, output $\perp$.

We require that $|\mathbb{R}|$ is sufficiently large, e.g., $\mathbb{R} = \{0,1\}^{80}$.

4.1 Security

Due to space limitation, we do not give a precise probability estimation. We will give it in the full version.

Notice that there are hash oracles H_1 and H_2 in the proof since our proposed scheme is secure in the random oracle model.

Theorem 1. *Suppose an IBE scheme is IND-ID-CPA secure, a PKE scheme has extensive γ_{PKE}-uniformity where γ_{PKE} is negligible and has the collision resistance of the encryption. Our proposed construction is then Type-I secure in the random oracle model.*

Sketch of Proof. Assume that there exists an adversary $\mathcal{A}$ that breaks Type-I security of our proposed scheme. Then we can also construct an adversary $\mathcal{B}_1$ that breaks IND-ID-CPA security of the IBE scheme.

On input *params*, $\mathcal{B}_1$ runs as follows.

Setup: $\mathcal{B}_1$ takes *params* as input. Then $\mathcal{B}_1$ chooses hash functions H_1 and H_2, sets $tpk = (params, H_1, H_2)$. In addition, $\mathcal{B}_1$ generates two lists $\mathcal{T}_{h_1} = \mathcal{T}_{h_2} = \phi$, and gives tpk to $\mathcal{A}$.

Hash queries: $\mathcal{B}_1$ responds to hash queries while maintaining two query-answer lists, $\mathcal{T}_{h_1}$ and $\mathcal{T}_{h_2}$, in the usual manner except if $\mathcal{A}$ issues $(T^*, \hat{c}_{\dot{b}})$ where $\dot{b} \in \{0, 1\}$, $\mathcal{B}_1$ immediately outputs $\dot{b}$ as a guess of the IND-ID-CPA game and halts. An input-output pair is recorded as (input, output) in these lists.

Release queries: $\mathcal{A}$ issues T. $\mathcal{B}_1$ passes it to its extraction oracle in the IND-ID-CPA game and gets d_T. Then $\mathcal{B}_1$ sets $s_T = (T, d_T)$ and responds with it.

Time-period decryption queries: $\mathcal{A}$ issues $(upk, T, CT = (\widetilde{T}, c))$. If $T \neq \widetilde{T}$, $\mathcal{B}_1$ immediately responds with $\perp$. If not, $\mathcal{B}_1$ searches $((T, \hat{c}), value2) \in \mathcal{T}_{h_2}$ such that $c = \mathsf{IBE.Enc}(params, T, \hat{c}; value2)$ holds, and also searches $((upk, m||r, T), value1) \in \mathcal{T}_{h_1}$ such that $\hat{c} = \mathsf{PKE.Enc}(upk, m||r; value1)$ holds. If both exist, $\mathcal{B}_1$ responds with m. Otherwise, responds with $\perp$.

Challenge: $\mathcal{A}$ issues (upk^*, T^*, m_0, m_1). $\mathcal{B}_1$ randomly chooses $r_i \in \mathbb{R}$ and computes $\hat{c}_i \leftarrow \mathsf{PKE.Enc}(upk^*, m_i||r_i; H_1(upk^*, m_i||r_i, T^*))$ for $i \in \{0, 1\}$. Next, $\mathcal{B}_1$ issues $(\hat{c}_0, \hat{c}_1)$ to the challenger of the IND-ID-CPA game and receives c^*. Then $\mathcal{B}_1$ sets $CT^* = (T^*, c^*)$ and responds with CT^*.

Output: Finally, $\mathcal{A}$ outputs a guess b'. $\mathcal{B}_1$ directly outputs b'.

If the above simulations succeed, the equation $b' = b$ holds with the probability $\mathbf{Adv}_{TRPKE\text{-}PC, \mathcal{A}}^{Type\text{-}I}$.

There are four special cases according to the activity of $\mathcal{A}$.

[Case 1]: $\mathcal{A}$ issues $(upk, T, CT = (T, c))$ to the time-period decryption oracle such that $|\{m \mid (CT, \cdot) = \mathsf{Enc}(tpk, T, upk, m)\}| \geq 2$.

[Case 2-1]: $\mathcal{A}$ issues $(upk, T, CT = (T, c))$ to the time-period decryption oracle where a issued ciphertext is valid and corresponding intermediate ciphertext was not issued to the H_2 hash oracle: For some r, m, these formulas $c = \mathsf{IBE.Enc}(params, T, \hat{c}; H_2(T, \hat{c}))$ and $(T, \hat{c}) \notin \mathcal{T}_{h_2}$ hold where $\hat{c} \leftarrow \mathsf{PKE.Enc}(upk, m||r; H_1(upk, m||r, T))$.

[**Case 2-2**]: The same event as Case 2-1 occurs except that corresponding intermediate ciphertext was issued to the H_2 hash oracle but corresponding plaintext was not issued to the H_1 hash oracle.

[**Case 3**]: $\mathcal{A}$ issues $(T^*, \hat{c}_{\dot{b}})$ where $\dot{b} \in \{0, 1\}$ to the H_2 hash oracle.

Case 1 contradicts the collision resistance of the PKE encryption. More precisely, the collision resistance of the PKE encryption guarantees that if a plaintext is different, an intermediate cipertext is also different. On the other hand, *params* is honestly generated so we use the completeness of the IBE scheme: if an intermediate ciphertext is different, a ciphertext is also different. In summary, a ciphertext does not collide, so the number of elements belonging to the group $|\{m \mid (CT, \cdot) = \mathsf{Enc}(tpk, T, upk, m)\}|$ is at most 1. Therefore, the Case 1 does not occur.

Case 2-1 occurs at most negligible probability. By definition of Case 2-1, a hash value $H_2(T, \hat{c})$ is not determined and we assume that H_2 is a random oracle. Therefore Case 2-1 occurs if $c = \mathsf{IBE.Enc}(params, T, \hat{c}; R)$ holds for randomly picked $R \in \mathbb{R}_{\mathrm{IBE}}$. Such a probability at most γ_{IBE}. Note that the adversary can choose user's public key, so we need to assume negligible γ_{PKE}-uniformity for any (possibly invalid) *pk*.

Case 2-2 occurs with the probability at most γ_{PKE}. We do not give the description since it is almost the same as Case 2-1.

When Case 3 occurs $\mathcal{B}_1$ fails to similate the H_2 hash oracle since $\mathcal{B}_1$ does not know $H_2(T^*, \hat{c}_b)$ and also does not know b. However, the probability that adversary issues $(T^*, \hat{c}_{\overline{b}})$ is at most to $1/|\mathbb{R}|$ at each query, where $\overline{b}$ denotes $1 - b$. This is because the adversary's view is independent to $\hat{c}_{\overline{b}}$. If the adversary issues $(T^*, \hat{c}_b)$, $\mathcal{B}_1$ wins the IND-ID-CPA game.

Consequently, those whole simulations succeeds in overwhelming probability.

$\square$

Theorem 2. *Suppose an IBE scheme has extensive γ_{IBE}-uniformity and the collision resistance of the encryption, and a PKE scheme is IND-CPA secure with extensive γ_{PKE}-uniformity where γ_{IBE} and γ_{PKE} are negligible. Our proposed scheme is then Type-II$^+$ secure in the random oracle model.*

Sketch of Proof. Assume that there exists an adversary $\mathcal{A}$ that breaks Type-II$^+$ security of our proposed scheme. Then we can also construct an adversary $\mathcal{B}_2$ that breaks the IND-CPA security of the PKE scheme.

On input pk, $\mathcal{B}_2$ runs as follows.

Setup: $\mathcal{B}_2$ gives security parameter 1^λ to $\mathcal{A}$ and gets tpk. $\mathcal{B}_2$ generates two lists $\mathcal{T}_{h_1} = \mathcal{T}_{h_2} = \phi$, sets $upk^* = pk$, and gives upk to $\mathcal{A}$

Hash queries: $\mathcal{B}_2$ responds to hash queries while maintaining two query-answer lists, $\mathcal{T}_{h_1}$ and $\mathcal{T}_{h_2}$, in the usual manner except that if $\mathcal{A}$ issues $(upk^*, m_{\dot{b}} \| r_{\dot{b}}, T^*)$ where $\dot{b} \in \{0, 1\}$ to the H_1 hash oracle, $\mathcal{B}_2$ immediately outputs $\dot{b}$ as a guess of the IND-CPA game and halts. An input-output pair is recorded as (input, output) in these lists.

Time-period decryption queries: $\mathcal{A}$ issues $(upk, T, CT = (\widetilde{T}, c))$. If $T \neq \widetilde{T}$, $\mathcal{B}_2$ immedeately responds with $\bot$. If not, $\mathcal{B}_2$ searches $((T, \hat{c}), value2) \in$

$\mathcal{T}_{h_2}$ such that c = IBE.Enc($params, T, \hat{c}; value2$) and also searches $((upk, m||r, T), value1) \in \mathcal{T}_{h_1}$ such that $\hat{c} =$ PKE.Enc($upk, m||r, T; value1$). If both exist, $\mathcal{B}_2$ responds with m. Otherwise, $\mathcal{B}_2$ responds with $\perp$.

Pre-open decryption queries: $\mathcal{A}$ issues $(upk, pok, CT = (T, c))$. $\mathcal{B}_2$ searches $((T, pok), value2) \in \mathcal{T}_{h_2}$ such that c = IBE.Enc($params, T, pok; value2$). and also searches $((upk, m||r, T), value1) \in \mathcal{T}_{h_1}$ such that $pok =$ PKE.Enc($upk, m||r, T; value1$). If both exist, $\mathcal{B}_2$ responds with m. Otherwise, $\mathcal{B}_2$ responds with $\perp$.

Challenge: $\mathcal{A}$ issues (T^*, m_0, m_1). $\mathcal{B}_2$ randomly chooses $r_0, r_1 \in \mathbb{R}$ and issues $(m_0||r_0, m_1||r_1)$ to the challenger in the IND-CPA game. After that, $\mathcal{B}_2$ receives $\hat{c}^*$, computes $c^* \leftarrow$ IBE.Enc($params, T^*, \hat{c}^*; H_2(T^*, \hat{c}^*)$), and responds $(CT^* = (T^*, c^*), pok^* = \hat{c}^*)$,

Output: Finally, $\mathcal{A}$ outputs a guess b'. $\mathcal{B}_2$ directly outputs b'.

If above simulations succeed, the equation $b' = b$ holds with the probability $\mathbf{Adv}_{\mathrm{TRPKE\text{-}PC}, \mathcal{A}}^{\mathrm{Type\text{-}II}+}$.

There are seven special cases according to the activity of $\mathcal{A}$.

[Case 1-1]: $\mathcal{A}$ issues $(upk, T, CT = (T, c))$ to the time-period decryption oracle such that $|\{m \mid (CT, \cdot) = \mathsf{Enc}(tpk, T, upk, m)\}| \geq 2$.

[Case 1-2]: $\mathcal{A}$ issues $(upk, pok, CT = (T, c))$ to the pre-open decryption oracle such that $|\{m \mid (CT, pok) = \mathsf{Enc}(tpk, \cdot, upk, m)\}| \geq 2$.

[Case 2-1]: This case is the same as Case 2-1 in the proof of Type-I security.

[Case 2-2]: The same as Case 2-2 in the proof of Type-I security.

[Case 3-1]: Above cases do not occur and $\mathcal{A}$ issues $(upk, pok, CT = (T, c))$ to the pre-open decryption oracle where a issued ciphertext is valid and corresonding intermediate ciphertext (pre-open key) was not queried to the H_2 hash oracle: For certain r, m, the equations $pok =$ PKE.Enc($upk, m||r; H_1(upk, m||r, T)$), and $c =$ IBE.Enc($params, T, pok;$

$H_2(T, pok)$) hold and $(T, pok) \notin \mathcal{T}_{h_2}$.

[Case 3-2]: Above cases do not occur and the same event as Case 3-1 occurs except that corresponding intermediate ciphertext (pre-open key) was issued to the H_2 hash oracle but corresponding plaintext was not issued to the H_1 hash oracle.

[Case 4]: Above cases do not occur and $\mathcal{A}$ issues $(upk^*, m_{\hat{b}}||r_{\hat{b}}, T^*)$ where $\hat{b} \in \{0, 1\}$ to the H_1 hash oracle.

Cases 1-1 and 1-2 do not occur as almost the same discussions as Case 1 in Type-I security proof except that pk is generated honestly and $params$ is generated maliciously. Therefore we needs the collision resistance of the IBE's encryption.

Cases 2-1 and 2-2 occurs at most negligible probability. These case is the same as Cases 2-1 and 2-2 in Type-I security except that a $params$ is chosen by the adversary, so we need extensive γ_{IBE}-uniformity.

Cases 3-1 and 3-2 also occurs at most negligible probability since the same as Cases 2-1 and 2-2.

When Case 4 occurs $\mathcal{B}_2$ fails to similate the H_1 hash oracle since $\mathcal{B}_2$ doesn't know $H_1(upk^*, m_b||r_b, T^*)$ and also doesn't know b. However, the probability that the adversary issues $(upk^*, m_{\bar{b}}||r_{\bar{b}}, T^*)$, is at most $1/|\mathbb{R}|$ at each query, where $\bar{b}$ denotes $1 - b$. This is because the adversary's view is independent to $m_{\bar{b}}||r_{\bar{b}}$. If the adversary issues $(upk^*, m_b||r_b, T^*)$, $\mathcal{B}_1$ wins the IND-CPA game.

Consequently, those whole simulations succeeds in overwhelming probability.

$\square$

It should be noted that in the above proof hash functions H_1 and H_2 are regarded as the random oracle in spite of the fact that they are chosen by the adversary. However, this is not a matter in the theoretical proof in the random oracle model, and is practically natural since the hash function is typically chosen according to the standard, as SHA-1.

Theorem 3. *Suppose an IBE scheme has the collision resistance of the encryption. Our proposed scheme then has the strong binding property.*

Sketch of Proof. Suppose an adversary $\mathcal{A}$ wins the strong binding game with outputs $(CT^* = (\tilde{T}^*, c^*), s_T^*, pok^*)$. By the definition of the game, $\perp \neq m_{tr} \neq m_{po} \neq \perp$ holds where $m_{tr} \leftarrow \mathsf{Dec_{TR}}(params, s_T^*, usk^*, c^*)$, and $m_{po} \leftarrow \mathsf{Dec_{PO}}(params, pok^*, usk^*, c^*)$. Please notice that in the decryption, our construction conducts a re-encryption check.

First, $\perp \neq m_{tr}$ means $T = \tilde{T}$.

Second, $\perp \neq m_{tr} \neq m_{po} \neq \perp$ and the PKE encryption being injective (notice that upk is generated honestly) mean $\hat{c}^* \neq pok^*$ where $pk^* = upk^*$, $\hat{c}^* = \mathsf{PKE.Enc}(pk^*, m_{tr}||r_{tr}; H_1(pk, m_{tr}||r_{tr}, T^*))$, $pok^* = \mathsf{PKE.Enc}(pk^*, m_{po}||r_{po}; H_1(pk, m_{po}||r_{po}, T^*))$ for some r_{tr} and some r_{po}.

Third, $\hat{c}^* \neq pok^*$ and the IBE encryption has collision resistance of the encryption also mean $\mathsf{IBE.Enc}(params, T^*, \hat{c}^*; H_2(T^*, \hat{c}^*)) \neq \mathsf{IBE.Enc}(params, T^*, pok^*; H_2(T^*, pok^*))$.

Table 2. Scheme Comparison

	Generic or concrete	Security notion	Model
HYL1 [22]	concrete	IND-TR-CPA$_{\mathsf{IS}}^-$ & *no estimation* & *no estimation*	ROM
HYL2 [22]	concrete	IND-TR-CCA$_{\mathsf{IS}}^-$ & *no estimation* & *no estimation*	ROM
NMKM [27]	generic	IND-TR-CPA$_{\mathsf{IS}}$ & IND-TR-CCA$_{\mathsf{TS}}$ & Binding	SM
MNM1 [26]	generic	IND-TR-CPA$_{\mathsf{IS}}$ & IND-TR-CCA$_{\mathsf{TS}}$ & Binding	SM
MNM2 [26]	generic	IND-TR-CPA$_{\mathsf{IS}}$ & IND-TR-CCA$_{\mathsf{TS}}$ & Binding	ROM
DT [15]	concrete	IND-TR-CPA$_{\mathsf{IS}}$ & IND-TR-CCA$_{\mathsf{TS}}$ & Binding	ROM
CY [11]	concrete	Type-I & Type-II & Binding	SM
ours	generic	Type-I & Type-II$^+$ & Strong binding	ROM

In the security notion & model columns, the left side is the security against malicious receivers, the middle is the security against a malicious time-server, and the right is the security against malicious senders. The compared schemes are ranked lower as the security notion is strong. Note that IND-TR-CCA$_{\mathsf{IS}}^-$ is weaker than IND-TR-CPA$_{\mathsf{IS}}$, and "*no estimation*" means there is no adversarial model where the adversary gets a time-server's secret key. "ROM" indicates that security is proved in the random oracle model, and "SM" indicates that security is proved in the standard model.

This contradicts to the definition when $\mathcal{A}$ wins. So our proposed construction has the strong binding property.   □

4.2 Comparison

We now compare the existing TRPKE-PC schemes [22,11,15,27,26] in the respect of the general/concrete construction, security, and model. Our construction satisfies the strongest security, and is generic construction in the random oracle model.

5 Conclusion

We showed that previous security definitions are incomplete or insufficient. Therefore, we defined new security definitions. We gave a precise definition of a strong decryption oracle against malicious receivers, strong security definition against a malicious time-server cheating under adversarial chosen keys, and strong security definition against senders cheating even if the adversary colludes with the time-server.

We also proposed a generic construction of a TRPKE-PC scheme satisfying stronger security notions. Our proposed construction is the first generic TRPKE(-PC) construction secure against a malicious time-server with malicious key generations.

Acknowledgment. We thank Keita Xagawa for useful discussions, and also thank to the anonymous reviewers for their helpful comments.

References

1. Au, M.H., Chen, J., Liu, J.K., Mu, Y., Wong, D.S., Yang, G.: Malicious KGC attack in certificateless cryptography. In: Proc. ACM Symposium on Information, Computer and Communications Security. ACM Press (2007)
2. Al-Riyami, S.S., Paterson, K.G.: Certificateless Public Key Cryptography. In: Laih, C.-S. (ed.) ASIACRYPT 2003. LNCS, vol. 2894, pp. 452–473. Springer, Heidelberg (2003)
3. Barbosa, M., Farshim, P.: Relations among Notions of Complete Non-malleability: Indistinguishability Characterisation and Efficient Construction without Random Oracles. In: Steinfeld, R., Hawkes, P. (eds.) ACISP 2010. LNCS, vol. 6168, pp. 145–163. Springer, Heidelberg (2010)
4. Bellare, M., Rogaway, P.: Random Oracles are Practical: A Paradigm for Designing Efficient Protocols. In: 1st ACM Conference on Computer and Communications Security, pp. 62–73. ACM, New York (1993)
5. Boneh, D., Franklin, M.: Identity-Based Encryption from the Weil Pairing. In: Kilian, J. (ed.) CRYPTO 2001. LNCS, vol. 2139, pp. 213–229. Springer, Heidelberg (2001)
6. Cathalo, J., Libert, B., Quisquater, J.-J.: Efficient and Non-Interactive Timed-Release Encryption. In: Qing, S., Mao, W., López, J., Wang, G. (eds.) ICICS 2005. LNCS, vol. 3783, pp. 291–303. Springer, Heidelberg (2005)

7. Chan, A.C.-F., Blake, I.F.: Scalable, Server-Passive, User-Anonymous Timed Release Public Key Encryption from Bilinear Pairing. In: 25th International Conference on Distributed Computing Systems, pp. 504–513. IEEE (2005), Full version of this paper is available at http://eprint.iacr.org/2004/211

8. Cheon, J.H., Hopper, N., Kim, Y., Osipkov, I.: Timed-Release and Key-Insulated Public Key Encryption. In: Di Crescenzo, G., Rubin, A. (eds.) FC 2006. LNCS, vol. 4107, pp. 191–205. Springer, Heidelberg (2006), Full version of this paper is available at http://eprint.iacr.org/2004/231

9. Cheon, J.H., Hopper, N., Kim, Y., Osipkov, I.: Provably Secure Timed-Release Public Key Encryption. ACM Trans. Inf. Syst. Secur. 11(2), 1–44 (2008)

10. Chow, S.S.M., Roth, V., Rieffel, E.G.: General Certificateless Encryption and Timed-Release Encryption. In: Ostrovsky, R., De Prisco, R., Visconti, I. (eds.) SCN 2008. LNCS, vol. 5229, pp. 126–143. Springer, Heidelberg (2008)

11. Chow, S.S.M., Yiu, S.M.: Timed-Release Encryption Revisited. In: Baek, J., Bao, F., Chen, K., Lai, X. (eds.) ProvSec 2008. LNCS, vol. 5324, pp. 38–51. Springer, Heidelberg (2008)

12. Dent, A.W.: A Survey of Certificateless Encryption Schemes and Security Models. Int. J. Inf. Sec. 7(5), 349–377 (2008)

13. Dent, A.W., Libert, B., Paterson, K.G.: Certificateless Encryption Schemes Strongly Secure in the Standard Model. In: Cramer, R. (ed.) PKC 2008. LNCS, vol. 4939, pp. 344–359. Springer, Heidelberg (2008)

14. Dodis, Y., Katz, J.: Chosen-Ciphertext Security of Multiple Encryption. In: Kilian, J. (ed.) TCC 2005. LNCS, vol. 3378, pp. 188–209. Springer, Heidelberg (2005)

15. Dent, A.W., Tang, Q.: Revisiting the Security Model for Timed-Release Encryption with Pre-open Capability. In: Garay, J., Lenstra, A.K., Mambo, M., Peralta, R. (eds.) ISC 2007. LNCS, vol. 4779, pp. 158–174. Springer, Heidelberg (2007)

16. Fischlin, M.: Completely Non-malleable Schemes. In: Caires, L., Italiano, G.F., Monteiro, L., Palamidessi, C., Yung, M. (eds.) ICALP 2005. LNCS, vol. 3580, pp. 779–790. Springer, Heidelberg (2005), Full version of this paper is available at http://www.cdc.informatik.tu-darmstadt.de/~fischlin/publications/fischlin.completely-non-malleable.2005.pdf

17. Fujioka, A., Okamoto, Y., Saito, T.: Security of Sequential Multiple Encryption. In: Abdalla, M., Barreto, P.S.L.M. (eds.) LATINCRYPT 2010. LNCS, vol. 6212, pp. 20–39. Springer, Heidelberg (2010)

18. Fujioka, A., Okamoto, Y., Saito, T.: Generic Construction of Strongly Secure Timed-Release Public-Key Encryption. In: Parampalli, U., Hawkes, P. (eds.) ACISP 2011. LNCS, vol. 6812, pp. 319–336. Springer, Heidelberg (2011)

19. Fujisaki, E., Okamoto, T.: How to Enhance the Security of Public-Key Encryption at Minimum Cost. In: Imai, H., Zheng, Y. (eds.) PKC 1999. LNCS, vol. 1560, pp. 53–68. Springer, Heidelberg (1999)

20. Fujisaki, E., Okamoto, T.: Secure Integration of Asymmetric and Symmetric Encryption Schemes. In: Wiener, M. (ed.) CRYPTO 1999. LNCS, vol. 1666, pp. 537–554. Springer, Heidelberg (1999)

21. Goldwasser, S., Micali, S.: Probabilistic Encryption. J. Comput. Syst. Sci. 28(2), 270–299 (1984)

22. Hwang, Y.H., Yum, D.H., Lee, P.J.: Timed-Release Encryption with Pre-open Capability and Its Application to Certified E-mail System. In: Zhou, J., López, J., Deng, R.H., Bao, F. (eds.) ISC 2005. LNCS, vol. 3650, pp. 344–358. Springer, Heidelberg (2005)

23. Hwang, Y.H., Liu, J.K., Chow, S.S.M.: Certificateless Public Key Encryption Secure against KGC Attacks in the Standard Model. Journal of Universal Computer Science, Special Issue on Cryptography in Computer System Security 14(3), 463–480 (2008)
24. Kawai, Y., Sakai, Y., Kunihiro, N.: On the (Im)possibility Results for Strong Attack Models for Public Key Cryptsystems. Journal of Internet Services and Information SecurityJISISj
25. May, T.: Timed-Release Crypto (1993) (manuscript).
26. Matsuda, T., Nakai, Y., Matsuura, K.: Efficient Generic Constructions of Timed-Release Encryption with Pre-open Capability. In: Joye, M., Miyaji, A., Otsuka, A. (eds.) Pairing 2010. LNCS, vol. 6487, pp. 225–245. Springer, Heidelberg (2010)
27. Nakai, Y., Matsuda, T., Kitada, W., Matsuura, K.: A Generic Construction of Timed-Release Encryption with Pre-open Capability. In: Takagi, T., Mambo, M. (eds.) IWSEC 2009. LNCS, vol. 5824, pp. 53–70. Springer, Heidelberg (2009)
28. Paterson, K.G., Quaglia, E.A.: Time-Specific Encryption. In: Garay, J.A., De Prisco, R. (eds.) SCN 2010. LNCS, vol. 6280, pp. 1–16. Springer, Heidelberg (2010)
29. Rivest, R.L., Shamir, A., Wagner, D.A.: Time-lock puzzles and timed-release crypto. Technical Report MIT/LCS/TR-684, Massachusetts Institute of Technology (1996)
30. Ventre, C., Visconti, I.: Completely Non-malleable Encryption Revisited. In: Cramer, R. (ed.) PKC 2008. LNCS, vol. 4939, pp. 65–84. Springer, Heidelberg (2008)
31. Yang, P., Kitagawa, T., Hanaoka, G., Zhang, R., Matsuura, K., Imai, H.: Applying Fujisaki-Okamoto to Identity-Based Encryption. In: Fossorier, M.P.C., Imai, H., Lin, S., Poli, A. (eds.) AAECC 2006. LNCS, vol. 3857, pp. 183–192. Springer, Heidelberg (2006)
32. Zhang, R., Hanaoka, G., Shikata, J., Imai, H.: On the Security of Multiple Encryption or CCA-security+CCA-security=CCA-security? In: Bao, F., Deng, R., Zhou, J. (eds.) PKC 2004. LNCS, vol. 2947, pp. 360–374. Springer, Heidelberg (2004)

A   More Related Works

Timed-release encryption (TRE) has a mechanism whereby a ciphertext can be decrypted after a time-period. (We distinguish TRE and TRPKE by whether a sender specifies a receiver.) There are two approaches to constructing a TRE scheme. One uses *time-lock puzzles* [29], in which a receiver has to solve "puzzles" to decrypt the ciphertext; and it is sufficiently difficult that the receiver cannot solve it within a specific time. The other uses a trusted agent (i.e., time-server) that periodically, or as needed, generates time-specific information.

Chan and Blake [7] proposed the first TRPKE scheme, and after that several additional functionalities, security aspects, and generalizations have been introduced. Cheon et al. [8,9] introduced a TRE scheme with *authentication*. Cathalo et al. [6] defined a notion of *release time confidentiality* for TRPKE and Chow and Yiu [11] defined it for TRPKE-PC. Chow et al. [10] generalized TRPKE so that a time-period can be *hierarchical*, and Paterson et al. [28] generalized TRE for *time-specific encryption* (TSE), in which a sender can specify a "interval" of a time-period, such as $[T_{from}, T_{to}]$, and a receiver can decrypt a ciphertext with a time instant key corresponding to a time-period $T \in [T_{from}, T_{to}]$.

Security notions for TRPKE-PC were first defined by Hwang et al. [22] and later adjusted by Dent and Tang [15]. Essentially, three security notions were defined: *time-server security*, *insider security* and *sender security*, respectively named IND-TR-CCA$_{TS}$, IND-TR-CPA$_{IS}$ and Binding. IND-TR-CCA$_{TS}$ security is defined in an indistinguishability game between a challenger and an adversary where the adversary can issue a time-period decryption query consisting of a ciphertext and a time-period to obtain a plaintext and can issue a pre-open decryption query consisting of a ciphertext and a pre-open key to obtain a corresponding plaintext. IND-TR-CPA$_{IS}$ security is similar to IND-TR-CCA$_{TS}$, except that the adversary can only issue a release query consisting of a time-period to obtain corresponding time signal. The Binding notion means that the adversary cannot make a ciphertext, time-period and pre-open key where a plaintext that is decrypted with time-period decryption differs from one decrypted with pre-open decryption.

B   IND-CPA Security

We describe IND-CPA security [21] for a PKE scheme based on the following IND-CPA game between a challenger C and an adversary A. At the beginning of the game, C runs the key generation algorithm $(pk, sk) \leftarrow \mathsf{PKE.KG}(1^\lambda)$ and gives the public key pk to A. A gives two messages m_0, m_1 to C. C randomly chooses $b \in \{0, 1\}$ and gives a challenge ciphertext $\hat{c}^* \leftarrow \mathrm{Penc}(pk, m_b)$ to A. A finally outputs a guess $b' \in \{0, 1\}$. We define the advantage of A for PKE in the IND-CPA game as $\mathbf{Adv}_{\mathrm{PKE}, A}^{\mathrm{IND\text{-}CPA}}(\lambda) = |\, 2 \Pr[b = b'] - 1\,|$.

Definition 4. *A public-key encryption scheme is IND-CPA secure if no PPT adversary A has non-negligible advantage $\mathbf{Adv}_{PKE, A}^{IND\text{-}CPA}(\lambda)$.*

C   IND-ID-CPA Security

We describe the IND-ID-CPA security [5] for an IBE scheme based on the following IND-ID-CPA game between a challenger C and an adversary A. At the beginning of the game, C takes a security parameter 1^λ, runs the setup algorithm $(params, msk) \leftarrow \mathsf{IBE.Setup}(1^\lambda)$, and gives $params$ to A. A gives two messages m_0, m_1 and an identity ID* to C. Then C randomly chooses $b \in \{0, 1\}$ and gives a challenge ciphertext $c^* \leftarrow \mathsf{IBE.Enc}(params, \text{ID}^*, m_b)$ to A. A finally outputs a guess $b' \in \{0, 1\}$.

During the game, the adversary A can issue extraction queries ID to the challenger C to obtain the decryption key d_{ID} except the challenge identity ID*.

We define the advantage of A in the IND-ID-CPA game as $\mathbf{Adv}_{\mathrm{IBE}, A}^{\mathrm{IND\text{-}ID\text{-}CPA}}(\lambda) = |\, 2 \Pr[b = b'] - 1\,|$.

Definition 5. *An identity-based encryption is IND-ID-CPA secure if no PPT adversary A has non-negligible advantage $\mathbf{Adv}_{IBE, A}^{IND\text{-}ID\text{-}CPA}(\lambda)$.*

D   Analysis of Chow and Yiu's Scheme

We show here a previous Type-II secure scheme [11] is vulnerable to a malicious time-server with malicious key generation. We first review how a time-server's/user's public key and a ciphertext are generated in the Chow and Yiu's scheme. We describe algorithms only Setup, KeyGen, Enc that are sufficient to analyze, and omit detail definitions (e.g., the definition of bilinear map) due to the space limitation.

- Setup(1^λ): Let $\mathbb{G}, \mathbb{G}_T$ be two multiplicative groups with a bilinea map $\hat{e} : \mathbb{G} \times \mathbb{G} \to \mathbb{G}_T$. They are of the same order p, which is a prime and $2^\lambda < p < 2^{\lambda+1}$. Pick the following components:
  - **Encryption key:** choose two generators $g, g_2 \leftarrow \mathbb{G}$.
  - **Master public key:** choose an exponent $\alpha \leftarrow \mathbb{Z}_p$ and set $g_1 = g^\alpha$.
  - **Hash key for time-identifier:** randomly pick $(\ell + 1)$ $\mathbb{G}$ elements $\vec{U} = (u', u_1, \ldots, u_\ell)$. Let $T = t_1 t_2 \cdots t_\ell$. Define $F_u(T) = u' \prod_{j=1}^{\ell} u_j^{t_j}$.
  - **Hash key for ciphertext validity:** randomly pick $\vec{V} = (v', v_1, \ldots, v_\ell) \in \mathbb{G}^{\ell+1}$. This vector defines $F_v(w) = v' \prod_{j=1}^{\ell} v_j^{b_j}$ where w is an ℓ-bits string $b_1 b_2 \cdots b_\ell$.
  - **Key-derivation function (KDF):** K is a KDF such that $K : \mathbb{G}_t \to \{0,1\}^{n+k+1}$, in which assuming that the output of K is computationally indistinguishable from a random distribution when the input comes from a uniformly distribution. Also assuming an implicit one-to-one mapping between $\mathbb{G}$ and $\{0,1\}^{k+1}$.

  Let H be a collision resistance hash function. The output is: ($params = (\lambda, p, \mathbb{G}, \mathbb{G}_T, \hat{e}(\cdot, \cdot), \ell, H, K, g, g_1, g_2, \vec{U}, \vec{V}), msk = g_2^\alpha$).
- KeyGen(tpk): Pick $usk \leftarrow \mathbb{Z}_p^*$ and set $upk = (g^{usk}, g_1^{usk})$.
- Enc(tpk, T, upk, m): See upk as (X, Y), If $e(X, g_1) = e(g, Y)$ holds, pick $s \leftarrow \mathbb{Z}_p$, set $k = K(\hat{e}(X, g_2)^s)$, compute $CT = (C_1, C_2, \tau, \sigma) = (m \cdot \hat{e}(Y, g_2)^s, (T || F_u(T)^s) \oplus k, g^s, F_v(w)^s)$ where $w = H(C_1 || C_2 || k || upk)$ and set $pok = g_1^s$. Otherwise, return $\bot$.

We next show how an adversary, i.e., malicious time-server with malicious key generation, decrypt a ciphertext without a receiver's secret key.

The adversary generate $params$ as follows. The adversary honestly chooses $H, K, g, g_1, \vec{U}, \vec{V}$. Then the adversary picks $x \leftarrow \mathbb{Z}_p^*$, sets $g_2 = g^x$ and then publishes $params$. To decrypt a ciphertext (C_1, C_2, τ, σ) encrypted with a public key (X, Y), the adversary computes $\frac{C_1}{\hat{e}(Y, \tau)^x}$.

A distribution of maliciously generated $params$ is identical to honestly generated one, so no one can detect the malicious key generation. Chow and Yiu [11] insists that in practice these elements of $params$ can be generated by using a pseudorandom function with a public seed. If so, we should proof the security under such condition but they did not: in the proof of Type-I security, a simulator maliciously generates $params$ and embeds a trapdoor in it without a public seed.

Fully Secure Unidirectional Identity-Based Proxy Re-encryption*

Song Luo[1,3,4], Qingni Shen[2,**], and Zhong Chen[2,3,4]

[1] College of Computer Science and Engineering,
Chongqing University of Technology, Chongqing, China
[2] School of Software and Microelectronics & MoE Key Lab of Network and
Software Assurance, Peking University, Beijing, China
[3] Institute of Software, School of Electronics Engineering and Computer Science,
Peking University
[4] Key Laboratory of High Confidence Software Technologies (Peking University),
Ministry of Education
{luosong,shenqn,chen}@infosec.pku.edu.cn

Abstract. Proxy re-encryption (PRE) allows the proxy to translate a ciphertext encrypted under Alice's public key into another ciphertext that can be decrypted by Bob's secret key. Identity-based proxy re-encryption (IB-PRE) is the development of identity-based encryption and proxy re-encryption, where ciphertexts are transformed from one identity to another. In this paper, we propose two novel unidirectional identity-based proxy re-encryption schemes, which are both non-interactive and proved secure in the standard model. The first scheme is a single-hop IB-PRE scheme and has master secret security, allows the encryptor to decide whether the ciphertext can be re-encrypted. The second scheme is a multi-hop IB-PRE scheme which allows the ciphertext re-encrypted multiple times but without the size of ciphertext growing linearly as previous multi-hop IB-PRE schemes.

Keywords: Proxy Re-encryption, Identity-Based Encryption, Single-hop, Multi-hop.

1 Introduction

The primitive of proxy re-encryption (PRE) is first proposed by Blaze et al. [2] which involves three parties: Alice, Bob, and a proxy. PRE allows allows Alice to temporarily delegate the decryption rights to Bob via a proxy, i.e., the proxy with proper re-encryption key can translate a ciphertext encrypted under Alice's public key into another ciphertext that can be decrypted by Bob's secret key. Unlike the traditional proxy decryption scheme, PRE doesn't need users to store any additional decryption key, in other words, any decryption would be finished

---

* Supported by National Natural Science Foundation of China (No.60873238, 61073156, 60970135, 60821003, 61170263).

** Corresponding author.

H. Kim (Ed): ICISC 2011, LNCS 7259, pp. 109–126, 2012.
© Springer-Verlag Berlin Heidelberg 2012

using only his own secret keys. PRE can be used in many scenarios, such as email forwarding, distributed file system, and the DRM of Apple's iTunes.

The concept of identity-based encryption (IBE) was first introduced by Shamir [16]. In an IBE system, arbitrary strings such as e-mail addresses or IP addresses can be used to form public keys for users. After Boneh and Franklin [5] proposed a practical identity-base encryption scheme, Green and Ateniese [11] proposed the first identity-based proxy re-encryption (IB-PRE). It allows the proxy to convert an encryption under Alice's identity into the encryption under Bob's identity. Due to the simplification of public-key infrastructure in identity-based framework, IB-PRE schemes are more desirable than non-identity-based ones.

According to the direction of transformation, IB-PRE schemes can be classified into two types, one is bidirectional, i.e., the proxy can transform from Alice to Bob and vice versa; the other is unidirectional, i.e., the proxy can only convert in one direction. Blaze et al. [2] also gave another method to classify IB-PRE schemes: single-hop, where the ciphertext can be transformed only once; and multi-hop, where the ciphertext can be transformed from Alice to Bob to Charlie and so on.

IB-PRE schemes are different from PRE schemes in which there exists a trusted private key generator (PKG) to generate all secret keys for identities. If Alice can compute re-encryption keys without the participation of Bob or PKG, the scheme is called non-interactive, or else called interactive. Obviously, it would be a hard work if all re-encryption keys are computed by the PKG. Therefore, it is more desirable to find non-interactive IB-PRE schemes. However, when generating secret keys, PKG insert the master key to users' secret keys. Obviously, re-encryption must involve some information of master key. But it is always hard to extract the part of master key from secret key to generate re-encryption, since elements of secret keys are always group elements and hard to get the discrete log based on a random generator.

Up to now, there are two ways to generate the re-encryption keys. One is proposed by Green and Ateniese [11]. In Green-Ateniese paradigm, to form a re-encryption key from Alice to Bob, a token is inserted in Alice's secret key and the token is encrypted to Bob, then these two parts form the re-encryption key. It is non-interactive in the generation of the re-encryption key and the re-encryption can be multi-hop where the ciphertext can be re-encrypted again and again. But the drawback of this method is that after one re-encryption, the encryption of the token would be attached to the ciphertext. So the ciphertext will grow linearly with the re-encryption times. The other is interactive proposed by Matsuo [15] in which the re-encryption key is generated by the private key generator or an extra re-encryption key generator which also owns the master key. This type of IB-PRE schemes are always single-hop where the re-encrypted ciphertext cannot be re-encrypted again.

1.1   Our Contribution

We present two novel unidirectional identity-based proxy re-encryption schemes. The first scheme is a single-hop scheme with master secret security. To make this scheme be unidirectional, we present two kinds of ciphertexts, the original ciphertext is called the second level ciphertext which is nearly the same as Lewko-Waters IBE scheme's ciphertext, the transformed ciphertext is called the first level ciphertext and cannot be re-encrypted more. Our way of generating re-encryption keys are different from Green-Ateniese and Matsuo. To make the re-encryption key be generated by the user itself, we introduce non-group elements containing part information of master keys in user's secret keys and provide re-randomization to avoid collusion of proxy and users.

Based on our single-hop scheme, we present a multi-hop scheme in which the decryption cost and size of ciphertext do not grow linearly with the re-encryption times. To the best of our knowledge, this scheme is the first unidirectional IB-PRE scheme without growing linearly in the size of ciphertext as the re-encryption times increasing. Both schemes are non-interactive, which means the re-encryption key can be generated by Alice without the participation of Bob or the private key generator. We construct our schemes in composite order groups and use dual system encryption to prove the security of proposed schemes.

1.2   Related Works

Identity-Based Encryption. The first practical IBE scheme, proposed by Boneh and Franklin [5], was proven secure in the random oracle model. To remove random oracles, Canetti, Halevi, and Katz [7] suggested a weaker security notion for IBE, known as selective identity (selective-ID) security, relative to which they were able to build an inefficient but secure IBE scheme without using random oracles. Boneh and Boyen [3] proposed two new efficient selective-ID secure IBE schemes without random oracles. Later Boneh and Boyen [4], Waters [20] proposed new IBE schemes with full security. In Eurocrypt'06, Gentry [10] proposed an efficient identity based encryption with tight security reduction in the standard model but based on a stronger assumption.

By using dual system encryption, Waters [21] proposed the first fully secure IBE and HIBE schemes with short parameters under simple assumptions. But Waters's HIBE scheme does not have constant ciphertext size. Afterwards, another two fully secure HIBE schemes with constant size ciphertexts were proposed in composite order groups [8, 13].

Identity-Based Proxy Re-encryption. Ateniese et al. [1] presented the first unidirectional and single-use proxy re-encryption scheme. In 2007, Green and Ateniese [11] provided the first identity-based proxy re-encryption scheme but their scheme is secure in the random oracle model. Chu and Tzeng [9] proposed a new multi-hop unidirectional identity-based proxy re-encryption scheme in the standard model. However, their scheme is not chosen-ciphertext secure, Shao et al. [17] pointed out that its transformed ciphertext can be modified to another well-formed transformed ciphertext by anyone. Recently Lai et al. [12] gave new constructions on IB-PRE based on identity-based mediated encryption. Luo et al. [14] also gave a new generic IB-PRE construction based on an existing IBE scheme. Wang et al. [18] proposed the first multi-use CCA-secure unidirectional IB-PRE scheme. All these schemes follow Green-Ateniese token paradigm, which makes the decryption cost and size of ciphertext grow linearly with the re-encryption times. In addition, Matsuo [15] proposed a new proxy re-encryption system for identity-based encryption, but his solution needs a re-encryption key generator (RKG) to generate re-encryption keys. Wang et al. [19] followed the route of Matsuo and proposed new secure IB-PRE schemes which let the PKG take part in generating the re-encryption keys.

1.3   Organization

The remaining paper is organized as follows. In Section 2, we review the definitions related to our proposals. In what follows, we present the single-hop scheme and its security analysis, and the multi-hop scheme and its security analysis, in Section 3 and Section 4, respectively. In Section 5 we discuss some extensions of the two schemes. Finally, we conclude the paper in Section 6.

2   Backgroud

2.1   Multi-hop Identity-Based Proxy Re-encryption

Definition 1. *A multi-hop unidirectional IB-PRE scheme consists of the following six algorithms:* **Setup, KeyGen, ReKeyGen, Enc, ReEnc,** *and* **Dec.**

Setup(1^λ). *This algorithm takes the security parameter λ as input and generates a public key* PK, *a master secret key* MK.

KeyGen(MK,$\mathcal{I}$). *This algorithm takes* MK *and an identity* $\mathcal{I}$ *as input and generates a secret key $SK_\mathcal{I}$ associated with $\mathcal{I}$.*

ReKeyGen($SK_\mathcal{I},\mathcal{I}'$). *This algorithm takes a secret key $SK_\mathcal{I}$ and an identity $\mathcal{I}'$ as input and generates a re-encryption key $RK_{\mathcal{I}\to\mathcal{I}'}$.*

Enc(PK,$M,\mathcal{I}$). *This algorithm takes* PK, *a message M, and an identity $\mathcal{I}$ as input, and generates a ciphertext $CT_\mathcal{I}$.*

ReEnc($CT_\mathcal{I},RK_{\mathcal{I}\to\mathcal{I}'}$). *This algorithm takes a a ciphertext $CT_\mathcal{I}$ encrypted to $\mathcal{I}$ and a re-encryption key $RK_{\mathcal{I}\to\mathcal{I}'}$ as input, generates a ciphertext $CT_{\mathcal{I}'}$ encrypted to $\mathcal{I}'$.*

Dec($CT_\mathcal{I},SK_\mathcal{I}$). *This algorithm takes a ciphertext $CT_\mathcal{I}$ and $SK_\mathcal{I}$ associated with $\mathcal{I}$ as input and returns the message M or the error symbol $\perp$ if $CT_\mathcal{I}$ is invalid.*

Correctness. A multi-hop unidirectional IB-PRE scheme should satisfy the following requirements:

1. **Dec(Enc(PK,$M,\mathcal{I}$),$SK_\mathcal{I}$)** $= M$;
2. **Dec(ReEnc(($\cdots$ReEnc(Enc(PK,$M,\mathcal{I}$),$RK_{\mathcal{I}\to\mathcal{I}_1}$)$\cdots$),$RK_{\mathcal{I}_{n-1}\to\mathcal{I}_n}$),$SK_{\mathcal{I}_n}$)** $= M$;

We describe the game-based security definitions for multi-hop unidirectional IB-PRE systems as follows.

Definition 2. *The security of a multi-hop unidirectional IB-PRE scheme is defined according to the following* IND-PrID-ATK *game, where* ATK $\in \{\text{CPA}, \text{CCA}\}$.

Setup. *Run the* **Setup** *algorithm and give* PK *to the adversary* $\mathcal{A}$.

Phase 1. $\mathcal{A}$ *makes the following queries.*

- *Extract($\mathcal{I}$):* $\mathcal{A}$ *submits an identity* $\mathcal{I}$ *for a* **KeyGen** *query, return the corresponding secret key* $SK_{\mathcal{I}}$.
- *RKExtract($\mathcal{I}, \mathcal{I}'$):* $\mathcal{A}$ *submits an identity pair* $(\mathcal{I}, \mathcal{I}')$ *for a* **ReKeyGen** *query, return the re-encryption key* $RK_{\mathcal{I} \to \mathcal{I}'}$.

If ATK = CCA, $\mathcal{A}$ *can make the additional queries:*

- *Reencrypt($\text{CT}_{\mathcal{I}}, \mathcal{I}, \mathcal{I}'$):* $\mathcal{A}$ *submits a ciphertext* $\text{CT}_{\mathcal{I}}$ *encrypted for* $\mathcal{I}$ *and an identity* $\mathcal{I}'$ *for a* **ReEnc** *query, return the re-encrypted ciphertext* $\text{CT}_{\mathcal{I}'} = \textbf{ReEnc}\,(\text{CT}_{\mathcal{I}}, RK_{\mathcal{I} \to \mathcal{I}'})$ *where* $RK_{\mathcal{I} \to \mathcal{I}'} = \textbf{ReKeyGen}(SK_{\mathcal{I}}, \mathcal{I}')$ *and* $SK_{\mathcal{I}} = \textbf{KeyGen}(\text{MK}, \mathcal{I})$.
- *Decrypt($\text{CT}_{\mathcal{I}}, \mathcal{I}$):* $\mathcal{A}$ *submits a ciphertext* $\text{CT}_{\mathcal{I}}$ *encrypted for* $\mathcal{I}$ *for a* **Dec** *query, return the corresponding plaintext* $M = \textbf{Dec}(\text{CT}_{\mathcal{I}}, SK_{\mathcal{I}})$, *where* $SK_{\mathcal{I}} = \textbf{KeyGen}(\text{MK}, \mathcal{I})$.

Note that $\mathcal{A}$ *is not permitted to choose* $\mathcal{I}^*$ *which will be submitted in* **Challenge** *phase such that trivial decryption is possible using keys extracted during this phase (e.g., by using extracted re-encryption keys to translate from* $\mathcal{I}^*$ *to some identity for which* $\mathcal{A}$ *holds a decryption key).*

Challenge. $\mathcal{A}$ *submits a challenge identity* $\mathcal{I}^*$ *and two equal length messages* M_0, M_1 *to* $\mathcal{B}$. $\mathcal{B}$ *flips a random coin* b *and passes the ciphertext* $\text{CT}^* = \text{Enc}(PK, M_b, \mathcal{I}^*)$ *to* $\mathcal{A}$.

Phase 2. *Phase 1 is repeated with the following restrictions. Let* $\mathcal{C}$ *be a set of ciphertext/identity pairs, initially containing the single pair* $\langle \mathcal{I}^*, \text{CT}^* \rangle$. *For all* $\text{CT} \in \mathcal{C}$ *and for all* RK *given to* $\mathcal{A}$, *let* $\mathcal{C}'$ *be the set of all possible values derived via (one or more) consecutive calls to* **Reencrypt**:

- $\mathcal{A}$ *is not permitted to issue any query* **Decrypt**$(\text{CT}, \mathcal{I})$ *where* $\langle \text{CT}, \mathcal{I} \rangle \in (\mathcal{C} \cap \mathcal{C}')$;
- $\mathcal{A}$ *is not permitted to issue any query* **Extract**$(\mathcal{I})$ *or* **RKExtract**$(\mathcal{I}, \mathcal{I}')$ *that would permit trivial decryption of any ciphertext in* $(\mathcal{C} \cap \mathcal{C}')$;
- $\mathcal{A}$ *is not permitted to issue any query* **Reencrypt**$(\text{CT}, \mathcal{I}, \mathcal{I}')$ *where* $\mathcal{A}$ *possesses the keys to trivially decrypt ciphertexts under* $\mathcal{I}'$ *and* $\langle \text{CT}, \mathcal{I} \rangle \in (\mathcal{C} \cap \mathcal{C}')$. *On successful execution of any re-encrypt query, let* CT' *be the result and add the pair* $\langle \text{CT}', \mathcal{I}' \rangle$ *to the set* $\mathcal{C}$.

Guess. $\mathcal{A}$ *outputs its guess* b' *of* b.

The advantage of $\mathcal{A}$ *in this game is defined as* $Adv_{\mathcal{A}} = |\Pr[b' = b] - \frac{1}{2}|$ *where the probability is taken over the random bits used by the challenger and the adversary. We say that a multi-hop unidirectional IB-PRE scheme is* IND-PrID-ATK *secure, where* ATK $\in \{\text{CPA}, \text{CCA}\}$, *if no probabilistic polynomial time adversary* $\mathcal{A}$ *has a non-negligible advantage in winning the* IND-PrID-ATK *game.*

2.2   Single-hop Identity-Based Proxy Re-encryption

Single-hop IB-PRE can be viewed as a weaker concept than multi-hop IB-PRE, in which the ciphertext can be re-encrypted only once or not. According to the re-encryption time, its ciphertext is divided into two levels: second level ciphertext and first level ciphertext. A second ciphertext can be re-encrypted into a first level one (intended for a possibly different receiver) using the suitable re-encryption key and a first level ciphertext cannot be re-encrypted for another party. So the algorithms **Enc** and **Dec** are divided into two sub-algorithms $\mathbf{Enc}_2$ and $\mathbf{Enc}_1$, $\mathbf{Dec}_2$ and $\mathbf{Dec}_1$, respectively. The other algorithms are similar to multi-hop IB-PRE schemes. Furthermore, a single-hop unidirectional IB-PRE scheme should satisfy the following requirements:

1. $\mathbf{Dec}_2(\mathbf{Enc}_2(\mathrm{PK}, M, \mathcal{I}), SK_\mathcal{I}) = M$;
2. $\mathbf{Dec}_1(\mathbf{Enc}_1(\mathrm{PK}, M, \mathcal{I}), SK_\mathcal{I}) = M$;
3. $\mathbf{Dec}_1(\mathbf{ReEnc}(\mathbf{Enc}_2(\mathrm{PK}, M, \mathcal{I}), RK_{\mathcal{I}\to\mathcal{I}'}), SK_{\mathcal{I}'}) = M$.

The game-based security definitions for single-hop unidirectional IB-PRE systems are derived from previous multi-hop IB-PRE systems. Since single-hop unidirectional IB-PRE system has two level ciphertexts, there are two level securities called IND-2PrID-CPA(CCA) security and IND-1PrID-CPA(CCA) security.

Definition 3. *The security of a single-hop unidirectional IB-PRE scheme at the second level is defined according to the following* IND-2PrID-ATK *game, where* ATK $\in \{$CPA, CCA$\}$.

Setup. *Run the* **Setup** *algorithm and give* PK *to the adversary* $\mathcal{A}$.
Phase 1. $\mathcal{A}$ *makes the following queries.*
   – *Extract(*$\mathcal{I}$*):* $\mathcal{A}$ *submits an identity* $\mathcal{I}$ *for a* **KeyGen** *query, return the corresponding secret key* $SK_\mathcal{I}$.
   – *RKExtract(*$\mathcal{I}, \mathcal{I}'$*):* $\mathcal{A}$ *submits an identity pair* $(\mathcal{I}, \mathcal{I}')$ *for a* **ReKeyGen** *query, return the re-encryption key* $RK_{\mathcal{I}\to\mathcal{I}'}$.
   If ATK $=$ CCA, $\mathcal{A}$ *can make the additional queries:*
   – *Reencrypt(*$\mathrm{CT}_\mathcal{I}, \mathcal{I}, \mathcal{I}'$*):* $\mathcal{A}$ *submits a second level ciphertext* $\mathrm{CT}_\mathcal{I}$ *encrypted for* $\mathcal{I}$ *and an identity* $\mathcal{I}'$ *for a* **ReEnc** *query, the challenger gives the adversary the re-encrypted ciphertext* $\mathrm{CT}_{\mathcal{I}'} = \mathbf{ReEnc}(\mathrm{CT}_\mathcal{I}, RK_{\mathcal{I}\to\mathcal{I}'})$ *where* $RK_{\mathcal{I}\to\mathcal{I}'} = \mathbf{ReKeyGen}(SK_\mathcal{I}, \mathcal{I}')$ *and* $SK_\mathcal{I} = \mathbf{KeyGen}(\mathrm{MK}, \mathcal{I})$.
   – *Decrypt(*$\mathrm{CT}_\mathcal{I}, \mathcal{I}$*):* $\mathcal{A}$ *submits a first level ciphertext* $\mathrm{CT}_\mathcal{I}$ *encrypted for* $\mathcal{I}$ *for a* **Dec**$_1$ *query, return the corresponding plaintext* $M = \mathbf{Dec}_1(\mathrm{CT}_\mathcal{I}, SK_\mathcal{I})$, *where* $SK_\mathcal{I} = \mathbf{KeyGen}(\mathrm{MK}, \mathcal{I})$.
Challenge. $\mathcal{A}$ *submits a challenge identity* $\mathcal{I}^*$ *and two equal length messages* M_0, M_1 *to* $\mathcal{B}$. *If the queries*
   – *Extract(*$\mathcal{I}^*$*); and*
   – *RKExtract(*$\mathcal{I}^*, \mathcal{I}'$*) and* *Extract(*$\mathcal{I}'$*) for any identity* $\mathcal{I}'$
are never made, then flip a random coin b *and pass the ciphertext* $\mathrm{CT}^* = \mathbf{Enc}_2(PK, M_b, \mathcal{I}^*)$ *to* $\mathcal{A}$.
Phase 2. *Phase 1 is repeated with the restriction that* $\mathcal{A}$ *cannot make the following queries:*

- **Extract($\mathcal{I}^*$)**;
- **RKExtract($\mathcal{I}^*, \mathcal{I}'$)** and **Extract($\mathcal{I}'$)** for any identity $\mathcal{I}'$;
- **Reencrypt(CT$^*, \mathcal{I}^*, \mathcal{I}'$)** and **Extract($\mathcal{I}'$)** for any identity $\mathcal{I}'$;
- **Decrypt(CT$_{\mathcal{I}'}, \mathcal{I}'$)** for any identity $\mathcal{I}'$, where CT$_{\mathcal{I}'} = \mathbf{ReEnc}(\text{CT}^*, \mathcal{I}^*, \mathcal{I}')$.

Guess. $\mathcal{A}$ outputs its guess b' of b.

The advantage of $\mathcal{A}$ in this game is defined as $Adv_{\mathcal{A}} = |\Pr[b' = b] - \frac{1}{2}|$ where the probability is taken over the random bits used by the challenger and the adversary. We say that a single-hop unidirectional IB-PRE scheme is IND-2PrID-ATK secure, where ATK $\in \{\text{CPA}, \text{CCA}\}$, if no probabilistic polynomial time adversary $\mathcal{A}$ has a non-negligible advantage in winning the IND-2PrID-ATK game.

Note that in the **Decrypt** query, we only provide the first level ciphertext decryption because any second level ciphertext can be re-encrypted to a first level ciphertext and then be queried for decryption.

Definition 4. The security of a single-hop unidirectional IB-PRE scheme at the first level is defined according to the following IND-1PrID-ATK game, where ATK $\in \{\text{CPA}, \text{CCA}\}$.

Setup. Run the **Setup** algorithm and give PK to the adversary $\mathcal{A}$.

Phase 1. $\mathcal{A}$ makes the following queries.

- **Extract($\mathcal{I}$)**: $\mathcal{A}$ submits an identity $\mathcal{I}$ for a **KeyGen** query, return the corresponding secret key $SK_{\mathcal{I}}$.
- **RKExtract($\mathcal{I}, \mathcal{I}'$)**: $\mathcal{A}$ submits an identity pair $(\mathcal{I}, \mathcal{I}')$ for a **ReKeyGen** query, return the re-encryption key $RK_{\mathcal{I} \to \mathcal{I}'}$.

If ATK = CCA, $\mathcal{A}$ can make the additional queries:

- **Decrypt(CT$_{\mathcal{I}}, \mathcal{I}$)**: $\mathcal{A}$ submits a first level ciphertext CT$_{\mathcal{I}}$ encrypted to $\mathcal{I}$ for a **Dec$_1$** query, return the corresponding plaintext $M = \mathbf{Dec}_1(\text{CT}_{\mathcal{I}}, SK_{\mathcal{I}})$, where $SK_{\mathcal{I}} = \mathbf{KeyGen}(\text{MK}, \mathcal{I})$.

Challenge. $\mathcal{A}$ submits a challenge identity $\mathcal{I}^*$ and two equal length messages M_0, M_1 to $\mathcal{B}$. If the query **Extract($\mathcal{I}^*$)** is never made, then $\mathcal{C}$ flips a random coin b and passes the ciphertext CT$^* = \mathbf{Enc}_1(PK, M_b, \mathcal{I}^*)$ to $\mathcal{A}$.

Phase 2. Phase 1 is repeated with the restriction that $\mathcal{A}$ cannot make the following queries:

- **Extract($\mathcal{I}^*$)**;
- **Decrypt(CT$^*, \mathcal{I}^*$)**.

Guess. $\mathcal{A}$ outputs its guess b' of b.

The advantage of $\mathcal{A}$ in this game is defined as $Adv_{\mathcal{A}} = |\Pr[b' = b] - \frac{1}{2}|$ where the probability is taken over the random bits used by the challenger and the adversary. We say that a single-hop unidirectional IB-PRE scheme is IND-1PrID-ATK secure, where ATK $\in \{\text{CPA}, \text{CCA}\}$, if no probabilistic polynomial time adversary $\mathcal{A}$ has a non-negligible advantage in winning the IND-1PrID-ATK game.

2.3   Master Secret Security

Master secret security is an important property for unidirectional PRE defined by Ateniese et al. [1]. Roughly speaking, even if the dishonest proxy colludes with the delegatee, it is still impossible for them to derive the delegator's secret key in full.

Definition 5. *The master secret security of a single-hop or multi-hop unidirectional IB-PRE scheme is defined according to the following master secret security game.*

Setup. *Run the* **Setup** *algorithm and give* PK *to the adversary* $\mathcal{A}$.
Phase 1. $\mathcal{A}$ *makes the following queries.*
  - **Extract($\mathcal{I}$):** $\mathcal{A}$ *submits an identity* $\mathcal{I}$ *for a* **KeyGen** *query, return the corresponding secret key* $SK_{\mathcal{I}}$.
  - **RKExtract($\mathcal{I}, \mathcal{I}'$):** $\mathcal{A}$ *submits an identity pair* $(\mathcal{I}, \mathcal{I}')$ *for a* **ReKeyGen** *query, return the re-encryption key* $RK_{\mathcal{I} \to \mathcal{I}'}$.
Challenge. $\mathcal{A}$ *submits a challenge identity* $\mathcal{I}^*$ *and query* **Extract($\mathcal{I}^*$)** *is never made.*
Phase 2. *Phase 1 is repeated with the restriction that* $\mathcal{A}$ *cannot make query* **Extract($\mathcal{I}^*$)**.
Output. $\mathcal{A}$ *outputs the secret key* $SK_{\mathcal{I}^*}$ *for the challenge identity* $\mathcal{I}^*$.

The advantage of $\mathcal{A}$ in this game is defined as $Adv_{\mathcal{A}} = \Pr[\mathcal{A} \ succeeds]$. A single-hop or multi-hop IB-PRE scheme has master secret security if no probabilistic polynomial time adversary $\mathcal{A}$ has a non-negligible advantage in winning the master secret security game.

For single-hop unidirectional IB-PRE schemes, it is easy to see that the master secret security is implied by the first level plaintext security. We have the following result.

Lemma 1. *For a single-hop unidirectional IB-PRE scheme, the master secret security is implied by the first level plaintext security. That is, if there exists an adversary* $\mathcal{A}$ *who can break the master secret security of a single-hop unidirectional IB-PRE scheme* $\mathcal{E}$, *then there also exists an adversary* $\mathcal{B}$ *who can also break* $\mathcal{E}$*'s IND-1PrID-CPA security.*

Lemma 1 is obvious, so we omit its proof here.

2.4   Composite Order Bilinear Groups

Composite order bilinear groups were first introduced by Boneh, Goh and Nissim in [6].

Definition 6. *Let* $\mathcal{G}$ *be an algorithm called a* **bilinear group generator** *that takes as input a security parameter* λ *and outputs a tuple* $(N = p_1 p_2 p_3, G, G_T, e)$ *where* p_1, p_2 *and* p_3 *are three distinct primes,* G *and* G_T *are two multiplicative abelian groups of order* N, *and* $e : G \times G \to G_T$ *is an efficiently computable map (or "pairing") satisfying the following properties:*

- *(Bilinear)* $\forall g, h \in G, a, b \in \mathbb{Z}_N, e(g^a, h^b) = e(g, h)^{ab}$.
- *(Non-degenerate)* $\exists g \in G$ *such that* $e(g, g)$ *has order* N *in* G_T.

We assume that the group action in G and G_T as well as the bilinear map e are all polynomial time computable in λ. Furthermore, we assume that the description of G and G_T includes a generator of G and G_T respectively.

We say that G, G_T are bilinear groups *if the group operation in* G *and the bilinear map* $e : G \times G \to G_T$ *are both efficiently computable*.

We let G_{p_1}, G_{p_2} and G_{p_3} denote the subgroups of order p_1, p_2 and p_3 in G respectively. There is an important property called "orthogonality" between two different order subgroups under the bilinear map e, i.e., if $g \in G_{p_i}$ and $h \in G_{p_j}$ where $i \neq j$, then $e(g, h) = 1$. If g_1 generates G_{p_1}, g_2 generates G_{p_2} and g_3 generates G_{p_3}, then every element h of G can be expressed as $g_1^x g_2^y g_3^z$ for some values $x, y, z \in \mathbb{Z}_N$.

2.5 Complexity Assumptions

We use the notation $X \xleftarrow{\text{R}} S$ to express that X is chosen uniformly randomly from the finite set S.

Assumption 1. Given a bilinear group generator $\mathcal{G}$, we define the following distribution:

$$\mathbb{G} = (N = p_1 p_2 p_3, G, G_T, e) \xleftarrow{\text{R}} \mathcal{G}(\lambda),$$
$$g \xleftarrow{\text{R}} G_{p_1}, X_3 \xleftarrow{\text{R}} G_{p_3},$$
$$D = (\mathbb{G}, g, X_3),$$
$$T_1 \xleftarrow{\text{R}} G_{p_1}, T_2 \xleftarrow{\text{R}} G_{p_1 p_2}.$$

We define the advantage of an algorithm $\mathcal{A}$ in breaking Assumption 1 to be

$$Adv_{\mathcal{A},\mathcal{G}}^{A1}(\lambda) := \left| \Pr[\mathcal{A}(D, T_1) = 1] - \Pr[\mathcal{A}(D, T_2) = 1] \right|.$$

Definition 7. *We say that* $\mathcal{G}$ *satisfies Assumption 1 if* $Adv_{\mathcal{A},\mathcal{G}}^{A1}(\lambda)$ *is a negligible function of* λ *for any probabilistic polynomial-time algorithm* $\mathcal{A}$.

Assumption 2. Given a bilinear group generator $\mathcal{G}$, we define the following distribution:

$$\mathbb{G} = (N = p_1 p_2 p_3, G, G_T, e) \xleftarrow{\text{R}} \mathcal{G}(\lambda),$$
$$g, X_1 \xleftarrow{\text{R}} G_{p_1}, X_2, Y_2 \xleftarrow{\text{R}} G_{p_2}, X_3, Y_3 \xleftarrow{\text{R}} G_{p_3},$$
$$D = (\mathbb{G}, g, X_1 X_2, X_3, Y_2 Y_3),$$
$$T_1 \xleftarrow{\text{R}} G_{p_1 p_3}, T_2 \xleftarrow{\text{R}} G.$$

We define the advantage of an algorithm $\mathcal{A}$ in breaking Assumption 2 to be

$$Adv_{\mathcal{A},\mathcal{G}}^{A2}(\lambda) := \left| \Pr[\mathcal{A}(D, T_1) = 1] - \Pr[\mathcal{A}(D, T_2) = 1] \right|.$$

Definition 8. *We say that $\mathcal{G}$ satisfies Assumption 2 if $Adv_{\mathcal{A},\mathcal{G}}^{A2}(\lambda)$ is a negligible function of λ for any probabilistic polynomial-time algorithm $\mathcal{A}$.*

Assumption 3. Given a bilinear group generator $\mathcal{G}$, we define the following distribution:

$$\mathbb{G} = (N = p_1p_2p_3, G, G_T, e) \xleftarrow{R} \mathcal{G}(\lambda), \alpha, s \xleftarrow{R} \mathbb{Z}_N,$$
$$g \xleftarrow{R} G_{p_1}, X_2, Y_2, Z_2 \xleftarrow{R} G_{p_2}, X_3 \xleftarrow{R} G_{p_3}$$
$$D = (\mathbb{G}, g, g^\alpha X_2, X_3, g^s Y_2, Z_2),$$
$$T_1 = e(g,g)^{\alpha s}, T_2 \xleftarrow{R} G_T.$$

We define the advantage of an algorithm $\mathcal{A}$ in breaking Assumption 3 to be

$$Adv_{\mathcal{A},\mathcal{G}}^{A3}(\lambda) := \left| \Pr[\mathcal{A}(D, T_1) = 1] - \Pr[\mathcal{A}(D, T_2) = 1] \right|.$$

Definition 9. *We say that $\mathcal{G}$ satisfies Assumption 3 if $Adv_{\mathcal{A},\mathcal{G}}^{A3}(\lambda)$ is a negligible function of λ for any probabilistic polynomial-time algorithm $\mathcal{A}$.*

3   Single-hop IB-PRE Scheme

In this section, we present a single-hop IB-PRE scheme. Our construction is based on Lewko-Waters IBE scheme [13] with small modification. We use its ciphertext as the second level ciphertext and add an extra element to make the re-encryption feasible. The scheme is constructed as follows.

3.1   Construction

Setup(1^λ). Given the security parameter λ, this algorithm first gets a bilinear group G of order $N = p_1p_2p_3$ from $\mathcal{G}(\lambda)$ where p_1 and p_2 are distinct primes. Let G_{p_i} denote the subgroup of order p_i in G. It then chooses $a, b, c, d, \alpha, \beta, \gamma \in \mathbb{Z}_N$ and $g \in G_{p_1}$ randomly. Next it computes $u_1 = g^a$, $h_1 = g^b$, $u_2 = g^c$, $h_2 = g^d$, $w = g^\beta$, and $v = g^\gamma$. The public parameters are published as

$$PK = \{N, g, u_1, h_1, u_2, h_2, w, v, e(g,g)^\alpha\}.$$

The master secret key MK is $\{\alpha, \beta, \gamma, a, b, c, d\}$ and a generator of G_{p_3}.
   The identity space is $\mathbb{Z}_N$ and the message space is G_T.

KeyGen(MK, $\mathcal{I}$). Given an identity $\mathcal{I} \in \mathbb{Z}_N$, this algorithm chooses r, t, t', $x, y, z \in \mathbb{Z}_N$ and $R_3, R_3', \hat{R}_3, \hat{R}_3' \in G_{p_3}$ randomly, and computes $D_1 = g^\alpha(u_1^{\mathcal{I}}h_1)^r R_3$, $D_2 = g^r R_3'$, $E_1 = \frac{c+x}{a\mathcal{I}+b}$, $E_2 = g^{\beta x}$, $F_1 = \frac{d+y}{a\mathcal{I}+b}$, $F_2 = g^{\beta y}$, $Z_1 = \frac{z}{a\mathcal{I}+b}$, $Z_2 = g^{\beta z}$, $K_1 = \frac{t}{\beta(c\mathcal{I}+d)}$, $K_2 = g^\alpha g^{t+\gamma t'} \hat{R}_3$, $K_3 = g^{t'} \hat{R}_3'$. We require that the PKG always use the same random value t for $\mathcal{I}$. This can be accomplished by using a pseudo-random function (PRF) or an internal log to ensure consistency.
   The secret key is $SK_{\mathcal{I}} = (D_1, D_2, E_1, E_2, F_1, F_2, Z_1, Z_2, K_1, K_2, K_3)$.

ReKeyGen$(SK_{\mathcal{I}}, \mathcal{I}')$. Given a secret key $SK_{\mathcal{I}} = (D_1, D_2, E_1, E_2, F_1, F_2, Z_1, Z_2, K_1, K_2)$ for $\mathcal{I}$ and an identity $\mathcal{I}' \neq \mathcal{I}$, this algorithm chooses $k_1, k_2 \in \mathbb{Z}_N$ randomly and computes $rk_1 = (E_1 + k_1 \cdot Z_1) \cdot \mathcal{I}' + (F_1 + k_2 \cdot Z_1)$, $rk_2 = (E_2 \cdot Z_2^{k_1})^{\mathcal{I}'} \cdot (F_2 \cdot Z_2^{k_2})$.

The re-encryption key is $RK_{\mathcal{I} \to \mathcal{I}'} = (rk_1, rk_2)$.

Enc$_2$ $(PK, M, \mathcal{I})$. To encrypt a message $M \in G_T$ for an identity $\mathcal{I}$, this algorithm chooses $s \in \mathbb{Z}_N$ randomly and computes $C = M \cdot e(g,g)^{\alpha s}$, $C_1 = (u_1^{\mathcal{I}} h_1)^s$, $C_2 = g^s$, $C_3 = v^s$.

The second level ciphertext is $CT_{\mathcal{I}} = (C, C_1, C_2, C_3)$.

Enc$_1$ $(PK, M, \mathcal{I})$. To encrypt a message $M \in G_T$ for an identity $\mathcal{I}$, this algorithm chooses $s \in \mathbb{Z}_N$ randomly and computes $C = M \cdot e(g,g)^{\alpha s}$, $C_1' = e(u_2^{\mathcal{I}} h_2, w)^s$, $C_2 = g^s$, $C_3 = v^s$.

The first level ciphertext is $CT_{\mathcal{I}} = (C, C_1', C_2, C_3)$.

ReEnc$(CT_{\mathcal{I}}, RK_{\mathcal{I} \to \mathcal{I}'})$. Given a second level ciphertext $CT_{\mathcal{I}} = (C, C_1, C_2, C_3)$ and a re-encryption key $RK_{\mathcal{I} \to \mathcal{I}'} = (rk_1, rk_2)$, this algorithm computes $C_1' = e(C_1, w)^{rk_1} e(C_2, rk_2)^{-1}$.

The re-encrypted ciphertext is $CT_{\mathcal{I}'} = (C, C_1', C_2, C_3)$.

Dec$_2$ $(CT_{\mathcal{I}}, SK_{\mathcal{I}})$. Let $CT_{\mathcal{I}} = (C, C_1, C_2, C_3)$ be a second level ciphertext for identity $\mathcal{I}$, it can be decrypted as

$$M = C \cdot \frac{e(D_2, C_1)}{e(D_1, C_2)}.$$

Dec$_1$ $(CT_{\mathcal{I}}, SK_{\mathcal{I}})$. Let $CT_{\mathcal{I}} = (C, C_1', C_2)$ be a first level ciphertext for identity $\mathcal{I}$, it can be decrypted as

$$M = C \cdot (C_1')^{K_1} \cdot \frac{e(K_3, C_3)}{e(K_2, C_2)}.$$

Correctness at Second Level

$$\frac{e(D_2, C_1)}{e(D_1, C_2)} = \frac{e(g^r R_3', (u_1^{\mathcal{I}} h_1)^s)}{e(g^\alpha (u_1^{\mathcal{I}} h_1)^r R_3, g^s)} = e(g,g)^{-\alpha s}.$$

Correctness at First Level

$$(C_1')^{K_1} \cdot \frac{e(K_3, C_3)}{e(K_2, C_2)} = e(u_2^{\mathcal{I}} h_2, w)^{s \cdot \frac{t}{\beta(c\mathcal{I}+d)}} \cdot \frac{e(g^{t'} \hat{R}_3', g^{\gamma s})}{e(g^\alpha g^{t+\gamma t'} \hat{R}_3, g^s)} = e(g,g)^{-\alpha s}.$$

3.2 Security

We have the following results for our proposed single-hop IB-PRE scheme.

Theorem 1. *If Assumptions 1, 2, 3 hold, then our single-hop IB-PRE scheme is IND-2PrID-CPA secure.*

Theorem 2. *If Assumptions 1, 2, 3 hold, then our single-hop IB-PRE scheme is IND-1PrID-CPA secure.*

It is easy to get the following result from Lemma 1 and Theorem 2.

Corollary 1. *Our single-hop IB-PRE scheme has master secret security.*

We use the dual system encryption technique to prove Theorem 1 and Theorem 2. First we define two additional structures: semi-functional keys and semi-functional ciphertexts. According to the encryption algorithms, there are two types of semi-functional ciphertext: second level semi-functional ciphertext and first level semi-functional ciphertext. These will not be used in the real system, but they will be used in our proof.

Second Level Semi-functional Ciphertext. Let g_2 denote a generator of the subgroup G_{p_2}. A second level semi-functional ciphertext is created as follows. The algorithm first runs the **Enc$_2$** algorithm to generate a normal second level ciphertext $\hat{C}, \hat{C}_1, \hat{C}_2, \hat{C}_3$, chooses $x, y \in \mathbb{Z}_N$ randomly and sets $C = \hat{C}, C_1 = \hat{C}_1 g_2^{xy}, C_2 = \hat{C}_2 g_2^{y}, C_3 = \hat{C}_3 g_2^{\gamma y}$.

First Level Semi-functional Ciphertext. A first level semi-functional ciphertext is created as follows. The algorithm first runs the **Enc$_1$** algorithm to generate a normal first level ciphertext $\hat{C}, \hat{C}_1', \hat{C}_2, \hat{C}_3$, chooses $y \in \mathbb{Z}_N$ randomly and sets $C = \hat{C}, C_1' = \hat{C}_1', C_2 = \hat{C}_2 g_2^{y}, C_3 = \hat{C}_3 g_2^{\gamma y}$.

Semi-functional Key. A semi-functional key is created as follows. The algorithm first runs the **KeyGen** algorithm to generate a normal secret key $\hat{D}_1, \hat{D}_2$, $\hat{E}_1, \hat{E}_2, \hat{F}_1, \hat{F}_2, \hat{Z}_1, \hat{Z}_2, \hat{K}_1, \hat{K}_2, \hat{K}_3$, chooses $\eta, \delta, z_1, z_2 \in \mathbb{Z}_N$ randomly and sets $D_1 = \hat{D}_1 g_2^{\eta z_1}, D_2 = \hat{D}_2 g_2^{\eta}, E_1 = \hat{E}_1, E_2 = \hat{E}_2, F_1 = \hat{F}_1, F_2 = \hat{F}_2, Z_1 = \hat{Z}_1$, $Z_2 = \hat{Z}_2, K_1 = \hat{K}_1, K_2 = \hat{K}_2 g_2^{\delta z_2}, K_3 = \hat{K}_3 g_2^{\delta}$.

We will prove the security of our system from Assumptions 1, 2, 3 using a hybrid argument over a sequence of games. We let q denote the number of key queries made by the attacker. We define these games as follows:

Game$_{2,Real}$: The IND-2PrID-CPA game defined previously in which the ciphertext and all the keys are normal.

Game$_{2,Restricted}$: This is like the real IND-2PrID-CPA game except that the attacker cannot ask for keys for identities which are equal to the challenge identity modulo p_2.

Game$_{2,i}$, $0 \le i \le q$: This is like **Game$_{2,Restricted}$** except that the challenge ciphertext is semi-functional and the first i private key is semi-functional. The rest of the keys are normal.

Game$_{2,Final}$: This is like **Game$_{2,q}$** except that the ciphertext is a semi-functional encryption of a random message, independent of the two messages provided by the attacker.

Game$_{1,Real}$: The IND-1PrID-CPA game defined previously in which the ciphertext and all the keys are normal.

Game$_{1,Restricted}$: This is like the real IND-1PrID-CPA game except that the attacker cannot ask for keys for identities which are equal to the challenge identity modulo p_2.

Game$_{1,i}$, $0 \leq i \leq q$**:** This is like **Game**$_{1,Restricted}$ except that the challenge ciphertext is semi-functional and the first i private key is semi-functional. The rest of the keys are normal.

Game$_{1,Final}$**:** This is like **Game**$_{1,q}$ except that the ciphertext is a semi-functional encryption of a random message, independent of the two messages provided by the attacker.

Game$_{2,Restricted}$ and **Game**$_{1,Restricted}$ are introduced in our proofs due to the same reason explained in Lewko-Waters IBE scheme's proof [13]. We note that **Game**$_{2,*}$ and **Game**$_{1,*}$ are defined differently due to the two base games IND-2PrID-CPA game and IND-1PrID-CPA game are different. In **Game**$_{*,0}$ the challenge ciphertext is semi-functional, but all keys are normal and in **Game**$_{*,q}$ all private keys are semi-functional. We will prove **Game**$_{2,*}$ type games and **Game**$_{1,*}$ type games are indistinguishable respectively.

Lemma 2. *Suppose there exists a polynomial time algorithm $\mathcal{A}$ where* **Game**$_{2,Real}$ $Adv_{\mathcal{A}}-$ **Game**$_{2,Restricted}Adv_{\mathcal{A}} = \epsilon$. *Then we can construct a polynomial time algorithm $\mathcal{B}$ with advantage $\geq \frac{\epsilon}{2}$ in breaking either Assumption 1 or Assumption 2.*

Proof. With probability ϵ, $\mathcal{A}$ produces identities $\mathcal{I}$ and $\mathcal{I}^*$ such that $\mathcal{I} \neq \mathcal{I}^*$ modulo N and p_2 divides $\mathcal{I} - \mathcal{I}^*$. Let $a = \gcd(\mathcal{I} - \mathcal{I}^*, N)$ and $b = \frac{N}{a}$. We have $p_2 \mid a$ and $a < N$. Note that $N = p_1 p_2 p_3$, so there are two cases:

1. $p_1 \mid b$ which means $a = p_2, b = p_1 p_3$ or $a = p_2 p_3, b = p_1$.
2. $p_1 \nmid b$ which means $a = p_1 p_2, b = p_3$.

At least one of these cases must occur with probability $\geq \frac{\epsilon}{2}$. In case 1, $\mathcal{B}$ will break Assumption 1. Given g, X_3, T, $\mathcal{B}$ can confirm that it is case 1 by checking whether $g^b = 1$. Then $\mathcal{B}$ can test whether $T^b = 1$. If yes, then $T \in G_{p_1}$. If not, then $T \in G_{p_1 p_2}$.

In case 2, $\mathcal{B}$ will break Assumption 2. Given $g, X_1 X_2, X_3, Y_2 Y_3$, $\mathcal{B}$ can confirm that it is case 2 by checking whether $g^a = 1$. Then $\mathcal{B}$ can test whether $e((Y_2 Y_3)^b, T) = 1$. If yes, then $T \in G_{p_1 p_3}$. If not, then $T \in G$. □

Lemma 3. *Suppose there exists a polynomial time algorithm $\mathcal{A}$ where* **Game**$_{2,Restricted}Adv_{\mathcal{A}}-$ **Game**$_{2,0}Adv_{\mathcal{A}} = \epsilon$. *Then we can construct a polynomial time algorithm $\mathcal{B}$ with advantage ϵ to Assumption 1.*

Proof. $\mathcal{B}$ receives g, X_3 and T to simulate **Game**$_{2,Restricted}$ or **Game**$_{2,0}$ with $\mathcal{A}$ depending on whether $T \in G_{p_1}$ or $T \in G_{p_1 p_2}$.

$\mathcal{B}$ sets the public parameters as follows. $\mathcal{B}$ chooses random exponents α, β, γ, a, b, c, d and computes $u_1 = g^a$, $h_1 = g^b$, $u_2 = g^c$, $h_2 = g^d$, $w = g^\beta$, and $v = g^\gamma$. It sends these public parameters $N, g, u_1, h_1, u_2, h_2, w, v, e(g,g)^\alpha$ to $\mathcal{A}$. And $\mathcal{B}$ uses X_3 as a generator of G_{p_3}. Note that $\mathcal{B}$ has the actual master secret key, it simply runs the key generation to generate the normal keys to $\mathcal{A}$ for any identity $\mathcal{I}$.

At the challenge phase, $\mathcal{A}$ submits two equal-length messages M_0, M_1 and the challenge identity $\mathcal{I}^*$ to $\mathcal{B}$. It then flips a coin μ and computes the challenge ciphertext as follows:

$$C = M_\mu e(g,T)^\alpha, C_1 = T^{a\mathcal{I}^*+b}, C_2 = T, C_3 = T^\gamma.$$

If $T \in G_{p_1}$, this is a normal ciphertext. If $T \in G_{p_1 p_2}$, then it can be written as $g^{s_1} g_2^{s_2}$ and the ciphertext is a semi-functional ciphertext with randomness $s = s_1$, $x = a\mathcal{I}^* + b$, $y = s_2$.

We can thus conclude that, if $T \in G_{p_1}$, then $\mathcal{B}$ has properly simulated **Game**$_{2,Restricted}$. If $T \in G_{p_1 p_2}$, then $\mathcal{B}$ has properly simulated **Game**$_{2,0}$. Hence, $\mathcal{B}$ can use the output of $\mathcal{A}$ to distinguish between these possibilities for T. □

Lemma 4. *Suppose there exists a polynomial time algorithm* $\mathcal{A}$ *where* **Game**$_{2,k-1}$ $Adv_\mathcal{A}$ − **Game**$_{2,k} Adv_\mathcal{A} = \epsilon$. *Then we can construct a polynomial time algorithm* $\mathcal{B}$ *with advantage* ϵ *to Assumption 2.*

Proof. $\mathcal{B}$ receives g, $X_1 X_2$, X_3, $Y_2 Y_3$, T to simulate **Game**$_{2,k-1}$ or **Game**$_{2,k}$ with $\mathcal{A}$ depending on whether $T \in G_{p_1 p_3}$ or $T \in G$.

$\mathcal{B}$ sets the public parameters as follows. $\mathcal{B}$ chooses random exponents α, β, γ, a, b, c, d and computes $u_1 = g^a$, $h_1 = g^b$, $u_2 = g^c$, $h_2 = g^d$, $w = g^\beta$ and $v = g^\gamma$. And $\mathcal{B}$ uses X_3 as a generator of G_{p_3}. It sends these public parameters N, g, u_1, h_1, u_2, h_2, w, v, $e(g,g)^\alpha$ to $\mathcal{A}$.

When $\mathcal{A}$ requests the i-th key for $\mathcal{I}_i$ where $i < k$, $\mathcal{B}$ returns a semi-functional key as follows. It chooses $r_i, \hat{r}_i, \hat{r}'_i, t_i, t'_i, \hat{t}_i, \hat{t}'_i, x_i, y_i, z_i \in \mathbb{Z}_N$ randomly and computes $D_1 = g^\alpha (u_1^{\mathcal{I}_i} h_1)^{r_i} (Y_2 Y_3)^{\hat{r}_i}$, $D_2 = g^{r_i} (Y_2 Y_3)^{\hat{r}'_i}$, $E_1 = \frac{c+x_i}{a\mathcal{I}_i + b}$, $E_2 = g^{\beta x_i}$, $F_1 = \frac{d+y_i}{a\mathcal{I}_i + b}$, $F_2 = g^{\beta y_i}$, $Z_1 = \frac{z_i}{a\mathcal{I}_i + b}$, $Z_2 = g^{\beta z_i}$, $K_1 = \frac{t_i}{\beta(c\mathcal{I}+d)}$, $K_2 = g^\alpha g^{t_i} g^{\gamma t'_i} (Y_2 Y_3)^{\hat{t}_i}$, $K_3 = g^{t'_i} (Y_2 Y_3)^{\hat{t}'_i}$.

When $i = k$, to response the key query for identity $\mathcal{I}_k$, $\mathcal{B}$ chooses r_k, r'_k, t_k, t'_k, $\hat{t}_k, x_k, y_k, z_k \in \mathbb{Z}_N$ randomly and computes $D_1 = g^\alpha T^{r_k(a\mathcal{I}_k + b)} X_3^{r'_k}$, $D_2 = T^{r_k}$, $E_1 = \frac{c+x_k}{a\mathcal{I}_k + b}$, $E_2 = g^{\beta x_k}$, $F_1 = \frac{d+y_k}{a\mathcal{I}_k + b}$, $F_2 = g^{\beta y_k}$, $Z_1 = \frac{z_k}{a\mathcal{I}_k + b}$, $Z_2 = g^{\beta z_k}$, $K_1 = \frac{t_k}{\beta(c\mathcal{I}+d)}$, $K_2 = g^\alpha g^{t_k} T^{\gamma \hat{t}_k} X_3^{t'_k}$, $K_3 = T^{\hat{t}_k}$. If $T \in G_{p_1 p_3}$, this is a normal key. If $T \in G$, then it is a semi-functional key.

For $i > k$, we note that $\mathcal{B}$ has the actual master secret key, so it only need to run the key generation algorithm to generate the normal keys to $\mathcal{A}$ for any identity $\mathcal{I}$.

At the challenge phase, $\mathcal{A}$ submits two equal-length messages M_0, M_1 and the challenge identity $\mathcal{I}^*$ to $\mathcal{B}$. It then flips a coin μ and computes the challenge semi-functional ciphertext as follows:

$$C = M_\mu e(g, X_1 X_2)^\alpha, C_1 = (X_1 X_2)^{a\mathcal{I}^* + b}, C_2 = X_1 X_2, C_3 = (X_1 X_2)^\gamma.$$

We can thus conclude that, if $T \in G_{p_1 p_3}$, then $\mathcal{B}$ has properly simulated **Game**$_{2,k-1}$. If $T \in G$, then $\mathcal{B}$ has properly simulated **Game**$_{2,k}$. Hence, $\mathcal{B}$ can use the output of $\mathcal{A}$ to distinguish between these possibilities for T. □

Lemma 5. *Suppose there exists a polynomial time algorithm* $\mathcal{A}$ *where* **Game**$_{2,q}$ $Adv_\mathcal{A}$ − **Game**$_{2,Final} Adv_\mathcal{A} = \epsilon$. *Then we can construct a polynomial time algorithm* $\mathcal{B}$ *with advantage* ϵ *to Assumption 3.*

Proof. $\mathcal{B}$ receives $g, g^\alpha X_2, X_3, g^s Y_2, Z_2, T$ to simulate **Game**$_{2,q}$ or **Game**$_{2,Final}$ with $\mathcal{A}$ depending on whether $T = e(g,g)^{\alpha s}$ or T is a random element of G_T.

$\mathcal{B}$ sets the public parameters as follows. $\mathcal{B}$ chooses random exponents β, γ, a, b, c, d and computes $u_1 = g^a$, $h_1 = g^b$, $u_2 = g^c$, $h_2 = g^d$, $w = g^\beta$ and $v = g^\gamma$. And $\mathcal{B}$ uses X_3 as a generator of G_{p_3}. It sends these public parameters N, g, u_1, h_1, u_2, h_2, w, v, $e(g, g^\alpha X_2) = e(g, g)^\alpha$ to $\mathcal{A}$. Note that α is unknown to $\mathcal{B}$.

When responding a key query from $\mathcal{A}$ for identity $\mathcal{I}_i$, $\mathcal{B}$ returns a semi-functional key as follows. It chooses r_i, t_i, $\hat{t}_i$, x_i, y_i, $\hat{x}_i$, w_i, w_i', $\hat{w}_i$, $\hat{w}_i'$, z_i, z_i', $\hat{z}_i$, $\hat{z}_i' \in \mathbb{Z}_N$ randomly and computes $D_1 = g^\alpha X_2 (u_1^{\mathcal{I}_i} h_1)^{r_i} Z_2^{w_i} X_3^{z_i}$, $D_2 = g^{r_i} Z_2^{w_i'} X_3^{z_i'}$, $E_1 = \frac{c + x_i}{a\mathcal{I}_i + b}$, $E_2 = g^{\beta x_i}$, $F_1 = \frac{d + y_i}{a\mathcal{I}_i + b}$, $F_2 = g^{\beta y_i}$, $Z_1 = \frac{\hat{x}_i}{a\mathcal{I}_i + b}$, $Z_2 = g^{\beta \hat{x}_i}$, $K_1 = \frac{t_i}{\beta(c\mathcal{I} + d)}$, $K_2 = g^\alpha X_2 g^{t_i} g^{\gamma \hat{t}_i} Z_2^{\hat{w}_i} X_3^{\hat{z}_i}$, $K_3 = g^{\hat{t}_i} Z_2^{\hat{w}_i'} X_3^{\hat{z}_i'}$.

At the challenge phase, $\mathcal{A}$ submits two equal-length messages M_0, M_1 and the challenge identity $\mathcal{I}^*$ to $\mathcal{B}$. It then flips a coin μ and computes the challenge semi-functional ciphertext as follows:

$$C = M_\mu T, C_1 = (g^s Y_2)^{a\mathcal{I}^* + b}, C_2 = g^s Y_2, C_3 = (g^s Y_2)^\gamma.$$

If $T = e(g, g)^{\alpha s}$, then this is a properly distributed semi-functional ciphertext with message M_μ. If T is a random element of G_T, then this is a semi-functional ciphertext with a random message. Hence, $\mathcal{B}$ can use the output of $\mathcal{A}$ to distinguish between these possibilities for T. □

Proof of Theorem 1. If Assumptions 1, 2, 3 hold then we have proved by Lemma 2, 3, 4, 5 that the real security game is indistinguishable from $\mathbf{Game}_{2, Final}$, in which the value of μ is information-theoretically hidden from the attacker. So there is no attacker that can obtain non-negligible advantage in winning the IND-2PrID-CPA game. □

Proof of Theorem 2 is similar but uses the games $\mathbf{Game}_{1,*}$, so the concrete proof is omitted here and provided in the full version of our paper due to similarity and space limitation.

4 Multi-hop IB-PRE Scheme

Now we construct a multi-hop IB-PRE scheme based on our single-hop IB-PRE scheme proposed in previous section. We observe that if we set $a = c$ and $b = d$, then the first level ciphertext can be re-encrypted using the same re-encryption key and has the same form. This means from the first level ciphertext, we can get a new multi-hop IB-PRE scheme. The new scheme is constructed as follows.

4.1 Construction

Setup(1^λ). Given the security parameter λ, this algorithm first gets a bilinear group G of order $N = p_1 p_2 p_3$ from $\mathcal{G}(\lambda)$ where p_1, p_2 and p_3 are distinct primes. Let G_{p_i} denote the subgroup of order p_i in G. It then chooses $a, b, \alpha, \beta \in \mathbb{Z}_N$ and $g \in G_{p_1}$ randomly. Next it computes $u = g^a$, $h = g^b$, $w = g^\beta$ and $v = g^\gamma$. The public parameters are published as

$$\text{PK} = \{N, g, u, h, w, v, e(g, g)^\alpha\},$$

the master secret key is MK $= \{\alpha, \beta, \gamma, a, b\}$ and a generator of G_{p_3}.
The identity space is $\mathbb{Z}_N$ and the message space is G_T.

KeyGen(MK, $\mathcal{I}$). Given an identity $\mathcal{I} \in \mathbb{Z}_N$, this algorithm chooses $t, r, x, y, z \in \mathbb{Z}_N$ and $R_3, R'_3 \in G_{p_3}$ randomly, and computes $D_1 = \frac{t}{\beta(a\mathcal{I}+b)}$, $D_2 = g^{\alpha} g^{t+\gamma r} R_3$, $D_3 = g^r R'_3$, $E_1 = \frac{a+x}{a\mathcal{I}+b}$, $E_2 = g^{\beta x}$, $F_1 = \frac{b+y}{a\mathcal{I}+b}$, $F_2 = g^{\beta y}$, $Z_1 = \frac{z}{a\mathcal{I}+b}$, $Z_2 = g^{\beta z}$. We also require that the PKG always use the same random value t for $\mathcal{I}$.

The secret key is $SK_{\mathcal{I}} = (D_1, D_2, D_3, E_1, E_2, F_1, F_2, Z_1, Z_2)$.

ReKeyGen($SK_{\mathcal{I}}, \mathcal{I}'$). Given a secret key $SK_{\mathcal{I}} = (D_1, D_2, E_1, E_2, F_1, F_2, Z_1, Z_2)$ for $\mathcal{I}$ and an identity $\mathcal{I}' \neq \mathcal{I}$, this algorithm chooses $k_1, k_2 \in \mathbb{Z}_N$ randomly and computes $rk_1 = (E_1 + k_1 \cdot Z_1) \cdot \mathcal{I}' + (F_1 + k_2 \cdot Z_1)$, $rk_2 = (E_2 \cdot Z_2^{k_1})^{\mathcal{I}'} \cdot (F_2 \cdot Z_2^{k_2})$.

The re-encryption key is $RK_{\mathcal{I} \to \mathcal{I}'} = (rk_1, rk_2)$.

Enc(PK, $M, \mathcal{I}$). To encrypt a message $M \in G_T$ for an identity $\mathcal{I}$, this algorithm $s \in \mathbb{Z}_N$ randomly and computes $C = M \cdot e(g, g)^{\alpha s}$, $C_1 = e(u^{\mathcal{I}} h, w)^s$, $C_2 = g^s$, $C_3 = v^s$.

The ciphertext is $CT_{\mathcal{I}} = (C, C_1, C_2, C_3)$.

ReEnc($CT_{\mathcal{I}}, RK_{\mathcal{I} \to \mathcal{I}'}$). Given a second level ciphertext $CT_{\mathcal{I}} = (C, C_1, C_2, C_3)$ and a re-encryption key $RK_{\mathcal{I} \to \mathcal{I}'} = (rk_1, rk_2)$, this algorithm computes $C'_1 = (C_1)^{rk_1} \cdot e(C_2, rk_2)^{-1}$.

The re-encrypted ciphertext is $CT_{\mathcal{I}'} = (C, C'_1, C_2, C_3)$.

Dec($CT_{\mathcal{I}}, SK_{\mathcal{I}}$). Let $CT_{\mathcal{I}} = (C, C_1, C_2, C_3)$ be a ciphertext for identity $\mathcal{I}$, it can be decrypted as

$$M = C \cdot (C_1)^{D_1} \cdot \frac{e(D_3, C_3)}{e(D_2, C_2)}.$$

The correctness of decryption process is easily observable.

4.2   Security

We have the following result for our proposed multi-hop IB-PRE scheme.

Theorem 3. *If Assumptions 1, 2, 3 hold, then our multi-hop IB-PRE scheme is IND-PrID-CPA secure.*

Proof of Theorem 3 is similar to proofs of Theorem 1 and Theorem 2, so we give the concrete proof in the full version due to similarity and space limitation.

5   Discussion

5.1   Re-encryption Control

In the single-hop proxy re-encryption scheme, we can see that the element $C_3 = v^s$ is of no use in the Dec_2 algorithm and it is only used in the Dec_1 algorithm. If the encryptor doesn't provide v^s in the second level ciphertext, the second level decryption is not affected but the decryption of re-encrypted ciphertext cannot go on. So the encryptor can decide whether the second level ciphertext can be re-encrypted (in fact he can decide whether the re-encrypted ciphertext can be decrypted).

5.2 Transitivity and Transferability

Transitivity means the proxy can redelegate decryption rights. For example, from $RK_{\mathcal{I}_1 \to \mathcal{I}_2}$ and $RK_{\mathcal{I}_2 \to \mathcal{I}_3}$, he can produce $RK_{\mathcal{I}_1 \to \mathcal{I}_3}$. Transferability means the proxy and a set of delegatees can redelegate decryption rights. For example, from $RK_{\mathcal{I}_1 \to \mathcal{I}_2}$ and $SK_{\mathcal{I}_2}$, they can produce $RK_{\mathcal{I}_1 \to \mathcal{I}_3}$. Note that the user $\mathcal{I}_2$ can produce the re-encryption key $RK_{\mathcal{I}_2 \to \mathcal{I}_3}$, so transferability is implied by transitivity. Our multi-hop scheme has such transitivity that the proxy can produce $RK_{\mathcal{I}_1 \to \mathcal{I}_3}$ by $RK_{\mathcal{I}_1 \to \mathcal{I}_2}$ and $RK_{\mathcal{I}_2 \to \mathcal{I}_3}$ as follows:

Let $RK_{\mathcal{I}_1 \to \mathcal{I}_2} = (rk_1, rk_2)$ and $RK_{\mathcal{I}_2 \to \mathcal{I}_3} = (rk_1', rk_2')$. It computes $rk_1'' = rk_1 \cdot rk_1'$ and $rk_2'' = (rk_2)^{rk_1'} \cdot rk_2'$. Then $RK_{\mathcal{I}_2 \to \mathcal{I}_3}$ is (rk_1'', rk_2'').

6 Conclusion

In this paper, we propose two novel unidirectional identity-based proxy re-encryption schemes, which are both non-interactive and proved secure in the standard model. The first scheme is a single-hop IB-PRE scheme and has master secret security, allows the encryptor to decide whether the ciphertext can be re-encrypted. The second scheme is a multi-hop IB-PRE scheme which allows the ciphertext re-encrypted many times but without the cost of ciphertext size growing linearly as previous multi-hop IB-PRE schemes.

References

1. Ateniese, G., Fu, K., Green, M., Hohenberger, S.: Improved proxy re-encryption schemes with applications to secure distributed storage. In: Proceedings of the Network and Distributed System Security Symposium, NDSS 2005. The Internet Society (2005)
2. Blaze, M., Bleumer, G., Strauss, M.J.: Divertible Protocols and Atomic Proxy Cryptography. In: Nyberg, K. (ed.) EUROCRYPT 1998. LNCS, vol. 1403, pp. 127–144. Springer, Heidelberg (1998)
3. Boneh, D., Boyen, X.: Efficient Selective-ID Secure Identity-Based Encryption Without Random Oracles. In: Cachin, C., Camenisch, J. (eds.) EUROCRYPT 2004. LNCS, vol. 3027, pp. 223–238. Springer, Heidelberg (2004)
4. Boneh, D., Boyen, X.: Secure Identity Based Encryption Without Random Oracles. In: Franklin, M. (ed.) CRYPTO 2004. LNCS, vol. 3152, pp. 443–459. Springer, Heidelberg (2004)
5. Boneh, D., Franklin, M.: Identity-Based Encryption from the Weil Pairing. In: Kilian, J. (ed.) CRYPTO 2001. LNCS, vol. 2139, pp. 213–229. Springer, Heidelberg (2001)
6. Boneh, D., Goh, E.-J., Nissim, K.: Evaluating 2-DNF Formulas on Ciphertexts. In: Kilian, J. (ed.) TCC 2005. LNCS, vol. 3378, pp. 325–341. Springer, Heidelberg (2005)
7. Canetti, R., Halevi, S., Katz, J.: A Forward-secure Public-key Encryption Scheme. In: Biham, E. (ed.) EUROCRYPT 2003. LNCS, vol. 2656, pp. 255–271. Springer, Heidelberg (2003)

8. Caro, A.D., Iovino, V., Persiano, G.: Fully secure anonymous hibe and secret-key anonymous ibe with short ciphertexts. Cryptology ePrint Archive, Report 2010/197 (2010), http://eprint.iacr.org/

9. Chu, C.-K., Tzeng, W.-G.: Identity-Based Proxy Re-encryption Without Random Oracles. In: Garay, J.A., Lenstra, A.K., Mambo, M., Peralta, R. (eds.) ISC 2007. LNCS, vol. 4779, pp. 189–202. Springer, Heidelberg (2007)

10. Gentry, C.: Practical Identity-Based Encryption Without Random Oracles. In: Vaudenay, S. (ed.) EUROCRYPT 2006. LNCS, vol. 4004, pp. 445–464. Springer, Heidelberg (2006)

11. Green, M., Ateniese, G.: Identity-Based Proxy Re-encryption. In: Katz, J., Yung, M. (eds.) ACNS 2007. LNCS, vol. 4521, pp. 288–306. Springer, Heidelberg (2007)

12. Lai, J., Zhu, W., Deng, R., Liu, S., Kou, W.: New constructions for identity-based unidirectional proxy re-encryption. Journal of Computer Science and Technology, 793–806 (2010)

13. Lewko, A., Waters, B.: New Techniques for Dual System Encryption and Fully Secure HIBE with Short Ciphertexts. In: Micciancio, D. (ed.) TCC 2010. LNCS, vol. 5978, pp. 455–479. Springer, Heidelberg (2010)

14. Luo, S., Hu, J., Chen, Z.: New construction of identity-based proxy re-encryption. In: Proceedings of the Tenth Annual ACM Workshop on Digital Rights Management, DRM 2010, pp. 47–50. ACM, New York (2010)

15. Matsuo, T.: Proxy Re-encryption Systems for Identity-Based Encryption. In: Takagi, T., Okamoto, T., Okamoto, E., Okamoto, T. (eds.) Pairing 2007. LNCS, vol. 4575, pp. 247–267. Springer, Heidelberg (2007)

16. Shamir, A.: Identity-Based Cryptosystems and Signature Schemes. In: Blakely, G.R., Chaum, D. (eds.) CRYPTO 1984. LNCS, vol. 196, pp. 47–53. Springer, Heidelberg (1985)

17. Shao, J., Cao, Z.: CCA-Secure Proxy Re-encryption without Pairings. In: Jarecki, S., Tsudik, G. (eds.) PKC 2009. LNCS, vol. 5443, pp. 357–376. Springer, Heidelberg (2009)

18. Wang, H., Cao, Z., Wang, L.: Multi-use and unidirectional identity-based proxy re-encryption schemes. Information Sciences (2010)

19. Wang, L., Wang, L., Mambo, M., Okamoto, E.: New Identity-Based Proxy Re-encryption Schemes to Prevent Collusion Attacks. In: Joye, M., Miyaji, A., Otsuka, A. (eds.) Pairing 2010. LNCS, vol. 6487, pp. 327–346. Springer, Heidelberg (2010)

20. Waters, B.: Efficient Identity-Based Encryption Without Random Oracles. In: Cramer, R. (ed.) EUROCRYPT 2005. LNCS, vol. 3494, pp. 114–127. Springer, Heidelberg (2005)

21. Waters, B.: Dual System Encryption: Realizing Fully Secure IBE and HIBE under Simple Assumptions. In: Halevi, S. (ed.) CRYPTO 2009. LNCS, vol. 5677, pp. 619–636. Springer, Heidelberg (2009)

Detecting Parasite P2P Botnet in *eMule*-like Networks through Quasi-periodicity Recognition

Yong Qiao, Yuexiang Yang, Jie He, Bo Liu, and Yingzhi Zeng

School of Computer, National University of Defense Technology,
Changsha, 410073, China
josayqiao@gmail.com, yyx@nudt.edu.cn, jack.237@163.com,
boliu615@yahoo.com.cn, zengyingzhi@nudt.edu.cn

Abstract. It's increasingly difficult to detect botnets since the introduction of P2P communication. The flow characteristics and behaviors can be easily hidden if an attacker exploits the common P2P applications' protocol to build the network and communicate. In this paper, we analyze two potential command and control mechanisms for *Parasite P2P Botnet*, we then identify the quasi periodical pattern of the request packets caused by *Parasite P2P Botnet* sending requests to search for the Botmaster's commands in PULL mode. Considering our observation, a *Parasite P2P Botnet* detection framework and a mathematical model are proposed, and two algorithms named *Passive Match Algorithm* and *Active Search Algorithm* are developed. Our experimental results are inspiring and suggest that our approach is capable of detecting the P2P botnet leeching in *eMule*-like networks.

Keywords: Quasi-periodicity, *Parasite P2P Botnet*, Pull, *eMule*, Active Search Algorithm.

1 Introduction

Botnet, as a special overlay network, is becoming one of the major threats to Internet security. It's commonly agreed that botnet is a malicious network constituted by a large number of compromised hosts which are also known as bots. Bots can be remotely controlled by a unified command from the Botmaster to launch DDOS attacks, send out spam messages, and conduct other group malicious activities. Currently, botnets are usually classified into three categories: centralized IRC-based botnets, distributed P2P-based botnets, and HTTP-based botnets.

Among those three types of botnets, the most widespread one is the IRC botnet. It has low complexity, simple structure and high efficiency in launching attacks. However, considering the congenital "single point of failure" defect brought by the centralized control structure, researchers have developed plenty of detection methods for the widespread IRC botnets [1]. In 2007, the outbreak of the *peacomm* [2] worm and the *storm* botnet [3] in Europe made the world begin to realize the possibility and significant threats of P2P botnets. Therefore the P2P botnets have been considered as the most promising next-generation botnets [4]. A series of P2P Botnets such as *Hybrid P2P Botnet* [5], *Super P2P Botnet* [6] and *Overbot* [7] were proposed shortly afterwards.

H. Kim (Ed): ICISC 2011, LNCS 7259, pp. 127–139, 2012.
© Springer-Verlag Berlin Heidelberg 2012

From current point of view in the research communities, a noteworthy phenomenon is that any P2P botnet based on a private protocol will be inevitably detected because of its individual flow characteristics or behaviors. So the researchers believe that the attackers may launch new P2P botnets by using the public P2P protocols and let P2P botnets leeching into the widely used P2P networks such as *eMule* or *BT* network. *Overbot* is such a P2P botnet designed to work on *eMule* or *BT* network.

In this paper, we propose a detection model to deal with these new challenges. Main contributions of our paper are shown as follows:

1. We analyze the two potential Command and Control (C&C) mechanisms for *Parasite P2P Botnet* and introduce the quasi-periodicity characteristic.
2. We propose a detection framework with a mathematical model based on the quasi periodical pattern of the request packets caused by Parasite P2P Botnet sending requests to search for the Botmaster's command, two algorithms called Passive *Match Algorithm* (PMA) and *Active Search Algorithm* (ASA) are developed to implement the model. During them, ASA can reduce the time complexity significantly.
3. We introduce the approach to identify the search requests from the *eMule* network traffic and some features of the packets *eMule* used to send requests.

2   Related Work

In terms of P2P botnets detection, Gu [8] *et al.* proposed a detection scheme named *Bothunter* to integrate the feedback information about different IDSs and perform a clustering analysis about the malicious activities. Others [9,10,11] also use the clustering methods to justify the flow similarities for botnets detection. *BotMiner* [12], another tool by Gu, can carry out detection independent of the structure and protocol of the botnet, which is an advanced method compared with its counterparts. Different from BotMiner, the proposed model in our research focuses on a specific type of P2P botnet which uses PULL mode to communicate and parasitizes in *eMule*-like networks. [13,14] use the Honeypots or Honeynet to detect the existing botnets, which fail in prediction botnets similarly to the one considered in this paper. Similar to our work, [15,16,17] also take periodic characteristics into account to detect botnets. However, their focal points and contexts are quite different from ours, we will analyze those differences in *Section* 5.

3   Background and Motivation

3.1   C&C of the Parasite P2P Botnet

Ping W. *et al.* [4] have pointed out, due to the structures of P2P network have evolved from early centralized form like *Napster* [18] to distributed or hybrid form such as *eMule* and *BT* network, current P2P applications are more stable in propagation and enjoy a higher bandwidth. Therefore, it attracts attacker to build P2P botnet based on existing P2P networks, which is named as *Parasite P2P Botnet* in this paper. There are two potential C&C mechanisms for Parasite P2P Botnet: PUSH mode and PULL mode.

In the PUSH mode, Botmaster needs to send commands to a certain number of bots, which will propagate these commands to the other bots. Due to the difficulties of using benign nodes to transmit commands actively in *Parasite P2P Botnet*, PUSH pattern is not a feasible choice for *Parasite P2P Botnet*.

Another C&C mechanism, PULL mode, is the main research object in *Parasite P2P Botnet*, which is called *Command Publishing/Subscribing* mechanism. In this mode, Botmaster randomly selects one or more bots to publish its commands. Here, this so-called publishing commands in P2P file-sharing systems is to allow a node to claim that it has a special shared file. In this way, the other bots could find these commands by searching for files periodically. If the formats of such search requests are consistent with the parasitized network, the benign nodes in the parasitized network can provide requests querying and store services for P2P botnet, which is exactly the same way as searching for normal resources and provided as the original functions of the benign nodes. Large-scale P2P application networks such as *eMule*, possess numerous nodes, so the commands searching will become efficient and accurate in these networks while adopting PULL pattern. Subsequently, the PULL mode is very suitable for *Parasite P2P Botnet* and more readily to be used by attackers to build such botnet. We assume that the PULL mode is used by *Parasite P2P Botnet* in this paper.

3.2 Quasi-periodicity Characteristic

Existing pull-model botnets usually perform periodic searches to look for commands. For example, Grizzard [2] mentioned in his analysis of *Storm* that nodes should *periodically* search for their own IDs in order to make sure they know the closest nodes near themselves. In addition, most theoretical P2P botnets proposed are also featured by search packets which are periodically sent. For instance, Ping W. et al [5], the authors of the famous *Hybrid P2P Botnet*, mentioned that both client and server bots actively and *periodically* connect to the server bots in their peer lists in order to retrieve commands issued by their Botmaster. Another example is *overbot* [7], which illustrates that nodes issue search requests *at regular intervals*. According to the analysis of representative P2P botnets above, we can summarize that it's universal for PULL model P2P botnets to send search packets periodically, and we believe that there are two reasons as follows:

1. This feature comes mainly from the unpredictability of the time when Botmaster send commands and the real-time demand of the bots. For convenience, botnets often adopt periodical method directly to search for commands punctually.
2. The pre-programmed behavior [16] of bots also leads to the periodic behaviors of sending search requests.

We note that this periodicity often comes with small fluctuations due to factors like network latency and packets timeout.It's not a strict mathematical periodicity, so we call it the *quasi-periodicity* feature.

Our detection-model focuses on the *Parasite P2P Botnet*, which is one of the high possibility predictive botnets. Based on our analysis of the development of P2P botnets, this botnet has the following characteristics:

1. This botnet builds C&C channels according to the communication protocol of *eMule*.

2. Bots send quasi-periodic packets to search for commands depending on the search protocol of *eMule*.

Here we take *eMule*[1] network as an example to illustrate our idea. Under above assumptions, our goal is to distinguish the malicious botnet search requests from benign *eMule* search requests automatically. In terms of the periodicity feature, what we need to do is to search for a periodic subsequence in hybrid ones.

4 Detection Framework

As demonstrated in Figure 1, our *Parasite P2P Botnet* detection framework is composed of five components: *Sequence Collector, Periodicity Recognition Detection Mathematical Model, Algorithms, Botnet Traffic Simulator, Parameters Trainer*, in which the *Sequence Collector Module* will complete the acquisition of the search requests' sequences, then outputs the results to the *mathematical model*. Two algorithms used to solve the model which we called *PMA* and *ASA* will be introduced in the *Algorithms* section. However, before that, we need to use *Parameters Trainer* module to get the recommended value of the parameters used in our model. *Botnet Traffic Simulator* will be used simultaneously. The last two modules will be explained in Section 5.

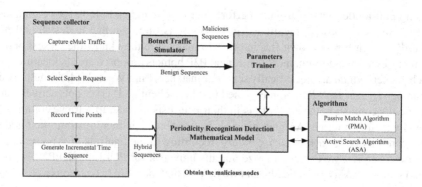

Fig. 1. Framework Overview

4.1 Sequence Collector

The procedure of the Sequence Collector shown in Figure 1 is straightforward. Next we will just introduce the core part about how to analyze the network output packets in order to decode the hybrid *eMule* search requests.

---

[1] *eMule* began to support *KAD* network which is based on *Kademlia* protocol since Version 0.42, Because *eDonkey*, another network in *eMule*, is based on centralized P2P protocol which is obviously unsuitable for the construction of a P2P botnet. *eMule* mentioned in this paper refers to the *KAD* network.

From [19] we know, *eMule* search requests have three action types: *KADEM-LIA_FIND_VALUE; KDEMLIA_STORE; KADEMLIA_FIND_NODE*. However, according to [7], *KADEMLIA_STORE* is not suitable for use in P2P botnets commands transmission, so we will focus on only another two types in the following part. The manual [20] has introduced the search types corresponding to the two action types in Table 1. We will not distinguish those different search types here, because all types can theoretically be used as carriers for P2P botnets to search for commands. With *Wireshark 1.6*, we have done numerous experiments on capturing the *eMule* search requests, and we have got a series of features as follows:

- They are all *UDP* packets;
- The size of each packet is 77 bytes, and the *UDP* part has 43 bytes;
- The payload part has 35 bytes, including the first three bytes of identifications as shown in the last column of Table 1. In these identifications, *E4* indicates that the packet is sent by *eMule*, *21* shows the packet is used to search for requests, *02* and *0B* can represent different action types. The rest of the 32 bytes are filled with encrypted MD4 value of the search contents and recipient's ID.

Table 1. The search Request's classification and identifications

Action Type	Search Type	Identification
	StoreFile	
FIND_VALUE	StoreKeyword	*E4 21 02*
	StoreNotes	
	FindBuddy	
FIND_NODE	Node	*E4 21 0B*
	NodeComplete	

Beside the three features above, we have noticed another interesting phenomenon. *eMule* does not put the MD4 values into a *UDP* packet directly, but changes the sequence of the MD4 characters as the way followed. Take the string "abc" for example, its MD4 value is *a448017a af21d852 5fc10ae8 7aa6729d*.

1. The 32 hexadecimal characters are divided into 4 big groups at the first time, for instance: *a448017a/af21d852/5fc10ae8/7aa6729d*.
2. Divides each big group into 4 small groups evenly, and reverse the 4 small groups' order. A case for the first group: *a448017a → a4/48/01/7a → 7a/01/48/a4*.
3. Recomposes the 32 hexadecimal characters, and the original MD4 value changes to : *7a0148a4 52d821af e80ac15f 9d72a67a*.

With the above knowledge, it is easy to identify and capture various types of search requests of *eMule* network.

4.2 Mathematical Model

Considering the analysis above, the result of the hybrid sequences is represented as Tn. Our target is to justify whether exists a subsequence of Tn fulfilled formula (1) and (2) showed below. represents the length limitation of the periodic sequence, α is proposed to relax the limitation of the periodic interval with the consideration of local volatility in the periodic sequences, ω can avoid the false judgment when very few parts of the sequences are lost or deviate from the regular time. We will discuss the recommended value of the parameters in *section* 5.

Table 2. The definition and introduction of parameters

Parameters	Introduction
$Tn=\{T_1,T_2,T_3,\ldots,T_i,\ldots,T_n\}, \forall i \in (1,n) \rightarrow T_{i+1} > T_i$	the ascending time sequence of the search requests
$S=\{t_1,t_2,t_3,\ldots,t_j,\ldots,t_m\}, j \in (1,m), t_j \in Tn$	S is one of Tn 's non-empty sequence
$\Delta S=\{\Delta t_1, \Delta t_2, \Delta t_3, \ldots, \Delta t_k, \ldots, \Delta t_{m-1}\}, \Delta t_k = t_{k+1} - t_k, k \in (1,m-1)$	ΔS is S's difference sequence
$Avg = \frac{\sum_1^{m-1} \Delta t_k}{|\Delta S|} = \frac{\sum_1^{m-1} \Delta t_k}{m-1}$	The average value of ΔS
$\alpha \in [0,1]$	The adjustment ratio
$\omega \in [0,1]$	The identification ratio
$K \in \mathbb{N}^+$	The minimum length of the periodic sequence

Target Problem: whether $\exists S \subseteq Tn$, Satisfies:

$$|S| \geq K \tag{1}$$

$$\frac{\sum_1^{m-1} f(\alpha, k) = \begin{cases} 1 & \text{if}: \Delta t_k \in [Avg(1-\alpha), Avg(1+\alpha)]; \\ 0 & \text{else.} \end{cases}}{|\Delta S|} \geq \omega. \tag{2}$$

4.3 Algorithms

It's not necessary for us to find out all the periodic sequences satisfied the conditions. One valid periodic sequence is enough to validate whether the hybrid sequence is anomalous.

Passive Match **Algorithm(PMA).** A simple method is to traverse through every sequence whose length is greater than or equal to K, then judge it by the constraints of the target function. Set the total length of the hybrid sequence as N, the amount of all possible subsequences will be:

$$Num = C_N^K + C_N^{K+1} + \ldots + C_N^N = 2^N - C_N^1 - C_N^2 - \ldots - C_N^{K-1} - 1 \tag{3}$$

Compared with N, K is a smaller natural number based on experience. So that the time complexity is approximately equal to $O(2^N)$, which is unacceptable in real context. It is necessary to design a fast approximation algorithm to replace the passive match method.

Active Search **Algorithm(ASA).** This algorithm will not traverse through all possible sequences passively. It starts from a certain node or a group of nodes, and finds the next possible node actively in the sequence according to the periodic conditions actively.

Active Search Algorithm (ASA)

Input: Tn, α, ω, K, S

Output: $SuccessS$: Have Got the periodic sequence if $SuccessS \neq \emptyset$

Begin:

    $SuccessS = \emptyset$;

    Foreach $T_i \in Tn$ **do**

      $S = \emptyset, t_1 = Ti$;

    GetSecondNode:

      $S = \{t_1\}$;

      If exists the nearest node t_2 ahead of t_1 **in** Tn except the nodes

      used before in the loop, **then:**

        Add t_2 to S

      else: goto **Begin**;

    Addnodes:

      If existed node t_j in Tn that satisfies formula (4),**then**

        Add t_j to S;

        if $S.length \geq K$ **then:**

        Return $Success = S$;

        else: goto Addnodes;

      else: goto GetSecondNode;

    Return $SuccessS$;

End

The judging rule which determines whether a suitable node exists in the hybrid sequence or not is: If the subsequence $S = \{t_1, t_2, t_3, \ldots, t_m\}(m < K)$ satisfies all the periodic conditions but the length condition, then do this:

Define: $Avg = \dfrac{\sum_1^{m-1} \Delta t_k}{m}, \Delta t_k = t_{k+1} - t_k$, whether $\exists t_j \subseteq Tn$,and satifies:

$$Avg(1 - \alpha) \leq t_j - t_m \leq Avg(1 + \alpha) \tag{4}$$

if a satisfied node t_j can be found in Tn, then add it to the S, and a new sequence formed:$S = \{t_1, t_2, t3, \ldots, t_m, t_j\}$,or else quit the procedure.

A: Time Complexity analysis

In ASA, the length of a periodic subsequence is gradually increased up to K. So the max beginning node is $T_{(N-K+1)}$. If T_i is the first node in S, the potential amount of the second node is $(N-K-i+2)$, and search steps of each progress are $(K-1)$ at most, so the total steps at this stage will be $(N-K-i+2)(K-1)$, and the total potential steps of all stages will be:

$$Num = (K-1)[1 + 2 + \ldots + (N - K + 1)]$$

$$= (K-1)\frac{(N-K+1)[1+(N-K+1)]}{2} \tag{5}$$

$$= \frac{K-1}{2} \times [(N - K + 1) + (N - K + 1)^2]$$

From the formula (5) we can see the time complexity here is $O(N^2)$, which reduces the computation cost effectively compared with the PMA.

B: Complementary Conditions

In ASA, if the current subsequence's first node is T_m, then the biggest interval of this subsequence is $(T_N - T_m)$. As the max length of the periodic subsequence is known, so we get a complementary condition:

$$\Delta t_m \leq (1 + \alpha)\frac{T_N - T_m}{K - 1} \quad (\Delta t_m = t_{m+1} - t_m) \tag{6}$$

The formula above can apply to every judgment step in ASA, which will further reduce the searching space and the computation costs.

5 Experiment and Analysis

5.1 Datasets

We captured real traffic from campus network, and tracked 100 *eMule* nodes through port identification. From the user behavior analysis of *eMule*, we know that the period between $18pm$ and $24pm$ is the peak of *eMule* usage, so the datasets used below are all captured in this period.

Botnet Traffic Simulator: In addition to real traffic collection of *eMule*, we developed a P2P botnet traffic simulator, which can periodically send packets in the format of standard *eMule* search packets. The simulator can make the period fluctuate in a certain degree or generate some controllable noise to correspond with real circumstances.

5.2 Determining the Empirical Value of Parameters

Those three parameters are interdependent, thus cannot be determined through separate experiments. As we have abandoned ω in our implementation of ASA, we only consider the determination of α and K here. According to the initial definition of α and K, we performed experiments on some possible combinations through crossing and rotating. Table 3 shows the recommended empirical value range of the two parameters. Because

Table 3. The Parameters' Range

Parameter	Range	Unit	Counts
α	$(0-0.3]$	0.02	15
K	$[5-20)$	1	15

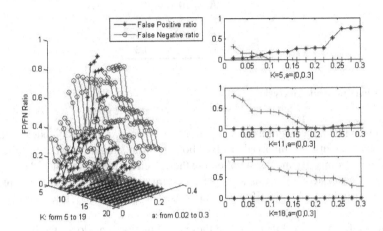

Fig. 2. False Positive ratio(FP) and False Negative ratio (FN)

the value of α lies in a continuous range, it's impossible to traverse all its value, we averagely divided the range into 15 pieces of units. As K has 15 possible values, the total number of potential combinations of the two parameters is 225.

Our experiments have separately collected traffic data from 100 hosts when they run normal *eMule* exclusively and when they run malicious periodical program as well in two days. Thus, we extracted 100 normal sequences containing search requests time points and another 100 anomalous sequences contain periodical malicious requests time points from the traffic data. Having known the true role of every node, we test all 225 combinations of α and K, and calculated their False Positive ratio (FP) and False Negative ratio (FN) shown in Figure 2, with 3 groups of typical values of parameters are illustrated together.

From Figure 2 we can see that both α and K can affect FP and FN significantly. A better combination should set formula (7) with the lowest value. Here, we assume that $\theta_1 = \theta_2 = 0.5$ empirically, in this situation, we just need to get the combinations with the lowest $(FN + FP)$.

$$F = \theta_1 \cdot FN + \theta_2 \cdot FP \tag{7}$$

From the experimental results above, we have got the optimal groups of parameters in Table 4, which convince us that $K \in [10, 12]$ and $\alpha \in [0.18, 0.22]$ is a suitable range for the justification of periodic sequences, and we can use them for verification in live circumstances.

Table 4. The Parameters' Range

$FN + FP$	Combinations
0	$K = 11, \alpha = 0.22$
	$K = 12, \alpha = 0.22$
1%	$K = 10, \alpha = 0.18$
	$K = 10, \alpha = 0.2$
	$K = 11, \alpha = 0.2$
2%	$K = 10, \alpha = 0.22$
	$K = 12, \alpha = 0.2$

5.3 Verification and Analysis

Accuracy

We run a test with all the combinations recommended above. A hundred nodes installed with the *eMule* clients are involved, 40 P2P botnet traffic simulators were installed on those nodes randomly. By means of collecting the search requests, we got 100 groups of requests, and generated 100 sequences by abstracting the times points of the requests' packets. We use the ASA to justify the sequences. The FP and FN rates were shown in Figure 3.

We can see from Figure 3 that, none of these combinations can do perfectly to justify the hosts absolutely right, but they have already fulfilled the detection task successfully to a great extent and the empirical values of the parameters exhibit a good robustness. In the future we still need to utilize the ω parameter, which can adjust the sequence when there is a partial fluctuation appearance.

Comparions

Similar to our work, [15,16,17] take the periodic characteristics into account when detecting botnet. Table 5 gives an analysis of relevant characteristics. In the table, **PO** indicates objects with periodicity in botnets, **DA** indicates the detection algorithm, **DOC**

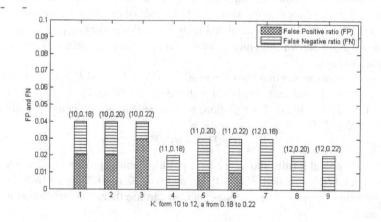

Fig. 3. The FP/FN with the change of (K, α)

is the degree of vertical-cross confusion[2] between periodic sequence and non-periodic sequence, and ***CUB*** indicates whether the method can be used in our work context without considering the detection effect. According to [17], there is no vertical cross but horizontal connections between periodic sequence and non-periodic sequence. ***AR*** represents the accuracy of detection when the algorithm is applied to our detection context using certain parameters.

Table 5. *Comparision of Four Detection Models*[*]

	PO	*DA*	*DOC*	*CUB*	*AR*
[15]	Connection intervals	Perceptual intuition	none	N	
[16]	Number of packets in C&C	Periodogram	weak	Y	2%
[17]	Packet size Sequence	Ukkonen	none	Y	16%
Our	Search Request's intervals	ASA	strong	Y	98%

PO:Periodic Object *DA*: Detection Algorithm

DOC:Degree of Confusion *CUB*: can be used here *AR*: Accuracy Rate

*: *Periodogram algorithm used here is under the same parameter definition in* [16]

and we define the ($\phi = 0.75$) *in algorithms* [17] *used here,* ($K = 11, \alpha = 0.18$) *is defined in ASA*

Table 5 also shows the detection effect of [16,17] and ASA when adopt specified parameters. As we can see, *Periodogram* algorithm [16] cannot achieve an effective detection, however, as some sequences here have low *DOC*. *Ukkonen* algorithm [17] is lucky to get a 16% detection accuracy rate here, which is still quite low. The low detection rate of the two algorithms can be mainly attributed to the huge differences of contexts, rather than the algorithms themselves. These results further demonstrate that our ASA algorithm is more suitable and effective in our detection context and the ASA also can be used to solve the similar problems theoretically.

6   Conclusion and Future Work

A detection framework with a mathematical model based on the *quasi-periodicity* characteristic is proposed in this paper after analyzing the potential C&C mechanisms in *Parasite P2P Botnet*. Two algorithms are proposed to solve the model. We verify the model and algorithms by doing a series of experiments. Compared with other detection methods in [8,9,10,11,12,13,14], our research focuses on a special predictive P2P botnet, which has been demonstrated as a devastating threat on the security of the future Internet security. The detection framework we proposed against such botnet is promising to be one step ahead of the attackers to discover and prevent this upcoming threat. We will further analyze possible communication modes of P2P botnets in the future, and provide better detection methods and suppression measures.

---

[2] Vertical-cross refers to situations where several time series get crossed in time domain. For example, for two series $\{1, 3, 5\}$ and $\{4, 5, 5.5\}$, a vertical-cross is $\{1, 3, 4, 5, 5.5\}$. On the contrary, the horizontal-cross refers to time series concatenated end to end: $\{1, 2, 3\}$ and $\{4, 5, 7\}$ connected as $\{1, 2, 3, 4, 5, 7\}$.

Acknowledgment. This work was supported by NSFC under grants NO. 61170286.

References

1. Feily, M., Shahrestani, A., Ramadass, S.: A survey of botnet and botnet detection. In: 2009 Third International Conference on Emerging Security Information, Systems and Technologies, pp. 268–273. IEEE (2009)
2. Grizzard, J.B., Sharma, V., Nunnery, C., Kang, B.B.H., Dagon, D.: Peer-to-peer botnets: Overview and case study. In: 1st USENIX Workshop on Hot Topics in Understanding Botnets, HostBots 2007 (2007)
3. Holz, T., Steiner, M., Dahl, F., Biersack, E., Freiling, F.: Measurements and mitigation of peer-to-peer-based botnets: a case study on storm worm. In: LEET 2008 Proceedings of the 1st Usenix Workshop on Large-Scale Exploits and Emergent Threats, p. 9. USENIX Association (2008)
4. Wang, P., Aslam, B., Zou, C.C.: Peer-to-Peer Botnets: The Next Generation of Botnet Attacks. Electrical Engineering, 1–25 (2010)
5. Wang, P., Sparks, S., Zou, C.C.: An advanced hybrid peer-to-peer botnet. In: Proceedings of the First Workshop on Hot Topics in Understanding Botnets, p. 2. USENIX Association (2007)
6. Vogt, R., Aycock, J., Jacobson, M.: Army of botnets. In: Proceedings of NDSS 2007, Citeseer, pp. 111–123 (2007)
7. Starnberger, G., Kruegel, C., Kirda, E.: Overbot: a botnet protocol based on Kademlia. In: 4th Int. Conf. on Security and Privacy in Communication Networks (SecureComm 2008), pp. 1–9. ACM (2008)
8. Gu, G., Porras, P., Yegneswaran, V., Fong, M., Lee, W.: Bothunter: Detecting malware infection through ids-driven dialog correlation. In: 16th USENIX Security Symp. (Security 2007), pp. 167–182. USENIX Association (2007)
9. Kang, J., Song, Y.-Z., Zhang, J.-Y.: Accurate detection of peer-to-peer botnet using Multi-Stream Fused scheme. Journal of Networks 6, 807–814 (2011)
10. Villamarin-Salomon, R., Brustoloni, J.C.: Bayesian bot detection based on DNS traffic similarity. In: 24th Annual ACM Symposium on Applied Computing, pp. 2035–2041. Association for Computing Machinery (2009)
11. Huang, Z., Zeng, X., Liu, Y.: Detecting and blocking P2P botnets Through contact tracing chains. International Journal of Internet Protocol Technology 5, 44–54 (2010)
12. Gu, G., Perdisci, R., Zhang, J., Lee, W.: BotMiner: clustering analysis of network traffic for protocol and structure independent botnet detection. In: 17th USENIX Security Symp., pp. 139–154. USENIX Association (2008)
13. Freiling, F.C., Holz, T., Wicherski, G.: Botnet Tracking: Exploring a Root-Cause Methodology to Prevent Distributed Denial-of-Service Attacks. In: de Capitani di Vimercati, S., Syverson, P., Gollmann, D. (eds.) ESORICS 2005. LNCS, vol. 3679, pp. 319–335. Springer, Heidelberg (2005)
14. Wang, P., Wu, L., Cunningham, R., Zou, C.C.: Honeypot detection in advanced botnet attacks. International Journal of Information and Computer Security 4, 30–51 (2010)
15. Lee, J.S., Jeong, H.C., Park, J.H., Kim, M., Noh, B.-N.: The activity analysis of malicious http-based botnets using degree of periodic repeatability. In: 2008 International Conference on Security Technology, pp. 83–86. Inst. of Elec. and Elec. Eng. Computer Society (2008)
16. AsSadhan, B., Moura, J.M.F., Lapsley, D.: Periodic behavior in botnet command and control channels traffic. In: 2009 IEEE Global Telecommunications Conference. Institute of Electrical and Electronics Engineers Inc. (2009)

17. Ma, X., Guan, X., Tao, J., Zheng, Q., Guo, Y., Liu, L., Zhao, S.: A novel IRC botnet detection method based on packet size sequence. In: 2010 IEEE International Conference on Communications. Institute of Electrical and Electronics Engineers Inc. (2010)
18. Saroiu, S., Gummadi, K.P., Gribble, S.D.: Measuring and analyzing the characteristics of Napster and Gnutella hosts. Multimedia Systems 9, 170–184 (2003)
19. eMule 0.47 code, eMule project (2011), http://www.emule-project.net/home/perl/general.cgi?l=42&rm=download
20. Kernel, H.: Emule Kad protocol Manual (2009), http://easymule.googlecode.com/files/Emule

AutoDunt: Dynamic Latent Dependence Analysis for Detection of Zero Day Vulnerability

Kai Chen[1], Yifeng Lian[1,2], and Yingjun Zhang[1]

[1] Institute of Software, Chinese Academy of Sciences, Beijing 100190, China
{chenk,lianyf,yjzhang}@is.iscas.ac.cn
[2] National Engineering Research Center for Information Security,
Beijing 100190, China

Abstract. Zero day vulnerabilities have played an important role in cyber security. Since they are unknown to the public and patches are not available, hackers can use them to attack effectively. Detecting software vulnerabilities and making patches could protect hosts from attacks that use these vulnerabilities. But this method cannot prevent all vulnerabilities. Some methods such as address space randomization could defend against vulnerabilities, but they cannot find them in software to help software vendors to generate patches for other hosts. In this paper, we design and develop a proof-of-concept prototype called AutoDunt (AUTOmatical zero Day vUlNerability deTector), which can detect vulnerable codes in software by analyzing attacks directly in virtual surroundings. It does not need any source codes or care about polymorphic/metamorphic shellcode (even no shellcode). We present a new kind of dependence between variables called *latent dependence* and use it to save necessary states for virtual surrounding replaying. In this way, AutoDunt does not need to use slicing or taint analysis method to find the vulnerable code in software, which saves managing time. We verify the effectiveness and evaluate the efficiency of AutoDunt by testing 81 real exploits and 7 popular applications at the end of this paper.

Keywords: AutoDunt, Latent dependence, Zero day vulnerability, Debug, Arbitrary code execution.

1 Introduction

Zero day vulnerabilities play a very important role in cyber security [1,2]. Many well known cyber security problems, such as worms, zombies and botnets, are rooted in that. Before a zero day vulnerability is disclosed, attacks based on it will be almost always effective since there is no patch for it. Moreover, a zero day vulnerability can usually be popular without being disclosed for many years. For example in 2008 Microsoft confirmed a vulnerability in Internet Explorer, which affected some versions that were released in 2001 [3]. This is very dangerous since hackers can use this vulnerability to attack again and again before its patch is published. New methods such as return-oriented programming [4] make those attacks even more difficult to detect.

H. Kim (Ed): ICISC 2011, LNCS 7259, pp. 140–154, 2012.
© Springer-Verlag Berlin Heidelberg 2012

People usually defend against zero day attacks in the following three ways. 1) Detecting all vulnerabilities in software and patching them seem to be a reasonable way [5,6,7]. But vulnerability detection is a challenging task. It is almost impossible for current detection techniques to find all the vulnerabilities. Thus, hackers could still use zero day vulnerabilities to attack. 2) Some methods such as LibSafe [8] and address space randomization [9,10] are used to prevent attacks without prior knowledge of vulnerabilities. But these methods cannot protect other hosts that do not use these methods. 3) Some methods such as Memsherlock [11] can point out the corruption position in software by slicing source codes [12] or using taint analysis method [13]. These methods usually need source codes. Unfortunately, not all of the software provide their source codes, especially the commercial software. It is not easy for them to perform operating system kernel protection. Moreover, slicing and taint analysis methods are not efficient since they need to record program execution traces or to instrument lots of instructions.

In this paper, we have designed and developed an end-to-end approach to automatically prevent attacks and point vulnerable positions in software by using attack information without source codes. At a high level, our approach is a four-step process: saving, detecting, rolling back and re-running. When examining a software sample, we first load it into our analysis environment. We save some snapshots of the CPU and memory when the sample is running. When an attack is detected, we roll back the running thread to a certain saved state. In the end, we re-run the thread from the checkpoint to find the position of the vulnerability. We present a new dependence between variables called *latent dependence* to decrease the number of saved states. To explore the feasibility of our approach, we design and develop a proof-of-concept prototype called AutoDunt and make several experiments on it. The results show that AutoDunt successfully detects arbitrary code execution vulnerabilities and points out the right vulnerable positions in the binary codes even if the attacks are polymorphic/metamorphic.

In summary, this paper makes the following contributions:

– the concept of *latent dependence* between variables represented by *medium instructions*, which is essential to arbitrary code execution vulnerabilities. By recognizing this type of relationship between variables and saving the states when *medium instructions* are executed, we decrease the number of saved states to about 0.232% of all the running blocks in average.

– an automatic system AutoDunt that performs kernel-level analysis and automatically points out vulnerable codes in binary programs without any source codes from zero day attacks. It avoids using slicing and taint analysis methods to increase managing speed. It does not care about the form of the shellcode even it is polymorphic/metamorphic.

– a description of how the malicious input exploits the unknown vulnerabilities. By re-running threads from checkpoints in virtual surroundings, we can replay the whole process of exploiting the vulnerability.

The rest of the paper is organized as follows. In Section 2, we summarize the work related to ours. In Section 3, we give an overview of our work. In Section 4, some definitions of *latent dependence* are shown. We give the implementation in Section 5 and show our experimental results in Section 6. Finally, we conclude the paper in Section 7.

2 Related Work

Our work draws on techniques from several areas, including: A) worm detection and signature generation, B) vulnerability prevention and C) virtual-machine replaying.

A) Generating worm signatures is an effective way to detect worms. Autograph [14] automatically generates signatures for novel Internet worms that propagate using TCP transport. Polygraph [15] can produce signatures that match polymorphic worms. Hamsa [16] enhances the efficiency, accuracy and attack resilience. SigFree [17] can block new and unknown buffer overflow attacks without signature. All the methods in this class need to analyze TCP traffic. Thus, if the attack is only performed once, it cannot be detected. Our work uses host-based method to detect such attacks without any need for network traffic. Most methods in this class rely on analyzing shellcode to detect attacks. New techniques such as return-oriented programming [4] make it possible that there is no shellcode in the malicious payload, which increases the difficulty to detect attacks.

B) Different methods are proposed to prevent vulnerabilities. Modifications of OS are made to prevent buffer overflow such as Pax [18], LibSafe [8] and e-NeXsh [19]. These methods need to re-compile the operating system. Mcgregor [20] proposed to store return address on the processor that no input can change any return address. This method needs hardware support. Our method supplies a cheap solution that we do not need to modify any operating system or hardware. Castro [21] proposed a new approach called data-flow integrity detection to prevent the system from being attacked. It needs to analyze source codes firstly and hence cannot handle indirect control flow precisely. Kiriansky etc. presented program shepherding to monitor control flow transfers to enforce a security policy [22]. Bhatkar etc. proposed address space randomization to randomize all code and data objects and their relative distances to protect against memory error exploits [23]. But it also needs source codes and it can only protect the software using randomization techniques while AutoDunt could detect the vulnerability and alarm other computers even without AutoDunt. Vigilant [24] can control the spread of worm by automatically generating SCAs. But if the worm is polymorphic, Vigilante will generate too many SCAs. MemSherlock [11] is very similar to our work. It can identify unknown memory corruption vulnerabilities upon the detection of malicious payloads that exploit such vulnerabilities, but it needs source codes and needs to analyze libraries files in advance. Furthermore, neither Vigilant nor MemSherlock can analyze operating system kernel while AutoDunt performs a whole-system, fine-grained analysis. Brumley and Newsome etc. focus on signature generation in their work [25,26] but they do not care about how to find the vulnerabilities. Panorama [27] and TTAnalyze

[28] can detect and analyze malware by capturing its behavior. They aim to comprehend the behavior and innerworking of unknown samples while our work aims to protect system and detect vulnerabilities. Most of the methods above use dynamic taint analysis [13,29], which suffers a lot from high overhead.

C) In our work, we use the method of virtual-machine replay. TTVM [30] is very similar to AutoDunt from the first view. It also saves the states of operating system and re-runs the system from checkpoints. Since we aim to find vulnerabilities while TTVM aims to debug operating system, the time when to save states differs a lot. TTVM saves the states when processing external inputs while AutDunt only saves the states related to vulnerabilities. In this way, the number of AutoDunt's checkpoints is much less than that of TTVM's checkpoints. Since different operating systems may have different input functions, TTVM needs to know all of them. AutoDunt is based on *latent dependence* which does not care about different platforms and operating systems. Moreover, we distinguish processes and threads to supply more precise results.

3   Overview of AutoDunt

Our approach is a four-step process: saving, detecting, rolling back and re-running. Figure 1 depicts the overview of AutoDunt. We first build virtual surrounding to load a program and monitor its behavior. In this way, AutoDunt saves necessary states without any interference with the programs running inside virtual machine. The next step is to detect the attack. For the sake of simplicity, in our current proof-of-concept implementation, we only detect arbitrary code execution vulnerability. However, using the method in Vigilante [24] to detect other types of vulnerabilities is straightforward. When an attack happens, we roll back and re-run the system from the saved states to find the vulnerable code.

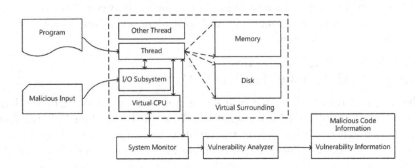

Fig. 1. An Overview of AutoDunt

Process is not the scheduled unit but the thread is. So we need to distinguish different threads. System Monitor can extract thread information by using operating system kernel data structure such as ETHREAD. When the vulnerability is triggered, *eip* (instruction pointer) will point to special position in memory such as stack or heap. Vulnerability Analyzer saves the thread states of the target program using *latent dependence* between variables. From the saved states, AutoDunt points out the vulnerable code in software.

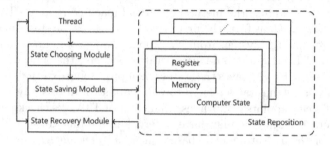

Fig. 2. Design of Vulnerability Analyzer

Figure 2 is the structure of Vulnerability Analyzer, which seems like a time machine. To re-run the target thread, it is necessary to save its running states. However, the number of the states may be too large even infinite to save. So proper states should be chosen by State Choosing Module. State Saving Module saves the chosen states into the State Repository according to the criteria in the next section. When an attack happens, State Recovery Module will pick out suitable state from the State Repository to re-run the target thread for vulnerability detection. In this way, complex methods such as slicing and taint analysis can be avoided. What AutoDunt needs to do is to compare the thread's states and to re-run the thread from those states.

4   Latent Dependence

In this section, we propose a new kind of dependence between variables in program. We also make some conclusions and implement AutoDunt based on them. Let's start from an example:

```
int a = 10 ; char b[10]; int i = 0;
do{cin>>b[i];} while(b[i++] != '0');
```

In the upper example, the value of a seems not to be changed after the loop. This can only be guaranteed if the length of the input is less than 10 bytes. Note that the loop exit condition is that the input value equals '0'. So if the input is not equal to '0' for 10 times, it will overflow buffer b and overwrite a. In this case, the value of a may not be '10' at the end of the program. So a may be data dependent on input. This kind of dependence is very special. It cannot be caught by regular analysis method. We refer to it as *latent dependence*. We define *static definition set* and *dynamic definition set* first.

Definition 1. [Static Definition Set]: Static definition set of instruction I includes variable(s) in I whose value(s) will always be changed after I is executed. It can be referred to as $def_s(I)$. For example, in instruction '$a = b$', $def_s(I) = a$.

Definition 2. [**Dynamic Definition Set**]: Dynamic definition set of instruction I includes variable(s) whose value(s) will possibly be changed after I is executed. It can be referred to as $def_d(I)$.

Definition 3. [**Latent Dependence**]: If $v \in def_d(I)$, then v *latent dependent* on $use(I)$. It can be referred to as $v \xrightarrow{LD(I)} use(I)$. $use(I)$ is the set of variable(s) that I uses. I can be called *medium instruction*.

Suppose instruction I is in the form of '$\text{op } opr_1, opr_2$'. op is opcode, and opr_1 and opr_2 are operands. **If** $def_d(I) \neq \emptyset$, 'opr_1' **must be a piece of memory with indirect index.** Otherwise, $opr_1 \in def_s(I)$ and $def_d(I) = \emptyset$. Suppose the indirect index is 'r'. There are two situations here: S1) $r \xrightarrow{DD} input$ (DD means *data dependence*), which means input data could change arbitrary memory locations. S2) $r \xcancel{\xrightarrow{DD}} input$. In this situation, I should be in a loop. Otherwise, r has a fixed value and $def_d(I) = \emptyset$.

S1 is not very common since a normal program will not let input data change arbitrary memory locations. Even if this situation happens, there are some limitations on 'r', which is less likely to be a vulnerability. So we focus our attention on S2. In this situation, we try to identify *medium instructions* and *latent dependence* between variables by recognizing loops. There are lots of algorithms to find loops in programs such as [31]. We use DJ graphs to identify loops [32] in this paper. In this way, we can get candidate *medium instructions* easily in static analysis. But not all the instructions in loops can become *medium instructions* even if their left values are with indirect indexes. More limitations should be added to narrow the scope of candidate *medium instructions*.

Definition 4. [**Super Variable**]: The *super variable* of a variable v is the continuous objects of the same type which v belongs to. It can be referred to as $sv(v)$. For example, the super variable of an element in an array is the array itself.

If an arbitrary code execution vulnerability is triggered in a loop l, there exists i which meets $sv(v^{(i)}) \neq sv(v^{(i+1)})$. $v \in def_s(I) \cup def_d(I)$ and $sv(v^{(i)})$ means the *super variable* of v in the ith execution of I in l. This can be used to detect *latent dependence* in dynamic analysis. Since we analyze binary codes directly without source codes, we do not know type information of variables. So it is difficult to judge whether $sv(v^{(i)}) \neq sv(v^{(i+1)})$.

We generate some rules to distinguish possibly different types. Table 1 shows those rules. Three classes of marks are used here: *read/write mark*, *loop mark* and *memory block mark*. *Read/write mark* indicates that a piece of memory has been read/written. We distinguish them since they can be used to detect other types vulnerabilities such as reading uninitialized memory in the future. *Loop mark* indicates that a piece of memory is read or written in a loop. *Memory block mark* is used to distinguish different blocks of memory. Any two blocks of memory have different *memory block marks*.

Table 1. Rules to Distinguish Possibly Different Types

Memory operations:	Rules:
Read $mem[i]$ (outside loop)	$smem[i] \leftarrow smem[i] + r$
Write $mem[i]$ (outside loop)	$smem[i] \leftarrow smem[i] + w$
Allocate $mem[i \sim j]$	$smem[i \sim j] \leftarrow mb$
Release $mem[i \sim j]$	$smem[i \sim j] \leftarrow \emptyset$
Read/Write $mem[i]$ (inside loop)	Check marks in $smem[i]$ and $smem[i] \leftarrow l$

In table 1, $mem[i]$ is the piece of memory indexed by i. $smem$ is the shadow memory of mem. If $mem[i]$ is read/written outside loops, we add *read/write mark* 'r/w' to $smem[i]$. When a new block of memory $mem[i \sim j]$ is allocated, we set *memory block mark* 'mb' to every piece of memory in $smem[i \sim j]$. mb is different from the *memory block mark* of $smem[i-1]$ and that of $smem[j+1]$. When a block of memory is released, we clear the marks. When a read/write operation I in loop l is executed, we can test whether $v^{(i)}$ and $v^{(i+1)}$ have the same mark ($v \in def_s(I) \cup def_d(I)$). If not, we will get $v \in def_d(I)$. We also add *loop mark* 'l' to it. We do not set 'r/w' to the piece of memory with l. Based on it, we design and develop a proof-of-concept prototype called AutoDunt. Detailed implementation will be shown in the next section.

5 Implementation

We choose QEMU [33] as the virtual surrounding because of its efficiency when compared to previous processor emulators such as Bochs. QEMU works in the following way. It first picks a basic block of executable codes from the program. A basic block in a procedure is a sequence of consecutive instructions with a single entry and a single exit point. Then QEMU translates the codes and runs it at last. AutoDunt plays as a middleware to analyze the codes and saves the necessary states of the guest OS. In this way, AutoDunt can be easily integrated with other virtual surroundings.

5.1 State Saving Module

In order to re-run a thread from a certain checkpoint, proper states should be saved when the thread is running. In our current implementation, AutoDunt only saves memory states and CPU states of the running thread, which are enough for vulnerability detection. Saving the state of every basic block works well. But most states are unnecessary for vulnerability detection. We use the results of *latent dependence analysis* to decrease the number of saved states. When we find $sv(v^{(i)}) \neq sv(v^{(i+1)})$ in a loop, we save the current state for the loop (only once). We use QEMU to instrument memory operation codes in table 1 to distinguish possibly different types. Some memory operations are implicit. For example, PUSH, POP, CALL and RET instructions can change stacks. Since CALL instruction will write return address in the stack, AutoDunt sets 'w' mark to the corresponding memory in *smem*. In this way, this type of vulnerability can be detected by *latent dependence analysis*. Overflowing SEH could also be detected since the address of exception handling function is marked in *smem*. Heap and stack have different kinds of memory allocation operations. Heap is usually allocated by system call 'HeapAlloc' while stack is allocated by any function call.

When AutoDunt manages nested loops, the states of inner loop may be saved many times. For example, suppose there are three loops. The outer-most one is l_1, the inner loop of l_1 is l_2, and the inner loop of l_2 is l_3. Suppose l_1 runs t_1 times and l_2 runs t_2 times. Then the states may be saved $t_1 \times t_2$ times. Most of the saved states are unnecessary. So before AutoDunt saves a state, it should firstly try to identify nested loops and only saves the state once. In this way, we can avoid saving unnecessary states. Evaluations about the number of saved states will be shown in Section 6.

Sometimes we do not need to analyze some trusted codes such as system calls and library functions. This saves a lot time since the number of these codes is not small. According to experiments in Section 6, we find kernel codes are about 55.6% of the overall codes in average. We define a set of instruction addresses which do not need to be analyzed. AutoDunt ignores the instructions whose addresses are in the set.

5.2   State Recovery Module

If a vulnerability is exploited, the vulnerable code can be found by comparing the saved states and re-running the thread. QEMU supplies an interface 'cpu_memory_rw_debug' to access virtual memory of guest operating system. So it is easy to re-run the thread from a saved state. There is no need to re-run every saved state to perform vulnerability detection. Suppose $eip \xrightarrow{DD} v \xrightarrow{LD} Input$ and the value of eip is v_{eip} when an attack happens. We choose state s_i as starting state for vulnerability analysis. s_i meets the following conditions: the value of v in s_{i+1} equals to v_{eip} and the value of v in s_i does not equals to v_{eip}. s_{i+1} is the next state after s_i. In this way, we can find the *medium instruction* that changes v. This information is valuable for software vendors to generate patches. By restarting the process from s_i, we avoid analyzing the codes from program start point, which improves the overall efficiency, especially when the program has executed for a long time. We analyzed 81 exploits and found this method could point out almost all of the vulnerable codes. Detailed results are shown in Section 6.

5.3   Distinguishing Different Processes and Threads

In windows operating system, different processes have different CR3 values. CR3 is a register in CPU to save the page directory address of current process, which can be used to distinguish different processes. When using QEMU, it is easy to read CR3 value in virtual CPU. We can use EPROCESS structure to get process names. In this way, we can choose the target program to analyze by names.

As different threads have different ID numbers, thread ID can be used to distinguish them. AutoDunt can get thread ID easily from 'ETHREAD structure by using the interface 'cpu_memory_rw_debug'. Although the structure of ETHREAD differs in different operating system, it is not a big task to acquire its structure information. In fact, most rootkits use ETHREAD to get thread id in the same way.

6   Evaluation

In order to evaluate the effectiveness and efficiency of AutoDunt, we deploy it with QEMU 0.9.0 and do some evaluations. Our testing computer is one IBM server which is equipped with two 3GHz Intel Pentium IV processors, 4GB memory cards and the operating system is Linux 2.6.11.

6.1   Effectiveness

To evaluate the effectiveness of AutoDunt, we test 81 real exploits. 35 of them are stack buffer overflow exploits, 23 of them are heap overflow exploits, 6 of them are integer overflow exploits and the rest are format string exploits. Most of the exploits are from Milw0rm [34] while the rest are collected from Internet.

We first choose MS06-055 (Vulnerability in Vector Markup Language Could Allow Remote Code Execution) [35] as a case study to illustrate AutoDunt. We select an unpatched Windows XP SP1 as guest operating system and run IE on it. Then we load a malicious HTML webpage (MS06-055.htm) which is shown as the left part of figure 3. AutoDunt successfully points out the *medium instruction* when the vulnerability is exploited. The code is 'MOV [EDI], DX' (0x6FF3ED1E) which copies malicious input to stack and overwrites the return address. This instruction is in a loop and $eip \xrightarrow{LD(MOV[EDI],DX)} Input$. AutoDunt also figures out the position of input data which changes the value of eip. To verify its correctness, we change the data at this position from 'AA' to 'MN' (as shown in the right part of figure 3) and load the webpage again. After the overflow, eip pointers to 0x004E004D (which is the Unicode form of 'MN' in memory).

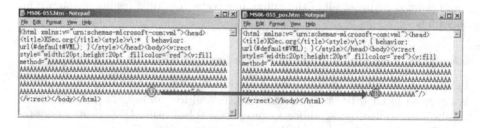

Fig. 3. A Malicious Webpage for MS06-055

In 81 exploits we have tested, we find AutoDunt could point out 77 right positions of the vulnerable codes successfully. The rest four vulnerabilities are all format string ones. Format string vulnerability falls into three categories: denial of service, reading and writing [36]. The four vulnerabilities are all in category three, which is the combination of arbitrary code execution attack and arbitrary function argument attack. We do not implement this kind of detection in our current prototype. It is not difficult to achieve this by using the method in [24]. Note that there may be some false positives in our system if the target program modifies the return address in stack itself. This exists in some self-modified codes and packed binaries. We will manage this problem in the future.

6.2   Obfuscated Shellcode Detection

Obfuscated shellcode usually has different signatures from its original one, which makes it difficult to detect. So attackers like to obfuscate codes to evade the detection of antivirus software and firewall [37,38]. In this part, several methods are chosen to obfuscate the shellcodes of five popular worms (including CodeRed, Slammer, Blaster, Sasser, MyDoom) to test the robustness of AutoDunt.

Table 2. Obfuscated Shellcode Detection

Obfuscation	Result
Code Reordering	Yes
Garbage Insertion	Yes
Equivalent Code Replacement	Yes
Jump Insertions	Yes
Code and Data Encapsulation	Yes
Register Renaming	Yes
Branch function	Yes
Opaque Predict	Yes
No Shellcodes	Yes

Table 2 shows that no obfuscated shellcode could evade the detection of AutoDunt. It is easy to understand since AutoDunt does not rely on signatures to detect shellcode. What AutoDunt cares is the change of *super variables*. Thus, no matter how different the shellcode seems, AutoDunt can still detect it and expose the vulnerable code in software.

6.3   Efficiency

We measured the efficiency of AutoDunt using several popular software in Windows in different ways. Table 3 compares the basic block number of user codes and kernel codes of several popular applications. It also shows the number of saved states and their proportion to all blocks. The result shows that AutoDunt chooses only a few states to save, which increases the efficiency of state saving and recovery. The peak memory usage of AutoDunt is also recorded. AutoDunt only saves the changed memory instead of the whole memory of each state, which saves a lot of memory and time. A comparison of these two methods is shown in figure 4. The first method is to record the whole memory of user stack in each state. We do not include heaps and global memory here since the memory usage is too high for AutoDunt to manage. The other method is to record changed memory (including stack, heap and all the other global memory). It is obvious that the memory usage of AutoDunt is extremely low. We also compare the memory usage between AutoDunt and TTVM [30], which also saves program states and replays them for program analysis. The memory usage (per second) of AutoDunt is about 0.5 MB/s while the usage of TTVM is about 4~7 MB/s [30]. The travel time to go to a desired point is about 10 seconds for TTVM while it is less than 1 second for AutoDunt.

To measure the overhead of state saving process, we record the time to open a document or a mp3 music according to different programs. Figure 5 compares the running time of those programs in native system and in virtual surroundings (with and without AutoDunt). The overhead of AutoDunt is about two times as

Table 3. Blocks number, States number and Memory Usage of Different Programs

	Microsoft Word 2003	Acrobat Reader 7	Media Player 10	Microsoft IE 6	Microsoft Live Messenger 8.1	Outlook 2003	Excel 2003
Block (user)	38251073 (42.108%)	13167959 (68.262%)	16601976 (31.926%)	20316738 (30.115%)	43028837 (62.143%)	36883635 (37.210%)	21935884 (39.350%)
Block (kernel)	52589922 (57.892%)	6122278 (31.738%)	35398642 (68.074%)	47146661 (69.885%)	26212978 (37.857%)	62240349 (62.790%)	33809681 (60.650%)
State Number	135972 (0.150%)	123907 (0.642%)	179815 (0.346%)	56864 (0.084%)	138585 (0.200%)	125420 (0.127%)	40545 (0.073%)
Memory (KB)	13,145	16,698	3,408	12,988	10,975	14,647	13,820

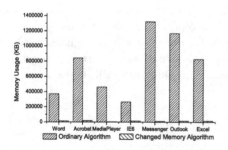

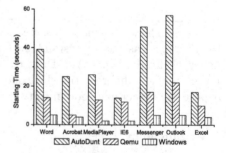

Fig. 4. Memory usage of two methods, one is to save all the memory of user stack and the other is to save changed memory

Fig. 5. Several programs' runtime performance in AutoDunt, QEMU and native system (Windows)

much as QEMU in average. As the performance of AutoDunt depends much on virtual surroundings, it increases with the improvement of virtual technologies. Moreover, some other methods [39,40,41] are proposed to decrease the overhead in dynamic analysis. We may use them in the future implementation of AutoDunt. Figure 6 and figure 7 show the results of omitting the trusted code and nested loop. They save as much as 2.95x numbers of states and 3.94x running time in average.

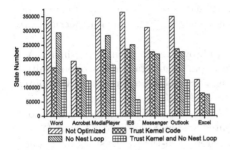

Fig. 6. The number of states with/without trusted codes and nested loops

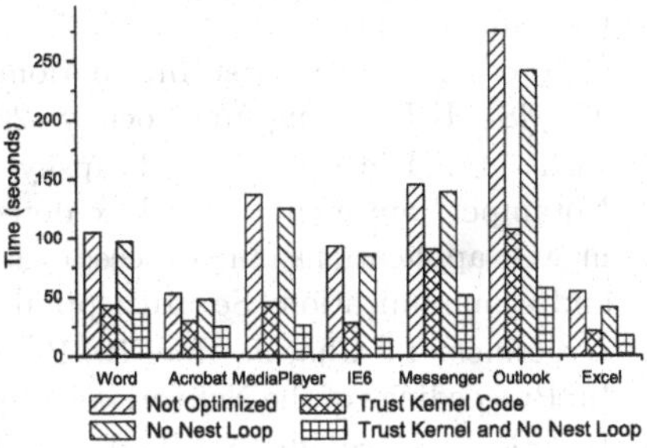

Fig. 7. The running time of different programs with/without trusted codes and nested loops

7 Conclusion and Future Work

In this paper, we design and develop a proof-of-concept prototype called Auto-Dunt to detect arbitrary code execution vulnerabilities in software by analyzing attacks directly in virtual surroundings. It could analyze both user codes and kernel codes without any source codes. It does not care about polymorphic/metamorphic shellcode (even no shellcode). We also propose a new kind of dependence between variables called *latent dependence* to decrease the number of saved states for virtual surrounding replaying. In this way, there is no need to use slicing method or taint analysis method, which increases the efficiency.

AutoDunt is a young system. Finding vulnerable codes greatly relies on attacks detection. In future work, we will enhance the detection part of AutoDunt. We will also try to decrease the number of saved states and increase the managing speed further. Moreover, we try to implement AutoDunt in a non-emulated environment to improve efficiency.

Acknowledgement. This work was supported by the National Natural Science Foundation of China (Grant No.61100226, No.60970028).

References

1. Cowan, C., Wagle, P., Pu, C., Beattie, S., Walpole, J.: Buffer overflows: attacks and defenses for the vulnerability of the decade. In: Foundations of Intrusion Tolerant Systems (Organically Assured and Survivable Information Systems), pp. 227–237 (2003)

2. Kuperman, B.A., Brodley, C.E., Ozdoganoglu, H., Vijaykumar, T.N., Jalote, A.: Detection and prevention of stack buffer overflow attacks. Communications of the ACM 48(11), 50–56 (2005)

3. BBC: Serious security flaw found in ie (2011), http://news.bbc.co.uk/2/hi/technology/7784908.stm

4. Hund, R., Holz, T., Freiling, F.: Return-oriented rootkits: Bypassing kernel code integrity protection mechanisms. In: Proceedings of the 18th Conference on USENIX Security Symposium, pp. 383–398. USENIX Association (2009)

5. Ganesh, V., Leek, T., Rinard, M.: Taint-based directed whitebox fuzzing. In: Proceedings of the 31st International Conference on Software Engineering, pp. 474–484. IEEE Computer Society (2009)

6. Bisht, P., Hinrichs, T., Skrupsky, N., Bobrowicz, R., Venkatakrishnan, V.: Notamper: automatic blackbox detection of parameter tampering opportunities in web applications. In: Proceedings of the 17th ACM Conference on Computer and Communications Security, pp. 607–618. ACM (2010)

7. Avgerinos, T., Cha, S., Hao, B., Brumley, D.: Aeg: Automatic exploit generation. In: Proceedings of the Network and Distributed System Security Symposium (2011)

8. Baratloo, A., Singh, N., Tsai, T.: Transparent run-time defense against stack smashing attacks. In: Proceedings of the USENIX Annual Technical Conference, pp. 251–262 (2000)

9. Shacham, H., Page, M., Pfaff, B., Goh, E., Modadugu, N., Boneh, D.: On the effectiveness of address-space randomization. In: Proceedings of the 11th ACM Conference on Computer and Communications Security, pp. 298–307. ACM (2004)

10. Kil, C., Jun, J., Bookholt, C., Xu, J., Ning, P.: Address space layout permutation (aslp): Towards fine-grained randomization of commodity software. In: 22nd Annual Computer Security Applications Conference, ACSAC 2006, pp. 339–348. IEEE (2006)

11. Sezer, E.C., Ning, P., Kil, C., Xu, J.: Memsherlock: An automated debugger for unknown memory corruption vulnerabilities (2007)

12. Weiser, M.: Programmers use slices when debugging. Communications of the ACM 25(7), 446–452 (1982)

13. Newsome, J., Song, D.: Dynamic taint analysis for automatic detection, analysis, and signature generation of exploits on commodity software. In: Proceedings of the 12th Annual Network and Distributed System Security Symposium (2005)

14. Kim, H.A., Karp, B.: Autograph: Toward automated, distributed worm signature detection. In: USENIX Security Symposium, vol. 286 (2004)

15. Newsome, J., Karp, B., Song, D.: Polygraph: automatically generating signatures for polymorphic worms. In: 2005 IEEE Symposium on Security and Privacy, pp. 226–241 (2005)

16. Li, Z., Sanghi, M., Chen, Y., Kao, M.Y., Chavez, B.: Hamsa: Fast signature generation for zero-day polymorphic worms with provable attack resilience. In: Proceedings of the 2006 IEEE Symposium on Security and Privacy, pp. 32–47 (2006)

17. Wang, X., Pan, C.C., Liu, P., Zhu, S.: Sigfree: A signature-free buffer overflow attack blocker. In: Proceedings of the 15th conference on USENIX Security (2006)

18. PaX.Team: Pax documentation (2003), http://pax.grsecurity.net/docs/pax.txt

19. Kc, G.S., Keromytis, A.D.: e-nexsh: Achieving an effectively non-executable stack and heap via system-call policing. In: 21st Annual Computer Security Applications Conference, pp. 286–302 (2005)

20. McGregor, J.P., Karig, D.K., Shi, Z., Lee, R.B.: A processor architecture defense against buffer overflow attacks. In: Proceedings of International Conference on Information Technology: Research and Education, ITR 2003, pp. 243–250 (2003)

21. Castro, M., Costa, M., Harris, T.: Securing software by enforcing data-flow integrity (2006)

22. Kiriansky, V., Bruening, D., Amarasinghe, S.: Secure execution via program shepherding. In: Proceedings of the 11th USENIX Security Symposium, pp. 191–205 (2002)

23. Bhatkar, S., Sekar, R., DuVarney, D.C.: Efficient techniques for comprehensive protection from memory error exploits. In: Proceedings of the 14th Conference on USENIX Security Symposium, vol. 14 table of contents, p. 17 (2005)

24. Costa, M., Crowcroft, J., Castro, M., Rowstron, A., Zhou, L., Zhang, L., Barham, P.: Vigilante: end-to-end containment of internet worms. In: Proceedings of the Twentieth ACM Symposium on Operating Systems Principles, pp. 133–147 (2005)

25. Brumley, D., Wang, H., Jha, S., Song, D.: Creating vulnerability signatures using weakest preconditions. In: 20th IEEE Computer Security Foundations Symposium, pp. 311–325 (2007)

26. Brumley, D., Newsome, J., Song, D., Wang, H., Jha, S.: Towards automatic generation of vulnerability-based signatures. In: Proceedings of the 2006 IEEE Symposium on Security and Privacy, pp. 2–16 (2006)

27. Yin, H., Song, D., Egele, M., Kruegel, C., Kirda, E.: Panorama: capturing system-wide information flow for malware detection and analysis. In: Proceedings of the 14th ACM Conference on Computer and Communications Security, pp. 116–127. ACM, New York (2007)

28. Bayer, U., Kruegel, C., Kirda, E.: Ttanalyze: A tool for analyzing malware. In: 15th Annual Conference of the European Institute for Computer Antivirus Research, EICAR (2006)

29. Suh, G.E., Lee, J.W., Zhang, D., Devadas, S.: Secure program execution via dynamic information flow tracking. In: Proceedings of the 11th International Conference on Architectural Support for Programming Languages and Operating Systems, pp. 85–96 (2004)

30. King, S.T., Dunlap, G.W., Chen, P.M.: Debugging operating systems with time-traveling virtual machines. In: Proceedings of the USENIX 2005 Annual Technical Conference (2005)

31. Ramalingam, G.: Identifying loops in almost linear time. ACM Transactions on Programming Languages and Systems (TOPLAS) 21(2), 175–188 (1999)

32. Sreedhar, V.C., Gao, G.R., Lee, Y.F.: Identifying loops using dj graphs. ACM Transactions on Programming Languages and Systems (TOPLAS) 18(6), 649–658 (1996)

33. Bellard, F.: Qemu, a fast and portable dynamic translator. In: Proceedings of the USENIX Annual Technical Conference, FREENIX Track, pp. 41–46 (2005)

34. Milw0rm: milw0rm-exploits: vulnerabilities: videos: papers: shellcode (2008), http://www.milw0rm.com/

35. Microsoft: Microsoft security bulletin ms06-055 (2006), http://www.microsoft.com/technet/security/Bulletin/MS06-055.mspx

36. techFAQ: What is a format string vulnerability? (2011), http://www.tech-faq.com/format-string-vulnerability.shtml

37. Collberg, C., Thomborson, C., Low, D.: A taxonomy of obfuscating transformations. University of Auckland Technical Report 170 (1997)

38. Linn, C., Debray, S.: Obfuscation of executable code to improve resistance to static disassembly. In: Proceedings of the 10th ACM Conference on Computer and Communications Security, pp. 290–299 (2003)
39. Ho, A., Fetterman, M., Clark, C., Warfield, A., Hand, S.: Practical taint-based protection using demand emulation. In: Proceedings of the 2006 EuroSys Conference, pp. 29–41 (2006)
40. Qin, F., Lu, S., Zhou, Y.: Safemem: exploiting ecc-memory for detecting memory leaks and memory corruption during production runs. In: 11th International Symposium on High-Performance Computer Architecture, HPCA-11, pp. 291–302 (2005)
41. Qin, F., Wang, C., Li, Z., Kim, H., Zhou, Y., Wu, Y.: Lift: A low-overhead practical information flow tracking system for detecting security attacks. In: Proceedings of the Annual IEEE/ACM International Symposium on Microarchitecture (Micro 2006), Orlando, Florida, USA (December 2006)

Weaknesses in Current RSA Signature Schemes

Juliane Krämer, Dmitry Nedospasov, and Jean-Pierre Seifert

Security in Telecommunications
Technische Universität Berlin and Deutsche Telekom Innovation Laboratories
Germany
{juliane,dmitry,jpseifert}@sec.t-labs.tu-berlin.de

Abstract. This work presents several classes of messages that lead to data leakage during modular exponentiation. Such messages allow for the recovery of the entire secret exponent with a single power measurement. We show that padding schemes as defined by industry standards such as PKCS#1 and ANSI x9.31 are vulnerable to side-channel attacks since they meet the characteristics defined by our classes. Though PKCS#1 states that there are no known attacks against RSASSA-PKCS1-v1.5, the EMSA-PKCS1-v1.5 encoding in fact makes the scheme vulnerable to side-channel analysis. These attacks were validated against a real-world smartcard system, the Infineon SLE78, which ran our proof of concept implementation. Additionally, we introduce methods for the elegant recovery of the full RSA private key from blinded RSA CRT exponents.

Keywords: RSA, PKCS#1, ANSI x9.31, Side-Channel Attacks, Simple Power Analysis, CRT, Exponent Blinding.

1 Introduction

Side-channel attacks exploit information leaked by the physical characteristics of a cryptosystem [8,9,17]. A common side-channel attack is power analysis. Power analysis can be categorized into two subcategories, *simple power analysis (SPA,* methods requiring few measurements) and *differential power analysis (DPA,* methods requiring many measurements) [12]. Since it is commonly impossible to recover the data being leaked in a single measurement, adversaries are often forced to perform DPA to recover data in its entirety. In turn, countermeasures, e.g., blindings, are implemented to counteract the attacks and to thwart DPA.

In this work, we attack the RSA signature process by performing simple power analysis to recover the potentially blinded secret exponent. We present several classes of messages that lead to data leakage during modular exponentiation. Specifically, we show that several properly formatted standardized input messages, including the message encodings of PKCS#1 [18] and ANSI x9.31 [1], meet the criteria defined by these classes. Thus, we show that compliance with industry standards can in fact lead to data leakage, although these standards are considered to be secure message encodings.

The analysis was performed against our proof of concept (POC) implementation running on an Infineon smartcard system, which performed ZDN-based

H. Kim (Ed): ICISC 2011, LNCS 7259, pp. 155–168, 2012.
© Springer-Verlag Berlin Heidelberg 2012

modular multiplication [7]. This setup allowed us to test all classes of input messages presented in this work. For all classes of input messages, the SPA yielded the entire private exponent or the entire blinded private exponent that was used for the signature process.

Recovering a potentially blinded exponent is sufficient to sign messages in the RSA signature scheme. However, this is not true for RSA CRT. In the case of RSA CRT, the attacker must instead recover the full private key. Methods for recovering the full RSA private key have been known since 1978, whereas we present two specific methods for RSA CRT, with and without exponent blinding. Since several methods for recovering a secret exponent fail to recover all of the exponent bits, we present the approach *exponent un-blinding*, which enables an attacker to compute the full private key more efficiently in such cases. This method can cope with more noise, more efficiently than other known methods.

The main contributions of this paper are:

1. Categorization of common vulnerable message classes and the corresponding attack scenarios.
2. Demonstrating that constant padding makes RSA signature schemes such as RSASSA-PKCS1-v1_5 and ANSI x9.31 vulnerable to side-channel analysis.
3. Practical validation of the attacks and the proposed attack scenarios against a proof of concept implementation on an Infineon smartcard system.
4. More efficient methods for recovering the full RSA CRT private key from blinded private exponents.

The paper is organized as follows: Section 2 presents necessary background information. In Section 3 we present several common types of input messages and explain attack vectors and scenarios which arise from certain characteristics of these messages. We categorize these characteristics into classes of messages which lead to data leakage. Any valid PKCS#1 and ANSI x9.31 message meets the criteria defined in one of these classes. We then demonstrate these attacks against an Infineon smartcard system running our POC implementation in Section 4. In Section 5, novel methods for the recovery of the full private key are explained for RSA CRT. Finally, we summarize the implications of our research in Section 6.

2   Background

In this section, we first give a brief introduction of the RSA and the RSA CRT signature scheme. We then explain the square-and-multiply algorithm for modular exponentiation and the ZDN algorithm for modular multiplication. Finally, we explain blinding techniques, which are used to thwart statistical attacks.

2.1   RSA CRT

Let (N, e) be the public RSA modulus and exponent, and $(p, q, d, \varphi(N))$ be the private key, satisfying $N = pq$ and $ed \equiv 1 \mod \varphi(N)$. As the modulus N is the product of two different primes, the *Chinese Remainder Theorem* (CRT) can be used to speed up the time intensive process of message signing by a factor of four [13,16]. Instead of computing the RSA signature $s = m^d \bmod N$ with an exponent of the order of $n = \log_2(N)$ bits (assuming a small e), two modular exponentiations with $n/2$-bit exponents are performed. In this setting and without loss of generality $q < p$, we precompute $d_p = d \bmod (p-1)$, $d_q = d \bmod (q-1)$ and $q_{inv} = q^{-1} \bmod p$. These constants are also part of the private key [18]. They are used for the computations

$$s_p = m^{d_p} \bmod p \qquad \text{and} \qquad s_q = m^{d_q} \bmod q. \qquad (1)$$

Subsequently, Garner's algorithm is used to yield the signature s of m:

$$s = s_q + (q_{inv} \cdot (s_p - s_q) \bmod p) \cdot q. \qquad (2)$$

2.2 Square-and-Multiply for Modular Exponentiation

A commonly used algorithm for modular exponentiation is the modular *square-and-multiply* algorithm, which exploits the binary representation of the exponent, see Figure 1. The input for this algorithm is (m, d, N) and its output, $s = m^d \bmod N$, is the signature of m. Let d_i, $i \in \{0, \ldots, l-1\}$, denote the ith bit of d, i.e., d_0 is the least significant bit. Thus, l is the bit length of d and we have $l = \lfloor \log_2(d) \rfloor + 1$. Performing a modular exponentiation with this algorithm needs $\mathcal{O}(\log_2(d))$ operations, i.e., it has logarithmic complexity.

2.3 The ZDN Algorithm for Modular Multiplication

Whithin the square-and-multiply algorithm, modular multiplications are perfomed. *ZDN*-based modular multiplications consist of three major parts, computation of the "look-ahead" multiplication (LABooth) [5,21], computation of the "look-ahead" reduction (LARed) [5,21], and a subsequent 3-operand addition, which finally yields the resulting partial product, see Figure 2. LABooth is optimized to shift across constant bit strings, whereas LARed requires only several significant bits to compute the reduction. The look-ahead reduction is designed so that its average reduction is approximately the same as the one of the look-ahead multiplication.[1] Thanks to this high level of optimization, the three parts are executed in parallel and require just a single clock cycle [5]. The algorithm ensures that the intermediate result Z fulfills $|Z| \leq \frac{1}{3}N$, hence the name (two-thirds N is *zwei Drittel N* in German).

2.4 Blinding Techniques to Thwart Statistical Attacks

Both RSA and RSA CRT are vulnerable to differential side-channel attacks [8,9,17]. To prevent these statistical side-channel attacks, randomized *blinding* is used to disguise intermediate results and to decouple the leaked information

---

[1] Both look-ahead sub-operations are explained in detail in [20,21].

```
1   input: m, d, N                    1   input: t, m, N
2   output: m^d mod N                 2   output: m·t mod N
3   k := log2(d) - 1, t := 1          3   Z := 0, C := m
4   while k >= 0                      4   l := log2(t) + 1, c := 0
5       // square                     5   while l > 0 or c > 0 do
6       // ZDN Mod-Mult               6       LABooth(t, &l, &s_t, &v_C)
7       t = (t·t) mod N               7       LARed(Z, N, c, &s_z, &v_N)
8       if d_k = 1 then               8       s_C := s_Z - s_t
9           // multiply               9       C := C·2^{s_C}
10          // ZDN Mod-Mult           10      Z := Z·2^{s_Z} + v_C·C + v_N·N
11          t = (t·m) mod N           11      c := c - s_C
12      k := k - 1                    12  endwhile
13  endwhile                         13  if Z < 0 then Z := Z + N
14  return t                         14  return Z
```

Fig. 1. Modular Square-and-Multiply   **Fig. 2.** ZDN-Based Modular Multiplication

from the processed data. We explain three different blinding techniques of which two are vulnerable to our attack. In all cases, the integers r and r_1, r_2, respectively, are random λ-bit numbers, commonly $\lambda = 32$ [22]. A new r is chosen independently for every operation.

The first of these blinding techniques is called *exponent blinding* [22]. The blinded exponent is $d' := d + r \cdot \varphi(N)$. Due to Euler's theorem, the following equation holds true: $s = m^d \bmod N = m^{d+r \cdot \varphi(N)} \bmod N$.[2] The same blinding can be applied to both exponents when using RSA CRT. In this case, the blinded exponents are $d'_p = d_p + r_1 \cdot (p - 1)$ and $d'_q = d_q + r_2 \cdot (q - 1)$, respectively.

The second blinding technique is called *base blinding* [9] or *message blinding*. Base blinding decouples the side channel leakage from the input m. For a random λ-bit integer r its inverse modulo N is calculated, i.e., $r \cdot r^{-1} \equiv 1 \bmod N$. The blinded message is $m' := r^e \cdot m$. Instead of m, the blinded message is signed, yielding the blinded signature s'. The blinding is reversed by computing $s = (r^{-1} \cdot s') \bmod N$. Since this form of base blinding includes a computationally expensive inverse calculation in $\mathbb{Z}_N{}^*$, it is relatively unattractive for embedded systems. Alternatively, another form of base blinding can be used. Given two random λ-bit integers r_1, r_2 where $r_1 < r_2$, the exponentiation can be computed as follows, $s = \left[(r_1 \cdot N + m)^d \bmod (r_2 \cdot N) \right] \bmod N$. This form of base blinding is far less computationally expensive.

3   SPA-Based Secret Exponent Recovery

In this section, we present several side-channel attacks which obtain the exponent of a modular exponentiation with a single-trace analysis. Most known methods

---

[2] It is very unlikely that the necessary condition $\gcd(m, N) = 1$ is not fulfilled.

for recovering exponents rely on statistical analysis [4,8,9,19,23]. As such, these methods require multiple exponentiations with the same exponent, and thus, can be prevented by exponent blinding. Instead, we consider approaches that are able to recover the entire exponent from a single modular exponentiation operation. These approaches also work whenever exponent blinding is used. First, we recall several known methods and then we present a list of criteria for the message that, when met, allow us to recover an exponent in a single trace.

3.1 Known Methods

In 2005, an attack that uses the specific input $m = -1$, i.e., $N - 1 \bmod N$, was presented [24]. This attack exploits the fact that whenever square-and-multiply is used, there are just three distinct pairs of operations, which are performed during exponentiation [14]. Due to the special input message, these distinct pairs result in three distinct power dissipation states, which can be identified within the power trace. Therefore, the bit pattern of the private exponent can be obtained by performing a single-trace SPA [14,24]. The same approach was extended for RSA CRT.

In 2010, an additional method for recovering secret data via a single-trace SPA emerged [3]. The authors consider systems, which utilize t-bit multipliers for performing long integer arithmetic, i.e., $t = 32$ or $t = 64$. If one or more t-bit strings of a message are equal to 0, i.e., 0-strings, the message will lead to data leakage in the power trace. The authors also describe several possible messages, which lead to data leakage, such as messages with a low Hamming weight, i.e., $m = 2^x$ where $x \leq \log_2(N)$. The authors mention that multiple constant t-bit strings, or constant strings that are longer in length, only increase the leakage even more. We demonstrate in this work that certain aspects of this attack are also applicable to systems that do not use t-bit multipliers of a certain length t, and consider systems that perform full length integer multiplication directly.

3.2 Classes of Input Messages

We present several classes of input messages and corresponding attack scenarios that lead to differences in the power consumption depending on the value of the exponent bit. As a result, whenever a cryptosystem performs a modular exponentiation with a message from one of these classes, the exponent bit sequence can be recovered. We validate these claims by performing an SPA against a POC implementation in Section 4.

Our attacks are based on messages that have constant bit strings, which can lead to data leakage. Specifically, we focus on two message types. The first class includes standardized messages, which consist largely of constant padding. In such cases, the constant padding of the leading bits constitutes a Leading Constant Bit String (LCBS). In the second class of message we consider regions of the message, which are set or affected by user input. Usually this is a region of the least significant bits or trailing bits. Hence we refer to this class of message as Trailing Constant Bit String (TCBS).

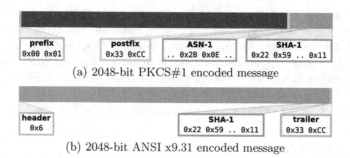

(a) 2048-bit PKCS#1 encoded message

(b) 2048-bit ANSI x9.31 encoded message

Fig. 3. Figures 3(a) and 3(b) are drawn to scale. The dark gray area of Figure 3(a) corresponds to the "heavier" Hamming weight of the leading 0xFF of the RSASSA-PKCS1-v1_5 padding and the lighter gray of Figure 3(b) to the "lighter" Hamming weight of the leading 0xBB padding of ANSI x9.31.

The message classes described in this section allow us to distinguish between square and multiply operations. Once we are able to distinguish a square from a multiply, the bit pattern of the exponent can be recovered from a single power trace, as is demonstrated in Section 4.

Leading Constant Bit String: The first class of message we consider, is the *Leading Constant Bit String* (LCBS). These are messages in which the most significant bits consist mostly of constant 0- or 1-strings. LCBS messages are particularly interesting because many valid messages utilizing non-random padding schemes constitute LCBS messages. For example, we classify both RSASSA-PKCS1-v1_5 of PKCS#1 [18] because of the leading 0xFF (11111111_2), and to a lesser extent ANSI x9.31 [1] with leading 0xBB(10111011_2) as LCBS messages. Thus, attacks that utilize LCBS messages are potentially harder to defend against because such attacks do not necessarily rely on the chosen message attack model. In such a scenario, the data is leaked by **any** valid message. Though PKCS#1 states that there are no known attacks against RSASSA-PKCS1-v1_5 [18], we demonstrate that the EMSA-PKCS1-v1_5 encoding in fact makes the scheme vulnerable to side-channel analysis, see Section 4. In the case of the exemplary 2048-bit PKCS#1 message, over 84% of the message is padding, see Figure 3(a). In the case of the exemplary 2048-bit ANSI x9.31 message, over 91% of the message is padding, see Figure 3(b).

LCBS messages do not necessarily reduce the workload of the modular reduction on systems that do not perform multiplication and reduction in parallel. However, the consistent structure of the leading 0- or 1-strings ensures a reduced workload on highly optimized systems implementing algorithms like ZDN [5].

Trailing Constant Bit String: The second class of message we consider is the *Trailing Constant Bit String* (TCBS). This is an important classification because many cryptographic schemes operate on messages that contain only a relatively small variable region set or affected by user input. In most cases, this region is a relatively small portion of the least significant bits and the bulk of the message consists of padding. However, if an attacker is able to set as little as 5%-10% of the least significant bits by, for example, supplying the specified hash to the signature scheme directly, then the attacker would be able to recover the secret data independent of the padding scheme being used. In this scenario, even if randomized padding is used, a very small region of trailing bits is sufficient to leak the entire secret data. Note, standards such as PKCS#1 also define multiple hash algorithms that can be used. Potentially, an attacker could even increase the region affected by user input to be as large as 512 bits if he is allowed to provide, for example, SHA-512-based messages instead.

As with LCBS, TCBS messages can be as long as the modulus in bits and, as such, TCBS messages do not necessarily reduce the workload of the modular reduction. This is generally the case whenever multiplication and reduction are not computed in parallel. However, on highly optimized systems, i.e. those which implement ZDN [5], the constant trailing 0- or 1-strings ensure a reduced workload.

Short Messages: Though of little interest if the implementation enforces padding, the third class of messages we consider is the *short message*. Short messages are messages $m \ll N$, where N is the modulus of the modular exponentiation operation in question. Short messages can be considered LCBS messages with leading 0-strings. We consider short messages in this work, primarily because they exploit both the multiplication and reduction step of modular multiplication and achieve the greatest difference in the power consumption of squares and multiplies, respectively. This was also validated against our POC implementation, where padding checks could be disabled, see Section 4. Note that efficient implementations generally ignore, or shift across any leading 0-strings, which in conjunction with the relatively low Hamming weight of the entire message greatly reduces the workload of the multiplication step. Additionally, because of the short length in bits of the message m, the intermediate result of the multiplication step of the square-and-multiply algorithm increases by only $\log_2(m)$ bits in length prior to reduction. In comparison, during the square operation, the bit length of the intermediate result approximately doubles. As a result, in addition to the lower computational workload of the multiplication, such messages also reduce the computational workload of the modular reduction after a multiplication and potentially eliminate reduction completely, further lowering the power consumption of the multiply operation.

4   Proof of Concept

In this section we present the practical evaluation of the classes of messages described in Section 3 on a real-world system.

The cryptosystem analyzed in this work is an Infineon SLE78-based [7] smart-card system. The SLE78 features a cryptographic coprocessor known as the Crypto@2304T, which provides 2304-bit registers and ZDN-based modular multiplication [5]. In contrast to previous works such as [3], which focus on cryptosystems that use "short" bit length multipliers (i.e. 32 or 64-bit multipliers), the SLE78 performs full-length arithmetic operations by utilizing registers and logic capable of 1024-bit and 2048-bit modular multiplication. With ZDN-based modular multiplication, multiplication and reduction are computed in parallel in multiple iterations of the modular multiplication loop, see Section 2.3. This improves performance and further reduces register length requirements by ensuring $|Z| \leq \frac{1}{3}N$ for the partial product Z and the RSA modulus N.

When used in conjunction with algorithms like square-and-multiply, the highly optimized nature of this modular multiplication introduces additional weaknesses. The two important characteristics of ZDN-based modular multiplication, which were exploited in this work are:

– LABooth ensures that the cryptosystem can shift across 0- (i.e., 00..00) and 1-strings (11..11) as well as 0- and 1-strings with isolated 1's and 0's, respectively, i.e., (0..010..0) and (1..101..1).
– LARed ensures that partial products are only actually reduced whenever they become too large, i.e. $|Z| > \frac{1}{3}N$.

By selecting messages, which meet the criteria outlined in Section 3, we exploit all of these characteristics of the algorithm. However, exploiting even any one characteristic of the algorithm allows for the recovery of the sequence of square and multiply operations, and thus, for the recovery of the secret exponent.

The system ran a proof of concept software implementation, which performed RSA signing. This implementation used square-and-multiply for modular exponentiation, the ZDN algorithm for modular multiplication, and it performed exponent blinding, as described in Section 2.2, 2.3 and 2.4. The system did not enforce padding, which allowed us to test all the message types described in Section 3, including short messages. The system was connected to a PC, which ran the client software, via a standard USB smartcard reader. The client software allowed us to select input messages and enable or disable additional software and hardware countermeasures.

Figures 4(a) and 4(c) show the first 3ms of the computation for a common exponent, but with the different classes of messages introduced in Section 3 as the input. The modulus of the modular exponentiation was 1024 bits in length for all the input message classes. For comparison since truly random messages do not produce data leakage, we provide a trace of a random message in the extended version of this paper (see [10]).

The data leakage is clearly visible in Figures 4(a), 4(c) and for the short message (see [10]). The attacks failed to recover a few of the leading bits depending on the class of input message, as described in Section 5. The system ran at 32MHz with no current limit and timing jitter enabled. Our experimental setup allowed us to capture the entire computation at this resolution. The system current was measured with a LeCroy 7-zi oscilloscope by performing a low-side shunt measurement over a 10Ω resistor.

Leading Constant Bit String: Figures 4(a) and 4(b) show the data leakage of the system while processing an LCBS input message. The LCBS message is the most important message class analyzed in this work, because any valid

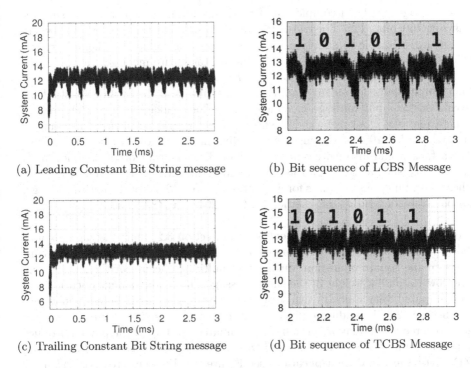

Fig. 4. First 3ms of the exponentiation for LCBS and TCBS input messages. The system current was measured with a LeCroy 7-Zi digital oscilloscope [11] via low-side current shunt insertion. Figures 4(b) and 4(d) are a magnification of the time 2ms - 3ms for the respective input message.

RSASSA-PKCS1-v1_5 message is a candidate LCBS message. On systems implementing highly optimized algorithms like ZDN, such as the smartcard system we analyzed, LCBS messages can also lead to data leakage despite the leading non-zero padding, i.e., leading 0xFF (11111111_2) and 0xBB(10111011_2). With ZDN, the look-ahead algorithm's sub-operations, LABooth and LARed, run in parallel and ensure that the system simply shifts across any leading 0- or 1-strings, deferring the bulk of the arithmetic operations, see Section 2.3. As a result, the constant structure of the leading bits of the message ensures a lower workload and lower power dissipation during the multiply operation. These effects are clearly visible for the LCBS input message in Figure 4(a). We chose the message according to the scenario described in Section 3.2, i.e., we used the constant RSASSA-PKCS1-v1_5 padding and added a random 160-bit string as hash value. Messages coded in the ANSI x9.31 format resulted in very similar data leakage.

Trailing Constant Bit String: Figures 4(c) and 4(d) show the data leakage of the system while processing a TCBS input message. The TCBS message succeeds in inducing leakage despite the random padding used in the input message.

The message consisted completely of random padding except for the least significant 160 bits. This illustrates the scenario described in Section 3.2, where the attacker is able to supply a hash value into the signature scheme directly. The 160 bits of SHA-1 make up only 16% of the entire message. However, the look-ahead algorithm's sub-operations, LABooth and LARed, ensure that the system simply shifts across any leading 0- or 1-strings and the trailing constant bit string of the input message is sufficient to induce leakage on our POC implementation. Note, our experiments show that a constant bit string consisting of as little as 5%-10% of the message is sufficient to induce data leakage. Thus, the attack would also work against 2048-bit RSA and for other common hash algorithms, such as SHA-256/384/512 or MD2 and MD5, respectively. Each of these hash functions accounts for at least 6.25% and 12.5% of the entire message, respectively, depending on whether 1024- or 2048-bit RSA is used.

Short Messages: As already mentioned in Section 3.2, short messages exploit both parts of parallel modular multiplication algorithms such as ZDN. In contrast to the square operation, during the multiply, because of the small value and low Hamming weight of the short message, the modular multiplication can be computed very quickly with very few iterations of the LABooth algorithm, see Section 2.3. In addition, since the intermediate result only grows by very few bits, reduction may potentially be eliminated entirely. If the partial product must be reduced, it can be computed with very few iterations of the loop during ZDN-based modular multiplication, see Figure 2.3. For these reasons the short message achieves the greatest difference in power consumption between squares and multiplies on the SLE78.

Potential Countermeasures: Because the attacks presented in this section require a particular structure, i.e., constant bit strings within the message, base blinding can defeat such attacks. However, it is worth noting that the "classical" blinding method as described by [9] actually fails in disrupting the constant bit string structure within the message. In this case, the message $m' := r^e \cdot m$ is used for the exponentiation instead of m, see Section 2.4. For common values of λ, i.e., 32-bit randoms, and small exponents, i.e., 3 or 17, the randomization introduced into the message is actually quite minimal. Additionally, the computation of the blinded message and its inverse becomes increasingly difficult for increasing λ's and exponents. For these reasons, an alternative form of base blinding should be used, which ensures randomization of the entire message, namely $s = [(r_1 \cdot N + m)^d \bmod (r_2 \cdot N)] \bmod N$.

Exponent blinding could be used to obfuscate the exponent, however, blinded exponents can also be used to sign messages in the RSA signature scheme. However, its worth noting that RSA CRT exponents cannot be used to forge signatures, and for this reason we present several methods for recovering the full RSA private key from potentially blinded exponents in Section 5.

Techniques that decouple the execution from the data being processed, such as square-and-multiply-always, were able to prevent our attacks. However, other DPA countermeasures, such as timing jitter, had no affect in our analysis.

5 Full RSA Private Key Recovery

If an attacker can obtain a CRT exponent, which also might be blinded, he can not generate valid signatures with it since the CRT computation of signatures requires both p and q. Thus, the attacker must factorize the modulus N. We present three methods for the factorization of N of which only the first is known.

Lemma 1. *Let $N = pq$ be an RSA modulus, e the public exponent, and d the private exponent with $ed \equiv 1 \mod \varphi(N)$. Let $d_p = d \mod (p-1)$ and $d_q = d \mod (q-1)$ be the RSA private CRT exponents. Then given d_p or d_q, N can be factorized [2].*

Proof. Let $m < N$ be an arbitrary message. Without loss of generality, let d_p be known. Given (N, e) and d_p, we can compute $c = m^e \mod N$ and $m_p = c^{d_p} \mod N$. Since $p \mid N$ and $d_p = d \mod (p-1)$, $m \equiv c^{d_p} \mod p$ and $m_p \equiv c^{d_p} \mod p$. Then $p = \gcd(N, m - m_p)$ [2].

Lemma 2. *Let $N = pq$ be an RSA modulus, e the public exponent, and d the private exponent with $ed \equiv 1 \mod \varphi(N)$. Let $d_p = d \mod (p-1)$ and $d_q = d \mod (q-1)$ be the RSA private CRT exponents. Then given a single blinded private CRT exponent $d_p' = d_p + r \cdot (p-1)$ or $d_q' = d_q + r \cdot (q-1)$, $r \in \mathbb{Z}$, N can be factorized.*

Proof. Let $m < N$ be an arbitrary message. Without loss of generality, let $d_p' = d_p + r \cdot (p-1)$, $r \in \mathbb{Z}$, be known. Given (N, e) and d_p', we can compute $c = m^e \mod N$ and $m_p = c^{d_p'} \mod N$. Since $p \mid N$ and $d_p' = d + x \cdot (p-1)$ for some $x \in \mathbb{Z}$, $m \equiv c^{d_p'} \mod p$ and $m_p \equiv c^{d_p'} \mod p$. Then $p = \gcd(N, m - m_p)$.

We propose an elegant method for recovering the full RSA private key from blinded CRT exponents, which we call *exponent un-blinding*. This method spares the expensive modular exponentiations necessary in Lemma 2.

Lemma 3. *Let $N = pq$ be an RSA modulus, e the public exponent, and d the private exponent with $ed \equiv 1 \mod \varphi(N)$. Let $d_p = d \mod (p-1)$ and $d_q = d \mod (q-1)$ be the RSA CRT private exponents. Then given at least $k \geq 3$ blinded exponents $d_{p_i}' = d_p + r_i \cdot (p-1)$ or $d_{q_i}' = d_q + r_i \cdot (q-1)$, $r_i \in \mathbb{N}$, $i \in \{1, \ldots, k\}$, N can be factorized.*

Proof. Without loss of generality, let k blinded exponents d_{p_i}' be known. We calculate the pairwise differences $d_{p_{i,j}}' = |d_{p_i}' - d_{p_j}'| = |d_p + r_i \cdot \varphi(p) - (d_p + r_j \cdot \varphi(p))| = |r_i - r_j| \cdot \varphi(p) = r_{i,j} \cdot \varphi(p) = r_{i,j} \cdot (p-1)$, since $\varphi(p) = p-1$. Subsequently, we get $G = \gcd\{d_{p_{i,j}}'\} = \gcd\{r_{i,j}\} \cdot (p-1) = g \cdot (p-1)$. Thus, we can test whether $g = 1$, i.e., if $G = p-1$, by testing whether $N \equiv 0 \mod (G+1)$. If fulfilled, we have found p and thus know the private key. Otherwise, we test whether $N \equiv 0 \mod (G/g + 1)$ for $g = 2, 3, 4 \ldots$.

Alternatively, if blinded exponents are easily obtainable, we simply obtain an additional blinded exponent $d'_{p_{k+1}}$ and perform the same calculations again. The higher k is, the higher is the probability of having a very small g. We provide experimental results and further information about the success probability in the extended version of this paper (see [10]).

The main advantage of the exponent un-blinding approach lies in the cheap computation costs. Many methods to obtain exponents, in fact, do not reveal the whole exponent [6,15]. This also holds true for an attack using the short message presented in this work, since such attack cannot distinguish square and multiply operations during the computation of the first 7 - 8 bits, i.e., before the first reduction. This means that we cannot determine the value of these bits, apart from the most significant bit, which is always 1. Thus, an attacker has to test all the resulting possibilities. The complexity of these computations can be heavily reduced by applying exponent un-blinding. Although it requires at least three blinded exponents, the method is very efficient. The efficiency of the algorithm is two-fold: First, the required computations are cheap, especially compared to modular exponentiation, which is needed when the method presented in Lemma 2 is used. Second, a false guess for one of the exponents will quickly lead to an obvious false intermediate result, i.e., $G \ll N/2$. Thus, in contrast to the other methods, exponent un-blinding is able to handle more noise, more efficiently.

6   Conclusion

In this work, we presented two classes of messages that lead to data leakage. These classes are referred to as *leading constant bit string* (LCBS) and *trailing constant bit string* (TCBS). Valid input messages of common signature schemes, including PKCS#1 and ANSI x9.31, meet the criteria for these classes. Both classes and the *short message*, a specific LCBS message, were validated against an advanced smartcard system from Infineon, which ran our POC implementation of the RSA signature scheme. In all cases the input messages allowed for the recovery of the RSA private exponent in a single-trace SPA.

The short message exploits both multiplication and reduction of the modular multiplication in RSA signing. However, short messages can be prevented by means such as message length checks or padding. TCBS messages reiterate the importance of restricting direct user input. Our analysis showed that even if the most significant 95% of the message bits consist of random padding, if the attacker is able to set the least significant 5%, he will be able to recover the secret data. Most importantly, LCBS messages demonstrate that even properly formatted messages can lead to distinct data leakage because of constant padding. For these reasons, we consider the constant paddings used by RSASSA-PKCS1-v1.5 and ANSI x9.31 to present a substantial security risk to modern cryptosystems that implement highly optimized algorithms, such as ZDN-based modular multiplication.

Our experimental results show that 0-strings result in far more distinctive data leakage than 1-strings. For this reason, zero padding should be avoided at all costs. In addition to non-constant padding, there are several countermeasures that can thwart such attacks, including square-and-multiply-always and a certain kind of base blinding. The initial reduction of RSA CRT also destroys any constant bit strings in the input message if the input message is larger than the modulus of the operation, i.e., larger than one of the prime factors of the modulus N. For these reasons, unless RSA CRT with initial reduction is used, we recommend that message blinding always be used on systems that implement a constant bit string padding scheme, such as the padding schemes of RSASSA-PKCS1-v1.5 and ANSI x9.31.

Additionally, a specific method for private key recovery when RSA CRT is used was presented, *exponent un-blinding*. This is substantially faster than the known methods if at least three distinct blinded exponents can be obtained. Most importantly, this method can cope with more noise, more efficiently and elegantly than any other known method.

In conclusion, this paper demonstrates that even an advanced cryptosystem, which implements recommended industry standards, can introduce additional unexpected side-channels. We believe that the claim "no attacks are known against RSASSA-PKCS1-v1.5" [18] is no longer true.

Acknowledgements. This work was supported by the German Federal Ministry of Education and Research, and by the Helmholtz Research School on Security Technologies.

The authors of this paper would like to thank all of their colleagues for their support. In particular we would like to thank Collin Mulliner and Christoph Bayer for their helpful and insightful input while writing the paper. We would also like to thank LeCroy Europe for their excellent technical support.

References

1. American National Standards Institute: ANSI X9.31-1998: Public Key Cryptography Using Reversible Algorithms for the Financial Services Industry (rDSA) (1998)
2. Campagna, M., Sethi, A.: Key recovery method for CRT implementation of RSA (2004)
3. Courrège, J.-C., Feix, B., Roussellet, M.: Simple Power Analysis on Exponentiation Revisited. In: Gollmann, D., Lanet, J.-L., Iguchi-Cartigny, J. (eds.) CARDIS 2010. LNCS, vol. 6035, pp. 65–79. Springer, Heidelberg (2010)
4. Dhem, J.F., et al.: A Practical Implementation of the Timing Attack. In: Working Conference on Smart Card Research and Advanced Application, pp. 167–182 (1998)
5. Fischer, W., Seifert, J.-P.: High-Speed Modular Multiplication. In: Okamoto, T. (ed.) CT-RSA 2004. LNCS, vol. 2964, pp. 264–277. Springer, Heidelberg (2004)
6. Halderman, J.A., et al.: Lest we remember: cold-boot attacks on encryption keys. Commun. ACM 52(5), 91–98 (2009)
7. Infineon Technologies AG: Contactless SLE 78 family: Next Generation Security, http://goo.gl/qbQ30
8. Kocher, P.C.: Timing Attacks on Implementations of Diffie-Hellman, RSA, DSS, and Other Systems. In: Koblitz, N. (ed.) CRYPTO 1996. LNCS, vol. 1109, pp. 104–113. Springer, Heidelberg (1996)

9. Kocher, P.C., Jaffe, J., Jun, B.: Differential Power Analysis. In: Wiener, M. (ed.) CRYPTO 1999. LNCS, vol. 1666, pp. 388–397. Springer, Heidelberg (1999)

10. Krämer, J., Nedospasov, D., Seifert, J.P.: Weaknesses in Current RSA Signature Schemes, Extended Version (2011), http://goo.gl/bu5MS

11. LeCroy Corporation: WavePro 7 Zi Oscilloscope, http://www.lecroy.com/Oscilloscope/OscilloscopeSeries.aspx?mseries=39

12. Mangard, S., Oswald, E., Popp, T.: Power Analysis Attacks: Revealing the Secrets of Smart Cards (Advances in Information Security). Springer New York, Inc. (2007)

13. Menezes, A., van Oorschot, P., Vanstone, S.: Handbook of Applied Cryptography. CRC Press (1997)

14. Miyamoto, A., Homma, N., Aoki, T., Satoh, A.: Enhanced power analysis attack using chosen message against RSA hardware implementations. In: ISCAS, pp. 3282–3285 (2008)

15. Percival, C.: Cache missing for fun and profit. In: Proc. of BSDCan 2005 (2005)

16. Quisquater, J.J., Couvreur, C.: Fast decipherment algorithm for RSA public-key cryptosystem. Electronic Letters 18(21), 905–907 (1982)

17. Quisquater, J.-J., Samyde, D.: ElectroMagnetic Analysis (EMA): Measures and Counter-Measures for Smart Cards. In: Attali, S., Jensen, T. (eds.) E-smart 2001. LNCS, vol. 2140, pp. 200–210. Springer, Heidelberg (2001)

18. RSA: PKCS #1 v2.1: RSA Cryptography Standard (2002), ftp://ftp.rsasecurity.com/pub/pkcs/pkcs-1/pkcs-1v2-1.pdf

19. Schindler, W.: A Timing Attack against RSA with the Chinese Remainder Theorem. In: Paar, C., Koç, Ç.K. (eds.) CHES 2000. LNCS, vol. 1965, pp. 109–124. Springer, Heidelberg (2000)

20. Sedlak, H.: Konzept und Entwurf eines Public-Key-Code Kryptographie-Prozessors (1985)

21. Sedlak, H.: The RSA Cryptography Processor. In: Price, W.L., Chaum, D. (eds.) EUROCRYPT 1987. LNCS, vol. 304, pp. 95–105. Springer, Heidelberg (1988)

22. Shamir, A.: Method and Apparatus for Protecting Public Key Schemes from Timing and Fault Attacks. US Patent 5991415 (November 23, 1999)

23. Walter, C., Thompson, S.: Distinguishing Exponent Digits by Observing Modular Subtractions. In: Naccache, D. (ed.) CT-RSA 2001. LNCS, vol. 2020, pp. 192–207. Springer, Heidelberg (2001)

24. Yen, S.-M., Lien, W.-C., Moon, S.-J., Ha, J.C.: Power Analysis by Exploiting Chosen Message and Internal Collisions – Vulnerability of Checking Mechanism for RSA-Decryption. In: Dawson, E., Vaudenay, S. (eds.) Mycrypt 2005. LNCS, vol. 3715, pp. 183–195. Springer, Heidelberg (2005)

Back Propagation Neural Network Based Leakage Characterization for Practical Security Analysis of Cryptographic Implementations

Shuguo Yang[1,2], Yongbin Zhou[1], Jiye Liu[1,2], and Danyang Chen[1,3]

[1] State Key Laboratory of Information Security,
Institute of Software Chinese Academy of Sciences
P.O. Box 8718, Beijing 100190, P.R. China
[2] Graduate University of Chinese Academy of Sciences,
19A Yuquan Lu, Beijing, 100049, P.R. China
[3] School of Mathematics Sciences, Beijing Normal University,
No.19, XinJieKouWai St., Beijing, 100875, P.R. China
{yangshuguo,zyb,qsgh,chendanyang}@is.iscas.ac.cn

Abstract. Side-channel attacks have posed serious threats to the physical security of cryptographic implementations. However, the effectiveness of these attacks strongly depends on the accuracy of underlying side-channel leakage characterization. Known leakage characterization models do not always apply into the real scenarios as they are working on some unrealistic assumptions about the leaking devices. In light of this, we propose a back propagation neural network based power leakage characterization attack for cryptographic devices. This attack makes full use of the intrinsic advantage of neural network in profiling non-linear mapping relationship as one basic machine learning tool, transforms the task of leakage profiling into a neural-network-supervised study process. In addition, two new attacks using this model have also been proposed, namely BP-CPA and BP-MIA.

In order to justify the validity and accuracy of proposed attacks, we perform a series of experiments and carry out a detailed comparative study of them in multiple scenarios, with twelve typical attacks using mainstream power leakage characterization attacks, the results of which are measured by quantitative metrics such as SR, GE and DL. It has been turned out that BP neural network based power leakage characterization attack can largely improve the effectiveness of the attacks, regardless of the impact of noise and the limited number of power traces.

Taking CPA only as one example, BP-CPA is 16.5% better than existing non-linear leakage characterized based attacks with respect to DL, and is 154% better than original CPA.

Keywords: Back Propagation Neural Network, Leakage Characterization, Side Channel Attack, Practical Security, Cryptographic Implementation.

H. Kim (Ed): ICISC 2011, LNCS 7259, pp. 169–185, 2012.
© Springer-Verlag Berlin Heidelberg 2012

1   Introduction

Power analysis attack takes instantaneous power consumptions of a device to be its side-channel leakages. In power analysis attack, power leakage model characterizes the correlation between intermediate values and corresponding power consumptions. The accuracy of power model determines not only whether the power analysis attack is feasible, but also how complex the attack will be. Therefore, cryptanalyst usually tries to model power leakages as precisely as possible before mounting an attack. We performed Hamming Weight(HW for short) based DPAs[8][2] against one software AES prototype implementation on 8-bit C51-compatible micro-controllers and have acquired the following two important observations:

Observation 1: HW-based DPA targeting different bits in one intermediate value results in DPA peaks with different heights.

Observation 2: HW-based DPA has a very low tolerance for noises.

These observations reflect that traditional HW based linear leakage approaches are not capable of characterizing the real leakage precisely. [15] also claims that some simple linear power models are no longer valid as the size of transistors shrinks, even for standard CMOS circuits.

As of today, several approaches have been proposed to model non-linear power leakage, e.g. SM(namely stochastic model)[17], VPA(namely variance power analysis)[14] and BWC(namely bit-wised characterizing)[10]. These three approaches use combination of independent components to model the power leakage. Specifically, SM approach uses basis function on predefined vector subspace, VPA approach uses predefined events(usually bit-flip event), and BWC uses DPA peak heights of different bits to approximate the leakage function. All these non-linear power leakage characterizing approaches are based upon one basic assumption: the basic vector(SM), events(VPA)or power consumption of each bit(BWC) are independent from one another. **This assumption is commonly referred to as Bit Independent Leakage Assumption.** Unfortunately, this assumption does not always hold in practice, thanks to the equivalent inductance among circuits. Of course, this inevitably jeopardize the accuracy of underlying characterization approaches.

Generally speaking, our work in this paper are motivated by the following two factors. First of all, traditional linear models could no longer precisely characterize the actual power leakage of circuits. Secondly, nearly all existing non-linear modeling approaches are based upon relatively strong and unrealistic assumption, which would lead to more errors when are used to deal with noisy scenario.

Taking two above-mentioned factors into consideration, we present a new power leakage modeling approach based on BP neural network. As a supervised learning method for capturing power leakage, BP neural network based modeling approach does NOT rely on any specific assumptions, thus having better compatibility and higher precision.

The main contributions of this paper are twofold. On one hand, we proposed one side-channel leakage modeling approach based upon BP neural network, which is applicable to non-linear power leakages and does NOT rely on any specific assumptions. By making use of intrinsic capability of BP neural network to model arbitrary mapping relationship, the proposed model is capable of profiling power leakages of unknown devices. On the other hand, we have constructed two new side-channel attacks based on BP neural network, namely BP-CPA and BP-MIA. The results of a series of experiments show that, BP-CPA shows compelling effectiveness and strong noise tolerance with respect to three quantitative metrics, including Success Rate(SR[2]), Guessing Entropy(GE[18]) and Distinctive Level(DL[5]).

The rest of this paper is organized as follows. Section 2 briefly introduces power leakage decomposition, existing attacks for leakage modeling and one generic leakage function. Section 3 introduces the elementary knowledge of BP neural network. Section 4 describes our BP Neural Network based leakage characterizing approach, and then proves its soundness, after which two new attacks are constructed. Section 5 presents a comprehensive and systematic comparison study of a set of typical attacks, in order to demonstrate the effectiveness of proposed BP neural network based power modeling approach. Section 6 concludes the whole paper.

2   Preliminaries

Modeling the power leakage is the basis for launching a power analysis attack, and its accuracy determines the final result of the attack. This section will first introduce the composition of power trace. Then we will introduce several classic linear power leakage modeling approaches(e.g HW model, HD model), followed by non-linear power leakage modeling approaches including BWC, VPA and SM. Finally, we will provide the formal description of generic power leakage model.

2.1   Power Leakage Decomposition

According to[12], any single point in power traces can be considered to be the sum of four independent components, namely signal, algorithm noise, electronic noise and constant. Let P be the overall power consumption, P_{op} the signal component(caused by operations), P_{data} the signal component(caused by targeted intermediate values), P_{noise} the electronic noise, and P_{const} the constant. Then, the power leakage decomposition can be represented as shown in Eq.(1).

$$P = P_{op} + P_{data} + P_{const} + P_{noise} \qquad (1)$$

where P_{data} is regarded as the power leakage. The main purpose of power modeling is to precisely model the mapping between intermediate value and P_{data}.

2.2   Linear Power Leakage Modeling

The most representative linear power model is Hamming Weight and Hamming Distance model. Hamming Weight model assumes that power leakage is proportional to the number of "1" in intermediate value[12]. Similarly, Hamming

Distance model assumes that power leakage is proportional to the number of "1"→"0" or "0"→"1" transition[12]. The relationship between HW and HD model is given as follow.

$$HD(v_0, v_1) = HW(v_0 \oplus v_1)$$

2.3 Nonlinear Power Leakage Modeling

SM. SM approach is, in essence, a two-step attack. In profiling phase, one suitable vector subspace is used to approximate the leakage function, while some basis functions are used to approximate various components of power leakages. Note that the selection of vector subspace mainly relay on experiences and intuitions. In subsequence analysis phase, minimum distance attack or maximum likelihood is used to evaluate key candidates.

VPA. VPA approach needs to predefine the events that caused the power consumption, assuming that different events are independent. Then, it calculates the weight of each event in order to characterize the overall power consumption. For instance, if Bit-flip is selected as the event, VPA is equivalent to weighted HD model.

BWC. BWC attack assumes that different bits in a intermediate value are independent from each other, and uses the DPA results of different bits as the weights.

2.4 Generic Leakage Function

Power leakage model is essentially a mapping from particular intermediate value to real power consumption. [3] gives the formal definition of power model.

$$P_{data}(v) = \alpha_{-1} + \sum_{i=0}^{n-1} \alpha_i v_i + \sum_{i_1,i_2=0}^{n-1} \alpha_{i_1,i_2} v_{i_1} v_{i_2} + ... + \sum_{i_1,...,i_d=0}^{n-1} \alpha_{i_1,i_2,...,i_d} v_{i_1} v_{i_2} \cdots v_{i_d}$$

(2)

where P_{data} is the signal(caused by targeted intermediate values), v is the intermediate value, n is the number of bits of the intermediate value, v_i is i^{th} bit of the the intermediate value, all the α's are the parameters which need to be characterized.

When all α_i are equal and the parameters of higher order terms are zero, Eq.(2) equivalently is HW model. Actually, SM, VPA and BWC consider only the constant part α_{-1}, and linear part α_i. They ignore the higher order terms, which is due to the assumed bit-wise dependence of intermediate value.

3   Introduction to Back Propagation Neural Network

According to [6][7], artificial neural network is a mathematic model for processing intelligent information by simulating the connections and activities of neurons.

It is also a computing model composed of numerous nodes and their connections. Each node has an unique activation function and a bias. Each connection between two nodes has a weight. These parameters stores the memory of the network. The output of a network is determined mainly by the activation function and bias of each node, the weight and mode of each connection, and the topological structure of the entire network. When all these parameters are determined, the output is fixed.

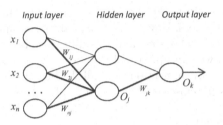

Fig. 1. A multi-layer BP neural network

BP neural network is a popular multi-layer network which uses the error back-propagation algorithm as its learning algorithm (in Fig.1). A BP neural network works with the followings five steps:

Step 1: Determination of Topological Structure. Specify the number of nodes in input layer, the number of hidden layers, the number of nodes in each hidden layer, and the number of nodes in output layer.

Step 2: Initialization of Weights and Biases. Generally, initial values for weights and biases set to be small random real numbers within [-1,1].

Step 3: Forward Propagation. Forward Propagation goes as follows.

(a) Initially, preprocess input data and transform them into the value scope of nodes in input layer.
(b) For input node i, set its output to be its input. So $O_i = I_i$ holds.
(c) For each node j not in input layer, compute its input as shown in Eq.3.

$$I_j = \sum_i W_{ij}O_i + \Theta_j \tag{3}$$

where W_{ij} means the weight of the connection between node i and node j, Θ_j is the bias of node j.

(d) Compute the output of nodes in hidden or output layer by $O_j = f(I_j)$, where f is the activation function of node j. BP neural network algorithm requests that the activation function be a nonconstant, bounded, and monotone-increasing continuous function. The activation function always uses the following form.

$$f(x) = \frac{1}{1 + e^{-x}} \qquad (4)$$

Step 4: Back Propagation. We can obtain an output value of the network by Forward Propagation process. However, there is always a difference between the actual output and the expected one, and this difference is called error. By back-propagating the error, we can update all the parameters stored in the network, including all the weights of connections and all the biases of nodes, thus bringing the output of the network closer to the expected value. It is the core of the whole learning algorithm.

This process goes as follows.

(a) For node k in output layer, compute the Err_k corresponding

$$Err_k = O_k(1 - O_k)(T_k - O_k) \qquad (5)$$

where O_k is the output of node k and T_k is the expected output value according to the training data.

(b) The error of a hidden layer node j is the weighted sum of all the errors of the nodes connected to node j in next layer. Compute the error of node j is shown in Eq.6.

$$Err_j = O_j(1 - O_j) \sum_k Err_k W_{jk} \qquad (6)$$

where W_{jk} is the weight of the connection between node j and node k in next layer, and Err_k is the error of node k.

(c) Update weights and biases according to the following two equations.

$$W_{ij} = W_{ij} + L \cdot Err_j O_i \qquad (7)$$

$$\Theta_j = \Theta_j + L \cdot Err_j \qquad (8)$$

where L is the learning rate, a constant range from 0 to 1. This constant is used for adjusting the behavior of the network. It helps avoid getting stuck at a locally optimum solution and encourages the search for a global optimum one. The bigger L is, the faster the network converges, and vice versa. Generally we can set L to $\frac{1}{t}$, where t is the number of iterations during the training process.

Step 5: Termination. Training process will stop either all parameters have converged, or the number of iterations exceeds its maximum, or the output of the network meets the desired result.

4 Our BP Neural Network based Leakage Characterizing Approach

In this section, a detailed description of the proposed approach to side-channel leakage characterization will be give firstly, with power leakage being a concrete example. Then we will prove that the BP network can precisely characterize the general leakage model (in Eq.2). Finally, two new attacks are constructed using our leakage model.

4.1 BP Neural Network Based Leakage Model

Essentially, BP neural network based leakage model is a BP neural network well-trained for the mapping from intermediate values to real power consumptions. The main steps of building BP Neural Network based leakage model is as follows.

Step 1: Definition of Topological Structure. The number of nodes in input layer is related to the bit-width of intermediate value. Each input value of nodes in input layer is derived from the bit value of the target intermediate value. We take AES as case of study. The implementation is an unprotected software implementation of AES, and the target intermediate value is the output of 1st S-box in 1st AES round. So the size of intermediate value is 8 and we set the number of nodes in the input layer to be 8. We use only one single hidden layer with 16 nodes. The output layer contains only one output node.

Step 2: Preparation of Training Data. We use V to denote one 8-bit intermediate value vector. Each element v in V is represented by $v(v_0, v_1, v_2, v_3, v_4, v_5, v_6, v_7)$ in binary form. Let R be real power consumption vector, each element r is according to intermediate value v. So the training set is pairs of (V, R). Since the total number of possible intermediate value is 256, so we only need 256 pairs of (v, r). We can use trace averaging to reduce Gaussian noise.

First, for each intermediate value $v(v_0, v_1, v_2, v_3, v_4, v_5, v_6, v_7)$, compute each component $I_{v,i}$ of the network input $I_v(I_{v,0} I_{v,1} I_{v,2} I_{v,3} I_{v,4} I_{v,5} I_{v,6} I_{v,7})$ by

$$I_{v,i} = GetInputValue(v_i) = \begin{cases} 1 & \text{if } v_i = 1 \\ -1 & \text{if } v_i = 0 \end{cases} \tag{9}$$

Then, for each real power consumption r, normalize it into $[-1, 1]$ by

$$T_r = GetTargetValue(r) = \frac{r - Min(R)}{Max(R) - Min(R)} * 2 - 1 \tag{10}$$

When one gets 256 pairs of (I_v, T_r), the training data set is ready.

Step 3: Training of BP Neural Network based Leakage Model. Since the number of training cases is limited, the training process will be iterative. Whether or not to continue the training epoch mainly relies on the precision of BP neural network output. We take Pearson Correlation Coefficient γ (in Eq.11) between network output vector O and real power consumption vector R to be the standard to judge when to stop training. Training will terminate when $\gamma > 0.95$.

$$\gamma = Pearson(X, Y) = \frac{\sum_{i=1}^{n}(x_i - \bar{x}) \cdot (y_i - \bar{y})}{\sqrt{\sum_{i=1}^{n}(x_i - \bar{x})^2} \cdot \sqrt{\sum_{i=1}^{n}(y_i - \bar{y})^2}} \qquad (11)$$

Training a BP neural network based leakage model is described in Algorithm 1. So far, we get a well-trained BP network to be our new BP neural network based leakage model.

> **Input**: Intermediate value vector:V, real power consumption vector:R
> **Output**: Trained BP Neural Network:network as the leakage model
> 1 Create BP neural network: network (8 nodes in input layer,16 nodes in hidden layer,1 node in output layer);
> 2 Initialize weights and biases;
> 3 **foreach** *Intermediate value v in* V **do**
> 4 $I_v = 0$;
> 5   **foreach** *inputnode $I_{v,i}$ in* I_v **do**
> 6 $I_{v,i} = $ GetInputValue(v_i) (Eq.9);
> 7   **end**
> 8 **end**
> 9 **foreach** *realvalue r in* R **do**
> 10 $T_r = $ GetTargetValue(r) (Eq.10);
> 11 **end**
> 12 $\gamma = 0$;
> 13 **while** $\gamma < 0.95$ **do**
> 14   **foreach** I_v **do**
> 15 $O_v = $ Forward-Propagation(network,I_v) ;
> 16   Back-Propagation(network,O_v,T_v);
> 17   **end**
> 18   Compute $\gamma = Pearson(O, T)$ (Eq.11);
> 19 **end**
> 20 return network;

Algorithm 1. The BP neural network based leakage model training algorithm

Step 4: Prediction of Hypothetical Leakages. Given an intermediate value, evaluation of hypothetic leakages by BP neural network based leakage model could be done by calling Forward-Propagation process of BP neural network. Algorithm 2 describes this process.

Input: Trained BP Neural Network:network, intermediate value vector:V
Output: The hypothetical leakage vector: H (according to V)

1 **foreach** *intermediate value v in V* **do**
2     $I_v = 0$;
3   **foreach** *inputnode $I_{v,i}$ in I_v* **do**
4        $I_{v,i}$ = GetInputValue(v_i) (Eq.9);
5   **end**
6   H_v = Forward-Propagation(network,I_v);
7 **end**
8 **return** H;

Algorithm 2. The BP neural network based leakage model prediction algorithm

4.2 Soundness of BP Neural Network for Leakage Characterization

In this subsection, we will prove the soundness of BP BP-Network in characterizing leakage model. We begin with this subsection with the following two theorems.

Theorem 1 (Weierstrass[9]). *Suppose f is a continuous complex-valued function defined on the real interval $[a, b]$. For every $\epsilon > 0$, there exists a polynomial function p over C such that for all x in $[a, b]$, we have $|f(x) - p(x)| < \epsilon$, or equivalently, the supremum norm $\|f(x) - p(x)\| < \epsilon$. If f is real-valued, the polynomial function can be taken over R.*

Theorem 2 (Universal Approximation[16]). *Let $\varphi(\cdot)$ be a nonconstant, bounded, and monotone-increasing continuous function. Let I_{m_0} denote the m_0-dimensional unit hypercube $[0, 1]^{m_0}$. The space of continuous function on I_{m_0} is denoted by $C(I_{m_0})$. Then, given any function $f \ni C(I_{m_0})$ and $\varepsilon > 0$, there exist an integer m_1 and sets of real constants α_i, b_i and ω_{ij}, where $i = 1, ..., m_1$ and $j = 1, ..., m_0$ such that we may define*

$$F(x_1, \cdots, x_{m_0}) = \sum_{i=1}^{m_1} \alpha_i \varphi(\sum_{j=1}^{m_0} \omega_{ij} x_j + b_i) \qquad (12)$$

as an approximate realization of the function $f(\cdot)$;that is,

$$|F(x_1, \cdots, x_{m_0}) - f(x_1, \cdots, x_{m_0})| < \varepsilon$$

for all $x_1, x_2, \cdots, x_{m_0}$ that lie in the input space.

Theorem 2 may be viewed as the natural extension of Weierstrass Theorem. It can be applied directly to multi-layer neural network[6]. Notice that the activation function in *Eq*.4 of each node in a BP neural network is required to be a nonconstant, bounded, and monotone-increasing continuous function, so it meets the condition of function $\varphi(\cdot)$. In fact, Eq.12 describes such a multi-layer neural network:

(a) Network has m_0 input nodes and a single hidden layer which is composed of m_1 nodes. The input of the network is denoted as $x_1, x_2, \cdots, x_m$.

(b) The node in hidden layers has its weights denoted by $\omega_{1i}, \cdots, \omega_{m_0i}$ and a bias denoted by Θ_i

(c) The output of network is the linear combination of the output of nodes in hidden layer, and weights of the node in the output layer are $\alpha_1, \cdots, \alpha_{m_1}$.

As we can see, Universal Approximation Theorem shows that for computing the ϵ uniform approximation of a fixed training set from input $x_1, x_2, \cdots, x_{m_0}$ to expected output $f(x_1, \cdots, x_{m_0})$, a single hidden layer is enough. The leakage characterizing is to map intermediate value to real leakage. In Eq.2, the leakage function can be represented as a high order polynomial function. Therefor, using BP neural network to training the relationship between intermediate value and real leakage is sound and practical.

4.3   Constructions of BP Neural Network Based Attacks

This subsection will briefly introduce two BP neural network based attacks we constructed, namely BP-CPA and BP-MIA, which are variants of CPA[1][11] and MIA[4][13][19] respectively. CPA and MIA are two popular and effective distinguishers, and they use different mathematic tools, i.e. Pearson Correlation Coefficient and Mutual Information to compare hypothetic power consumption and real power consumption. In order to construct new attacks, we only need to replace the original HW model with our BP neural network based leakage model. Note that performing BP-CPA and BP-MIA requires only one well-trained BP neural network based leakage model. In this way, CPA-like and MIA-like attacks can be built over any specific leakage model.

5   Experiments

A series of typical experiments have been conducted to test the reliability of our new approach. Specially, we performed simulated experiments on a 8-bit software implementation of an unprotected AES. The leakage model for simulation adopted the function in Eqa.2, where $d = 3$. It means that the function $P_{data}(v)$ has only four terms: one constant, one order term, one quadratic term and one three order term. These parameters α_{i_1,i_2} (or α_{i_1,i_2,i_3}) will be set to small random real numbers when values of corresponding adjacent 2(or 3) bits in the intermediate value are 1 simultaneously, otherwise the parameters will be set to 0. This takes into consideration the interactions between multi-bits themselves in an intermediate value. This leakage function describes the superimposed situation of both linear and non-linear scenarios. Therefore, it is very close to the scenario of a real device[15].

We divide the considered contexts with different noise levels into three categories: ideal scenario with a noise level $\sigma \leqslant 10$, realistic scenario with a noise level $\sigma \in (10, 30)$, challenging scenario with a noise level $\sigma \geqslant 30$. Three quantitative metrics, including SR, GE and DL are used to measure the effectiveness.

Each attack in this section is viewed as a combination of power leakage model and distinguisher. We considered five power leakage models and four distinguishers. The power leakage models are HW, BP, BWC, VPA and SM. And the distinguishers are CPA, MIA, VPA and SM. Note that all attacks are denoted by "A-B", where "A" denotes power leakage model of the attack and "B" denotes its distinguisher. For example, BP-CPA stands for a attack that uses BP neural network based leakage model and Pearson Correlation Coefficient.

Considering that CPA and MIA are two mainstream distinguishers most widely used, we take them as two standard distinguishers to evaluate the effectiveness of different leakage models. The results thus obtained will be reliable and convincing. All results in this paper are obtained by averaging the results of 100 times of repeated tests.

The experiments have been preformed from following three aspects: evaluation and comparison with CPA-like attacks, evaluation and comparison with MIA-like attacks, evaluation and Comparison with attacks which has a modeling phase.

5.1 Comparison with CPA-like Attacks

This subsection compares 5 CPA-like attacks in three scenarios, including HW-CPA, BP-CPA, BWC-CPA, VPA-CPA and SM-CPA. Three scenarios are ideal scenario($\sigma = 1$), realistic scenario($\sigma = 25$) and challenging scenario($\sigma = 50$) respectively. Additionally, three metrics(SR, GE and DL) are used in our evaluation. The results are showed in Fig.2. As we can see, in all three scenarios, SR values of tested attacks present the following pattern: HW-CPA < SM-CPA < BWC-CPA $\approx$ VPA-CPA < BP-CPA. The same results also holds for GE and DL.

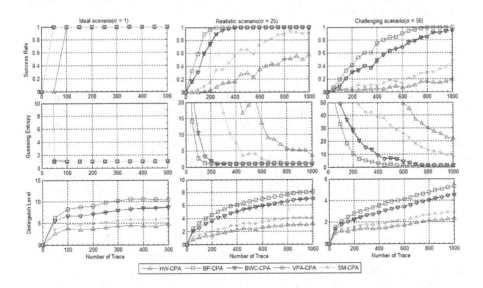

Fig. 2. Results of CPA with different leakage models in three scenarios

Specifically, in ideal scenario, all attacks except HW-CPA need only 50 traces to reach 100% SR, while HW-CPA needs 100 traces. In realistic scenario, BP-CPA needs 250 traces to achieve 100%SR, while BWC-CPA and VPA-CPA need 400 traces. The SM-CPA and HW-CPA have performed a little worse; with 1000 traces they achieve 91% and 57% respectively. In challenging scenario, BP-CPA needs 800 traces to get SR 100%, while BWC-CPA and VPA-CPA get only 95% with 1000 traces.

Table 1 shows DLs in three scenarios when trace number is 1000.

Table 1. Distinctive Levels of five CPA-like attacks in three scenarios

N=1000	Ideal($\sigma = 1$)		Realistic($\sigma = 25$)		Challenging ($\sigma = 50$)	
	DL	(*/HW-CPA)%	DL	(*/HW-CPA)%	DL	(*/HW-CPA)%
HW-CPA	5.17800841	-	3.01123095	-	2.3206056	-
BP-CPA	10.8724881	209.9743	7.65685814	254.2767	5.3341048	229.8583
BWC-CPA	9.21271198	177.9200	6.51802617	216.4572	4.51194033	194.4294
VPA-CPA	9.33850383	180.3493	6.56984784	218.1781	4.51801854	194.6914
SM-CPA	6.62982683	128.0382	4.11803832	136.7560	2.96579518	127.8026

As Table.1 shows, BP-CPA outperforms other attacks in all three scenarios. The difference is most significant in realistic scenario, where the DL of BP-CPA exceeds that of HW-CPA by 154%. The DL of BWC-CPA, VPA-CPA and SM-CPA is higher than that of HW-CPA 116%, 118% and 37% respectively. So in terms of DL, BP-CPA is better than the second best CPA-like attacks 16.5%(refer to VPA-CPA).

5.2   Comparison with MIA-like Attacks

This subsection compares 5 MIA-like attacks in three scenarios, including HW-MIA, BP-MIA, BWC-MIA, VPA-MIA and SM-MIA. Three scenarios are ideal scenario($\sigma = 4$), realistic scenario($\sigma = 16$) and challenging scenario($\sigma = 30$) respectively. Additionally, two metrics(SR and GE) are used. The results are showed in Fig.3. The results are showed in Fig.2. **Note:** Due to the poor noise tolerance of MIA-like attacks, we reduce the noise levels of three scenarios respectively. As illustrated in Fig.3, in all three scenarios, SR values of tested attacks present the following pattern: HW-MIA < SM-MIA < BWC-MIA ≈ VPA-MIA < BP-MIA. The HW-MIA performs the worst due to an rough leakage model.

In ideal scenario, BP-MIA only needs 150 traces to get 100% SR, while BWC-MIA and VPA-MIA need 200 traces. In realistic scenario, BP-MIA needs 800 traces to get 100%SR. Meanwhile, it takes 1000 traces for BWC-MIA and VPA-MIA to reach 82% and 84% respectively.

In challenging scenario, none of attacks achieved any high SR. However, we can see that BP-MIA still outperforms others significantly.

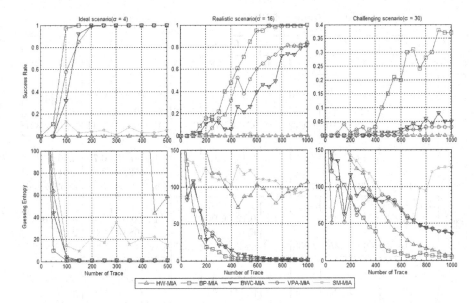

Fig. 3. Results of MIA with different leakage models in three scenarios

Results show that all MIA-like attacks have a low noise tolerance. Even though BP-MIA does not performs well in challenging scenario even when the noise level reduces to $\sigma = 4$, BP-MIA is still the best among all MIA-like attacks.

5.3   Comparison among Attacks Using Different Leakage Characterizations

The results of above experiments have proved that BP-CPA is the best among CPA-like attacks, and BP-MIA best among MIA-like attacks. The following part

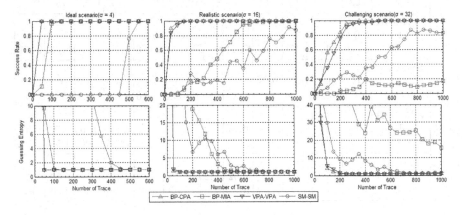

Fig. 4. Results of attacks using different leakage characterization in three scenarios

Table 2. Results of 12 Attacks in Three Scenarios

Scenarios		Ideal(σ=1)				Realistic(σ=30)				Challenging(σ=50)			
#Trace		50	400	700	1000	50	400	700	1000	50	400	700	1000
CPA	HW-CPA SR	0	1	1	1	0	0.11	0.32	0.5	0.02	0.05	0.1	0.19
	GE	1.25	1	1	1	115.8	44.66	17.9	6.6	97.78	61.57	36.83	22.35
	HL	2.591	4.316	4.985	5.178	0.749	1.784	2.417	3.011	0.991	1.44	1.944	2.321
	BP-CPA SR	1	1	1	1	0.19	1	1	1	0.07	0.75	0.96	1
	GE	1	1	1	1	22.37	1	1	1	52.34	3.18	1.01	1
	HL	6.298	10.54	10.8	10.87	2.221	5.319	6.707	7.657	1.554	3.515	4.459	5.334
	BWC-CPA SR	1	1	1	1	0.12	0.95	1	1	0.04	0.49	0.79	0.95
	GE	1	1	1	1	45.06	1.06	1	1	68.06	8.97	2.23	1.05
	HL	5.609	8.455	9.058	9.213	1.846	4.368	5.625	6.518	1.399	2.883	3.764	4.512
	VPA-CPA SR	1	1	1	1	0.11	0.95	1	1	0.04	0.47	0.78	0.95
	GE	1	1	1	1	45.11	1.04	1	1	69	9.13	2.25	1.03
	HL	5.621	8.561	9.179	9.339	1.838	4.403	5.667	6.57	1.397	2.88	3.768	4.518
	SM-CPA SR	1	1	1	1	0.04	0.29	0.66	0.88	0.04	0.1	0.24	0.49
	GE	1	1	1	1	98.15	15.25	3.49	1.27	88.56	38.62	18.17	8.15
	HL	3.515	5.797	6.488	6.63	1.023	2.546	3.398	4.118	1.128	1.818	2.449	2.966
MIA	HW-MIA SR	0	0	0	0	0	0	0	0	0	0	0	0
	GE	226.9	18.13	167.8	106	122.8	81.31	21.39	6.92	220.2	180.5	120.5	123.2
	BP-MIA SR	0.69	1	1	1	0	0.03	0.31	0.37	0	0	0	0
	GE	1.09	1	1	1	121.1	39.42	4.49	4.99	103.9	124.9	93.55	82.04
	BWC-MIA SR	0	1	1	1	0	0.01	0.04	0.05	0	0	0	0
	GE	60.23	1	1	1	137	82.96	54.03	35.81	68.52	136.7	113.5	93.5
	VPA-MIA SR	0	1	1	1	0	0	0.02	0.03	0	0	0	0
	GE	69.19	1	1	1	51.11	81.5	54.31	36.46	94.68	119.1	117.2	94.76
	SM-MIA SR	0	0	0.01	0.08	0	0.01	0	0	0	0	0	0
	GE	203.6	26.28	17.96	3.7	142.5	105.4	50.74	126.6	156.7	157.4	153.3	140.3
VPA	VPA-VPA SR	1	1	1	1	0.13	0.99	1	1	0.12	0.56	0.79	0.97
	GE	1	1	1	1	22.49	1	1	1	42.83	5.06	1.56	1
SM	SM-SM SR	0	1	1	1	0.02	0.29	0.73	0.93	0	0	0.09	0.22
	GE	5.05	1	1	1	100.3	7.44	1.35	1.01	104.6	57.4	30.46	15.29

will compare BP-CPA and BP-MIA with non CPA-like or MIA-like attacks that use a characterized leakage model.

In this subsection, BP-CPA, BP-MIA, VPA-VPA and SM-SM are compared under three scenarios, namely ideal scenario($\sigma = 4$), realistic scenario($\sigma = 16$) and challenging scenario($\sigma = 32$), with SR and GE as metrics. The attack results are showed in Fig.4. As illustrated in Fig.4,BP-CPA still performs best among all attacks. In ideal scenario, all attacks can get 100%SR with a small mount of traces. In realistic scenario, BP-CPA only needs 150 traces to get 100%SR, while VPA-VPA needs 200 traces and BP-MIA needs 850 traces. However, SM-SM only gets 87% with 1000 traces. BP-MIA outperforms SM-SM in this scenario.

It is worth noticing that the performance of BP-MIA reduces significantly from realistic scenario to challenging scenario, while other non-MIA-like attacks show a better tolerance to the increasing noise.

So far, we have obtained performance data of all 12 attacks in different scenarios with different metrics. Due to the limit of space, we only present part of results in Table.2.

6 Conclusions

In summary, this paper has proposed a new leakage characterizing method based upon BP neural network. This method makes full use of the intrinsic advantage of the machine learning method in profiling non-linear mapping relationship, and does NOT rely on any specific assumptions. Leakage characterizing phase of BP neural network based leakage model is realized in the training process of one BP neural network, and hypothetical leakage prediction phase is realized in the forward propagation process of a well-trained BP neural network.

Two new BP Neural Network based side channel attacks have been proposed, namely BP-CPA and BP-MIA. They have been validated with a series of simulated experiments on a 8-bit software implementation of an unprotected AES. Results show that BP Neural Network based attacks require fewer traces to reach an acceptable level of success rate, and they outperform other attacks under some harsh conditions. Under consideration by the data, we believe that BP based model can more accurately characterize correlations (linear as well as non-linear) between intermediate value and side channel leakages of cryptographic implementations.

However, we admit that the proposed approach also has some disadvantages. For example, the training process of a back propagation is inefficient. Each iteration of a pair of intermediate value and real leakage value has to go through the entire network twice, forward computing the output of every nodes and backward updating parameters of every nodes and connections. As shown in line 13 of Algorithm 1, we used 0.95 as our termination target precision. The γ value grows rapidly when it is below 0.85, but when it exceeds 0.90 the increasing speed shrinks quickly. The training process lasts almost several hours to meet the termination condition. Considering that the training process for each target device needs to be done only once, we consider it reasonable and acceptable to sacrificed some time penalty in training process for a higher predicting precision. In the future, we will focus on improving the efficiency, e.g. by changing the topological structure of network or using a new machine learning algorithm.

Acknowledgements. This work is supported in part by the National Natural Science Foundation of China (No.61073178) and Beijing Natural Science Foundation (No.4112064).

References

1. Brier, E., Clavier, C., Olivier, F.: Correlation Power Analysis with a Leakage Model. In: Joye, M., Quisquater, J.-J. (eds.) CHES 2004. LNCS, vol. 3156, pp. 16–29. Springer, Heidelberg (2004)
2. Agrawal, D., Archambeault, B., Rao, J.R., Rohatgi, P.: The EM Side-Channel(s). In: Kaliski Jr., B.S., Koç, Ç.K., Paar, C. (eds.) CHES 2002. LNCS, vol. 2523, pp. 29–45. Springer, Heidelberg (2003)
3. Doget, J., Prouff, E., Rivain, M., Standaert, F.-X.: Univariate side channel attacks and leakage modeling. Journal Cryptographic Engineering 1(2), 123–144 (2011)
4. Gierlichs, B., Batina, L., Tuyls, P., Preneel, B.: Mutual Information Analysis - A Generic Side-Channel Distinguisher. In: Oswald, E., Rohatgi, P. (eds.) CHES 2008. LNCS, vol. 5154, pp. 426–442. Springer, Heidelberg (2008)
5. Huang, J., Zhou, Y., Liu, J.: Measuring the effectiveness of dpa attacks - from the perspective of distinguishers statistical characteristics. In: IEEE ICCSIT 2010, pp. 161–168 (2010)
6. Haykin, S.: Neural Networks and Learning Machines, 3rd edn. Pearson (2009)
7. Han, J., Kamber, M.: Data Mining: Concepts and Techniques, 2nd edn. Elseview Inc. (2006)
8. Kocher, P.C., Jaffe, J., Jun, B.: Differential Power Analysis. In: Wiener, M. (ed.) CRYPTO 1999. LNCS, vol. 1666, pp. 388–397. Springer, Heidelberg (1999)
9. Weierstrass, K.: Über die analytische Darstellbarkeit sogenannter willkürlicher Functionen einer reellen Veränderlichen. Sitzungsberichte der Kniglich Preuischen Akademie der Wissenschaften zu Berlin (1885) II
10. Liu, J., Zhou, Y., Han, Y., Li, J., Yang, S., Feng, D.: How to Characterize Side-Channel Leakages More Accurately? In: Bao, F., Weng, J. (eds.) ISPEC 2011. LNCS, vol. 6672, pp. 196–207. Springer, Heidelberg (2011)
11. Le, T.-H., Clédière, J., Canovas, C., Robisson, B., Servière, C., Lacoume, J.-L.: A Proposition for Correlation Power Analysis Enhancement. In: Goubin, L., Matsui, M. (eds.) CHES 2006. LNCS, vol. 4249, pp. 174–186. Springer, Heidelberg (2006)
12. Mangard, S., Oswald, E., Popp, S.: Power Analysis Attacks: Revealing the Secrets of Smart Cards. Springer (2007)
13. Prouff, E., Rivain, M.: Theoretical and Practical Aspects of Mutual Information Based Side Channel Analysis. In: Abdalla, M., Pointcheval, D., Fouque, P.-A., Vergnaud, D. (eds.) ACNS 2009. LNCS, vol. 5536, pp. 499–518. Springer, Heidelberg (2009)
14. Hoogvorst, P.: The Variance Power Analysis. In: Proceeding of COSADE 2010, pp. 4–9 (2010)
15. Renauld, M., Standaert, F.-X., Veyrat-Charvillon, N., Kamel, D., Flandre, D.: A Formal Study of Power Variability Issues and Side-Channel Attacks for Nanoscale Devices. In: Paterson, K.G. (ed.) EUROCRYPT 2011. LNCS, vol. 6632, pp. 109–128. Springer, Heidelberg (2011)

16. Hecht-Nielsen, R.: Kolmogorov's Mapping Neural Network Existence Theorem. In: Proceedings of First IEEE International Conference on Neural Networks, San Diego, CA, pp. 11–14 (1987)
17. Schindler, W., Lemke, K., Paar, C.: A Stochastic Model for Differential Side Channel Cryptanalysis. In: Rao, J.R., Sunar, B. (eds.) CHES 2005. LNCS, vol. 3659, pp. 30–46. Springer, Heidelberg (2005)
18. Standaert, F.-X., Gierlichs, B., Verbauwhede, I.: Partition *vs.* Comparison Side-Channel Distinguishers: An Empirical Evaluation of Statistical Tests for Univariate Side-Channel Attacks against Two Unprotected CMOS Devices. In: Lee, P.J., Cheon, J.H. (eds.) ICISC 2008. LNCS, vol. 5461, pp. 253–267. Springer, Heidelberg (2009)
19. Veyrat-Charvillon, N., Standaert, F.-X.: Mutual Information Analysis: How, When and Why? In: Clavier, C., Gaj, K. (eds.) CHES 2009. LNCS, vol. 5747, pp. 429–443. Springer, Heidelberg (2009)

A Revocable Group Signature Scheme with the Property of Hiding the Number of Revoked Users

Keita Emura[1], Atsuko Miyaji[2], and Kazumasa Omote[2]

[1] Center for Highly Dependable Embedded Systems Technology
[2] School of Information Science
Japan Advanced Institute of Science and Technology, 1-1, Asahidai, Nomi, Ishikawa, 923-1292, Japan
{k-emura,miyaji,omote}@jaist.ac.jp

Abstract. If there are many displaced workers in a company, then a person who goes for job hunting might not select this company. That is, the number of members who quit is quite negative information. Similarly, in revocable group signature schemes, if one knows (or guesses) the number of revoked users (say r), then one may guess the reason behind such circumstances, and it may lead to harmful rumors. However, no previous revocation procedure can achieve to hide r. In this paper, we propose the first revocable group signature scheme, where r is kept hidden. To handle these properties, we newly define the security notion called anonymity w.r.t. the revocation which guarantees the unlinkability of revoked users.

Keywords: Group signature, Revocation, Hiding the Number of Revoked Users.

1 Introduction

Imagine that there are many users who have stopped using a service. If this fact is published, then how would the newcomers feel about this? One may guess the reason behind such circumstances, and may judge that those users did not find the service attractive or the service fee is expensive. The same thing may occur in other cases, e.g., if there are many displaced workers in a company, then a person who goes for job hunting might not select this company. That is, the number of members who quit is quite negative information.

Many cryptographic attempts for the revocation of rights of users have been considered so far, especially, in group signature [12][1], anonymity revocation has

---

[1] The concept of group signature was investigated by Chaum and Heyst [12], and its typical usage is described as follows: The group manager (GM) issues a membership certificate to a signer. A signer makes a group signature by using its own membership certificate. A verifier anonymously verifies whether a signer is a member of a group or not. In order to handle some special cases (e.g., an anonymous signer behaves maliciously), GM can identify the actual signer through the open procedure. Since verifiers do not have to identify individual signers, group signature is a useful and powerful tool for protecting signers' privacy.

H. Kim (Ed): ICISC 2011, LNCS 7259, pp. 186–203, 2012.
© Springer-Verlag Berlin Heidelberg 2012

been introduced [7,8,14,25,27,28,31][2]. However, the number of revoked users (say r) is revealed in all previous revocable group signature schemes. As mentioned previously, the number of revoked users r is quite a negative information. As a concrete example, we introduce an application of revocable group signature for outsourcing businesses [20]. By applying group signature, the service authentication server (outsourcee) has only to verify whether a user is a legitimate member or not, and does not have to manage the list of identities of users. Therefore, the risk of leaking the list of identities of users can be minimized, and this is the merit of using group signature in identity management. After a certain interval, the service provider charges the users who have already used the service, by using the opening procedure of group signature. When a user would like to leave the group, or when a user have not paid, the service provider revokes this user. In this system, if r is revealed, then one may think that there might be many users who have stopped using the service, i.e., this service may not be interesting, or he/she have not paid the service fee, namely, the service fee may be expensive, and so on.

So, our main target is to propose a revocable group signature scheme with the property of hiding the number of revoked users r. Then, we need to investigate the methodology for achieving the following:

1. The size of any value does not depend on r.
2. The costs of any algorithm do not depend on r, except the revocation algorithm executed by GM.
3. Revoked users are unlinkable.

In particular, if revoked users are linkable, then anyone can guess (i.e., not exactly obtain) r by linking and counting revoked users. Although we assume that an adversary can obtain the polynomial (of the security parameter) number of group signatures, this assumption is not unreasonable (actually, the adversary can issue the polynomial times queries of the signing oracle). In addition, r is also a polynomial-size value. That is, this guessing attack works given that revoked users are linkable.

However, no previous revocable signature scheme satisfying all requirements above has been proposed. For example, in revocable group signatures [7,11,14,31] (which are based on updating the group public values, e.g., using accumulators), either the size of public value or the costs of updating membership certificate depend on r. Nakanishi et al. [27] proposed a novel technique of group signature, where no costs of the GSign algorithm (or the Verify algorithm also) depend on r. However, their methodology requires that r signatures are published to make a group signature, and therefore r is revealed. In [8,13,25,28] (which are verifier-local revocation (VLR) type group signature), revoked users are linkable. In this case, anyone can guess r by executing the verification procedure. For the sake of clarity, we introduce the Nakanishi-Funabiki methodology [28] as follows: let

---

[2] Since a long RSA modulus might lead to certain inefficiency aspects (e.g., long signatures, heavy complexity costs, and so on), we exclude RSA-based revocable group signatures (e.g., [29,30]) in this paper.

$RL = \{h^{x_1}, h^{x_2}, \ldots, h^{x_r}\}$ be the revocation list, where x_i is the secret value of revoked user U_i. Note that by adding dummy values, we can easily expand $|RL|$. So, we can assume that r is not revealed from the size of RL. But, r is revealed (or rather, guessed) as follows. Each group signature σ (made by U_j) contains $f^{x_j+\beta}$ and h^β for some random β and some group elements f and h. If U_j has been revoked, then there exists h^{x_i} such that $e(f^{x_j+\beta}, h) = e(h^{x_i}h^\beta, f)$ holds. By counting such i, one can easily guess r even if RL is expanded by dummy values. Since each value in RL is linked to a user (i.e., h^{x_i} is linked to U_i), even if values in RL are randomized (e.g., $(h^{x_i})^{r_i}$ for some random r_i), this connection between a user and a value in RL is still effective. So, one can easily guess r even if RL is randomized.

From the above considerations, no previous revocation procedure can be applied for hiding r. One solution has been proposed in [16], where only the designated verifier can verify the signature. By preventing the verification of signature from the third party, r is not revealed from the viewpoint of the third party. However, this scheme (called anonymous designated verifier signature) is not group signature any longer. Next, as another methodology, consider the multi group signature [1] with two groups (valid user group and revoked user group). However, this attempt does not work, since each user is given his/her membership certificate (corresponding the group he/she belongs to) in the initial setup phase, and the revocation procedure is executed after the setup phase.

Our Contribution: In this paper, we propose the first group signature scheme with the property of hiding the number of revoked users r, by applying attribute-based group signature (ABGS) [15,18,21,22]. By considering two attributes: (1) valid group user and (2) the user's identity, we can realize the property of hiding r. To handle this property, we newly define the security notion called anonymity w.r.t. the revocation. As the main difference among our anonymity definition and previous ones, to guarantee the unlinkability of revoked users, $\mathcal{A}$ can issue the revocation queries against the challenge users.

2   Bilinear Groups and Complexity Assumptions

Definition 1 (Bilinear Groups). *Let $\mathbb{G}_1$, $\mathbb{G}_2$, and $\mathbb{G}_T$ be cyclic groups with a prime order p, and $\mathbb{G}_1 = \langle g \rangle$ and $\mathbb{G}_2 = \langle h \rangle$. Let $e : \mathbb{G}_1 \times \mathbb{G}_2 \to \mathbb{G}_T$ be an (efficient computable) bilinear map with the following properties: (1) bilinearity: for all $(g, g') \in \mathbb{G}_1^2$ and $(h, h') \in \mathbb{G}_2^2$, $e(gg', h) = e(g, h)e(g', h)$ and $e(g, hh') = e(g, h)e(g, h')$ hold, and (2) non-degeneracy : $e(g, h) \neq 1_T$, where 1_T is the unit element over $\mathbb{G}_T$.*

Definition 2 (The Computational Diffie-Hellman (CDH) assumption). *We say that the CDH assumption holds if for all probabilistic polynomial time (PPT) adversary $\mathcal{A}$, $\Pr[\mathcal{A}(g_1, g_1^a, g_1^b) = g_1^{ab}]$ is negligible, where $g_1 \in \mathbb{G}_1$ and $(a, b) \in \mathbb{Z}_p^2$.*

Definition 3 (The Decision Diffie-Hellman (DDH) assumption). *We say that the DDH assumption holds if for all PPT adversary $\mathcal{A}$, $|\Pr[\mathcal{A}(g_1, g_1', g_1^x, g_1'^x) = 0] - \Pr[\mathcal{A}(g_1, g_1', g_1^x, g_1'^r) = 0]|$ is negligible, where $(g_1, g_1') \in \mathbb{G}_1^2$ and $(x, r) \in \mathbb{Z}_p^2$ with $x \neq r$.*

Definition 4 (The Decision Linear (DLIN) assumption [7]). *We say that the DLIN assumption holds if for all PPT adversary $\mathcal{A}$, $|\Pr[\mathcal{A}(u, v, h, u^a, v^b, h^{a+b}) = 0] - \Pr[\mathcal{A}(u, v, h, u^a, v^b, \eta) = 0]|$ is negligible, where $(u, v, h, \eta) \in \mathbb{G}_2^4$ and $(a, b) \in \mathbb{Z}_p^2$.*

Definition 5 (The Hidden Strong Diffie-Hellman (HSDH) assumption [9]). *We say that ℓ-HSDH assumption holds if for all PPT adversary $\mathcal{A}$,*
$\Pr[\mathcal{A}(g_1, h, h^\omega, (g_1^{\frac{1}{\omega + c_i}}, h^{x_i})_{i=1,\dots,\ell}) = (g_1^{\frac{1}{\omega + x}}, h^x) \wedge \forall x_i \neq x]$ *is negligible, where $(g_1, h) \in \mathbb{G}_1 \times \mathbb{G}_2$ and $(\omega, x, x_1, \dots, x_\ell) \in \mathbb{Z}_p^{\ell+2}$.*

Definition 6 (The Strong Diffie-Hellman (SDH) assumption [6]). *We say that q-SDH assumption holds if for all PPT adversary $\mathcal{A}$, $\Pr[\mathcal{A}(g_1, h, h^\omega, h^{\omega^2}, \dots, h^{\omega^q}) = (g_1^{\frac{1}{\omega + x}}, x)]$ is negligible, where $(g_1, h) \in \mathbb{G}_1 \times \mathbb{G}_2$ and $(\omega, x) \in \mathbb{Z}_p^2$.*

Definition 7 (The external Diffie-Hellman (XDH) assumption [14]). *Let $(\mathbb{G}_1, \mathbb{G}_2, \mathbb{G}_T)$ be a bilinear group. We say that the XDH assumption holds if for all PPT adversary $\mathcal{A}$, the DDH assumption over $\mathbb{G}_1$ holds.*

3   Definitions of Group Signature

Here, we define the system operations of revocable group signature and security requirements (anonymity w.r.t. the revocation and traceability) by adapting [27]. Note that our definition follows the static group settings [4]. However, we can easily handle the dynamic group settings [3] (and non-frameability) by applying an interactive join algorithm.

Definition 8. *System Operations of Group Signature*

Setup : *This probabilistic setup algorithm takes as input the security parameter 1^κ, and returns public parameters params.*

KeyGen : *This probabilistic key generation algorithm (for GM) takes as input the maximum number of users N and params, and returns the group public key gpk, GM's secret key msk, all user's secret key $\{usk_i\}_{i \in [1,N]}$, and the initial revocation-dependent value $\mathcal{T}_0$.*

GSign : *This probabilistic signing algorithm (for a user U_i) takes as input gpk, usk_i, a signed message M, and a revocation-dependent value (in the period t) $\mathcal{T}_t$, and returns a group signature σ.*

Verify : *This deterministic verification algorithm takes as input gpk, M, σ, and $\mathcal{T}_t$, and returns 1 if σ is a valid group signature, and 0 otherwise.*

Revoke : *This (potentially) probabilistic revocation algorithm takes as input gpk, msk, a set of revoked users $RL_{t+1} = \{U_i\}$, and $\mathcal{T}_t$, and returns $\mathcal{T}_{t+1}$.*

Open : *This deterministic algorithm takes as input msk and a valid pair (M, σ), and returns the identity of the signer of σ ID. If ID is not a group member, then the algorithm returns 0.*

In the Revoke algorithm, we set $RL_0 = \emptyset$, and assume that the non-revoked user in t is $\{U_1, \ldots, U_N\} \setminus RL_t$. Under this setting, boomerang users (who re-join the group) are available (i.e., U_i such that $U_i \in RL_{t-1}$ and $U_i \notin RL_t$). In addition, if an invalid pair (M, σ) is input to the Open algorithm, then the Open algorithm easily detect this fact by using the Verify algorithm. So, we exclude this case from the definition of the Open algorithm.

Next, we define anonymity w.r.t. the revocation and traceability. As the main difference among our anonymity definition and previous ones, to guarantee the unlinkability of revoked users, $\mathcal{A}$ can issue the revocation queries against the challenge users. Note that we do not handle the CCA-anonymity, where an adversary $\mathcal{A}$ can issue the open queries. So, we just handle the CPA-anonymity [7] only in this paper. However, as mentioned by Boneh et al. [7], the CCA-anonymity can be handled by applying a CCA secure public key encryption for implementing the open algorithm.

Definition 9 (Anonymity w.r.t. the Revocation)

Setup : *The challenger $\mathcal{C}$ runs the Setup algorithm and the KeyGen algorithm, and obtains params, gpk, msk, and all $\{usk_i\}_{i=1}^{N}$. $\mathcal{C}$ gives params and gpk to $\mathcal{A}$, and sets $t = 0$, $RU_0 = \emptyset$, and $CU = \emptyset$, where RU_0 denotes the (initial) set of ID's of revoked users, and CU denotes the set of ID's of corrupted users.*

Queries : *$\mathcal{A}$ can issue the following queries:*

   Revocation : *$\mathcal{A}$ can request the revocation of users $ID_{i_1}, \ldots, ID_{i_{k_{t+1}}}$ for some constant $k_{t+1} \in [1, N]$. $\mathcal{C}$ uns $\mathcal{T}_{t+1} \leftarrow$ Revoke$(msk, \{ID_{i_1}, \ldots, ID_{i_{k_{t+1}}}\}, \mathcal{T}_t)$ and adds $ID_{i_1}, \ldots, ID_{i_{k_{t+1}}}$ to RU_{t+1}.*

   Signing : *$\mathcal{A}$ can request a group signature on a message M for a user U_i where $ID_i \notin CU$. $\mathcal{C}$ runs $\sigma \leftarrow$ GSign$(gpk, usk_i, M, \mathcal{T}_t)$, where $\mathcal{T}_t$ is the current revocation-dependent value, and gives σ to $\mathcal{A}$.*

   Corruption : *$\mathcal{A}$ can request the secret key of a user U_i. $\mathcal{C}$ adds ID_i to CU, and gives usk_i to $\mathcal{A}$.*

Challenge : *$\mathcal{A}$ sends a message M^* and two users U_{i_0} and U_{i_1}, where $ID_{i_0}, ID_{i_1} \notin CU$. $\mathcal{C}$ chooses a bit $b \leftarrow \{0,1\}$, and runs $\sigma^* \leftarrow$ GSign$(gpk, usk_{i_b}, M^*, \mathcal{T}_{t^*})$, where $\mathcal{T}_{t^*}$ is the current revocation-dependent value, and gives σ^* to $\mathcal{A}$.*

Queries : *The same as the previous one (Note that no corruption query for the challenge users is allowed).*

Output : *$\mathcal{A}$ outputs a guessing bit $b' \in \{0, 1\}$.*

We say that anonymity holds if for all PPT adversaries $\mathcal{A}$, the advantage

$$\mathsf{Adv}_{\mathcal{A}}^{anon}(1^\kappa) := |\Pr[b = b'] - \frac{1}{2}|$$

is negligible.

There are two types of revocable group signature such that (1) any users can make a valid group signature, and anyone can distinguish whether a signer has been revoked or not [8,13,25,28], or (2) no revoked user can make a valid group signature without breaking traceability [7,11,14,27,31]. We implicitly require the second type revocable group signature, since clearly anonymity is broken if one of the challenge users is revoked in a first type scheme. We also require that the challenger C (that has msk) can break traceability to compute the challenge group signature σ^* for the case that a challenger user is revoked. Note that since msk is used for generating user's secret keys, obviously any entity with msk makes an "untraceable" group signature, and this fact does not detract the security of our group signature.

One may think that the above anonymity definition can be extended that A can issue the corruption query against the challenge users, as in the Full-Anonymity [4]. It might be desired that r is not revealed even if revoked users reveal their secret signing keys, since their signing keys are already meaningless (i.e., the rights of signing have been expired). For example, if users are not intentionally revoked (e.g., a user has not paid in the outsourcing businesses example [20]), then users might reveal their secret signing keys to compromise the systems. Or, even if users are intentionally revoked (e.g., they feel that this service is not interesting in the outsourcing businesses example), they might reveal their secret signing keys as a crime for pleasure. However, even if r is kept hidden when revoked users reveal their secret signing keys, one can easily guess r by counting the number of revealed secret keys. So, in our opinion such secret key leakage resilient property is too strong, and therefore our proposed group signature does not follow this leakage property. Next, we define traceability.

Definition 10 (Traceability)

Setup : *The challenger C runs the* Setup *algorithm and the* KeyGen *algorithm, and obtains params, gpk, msk, and all $\{usk_i\}_{i=1}^N$. C gives params and gpk to A, and sets $t = 0$, $RU_0 = \emptyset$, and $CU = \emptyset$, where RU_0 denotes the (initial) set of ID's of revoked users, and CU denotes the set of ID's of corrupted users.*

Queries : *A can issue the following queries:*

  Revocation : *A can request the revocation of users $ID_{i_1}, \ldots, ID_{i_{k_{t+1}}}$ for some constant $k_{t+1} \in [1, N]$. C runs $T_{t+1} \leftarrow$ Revoke$(msk, \{ID_{i_1}, \ldots, ID_{i_{k_{t+1}}}\}, T_t)$ and adds $ID_{i_1}, \ldots, ID_{i_{k_{t+1}}}$ to RU_{t+1}.*

  GSigning : *A can request a group signature on a message M for a user U_i where $ID_i \notin CU$. C runs $\sigma \leftarrow$ GSign(gpk, usk_i, M, T_t), where T_t is the current revocation-dependent value, and gives σ to A.*

  Corruption : *A can request the secret key of a user U_i. C adds ID_i to CU, and gives usk_i to A.*

  Opening : *A can request to a group signature σ on a message M. C returns the result of Open(msk, M, σ) to A.*

Output : *A outputs a past interval $t^* \leq t$ for the current interval t, and (M^*, σ^*).*

We say that A wins if $(1) \wedge (2) \wedge ((3) \vee (4))$ holds, where

1. Verify$(gpk, M^*, \sigma^*, \mathcal{T}_{t^*}) = 1$
2. $\mathcal{A}$ did not obtain σ^* by making a signing query at M^*.
3. for $ID_{i^*} \leftarrow$ Open(msk, M^*, σ^*), $ID_{i^*} \notin CU$
4. for $ID_{i^*} \leftarrow$ Open(msk, M^*, σ^*), $ID_{i^*} \in RU_{t^*}$

We say that traceability holds if for all PPT adversaries $\mathcal{A}$, the advantage

$$\mathsf{Adv}_{\mathcal{A}}^{trace}(1^\kappa) := \Pr[\mathcal{A} \ wins]$$

is negligible.

4   Other Cryptographic Tools

In this section, we introduce cryptographic tools applied for our construction.

BBS+ Signature [2,7,19,27]: Let L be the number of signed messages, and $(\mathbb{G}_1, \mathbb{G}_2, \mathbb{G}_T)$ be a bilinear group. Select $g, g_1, \ldots, g_L \xleftarrow{\$} \mathbb{G}_1$, $h \xleftarrow{\$} \mathbb{G}_2$, and $\omega \leftarrow \mathbb{Z}_p$, and compute $\Omega = g^\omega$. The signing key is ω and the verification key is $(p, \mathbb{G}_1, \mathbb{G}_2, \mathbb{G}_T, e, g, g_1, \ldots, g_{L+1}, h, \Omega)$. For a set of signed messages $(m_1, \ldots, m_L) \in \mathbb{Z}_p^L$, choose $r, y \xleftarrow{\$} \mathbb{Z}_p$, and compute $A = (g_1^{m_1} \cdots g_L^{m_L} g_{L+1}^r g)^{\frac{1}{\omega+y}}$. For a signature (A, r, y), the verification algorithm output 1 if $e(A, \Omega h^y) = e(g_1^{m_1} \cdots g_L^{m_L} g_{L+1}^r g, h)$ holds. The BBS+ signature scheme satisfies existential unforgeability against chosen message attack (EUF-CMA)[3] under the q-SDH assumption.

Linear Encryption [7]: A public key is $pk = (u, v, h) \in \mathbb{G}_2$ such that $u^{X_1} = v^{X_2} = h$ for $X_1, X_2 \in \mathbb{Z}_p$. The corresponding secret key is (X_1, X_2). For a plaintext $M \in \mathbb{G}_2$, choose $\delta_1, \delta_2 \xleftarrow{\$} \mathbb{Z}_p$, compute a ciphertext $C = (F_1, F_2, F_3)$, where $F_1 = M \cdot h^{\delta_1, \delta_2}$, $F_2 = u^{\delta_1}$, and $F_3 = v^{\delta_2}$. C can be decrypted as $M = F_1/(F_2^{X_1} F_3^{X_2})$. The linear encryption is IND-CPA secure[4] under the DLIN assumption.

Signature Based on Proof of Knowledge: In our group signature, we apply the conversion of the underlying interactive zero knowledge (ZK) proof into non-interactive ZK (NIZK) proof by applying the Fiat-Shamir heuristic [17]. We describe such converted signature based on proof of knowledge (SPK) as $SPK\{x : (y, x) \in R\}(M)$, where x is the knowledge to be proved, R is a relation (e.g., $y = g^x$

---

[3] First an adversary $\mathcal{A}$ is given vk from the challenger $\mathcal{C}$. Then $\mathcal{A}$ sends messages to $\mathcal{C}$ and obtains the corresponding signatures. Finally, $\mathcal{A}$ outputs a message/signature pair (M^*, σ^*). We say that $\mathcal{A}$ wins if (M^*, σ^*) is valid and $\mathcal{A}$ has not sent M^* as a signing query. The EUF-CMA security guarantees that the probability $\Pr[\mathcal{A} \ wins]$ is negligible.

[4] First an adversary $\mathcal{A}$ is given pk from the challenger $\mathcal{C}$. Then $\mathcal{A}$ sends the challenge message (M_0^*, M_1^*) to $\mathcal{C}$, and $\mathcal{C}$ chooses $\mu \xleftarrow{\$} \{0, 1\}$, and computes the challenge ciphertext C^* which is a ciphertext of M_μ^*. $\mathcal{A}$ is given C^*, and outputs a bit μ'. The IND-CPA security guarantees that $|\Pr[\mu = \mu'] - \frac{1}{2}|$ is negligible.

in the case of the knowledge of the discrete logarithm), and M is a signed message. The SPK has an extractor of the proved knowledge from two accepting protocol views whose commitments are the same but challenges are different.

5   Proposed Group Signature Scheme with Hiding of the Number of Revoked Users

In this section, we propose a group signature scheme hiding the number of revoked users by applying ABGS. Before explaining our scheme, we introduce ABGS as follows:

Attribute-Based Group Signature (ABGS): ABGS [15,18,21,22] is a kind of group signature, where a user with a set of attributes can prove anonymously whether he/she has these attributes or not. Anonymity means a verifier cannot identify who the actual signer is among group members. As a difference from attribute-based signature [23,24,26,32], there is an opening manager (as in group signatures) who can identify the actual signer (anonymity revocation), and a verifier can "explicitly" verify whether a user has these attributes or not [15,21,22]. By applying this explicitly attribute verification, anonymous survey for the collection of attribute statistics is proposed [15]. As one exception, the Fujii et al. ABGS scheme [18] achieves signer-attribute privacy, where a group signature does not leak which attributes were used to generate it, except that assigned attributes satisfy a predicate. As another property (applied for our construction), the dynamic property has been proposed in [15], where the attribute predicate can be updated without re-issuing the user's secret keys.

Our Methodology: We consider two attributes: (1) valid group user and (2) the user's identity (say U_i), and apply the dynamic property of ABGS [15] and the signer-attribute privacy of ABGS [18]. Here we explain our methodology. Let the initial access tree be represented as in Fig 1:

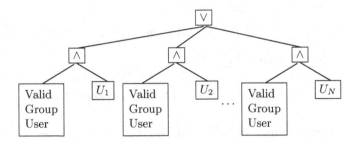

Fig. 1. Initial Access Tree

Due to the signer-attribute privacy, a user U_i can anonymously prove that he/she has attributes "valid group user" and "U_i". Namely, anyone can verify whether the signer's attributes satisfy the access tree, without detecting the actual attribute (i.e., the user's identity).

When a user (say U_1) is revoked, the tree structure is changed as in Fig 2.

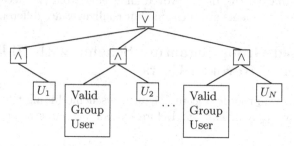

Fig. 2. Modified Access Tree

Due to the dynamic property of ABGS, this modification can be done without re-issuing the user's secret keys. By removing the attribute "valid group user" from the subtree of U_1, we can express the revocation of U_1, since U_1 cannot prove that his/her attributes satisfy the current access tree.

In addition, we propose a randomization and dummy attribute technique to implement the revocation procedure (Fig 3). We apply the Boldyreva multisignature [5], since it is applied for the computation of the membership certificate in the Fujii et al. ABGS. Let t be the time interval and v denote the attribute "valid group user".

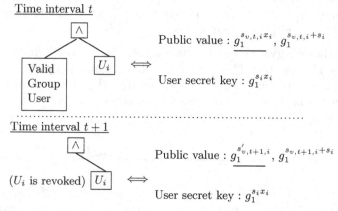

Fig. 3. Our Randomization and Dummy Attribute Technique

For a non-revoked user U_i, GM publishes the dummy value $g_1^{s_{v,t,i}x_i}$. Then U_i can compute $g_1^{(s_{v,t,i}+s_i)x_i}(= H_i)$ from $d_{T,t,i} = g_1^{s_{v,t,i}x_i}$ and U_i's secret key $B_i = g_1^{s_i x_i}$. Let U_i be revoked in the time interval $t+1$. Then, GM publishes a randomized dummy value $g_1^{s'_{v,t+1,i}}$ (instead of $g_1^{s_{v,t+1,i}x_i}$), and therefore U_i cannot compute $g^{(s_{v,t+1,i}+s_i)x_i}$ due to the CDH assumption. Note that $(g_1^{s_{v,t+1,i}+s_i}, g_1^{s_{v,t+1,i}x_i})$ and $(g_1^{s_{v,t+1,i}+s_i}, g_1^{s'_{v,t+1,i}})$ are indistinguishable, under the XDH assumption, where the DDH assumption holds in $\mathbb{G}_1$. Next, we give our group signature scheme.

Protocol 1 (Our revocable group signature)

Setup(1^κ) : *Select a bilinear group* $(\mathbb{G}_1, \mathbb{G}_2, \mathbb{G}_T)$ *with prime order* p, *a bilinear map* $e : \mathbb{G}_1 \times \mathbb{G}_2 \to \mathbb{G}_T$, $g, g_1, \ldots, g_4, \tilde{g} \xleftarrow{\$} \mathbb{G}_1$, $\tilde{h} \xleftarrow{\$} \mathbb{G}_2$. *Output params* = $(p, \mathbb{G}_1, \mathbb{G}_2, \mathbb{G}_T, e, H, g, g_1, g_2, g_3, g_4, \tilde{g}, \tilde{h})$

KeyGen($params, N$) : *Let* $(U_1, \ldots, U_N)$ *be all users. Set* $t = 0$. *Select* ω_1, ω_2, X_1, $X_2, x_1, \ldots, x_N, s_1, \ldots, s_N, s_{v,0,1}, \ldots, s_{v,0,N} \xleftarrow{\$} \mathbb{Z}_p^*$. *Compute*

- $u, v, h \in \mathbb{G}_2$ *with the condition* $u^{X_1} = v^{X_2} = h$ *(note that* (u, v, h) *is a public key of the linear encryption, and* (X_1, X_2) *is the corresponding secret key)*,

- $K_{i,1} = g_1^{\frac{1}{\omega_1 + x_i}}$, $K_{i,2} = h^{x_i}$, *and* $B_i = g_1^{s_i x_i}$ *for all* $i \in [1, N]$, *and*

- $\Omega_1 = h^{\omega_1}$ *and* $\Omega_2 = h^{\omega_2}$.

For all $i \in [1, N]$, *choose* $s_{v,0,i}, y_{0,i}, r_{0,i} \xleftarrow{\$} \mathbb{Z}_p^*$. *If* $s_{v,0,i} + s_i = 0 \bmod p$, *then choose* $s_{v,0,i}$ *again until* $s_{v,0,i} + s_i \neq 0 \bmod p$ *holds. Set* $s_{T,0,i} := s_{v,0,i} + s_i$, *and compute*

- $h_{T,0,i} = g_1^{s_{T,0,i}}$

- $A_{0,i} = (g_1^{s_{T,0,i}} g_2^t g_3^{r_{0,i}} g_4)^{\frac{1}{\omega_2 + y_{0,i}}}$ *(which is a BBS+ signature for signed messages* $(s_{T,0,i}, t)$*), and*

- $d_{T,0,i} := g_1^{s_{v,0,i} x_i}$.

Set $Sign(s_{T,0,i}, i) = (A_{0,i}, y_{0,i}, r_{0,i})$. *Output*

- $gpk = (params, \Omega_1, \Omega_2, u, v, \mathcal{H})$, *where* $\mathcal{H} : \{0, 1\}^* \to \mathbb{Z}_p^*$ *is a hash function which is modeled as a random oracle.*

- $msk = (X_1, X_2, s_1, \ldots, s_N, s_{v,0,1}, \ldots, s_{v,0,N}, x_1, \ldots, x_N, \mathsf{reg} := \{(K_{i,2}, i)\}_{i=1}^N)$,

- $usk_i = (K_{i,1}, K_{i,2}, B_i)$ *for all* $i \in [1, N]$, *and*

- $\mathcal{T}_0 = \{(Sign(s_{T,0,i}, i), h_{T,0,i}, d_{T,0,i})\}_{i=1}^N$.

GSign($gpk, usk_i, M, \mathcal{T}_t$) : *Let* U_i *be a non-revoked user in the current time interval* t. *That is, for* $(Sign(s_{T,t,i}, i), h_{T,t,i}, d_{T,t,i}) \in \mathcal{T}_t$, $h_{T,t,i} = g_1^{s_{v,t,i} + s_i} := g_1^{s_{T,t,i}}$ *and* $d_{T,t,i} = g_1^{s_{v,t,i} x_i}$ *hold for some unknown exponent* $s_{v,t,i} \in \mathbb{Z}_p^*$. U_i *chooses* $r_1, r_2, \ldots, r_{10}, \delta_1, \delta_2 \xleftarrow{\$} \mathbb{Z}_p^*$, *sets* $\alpha = -r_1 r_2$, $\beta = -r_2 r_4$, $\beta' = r_5 y_{t,i} - r_4$, $\gamma = r_2 r_6 + r_7$, $\gamma' = r_4 r_8 + r_9$, *and* $\gamma'' = r_{10} y_{t,i}$, *and computes*

$$H_i = B_i \cdot d_{T,t,i} = g_1^{s_i x_i + s_{v,t,i} x_i} = h_{T,t,i}^{x_i},$$

$$T_1 = K_{i,1} \tilde{g}^{r_1}, T_2 = K_{i,2} \tilde{h}^{r_2}, T_3 = H_i \tilde{g}^{r_3}, T_4 = h_{T,t,i} \tilde{g}^{r_4}, T_5 = A_{t,i} \tilde{g}^{r_5},$$

$$C_1 = g^{r_1} \tilde{g}^{r_6}, C_2 = g^\alpha \tilde{g}^{r_7}, C_3 = g^{r_2} \tilde{g}^{r_8}, C_4 = g^\beta \tilde{g}^{r_9}, C_5 = g^{r_{10}} \tilde{g}^{-r_5}, C_6 = g^{\gamma''} \tilde{g}^{-r_4},$$

$$F_1 = K_{i,2} h^{\delta_1 + \delta_2}, F_2 = u^{\delta_1}, \text{ and } F_3 = v^{\delta_2}, \text{ and}$$

$$V = SPK\{(r_1, r_2, r_3, r_4, r_5, r_6, r_7, r_8, r_9, r_{10}, y_{t,i}, r_{t,i}, \alpha, \beta, \beta', \gamma, \gamma', \gamma'', \delta_1, \delta_2):$$

$$\frac{e(T_1, \Omega_1 T_2)}{e(g_1, h)} = e(\tilde{g}, \Omega_1 T_2)^{r_1} e(T_1, \tilde{h})^{r_2} e(\tilde{g}, \tilde{h})^{\alpha}$$

$$\wedge \frac{e(T_4, T_2)}{e(T_3, h)} = \frac{e(\tilde{g}, T_2)^{r_4} e(T_4, \tilde{h})^{r_2} e(\tilde{g}, \tilde{h})^{\beta}}{e(\tilde{g}, h)^{r_3}}$$

$$\wedge \frac{e(T_5, \Omega_2)}{e(g_4, h) e(T_4, h) e(g_2, h)^t} = \frac{e(\tilde{g}, \Omega_2)^{r_5} e(g_3, h)^{r_{t,i}} e(\tilde{g}, h)^{\beta'}}{e(T_5, h)^{y_{t,i}}}$$

$$\wedge C_1 = g^{r_1} \tilde{g}^{r_6} \wedge C_2 = g^{\alpha} \tilde{g}^{r_7} \wedge C_2 = C_1^{-r_2} \tilde{g}^{\gamma}$$

$$\wedge C_3 = g^{r_2} \tilde{g}^{r_8} \wedge C_4 = g^{\beta} \tilde{g}^{r_9} \wedge C_4 = C_3^{-r_4} \tilde{g}^{\gamma'}$$

$$\wedge C_5 = g^{r_{10}} \tilde{g}^{-r_5} \wedge C_6 = g^{\gamma''} \tilde{g}^{-r_4} \wedge C_6 = C_5^{y_{t,i}} \tilde{g}^{\beta'}$$

$$\wedge \frac{T_2}{F_1} = \frac{\tilde{h}^{r_2}}{h^{\delta_1 + \delta_2}} \wedge F_2 = u^{\delta_1} \wedge F_3 = v^{\delta_2}\}(M)$$

Output $\sigma = (C_1, C_2, C_3, C_4, C_5, C_6, F_1, F_2, F_3, T_1, T_2, T_3, T_4, T_5, V)$[5].

Verify$(gpk, M, \sigma, \mathcal{T}_t)$: *Return 1 if σ is a valid group signature*[6], *and 0 otherwise.*

Revoke$(gpk, msk, \{U_i\}, \mathcal{T}_t)$: *Let $RL_{t+1} := \{U_i\}$ be a set of revoked users. Set $t \to t+1$. For all $i \in \{i | U_i \in RL_{t+1}\}$, choose $s'_{v,t+1,i} \xleftarrow{\$} \mathbb{Z}_p^*$. For all $i \in [1, N]$, choose $s_{v,t+1,i}, y_{t+1,i}, r_{t+1,i} \xleftarrow{\$} \mathbb{Z}_p^*$ (until $s_{v,t+1,i} + s_i \neq 0 \mod p$ holds), set $s_{T,t+1,i} := s_{v,t+1,i} + s_i$, and compute*

$$h_{T,t+1,i} = g^{s_{T,t+1,i}},$$

$$A_{t+1,i} = (g_1^{s_{T,t+1,i}} g_2^{t+1} g_3^{r_{t+1,i}} g_4)^{\frac{1}{\omega_2 + y_{t+1,i}}},$$

and compute $d_{T,t+1,i}$ such that:

$$d_{T,t+1,i} = \begin{cases} g_1^{s_{v,t+1,i} x_i} & (U_i \notin RL_{t+1}) \\ g_1^{s'_{v,t+1,i}} & (U_i \in RL_{t+1}) \end{cases}$$

and set $Sign(s_{T,t+1,i}, i) = (A_{t+1,i}, y_{t+1,i}, r_{t+1,i})$. Output $\mathcal{T}_{t+1} = \{(Sign(s_{T,t+1,i}, i), h_{T,t+1,i}, d_{T,t+1,i})\}_{i=1}^N$.

Open(gpk, msk, M, σ) : *Compute $\frac{F_1}{F_2^{X_1} F_3^{X_2}} = K$, and search i such that $(K_{i,2}, i) \in$ reg and $K = K_{i,2}$. If there is no such i, output 0. Otherwise, output i.*

In our scheme, no public values have size dependent on r, and no costs of the GSign algorithm (or the Verify algorithm) depend on r or N. In addition, our scheme satisfies anonymity w.r.t. the revocation which guarantees the unlinkability of revoked users. So, in our scheme, no r is revealed.

---

[5] We give the detailed form of SPK V in the appendix.

[6] We give the procedure of the verification algorithm in the appendix.

6 Discussion

One drawback of our scheme is that the number of public values depends on N, since no common attribute can be applied for implementing the revocation procedure of "each" user. So, one may think that there might be a more trivial construction (without applying ABGS) if such big-size public value is allowed. For example, as one of the most simple group signature construction, let $g^{x_1}, \ldots, g^{x_N}$ be users' public keys, and GM randomizes these value such that $y_1 := (g^{x_1})^{r_{GM}}, \ldots, y_N := (g^{x_N})^{r_{GM}}$, and publishes $y := g^{r_{GM}}$. Each user (say U_i) proves the knowledge of x_i for the relation $(g^{r_{GM}})^{x_i}$ using the OR relation such that $SPK\{x : y^x = y_1 \vee \cdots \vee y^x = y_N\}(M)$ to hide the identity $i \in [1, N]$. If a user (say U_j) is revoked, then GM publishes a random value R_j (instead of $(g^{x_j})^{r_{GM}}$). In this case, the number of revoked users is not revealed under the DDH assumption, since $(g, g^{x_j}, g^{r_{GM}}, (g^{x_j})^{r_{GM}})$ is a DDH tuple. However, this trivial approach requires N-dependent signing/verification cost, whereas our scheme achieves constant proving costs.'

As another candidate, Sudarsono et al. [33] proposed an attribute-based anonymous credential system by applying an efficient pairing-based accumulator proposed by Camenisch et al. [10]. Since the Sudarsono et al. construction follows AND/OR relations of attributes, a revocable group signature scheme with the property of hiding r might be constructed. However, it is not obvious whether 2-DNF formulae $\vee_{i=1}^{N}(\text{valid group user} \wedge U_i)$ can be implemented or not in the Sudarsono et al. attribute-based proof system. In addition, their construction also requires the N-dependent-size (N is the number of attributes in this context) public values to update the witness of users, as in our group signature scheme. So, we insist that proposing a revocable group signature scheme with both the property of hiding r and constant proving costs is not trivial if such large-size public key is allowed.

7 Security Analysis

The security proofs of following theorems are given in the appendix.

Theorem 1. *The proposed group signature scheme satisfies anonymity w.r.t. the revocation under the DLIN assumption and the XDH assumption.*

Theorem 2. *The proposed group signature scheme satisfies traceability under the N-HSDH assumption, the CDH assumption, and Nt-SDH assumption, where t is the final time interval that $\mathcal{A}$ outputs (M^*, σ^*).*

8 Conclusion

In this paper, we propose a revocable group signature scheme with the property of hiding r, by applying ABGS. Under a XDH-hard elliptic curve with 170 bits p (as in [14,28]), the size of signature is 7242 bits, where the size of an element of $\mathbb{G}_1$ is 171 bits, the size of an element of $\mathbb{G}_2$ is 513 bits, and the size of the challenge c is 80 bits. Since the size of signature in [14] (resp. in [28]) is 1444 (resp. 1533) bits, there is space for improvement the signature size. In addition, proposing a r-hiding revocable group signature with small-size public key is also interesting future work.

References

1. Ateniese, G., Tsudik, G.: Some Open Issues and New Directions in Group Signatures. In: Franklin, M.K. (ed.) FC 1999. LNCS, vol. 1648, pp. 196–211. Springer, Heidelberg (1999)
2. Au, M.H., Susilo, W., Mu, Y.: Constant-Size Dynamic k-TAA. In: De Prisco, R., Yung, M. (eds.) SCN 2006. LNCS, vol. 4116, pp. 111–125. Springer, Heidelberg (2006)
3. Bellare, M., Shi, H., Zhang, C.: Foundations of Group Signatures: The Case of Dynamic Groups. In: Menezes, A. (ed.) CT-RSA 2005. LNCS, vol. 3376, pp. 136–153. Springer, Heidelberg (2005)
4. Bellare, M., Micciancio, D., Warinschi, B.: Foundations of Group Signatures: Formal Definitions, Simplified Requirements, and a Construction Based on General Assumptions. In: Biham, E. (ed.) EUROCRYPT 2003. LNCS, vol. 2656, pp. 614–629. Springer, Heidelberg (2003)
5. Boldyreva, A.: Threshold Signatures, Multisignatures and Blind Signatures Based on the Gap-Diffie-Hellman-Group Signature Scheme. In: Desmedt, Y.G. (ed.) PKC 2003. LNCS, vol. 2567, pp. 31–46. Springer, Heidelberg (2002)
6. Boneh, D., Boyen, X.: Short signatures without random oracles and the SDH assumption in bilinear groups. J. Cryptology 21(2), 149–177 (2008)
7. Boneh, D., Boyen, X., Shacham, H.: Short Group Signatures. In: Franklin, M. (ed.) CRYPTO 2004. LNCS, vol. 3152, pp. 41–55. Springer, Heidelberg (2004)
8. Boneh, D., Shacham, H.: Group signatures with verifier-local revocation. In: ACM Conference on Computer and Communications Security, pp. 168–177 (2004)
9. Boyen, X., Waters, B.: Full-Domain Subgroup Hiding and Constant-Size Group Signatures. In: Okamoto, T., Wang, X. (eds.) PKC 2007. LNCS, vol. 4450, pp. 1–15. Springer, Heidelberg (2007)
10. Camenisch, J., Kohlweiss, M., Soriente, C.: An Accumulator Based on Bilinear Maps and Efficient Revocation for Anonymous Credentials. In: Jarecki, S., Tsudik, G. (eds.) PKC 2009. LNCS, vol. 5443, pp. 481–500. Springer, Heidelberg (2009)
11. Camenisch, J., Lysyanskaya, A.: Dynamic Accumulators and Application to Efficient Revocation of Anonymous Credentials. In: Yung, M. (ed.) CRYPTO 2002. LNCS, vol. 2442, pp. 61–76. Springer, Heidelberg (2002)
12. Chaum, D., van Heyst, E.: Group Signatures. In: Davies, D.W. (ed.) EUROCRYPT 1991. LNCS, vol. 547, pp. 257–265. Springer, Heidelberg (1991)
13. Chen, L., Li, J.: VLR group signatures with indisputable exculpability and efficient revocation. In: SocialCom/PASSAT, pp. 727–734 (2010)
14. Delerablée, C., Pointcheval, D.: Dynamic Fully Anonymous Short Group Signatures. In: Nguyên, P.Q. (ed.) VIETCRYPT 2006. LNCS, vol. 4341, pp. 193–210. Springer, Heidelberg (2006)
15. Emura, K., Miyaji, A., Omote, K.: A dynamic attribute-based group signature scheme and its application in an anonymous survey for the collection of attribute statistics. Journal of Information Processing 17, 216–231 (2009)

16. Emura, K., Miyaji, A., Omote, K.: An Anonymous Designated Verifier Signature Scheme with Revocation: How to Protect a Company"s Reputation. In: Heng, S.-H., Kurosawa, K. (eds.) ProvSec 2010. LNCS, vol. 6402, pp. 184–198. Springer, Heidelberg (2010)

17. Fiat, A., Shamir, A.: How to Prove Yourself: Practical Solutions to Identification and Signature Problems. In: Odlyzko, A.M. (ed.) CRYPTO 1986. LNCS, vol. 263, pp. 186–194. Springer, Heidelberg (1987)

18. Fujii, H., Nakanishi, T., Funabiki, N.: A proposal of efficient attribute-based group signature schemes using pairings. IEICE Technical Report 109(272), 15–22 (2009) (in Japanese), http://ci.nii.ac.jp/naid/110007520932/en/

19. Furukawa, J., Imai, H.: An efficient group signature scheme from bilinear maps. IEICE Transactions 89-A(5), 1328–1338 (2006)

20. Isshiki, T., Mori, K., Sako, K., Teranishi, I., Yonezawa, S.: Using group signatures for identity management and its implementation. In: Digital Identity Management, pp. 73–78 (2006)

21. Khader, D.: Attribute based group signature with revocation. Cryptology ePrint Archive, Report 2007/241 (2007)

22. Khader, D.: Attribute based group signatures. Cryptology ePrint Archive, Report 2007/159 (2007)

23. Li, J., Au, M.H., Susilo, W., Xie, D., Ren, K.: Attribute-based signature and its applications. In: ASIACCS, pp. 13–16 (2010)

24. Li, J., Kim, K.: Hidden attribute-based signatures without anonymity revocation. International Journal of Information Sciences 180(9), 1681–1689 (2010)

25. Libert, B., Vergnaud, D.: Group Signatures with Verifier-Local Revocation and Backward Unlinkability in the Standard Model. In: Garay, J.A., Miyaji, A., Otsuka, A. (eds.) CANS 2009. LNCS, vol. 5888, pp. 498–517. Springer, Heidelberg (2009)

26. Maji, H.K., Prabhakaran, M., Rosulek, M.: Attribute-Based Signatures. In: Kiayias, A. (ed.) CT-RSA 2011. LNCS, vol. 6558, pp. 376–392. Springer, Heidelberg (2011)

27. Nakanishi, T., Fujii, H., Hira, Y., Funabiki, N.: Revocable Group Signature Schemes with Constant Costs for Signing and Verifying. In: Jarecki, S., Tsudik, G. (eds.) PKC 2009. LNCS, vol. 5443, pp. 463–480. Springer, Heidelberg (2009)

28. Nakanishi, T., Funabiki, N.: A Short Verifier-Local Revocation Group Signature Scheme with Backward Unlinkability. In: Yoshiura, H., Sakurai, K., Rannenberg, K., Murayama, Y., Kawamura, S.-i. (eds.) IWSEC 2006. LNCS, vol. 4266, pp. 17–32. Springer, Heidelberg (2006)

29. Nakanishi, T., Funabiki, N.: Efficient revocable group signature schemes using primes. Journal of Information Processing 16, 110–121 (2008)

30. Nakanishi, T., Kubooka, F., Hamada, N., Funabiki, N.: Group Signature Schemes with Membership Revocation for Large Groups. In: Boyd, C., González Nieto, J.M. (eds.) ACISP 2005. LNCS, vol. 3574, pp. 443–454. Springer, Heidelberg (2005)

31. Nguyen, L.: Accumulators from Bilinear Pairings and Applications. In: Menezes, A. (ed.) CT-RSA 2005. LNCS, vol. 3376, pp. 275–292. Springer, Heidelberg (2005)

32. Shahandashti, S.F., Safavi-Naini, R.: Threshold Attribute-Based Signatures and Their Application to Anonymous Credential Systems. In: Preneel, B. (ed.) AFRICACRYPT 2009. LNCS, vol. 5580, pp. 198–216. Springer, Heidelberg (2009)

33. Sudarsono, A., Nakanishi, T., Funabiki, N.: Efficient Proofs of Attributes in Pairing-Based Anonymous Credential System. In: Fischer-Hübner, S., Hopper, N. (eds.) PETS 2011. LNCS, vol. 6794, pp. 246–263. Springer, Heidelberg (2011)

Appendix A: Detailed SPK

First, we explain the relations proved in SPK V. V proves that:

1. A signer has a valid $(K_{i,1}, K_{i,2})$ generated by the KeyGen algorithm.
   - $(K_{i,1}, K_{i,2})$ can be verified by using the public value Ω_1 such that:

$$e(K_{i,1}, \Omega_1 K_{i,2}) = e(g_1, h)$$

   - Since $K_{i,1}$ (resp. $K_{i,2}$) is hidden such that $T_1 = K_{i,1}\tilde{g}^{r_1}$, (resp. $T_2 = K_{i,2}\tilde{h}^{r_2}$), this relation is represented as:

$$\frac{e(T_1, \Omega_1 T_2)}{e(g_1, h)} = e(\tilde{g}, \Omega_1 T_2)^{r_1} e(T_1, \tilde{h})^{r_2} e(\tilde{g}, \tilde{h})^{\alpha}$$

   - We need to guarantee the relation $\alpha = -r_1 r_2$ in the relation above. To prove this, introduce an intermediate value $\gamma = r_2 r_6 + r_7$, and prove that:

$$C_1 = g^{r_1}\tilde{g}^{r_6} \wedge C_2 = g^{\alpha}\tilde{g}^{r_7} \wedge C_2 = C_1^{-r_2}\tilde{g}^{\gamma}$$

   Note that $C_2 = C_1^{-r_2}\tilde{g}^{\gamma} = (g^{r_1}\tilde{g}^{r_6})^{-r_2}\tilde{g}^{\gamma} = g^{-r_1 r_2}\tilde{g}^{-r_2 r_6 + \gamma} = g^{\alpha}\tilde{g}^{r_7}$ yields $\alpha = -r_1 r_2$ and $\gamma = r_2 r_6 + r_7$.

2. A signer has not been revoked.
   - A non-revoked signer can compute $H_i = h_{T,t,i}^{\log_h K_{i,2}} = (g_1^{s_{T,t,i}})^{x_i}$ from B_i and $d_{T,t,i}$, where $s_{T,t,i}$ is a signed message of $A_{t,i}$. These satisfy the relations

$$e(h_{T,t,i}, K_{i,2}) = e(H_i, h)$$
$$e(A_{t,i}, \Omega_2 h^{y_{t,i}}) = e(g_1^{s_{T,t,i}} g_2^t g_3^{r_{t,i}} g_4, h)$$

   - Since H_i, $h_{T,t,i}$, and A_i are hidden such that $T_3 = H_i\tilde{g}^{r_3}$, $T_4 = h_{T,t,i}\tilde{g}^{r_4}$, and $T_5 = A_{t,i}\tilde{g}^{r_5}$, these relations are represented as:

$$\frac{e(T_4, T_2)}{e(T_3, h)} = \frac{e(\tilde{g}, T_2)^{r_4} e(T_4, \tilde{h})^{r_2} e(\tilde{g}, \tilde{h})^{\beta}}{e(\tilde{g}, h)^{r_3}}$$

$$\frac{e(T_5, \Omega_2)}{e(g_4, h)e(T_4, h)e(g_2, h)^t} = \frac{e(\tilde{g}, \Omega_2)^{r_5} e(g_3, h)^{r_{t,i}} e(\tilde{g}, h)^{\beta'}}{e(T_5, h)^{y_{t,i}}}$$

   - We need to guarantee the relations $\beta = -r_2 r_4$ and $\beta' = r_5 y_{t,i} - r_4$ in the relations above. To prove these, introduce intermediate values $\gamma' = r_4 r_8 + r_9$ and $\gamma'' = r_{10} y_{t,i}$, and prove that:

$$C_3 = g^{r_2}\tilde{g}^{r_8} \wedge C_4 = g^{\beta}\tilde{g}^{r_9} \wedge C_4 = C_3^{-r_4}\tilde{g}^{\gamma'}$$
$$C_5 = g^{r_{10}}\tilde{g}^{-r_5} \wedge C_6 = g^{\gamma''}\tilde{g}^{-r_4} \wedge C_6 = C_5^{y_{t,i}}\tilde{g}^{\beta'}$$

   As in α and γ explained before, relations $\beta = -r_2 r_4$, $\beta' = r_5 y_{t,i} - r_4$, $\gamma' = r_4 r_8 + r_9$, and $\gamma'' = r_{10} y_{t,i}$ are obtained from the relations above.
   - Note that $(A_{t,i}, r_{t,i}, y_{t,i})$ is a BBS+ signature for signed messages $(s_{T,t,i}, t)$, and therefore V depends on the current time interval t.

3. A value for the Open algorithm is included in σ.
   - (F_1, F_2, F_3) is a ciphertext (of the linear encryption scheme) of the plaintext $K_{i,2}$, which can be computed by decrypting (F_1, F_2, F_3) using msk.

Next, we describe the detailed SPK of our scheme as follows.

1. Choose $r_{r_1}, r_{r_2}, r_{r_3}, r_{r_4}, r_{r_5}, r_{r_6}, r_{r_7}, r_{r_8}, r_{r_9}, r_{r_{10}}, r_{y_{t,i}}, r_{r_{t,i}}, r_\alpha, r_\beta, r_{\beta'}, r_\gamma, r_{\gamma'},$
   $r_{\gamma''}, r_{\delta_1}, r_{\delta_2} \xleftarrow{\$} \mathbb{Z}_p^*.$

2. Compute

$$R_1 = e(\tilde{g}, \Omega_1 T_2)^{r_{r_1}} e(T_1, \tilde{h})^{r_{r_2}} e(\tilde{g}, \tilde{h})^{r_\alpha}, \quad R_2 = \frac{e(\tilde{g}, T_2)^{r_{r_4}} e(T_4, \tilde{h})^{r_{r_2}} e(\tilde{g}, \tilde{h})^{r_\beta}}{e(\tilde{g}, h)^{r_{r_3}}},$$

$$R_3 = \frac{e(\tilde{g}, \Omega_2)^{r_{r_5}} e(g_3, h)^{r_{r_{t,i}}} e(\tilde{g}, h)^{r_{\beta'}}}{e(T_5, h)^{r_{y_{t,i}}}}, \quad R_4 = g^{r_{r_1}} \tilde{g}^{r_{r_6}}, \quad R_5 = g^{r_\alpha} \tilde{g}^{r_{r_7}}, \quad R_6 = C_1^{-r_{r_2}} \tilde{g}^{r_\gamma},$$

$$R_7 = g^{r_{r_2}} \tilde{g}^{r_{r_8}}, \quad R_8 = g^{r_\beta} \tilde{g}^{r_{r_9}}, \quad R_9 = C_3^{-r_{r_4}} \tilde{g}^{r_{\gamma'}}, \quad R_{10} = g^{r_{r_{10}}} \tilde{g}^{-r_{r_5}}, \quad R_{11} = g^{r_{\gamma''}} \tilde{g}^{-r_{r_4}},$$

$$R_{12} = C_5^{r_{y_{t,i}}} \tilde{g}^{r_{\beta'}}, \quad R_{13} = \frac{\tilde{h}^{r_{r_2}}}{h^{r_{\delta_1} + r_{\delta_2}}}, \quad R_{14} = u^{r_{\delta_1}}, \quad R_{15} = v^{r_{\delta_2}},$$

$$c = \mathcal{H}(gpk, M, C_1, C_2, C_3, C_4, C_5, C_6, F_1, F_2, F_3, T_1, T_2, T_3, T_4, T_5, R_1, \ldots, R_{15}),$$

$$s_{r_i} = r_{r_i} + cr_i \ (i \in [1, 10]), \quad s_{y_{t,i}} = r_{y_{t,i}} + cy_{t,i}, \quad s_{r_{t,i}} = r_{r_{t,i}} + cr_{t,i},$$

$$s_\alpha = r_\alpha + c\alpha, \quad s_\beta = r_\beta + c\beta, \quad s_{\beta'} = r_{\beta'} + c\beta', \quad s_\gamma = r_\gamma + c\gamma, \quad s_{\gamma'} = r_{\gamma'} + c\gamma',$$

$$s_{\gamma''} = r_{\gamma''} + c\gamma'', \quad s_{\delta_1} = r_{\delta_1} + c\delta_1, \quad \text{and} \quad s_{\delta_2} = r_{\delta_2} + c\delta_2,$$

3. Output $V = (c, \{s_{r_i}\}_{i=1}^{10}, s_{y_{t,i}}, s_{r_{t,i}}, s_\alpha, s_\beta, s_{\beta'}, s_\gamma, s_{\gamma'}, s_{\gamma''}, s_{\delta_1}, s_{\delta_2}).$

Next, we describe the verification of $\sigma = (C_1, C_2, C_3, C_4, C_5, C_6, F_1, F_2, F_3, T_1,$
$T_2, T_3, T_4, T_5, c, \{s_{r_i}\}_{i=1}^{10}, s_{y_{t,i}}, s_{r_{t,i}}, s_\alpha, s_\beta, s_{\beta'}, s_\gamma, s_{\gamma'}, s_{\gamma''}, s_{\delta_1}, s_{\delta_2})$ as follows.

1. Compute

$$\tilde{R}_1 = e(\tilde{g}, \Omega_1 T_2)^{s_{r_1}} e(T_1, \tilde{h})^{s_{r_2}} e(\tilde{g}, \tilde{h})^{s_\alpha} \left(\frac{e(T_1, \Omega_1 T_2)}{e(g_1, h)}\right)^{-c},$$

$$\tilde{R}_2 = \frac{e(\tilde{g}, T_2)^{s_{r_4}} e(T_4, \tilde{h})^{s_{r_2}} e(\tilde{g}, \tilde{h})^{s_\beta}}{e(\tilde{g}, h)^{s_{r_3}}} \left(\frac{e(T_4, T_2)}{e(T_3, h)}\right)^{-c},$$

$$\tilde{R}_3 = \frac{e(\tilde{g}, \Omega_2)^{s_{r_5}} e(g_3, h)^{s_{r_{t,i}}} e(\tilde{g}, h)^{s_{\beta'}}}{e(T_5, h)^{s_{y_{t,i}}}} \left(\frac{e(T_5, \Omega_2)}{e(g_4, h) e(T_4, h) e(g_2, h)^t}\right)^{-c},$$

$$\tilde{R}_4 = g^{s_{r_1}} \tilde{g}^{s_{r_6}} C_1^{-c}, \quad \tilde{R}_5 = g^{r_\alpha} \tilde{g}^{s_{r_7}} C_2^{-c}, \quad \tilde{R}_6 = C_1^{-s_{r_2}} \tilde{g}^{s_\gamma} C_2^{-c},$$

$$\tilde{R}_7 = g^{s_{r_2}} \tilde{g}^{s_{r_8}} C_3^{-c}, \quad \tilde{R}_8 = g^{s_\beta} \tilde{g}^{s_{r_9}} C_4^{-c}, \quad \tilde{R}_9 = C_3^{-s_{r_4}} \tilde{g}^{s_{\gamma'}} C_4^{-c},$$

$$\tilde{R}_{10} = g^{s_{r_{10}}} \tilde{g}^{-s_{r_5}} C_5^{-c}, \quad \tilde{R}_{11} = g^{s_{\gamma''}} \tilde{g}^{-s_{r_4}} C_6^{-c}, \quad \tilde{R}_{12} = C_5^{s_{y_{t,i}}} \tilde{g}^{s_{\beta'}} C_6^{-c},$$

$$\tilde{R}_{13} = \frac{\tilde{h}^{s_{r_2}}}{h^{s_{\delta_1} + s_{\delta_2}}} \left(\frac{T_2}{F_1}\right)^{-c}, \quad \tilde{R}_{14} = u^{s_{\delta_1}} F_2^{-c}, \quad \text{and}$$

$$\tilde{R}_{15} = v^{r_{\delta_2}} F_3^{-c}.$$

Note that a verifier computes $e(g_2, h)^t$ to check whether σ is made in the time interval t or not.

2. Check $c = \mathcal{H}(gpk, M, C_1, C_2, C_3, C_4, C_5, C_6, F_1, F_2, F_3, T_1, T_2, T_3, T_4, T_5, \tilde{R}_1,$
   $\ldots, \tilde{R}_{15})$. If it holds, then output 1, and 0, otherwise.

Appendix B: Security Analysis

Proof of Theorem 1

Proof. Let $\mathcal{C}$ be the challenger of the linear encryption, and $\mathcal{A}$ be the adversary who breaks anonymity w.r.t. the revocation of our scheme. We construct the algorithm $\mathcal{B}$ that breaks the IND-CPA security of the linear encryption. First, $\mathcal{C}$ gives the public key of the linear encryption (u, v, h). $\mathcal{B}$ chooses all values, except (u, v, h), and therefore $\mathcal{B}$ can answer all queries issued from $\mathcal{A}$.

In the challenge phase, $\mathcal{A}$ sends (M^*, U_{i_0}, U_{i_1}). Let $h^{x_{i_0}}$ and $h^{x_{i_1}}$ be (a part of) secret key of U_{i_0}, and U_{i_1}, respectively. $\mathcal{B}$ sets $M_0^* := h^{x_{i_0}}$ and $M_1^* := h^{x_{i_1}}$, and sends (M_0^*, M_1^*) to $\mathcal{C}$ as the challenge messages of the linear encryption. $\mathcal{C}$ sends the challenge ciphertext C^*. $\mathcal{B}$ sets $C^* = (F_1, F_2, F_3)$, and computes the challenge group signature σ^*. Note that $\mathcal{B}$ does not know the random number (δ_1^*, δ_2^*) and $\mu \in \{0, 1\}$ such that $C^* = (h^{x_{i_\mu}} h^{\delta_1^* + \delta_2^*}, u^{\delta_1^*}, v^{\delta_2^*})$, since $(\delta_1^*, \delta_2^*, \mu)$ are chosen by $\mathcal{C}$. So, $\mathcal{B}$ uses the backpatch of the random oracle $\mathcal{H}$ for computing σ^*, and includes C^* in σ^*. Then, all values (except C^*) is independent of μ. Note that even if U_{i_μ} is revoked in the challenge interval, $\mathcal{B}$ can compute σ^*, since $\mathcal{B}$ knows msk. If either U_{i_0} or U_{i_1} is revoked in the challenge interval, this fact is not used for guessing μ under the XDH assumption, since $(g_1^{s_{v,t+1,i_\mu} + s_{i_\mu}}$, $g_1^{s_{v,t+1,i_\mu} x_{i_\mu}})$ and $(g_1^{s_{v,t+1,i_\mu} + s_{i_\mu}}, g_1^{s'_{v,t+1,i_\mu}})$ are indistinguishable.

Finally, $\mathcal{A}$ outputs the guessing bit $\mu' \in \{0, 1\}$. $\mathcal{B}$ outputs μ' as the guessing bit of the IND-CPA game of the linear encryption. □

Proof of Theorem 2

Proof. Let $\mathcal{A}_1$ be an adversary who outputs (M^*, σ^*) where for $ID_{i^*} \leftarrow$ Open(msk, M^*, σ^*), $ID_{i^*} \notin CU$ holds. As a case of the first one, let $\mathcal{A}_2$ be an adversary who outputs (M^*, σ^*) where for $ID_{i^*} \leftarrow$ Open(msk, M^*, σ^*), $ID_{i^*} \notin CU$ and $U_{i^*} \notin \{U_1, \ldots, U_N\}$ holds. In addition, let $\mathcal{A}_3$ be an adversary who outputs (M^*, σ^*) where for $ID_{i^*} \leftarrow$ Open(msk, M^*, σ^*), $ID_{i^*} \in RU$ holds. We construct an algorithm $\mathcal{B}_1$ (resp. $\mathcal{B}_2$ and $\mathcal{B}_3$) that breaks the N-HSDH assumption (resp. q-SDH assumption, where q is the number of signing queries, and the CDH assumption) by using $\mathcal{A}_1$ (resp. $\mathcal{A}_2$ and $\mathcal{A}_3$).

First, we describe $\mathcal{B}_1$. Let $g_1, h, h^{\omega_1}, \{(g_1^{\frac{1}{\omega_1 + x_i}}, h^{x_i})\}_{i=1,\ldots,N}$ be an N-HSDH instance. $\mathcal{B}_1$ selects $U_{i^*} \in \{U_1, \ldots, U_N\}$, and choose all values, except g_1, h, and $\Omega_1 := h^{\omega_1}$. $\mathcal{B}_1$ answers queries issued by $\mathcal{A}_1$ as follows:

Revocation : $\mathcal{A}_1$ requests the revocation of users $ID_{i_1}, \ldots, ID_{i_{k_t}}$ for some constant $k_t \in [1, N]$. Since $\mathcal{B}_1$ knows ω_2, $\mathcal{B}_1$ adds $ID_{i_1}, \ldots, ID_{i_{k_t}}$ to RU_t, and simply returns the result of the Revoke algorithm.

GSigning : $\mathcal{A}_1$ requests a group signature on a message M for a user U_i where $ID_i \notin CU$. Since $\mathcal{B}_1$ does not know $g_1^{x_i}$, $\mathcal{B}_1$ computes σ by using the backpatch of the random oracle $\mathcal{H}$, and gives σ to $\mathcal{A}$.

Corruption : $\mathcal{A}_1$ requests the secret key of a user U_i. If $U_i = U_{i^*}$, then $\mathcal{B}_1$ aborts. Otherwise, $\mathcal{B}_1$ sets $(g_1^{\frac{1}{\omega_1 + x_i}}, h^{x_i}) = (K_{i,1}, K_{i,2})$, chooses $s_i' \xleftarrow{\$} \mathbb{Z}_p^*$, sets $s_i' =$

$s_i x_i$, and computes $B_i = g^{s'_i}$. $\mathcal{B}_1$ adds ID_i to CU, and gives $(K_{i,1}, K_{i,2}, B_i)$ to $\mathcal{A}_1$.

Opening : Since $\mathcal{B}_1$ has (X_1, X_2), $\mathcal{B}_1$ simply returns the result of the Open algorithm.

Finally, $\mathcal{A}_1$ outputs a past interval $t^* \le t$ for the current interval t, and a pair (M^*, σ^*). By using the extractor of SPK, $\mathcal{B}_1$ gets: $(K^*_{i,1}, K^*_{i,2}, H^*_i)$, where $e(K^*_{i,1}, \Omega_1 K^*_{i,2}) = e(g_1, h)$, $e(h_{T,t,i}, K^*_{i,2}) = e(H^*_i, h)$, $F_1 = K^*_{i,2} h^{\delta_1 + \delta_2}$, $F_2 = u^{\delta_1}$, and $F_3 = v^{\delta_2}$ hold. From (F_1, F_2, F_3), $\mathcal{B}_1$ obtains i by using the Open algorithm. If $i \ne i^*$, then $\mathcal{B}_1$ aborts. Otherwise, $\mathcal{B}_1$ outputs $(K^*_{i,1}, K^*_{i,2})$ as a solution of the N-HSDH problem.

Next, we describe $\mathcal{B}_2$ that outputs a forged BBS+ signature. Let $\mathcal{C}$ be the challenger of the BBS+ signature. $\mathcal{B}_2$ is given $(g, g_1, g_2, g_3, g_4, h, \Omega_2)$ from $\mathcal{C}$. $\mathcal{B}_2$ chooses all values, except $(g, g_1, g_2, g_3, g_4, h, \Omega_2)$. For each revocation query, $\mathcal{B}_2$ issues N signing queries to $\mathcal{C}$ for obtaining $A_{.,i}$. So, $\mathcal{B}_2$ needs to issue the signing query in Nt times. For other queries, $\mathcal{B}_2$ can answer since $\mathcal{B}_2$ knows all other secret values. Finally, $\mathcal{A}_3$ outputs a past interval $t^* \le t$ for the current interval t, and a pair (M^*, σ^*). By using the extractor of SPK, $\mathcal{B}_2$ gets: $(A_{t^*,i^*}, y_{t^*,i^*}, r_{t^*,i^*})$, where $e(A_{t^*,i^*}, \Omega_2 h^{y_{t^*,i^*}}) = e(g_1^{s_{T,t^*,i^*}} g_2^{t^*} g_3^{r_{t^*,i^*}} g_4, h)$. Note that, since $U_{i^*} \notin \{U_1, \ldots, U_N\}$, $\mathcal{B}_2$ does not obtain $(A_{t^*,i^*}, y_{t^*,i^*}, r_{t^*,i^*})$ from $\mathcal{C}$. So, $\mathcal{B}_2$ outputs a forged BBS+ signature $(A_{t^*,i^*}, y_{t^*,i^*}, r_{t^*,i^*})$.

Finally, we describe $\mathcal{B}_3$ that breaks the CDH assumption. Let (g_1, g_1^a, g_1^b) be an CDH instance. $\mathcal{B}_3$ selects $U_{i^*} \in \{U_1, \ldots, U_N\}$, sets $x_{i^*} := a$ and $s_{i^*} := b$, and choose all values, except g_1 and usk_{i^*}. $\mathcal{B}_3$ answers queries issued by $\mathcal{A}_3$ as follows:

Revocation : $\mathcal{A}_3$ requests the revocation of users $ID_{i_1}, \ldots, ID_{i_{k_t}}$ for some constant k_t. Since $\mathcal{B}_3$ knows ω_2, $\mathcal{B}_3$ adds $ID_{i_1}, \ldots, ID_{i_{k_t}}$ to RU_t, and simply returns the result of the Revoke algorithm.

GSigning : $\mathcal{A}_3$ requests a group signature on a message M for a user U_i where $ID_i \notin CU$. $\mathcal{B}_3$ computes σ by using the backpatch of the random oracle $\mathcal{H}$, and gives σ to $\mathcal{A}$.

Corruption : $\mathcal{A}_3$ requests the secret key of a user U_i. If $U_i = U_{i^*}$, then $\mathcal{B}_3$ aborts. Otherwise, $\mathcal{B}_3$ adds ID_i to CU, and gives $(K_{i,1}, K_{i,2}, B_i)$ to $\mathcal{A}_3$.

Opening : Since $\mathcal{B}_3$ has (X_1, X_2), $\mathcal{B}_3$ simply returns the result of the Open algorithm.

Finally, $\mathcal{A}_3$ outputs a past interval $t^* \le t$ for the current interval t, and a pair (M^*, σ^*). By using the extractor of SPK, $\mathcal{B}_3$ gets: H^*_i, where $e(K^*_{i,1}, \Omega_1 K^*_{i,2}) = e(g_1, h)$, $e(h_{T,t,i}, K^*_{i,2}) = e(H^*_i, h)$, $F_1 = K^*_{i,2} h^{\delta_1 + \delta_2}$, $F_2 = u^{\delta_1}$, and $F_3 = v^{\delta_2}$ hold. From (F_1, F_2, F_3), $\mathcal{B}_3$ obtains i by using the Open algorithm. If $i \ne i^*$, then $\mathcal{B}_3$ aborts. Otherwise, $\mathcal{B}_3$ solves the CDH problem as follows. Since $U_i \in RL_t$, $\mathcal{B}_3$ has computed $g_1^{s_{v,t,i^*}} \cdot g_1^b = g_1^{s_{v,t,i^* + s_{i^*}}}$ and $g_1^{s'_{v,t,i^*}}$. That is, $H^*_i = B_{i^*} \cdot g_1^{s_{v,t,i^* x_i}} = g_1^{ab + a s_{v,t,i^* x_i}}$ holds. So, $\mathcal{B}_3$ outputs $H^*_i / (g_1^a)^{s_{v,t,i^*}} = g_1^{ab}$ as the solution of the CDH problem. $\square$

Generic Constructions
for Verifiable Signcryption*

Laila El Aimani

Technicolor, 1 avenue de Belle Fontaine - CS17616, 35576 Cesson-Sévigné, France

Abstract. Signcryption is a primitive which simultaneously performs
the functions of both signature and encryption in a way that is more
efficient than signing and encrypting separately. We study in this paper constructions of signcryption schemes from basic cryptographic
mechanisms; our study concludes that the known constructions require
expensive encryption in order to attain confidentiality, however some
adjustments make them rest on cheap encryption without compromising
their security. Our constructions further enjoy verifiability which entitles
the sender or the receiver to prove the validity of a signcryption with/out
revealing the *signcrypted* message. They also allow the receiver to release
some information which allows anyone to publicly verify a signcryption
on a given message. Finally, our constructions accept efficient instantiations if the building blocks belong to a wide class of signature/encryption
schemes.

Keywords: signcryption, sign-then-encrypt paradigm, commit-then-
encrypt-and sign paradigm, encrypt-then-sign paradigm, (public) verifiability, homomorphic encryption.

1 Introduction

Cryptographic mechanisms that proffer both the functionalities of signature and
of encryption are becoming nowadays increasingly important. In fact, many real-
life applications entail both the confidentiality and the authenticity/integrity of
the transmitted data; an illustrative example is electronic elections where the
voter wants to encrypt his vote to guarantee privacy, and at the same time, the
voting center needs to ensure that the encrypted vote comes from the entity
that claims to be its provenance. To respond to this need, Zheng [27] introduced
the notion of *signcryption* which is a primitive that simultaneously performs the
functions of both signature and encryption in a way that is more efficient than
signing and encrypting separately.

Related work. Since the introduction of this primitive, many constructions which
achieve different levels of security have been proposed. On a high level, security

---

* This is an extended abstract. The full version [13] is available at the Cryptology
ePrint Archive.

H. Kim (Ed): ICISC 2011, LNCS 7259, pp. 204–218, 2012.
© Springer-Verlag Berlin Heidelberg 2012

of a signcryption scheme involves two properties; privacy and authenticity. Privacy is analogous to indistinguishability in encryption schemes, and it denotes the infeasibility to infer any information about the *signcrypted* message. Authenticity is similar to unforgeability in signature schemes and it denotes the difficulty to impersonate the *signcrypter*. Defining formally those two properties is a fundamental divergence in signcryption constructions as there are many issues which come into play:

- TWO-USER VERSUS MULTI-USER SETTING In the two-user setting, adopted for instance in [1], a single sender (the entity that creates the signcryption) interacts with a single receiver (the entity that recovers the message from the signcryption). Although such a setting is too simplistic to represent the reality, e.g. the case of electronic elections, it provides however an important preliminary step towards modeling and building schemes in the multi-user setting. In fact, many works have proposed simple tweaks in order to derive multi-user security from two-user security [1,22].

- INSIDER VERSUS OUTSIDER SECURITY Another consequential difference between security models is whether the adversary is external or internal to the entities of the system. The former case corresponds to outsider security, e.g. [10], whereas the latter denotes insider security which protects the system protagonists even when some of their fellows are malicious or have compromised/lost their private keys [1,22]. It is naturally possible to mix these notions into one single signcryption scheme, i.e. insider indistinguishability and outsider unforgeability [1,6], or outsider indistinguishability and insider unforgeability [2]. However, the most frequent mix is the latter as illustrated by the number of works in the literature, e.g. [1,19,2]; it is also justified by the necessity to protect the sender from anyone trying to impersonate him including entities in the system. Insider indistinguishability is by contrast needed in very limited applications; the typical example [1] is when the adversary happens to steal the private key of the sender, thus when it is able to send "fake" messages, but we still wish to protect the privacy of the recorded signcryptions sent by the genuine sender.

- VERIFIABILITY A further requirement on signcryption is verifiability which consists in the possibility to prove efficiently the validity of a given signcryption, or to prove that a signcryption has indeed been produced on a given message. In fact, if we consider the example of electronic elections, the voting center might require from the voter a proof of validity of the "signcrypted" vote. Also, the trusted party (the receiver) that decrypts the vote might be compelled, for instance to resolve some later disputes, to prove that the sender has indeed produced the vote in question; therefore, it would be desirable to support the prover with efficient means to provide such a proof without having to disclose his private input. This property is also needed in filtering out spams in a secure email system. Although a number of constructions [3,26,7,21,25] have tackled the notion of verifiability (this notion is often referred to in the literature as public verifiability, and it denotes the possibility to release (by the receiver) some information which allows to publicly verify a signcryption with/out revealing the message in question), most of these schemes do not allow the sender to prove the validity of the created signcryption, nor allow the receiver to prove *without revealing any information, ensuring consequently non-transferability*, to a third party, the validity of a signcryption w.r.t. a given message. It is worth noting that the former need, i.e. allowing the sender to prove the validity of a signcryption without revealing the message, already manifests in the IACR electronic voting scheme (The Helios voting scheme) where the sender proves the validity of the encrypted vote to the voting manager. The scheme nonetheless does not respond to the formal security requirements of a signcryption scheme.

Before ending this paragraph, we recall the main generic constructions of signcryption schemes that were proposed so far. In fact, building complex mechanisms from basic ones is customary in cryptography as it allows achieving easy-to-analyze schemes, compared to dedicated/monolithic constructions. The first constructions of signcryption were given and analyzed in [1], where the authors study how to derive signcryption schemes, mainly in the two-user setting, using the classical combinations "sign-then-encrypt", "encrypt-then-sign", and "commit-then-encrypt-and-sign". Subsequently, the work [22] presented several optimizations of these combinations that lead to signcryptions with multi-user security. The paper shows also how to use symmetric encryption in order to derive constructions in the outsider multi-user setting. Finally, there are the recent constructions [6] which achieve security in the insider multi-user setting without key registration assumptions (on the receiver's side). It is worth noting that none of these constructions treat verifiability.

To the best of our knowledge, there are no generic constructions which provide verifiability in a reasonable security model. The main contribution of this paper is to provide such constructions.

Our Contributions. We make the following contributions. First, we propose a new model for signcryption schemes which upgrades the existing models by three interactive protocols: 1. a protocol that allows the sender to prove, to a third party, the validity of the created signcryption, 2. and two protocols that allow the receiver to prove, to a third party, the validity of a given signcryption with/out revealing the message. All these protocols do not require the provers to reveal *any information*.

In Section 3, we show that the "sign-then-encrypt" (StE) and the "commit-then-encrypt-and-sign" (CtEaS) paradigms require expensive assumptions on the underlying encryption in order to derive signcryption with outsider indistinguishability. We do this by first proving the insufficiency of OW-CCA and NM-CPA secure encryption, then by exhibiting a simple attack if the system is instantiated from certain encryption schemes. Next, we propose simple tweaks of the paradigms that make the resulting constructions rest on cheap encryption.

In Section 4, we show that the "encrypt-then-sign" ("TagEncrypt-then-sign") paradigm provides efficient constructions which are proven secure in our adopted model. We demonstrate the efficiency of these schemes by explicitly describing the different verification protocols if the constructions are instantiated from a wide class of encryption (tag-based encryption) schemes.

Finally, in Section 5, we propose a new paradigm which combines the merits of both the "sign-then-encrypt" (StE) and "encrypt-then-sign" (EtS) paradigms while avoiding their drawbacks. In fact, the former (both the old and the new variant) suffers the problem of verifiability. The latter suffers the recourse to stronger security assumptions on the underlying signature. Moreover, the paradigm does not provide anonymity of the sender. In this section, we show that our new proposed paradigm, called "encrypt-then-sign-then-encrypt" (EtStE) circumvents these problems while accepting many efficient instantiations.

2 Model and Main Constructions

A verifiable signcryption scheme consists of the following algorithms/protocols:

Setup (setup(1^κ)). This probabilistic algorithm inputs a security parameter κ, and generates the public parameters param of the signcryption scheme.

Key generation (keygen$_U(1^\kappa$, param), $U \in \{S, R\}$). This probabilistic algorithm inputs the security parameter κ and the public parameters param, and outputs a key pair ($\mathsf{pk}_U, \mathsf{sk}_U$) for the system user U which is either the sender S or the receiver R.

Signcryption (signcrypt($m, \mathsf{sk}_S, \mathsf{pk}_S, \mathsf{pk}_R$)). This probabilistic algorithm inputs a message m, the key pair ($\mathsf{sk}_S, \mathsf{pk}_S$) of the sender, the public key pk_R of the receiver, and outputs the signcryption μ of the message m.

Proof of validity (proveValidity($\mu, \mathsf{pk}_S, \mathsf{pk}_R$)). This is an interactive protocol between the receiver or the sender who has just generated a signcryption μ on some message, and any verifier: the sender uses the randomness used to create μ (as private input) and the receiver uses his private key sk_R in order to convince the verifier that μ is a valid signcryption on some message. The common input to both the prover and the verifier comprise the signcryption μ in question, pk_S, and pk_R. At the end of the protocol, the verifier either accepts or rejects the proof.

Unsigncryption (unsigncrypt($\mu, \mathsf{sk}_R, \mathsf{pk}_R, \mathsf{pk}_S$)). This is a deterministic algorithm which inputs a putative signcryption μ on some message, the key pair ($\mathsf{sk}_R, \mathsf{pk}_R$) of the receiver, and the public key pk_S of the sender, and outputs either the message underlying μ or an error symbol $\perp$.

Confirmation/Denial ($\{\mathsf{confirm}, \mathsf{deny}\}(\mu, m, \mathsf{pk}_R, \mathsf{pk}_S$)). These are interactive protocols between the receiver and any verifier; the receiver uses his private key sk_R (as private input) to convince any verifier that a signcryption μ on some message m is/is not valid. The common input comprises the signcryption μ and the message m in question, in addition to pk_R and pk_S. At the end of the protocol, the verifier is either convinced of the validity/invalidity of μ w.r.t. m or not.

Public verification (publicVerify($\mu, m, \mathsf{sk}_R, \mathsf{pk}_R, \mathsf{pk}_S$)). This is an algorithm which inputs a signcryption μ, a message m, the key pair ($\mathsf{sk}_R, \mathsf{pk}_R$) of the receiver, and the public key pk_S of the sender, and outputs either an error symbol $\perp$ if μ is not a valid signcryption on m, or a string which allows to publicly verify the validity of μ on m otherwise.

It is natural to require the correctness of a signcryption scheme:

$$\text{unsigncrypt}(\text{signcrypt}(m, \text{sk}_S, \text{pk}_S, \text{pk}_R), \text{sk}_R, \text{pk}_R, \text{pk}_S) = m.$$

$$\text{publicVerify}(m, \text{signcrypt}(m, \text{sk}_S, \text{pk}_S, \text{pk}_R), \text{sk}_R, \text{pk}_R, \text{pk}_S) \neq \perp.$$

Moreover, the protocols proveValidity and {confirm, deny} must be complete, sound, and zero knowledge. We refer to [15] for details of these notions.

2.1 Unforgeability

This notion protects the sender's authenticity from *malicious insider* adversaries, i.e. the receiver. It is defined through a game between a challenger $\mathcal{C}$ and an adversary $\mathcal{A}$ where the latter gets the public key pk_S of the sender, generates the key pair $(\text{pk}_R, \text{sk}_R)$ of the receiver, and hands pk_R to the challenger. During the game, $\mathcal{A}$ is allowed to ask adaptively signcryption queries w.r.t. pk_R and pk_S on messages of his choice to $\mathcal{C}$. The scheme is said to be *Existentially Unforgeable against Chosen Message Attacks (EUF-CMA)* if the adversary is unable to produce a valid signcryption μ^* on a message m^* that he did not ask to the signcryption oracle.

Definition 1 (Unforgeability). *We consider a signcryption scheme sc given by the algorithms/protocols defined earlier in this section. Let $\mathcal{A}$ be a PPTM. We consider the following random experiment:*

Experiment $\mathbf{Exp}_{\text{sc},\mathcal{A}}^{\text{euf-cma}}(1^\kappa)$

$\text{param} \leftarrow \text{sc.setup}(1^\kappa)$
$(\text{pk}_S, \text{sk}_S) \leftarrow \text{sc.keygen}_S(1^\kappa, \text{param})$
$\text{pk}_R \leftarrow \mathcal{A}(\text{pk}_S)$
$\mu^* \leftarrow \mathcal{A}^{\mathfrak{S}}(\text{pk}_S, \text{pk}_R, \text{sk}_R)$
$\qquad \mathfrak{S} : m \longmapsto \text{sc.signcrypt}\{\text{sk}_S, \text{pk}_S, \text{pk}_R\}(m)$
return 1 if and only if the following properties are satisfied:
$\quad$ - $\text{sc.unsigncrypt}_{\{\text{sk}_R, \text{pk}_R, \text{pk}_S\}}[\mu^*] = m^*$
$\quad$ - m^* *was not queried to* $\mathfrak{S}$

We define the success *of $\mathcal{A}$ via:*

$$\mathbf{Succ}_{\text{sc},\mathcal{A}}^{\text{euf-cma}}(1^\kappa) = \Pr\left[\mathbf{Exp}_{\text{sc},\mathcal{A}}^{\text{euf-cma}}(1^\kappa) = 1\right].$$

Given $(t, q_s) \in \mathbb{N}^2$ and $\varepsilon \in [0, 1]$, $\mathcal{A}$ is called a (t, ε, q_s)-EUF-CMA adversary against sc if, running in time t and issuing q_s queries to the sc.signcrypt oracle, $\mathcal{A}$ has $\mathbf{Succ}_{\text{sc},\mathcal{A}}^{\text{euf-cma}}(1^\kappa) \geq \varepsilon$. The scheme sc is said to be (t, ε, q_s)-EUF-CMA secure if no (t, ε, q_s)-EUF-CMA adversary against it exists.

Remark 1. Note that $\mathcal{A}$ in the above definition is not given the oracles sc.proveValidity, sc.unsigncrypt, sc.publicVerify, and sc.{confirm, deny}. In fact, these oracles are useless for him as he has the receiver's private key sk_R at his disposal.

2.2 Indistinguishability

This notion protects the sender's privacy from *outsider adversaries*. It is defined through a game between a challenger $\mathcal{C}$ and an adversary $\mathcal{A}$; $\mathcal{C}$ generates the key pairs $(\mathsf{sk}_S, \mathsf{pk}_S)$ and $(\mathsf{sk}_R, \mathsf{pk}_R)$ for the sender and for the receiver respectively, and hands $(\mathsf{pk}_S, \mathsf{pk}_R)$ to $\mathcal{A}$. During the first phase of the game, $\mathcal{A}$ queries adaptively signcrypt and proveValidity (actually proveValidity is only invoked on inputs just obtained from the signcryption oracle), unsigncrypt, $\{\mathsf{confirm}, \mathsf{deny}\}$, and publicVerify for any input. Once $\mathcal{A}$ decides that this phase is over, he generates two messages $m_0^\star, m_1^\star$ and hands them to $\mathcal{C}$ who generates a signcryption $\mu^\star$ on $m_b^\star$ for $b \xleftarrow{R} \{0,1\}$ and gives it $(\mu^\star)$ to $\mathcal{A}$. The latter resumes querying the previous oracles adaptively on any input with the exception of not querying unsigncrypt on $\mu^\star$, and $\{\mathsf{confirm}, \mathsf{deny}\}$ and publicVerify on the pair $(\mu^\star, m_i^\star)$ for $i \in \{0,1\}$. At the end, the adversary outputs his guess b' for the message underlying the signcryption $\mu^\star$. He is considered successful if $b = b'$.

Definition 2 (Indistinguishability (IND-CCA)). *Let* sc *be a signcryption scheme, and let* $\mathcal{A}$ *be a PPTM. We consider the following random experiment for* $b \xleftarrow{R} \{0,1\}$:

Experiment $\mathbf{Exp}_{\mathsf{sc},\mathcal{A}}^{ind\text{-}cca\text{-}b}(1^\kappa)$

param $\leftarrow$ sc.setup(1^κ)
$(\mathsf{sk}_S, \mathsf{pk}_S) \leftarrow$ sc.keygen$_S(1^\kappa, \mathsf{param})$
$(\mathsf{sk}_R, \mathsf{pk}_R) \leftarrow$ sc.keygen$(1^\kappa, \mathsf{param})$
$(m_0^\star, m_1^\star, \mathcal{I}) \leftarrow \mathcal{A}^{\mathfrak{S}, \mathfrak{V}, \mathfrak{U}, \mathfrak{C}}(\textit{find}, \mathsf{pk}_S, \mathsf{pk}_R)$
$\qquad\Big|\;\; \mathfrak{S} : m \longmapsto$ sc.signcrypt$_{\{\mathsf{sk}_S, \mathsf{pk}_S, \mathsf{pk}_R\}}(m)$
$\qquad\Big|\;\; \mathfrak{V} : \mu \longmapsto$ sc.proveValidity$(\mu, \mathsf{pk}_S, \mathsf{pk}_R)$
$\qquad\Big|\;\; \mathfrak{U} : \mu \longmapsto$ sc.unsigncrypt$_{\mathsf{sk}_R, \mathsf{pk}_R, \mathsf{pk}_S}(\mu)$
$\qquad\Big|\;\; \mathfrak{C} : (\mu, m) \longmapsto$ sc.$\{\mathsf{confirm}, \mathsf{deny}\}(\mu, m, \mathsf{pk}_R, \mathsf{pk}_S)$
$\qquad\Big|\;\; \mathfrak{P} : (\mu, m) \longmapsto$ sc.publicVerify$(\mu, m, \mathsf{pk}_R, \mathsf{pk}_S)$
$\mu^\star \leftarrow$ sc.signcrypt$_{\{\mathsf{sk}_S, \mathsf{pk}_S, \mathsf{pk}_R\}}(m_b^\star)$
$d \leftarrow \mathcal{A}^{\mathfrak{S}, \mathfrak{V}, \mathfrak{U}, \mathfrak{C}}(\textit{guess}, \mathcal{I}, \mu^\star, \mathsf{pk}_S, \mathsf{pk}_C)$
$\qquad\Big|\;\; \mathfrak{S} : m \longmapsto$ sc.signcrypt$_{\{\mathsf{sk}_S, \mathsf{pk}_S, \mathsf{pk}_R\}}(m)$
$\qquad\Big|\;\; \mathfrak{V} : \mu \longmapsto$ sc.proveValidity$(\mu, \mathsf{pk}_S, \mathsf{pk}_R)$
$\qquad\Big|\;\; \mathfrak{U} : \mu(\neq \mu^\star) \longmapsto$ sc.unsigncrypt$_{\mathsf{sk}_R, \mathsf{pk}_R, \mathsf{pk}_S}(\mu)$
$\qquad\Big|\;\; \mathfrak{C} : (\mu, m)(\neq (\mu^\star, m_i^\star), i = 0, 1) \longmapsto$ sc.$\{\mathsf{confirm}, \mathsf{deny}\}(\mu, m, \mathsf{pk}_R, \mathsf{pk}_S)$
$\qquad\Big|\;\; \mathfrak{P} : (\mu, m)(\neq (\mu^\star, m_i^\star), i = 0, 1) \longmapsto$ sc.publicVerify$(\mu, m, \mathsf{pk}_R, \mathsf{pk}_S)$

Return d

We define the advantage *of* $\mathcal{A}$ *via:*

$$\mathbf{Adv}_{\mathsf{sc},\mathcal{A}}^{ind\text{-}cca}(1^\kappa) = \left| \Pr\left[\mathbf{Exp}_{\mathsf{sc},\mathcal{A}}^{ind-cca-b}(1^\kappa) = b \right] - \frac{1}{2} \right|.$$

Given $(t, q_s, q_v, q_u, q_{cd}, q_{pv}) \in \mathbb{N}^6$ *and* $\varepsilon \in [0, 1]$, $\mathcal{A}$ *is called a* $(t, \varepsilon, q_s, q_v, q_u, q_{cd}, q_{pv})$-IND-CCA *adversary against* sc *if, running in time* t *and issuing* q_s *queries to the* sc.signcrypt *oracle,* q_v *queries to the* sc.proveValidity *oracle,* q_u *queries to the* sc.unsigncrypt *oracle,* q_{cd} *queries to the* sc.$\{\mathsf{confirm}, \mathsf{deny}\}$

oracle, and q_{pv} *to the* publicVerify *oracle,* A *has* $\mathbf{Adv}_{sc,A}^{ind-cca}(1^\kappa) \geq \varepsilon$. *The scheme* sc *is said to be* $(t, \varepsilon, q_s, q_v, q_u, q_{cd}, q_{pv})$-*IND-CCA secure if no* $(t, \varepsilon, q_s, q_v, q_u, q_{cd}, q_{pv})$-*IND-CCA adversary against it exists.*

In the full version [13], we provide the above properties in the multi-user setting, namely the dM-EUF-CMA and the fM-IND-CCA security properties, where the adversary is further given all the private keys except those of the target sender and of the target receiver.

2.3   Main Constructions

Let Σ be a digital signature scheme given by Σ.keygen which generates a key pair $(\Sigma.\mathsf{sk}, \Sigma.\mathsf{pk})$, Σ.sign, and Σ.proveValidity. Let furthermore Γ denote a public key encryption scheme described by Γ.keygen that generates the key pair $(\Gamma.\mathsf{sk}, \Gamma.\mathsf{pk})$, Γ.encrypt, and Γ.decrypt. Finally, let Ω be a commitment scheme given by the algorithms Ω.commit and Ω.open. The most popular paradigms used to devise signcryption schemes from basic primitives are:

- The *"sign-then-encrypt" (StE) paradigm.* Given a message m, signcrypt first produces a signature σ on the message using $\Sigma.\mathsf{sk}$, then encrypts $m\|\sigma$ under $\Gamma.\mathsf{pk}$. The result forms the signcryption on m. To unsigncrypt, one first decrypts the signcryption using $\Gamma.\mathsf{sk}$ in $m\|\sigma$, then checks the validity of σ, using $\Sigma.\mathsf{pk}$, on m. Finally, publicVerify of a valid signcryption $\mu = \Gamma.\mathsf{encrypt}(m\|\sigma)$ on m outputs σ.
- The *"encrypt-then-sign" (EtS) paradigm.* Given a message m, signcrypt produces an encryption e on m using $\Gamma.\mathsf{pk}$, then produces a signature σ on e using $\Sigma.\mathsf{sk}$; the signcryption is the pair (e, σ). To unsigncrypt such a signcryption, one first checks the validity of σ w.r.t. e using $\Sigma.\mathsf{pk}$, then decrypts e using $\Gamma.\mathsf{sk}$ to get m. Finally, publicVerify outputs a zero knowledge non-interactive (NIZK) proof that m is the decryption of e; such a proof is possible since the statement in question is in NP ([16] and [4]).
- The *"commit-then-encrypt-and-sign" (CtEaS) paradigm.* This construction has the advantage of performing the signature and the encryption *in parallel* in contrast to the previous sequential compositions. Given a message m, one first produces a commitment c on it using some random nonce r, then encrypts $m\|r$ under $\Gamma.\mathsf{pk}$, *and* produces a signature σ on c using $\Sigma.\mathsf{sk}$. The signcryption is the triple (e, c, σ). To unsigncrypt such a signcryption, one first checks the validity of σ w.r.t. c, then decrypts e to get $m\|r$, and finally checks the validity of the commitment c w.r.t (m, r). publicVerify is achieved by releasing the decryption of e, namely $m\|r$.

The proofs of well (mal) formed-ness, namely proveValidity and {confirm, deny} can be carried out since the languages in question are in NP (co-NP) and thus accept zero knowledge proof systems [16]. Finally, it is possible to require a proof in the publicVerify algorithms of StE and CtEaS, that the revealed information is indeed a correct decryption of the encryption in question; such a proof is again possible to issue since the corresponding statement is in NP.

3   Analysis of the StE and CtEaS Paradigms

3.1   Insufficiency of OW-CCA and NM-CPA Secure Encryption

We proceed in this subsection as in [11] where the author shows the impossibility to derive secure confirmer signatures, using the StE and the CtEaS paradigms, from both OW-CCA and NM-CPA secure encryption; we first show the impossibility result for the so-called *key-preserving reductions*, i.e. reductions which launch the adversary on its challenge public key in addition to some freely chosen parameters, then we generalize the result to arbitrary reductions assuming new assumptions on the underlying encryption scheme.

Lemma 1. *Assume there exists a key-preserving reduction $\mathcal{R}$ that converts an IND-CCA adversary $\mathcal{A}$ against signcryptions from the StE (CtEaS) paradigm to a OW-CCA (NM-CPA) adversary against the underlying encryption scheme. Then, there exists a meta-reduction $\mathcal{M}$ that OW-CCA (NM-CPA) breaks the encryption scheme in question.*

Moreover, we can rule out the OW-CPA, OW-PCA, and IND-CPA notions by remarking that ElGamal's encryption meets all those notions (under different assumptions), but cannot be employed in StE and CtEaS as it is malleable. We refer again to [13] for further details.

In consequence of the above analysis, the used encrypted scheme has to satisfy at least IND-PCA security in order to lead to secure signcryption from StE or CtEaS. Since there are no known encryption schemes in the literature which separate the notions IND-PCA and IND-CCA, our result practically means that the encryption scheme underlying the previous constructions has to satisfy the highest security level (IND-CCA) in order to lead to secure signcryption. This translates in expensive operations, especially if verifiability is further required for the resulting signcryption.

3.2   Positive Results

Constructions from StE or CtEaS suffer the strong forgeability: given a signcryption on some message, one can create another valid signcryption on the same message without the sender's help. To circumvent this problem, we propose the following techniques which bind the digital signature to the resulting signcryption.

The New *"Sign-then-Encrypt"(StE) Paradigm.* Let Σ be a digital signature scheme given by Σ.keygen, which generates a key pair $(\Sigma.\text{sk}, \Sigma.\text{pk})$, Σ.sign, and Σ.proveValidity. Let furthermore $\mathcal{K}$ be a KEM given by $\mathcal{K}$.keygen, which generates a key pair $(\mathcal{K}.\text{pk}, \mathcal{K}.\text{sk})$, $\mathcal{K}$.encap, and $\mathcal{K}$.decap. Finally, we consider a DEM $\mathcal{D}$ given by $\mathcal{D}$.encrypt and $\mathcal{D}$.decrypt. We assume that the message space of $\mathcal{D}$ includes the concatenation of elements from the message space of Σ, and of signatures produced by Σ, and that the encapsulations generated by $\mathcal{K}$ are exactly κ-bit long, where κ is a security parameter.

A signcryption scheme sc is defined as follows: sc.setup invokes the setup algorithms of Σ, $\mathcal{K}$, and $\mathcal{D}$. sc.keygen$_S$ and sc.keygen$_R$ consist of Σ.keygen and $\mathcal{K}$.keygen respectively. To sc.signcrypt a message m, one first generates a key k with its encapsulation c using $\mathcal{K}$.encap, then produces a signature σ on $c\|m$, and finally outputs $\mu = (c, \mathcal{D}.\text{encrypt}_k(m\|\sigma))$ as a signcryption of m. Unsigncryption of some (μ_1, μ_2) is done by first recovering the key k from μ_1 using $\mathcal{K}$.decap, then using $\mathcal{D}$.decrypt and k to decrypt μ_2, and finally checking that the result is a valid digital signature on $\mu_1\|m$ where m is the retrieved message. The rest is similar to the original StE.

Theorem 1. *Given $(t, q_s) \in \mathbb{N}^2$ and $\varepsilon \in [0, 1]$, the above construction is (t, ϵ, q_s)-EUF-CMA secure if the underlying digital signature scheme is (t, ϵ, q_s)-EUF-CMA secure.*

Theorem 2. *Given $(t, q_s, q_v, q_u, q_{cd}, q_{pv}) \in \mathbb{N}^6$ and $(\varepsilon, \epsilon') \in [0, 1]^2$, the above construction is $(t, \epsilon, q_s, q_v, q_u, q_{cd}, q_{pv})$-IND-CCA secure if it uses a (t, ϵ', q_s)-EUF-CMA secure digital signature, an IND-OT secure DEM and an $(t + q_s(q_u + q_{cd} + q_{pv}), \epsilon \cdot (1 - \epsilon')^{q_u + q_{cd} + q_{pv}})$-IND-CPA secure KEM.*

The new "Commit-then-Encrypt-and-Sign" (CtEaS) paradigm The new "commit-then-encrypt-and-sign" (CtEaS) paradigm. The construction is similar to the basic one described earlier, with the exception of producing the digital signature on both the commitment c and the encryption e. The new construction looses the parallelism of the original one, i.e. encryption and signature can longer be carried out in parallel, however it has the advantage of resting on cheap encryption compared to the basic one.

Theorem 3. *Given $(t, q_s) \in \mathbb{N}^2$ and $(\varepsilon, \epsilon_b) \in [0, 1]^2$, the above construction is (t, ϵ, q_s)-EUF-CMA secure if it uses a uses a (t, ϵ_b) binding commitment scheme and a $(t, \epsilon(1 - \epsilon_b)^{q_s}, q_s)$-EUF-CMA secure digital signature scheme.*

Theorem 4. *Given $(t, q_s, q_v, q_u, q_{cd}, q_{pv}) \in \mathbb{N}^6$ and $(\varepsilon, \epsilon', \epsilon_h) \in [0, 1]^3$, the new CtEaS construction is $(t, \epsilon, q_s, q_v, q_u, q_{cd}, q_{pv})$-IND-CCA secure if it uses a (t, ϵ', q_s)-SEUF-CMA secure digital signature, a statistically binding, and (t, ϵ_h)-hiding commitment, and a $(t + q_s(q_u + q_{cd} + q_{pv}), \frac{1}{2}(\epsilon + \epsilon_h)(1 - \epsilon')^{q_u + q_{cd} + q_{pv}})$-IND-CPA secure encryption scheme.*

4   Efficient Verifiable Signcryption from the EtS Paradigm

The EtS paradigm turns out to provide efficient signcryptions schemes that are proven secure in the model we adhere to.

Theorem 5. *Given $(t, q_s) \in \mathbb{N}^2$ and $\varepsilon \in [0, 1]$, signcryption schemes from EtS are (t, ϵ, q_s)-EUF-CMA secure if the underlying digital signature scheme is (t, ϵ, q_s)-EUF-CMA secure.*

Theorem 6. *Given* $(t, q_s, q_v, q_u, q_{cd}, q_{pv}) \in \mathbb{N}^6$ *and* $(\varepsilon, \epsilon') \in [0,1]^2$, *signcryptions from EtS are* $(t, \epsilon, q_s, q_v, q_u, q_{cd}, q_{pv})$-*IND-CCA secure if they use a* (t, ϵ', q_s)-*SEUF-CMA secure digital signature and a* $(t + q_s(q_u + q_{cd} + q_{pv}), \epsilon(1 - \epsilon')^{q_u + q_{cd} + q_{pv}})$-*IND-CPA secure encryption scheme*

4.1   Efficient Instantiations

To allow efficient proveValidity, {confirm, deny}, and publicVerify protocols/algorithms, we propose to instantiate the encryption scheme from the class $\mathbb{E}$ defined in [12] which includes most homorphic encryption, e.g. ElGamal's encryption [14], the encryption scheme defined in [5], or Paillier's [23] encryption scheme.

We describe in the following proveValidity, {confirm, deny}, and publicVerify protocols/algorithms if the used encryption belongs to the class $\mathbb{E}$.

Proof of Validity. We depict the proveValidity protocol in Figure 1.

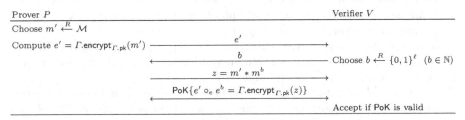

Fig. 1. Proof system for membership to the language $\{m : e = \Gamma.\mathsf{encrypt}_{\Gamma.\mathsf{pk}}(m)\}$ Common input: $(e, \Gamma.\mathsf{pk})$ and Private input: m and $\Gamma.\mathsf{sk}$ or randomness used to produce e.

Theorem 7. *Let Γ be a one-way encryption scheme from the class $\mathbb{E}$. The protocol depicted in Figure 1 is a ZK proof of knowledge of the decryption of e.*

Confirmation/Denial Protocols. The confirm protocol is nothing but the proof PoK which is in case of [14,5] a proof of equality of two discrete logarithms, and in case of [23] a proof of knowledge of an N-th root. We depict the deny protocol in Figure 2, where f denotes an arbitrary *homomorphic injective one way function*:

$$\forall m, m' : f(m \star m') = f(m) \circ f(m')$$

Theorem 8. *Let Γ be an IND-CPA encryption scheme from the above class $\mathbb{E}$. The protocol depicted in Figure 2 is a ZK proof of the decryption of e which is different from the message m.*

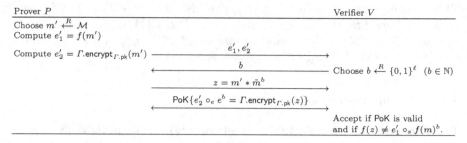

Fig. 2. Proof system for membership to the language $\{(m, e): \exists \tilde{m} : e = \Gamma.\mathsf{encrypt}(\tilde{m}) \wedge \tilde{m} \neq m\}$ Common input: $(m, e, \Gamma.\mathsf{pk})$ and Private input: $\Gamma.\mathsf{sk}$ or randomness encrypting $\tilde{m}$ in e

Public Verification. the publicVerify algorithm outputs a ZK non-interactive proof of the correctness of a decryption. We note the following three solutions according to the used encryption:

1. The case of Paillier's encryption [23]: this scheme belongs to *fully decryptable* encryption schemes, i.e. encryption schemes where decryption leads to the randomness used to produce the ciphertext. Thus, publicVerify will simply release the randomness used to generate the ciphertext.
2. The case of [5]'s encryption: Groth and Sahai [17] presented an efficient ZK non-interactive proof that a given encryption using this scheme encrypts a given message under a given public key.
3. The case of DL-based encryption schemes, e.g. [14,5,8]: the interactive proof of correctness of most such schemes reduces to a proof of equality of two discrete logarithms. The work [9] presented an efficient method to remove interaction using additively homomorphic encryption, e.g. Paillier [23].

4.2   Extension to Multi-user Security

The construction is the same provided in [22], namely the TagEncrypt-then-Sign paradigm (TEtS), which deviates from the standard EtS paradigm as follows:

1. It considers a tag-based encryption scheme where the tag is set to the public key of the sender pk_S.
2. The digital signature is produced on the resulting ciphertext and on the public key of the receiver.

Theorem 9. *Given $(t, q_s) \in \mathbb{N}^2$ and $\varepsilon \in [0, 1]$, signcryption schemes from the TEtS paradigm are (t, ϵ, q_s)-dM-EUF-CMA secure if the underlying digital signature scheme is (t, ϵ, q_s)-EUF-CMA secure.*

Theorem 10. *Given $(t, q_s, q_v, q_u, q_{cd}, q_{pv}) \in \mathbb{N}^6$ and $(\varepsilon, \epsilon') \in [0, 1]^2$, signcryption constructions from the TEtS paradigm are $(t, \epsilon, q_s, q_v, q_u, q_{cd}, q_{pv})$-IND-CCA secure if they use a (t, ϵ', q_s)-SEUF-CMA secure digital signature and a $(t + q_s(q_u + q_{cd} + q_{pv}), \epsilon(1 - \epsilon')^{q_u + q_{cd} + q_{pv}}, q_u + q_{cd} + q_{pv})$-IND-sTag-CCA secure tag-based encryption scheme.*

5 Efficient Verifiable Signcryption from the EtStE Paradigm

The EtS technique provides efficient verifiability but at the expense of the sender's anonymity, and of the security requirements on the building blocks. StE achieves better privacy using cheap constituents but at the expense of verifiability. It would be nice to have a technique that combines the merits of both paradigms while avoiding their drawbacks. This is the main contribution in this section; the core of the idea consists in first encrypting the message to be signcrypted using a public key encryption scheme, then applying the StE paradigm to the produced encryption. The result of this operation in addition to the encrypted message form the new signcryption of the message in question. In other terms, this technique can be seen as a merge between EtS and StE; thus we can term it the "encrypt-then-sign-then-encrypt" paradigm (EtStE).

5.1 The Construction

Setup. Consider a signature scheme Σ, an encryption scheme Γ, and another encryption scheme $(\mathcal{K}, \mathcal{D})$ derived from the KEM/DEM paradigm. Next, on input the security parameter $\kappa = (\kappa_1, \kappa_2, \kappa_3)$, generate the parameters param of these schemes. We assume that signatures issued with Σ can be written as (r, s), where r reveals no information about the signed message nor about the public signing key, and s represents the "significant" part of the signature.

Key generation. On input the security parameter κ and the public parameters param, invoke the key generation algorithms of the building blocks and set the sender's key pair to $(\Sigma.\mathsf{pk}, \Sigma.\mathsf{sk})$, and the receiver's key pair to $(\{\Gamma.\mathsf{pk}, \mathcal{K}.\mathsf{pk}\}, \{\Gamma.\mathsf{sk}, \mathcal{K}.\mathsf{sk}\})$.

Signcrypt. On a message m, produce an encryption $e = \Gamma.\mathsf{encrypt}_{\Gamma.\mathsf{pk}}(m)$ of m. Then fix a key k along with its encapsulation c using $\mathcal{K}.\mathsf{encrypt}_{\mathcal{K}.\mathsf{pk}}$, produce a signature (r, s) on $c\|e$, and finally encrypt s with k using $\mathcal{D}.\mathsf{encrypt}$. The signcryption of m is the tuple $(e, c, \mathcal{D}.\mathsf{encrypt}_k(s), r)$.

Prove Validity. Given a signcryption $\mu = (\mu_1, \mu_2, \mu_3, \mu_4)$ on a message m, the prover proves knowledge of the decryption of μ_1, and of the decryption of (μ_2, μ_3), which together with μ_4 forms a valid digital signature on $\mu_2\|\mu_1$. The private input is either the randomness used to create μ or $\{\Gamma.\mathsf{sk}, \mathcal{K}.\mathsf{sk}\}$.

Unsigncrypt. On a signcryption a $(\mu_1, \mu_2, \mu_3, \mu_4)$, compute $m = \Gamma.\mathsf{decrypt}_{\Gamma.\mathsf{sk}}(\mu_1)$ and $k = \mathcal{K}.\mathsf{decapsulate}_{\mathcal{K}.\mathsf{sk}}(\mu_2)$. Check whether $(\mathcal{D}.\mathsf{decrypt}_k(\mu_3), \mu_4)$ is valid signature on $\mu_2\|\mu_1$; if yes then output m, otherwise output $\bot$.

Confirm/Deny. On input a putative signcryption $\mu = (\mu_1, \mu_2, \mu_3, \mu_4)$ on a message m, use the receiver's private key to prove that m is/isn't the decryption of μ_1, and prove knowledge of the decryption of (μ_2, μ_3), which together with μ_4 forms a valid/invalid digital signature on $\mu_2\|\mu_1$.

Public Verify. On a valid signcryption $\mu = (\mu_1, \mu_2, \mu_3, \mu_4)$ on a message m, output a ZK non-interactive proof that μ_1 encrypts m, in addition to $(\mathcal{D}.\mathsf{decrypt}_{\mathcal{K}.\mathsf{decap}(\mu_2)}(\mu_3), \mu_4)$.

5.2   Analysis

Theorem 11. *Given* $(t, q_s) \in \mathbb{N}^2$ *and* $\varepsilon \in [0, 1]$, *the above construction is* (t, ϵ, q_s)-*EUF-CMA secure if the underlying signature is* (t, ϵ, q_s)-*EUF-CMA secure.*

Theorem 12. *Given* $(t, q_s, q_v, q_u, q_{cd}, q_{pv}) \in \mathbb{N}^6$ *and* $(\varepsilon, \epsilon') \in [0, 1]^2$, *the construction proposed above is* $(t, \epsilon, q_s, q_v, q_u, q_{cd}, q_{pv})$-*IND-CCA secure if it uses a* (t, ϵ', q_s)-*EUF-CMA secure signature, an IND-CPA secure encryption, an IND-OT secure DEM, and a* $(t + q_s(q_u + q_{cd} + q_{pv}), \frac{\epsilon(1-\epsilon')^{q_{cd}+q_u+q_{pv}}}{2})$-*IND-CPA secure KEM.*

Our aim in the rest of this paragraph consists in identifying suitable classes of encryption/signature schemes that renders the proveValidity and {confirm, deny} efficient. These protocols comprise the following sub-protocols:

1. Proving knowledge of the decryption of a ciphertext produced using the encryption scheme Γ.
2. Proving that a message is/isn't the decryption of a certain ciphertext produced using Γ.
3. Proving knowledge of the decryption of a ciphertext produced using $(\mathcal{K}, \mathcal{D})$, and that this decryption forms a valid/invalid digital signature, issued using Σ, on some known string.

It is natural to instantiate the encryption scheme Γ from the class $\mathbb{E}$. The first two sub-protocols can be efficiently carried out using the proofs depicted in Figure 1 and Figure 2. For the last sub-protocol, one can consider encryption schemes from the class $\mathbb{E}$ that are derived from the KEM/DEM paradigm, in addition to signature schemes that accept efficient proofs of knowledge. See the full version [13] for further details.

5.3   Extension to Multi-user Security

The above EtStE technique can be extended to achieve security in the multi-user setting by applying the standard techniques [1,22]. More specifically, one considers a tag-based encryption scheme Γ, a tag-based KEM $\mathcal{K}$, a DEM $\mathcal{D}$, an a signature scheme. The sender's key pair is the signature scheme key pair, whereas the receiver's key pair comprise both key pairs of Γ and $\mathcal{K}$. Signcryption on a message m w.r.t. a sender's public key Σ.pk and a receiver's public key $(\Gamma.\text{pk}, \mathcal{K}.\text{pk})$ is generated as follows. First compute an encryption e on m (with Γ) w.r.t. the tag Σ.pk, then generate a key k and its encapsulation c w.r.t. the same tag (with $\mathcal{K}$), then compute a digital signature on $c\|e\|\{\Gamma.\text{pk}, \mathcal{K}.\text{pk}\}$, and finally sign the "significant" part of this signature using k. The signcryption consists of the result of this encryption, the remaining part of the signature, and (e, c). The rest is similar to the paradigm in the two-user setting.

Theorem 13. *Given* $(t, q_s) \in \mathbb{N}^2$ *and* $\varepsilon \in [0, 1]$, *the above construction is* (t, ϵ, q_s)-*dM-EUF-CMA secure if the underlying digital signature scheme is* (t, ϵ, q_s)-*EUF-CMA secure.*

Theorem 14. *Given* $(t, q_s, q_v, q_u, q_{cd}, q_{pv}) \in \mathbb{N}^6$ *and* $(\varepsilon, \epsilon') \in [0,1]^2$, *the above construction is* $(t, \epsilon, q_s, q_v, q_u, q_{cd}, q_{pv})$-*fM-IND-CCA secure if it uses a* (t, ϵ', q_s)-*EUF-CMA secure digital signature, an IND-sTag-CCA secure encryption, an IND-OT secure DEM, and a* $(t + q_s(q_u + q_{cd} + q_{pv}), \frac{\epsilon(1-\epsilon')^{q_{cd}+q_u+q_{pv}}}{2}, q_{cd} + q_u + q_{pv})$-*IND-sTag-CCA secure KEM.*

References

1. An, J.H., Dodis, Y., Rabin, T.: On the Security of Joint Signature and Encryption. In: Knudsen, L.R. (ed.) EUROCRYPT 2002. LNCS, vol. 2332, pp. 83–107. Springer, Heidelberg (2002)
2. Baek, J., Steinfeld, R., Zheng, Y.: Formal Proofs for the Security of Signcryption. J. Cryptology 20(2), 203–235 (2007)
3. Bao, F., Deng, R.H.: A Signcryption Scheme with Signature Directly Verifiable by Public Key. In: Imai, H., Zheng, Y. (eds.) PKC 1998. LNCS, vol. 1431, pp. 55–59. Springer, Heidelberg (1998)
4. Blum, M., Feldman, P., Micali, S.: Non-Interactive Zero-Knowledge and Its Applications (Extended Abstract). In: Simon, J. (ed.) STOC, pp. 103–112. ACM Press (1988)
5. Boneh, D., Boyen, X., Shacham, H.: Short Group Signatures. In: Franklin, M. (ed.) CRYPTO 2004. LNCS, vol. 3152, pp. 41–55. Springer, Heidelberg (2004)
6. Chiba, D., Matsuda, T., Schuldt, J.C.N., Matsuura, K.: Efficient Generic Constructions of Signcryption with Insider Security in the Multi-user Setting. In: Lopez, J., Tsudik, G. (eds.) ACNS 2011. LNCS, vol. 6715, pp. 220–237. Springer, Heidelberg (2011)
7. Chow, S.M., Yiu, S.-M., Hui, L., Chow, K.P.: Efficient Forward and Provably Secure ID-Based Signcryption Scheme with Public Verifiability and Public Ciphertext Authenticity. In: Lim, J.-I., Lee, D.-H. (eds.) ICISC 2003. LNCS, vol. 2971, pp. 352–369. Springer, Heidelberg (2004)
8. Cramer, R., Shoup, V.: Design and Analysis of Practical Public-Key Encryption Schemes Secure Against Adaptive Chosen Ciphertext Attack. SIAM J. Comput. 33(1), 167–226 (2003)
9. Damgård, I., Fazio, N., Nicolosi, A.: Non-interactive Zero-Knowledge from Homomorphic Encryption. In: Halevi, S., Rabin, T. (eds.) TCC 2006. LNCS, vol. 3876, pp. 41–59. Springer, Heidelberg (2006)
10. Dent, A.W.: Hybrid Signcryption Schemes with Outsider Security. In: Zhou, J., López, J., Deng, R.H., Bao, F. (eds.) ISC 2005. LNCS, vol. 3650, pp. 203–217. Springer, Heidelberg (2005)
11. El Aimani, L.: On Generic Constructions of Designated Confirmer Signatures. In: Roy and Sendrier [24], Full version available at the Cryptology ePrint Archive, Report 2009/403, pp. 343–362
12. El Aimani, L.: Efficient Confirmer Signature from the "Signature of a Commitment" Paradigm. In: Heng, Kurosawa [18], Full version available at the Cryptology ePrint Archive, Report 2009/435, pp. 87–101
13. El Aimani, L.: Generic Constructions for Verifiable Signcryption (2011), Available at the Cryptology ePrint Archive. Report 2011/592
14. El Gamal, T.: A Public Key Cryptosystem and a Signature Scheme based on Discrete Logarithms. IEEE Trans. Inf. Theory 31, 469–472 (1985)

15. Goldreich, O.: Foundations of cryptography. Basic Tools. Cambridge University Press (2001)
16. Goldreich, O., Micali, S., Wigderson, A.: How to Prove All NP-Statements in Zero-Knowledge and a Methodology of Cryptographic Protocol Design. In: Odlyzko, A.M. (ed.) CRYPTO 1986. LNCS, vol. 263, pp. 171–185. Springer, Heidelberg (1987)
17. Groth, J., Sahai, A.: Efficient Non-interactive Proof Systems for Bilinear Groups. In: Smart, N.P. (ed.) EUROCRYPT 2008. LNCS, vol. 4965, pp. 415–432. Springer, Heidelberg (2008)
18. Heng, S.-H., Kurosawa, K. (eds.): ProvSec 2010. LNCS, vol. 6402. Springer, Heidelberg (2010)
19. Jeong, I., Jeong, H., Rhee, H., Lee, D., Lim, J.: Provably Secure Encrypt-then-Sign Composition in Hybrid Signcryption. In: Lee, Lim [20], pp. 16–34
20. Lee, P.J., Lim, C.H. (eds.): ICISC 2002. LNCS, vol. 2587. Springer, Heidelberg (2003)
21. Ma, C.: Efficient Short Signcryption Scheme with Public Verifiability. In: Lipmaa, H., Yung, M., Lin, D. (eds.) Inscrypt 2006. LNCS, vol. 4318, pp. 118–129. Springer, Heidelberg (2006)
22. Matsuda, T., Matsuura, K., Schuldt, J.: Efficient Constructions of Signcryption Schemes and Signcryption Composability. In: Roy, Sendrier [24], pp. 321–342
23. Paillier, P.: Public-Key Cryptosystems Based on Composite Degree Residuosity Classes. In: Stern, J. (ed.) EUROCRYPT 1999. LNCS, vol. 1592, pp. 223–238. Springer, Heidelberg (1999)
24. Roy, B., Sendrier, N. (eds.): INDOCRYPT 2009. LNCS, vol. 5922. Springer, Heidelberg (2009)
25. Selvi, S., Vivek, S., Pandu Rangan, P.: Identity Based Public Verifiable Signcryption Scheme. In: Heng, Kurosawa [18], pp. 244–260
26. Shin, J.-B., Lee, K., Shim, K.: New DSA-Verifiable Signcryption Schemes. In: Lee, Lim [20], pp. 35–47
27. Zheng, Y.: Digital Signcryption or How to Achieve Cost (Signature & Encryption) << Cost(Signature) + Cost(Encryption). In: Kaliski Jr., B.S. (ed.) CRYPTO 1997. LNCS, vol. 1294, pp. 165–179. Springer, Heidelberg (1997)

Non-delegatable Strong Designated Verifier Signature on Elliptic Curves

Haibo Tian[1], Xiaofeng Chen[2], Zhengtao Jiang[3], and Yusong Du[1]

[1] School of Information Science and Technology,
Sun Yat-Sen University, Guangzhou, 510275, China
sysutianhb@gmail.com, yud80h@163.com
[2] School of Telecommunications Engineering, Xidian University,
Xi'an, 710071, China
xfchen@xidian.edu.cn
[3] School Of Computer, Communication University of China,
Beijing, 100024, China
z.t.jiang@163.com

Abstract. We propose a non-delegatable strong designated verifier signature on elliptic curves. The size of the signature is less than 500 bits considering an 80 bits security strength. It provably satisfies the non-delegatability and signer ambiguity properties. The construction method is a combination of the Schnorr signature and the elliptic curve Diffie-Hellman problem.

Keywords: Signature Schemes, Strong Designated Verifier Signature, Non-delegatability, Signer Ambiguity.

1 Introduction

Jakobsson et al. [10] proposed the concept of designated verifier signature (DVS). A DVS consists of a proof that either "the signer has signed on a message" or "the signer has the verifier's secret key". If the designated verifier is confident that her/his private key is kept in secret, the verifier makes sure that the signer has signed on a message. No other parties can be convinced by the DVS since the designated verifier can generate it with her/his private key. It is useful in various commercial cryptographic applications, such as e-voting, copyright protection, etc.

A strong DVS (SDVS) is an extension of the DVS. In the appendix, Jakobsson et al. [10] gave a definition of SDVS. It means that the verifier needs to use her/his private key to verify the signature. It considers a situation where the signature is captured before reaching the verifier. In this case, an adversary can know who is the real signer as there are only two possibilities. Laguillaumie and Vergnaud [13], and Saeednia [20] both formalized the notion.

There are five properties of SDVS, three basic properties and two enhancements. The basic properties include the unforgeability and non-transferability [15], and privacy of signer's identity (PSI) [13]. Informally, the unforgeability means that if an adversary can forge a (strong) DVS, it solves some hard problems. The non-transferability means that the designated verifier cannot convince

H. Kim (Ed): ICISC 2011, LNCS 7259, pp. 219–234, 2012.
© Springer-Verlag Berlin Heidelberg 2012

a third party that an SDVS is produced by a signer. The PSI means that nobody can identify a real signer from an SDVS except the designated verifier. Generally, a PSI attacker does not know the private key of a signer. The enhancements are the non-delegatability (ND) [17] and signer ambiguity (SA) [14]. The ND property means that the only way to generate an SDVS is to own a private key of a signer or a verifier. The SA property means that an attacker equipped with the private key of a signer, cannot identify who is the real signer of an SDVS.

This paper focuses on the SDVS. We give a new approach to construct an SDVS and prove that the scheme satisfies all above properties. We emphasize the ND and SA properties that are detailed in Section 3. It is interesting to design schemes with good performance under stricter conditions.

1.1 Related Works

Jakobsson et al. [10] gave a DVS scheme and proposed a method to transform a DVS to an SDVS by an encryption layer. Saeednia et al. [20] proposed a DVS scheme and an SDVS scheme. Lee [14] showed that the SDVS construction [20] lost the SA property. Lipmaa et al. [17] showed that the construction in [20] was delegatable. Laguillaumie et al. [13] proposed an ID-based SDVS scheme by using the long term symmetric secret of two users. Tso et al. [22] proposed to construct SDVS schemes by using authenticated key agreement protocols. Zhang et al. [24] proposed an SDVS scheme where the verification needed a long term symmetric key. Huang et al. [5] proposed a short DVS scheme by using a long term symmetric key. Kancharla et al. [11] proposed an ID-based strong DVS by using a temporal symmetric secret. According to the analysis technique in [14], schemes lose the SA property if they used a symmetric secret. Sun et al. [21] showed that the scheme in [11] was delegatable.

Lipmaa et al. [17] proposed a non-delegatable DVS scheme based on the signature scheme in [12]. Huang et al. [4] proposed an universal DVS without delegatability. Shahandashti and Safavi-Naini [19] proposed a general approach to construct an universal DVS without delegatability. Liao and Jia [16] proposed a DVS with an argument about its non-delegatability. Wang [23] proposed a non-delegatable ID-based SDVS scheme. Huang et al. [7] proposed a non-delegatable SDVS scheme. The team also [6] proposed a non-delegatable ID-based DVS scheme and a non-delegatable ID-based SDVS scheme [8].

1.2 Contributions

- First of all, we propose a new method to construct non-delegatable schemes. Some literatures also proposed (S)DVS schemes with the non-delegatability property. These schemes show two construction ideas:
  - One idea is to use an approach in [10], such as schemes in [17,23], which uses a trap-door commitment and the Fiat-Shamir heuristic;
  - The other idea is to use the typical OR proofs of two three-round zero knowledge protocols, such as schemes in [4,6–8,19].

We give another idea to combine the Schnorr signature and a hard problem. The advantage of the idea is that it can produce a signature consisting of three elements. Comparatively, the two existing ideas lead to a signature with at least four elements.

- The scheme enjoys ND and SA properties at the same time. Other schemes usually enjoy one property and lose the other.

  - Lee et al. [14] showed that the SDVS in [20] lost the SA property. They proposed a scheme and argued that it satisfied the SA. However, their scheme is delegatable. A signer can generate a random group value r and compute a temporal value s in the same way as the signer in their signing algorithm. Then the signer can give (r, s) to an agent to sign any message.

  - Huang et al. [7] proposed a non-delegatable SDVS scheme. Their scheme loses the SA property. The value K in their scheme is a long term secrete between a signer and a designated verifier. With the private key of the signer, an attacker can compute the value K and verify an SDVS.

  - Wang [23] proposed a non-delegatable identity-based SDVS scheme. The scheme loses the SA property. The value V_S in the verification algorithm can be computed without using the private key of the designated verifier. Then with the private key of the signer, an attacker can check their verification equation.

  In fact, we only find that the scheme in [8] enjoys the ND and SA properties with high bandwidth and computation costs.

In summary, the features of this paper are as follows:

- It shows a new method to construct a non-delegatable SDVS;
- The scheme enjoys the ND and SA properties with a short signature size and a moderate computation cost.

1.3 Organizations

Section 2 includes elementary materials about assumptions and definitions of an SDVS. Section 3 presents the ND and SA properties. Section 4 is the new SDVS scheme. The scheme is proven secure in section 5. Section 6 compares our scheme and other non-delegatable (strong) DVS schemes.

2 Preliminaries

2.1 Assumptions

Let p be a large prime and $\mathbb{F}_p$ be a finite filed. Let $a, b \in_R \mathbb{F}_p$ be random elements in the field to define a curve $\mathbb{E}$. Let $P \in_R \mathbb{E}$ be a point in the curve and be a generator of a group $\mathbb{G}$ with an order q.

- **Elliptic Curve Diffie-Hellman Problem (ECDHP):** Given points αP, $\beta P \in \mathbb{G}$, find another point $\alpha \beta P$.

- **Elliptic Curve Decisional Diffie-Hellman Problem (ECDDHP):** Given points $\alpha P, \beta P, \gamma P \in \mathbb{G}$, check whether $\gamma = \alpha\beta \bmod q$.

The assumption is that there are no (t, ϵ) algorithms to solve the ECDHP (ECD-DHP) problem in time t with a non-negligible probability ϵ if q is big enough.

Knowledge Extractor Assumption version 1 (KEAv1) [9]: Let T denotes a polynomial time bounded algorithm which on input $(P, \alpha P)$, produces $(\beta P, \alpha\beta P)$ where β is chosen by T. The assumption is that there exists another polynomial time bounded algorithm T^*, which takes as the same input of T, uses the same coins of T, and produces $(\beta, \beta P, \alpha\beta P)$ with a probability $1 - \epsilon$ where ϵ is a negligible value.

Remark 1. The KEAv1 is proven in a generic group model [3]. This gives it an evidence about its plausibility. It is only used in the proof of the non-delegatability in this paper.

2.2   SDVS

We define an SDVS scheme as follows.

- **System Parameters Generation SP:** A probabilistic polynomial time algorithm that, on input a security parameter k, produces the system parameters sp.
- **Key Generation KG:** A probabilistic polynomial time algorithm that, on input the system parameters sp, produces key pairs (pk_s, sk_s) for a signer and (pk_v, sk_v) for a verifier.
- **Signature Generation SG:** A probabilistic polynomial time algorithm that, on input the signer's private key sk_s, the verifier's public key pk_v and a message m, produces a signature δ.
- **Signature Verification SV:** A deterministic polynomial time algorithm that, on input the public key pk_s of the signer, the private key sk_v of the verifier, the message m, and a signature δ, produces a verification decision $b \in \{True, False\}$.
- **Transcript Simulation TS:** A probabilistic polynomial time algorithm that, on input the public key pk_s of the signer, the private key sk_v of the verifier, and the message m, produces a signature δ.

Properties

There are three basic properties about SDVS, namely the non-transferability, unforgeability, and PSI.

- Unforgeability [15]: It is formally defined by using a game between an adversary $\mathcal{A}$ and a challenger $\mathcal{C}$:
  - $\mathcal{C}$ provides the system parameters sp to $\mathcal{A}$.
  - $\mathcal{C}$ provides $\mathcal{A}$ a public key pk_s of a signer and a public key pk_v of a designated verifier.

- At any time, $\mathcal{A}$ can send any message m_i to $\mathcal{C}$ for the signature of m_i. These singing queries are up to q_s times. $\mathcal{C}$ will answer a query by providing the signature $\delta = SG(sk_s, pk_v, m_i)$.
- Eventually, $\mathcal{A}$ will produce a new signature δ^* for a message m^*. $\mathcal{A}$ succeeds if $SV(pk_s, sk_v, m^*, \delta^*) = True$ and m^* has never been queried.

The success probability of $\mathcal{A}$ is defined by $Adv_{SDVS,\mathcal{A}}(k)$.

Definition 1. *An SDVS scheme is* (t, q_s, ϵ) *unforgeable, if no polynomial time bounded adversary* $\mathcal{A}$ *has a success probability* $Adv_{SDVS,\mathcal{A}}(k) \geq \epsilon$ *running in time* t *with* q_s *signing queries.*

- non-transferability [15]: It is formally defined by using a game between a distinguisher $\mathcal{D}$ and a challenger $\mathcal{C}$:
  - $\mathcal{C}$ provides the system parameters sp to $\mathcal{D}$.
  - $\mathcal{C}$ provides $\mathcal{D}$ a public key pk_s of a signer and a public key pk_v of a designated verifier.
  - $\mathcal{D}$ issues signing queries on any message m_i. $\mathcal{C}$ replies to $\mathcal{D}$ with $\delta_i = SG(sk_s, pk_v, m_i)$.
  - $\mathcal{D}$ submits a new message m^* to $\mathcal{C}$. $\mathcal{C}$ flips a fair coin $b \leftarrow \{0, 1\}$, and produces a signature $\delta^* = SG(sk_s, pk_v, m_i)$ if $b = 0$ or $\delta^* = TS(pk_s, sk_v, m_i)$ if $b = 1$. $\mathcal{C}$ replies to $\mathcal{D}$ with δ^*.
  - On receiving δ^*, $\mathcal{D}$ can issue new singing queries at will. These signing queries are up to q_s times.
  - Eventually, $\mathcal{D}$ produces a bit b' and succeeds if $b' = b$.

The advantage of $\mathcal{D}$ is defined as $Adv_{SDVS,\mathcal{D}}(k) = |Pr[b' = b] - 1/2|$.

Definition 2. *An SDVS scheme is* (t, q_s, ϵ) *non-transferable if no polynomial time bounded distinguisher* $\mathcal{D}$ *has an advantage* $Adv_{SDVS,\mathcal{D}}(k) \geq \epsilon$ *in time* t *with* q_s *signing queries.*

- Privacy of Signer's Identity [13]: It is formally defined by using a game between a distinguisher $\mathcal{D}$ and a challenger $\mathcal{C}$:
  - $\mathcal{C}$ provides the system parameters sp to $\mathcal{D}$.
  - $\mathcal{C}$ provides $\mathcal{D}$ two public keys pk_{s0} and pk_{s1} of two signers and a public key pk_v of a designated verifier.
  - $\mathcal{C}$ provides $\mathcal{D}$ two signing oracles Σ_{s0} and Σ_{s1}, and a verifying oracle Υ. On a signing query m_i of $\mathcal{D}$, Σ_{s0} replies to $\mathcal{D}$ with $\delta_{i0} = SG(sk_{s0}, pk_v, m_i)$, and Σ_{s1} with $\delta_{i1} = SG(sk_{s1}, pk_v, m_i)$. On a verifying query $(pk_{s\zeta}, m_j, \delta_j)$, $\zeta \in \{0, 1\}$, Υ replies to $\mathcal{D}$ with $True$ or $False$. $\mathcal{D}$ produces a message m^* and a state information I^* after enough queries. $\mathcal{D}$ sends the message m^* to $\mathcal{C}$.
  - $\mathcal{C}$ flips a fair coin $b \leftarrow \{0, 1\}$, and computes $\delta^* = SG(sk_{sb}, pk_v, m^*)$ as a reply to $\mathcal{D}$.
  - $\mathcal{D}$ continues to issue signing and verifying queries on any message with the restriction of not querying Υ about $(pk_{s\zeta}, m^*, \delta^*)$, $\zeta \in \{0, 1\}$. The state information is taken as input in this step. The signing queries are up to q_s times. The verifying queries are up to q_v times.

- Eventually, $\mathcal{D}$ produces a bit b' and succeeds if $b' = b$.

The advantage of $\mathcal{D}$ is defined as $Adv_{SDVS,\mathcal{D}}^{PSI}(k) = |Pr[b' = b] - 1/2|$.

Definition 3. *An SDVS scheme is (t, q_s, q_v, ϵ) secure about privacy of a signer's identity if no polynomial time bounded distinguisher $\mathcal{D}$ has an advantage $Adv_{SDVS,\mathcal{D}}^{PSI}(k) \geq \epsilon$ in time t with q_s signing queries and q_v verifying queries.*

3   Modified Definitions

3.1   Non-delegatability

A Brief Review

The notion of ND was initiated in 2005 by Lipmaa et al. [17]. A real life scenario is that an adversary interacts with a signer who is responsible for the generation of the public and secret key pair (pk_s, sk_s). After this interaction, the adversary may obtain something which can be used to create signatures for a particular designated verifier. It is intended to prevent a dishonest signer to sell a specifically constructed key which is capable of creating a signature for a particular designated verifier.

The requirement of ND is then given access to the adversary, an extractor should be able to extract the private key sk_s. This guarantees that any designated verifier signature must be created from an entity who is in possession of the secret key from either the signer or the verifier. The original definition of ND [17] is as follows:

Definition 4. *Let $\kappa \in [0, 1]$ be the knowledge error. A scheme is (τ, κ)-non-delegatable if there is a black-box knowledge extractor $\mathcal{K}$ that, for every algorithm $\mathcal{F}$ and for every valid signature δ, satisfies the following condition: For every $(pk_s, sk_s) \leftarrow KG(sp)$, $(pk_v, sk_v) \leftarrow KG(sp)$ and message m, if $\mathcal{F}$ produces a valid signature on m with probability $\varepsilon \geq \kappa$ then, on input m and on access to the oracle $\mathcal{F}_m$, $\mathcal{K}$ produces either sk_s or sk_v in expected time $\tau/(\varepsilon - \kappa)$ (without counting the time to make the oracle queries).*

In the proof of ND [17], it said that "$\mathcal{K}$ executes $\mathcal{F}_m$ step by step" and $\mathcal{K}$ answered hashing queries of $\mathcal{F}_m$. As $\mathcal{F}_m$ is a black-box, nobody knows what it has, and how it produces a signature. Luckily, in the random oracle model, $\mathcal{F}_m$ must request a hashing oracle. By providing the oracle, $\mathcal{K}$ uses the rewinding technique to extract the private keys. As their scheme is the Schnorr style, this technique works.

The Motivation. There are some subtle problems.

- As $\mathcal{F}$ is a black-box, nobody knows how it works. It may just need a hashing oracle if we are lucky. However, it may also need other oracles, such as signing oracles or verifying oracles. That is, a malicious signer may sell some partial secret so that $\mathcal{F}$ can produce a signature for a valuable message after some online signing queries for garbage messages (according to the agreement between the signer and $\mathcal{F}$).

- For the same reason, nobody knows what $\mathcal{F}$ has. It may directly buy the private key of the singer with a high price. It may buy a partial secret from a signer or a verifier with a moderate cost. And it may buy nothing, but can simply produce a signature after some oracle queries.
- It seems that there are four players in the original definition. Two of them are a signer and a designated verifier that produce key pairs, and the other two of them are $\mathcal{F}$ and $\mathcal{K}$. There is no specification about the interaction between $\mathcal{F}$ and the two key-pair producers.
- The random oracle is the only available tool for $\mathcal{K}$ to extract the possible private key of $\mathcal{F}$. This makes the ND proof be limited to use the rewinding of a hashing oracle.

These observations make the proof of ND uneasy. It is not easy to determine what $\mathcal{K}$ should provide to $\mathcal{F}$ so that $\mathcal{F}$ can produce a signature, and which key will be extracted by $\mathcal{K}$, and not easy to apply the definition to new constructions.

The Instinctive Version

However, instinctively, an (S)DVS is non-delegatable means that if a party can create a valid signature, this party must know either the secret key of the signer or the verifier. Then a direct model is that if $\mathcal{K}$ interacts with a signature producer $\mathcal{F}$, $\mathcal{K}$ can extract a private key.

To overcome the subtle problems, we let $\mathcal{F}$ provide a public key $pk_{s(v)}$ of a signer or a verifier as the target to be extracted. It doesn't matter whether $\mathcal{F}$ knows the private key $sk_{s(v)}$ corresponding to $pk_{s(v)}$. We do not care what $\mathcal{F}$ has as it is a black-box. We let $\mathcal{K}$ provide any possible help to $\mathcal{F}$ to produce a signature. That is, $\mathcal{K}$ provides the signing, verifying and hashing oracles if possible. $\mathcal{K}$ can provide them easily since it controls the other key pair $(sk_{v(s)}, pk_{v(s)})$.

We define the instinctive version in a game style as follows. Suppose a polynomial time bounded signature producer $\mathcal{F}$ and an extractor $\mathcal{K}$.

- $\mathcal{K}$ produces the system parameter sp, and sends it to $\mathcal{F}$.
- $\mathcal{F}$ produces a public key $pk_{s(v)}$, and sends it to $\mathcal{K}$.
- $\mathcal{K}$ produces the other public key $pk_{v(s)}$ and sends it to $\mathcal{F}$.
- $\mathcal{F}$ produces a valid SDVS for a message m queried by $\mathcal{K}$ with a non-negligible probability. $\mathcal{F}$ accesses to the signing and hashing oracles if $\mathcal{F}$ produces pk_v, or accesses to the signing, verifying, and hashing oracles if $\mathcal{F}$ produces pk_s.
- $\mathcal{K}$ produces $sk_{s(v)}$.

The success probability of $\mathcal{K}$ is defined as $Adv_{SDVS,\mathcal{K}}^{ND}(k)$.

Definition 5. *An SDVS scheme is $(t, \epsilon, t', \epsilon')$ non-delegatable if $\mathcal{K}$ can extract the private key in time t with a probability $Adv_{SDVS,\mathcal{K}}^{ND}(k) \geq \epsilon$ against $\mathcal{F}$ that can produce a signature in time t' with a probability ϵ', where $\epsilon > poly_1(\epsilon')$ and $t < poly_2(t')$ for two polynomial functions $poly_1$ and $poly_2$.*

The Relationship. If a scheme is non-delegatable in the original definition, so it is in the instinctive version. If a scheme can be proven secure in the new version, it may NOT be proven in the old one. In this way, we can view the new version as a relaxed version. While it captures the core of the ND directly, it allows more flexible constructions.

3.2 Signer Ambiguity

Lee and Chang [14] gave a comment on Saeednia et al.'s SDVS scheme [20]. They found that Saeednia et al.'s scheme would reveal the identity of the signer if the secret key of this signer is compromised. That is, if an adversary is equipped with a signer's private key, for any SDVS produced by the signer, the adversary can verify the SDVS. Then, if such an SDVS is captured before it reaches the designated verifier, the real signer can be revealed. It is deemed as a weakness for the signer ambiguity. There is no formal definition about it in [14]. They argued their new scheme for the signer ambiguity.

Huang et al. [6] defined a stronger definition for the PSI property. It allows an adversary to know the private key of a signer. The signer ambiguity property is integrated. However, a simulator should provide various oracles for an adversary, including the signing, verifying and hashing oracles in their definition. It is unnecessary to provide the signing oracle as the adversary knows the signer's private key. It provides more information than what the adversary needs in [14].

We define a tailored model for the attack in [14]. It is defined by using a game between a distinguisher $\mathcal{D}$ and a challenger $\mathcal{C}$:

- $\mathcal{C}$ provides the system parameter sp to $\mathcal{D}$.
- $\mathcal{C}$ provides $\mathcal{D}$ two key pairs (pk_{s0}, sk_{s0}) and (pk_{s1}, sk_{s1}) of two signers and a public key pk_v of a designated verifier.
- $\mathcal{D}$ produces a message m^* and sends it to $\mathcal{C}$.
- $\mathcal{C}$ flips a fair coin $b \leftarrow \{0, 1\}$, and computes $\delta^* = SG(sk_{sb}, pk_v, m^*)$ as a reply to $\mathcal{D}$.
- $\mathcal{D}$ produces a bit b' and succeeds if $b' = b$.

The advantage of $\mathcal{D}$ is defined as $Adv_{SDVS,\mathcal{D}}^{SA}(k) = |Pr[b' = b] - 1/2|$.

Definition 6. *An SDVS scheme is (t, ϵ) ambiguous about the signer if no polynomial time bounded distinguisher $\mathcal{D}$ has an advantage $Adv_{SDVS,\mathcal{D}}^{SA}(k) \geq \epsilon$ in time t.*

Remark 2. As the attack in [14] does not need any oracles, our definition provide no oracles too. It is reasonable to provide a verifying oracle as a distinguisher cannot do so. However, it is not easy to provide such a oracle in our proof, and that enhancement is beyond the attack technique in [14].

4 The SDVS Scheme

- SP: Let p be a large prime and $\mathbb{F}_p$ be a finite filed. Let $a, b \in_R \mathbb{F}_p$ be random elements in the field to define a curve $\mathbb{E}$. Let $P \in_R \mathbb{E}$ be a point in the curve and be a generator of a group $\mathbb{G}$ with an order q. Let $H_1 : \mathbb{G} \to \mathbb{Z}_q^*$ and $H_2 : \{0, 1\}^* \to \mathbb{G}$ be secure hashing functions. The system parameter sp is $(\mathbb{F}_p, a, b, P, q, H_1, H_2)$.
- KG: For a signer, randomly select $sk_s = s \in_R \mathbb{Z}_q^*$ and compute $pk_s = Q_s = -sP$. For a designated verifier, randomly select $sk_v = v \in_R \mathbb{Z}_q^*$ and compute $pk_v = Q_v = vP$.

- SG: Randomly select $r, l \in_R \mathbb{Z}_q^*$ and compute $A = lP$, $C_0 = rP$ and $C_1 = H_2(m, A)$, $C = C_0 + C_1 = (c_x, c_y)$, $z = l + c_x s \bmod q$, $R = rQ_v$, and $t = H_1(R)$. The signature is (C, z, t).
- SV: Compute $A' = zP + c_x Q_s$, $C_1' = H_2(m, A')$, $C_0' = C - C_1'$, $R' = vC_0'$, and $t' = H_1(R')$. Verify that $t = t'$.
- TS: Randomly select $C \in_R \mathbb{G}, z \in_R \mathbb{Z}_q^*$ and compute $A = zP + c_x Q_s$. Compute $C_1 = H_2(m, A)$, $C_0 = C - C_1$, $R = vC_0$ and $t = H_1(R)$.

Remark 3. The point C in the signature will be presented in a compressed fashion to shorten the size. That is, only the X coordinate and a compressed one-bit representation of the Y coordinate of the point will be transmitted.

The computation of values A and C_0 in the SG algorithm is taken as one scalar multiplication as they can be computed sequentially where they are sorted by the values of their indexes. The computation of $zP + c_x Q_s$ is approximated to one scalar multiplication thanks to the algorithm 15.2 recorded in a book [18].

5 Proof of Properties

We use the symbol τ_m to denote the time of a scalar multiplication on elliptic curves. Generally, we suppose the hash functions, H_1, H_2 are random oracles, and the q_h is the number of the hashing queries to the H_2 oracle.

Unforgeability

Proposition 1. *The SDVS scheme is (t, q_s, ϵ) unforgeable if the ECDHP problem is (t', ϵ') unsolvable, where $t' \leq 4(t + (q_h + 2q_s)\tau_m)$, and $\epsilon' > 1/3\epsilon(\epsilon/q_h - 1/q)$*

Proof. Assume a simulator $\mathcal{C}$ which tries to solve an ECDHP problem. Assume an ECDHP problem instance is $(\alpha P, \beta P)$ with group parameters $(\mathbb{F}_p, a, b, P, q)$. Suppose an adversary, $\mathcal{A}$, which claims a non-negligible success probability ϵ over the SDVS scheme in time t with q_s signing queries. $\mathcal{C}$ runs two games with $\mathcal{A}$.

- **Game 0:**
  - $\mathcal{C}$ sets $sp = (\mathbb{F}_p, a, b, P, q, H_1, H_2)$, and gives sp to $\mathcal{A}$.
  - $\mathcal{C}$ runs KG to produce a verifier's key pair (pk_v, sk_v), and gives $\mathcal{A}$ public keys $pk_s = \alpha P$ and pk_v.
  - $\mathcal{C}$ provides a signing oracle by running the TS algorithm.
  - $\mathcal{C}$ provides two hashing oracles by maintaining two hashing lists.
    1. $\mathcal{C}$ maintains an H_{list}^1 for hashing queries of H_1. The H_{list}^1 is empty at the beginning. When $\mathcal{A}$ provides an R_i for hashing, if there is a match in the H_{list}^1, $\mathcal{C}$ replies to $\mathcal{A}$ the value t_i in the match directly. Else $\mathcal{C}$ randomly selects $t_i \in_R \mathbb{Z}_q^*$ and replies to $\mathcal{A}$ with t_i and records an entry (R_i, t_i) in the H_{list}^1.
    2. $\mathcal{C}$ maintains an H_{list}^2 for hashing queries of H_2. The H_{list}^2 is empty at the beginning. When $\mathcal{A}$ provides (m_i, A_i) for hashing, if there is a match in the H_{list}^2, $\mathcal{C}$ replies to $\mathcal{A}$ with the value Q_i in the match directly. Else $\mathcal{C}$ randomly selects $d_i \in_R \mathbb{Z}_q^*$ and computes $Q_i = d_i P$. $\mathcal{C}$ replies to $\mathcal{A}$ with Q_i and records an entry (m_i, A_i, d_i, Q_i) in the H_{list}^2.

- Since the non-transferability property will show that the signatures produced by the TS algorithm are distributed the same as those produced by the SG algorithm, $\mathcal{A}$ should give an SDVS (C^*, z^*, t^*) for a new message m^*.

According to the general forking lemma [2], $\mathcal{A}$ should give another SDVS (C', z', t') for the same message m^* with fixed coins. The probability is $\theta = \epsilon(\epsilon/q_h - 1/q)$. Suppose the probability of $z^* \neq z'$ is $\xi \in [0, 1]$. Then with a probability $\xi\theta$, $\mathcal{C}$ can extract the private key α of the signer. Then $\mathcal{C}$ computes $\beta\alpha P$ as the answer of the ECDHP problem. The total run time of $\mathcal{C}$ in this case is less than $2(t + (q_h + 2q_s)\tau_m)$.

There is a probability $1 - \xi$ for the event $z^* = z'$. This leads to the design of Game 1.

- **Game 1:**
  - $\mathcal{C}$ gives the system parameter sp to $\mathcal{A}$.
  - $\mathcal{C}$ runs KG to produce a signer's key pair (pk_s, sk_s), and sets $pk_v = \beta P$, and gives public keys pk_s, pk_v to $\mathcal{A}$.
  - $\mathcal{C}$ provides a signing oracle by running the SG algorithm.
  - $\mathcal{C}$ provides the hashing oracles in the same way as it does in the Game 0.
  - Finally, $\mathcal{A}$ should give an SDVS (C^*, z^*, t^*) for a new message m^*.

Again, $\mathcal{A}$ should give another SDVS (C', z', t') for the same message m^* with fixed coins with a probability $\theta = \epsilon(\epsilon/q_h - 1/q)$. However, when $\mathcal{A}$ queries $(m^*, A^* = z^*P + c_x^*Q_s)$ at the second run, $\mathcal{C}$ sets $H_2(m^*, A^*) = C^* - \alpha P$ or $H_2(m^*, A^*) = -C^* - \alpha P$ with a probability $1/2$. After $\mathcal{A}$ gives (C', z', t'), $\mathcal{C}$ finds the R' in H_{list}^1 indexed by t'. The answer is R' for the ECDHP problem. Note that if $z^* = z'$, $c_x^* = c_x' \mod q$. Then $C' = C^*$ or $C' = -C^*$. Since $\alpha P = C^* - H_2(m^*, A^*)$ or $\alpha P = -C^* - H_2(m^*, A^*)$, and $pk_v = \beta P$, it should be $R' = \beta\alpha P$ with a probability $1/2$.

As we are in the random oracle, we omit the collision event that when $R' \neq R''$, $H_1(R') = H_1(R'')$. There is another error event that although $c_x^* \neq c_x'$, $c_x^* = c_x' \mod q$. Note that $p + 1 - 2\sqrt{p} < q < p + 1 + 2\sqrt{p}$ by the well-known Hasse's theorem. If $q > p$, there is no error probability. If $q < p$, the error probability is $(p - q)/p \cdot 1/(p - q) = 1/p$.

The success probability of $\mathcal{C}$ is at least $1/2(1 - 1/p)(1 - \xi)\theta > 1/3(1 - \xi)\theta$ to solve the ECDHP problem. The runtime of $\mathcal{C}$ in this case is similar to it in the Game 0.

So the total runtime of $\mathcal{C}$ is less than $4(t + (q_h + 2q_s)\tau_m)$. The total success probability of $\mathcal{C}$ is at least $1/3(1 - \xi)\theta + \xi\theta > 1/3\theta$. $\square$

Remark 4. The general forking lemma [2] can be used in contexts other than traditional signatures as it only considers the inputs and outputs of a function.

Non-transferability

Proposition 2. *The SDVS scheme is non-transferable in the context of Definition 2.*

Proof. According to the Definition 2, we assume a challenger $\mathcal{C}$. $\mathcal{C}$ runs algorithms SP and KG to generate the system parameter sp and key pairs. Then $\mathcal{C}$ sends the sp and public keys to a distinguisher $\mathcal{D}$. To answer the signing queries, $\mathcal{C}$ uses the private key of a signer directly to sign a query message according to the SG algorithm. When $\mathcal{D}$ submits a message m^* for a test, $\mathcal{C}$ flips a fair coin $b \in \{0,1\}$. If $b = 0$, $\mathcal{C}$ uses the private key of the signer to produce a signature $(\hat{C}_0, \hat{z}_0, \hat{t}_0)$. If $b = 1$, $\mathcal{C}$ uses the private key of the designated verifier to produce a signature $(\hat{C}_1, \hat{z}_1, \hat{t}_1)$ according to the TS algorithm.

The distinguisher $\mathcal{D}$ has no advantage to produce a b' such that $b' = b$ since the distribution of $(\hat{C}_0, \hat{z}_0, \hat{t}_0)$ is identical to that of $(\hat{C}_1, \hat{z}_1, \hat{t}_1)$.

Let $(\hat{C}, \hat{z}, \hat{t})$ be an SDVS that is randomly chosen in the set of all valid signatures of a signer intended to a designated verifier. Then we consider the probability of the event: $\hat{C}_0 = r_0 P + H_2(m, l_0 P) = \hat{C}$, $\hat{z}_0 = l_0 + \hat{c}_{0x}(sk_s) = \hat{z}$, and $t_0 = H_1(r_0 pk_v) = \hat{t}$. The randomness is over the variables $r_0, l_0 \in_R \mathbb{Z}_q^*$. The probability is about $1/q^2$.

Another event is $\hat{C}_1 = \hat{C}$, and $\hat{z}_1 = \hat{z}$, and $\hat{t}_1 = H_1((C_1 - H_2(m, \hat{c}_{1x}Q_s + z_1 P))^{sk_v}) = \hat{t}$. The randomness is over the variables $\hat{z}_1 \in_R \mathbb{Z}_q^*$ and $\hat{C}_1 \in_R \mathbb{G}$. The probability is also about $1/q^2$. □

Privacy of signer's identity (PSI)

Proposition 3. *If the SDVS scheme is (t', q'_s, ϵ') unforgeable, and the ECDDHP problem is (t'', ϵ'') unsolvable, the SDVS scheme is (t, q_s, q_v, ϵ) secure in the context of Definition 3.*

Proof. There is a challenger $\mathcal{C}$ and a distinguisher $\mathcal{D}$ as in the Definition 3. They play a serial of games.

Game 0

- $\mathcal{C}$ invokes SP to produce the system parameter sp and gives it to $\mathcal{D}$.
- $\mathcal{C}$ runs KG to set up two key pairs $(sk_{s\zeta}, pk_{s\zeta})$, $\zeta \in \{0,1\}$, for two signers and one key pair (pk_v, sk_v) for a designated verifier. Then $\mathcal{C}$ gives $\mathcal{D}$ public keys (pk_{s0}, pk_{s1}, pk_v).
- $\mathcal{C}$ installs the private key sk_{s0} in the signing oracle Σ_{s0}, and sk_{s1} in Σ_{s1}. $\mathcal{C}$ installs the private key sk_v in the verifying oracle Υ. $\mathcal{C}$ provides a hashing oracle for H_1 that is the same as it in the Game 0 of the Proposition 1. Then $\mathcal{C}$ provides $\mathcal{D}$ the signing, verifying and hashing oracles.
- When $\mathcal{D}$ provides the message m^*, $\mathcal{C}$ produces a challenge SDVS (C^*, z^*, t^*). $\mathcal{C}$ flips a fair coin $b^* \in \{0,1\}$, and uses the private key sk_{sb^*} to produce a challenge SDVS according to the SG algorithm.
- $\mathcal{C}$ provides the signing and verifying oracles as before except that the verifying oracle Υ has no response to the query $(pk_{s\zeta}, m^*, C^*, z^*, t^*)$, $\zeta \in \{0,1\}$.
- Eventually, $\mathcal{D}$ produces a bit b and succeeds if $b = b^*$.

This game is exactly the definition of the PSI. So $\mathcal{D}$ should succeed with a non-negligible advantage ϵ in time t with q_s signing queries and q_v verifying queries.

Game 1: This game is intended to show that $\mathcal{D}$ cannot tell whether a challenge signature is valid.

- $\mathcal{C}$ takes an ECDDHP problem instance $(\alpha P, \beta P, \gamma P)$.
- $\mathcal{C}$ sets $pk_v = \beta P$.
- $\mathcal{C}$ provides the signing oracles as before. However, $\mathcal{C}$ maintains a signing list S_{list} for all signatures produced by the signing oracles. The verifying oracle Υ uses the S_{list} to answer queries. If a query is in the list, Υ answers $True$, else $False$.
- When $\mathcal{D}$ provides the message m^*, $\mathcal{C}$ produces a challenge SDVS (C^*, z^*, t^*) as follows. $\mathcal{C}$ flips a fair coin $b^* \in \{0,1\}$, and uses the private key sk_{sb^*}. It randomly selects $l \in_R \mathbb{Z}_q^*$ and computes $A^* = lP$, and $C_1 = H_2(m^*, A^*)$. It sets $C_0 = \alpha P$ and $R = \gamma P$. It computes $C^* = C_0 + C_1$, and $z^* = l + c_x^* sk_{sb^*}$, and $t^* = H_1(R)$. The challenge signature is (C^*, z^*, t^*).
- Other steps keep unchanged.

The distinguisher $\mathcal{D}$ has two possible ways to distinguish Game 0 from Game 1. The first is about the two challenge signatures in the two games. However, If the input of $\mathcal{C}$ is an ECDDHP tuple, the signature is valid and indistinguishable from it in the Game 0. If it is not, the challenge SDVS is an invalid signature. If $\mathcal{D}$ can distinguish the Game 1 from Game 0 by the two signatures, $\mathcal{C}$ can solve the ECDDHP problem directly. So the probability is ϵ'' for $\mathcal{D}$ to distinguish the two games by the challenge signatures in time t''.

The second way is through the verifying oracle Υ. From the Proposition 1, we know the probability of successful forgery is a negligible probability ϵ' in time t' with q_s' signing queries. So all $True$ signatures queried to the oracle Υ can be verified rightly. The only questionable signature is the challenge signature, which is not allowed to be a query by the game rules. So the probability is ϵ' for $\mathcal{D}$ to use the oracle Υ to distinguish Game 1 from Game 0 in time t' in this case.

Game 2: The challenge SDVS (C^*, z^*, t^*) is produced without using any private keys.

- When $\mathcal{D}$ provides the message m^*, $\mathcal{C}$ produces a challenge SDVS (C^*, z^*, t^*) as follows. $\mathcal{C}$ randomly selects $t^*, z^* \in_R \mathbb{Z}_q^*$, and $C^* \in_R \mathbb{G}$. The challenge signature is (C^*, z^*, t^*).
- Other steps keep unchanged.

The values C^* and z^* are distributed the same as them in the Game 1 due to the non-transferability property. The value t^* is also distributed the same as it in the Game 1 since we are in the random oracle model. So $\mathcal{D}$ cannot distinguish the Game 2 from the Game 1.

Since no private keys are used to produce the challenge SDVS in Game 2, it is meaningless for the distinguisher $\mathcal{D}$ to claim that the message m^* is signed by the owner of a private key sk_0 or sk_1, or simulated by a designated verifier. There is no advantage for $\mathcal{D}$ in Game 2.   □

Non-delegatability

Proposition 4. *Suppose that the KEAv1 assumption holds with a probability $1 - \epsilon''$. Then the SDVS scheme is $(t, \epsilon, t', \epsilon')$ non-delegatable in the context of Definition 5.*

Proof. According to the Definition 5, there is an extractor $\mathcal{K}$ and a signature producer $\mathcal{F}$. $\mathcal{F}$ produces a valid signature on a message m with a non-negligible probability ϵ' in time t'.

- $\mathcal{F}$ produces a public key pk_s as the target.
  - $\mathcal{K}$ produces the sp and sends it to $\mathcal{F}$
  - $\mathcal{F}$ produces pk_s and sends it to $\mathcal{K}$.
  - $\mathcal{K}$ produces the key pair (pk_v, sk_v) and sends the pk_v to $\mathcal{F}$.
  - $\mathcal{K}$ provides a signing oracle by the TS algorithm, and a verifying oracle by the SV algorithm, and two hashing oracles for the H_1 and H_2 in the same way as them in the Game 0 of the Proposition 1.
  - $\mathcal{K}$ selects a message m^* and sends it to $\mathcal{F}$.
  - $\mathcal{F}$ produces a signature (C^*, z^*, t^*) for the message m^*.
  - $\mathcal{K}$ uses the same method as $\mathcal{C}$ in the Game 0 of the unforgeability proof to extract the private key sk_s. The successful probability is $\epsilon \geq \xi\epsilon'(\epsilon'/q_h - 1/q)$ where ξ is the same as it in the Game 0 of the Proposition 1. The runtime is $t < 2(t' + (q_h + 2q_s + 2q_v)\tau_m)$.
- $\mathcal{F}$ produces a public key pk_v as the target.
  - $\mathcal{K}$ produces the sp and sends it to $\mathcal{F}$
  - $\mathcal{F}$ produces pk_v and sends it to $\mathcal{K}$.
  - $\mathcal{K}$ produces the key pair (pk_s, sk_s) and sends pk_s to $\mathcal{F}$.
  - $\mathcal{K}$ provides a signing oracle by the SG algorithm, and two hashing oracles for the H_1 and H_2 in the same way as them in the Game 0 of the Proposition 1.
  - $\mathcal{K}$ selects a message m^* and sends it to $\mathcal{F}$.
  - $\mathcal{F}$ produces a signature (C^*, z^*, t^*) for the message m^*.
  - $\mathcal{K}$ runs $\mathcal{F}$ again with fixed coins. When $\mathcal{F}$ queries the H_2 with $(m^*, z^*P + c_x^* pk_s)$, $\mathcal{K}$ replies $C_1' = C^* - C_0''$ or $C_1' = -C^* - C_0''$ with a probability $1/2$, where $C_0'' = r'P \in_R \mathbb{G}$ for a random $r' \in_R \mathbb{Z}_q^*$. $\mathcal{F}$ produces another signature (C', z', t') for m^*. If $z' \neq z^*$, $\mathcal{K}$ fails. Else $\mathcal{K}$ finds the value $R' = r'pk_v$ in the H_{list}^1. We take the values (P, C_0'') as an input to $\mathcal{F}$ because the values are totally determined by $\mathcal{K}$. We take the values (pk_v, R') as the output of $\mathcal{F}$. Then according to the KEAv1 assumption, $\mathcal{K}$ can build another $\mathcal{F}^*$ with the same coins to produce an output (sk_v, pk_v, R').

    Similar to the analysis in the Game 1 of the Proposition 1, the success probability of $\mathcal{F}$ is $\epsilon > 1/3(1 - \epsilon'')(1 - \xi)(\epsilon'(\epsilon'/q_h - 1/q))$ considering the extra assumption of KEAv1. The runtime is $t < 2(t' + (q_h + 2q_s)\tau_m)$.   □

Signer Ambiguity

Proposition 5. *The SDVS scheme is (t, ϵ) ambiguous about the signer if the ECDDHP problem is (t', ϵ') unsolvable, where $t \approx t' - \tau_m$ and $\epsilon = \epsilon'$.*

Proof. Suppose a challenger $\mathcal{C}$ which tries to solve an ECDDHP problem instance $(\alpha P, \beta P, \gamma P)$. Suppose a distinguisher $\mathcal{D}$ which tries to break the SA property of the scheme with an advantage ϵ in time t.

- $\mathcal{C}$ provides the system parameters sp to $\mathcal{D}$.

- $\mathcal{C}$ provides $\mathcal{D}$ two key pairs (pk_{s0}, sk_{s0}) and (pk_{s1}, sk_{s1}) of two signers and a public key $pk_v = \beta P$ of a designated verifier.
- $\mathcal{D}$ produces a message m^* and sends it to $\mathcal{C}$.
- $\mathcal{C}$ flips a fair coin $b \leftarrow \{0,1\}$. It randomly selects $l \in_R \mathbb{Z}_q^*$, and computes $A^* = lP$. Then it sets $C_0 = \alpha P$ and $R = \gamma P$. Then $C^* = H_2(m^*, A^*) + C_0$, and $z^* = l + c_x^* sk_{sb} \bmod q$, and $t^* = H_1(R)$. The SDVS is (C^*, z^*, t^*).
- $\mathcal{C}$ produces whatever produced by $\mathcal{D}$.

If the tuple $(\alpha P, \beta P, \gamma P)$ is an ECDDHP tuple, $\mathcal{D}$ should have an advantage ϵ in time t. If it is not, the signature is invalid and $\mathcal{D}$ has no advantage. The advantage of $\mathcal{C}$ is the same as $\mathcal{D}$. The runtime of $\mathcal{C}$ is about $t + \tau_m$.   □

6 Performance

We compare our scheme with some current non-delegatable (strong) DVS schemes. The column "Type" shows the signature type of each scheme. The column "SA" shows whether a scheme enjoys the SA property. The "RO" column shows the proof model of a scheme. The "NPRO" means a non-programable random oracle model. The "RO" means a random oracle model. Other columns are about the signature size and computation cost.

There are three kinds of system parameters.

- Let p', q' be large primes such that $q'|p' - 1$. Let $\mathbb{G}'$ be a group of order q'.
- The parameters $(\mathbb{F}_p, a, b, P, q)$ is for a group $\mathbb{G}$ in a non-supersingular elliptic curve.
- Let $\mathbb{G}_e$ be a group derived from the curve defined by $y^2 = x^3 + 2x + 1$ in the filed $\mathbb{F}_{3^u}$. There is a pairing evaluation $e : \mathbb{G}_e \times \mathbb{G}_e \to \mathbb{G}_t$. The order is q_e of the two groups.

Let $|p|$ denote the bits length of the value p. Considering a cryptographic strength of approximate 80 security bits, $|p'| = 1024$ and $|q'| = 160$, $|q| = 160$, and $u = 97$ and $|q_e| = 151$. If $Q \in \mathbb{G}$, $|Q| = 161$. If $Q \in \mathbb{G}_e$, $|Q| = 154$. If $Q \in \mathbb{G}_t$, $|Q| \approx 923$.

We use symbols $\tau', \tau_m, \tau_p, \tau_h$ to denote the computation of modular exponentiation in $\mathbb{G}'$ or $\mathbb{G}_t$, the scalar multiplication in $\mathbb{G}$ or $\mathbb{G}_e$, the bilinear pairing computation and the hash function that maps arbitrary input strings to elliptic curve points (the *MapToGroup* function [1]).

From the Table 1, we observe the following points:

Table 1. Comparison among non-delegatable (strong) DVS schemes

Scheme	Type	SA	Signature-Size (bits)	Sign-Cost	Verify-Cost	RO
[17]	DVS	-	640 ($\mathbb{Z}_{q'}^4$)	$3\tau'$	$3\tau'$	NPRO
[6]	IBDVS	-	758 ($\mathbb{Z}_{q_e}^4 \times \mathbb{G}_e$)	$3\tau' + 3\tau_p + \tau_m + 2\tau_h$	$4\tau' + 4\tau_p + 3\tau_h$	RO
[7]	SDVS	No	640 ($\mathbb{Z}_{q'}^4$)	$3\tau'$	$3\tau'$	RO
[8]	IBSDVS	Yes	2607 ($\mathbb{Z}_{q_e}^3 \times \mathbb{G}_e^2 \times \mathbb{G}_t^2$)	$4\tau' + 4\tau_p + \tau_m + 3\tau_h$	$4\tau' + 5\tau_p + 4\tau_h$	RO
Ours	SDVS	Yes	481 ($\mathbb{Z}_q^2 \times \mathbb{G}$)	$2\tau_m + \tau_h$	$2\tau_m + \tau_h$	RO

- Only one IBSDVS scheme [8] enjoys the ND and SA properties, which has a big signature size and high signing and verifying costs. Comparatively, our scheme shows advantages in both the signature size and computation costs.
- The signature size of our scheme is short. It is the only scheme consists of three elements in a signature.
- The signing and verifying costs of our scheme are moderate due to the cost of the scalar multiplication on elliptic curves.

Acknowledgment. This work is supported by the National Natural Science Foundation of China (Nos. 60970144, 60803135, 61070168, 61003244, 61103199), Fundamental Research Funds for the Central Universities (Nos. 10lgpy31, 11lgpy71, 11lgzd06, 10lgzd14), Specialized Research Fund for the Doctoral Program of Higher Education for New Teachers (No. 20090171120006), and Beijing Municipal Natural Science Foundation(No. 4112052).

References

1. Boneh, D., Lynn, B., Shacham, H.: Short Signatures from the Weil Pairing. In: Boyd, C. (ed.) ASIACRYPT 2001. LNCS, vol. 2248, pp. 514–532. Springer, Heidelberg (2001)
2. Bellare, M., Neven, G.: Multi-Signatures in the Plain Public-Key Model and a General Forking Lemma. In: Proceedings of the 13th Association for Computing Machinery (ACM) Conference on Computer and Communications Security (CCS), pp. 390–399. ACM, Alexandria (2006)
3. Dent, A.W., Galbraith, S.D.: Hidden Pairings and Trapdoor DDH Groups. In: Hess, F., Pauli, S., Pohst, M. (eds.) ANTS 2006. LNCS, vol. 4076, pp. 436–451. Springer, Heidelberg (2006)
4. Huang, X., Susilo, W., Mu, Y., Wu, W.: Universal Designated Verifier Signature Without Delegatability. In: Ning, P., Qing, S., Li, N. (eds.) ICICS 2006. LNCS, vol. 4307, pp. 479–498. Springer, Heidelberg (2006)
5. Huang, X., Susilo, W., Mu, Y., Zhang, F.: Short Designated Verifier Signature Scheme and Its Identity-based Variant. International Journal of Network Security 6(1), 82–93 (2008)
6. Huang, Q., Susil, W., Wong, D.S.: Non-delegatable Identity-based Designated Verifier Signature. Cryptology ePrint Archive: Report 2009/367 (2009)
7. Huang, Q., Yang, G., Wong, D.S., Susilo, W.: Efficient Strong Designated Verifier Signature Schemes without Random Oracles or Delegatability. Cryptology ePrint Archive: Report 2009/518 (2009)
8. Huang, Q., Yang, G., Wong, D.S., Susilo, W.: Identity-based strong designated verifier signature revisited. Journal of Systems and Software 84(1), 120–129 (2011)
9. Damgård, I.B.: Towards Practical Public Key Systems Secure against Chosen Ciphertext Attacks. In: Feigenbaum, J. (ed.) CRYPTO 1991. LNCS, vol. 576, pp. 445–456. Springer, Heidelberg (1992)
10. Jakobsson, M., Sako, K., Impagliazzo, R.: Designated Verifier Proofs and Their Applications. In: Maurer, U.M. (ed.) EUROCRYPT 1996. LNCS, vol. 1070, pp. 143–154. Springer, Heidelberg (1996)
11. Kancharla, P.K., Gummadidala, S., Saxena, A.: Identity Based Strong Designated Verifier Signature Scheme. Journal of Informatica 18(2), 239–252 (2007)

12. Katz, J., Wang, N.: Efficiency Improvements for Signature Schemes with Tight Security Reductions. In: 10th ACM Conference on Computer and Communications Security, pp. 155–164. ACM Press (2003)
13. Laguillaumie, F., Vergnaud, D.: Designated Verifier Signatures: Anonymity and Efficient Construction from *Any* Bilinear Map. In: Blundo, C., Cimato, S. (eds.) SCN 2004. LNCS, vol. 3352, pp. 105–119. Springer, Heidelberg (2005)
14. Lee, J., Chang, J.: Comment on Saeednia et al.'s strong designated verifier signature scheme. Journal of Computer Standards & Interfaces - CSI 31(1), 258–260 (2009)
15. Li, Y., Susilo, W., Mu, Y., Pei, D.: Designated Verifier Signature: Definition, Framework and New Constructions. In: Indulska, J., Ma, J., Yang, L.T., Ungerer, T., Cao, J. (eds.) UIC 2007. LNCS, vol. 4611, pp. 1191–1200. Springer, Heidelberg (2007)
16. Liao, Y., Jia, C.: Designated verifier signature without random oracles. Communications, Circuits and Systems. In: IEEE International Conference on ICCCAS 2008, pp. 474–477 (2008)
17. Lipmaa, H., Wang, G., Bao, F.: Designated Verifier Signature Schemes: Attacks, New Security Notions and a New Construction. In: Caires, L., Italiano, G.F., Monteiro, L., Palamidessi, C., Yung, M. (eds.) ICALP 2005. LNCS, vol. 3580, pp. 459–471. Springer, Heidelberg (2005)
18. Mao, W.: Modern cryptography: theory and practice. Prentice Hall Professional Technical Reference (2003)
19. Shahandashti, S.F., Safavi-Naini, R.: Construction of Universal Designated-Verifier Signatures and Identity-Based Signatures from Standard Signatures. In: Cramer, R. (ed.) PKC 2008. LNCS, vol. 4939, pp. 121–140. Springer, Heidelberg (2008)
20. Saeednia, S., Kramer, S., Markovitch, O.: An Efficient Strong Designated Verifier Signature Scheme. In: Lim, J.-I., Lee, D.-H. (eds.) ICISC 2003. LNCS, vol. 2971, pp. 40–54. Springer, Heidelberg (2004)
21. Sun, X., Li, J., Hu, Y., Chen, G.: Delegatability of an Identity Based Strong Designated Verifier Signature Scheme. INFORMATICA 21(1), 117–122 (2010)
22. Tso, R., Okamoto, T., Okamoto, E.: Practical Strong Designated Verifier Signature Schemes Based on Double Discrete Logarithms. In: Feng, D., Lin, D., Yung, M. (eds.) CISC 2005. LNCS, vol. 3822, pp. 113–127. Springer, Heidelberg (2005)
23. Wang, B.: A non-delegatable identity-based strong designated verifier signature scheme, http://eprint.iacr.org/2008/507
24. Zhang, J., Mao, J.: A novel ID-based designated verifier signature scheme. Information Sciences 178(3), 766–773 (2008)

An Improved Known Plaintext Attack on PKZIP Encryption Algorithm

Kyung Chul Jeong, Dong Hoon Lee, and Daewan Han

The Attached Institute of ETRI,
P.O. Box 1, Yuseong-Gu, Daejeon, Korea
{jeongkc,dlee,dwh}@ensec.re.kr

Abstract. The PKZIP encryption algorithm has been widely used to protect the contents of compressed archives despite the known security weakness. Biham and Kocher proposed a known plaintext attack with the complexity 2^{40} when 12 plaintext bytes are given. Stay suggested a different way to attack and addressed an idea which makes the complexity be reduced if the information of additional files encrypted under the same password is provided. However, the complexity of Stay's attack is quite large when only one file is used.

In this paper, we propose a new attack based on Biham and Kocher's attack. We introduce a method to reduce the complexity using the information of multi-files, so our attack can have the both advantages of previous two attacks. As a result, our attack becomes about $(3.4)^l$ times faster than the attack of Biham and Kocher when l additional files are used. Our experiment shows that ours is at least 10 times faster than Stay's. In addition, our attack can be improved in the chosen ciphertext model. It is about $(21.3)^l$ times faster than Biham and Kocher's attack with chosen plaintext of l additional files.

Keywords: PKZIP encryption, known plaintext attack.

1 Introduction

Compression softwares are used for various reasons: to reduce the size of big files, to unify many files and folders into a single archive, to split a big file into several parts with a small size and to protect the contents of the files by the password-based encryption. Most compression softwares support the ZIP file format among several compression file formats such as ZIP, RAR, ARJ, 7z and etc. As one of their protection algorithms, they also support the *traditional* PKZIP encryption included in the ZIP format specification. Almost all softwares support more strong encryption algorithms like AES in addition.

The traditional PKZIP encryption algorithm (a.k.a standard Zip 2.0 encryption) was designed by Roger Schlafly [4]. It has been widely supported by most compression softwares despite the publicly known security weakness. Biham and Kocher presented a known plaintext attack on the PKZIP encryption in [1]. They described an algorithm which extracts the encryption key (initialized with

H. Kim (Ed): ICISC 2011, LNCS 7259, pp. 235–247, 2012.
© Springer-Verlag Berlin Heidelberg 2012

the password) with 2^{40} complexity providing 12 plaintext bytes (or 2^{38} complexity providing 13 plaintext bytes). Stay introduced a ciphertext-only attack on the PKZIP encryption for some of the compression softwares [3]. If 5 files in an archive are given, the first 10 plaintext bytes of all files could be derived in some softwares at that time. He proposed a new type of attack to utilize this additional information. His attack has the complexity of 2^{63} when only one file with 12 plaintext bytes is given, but it can be much more efficient using every plaintext of 5 files.

In this paper, we propose an attack which includes a new method to reduce the complexity using the additional file's plaintext. Our attack can be regarded as a generalization of Biham and Kocher's attack in the sense that ours is same as theirs when there is only one file. In the early state of our attack, a portion of key candidates can be filtered out by checking a certain condition induced from the relation between the plaintexts. As a result, we combine the two advantages: smaller complexity of [1] and the efficiency of utilizing multi-files in [3]. In the known plaintext attack, our attack becomes about $(3.4)^l$ times faster than the attack of [1] if plaintexts of l additional files are given. The experiment supports our claim and shows that ours is at least 10 times faster than Stay's attack.

The ratio of the reduced complexity depends on the relation between plaintext values. In the chosen ciphertext attack, we can determine plaintexts to satisfy the optimal relation. As a result, our attack can be improved to become $(21.3)^l$ times faster than the original attack when l additional files are given.

This paper is organized as follows. The preliminary and previous works are briefly described in the next section. It includes the overview of PKZIP encryption algorithm and the sketches of the attacks of [1] and [3]. We explain a new attack in Section 3, 4 and validate our result by comparing with other results by some experiments in Section 5. Finally we conclude in Section 6.

2   Previous Works

In this section, we briefly describe the PKZIP encryption which can be found in [4] and fix the notation. The attack of Biham and Kocher [1] is summarized in 2.2 and the recent attack of Stay [3] is summarized in 2.3.

2.1   The PKZIP Encryption

The PKZIP encryption is a stream cipher which encrypts one byte at a time. Three 32-bit keys K^0, K^1 and K^2 are used as an internal state. One byte information 'B' is used to update these 3 keys as follows.

$$\texttt{Key\_update(B)}: \quad K^0 = \text{CRC32}(K^0, B)$$
$$K^1 = \{K^1 + L(K^0)\} \times \texttt{0x08088405} + 1$$
$$K^2 = \text{CRC32}(K^2, M(K^1))$$

The definition of CRC32(,) is described in [1]. $L(X)$ and $M(X)$ are the least and the most significant byte of X, respectively. Each bit is numbered from right to left such that $(i+1)$-st bit is the (2^i)'s position. We denote consecutive k bits of X from $(i+1)$-st bit to $(i+k)$-th bit by $k(i+k-1..i)$ bits of X. For example, if X is a 32-bit value, L(X) consists of 8(7..0) bits of X and M(X) consists of 8(31..24) bits of X. The prefix 0x indicates that it is the hexadecimal representation. The 4 basic binary operations $+$, $-$, $\times$ and $()^{-1}$ in this paper are the modular arithmetics (mod 2^{32}) in almost all cases.

The characters of the password, denoted by PW[1], PW[2], $\cdots$, are firstly used to update the internal state. After initializing with the given password, the initial internal state is denoted by $K^0[0]$, $K^1[0]$ and $K^2[0]$. Plaintext is used to update the state during the encryption. We denote plaintext bytes by P[1], P[2], $\cdots$ and ciphertext bytes by C[1], C[2], $\cdots$. The i-th internal state $K^0[i-1]$, $K^1[i-1]$ and $K^2[i-1]$ are updated with P[i] and become $K^0[i]$, $K^1[i]$ and $K^2[i]$ for i=1, 2, $\cdots$.

Key stream bytes are denoted by S[1], S[2], $\cdots$. One key stream byte is generated using $K^2[]$ of each state. S[i]=Stream_gen($K^2[i-1]$) is computed as follows.

$$\texttt{Stream\_gen}(K^2[i-1]): \quad \texttt{tmp} = (K^2[i-1]\&\texttt{0xFFFC})|2$$
$$S[i] = L(\{\texttt{tmp}\times(\texttt{tmp}\oplus1)\}\gg 8),$$

where & is the bitwise logical 'and', | is the bitwise logical 'or', $\oplus$ is the bitwise 'exclusive or' and $\gg$ is the right shift as usual.

The algorithm 1 is the overall process of the PKZIP encryption.

Algorithm 1. The PKZIP encryption overview

Require:
    1. PW[m] - The password of length m
    2. P[n] - The plaintext of length n
    3. C[n] - The empty array for ciphertext
Ensure:
    1. K^0=0x12345678, K^1=0x23456789, K^2=0x34567890 at first

1: Init_Keys(PW[m])
2: Encryption(P[n], C[n])

3: **procedure** INIT_KEYS(PW[m])
4:     **for** i=1 to m **do**
5:         Key_update(PW[i])
6:     **end for**
7: **end procedure**

8: **procedure** ENCRYPTION(P[n], C[n])
9:     **for** i=1 to n **do**
10:        C[i] = P[i]$\oplus$Stream_gen($K^2[i-1]$)
11:        Key_update(P[i])
12:     **end for**
13: **end procedure**

The followings are another representation of CRC32(,) and the definition of CRC32^{-1}(,) which is introduced in [1].

$$Y = \text{CRC32}(X, C) = (X \gg 8) \oplus \text{crc}[L(X) \oplus C], \tag{1}$$
$$X = \text{CRC32}^{-1}(Y, C) = (Y \ll 8) \oplus \text{inv}[M(Y)] \oplus C, \tag{2}$$

where crc[] and inv[] are the tables same as [1]. $\ll$ is the left shift. The crc[] tables has the linear property such that crc[L(X)$\oplus$C]= crc[L(X)]$\oplus$crc[C].

2.2   The Attack of Biham and Kocher

Biham and Kocher suggested a known plaintext attack on the PKZIP encryption in [1]. They assumed that adjacent plaintext bytes of arbitrary position are known. Following is the sketch of the trade-off version of the attack which is mentioned in [3]. We are going to explain details about some computations in section 3 since our attack is based on this algorithm. We assume that P[1], P[2], $\cdots$, P[12] are given.

1. **Making a list of 2^{24} K^2[]'s candidates**
   Guess 16(31..16) bits of K^2[11] and get 64 values of 14(15..2) bits of K^2[11] using S[12].
   - Determine 30(31..2) bits of K^2[10] using K^2[11] and S[11]
     $\rightarrow$ Find 2(1,0) more bits of K^2[11]
   - Determine 30(31..2) bits of K^2[9] using K^2[10] and S[10]
     $\rightarrow$ Find 2(1,0) more bits of K^2[10] $\rightarrow$ Find M(K^1[11])
     $\vdots$
   - Determine 30(31..2) bits of K^2[0] using K^2[1] and S[1]
     $\rightarrow$ Find 2(1,0) more bits of K^2[1] $\rightarrow$ Find M(K^1[2])
   Guess 2(1,0) bits of K^2[0] $\rightarrow$ Find M(K^1[1]).

2. **Given K^2[] and M(K^1[]), making a list of 2^{16} K^1[]'s candidates**
   Get 2^{16} values of 24(23..0) bits of K^1[11] using M(K^1[11]) and M(K^1[10])
   - Find L(K^0[11]) using K^1[11] and M(K^1[9]) $\rightarrow$ Determine K^1[10]
   - Find L(K^0[10]) using K^1[10] and M(K^1[8]) $\rightarrow$ Determine K^1[9]
     $\vdots$
   - Find L(K^0[3]) using K^1[3] and M(K^1[1])

3. **Deriving K^0[8] from L(K^0[8]), L(K^0[9]), L(K^0[10]) and L(K^0[11])**
   A list of 2^{40} candidates for the 9-th internal state is made.

4. **Checking the validity of each candidate**
   Compute K^0[i−1] using K^0[i] and P[i] for i = 8,7,6,5,4 and compare the least significant bytes with the values found in step 2. Finally the unique and correct 4-th internal state is found.

This attack requires 12 known plaintext bytes and has the complexity of 2^{40} (the number of candidates in the list). The original attack presented in [1] requires 13 bytes with 2^{38} complexity. In fact, last 2 bits guess of K^2[0] in step 1 can be done after 4 bytes comparison of $L(K^0[])$ in step 4. So the actual running times of these 2 versions seem to be similar to each other.

The complexity can be reduced if there are more adjacent plaintext bytes of the target compressed file. For example, the complexity may be reduced to about 2^{27} using about 10,000 known plaintext bytes. If an internal state is given, we can find a corresponding password which makes that initial state much more efficiently than the brute-force search.

2.3   The Attack of Stay

Stay suggested another type of attack on the PKZIP encryption in [3]. This attack has the bigger complexity than the previous one, but the number of candidates of the key can be reduced naturally when multi-files encrypted under an identical password are given.

1. 2^{23} candidates (31 bits guess and 8 bits check)
    - 2^8: Guess $L(CRC32(K^0[0],0)) \rightarrow$ Determine $L(K^0[1])$ using $P[1]$
    - 2^9: Guess $M(K^1[0] \times 0x08088405)$ and 1 carry bit
      $\rightarrow$ Determine $M(K^1[1])$ using $L(K^0[1])$
    - 2^6: Guess $14(15..2)$ bits of $CRC32(K^2[0],0)$
      $\rightarrow$ Determine $14(15..2)$ bits of $K^2[1]$ using $M(K^1[1])$ and check 8 bits condition using $S[2]$
2. 2^{19} candidates (27 bits guess and 8 bits check) $\rightarrow 2^{42}$ candidates in total
    - 2^8: Guess $8(15..8)$ bits of $CRC32(K^0[0],0)$
      $\rightarrow$ Determine $L(K^0[2])$ using $P[2]$
    - 2^9: Guess $M(K^1[0] \times 0x08088405^2)$ and 1 carry bit
      $\rightarrow$ Determine $M(K^1[2])$ using $L(K^0[2])$
    - 2^2: Guess $8(23..16)$ bits and $2(1..0)$ bits of $CRC32(K^2[0],0)$
      $\rightarrow$ Determine $14(15..2)$ bits of $K^2[2]$ using $M(K^1[2])$ and check 8 bits condition using $S[3]$
3. 2^{13} candidates (29 bits guess and 16 bits check) $\rightarrow 2^{55}$ candidates in total
    - 2^8: Guess $8(23..16)$ bits of $CRC32(K^0[0],0)$
      $\rightarrow$ Determine $L(K^0[3])$ using $P[3]$
    - 2^{13}: Guess $K^1[0]$ using the previous conditions (about 2^{13} possibilities)
      $\rightarrow$ Determine $M(K^1[3])$ using $L(K^0[3])$
    - 2^{-8}: Guess $M(CRC32(K^0[0],0))$, determine $K^0[0]$ and check 8 bits condition using $S[1]$
      $\rightarrow$ Determine $K^2[3]$ using $M(K^1[3])$ and check 8 bits condition using $S[4]$
4. 2^8 candidates (8 bits guess) $\Rightarrow 2^{63}$ candidates of $K^0[0]$, $K^1[0]$, $K^2[0]$ are found.
    - 2^8: Guess $8(31..24)$ bits of $CRC32(K^0[0],0)$ and determine $K^0[0]$
5. Check 64 bits condition using $S[5]$, $S[6]$, $\cdots$ $S[12]$
$\Rightarrow$ The correct internal state is found.

This attack requires 12 known plaintext bytes and has the complexity(also the number of key candidates) of 2^{63}.

Stay introduced a ciphertext-only attack on the public compression softwares such as WinZip and NetZip. Due to the flaws of a little different encryption method and the random byte generator, if there were 5 files in an encrypted archive, the first 10 plaintext bytes of all 5 files can be extracted.

If we can use an external method to check the validity of the internal state, less number of plaintext bytes are required in both attacks shown above. The effect of this assumption seems to similar. So we consider no situation like that in this paper.

3   New Known Plaintext Attack

Our new attack is based on the algorithm of Biham and Kocher. The attack of Stay is suitable to utilize additional files but has larger complexity. So we take the advantage of the small complexity of [1] and mount a new idea to utilize the information of multi-files. The complexity of Biham and Kocher's attack mainly depends on the product of the number of K^2 candidates and the number of K^1 candidates. We are going to filter out some of the K^2 candidates, thus improve the complexity. Our attack can be regarded as a generalization of [1] when the known plaintexts of more than one encrypted files are given.

3.1   Main Idea

In each step of Stay's attack, every bit of the initial state required to update the state and to produce the keystream is guessed in order to update the state again with other plaintext bytes. So the basic complexity is large, whereas the way of using additional information is quite straightforward.

In the first step of Biham and Kocher's algorithm, the series of $K^2[\,]$ and $M(K^1[\,])$ can be determined using the keystream after some bits of $K^2[11]$ are guessed. The plaintext bytes influence this algorithm at the last step. On the other hand, the update of K^0 is basically the xor operation with $P[1]$ due to the linear property of $crc[\,]$ table and the K^1 update begins with the addition with $L(K^0)$. So the partial information about relation between $M(K^1)$'s updated with different plaintext bytes can be obtained from the plaintext values if we adopt the concept of *the additive differential of xor* whose formal definition is found in [2]. Then the key candidates can be sieved in the first step with no more bits guess.

3.2   Attack Using 2 Files

For the clear description of our concept we deal with the case of two known plaintexts. We assume that there are two zip files[1] encrypted under an identical password and their first 12 bytes of plaintext are given. The plaintext bytes are denoted by $P_0[1], \ldots, P_0[12]$ and $P_1[1], \ldots, P_1[12]$, respectively. The lower index indicates the file identity and is going to be applied to K^0, K^1, K^2 and $S[i]$ as well.

---

[1] Note that an encrypted archive containing two files is also allowed.

Since $P_0[i]$ and $P_1[i]$ are equal for all i with the probability of $1/2^{96}$, it is natural to assume that two 12-byte plaintexts are not all the same. It does not matter which i is the smallest index such that $P_0[i] \neq P_1[i]$, so we may assume that $P_0[1] \neq P_1[1]$. Some of the computations below can be found in [1] or [3]. We write them down again for the readability.

Find the Partial Information of $K_0^1[1]$ and $K_1^1[1]$. At first, the most significant $16(31..16)$ bits of $K_0^2[1]$ has to be guessed. One byte output $S_0[2]$ is produced with the input of $14(15..2)$ bits of $K_0^2[1]$ through Stream_gen( ). For each key stream byte, there are exactly 64 inputs of 14-bit value. So 64 values for $14(15..2)$ bits of $K_0^2[1]$ can be listed directly from $S_0[2]$ with the prepared table. Thus we have a list of 2^{22} candidates for $30(31..2)$ bits of $K_0^2[1]$.

The most significant $22(31..10)$ bits of the previous state's $K_0^2[0]$ are computed from $30(31..2)$ bits of $K_0^2[1]$ using the equation

$$K_0^2[0] = (K_0^2[1] \ll 8) \oplus \text{inv}\big[M\big(K_0^2[1]\big)\big] \oplus M\big(K_0^1[1]\big).$$

Independently, 64 values for $14(15..2)$ bits of $K_0^2[0]$ are derived from $S_0[1]$ in the same manner as above.

This can be presented as Figure 1.

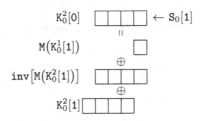

Fig. 1. The equation for $K_0^2[0]$

On average, one out of 64 values has the $6(15..10)$ bits same as the previous $22(31..10)$ bits at the overlapped location. So $30(31..2)$ bits of $K_0^2[0]$ are determined using $30(31..2)$ bits of $K_0^2[1]$ and $S_0[1]$.

The least significant $2(1,0)$ unknown bits of $K_0^2[1]$ can be computed using $2(9..8)$ bits of $K_0^2[0]$. $6(7..2)$ bits of $M\big(K_0^1[1]\big)$ are same as $6(7..2)$ bits of $K_0^2[0] \oplus \text{inv}\big[M\big(K_0^2[1]\big)\big]$. The complete value of $K_0^2[1]$, $30(31..2)$ bits of $K_0^2[0]$ and $6(7..2)$ bits of $M\big(K_0^1[1]\big)$ are included in each of the 2^{22} candidates.

Since the internal states of the two files are initialized with an identical password, their initial states are identical. This fact plays an essential role to gather more information from the second plaintext. From the equation (1) we have

$$K_1^2[1] = (K_1^2[0] \gg 8) \oplus \text{crc}\big[L\big(K_1^2[0]\big) \oplus M\big(K_1^1[1]\big)\big]$$

and already know $30(31..2)$ bits of $K_1^2[0](=K_0^2[0])$. On the other hand, 64 values for $14(15..2)$ bits of $K_1^2[1]$ are obtained from $S_1[2]$. Therefore 64 values of $14(15..2)$ bits of $\text{crc}\big[L\big(K_1^2[0]\big) \oplus M\big(K_1^1[1]\big)\big]$ can be obtained.

Since $\mathtt{crc}[\ ]$ table has the 2^8 elements among all the 32-bit values, any given 14-bit value can be one of the 14(15..2)-bit values of the table elements with the probability of 1/64. So a unique 14(15..2)-bit value of $\mathtt{crc}\big[\mathtt{L}(\mathtt{K}_1^2[0])\oplus\mathtt{M}(\mathtt{K}_1^1[1])\big]$ can survive on average. We can get the value of $\mathtt{L}(\mathtt{K}_1^2[0])\oplus\mathtt{M}(\mathtt{K}_1^1[1])$, then 6(7..2) bits of $\mathtt{M}(\mathtt{K}_1^1[1])$ come from 6(7..2) bits of $\mathtt{K}_1^2[0]$. In addition, the complete 32-bit value of $\mathtt{K}_1^2[1]$ can be computed.

Relation between $\mathtt{K}_0^1[1]$ and $\mathtt{K}_1^1[1]$. Now we will show that $\mathtt{M}(\mathtt{K}_0^1[1])-\mathtt{M}(\mathtt{K}_1^1[1])$ should satisfy a certain condition, provided that we guess $\mathtt{K}_0^2[1]$ correctly in the previous step. Thus we can filter out some of the wrong candidates from the list.

The staring point of the condition is $\mathtt{P}_0[1]\oplus\mathtt{P}_1[1]$. The exact value of $\mathtt{K}_0^0[1]\oplus\mathtt{K}_1^0[1]$ can be computed even without any knowledge of each value.

$$\begin{aligned}
\mathtt{K}_0^0[1] \oplus \mathtt{K}_1^0[1] &= \mathtt{CRC32}(\mathtt{K}_0^0[0],\mathtt{P}_0[1]) \oplus \mathtt{CRC32}(\mathtt{K}_1^0[0],\mathtt{P}_1[1]) \\
&= (\mathtt{K}_0^0[0] \gg 8) \oplus \mathtt{crc}\big[\mathtt{L}(\mathtt{K}_0^0[0])\oplus\mathtt{P}_0[1]\big] \\
&\quad \oplus (\mathtt{K}_1^0[0] \gg 8) \oplus \mathtt{crc}\big[\mathtt{L}(\mathtt{K}_1^0[0])\oplus\mathtt{P}_1[1]\big] \\
&= \mathtt{crc}[\mathtt{P}_0[1]\oplus\mathtt{P}_1[1]] \text{ (Because } \mathtt{K}_0^0[0] = \mathtt{K}_1^0[0]).
\end{aligned}$$

This *xor difference* of $\mathtt{K}_0^0[1]$ and $\mathtt{K}_1^0[1]$ affects on the *subtraction difference* of $\mathtt{K}_0^1[1]$ and $\mathtt{K}_1^1[1]$ through the key update process as follows.

$$\begin{aligned}
\mathtt{K}_0^1[1] - \mathtt{K}_1^1[1] &= \big\{(\mathtt{K}_0^1[0] + \mathtt{L}(\mathtt{K}_0^0[1])) \times \mathtt{0x08088405} + 1\big\} \\
&\quad - \big\{(\mathtt{K}_1^1[0] + \mathtt{L}(\mathtt{K}_1^0[1])) \times \mathtt{0x08088405} + 1\big\} \\
&= \big\{\mathtt{L}(\mathtt{K}_0^0[1]) - \mathtt{L}(\mathtt{K}_1^0[1])\big\} \times \mathtt{0x08088405}.
\end{aligned}$$

In summary, we have

$$\mathtt{K}_0^1[1]-\mathtt{K}_1^1[1]=\big\{\mathtt{L}(\mathtt{K}_0^0[1])-\mathtt{L}(\mathtt{K}_0^0[1]\oplus\mathtt{crc}[\mathtt{P}_0[1]\oplus\mathtt{P}_1[1]])\big\}\times\mathtt{0x08088405}. \tag{3}$$

For a fixed $\mathtt{P}_0[1]\oplus\mathtt{P}_1[1]$, not all the possible most significant 6(31..26) bits appear in the equation (3) even running it over all 256 values of $\mathtt{L}(\mathtt{K}_0^0[1])$. For example, if $\mathtt{P}_0[1]\oplus\mathtt{P}_1[1]=\mathtt{0x1d}$, then $\mathtt{L}(\mathtt{crc}[\mathtt{P}_0[1]\oplus\mathtt{P}_1[1]])=\mathtt{0xd9}$. In this case, there are only 18 values for 6(7..2) bits of $\mathtt{M}(\mathtt{K}_0^1[1])-\mathtt{M}(\mathtt{K}_1^1[1])$ considering the possibility of the unknown carry bit. The $-$ operation between 6(7..2) bits of $\mathtt{M}(\mathtt{K}_0^1[1])$ and $\mathtt{M}(\mathtt{K}_1^1[1])$ is calculated under the modulus 2^6.

On average over all $\mathtt{P}_0[1]\oplus\mathtt{P}_1[1]$, 18.7 out of 64 values for 6(7..2) bits of $\mathtt{M}(\mathtt{K}_0^1[1])-\mathtt{M}(\mathtt{K}_1^1[1])$ are possible. The table 2 in the appendix A shows the number of possible $\mathtt{M}(\mathtt{K}_0^1[1])-\mathtt{M}(\mathtt{K}_1^1[1])$ values for each $\mathtt{P}_0[1]\oplus\mathtt{P}_1[1]$.

We can filter out some of the candidates using the second file's information $\mathtt{P}_1[1]$ and $\mathtt{S}_1[2]$ by keeping track of the plaintext difference influence. The trace of the $\mathtt{P}[1]$ difference in the key updated process is depicted in figure 2. The number of $\mathtt{K}_0^2[1]$ candidates is reduced to $2^{22} \times (18.7/64) = 2^{22} \times (1/3.4) \approx 2^{20.2}$ on average.

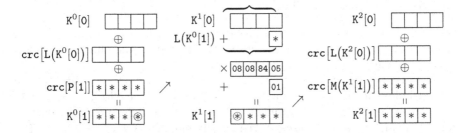

Fig. 2. Difference trace of P[1] in the key update process

Rest of the Attack. The rest of the attack is similar to [1]. Now we have $2^{20.2}$ candidates of $K_0^2[1]$, $K_1^2[1]$, 30(31..2) bits of $K_0^2[0]$, 6(7..2) bits of $M\bigl(K_0^1[1]\bigr)$ and 6(7..2) bits of $M\bigl(K_1^1[1]\bigr)$. By the similar way of finding $K_1^1[1]$ and 6(7..2) bits of $M\bigl(K_1^1[1]\bigr)$, we can find the forward state's $K_0^2[i]$ and $M\bigl(K_0^1[i]\bigr)$ using $S_0[i+1]$ in the sequel for i=2,...,7.

At this point, we have to increase the number of candidates by the factor of 2^{16} because of guessing the least significant 24-bit value of $K_0^1[7]$. We guess 24(23..0) bits of $K_0^1[7]$ and compute

$$(K_0^1[7] - 1) \times \text{0x08088405}^{-1} = K_0^1[6] + L\bigl(K_0^0[7]\bigr).$$

About 2^{16} values remain by comparing the most significant bytes of the both side. Through the pre-computed table, we can determine about 2^{16} values for 24(23..0) bits of $K_0^1[7]$ from $M\bigl(K_0^1[6]\bigr)$ and $M\bigl(K_0^1[7]\bigr)$ without excess guessing.

$L\bigl(K_0^0[7]\bigr)$ is determined by comparing the most significant bytes of following equation.

$$K_0^1[5] + L\bigl(K_0^0[6]\bigr) = (K_0^1[6] - 1) \times \text{0x08088405}^{-1}$$
$$= \bigl\{ (K_0^1[7] - 1) \times \text{0x08088405}^{-1} + L\bigl(K_0^0[7]\bigr) - 1 \bigr\} \times \text{0x08088405}^{-1}$$

Determining $L\bigl(K_0^0[7]\bigr)$ can also be done by reading the pre-computed table instead of comparing 8 bits for all 256 trials for $L\bigl(K_0^0[7]\bigr)$. In the same manners, we can compute the backward state's $K_0^1[i]$ and $L\bigl(K_0^0[i]\bigr)$ in the sequel for i=6,5,4.

By the linearity of CRC32(,) function, we can determine $K_0^0[7]$ with $L\bigl(K_0^0[4]\bigr)$, $L\bigl(K_0^0[5]\bigr)$, $L\bigl(K_0^0[6]\bigr)$ and $L\bigl(K_0^0[7]\bigr)$. Then a complete state which consists of $K_0^0[7]$, $K_0^1[7]$ and $K_0^2[7]$ is made for each element of $2^{36.2}$ lists.

Until this point, we utilize 10 known plaintext bytes, $P_0[1]$,...,$P_0[8]$, $P_1[1]$ and $P_1[2]$. We need 5 more plaintext bytes except $P_0[1]$,...,$P_0[8]$ to check whether the internal state is guessed correctly. We remark that $P_1[1]$ and $P_1[2]$ can be reused. The efficiency of using $P_1[1]$ is relatively low because it was already used for the same purpose at the beginning.

As a result, the attack has the complexity of $2^{36.2}$ and uses at most 14-byte known plaintext. It is aforementioned that the two versions of the Biham and Kocher's attack seem to have similar time complexity. So our new attack is about 3.4 times faster than the previous attack on average.

3.3   Attack with More Files

If some additional files encrypted under the same password are given with their forepart plaintext bytes, they also can be used to reduce the number of candidates at the early stage of the attack.

The exact position where we use the j-th additional file F_j is related to the smallest index, denoted by i_j, such that $P_0[i_j] \neq P_j[i_j]$. The effects of F_j and F_k are independent unless $i_j = i_k$ and $P_j[i_j] = P_k[i_k]$. The difference between $P_j[]$ and $P_k[]$ is also able to be used to reduce the number of key candidates. The effects of this additional filtering differ according to the positions of the discrepancy and the difference values. The algorithm of the attack with many files is a straightforward extension. As a result, we can expect the attack to be about $(3.4)^l$ times faster when l additional files with their plaintext are given.

4   Chosen Plaintext Attack

In the attack of the previous section, the ratio of the filtered wrong keys using the plaintext of j-th additional file varies with the value of $P_0[i_j] \oplus P_j[i_j]$. Unlike the Stay's attack, our attack's efficiency depends on the *relation* between the plaintexts, so we can take an advantage in the chosen ciphertext attack. The number's minimum of the possible values for $6(7..2)$ bits of $M(K_0^1[1]) - M(K_1^1[1])$ is 3 in the table 2 of the appendix A. If we choose the plaintext bytes to reach this minimum, our attack becomes $64/3 \approx 21.3$ times faster whenever one file is additionally given.

The number of candidate keys in the list right before the filtering is 2^{22}. In the known plaintext attack, each additional file can make the number decrease by the factor of about 3.4 on average. Since 13 is the smallest inter i satisfying $2^{22} \times (1/3.4)^i < 1$, it can be expected that there remains a unique key candidate when 13 additional files are given. Similarly, in the chosen plaintext attack, one key candidate can be obtained at this filtering stage with 5 additional files because 5 is the smallest i such that $2^{22} \times (1/21.3)^i < 1$.

5   Experimental Result and Comparison

In this section, we give the time measuring result about the implementation of our attack. It is briefly described how we can use the additional files in the Stay's attack. We implement the attack of Stay including the case of many files and compare these results.

5.1   The Implementation of Our Attack

Our attack is implemented on the PC equipped with the Intel(R) Core(TM) i7 CPU 870 @ 2.93GHz using the C language. Since 8 threads are available on this CPU, we have to employ a parallelizing technique. We use 'openmp' to work with all 8 threads and simply parallelize the first loop for 16 bits guessing of

Table 1. The experimental result of our attack

The number of files	1	2	3	4	5
Average attack time	124.5 min	24.1 min	4.2 min	37.8 sec	11.7 sec
Ratio to the left column	·	0.19	0.17	0.15	0.31
Ratio to the 1st col. per additional file	1	0.19	0.18	0.17	0.20

$K_0^2[1]$. We measure the worst case time complexity which means the key finding process goes till the end even after the correct key is found. The correct key is found around the middle on average.

The result is summarized in the table 1. Whenever one file is added, the ratio of time deceasing seems to be somewhat regular. To focus on measuring the ratios, we use the common F_0 for all cases, the common F_1 for the last 4 cases, and so on, in each iteration. This result is better than the expected. One additional file makes the attack about 5 times faster.

5.2   The Comparison with the Attack of Stay

The way to utilize an additional file in the Stay's attack is natural. We explicitly and briefly write down the case of 2 files. We denote temporarily 0 and 1 as $\star$, for example, $K_\star^0[1]$ means $K_0^0[1]$ and $K_1^0[1]$.

1. 2^{16} candidates (32 bits guess / 16 bits check)
   - Guess $L(CRC32(K_0^0[0],0)) \to$ Determine $L(K_\star^0[1])$ using $P_\star[1]$
   - Guess $M(K_0^1[0] \times 0x08088405)$ and 2 carries $\to$ Determine $M(K_\star^1[1])$
   - Guess $14(15..2)$ of $CRC32(K_0^2[0],0)$
     $\to$ Determine $14(15..2)$ of $K_\star^2[1]$ and check 16 bits using $S_\star[2]$
2. 2^{28} candidates (28 bits guess / 16 bits check)
   - Guess $8(15..8)$ of $CRC32(K_0^0[0],0) \to$ Determine $L(K_\star^0[2])$ using $P_\star[2]$
   - Guess $M(K_0^1[0] \times 0x08088405^2)$ and 2 carries $\to$ Determine $M(K_\star^1[2])$
   - Guess $10(23..16,1,0)$ of $CRC32(K_0^2[0],0)$
     $\to$ Determine $14(15..2)$ of $K_\star^2[2]$ and check 16 bits using $S_\star[3]$
3. 2^{30} candidates (26 bits guess / 24 bits check)
   - Guess $8(23..16)$ of $CRC32(K_0^0[0],0) \to$ Determine $L(K_\star^0[3])$ using $P_\star[3]$
   - Guess $K_0^1[0]$ (2^{10} possibilities) $\to$ Determine $M(K_\star^1[3])$
   - Guess $M(CRC32(K_0^0[0],0))$, determine $K_0^0[0]$ and check 8 bits using $S_0[1]$
     $\to$ Determine $K_\star^2[3]$ and check 16 bits using $S_\star[4]$
4. 2^{38} candidates of $K_0^0[0]$, $K_0^1[0]$, $K_0^2[0]$ (8 bits guess)
   - 2^8: Guess $8(31..24)$ of $CRC32(K_0^0[0],0)$ and determine $K_0^0[0]$
5. Check 40 bits using $S_\star[5]$, $S_\star[6]$ and $S_0[7] \Rightarrow$ The correct state is found.

The maximum number of candidates is 2^{38}, which is a great improvement from the single file case. One of the greatest advantages of this algorithm is that the number of lists decreases by the factor of 2^7 in every step whenever one additional file is given. However, it must be done in the way of one more bits guessing and 8 more bits check. It is not easy to reduce the number of intermediate candidates using some preparations.

We have to mention that finding the list of $K_0^1[0]$ efficiently in step 3 does not seem to be easy to us. The 32-bit brute force search is very inefficient. The pre-computation table including about 16GB information can be made if we do not apply the information about the carry bits. It is not easy to handle this table due to its size.

If 5 files are given, there remains a unique list in each step. So no more additional file is needed. Thus we implemented this attack of 5-file case in the same environment including the parallel computing with 8 threads as the previous subsection. We employed an inevitable cheating about $K_0^1[0]$ in step 3. We just use the stored right key value to make the list of appropriate size. The worst case average time of this attack is about 2 minutes. It is about 10 times slower than 5 file case of our new attack.

If there is only one file, the Biham and Kocher's algorithm is more efficient than the Stay's algorithm. The improvement of the complexity per additional file is much greater in the Stay's attack. The complexity of the Stay's attack keep decreasing until the number of files encrypted under an identical password grows to 5. Our attack's complexity decreases until 14 files are given on average. So we can conclude that our attack is always at least 10 times faster than the Stay's attack if the same number of files are given with their plaintext.

6   Conclusion

In this paper, we improved the known plaintext attack on the PKZIP encryption suggested in [1]. Our attack can utilize the additional information of multi-files encrypted under the same password. Comparing the single file case (original version of [1]), our attack can be about $(3.4)^l$ times faster when $(l+1)$ files with their plaintext bytes are given. In the chosen ciphertext model, ours becomes $(21.3)^l$ times faster when $(l+1)$ files are used.

References

1. Biham, E., Kocher, P.: A Known Plaintext Attack on the PKZIP Stream Cipher. In: Preneel, B. (ed.) FSE 1994. LNCS, vol. 1008, pp. 144–153. Springer, Heidelberg (1995)
2. Lipmaa, H., Wallén, J., Dumas, P.: On the Additive Differential Probability of Exclusive-Or. In: Roy, B., Meier, W. (eds.) FSE 2004. LNCS, vol. 3017, pp. 317–331. Springer, Heidelberg (2004)
3. Stay, M.: ZIP Attacks with Reduced Known Plaintext. In: Matsui, M. (ed.) FSE 2001. LNCS, vol. 2355, pp. 125–134. Springer, Heidelberg (2002)
4. PKWARE, Inc., APPNOTE.TXT - .ZIP File Format Specification. version 6.3.2, http://www.pkware.com/documents/casestudies/APPNOTE.TXT

A   The Subtraction Difference of $K^1[1]$

We are going to explain the detail about the process to find the possible values of $K^1[1]$ difference. Let assume that $P_0[1] \oplus P_1[1] = \text{0x1d}$, then $L(\text{crc}[P_0[1] \oplus P_1[1]])$ is 0xd9. The equation (3) can be written again with the specific value.

$$K_0^1[1] - K_1^1[1] = \left\{ L(K_0^0[1]) - L(K_0^0[1]) \oplus \text{0xd9} \right\} \times \text{0x08088405} \qquad (4)$$

For all the values of $L(K_0^0[1])$, 16 distinct 6-bit values appear as the most significant 6(31..26) bits of the righthand side of the equation (4). The entire values of this list are 0xc, 0xd, 0xe, 0xf, 0x10, 0x11, 0x12, 0x13, 0x2c, 0x2d, 0x2e, 0x2f, 0x30, 0x31, 0x32 and 0x33.

Assume that the actual value of the most significant 6 bits of equation (4) is 0x13. If 26(25..0) bits of $K_0^1[1]$ is greater than 26(25..0) bits of $K_1^1[1]$, 6(7..2) bits of $M(K_0^1[1])$ minus 6(7..2) bits of $M(K_1^1[1])$ equals to 0x14. So the number of all the possible values for the most significant 6-bit subtraction of $K^1[1]$ considering the carry possibility is 18, which consists of 0xc, 0xd, 0xe, 0xf, 0x10, 0x11, 0x12, 0x13, 0x14, 0x2c, 0x2d, 0x2e, 0x2f, 0x30, 0x31, 0x32, 0x33 and 0x34.

The numbers of the possible subtraction differences for all values of $P_0[1] \oplus P_1[1]$ are presented in Table 2. For example, when $P_0[1] \oplus P_1[1] = \text{0x1d}$, the number of subtraction difference is placed in the intersection of the row with index '1' and the column with index 'd' in the table. The result is 18 as the above description. The average value of all 255 elements is 18.7.

Table 2. The subtraction difference table of K^1

	0	1	2	3	4	5	6	7	8	9	a	b	c	d	e	f
0	·	20	12	20	12	56	12	17	6	10	24	8	24	36	16	9
1	6	12	6	40	28	34	6	33	12	5	14	20	48	18	12	20
2	10	24	10	6	9	24	40	24	24	12	20	3	17	12	64	12
3	20	12	4	12	18	12	18	32	40	4	10	6	25	6	34	24
4	5	8	20	12	16	48	18	12	10	6	32	6	34	24	33	4
5	12	6	10	24	36	28	9	24	20	3	20	14	64	12	19	12
6	6	40	6	10	6	33	24	36	12	24	12	5	12	18	48	18
7	12	20	3	16	12	17	8	64	16	10	6	10	24	7	24	36
8	3	20	8	20	12	64	12	13	6	8	24	10	16	36	24	9
9	6	10	6	48	24	34	6	33	12	5	12	20	48	20	12	18
a	10	28	10	6	11	24	36	24	20	12	24	3	17	12	64	14
b	16	12	5	12	18	8	18	48	40	6	10	4	33	6	30	24
c	5	12	20	8	18	48	14	12	8	6	40	6	34	16	33	6
d	10	6	12	24	36	24	9	24	20	3	20	12	64	12	17	12
e	6	40	6	10	6	35	24	34	12	20	12	5	12	18	55	18
f	12	16	3	20	8	17	12	64	24	10	4	10	24	9	24	30

Synthetic Linear Analysis: Improved Attacks on CubeHash and Rabbit

Yi Lu[1], Serge Vaudenay[2], Willi Meier[3],
Liping Ding[1], and Jianchun Jiang[1]

[1] National Engineering Research Center of Fundamental Software,
Institute of Software, Chinese Academy of Sciences, Beijing, China
[2] EPFL, Lausanne, Switzerland
[3] FHNW, Windisch, Switzerland

Abstract. It has been considered most important and difficult to analyze the bias and find a large bias regarding the security of crypto-systems, since the invention of linear cryptanalysis. The demonstration of a large bias will usually imply that the target crypto-system is not strong. Regarding the bias analysis, researchers often focus on a theoretical solution for a specific problem. In this paper, we take a first step towards the synthetic approach on bias analysis. We successfully apply our synthetic analysis to improve the most recent linear attacks on CubeHash and Rabbit respectively. CubeHash was selected to the second round of SHA-3 competition. For CubeHash, the best linear attack on 11-round CubeHash with 2^{470} queries was proposed previously. We present an improved attack for 11-round CubeHash with complexity $2^{414.2}$. Based on our 11-round attack, we give a new linear attack for 12-round CubeHash with complexity 2^{513}, which is sharply close to the security parameter 2^{512} of CubeHash. Rabbit is a stream cipher among the finalists of ECRYPT Stream Cipher Project (eSTREAM). For Rabbit, the best linear attack with complexity 2^{141} was recently presented. Our synthetic bias analysis yields the improved attack with complexity 2^{136}. Moreover, it seems that our results might be further improved, according to our ongoing computations.

Keywords: bias, linear cryptanalysis, synthetic analysis, conditional dependence, CubeHash, Rabbit.

1 Introduction

It has been considered most important and difficult to analyze the bias and find a large bias regarding the security of crypto-systems, since the invention of linear cryptanalysis [6] almost 20 years ago. The demonstration of a large bias will usually imply that the target crypto-system is not as strong as expected. Regarding the bias analysis, researchers often focus on a theoretical solution for a specific problem. Unfortunately, it does not help much to analyze the bias for a broad class of problems.

H. Kim (Ed): ICISC 2011, LNCS 7259, pp. 248–260, 2012.
© Springer-Verlag Berlin Heidelberg 2012

Most often, we need to study the combined bias of multiple Boolean functions (such as multiple linear approximations) with many input variables. Assuming that these Boolean functions are all independent pairwise, the problem reduces to the bias computation of each Boolean function separately. Apparently, if the terms involved in each Boolean function are statistically independent of the terms in the others, we are sure that all are independent pairwise and it is "safe" to concentrate on bias computation of each Boolean function. Further, it is worth pointing out that it is incorrect to conclude independence when the terms involved in each function "differ" from the terms occurring in the others. It is thus essential to conduct synthetic analysis to study these bias problems. In this paper, we take a first step towards the synthetic approach on bias analysis. We also propose a conditional dependent bias problem and we give an analysis to estimate the bias.

We apply our synthetic analysis to improve the most recent linear attack [1] on the hash function CubeHash [2]. CubeHash was selected to the second round of SHA-3 competition [8]. In [1], based on the bias analysis for 11-round CubeHash, the best linear attack on 11-round CubeHash with 2^{470} queries was proposed. Our results improve the bias analysis [1]. We show the largest bias $2^{-207.1}$ for 11-round CubeHash, and we present an improved linear attack for 11-round CubeHash with complexity $2^{414.2}$. Further, based on our 11-round attack, we give a new linear attack for 12-round CubeHash with complexity 2^{513}, which is sharply close to the security parameter 2^{512} of CubeHash.

Meanwhile, our synthetic analysis is applied to the recent linear attack [5] on stream cipher Rabbit [3]. Rabbit is a stream cipher among the finalists of ECRYPT Stream Cipher Project (eSTREAM). It has also been published as informational RFC 4503 with the Internet Engineering Task Force (IETF). In [5], the best linear attack with complexity 2^{141} was presented. As reference, Rabbit designers claim the security level 2^{128}. Our synthetic analysis applies to the main part of the bias analysis [5]. Our results yield the improved linear attack with complexity 2^{136}.

The rest of the paper is organized as follows. In Section 2, we give preliminary analysis on CubeHash. In Section 3, we introduce the synthetic approach to the bias analysis problem and discuss how to apply to CubeHash round function in details. In Section 4, we propose the synthetic bias analysis for the conditional dependent problem. In Section 5 and Section 6, we present our improved attacks on CubeHash and Rabbit. We conclude in Section 7.

2   Preliminary Analysis on CubeHash Round Function

The hash function CubeHash [2] was designed by Prof. Daniel J. Bernstein. It was one of the 14 candidates which were selected to the second round of SHA-3 competition [8]. SHA-3 was initiated by the U.S. National Institute of Standards and Technology to push forwards the development of a new hash standard, following the recent fruitful research work on the hash function cryptanalysis. CubeHash is a family of cryptographic hash functions, parameterized by the performance and security requirement. At the heart of it, CubeHash consists of an internal state of 1024 bits, round transformation T, round number r, between introduction of new message blocks. At the end, T is repeated $10r$ times before outputting h bits of its state as the final hash value. Security/performance tradeoffs are provided with different combinations h, r and the message block length b. The normal security parameters are $r = 16, b = 32$, according to [1].

Each Round of CubeHash consists of two half rounds. Each half round consists of five steps, and only one step out of five introduces nonlinearity to the internal state by performing the modular addition operations. We will investigate the largest bias [1] for CubeHash. It was shown that due to this largest bias, a non-trivial linear attack on 11-round CubeHash with 2^{470} queries exists. As reference, the security parameter is 2^{512}. We will improve the bias analysis of multiple linear approximations in [1]. Recall that the bias[1] of a binary random variable X is $\Pr[X = 0] - \Pr[X = 1]$. Our main focus is that, within each round, the linear approximations are *not* all independent pairwise. This can be justified by the fact that nonlinearity is introduced by two separate steps (Step 1 and Step 6) instead of one step within a round.

Let us start from a simple case of Round 7 first. We let 32 words $x_{00000}, x_{00001},$ $\ldots, x_{11111}$ to denote the internal states of 1024 bits (each word has 32 bits). The round transformation T can be described by the following ten steps of operations ('+' denotes modular addition):

$$\text{Step 1:} \quad x_{0n} + x_{1n} \to x_{1n} \text{ for all 4-bit } n$$
$$\text{Step 2:} \quad x_{0n} \lll 7 \to x_{0n} \text{ for all 4-bit } n$$
$$\text{Step 3:} \quad x_{00n} \leftrightarrow x_{01n} \text{ for all 3-bit } n$$
$$\text{Step 4:} \quad x_{0n} \oplus x_{1n} \to x_{0n} \text{ for all 4-bit } n$$
$$\text{Step 5:} \quad x_{1jk0m} \leftrightarrow x_{1jk1m} \text{ for all 1-bit } j, k, m$$
$$\text{Step 6:} \quad x_{0n} + x_{1n} \to x_{1n} \text{ for all 4-bit } n$$
$$\text{Step 7:} \quad x_{0n} \lll 11 \to x_{0n} \text{ for all 4-bit } n$$
$$\text{Step 8:} \quad x_{0j0km} \leftrightarrow x_{0j1km} \text{ for all 1-bit } j, k, m$$
$$\text{Step 9:} \quad x_{0n} \oplus x_{1n} \to x_{0n} \text{ for all 4-bit } n$$
$$\text{Step 10:} \quad x_{1jkm0} \leftrightarrow x_{1jkm1} \text{ for all 1-bit } j, k, m$$

For our purpose, we use superscripts to represent the step number within the round. We let the states without superscripts to represent the states right at beginning of the round. The round number of the internal states which we study is clear from the context, and we omit it from the notations. At Step 1 of Round 7, the step operation allows us to deduce:

$$0x300 \cdot x_{10100} \oplus 0x300 \cdot x_{10110} \tag{1}$$
$$= 0x300 \cdot x_{10100} \oplus 0x300 \cdot x_{00100} \oplus 0x300 \cdot x_{00100} \oplus \tag{2}$$
$$0x300 \cdot x_{10110} \oplus 0x300 \cdot x_{00110} \oplus 0x300 \cdot x_{00110} \tag{3}$$
$$\approx 0x300 \cdot x_{10100}^{1} \oplus 0x300 \cdot x_{10110}^{1} \oplus 0x300 \cdot x_{00100}^{1} \oplus 0x300 \cdot x_{00110}^{1} \tag{4}$$

---

[1] Our definition of bias is slightly different from [1], it was defined as $\Pr[X = 0] - 1/2$ in [1].

We note that two linear approximations are introduced into (4):

$$0x300 \cdot x^1_{10100} \oplus 0x300 \cdot x^1_{00100} \approx 0x300 \cdot (x^1_{10100} - x^1_{00100}) \tag{5}$$

$$0x300 \cdot x^1_{10110} \oplus 0x300 \cdot x^1_{00110} \approx 0x300 \cdot (x^1_{10110} - x^1_{00110}) \tag{6}$$

We continue on (4) from Step 2 through Step 5:

$$= 0x300 \cdot x^2_{10100} \oplus 0x300 \cdot x^2_{10110} \oplus 0x18000 \cdot x^2_{00100} \oplus 0x18000 \cdot x^2_{00110}$$

$$= 0x300 \cdot x^3_{10100} \oplus 0x300 \cdot x^3_{10110} \oplus 0x18000 \cdot x^3_{01100} \oplus 0x18000 \cdot x^3_{01110}$$

$$= 0x300 \cdot x^4_{10100} \oplus 0x300 \cdot x^4_{10110} \oplus 0x18000 \cdot x^4_{01100} \oplus 0x18000 \cdot x^4_{11100} \oplus$$
$$0x18000 \cdot x^4_{01110} \oplus 0x18000 \cdot x^4_{11110}$$

$$= 0x300 \cdot x^5_{10110} \oplus 0x300 \cdot x^5_{10100} \oplus 0x18000 \cdot x^5_{01100} \oplus 0x18000 \cdot x^5_{11110}$$
$$\oplus 0x18000 \cdot x^5_{01110} \oplus 0x18000 \cdot x^5_{11100} \tag{7}$$

At Step 6, (7) can be rewritten as,

$$= 0x300 \cdot x^5_{10110} \oplus 0x300 \cdot x^5_{00110} \oplus 0x300 \cdot x^5_{00110} \oplus$$
$$0x300 \cdot x^5_{10100} \oplus 0x300 \cdot x^5_{00100} \oplus 0x300 \cdot x^5_{00100} \oplus$$
$$0x18000 \cdot x^5_{01100} \oplus 0x18000 \cdot x^5_{11100} \oplus 0x18000 \cdot x^5_{01110} \oplus 0x18000 \cdot x^5_{11110}$$
$$\approx 0x300 \cdot x^6_{10110} \oplus 0x300 \cdot x^6_{00110} \oplus 0x300 \cdot x^6_{10100} \oplus 0x300 \cdot x^6_{00100} \oplus$$
$$0x18000 \cdot x^6_{11100} \oplus 0x18000 \cdot x^6_{11110}$$

Four linear approximations are introduced in this step:

$$0x300 \cdot x^5_{10110} \oplus 0x300 \cdot x^5_{00110} \approx 0x300 \cdot (x^5_{10110} + x^5_{00110}) \tag{8}$$

$$0x300 \cdot x^5_{10100} \oplus 0x300 \cdot x^5_{00100} \approx 0x300 \cdot (x^5_{10100} + x^5_{00100}) \tag{9}$$

$$0x18000 \cdot x^5_{01100} \oplus 0x18000 \cdot x^5_{11100} \approx 0x18000 \cdot (x^5_{01100} + x^5_{11100}) \tag{10}$$

$$0x18000 \cdot x^5_{01110} \oplus 0x18000 \cdot x^5_{11110} \approx 0x18000 \cdot (x^5_{01110} + x^5_{11110}) \tag{11}$$

It is clear that the bias for the linear approximation at Round 7,

$$0x300 \cdot x_{10100} \oplus 0x300 \cdot x_{10110} \tag{12}$$

$$\approx 0x180000 \cdot (x^{10}_{00000} \oplus x^{10}_{00010} \oplus x^{10}_{10001} \oplus x^{10}_{10011}) \oplus$$
$$0x300 \cdot (x^{10}_{10101} \oplus x^{10}_{10111}) \oplus 0x18000 \cdot (x^{10}_{11101} \oplus x^{10}_{11111})$$

equals the combined bias of the six approximations (5), (6), (8), (9), (10), (11) holding simultaneously. Furthermore, if these approximations are independent, we apply Piling-up lemma [6] to deduce that the total bias is equal to the product of the six individual biases for each linear approximation. Unfortunately, as we demonstrate below, this independence assumption is not true.

Obviously, we can easily justify that Approximations (5), (6) are independent, because the involved states x_{10100}, x_{00100} in (5) are independent of the involved states x_{10110}, x_{00110} in (6); similarly, Approximations (8), (9), (10), (11) are independent pairwise. Our main focus here is to show below that these two groups of approximations are, however, *not* independent. The internal states are invertible with the CubeHash round function T, as each step operation is invertible. Thus, we can rewrite Approximations (8), (9), (10), (11) in terms of states right after step one as follows respectively,

$$0x300 \cdot (x^1_{10100} + (x^1_{01110} \lll 7 \oplus x^1_{10110})) \approx 0x300 \cdot (x^1_{10100} \qquad (13)$$
$$\oplus x^1_{01110} \lll 7 \oplus x^1_{10110})$$

$$0x300 \cdot (x^1_{10110} + (x^1_{01100} \lll 7 \oplus x^1_{10100})) \approx 0x300 \cdot (x^1_{10110} \qquad (14)$$
$$\oplus x^1_{01100} \lll 7 \oplus x^1_{10100})$$

$$0x18000 \cdot (x^1_{11110} + (x^1_{00100} \lll 7 \oplus x^1_{11100})) \approx 0x18000 \cdot (x^1_{11110} \qquad (15)$$
$$\oplus x^1_{00100} \lll 7 \oplus x^1_{11100})$$

$$0x18000 \cdot (x^1_{11100} + (x^1_{00110} \lll 7 \oplus x^1_{11110})) \approx 0x18000 \cdot (x^1_{11100} \qquad (16)$$
$$\oplus x^1_{00110} \lll 7 \oplus x^1_{11110})$$

In the next section, we will discuss the synthetic bias analysis and apply it for CubeHash Round 7 to analyze (5), (6), (13), (14), (15) and (16).

3　The Synthetic Approach

When we study the combined bias of multiple Boolean functions, such as multiple linear approximations, it is common to assume that they are all independent pairwise. This way, the problem reduces to the bias computation of each Boolean function separately. Apparently, if the terms involved in each function are statistically independent of the terms in the other functions, we are sure that all function outputs are independent pairwise and it is "safe" to concentrate on bias computation of each Boolean function. Further, it is worth pointing out that it is incorrect to conclude independence when the terms involved in each function "differ" from the terms occurring in the other functions. For example, one might take it for granted that (5), (6), (8), (9), (10), (11) are all independent, as the terms occurring in any linear approximation never occurs in other approximations. As a matter of fact, as we will see later, after re-writing (8), (9), (10), (11) equivalently by (13), (14), (15), (16) respectively, they are *not* all independent.

It thus leads us naturally to the "Divide-and-Conquer" method to the bias analysis involving multiple Boolean functions. That is, we try to group multiple possibly dependent Boolean functions (eg. linear approximations with regards to CubeHash). The aim is that the functions in each group are dependent and the functions in different groups are independent. When grouping, it is desirable to make each group size as small as possible. The group size is referred to the number of functions contained in the group. The rationale behind grouping is that, we are dividing originally one (big) group of a larger number of functions into multiple independent groups; once grouping is done, we just need to study each group of smaller size individually. This helps make the task of bias analysis easier by reducing the number of the functions, which have to be studied simultaneously. We will explain in details next on CubeHash.

3.1   Our Analysis on CubeHash Round Function

We will first see how to group the six approximations (5), (6), (13), (14), (15) and (16) for CubeHash Round 7. We look at (13) and (14) first. At first glance, it seems that they are dependent as both have x^1_{10110} and x^1_{10100}. However, we note that x^1_{01110}, x^1_{01100} only occurs once in (5), (6), (13), (14), (15), (16), i.e., neither occurs in (5), (6), (15), (16). From the fact that x^1_{01110}, x^1_{01100} are independent, we deduce that $x^1_{10100}, x^1_{01110} \lll 7$ are independent of $x^1_{10110}, x^1_{01100} \lll 7$. Thus, $x^1_{10100}, x^1_{01110} \lll 7 \oplus x^1_{10110}$ are independent of $x^1_{10110}, x^1_{01100} \lll 7$; and $x^1_{10100}, x^1_{01110} \lll 7 \oplus x^1_{10110}$ are independent of $x^1_{10110}, x^1_{01100} \lll 7 \oplus x^1_{10100}$. Consequently, we know that (13) and (14) are independent. As x^1_{10100} occurs in both (5) and (13), we group (5) and (13) together. Likewise, as x^1_{10110} occurs in both (6) and (14), we group (6) and (14) together.

Both (15), (16) involve x^1_{11110}, x^1_{11100}, and it thus seems that (15), (16) are dependent. We can use the fact that x^1_{00100}, x^1_{00110} are independent to show that $x^1_{11110}, x^1_{00100} \lll 7 \oplus x^1_{11100}$ are independent of $x^1_{11100}, x^1_{00110} \lll 7 \oplus x^1_{11110}$. So, we deduce (15) and (16) are independent. As (15), (16) relates to x^1_{00100}, x^1_{00110} respectively, x^1_{00100} is related to (5), (13), and x^1_{00110} is related to (6) and (14). Therefore, we are able to make two groups. Group One contains (5), (13), (15). Group Two contains (6), (14), (16). These two groups are independent as we have just explained above.

When it comes to the joint bias computation of a group of dependent linear approximations, in general, it is a computationally hard problem, although in certain cases it might be feasible to calculate the bias for a single linear approximation. For example, [4] is applicable to analyze the bias of a single linear approximation in our above CubeHash problem. Nevertheless, when the bias is large, we can always compute it empirically, as successfully showed with recent results on RC4 biases (eg. [9]). The direct bias computation when the bias is small is beyond the scope of this paper.

Our computations show that the joint bias for the group of approximations (5), (13), (15) holding simultaneously is around $2^{-2.5}$ and the joint bias for (6), (14), (16) is around $2^{-2.5}$.

Consequently, the total bias for the linear approximation (12) at Round 7, is calculated as $2^{-2.5} \times 2^{-2.5} = 2^{-5}$. In contrast, if the dependency within the round is ignored, we would have a smaller bias 2^{-6} at Round 7.

For CubeHash Round 8, we can show that six pairwise independent approximations arise at Step 1:

$$0x180000 \cdot (x^1_{00001} \oplus x^1_{10001}) \approx 0x180000 \cdot (x^1_{00001} - x^1_{10001}) \tag{17}$$

$$0x180000 \cdot (x^1_{00011} \oplus x^1_{10011}) \approx 0x180000 \cdot (x^1_{00011} - x^1_{10011}) \tag{18}$$

$$0x300 \cdot (x^1_{00101} \oplus x^1_{10101}) \approx 0x300 \cdot (x^1_{00101} - x^1_{10101}) \tag{19}$$

$$0x300 \cdot (x^1_{00111} \oplus x^1_{10111}) \approx 0x300 \cdot (x^1_{00111} - x^1_{10111}) \tag{20}$$

$$0x18000 \cdot (x^1_{01101} \oplus x^1_{11101}) \approx 0x18000 \cdot (x^1_{01101} - x^1_{11101}) \tag{21}$$

$$0x18000 \cdot (x^1_{01111} \oplus x^1_{11111}) \approx 0x18000 \cdot (x^1_{01111} - x^1_{11111}) \tag{22}$$

Eight pairwise independent approximations arise at Step 6. They are presented in terms of states right after step one of the round:

$$0x180000 \cdot (x^1_{10011} + (x^1_{01001} \lll 7 \oplus x^1_{10001})) \approx$$
$$0x180000 \cdot (x^1_{10011} \oplus x^1_{01001} \lll 7 \oplus x^1_{10001}) \tag{23}$$

$$0x180000 \cdot (x^1_{10001} + (x^1_{01011} \lll 7 \oplus x^1_{10011})) \approx$$
$$0x180000 \cdot (x^1_{10001} \oplus x^1_{01011} \lll 7 \oplus x^1_{10011}) \tag{24}$$

$$0xc00300 \cdot (x^1_{10111} + (x^1_{01101} \lll 7 \oplus x^1_{10101})) \approx$$
$$0xc00300 \cdot (x^1_{10111} \oplus x^1_{01101} \lll 7 \oplus x^1_{10101}) \tag{25}$$

$$0xc00300 \cdot (x^1_{10101} + (x^1_{01111} \lll 7 \oplus x^1_{10111})) \approx$$
$$0xc00300 \cdot (x^1_{10101} \oplus x^1_{01111} \lll 7 \oplus x^1_{10111}) \tag{26}$$

$$0xc00000 \cdot (x^1_{11010} + (x^1_{00000} \lll 7 \oplus x^1_{11000})) \approx$$
$$0xc00000 \cdot (x^1_{11010} \oplus x^1_{00000} \lll 7 \oplus x^1_{11000}) \tag{27}$$

$$0xc00000 \cdot (x^1_{11011} + (x^1_{00001} \lll 7 \oplus x^1_{11001})) \approx$$
$$0xc00000 \cdot (x^1_{11011} \oplus x^1_{00001} \lll 7 \oplus x^1_{11001}) \tag{28}$$

$$0xc00000 \cdot (x^1_{11000} + (x^1_{00010} \lll 7 \oplus x^1_{11010})) \approx$$
$$0xc00000 \cdot (x^1_{11000} \oplus x^1_{00010} \lll 7 \oplus x^1_{11010}) \tag{29}$$

$$0xc00000 \cdot (x^1_{11001} + (x^1_{00011} \lll 7 \oplus x^1_{11011})) \approx$$
$$0xc00000 \cdot (x^1_{11001} \oplus x^1_{00011} \lll 7 \oplus x^1_{11011}) \tag{30}$$

Thus, we have 6+8=14 linear approximations involved in this round. As was done for Round 7, we can demonstrate that these 14 approximations fall into four independent groups.

Group One: (17), (18), (23), (24), (28), (30).
Group Two: (19), (20), (21), (22), (25), (26).
Group Three: (27).
Group Four: (29).

As Group Three and Group Four each contains only one approximation, we easily know the bias is 2^{-1} for each group directly from [4]. Group One and Group Two each contains six approximations. We compute the total bias for each group separately. Our results show that the bias for Group One is 2^{-5} and the bias for Group Two is $2^{-6.8}$. Note that the independence assumption would yield a smaller bias $2^{-6}, 2^{-8}$ for Group One, Group Two respectively. Consequently, we deduce the total bias $2^{-5} \times 2^{-6.8} \times 2^{-1} \times 2^{-1} = 2^{-13.8}$ for Round 8, by considering the dependence within the round. Note that, if the dependency within the round is ignored, we would have a smaller bias 2^{-16}.

4 Synthetic Bias Analysis on the Conditional Dependent Problem

When analyzing CubeHash round function, we note a new bias problem, which we shall call conditional dependence from now on. This is in contrast to the well-known concept of conditional independence in statistics. Let X, Y, Z be random variables. Recall that X, Y are conditional independent given Z, if X, Y, Z satisfy $\Pr(X = x, Y = y|Z) = \Pr(X = x|Z)\Pr(Y = y|Z)$ for all x, y. In our problem, X, Y are statistically independent variables, but X, Z are dependent as well as Y, Z. We say that X, Y are conditional dependent given Z. We are concerned with the bias of $f_1(X) \oplus f_2(Y) \oplus f_3(Z)$ for Boolean functions f_1, f_2, f_3. For convenience, we say $f_1(X), f_2(Y)$ are conditional dependent given $f_3(Z)$ rather than that X, Y are conditional dependent given Z.

Formally speaking, we consider that u_0, u_1, u_2, v_1, v_2 are independent variables of binary strings (of fixed length). Three Boolean functions $f_A(u_0, u_1, u_2)$, $f_B(u_2, v_2)$, $f_C(u_1, v_1)$ are defined over those variables. For simplicity, they are denoted by A, B, C in shorthand. We assume that we already know the bias for A, B, C respectively. The main question is, we want to estimate the bias for $A \oplus B \oplus C$, and due to the dependence we do not want to use the Piling-up approximation. We assume it infeasible to compute directly. Our first solution is to obtain the bias for $A \oplus C$ (or $A \oplus B$) first. Then, estimate the bias for $A \oplus B \oplus C$ by taking either of the two

$$\mathrm{Bias}(A \oplus C) \cdot \mathrm{Bias}(B), \ \mathrm{Bias}(A \oplus B) \cdot \mathrm{Bias}(C).$$

Here, we consider only one dependence relation and ignore another dependence relation.

By considering the functions as black-boxes (of random functions), we propose to use the heuristics and make a more delicate estimate as follows. As u_1 affects both $A \oplus B$ and C, we make a simple assumption about the two distributions of the bias for $A \oplus B$ and for C over u_1: the absolute value of the bias is (almost) a constant and can only take values in a set of two elements. Thus, it leads us to compute the average p_+ (resp. p_-) of the positive (resp. negative) biases for $A \oplus B$ over randomly chosen u_1 and the percentage q of the positive biases for $A \oplus B$ over randomly chosen u_1. Similarly, we also compute the average of the positive p'_+ (resp. negative p'_-) biases for C over randomly chosen u_1 and the percentage q' of the positive biases.

The distribution of the bias for $A \oplus B$ over u_1 is independent of the distribution of the bias for C over u_1, so we combine the results and give an estimate on the bias of $A \oplus B \oplus C$ by

$$qq'p_+p'_+ + (1-q)(1-q')p_-p'_- - q(1-q')p_+p'_- - (1-q)q'p_-p'_+ \quad (31)$$

5   Improved Attacks on CubeHash

Using our synthetic analysis, we analyzed all the 11 rounds for CubeHash. We give our results in Table 1. Note that we can show that all the linear approximations for Round 5 are independent and for Round 6, so no bias improvement is possible for Round 5 as well as Round 6. Due to the dependence within each round, we are able to improve the bias estimate for 11-round CubeHash from 2^{-234} in [1] to $2^{-207.1}$. This gives an improved attack for 11-round CubeHash with complexity $2^{414.2}$.

Table 1. Our analysis results on 11-round linear approximations of CubeHash

round	1	2	3	4	5	6	7	8	9	10	11	total
our bias	2^{-29}	$2^{-35.7}$	$2^{-16.9}$	2^{-13}	2^{-4}	2^{-2}	2^{-5}	$2^{-13.8}$	$2^{-18.7}$	$2^{-36.5}$	$2^{-32.5}$	$2^{-207.1}$
paper [1]	2^{-34}	2^{-40}	2^{-18}	2^{-14}	2^{-4}	2^{-2}	2^{-6}	2^{-16}	2^{-22}	2^{-42}	2^{-36}	2^{-234}

We can extend our above results to attack 12-round CubeHash. Our analysis shows that by choosing the same output masks from the set $\{0x600, 0x18000, 0x180000, 0xc000000, 0xc0000000\}$ for x_{01101} and x_{01111} at the end of Round 5, going backwards 6 rounds, forwards 6 rounds, we get[2] five new linear approximations (given in Appendix A) on 12-round CubeHash. They all have the same bias of around $2^{-261.1}$. In particular, with this construction, the last 11 rounds all have the same bias as our 11-round CubeHash above. The bias for its first round is 2^{-54}, assuming all linear approximations are independent[3].

In above analysis, the analysis is focused within each round. According to the specification of CubeHash, no randomization is introduced between consecutive rounds, and the biases of consecutive rounds of CubeHash are likely to be dependent. Our current quick results show that the bias for round 6 and round 7 can be improved to 2^{-6}, and the bias for round 4 and round 5 can be improved to 2^{-16}. Thus, we have the improved bias estimate $2^{-259.1}$ for 12-round CubeHash. By using the five equal biases, we have an attack complexity $2^{(-256.5)\times(-2)}$, ie. $O(2^{513})$.

6   Our Improved Analysis on Stream Cipher Rabbit

Rabbit [3] is a stream cipher among the finalists of EU-funded ECRYPT Stream Cipher Project (eSTREAM). Rabbit encryption algorithm has been published as informational RFC 4503 with the Internet Engineering Task Force (IETF), the standardization body for Internet technology. We give a brief description on Rabbit in Appendix B. Recently, the bias for Rabbit keystream outputs, $0x606 \cdot s_{i+1}^{[47..32]} \oplus 0x606 \cdot s_{i+1}^{[79..64]} \oplus 0x606 \cdot s_{i+1}^{[111..96]}$ was estimated to be $2^{-70.5}$ in [5]. It yields the best distinguishing attack with complexity 2^{141}, which is still above the claimed security level 2^{128}.

In this section, we apply our synthetic approach to analyze the main part of the bias analysis, i.e., the total combined bias of the six linear approximations below for $m = 0x606, m' = 0x6060000$ (for simplicity we omit the irrelevant subscripts i from the variables g):

---

[2] Note that in above analysis on 11-round CubeHash, the 11-round linear approximation can be obtained by going backwards 5 rounds and forwards 6 rounds with mask $0x6$ for x_{01101} and x_{01111} at the end of Round 5.

[3] As our computation is going on, we expect that our previous analysis on the internal round dependence would further improve it.

$$m \cdot (g_2 + g_1 \lll 16 + g_0 \lll 16) \approx m \cdot (g_2 \oplus g_1 \lll 16 \oplus g_0 \lll 16) \tag{32}$$

$$m \cdot (g_4 + g_3 \lll 16 + g_2 \lll 16) \approx m \cdot (g_4 \oplus g_3 \lll 16 \oplus g_2 \lll 16) \tag{33}$$

$$m \cdot (g_6 + g_5 \lll 16 + g_4 \lll 16) \approx m \cdot (g_6 \oplus g_5 \lll 16 \oplus g_4 \lll 16) \tag{34}$$

$$m' \cdot (g_1 + g_0 \lll 8 + g_7) \approx m' \cdot (g_1 \oplus g_0 \lll 8 \oplus g_7) \tag{35}$$

$$m' \cdot (g_3 + g_2 \lll 8 + g_1) \approx m' \cdot (g_3 \oplus g_2 \lll 8 \oplus g_1) \tag{36}$$

$$m' \cdot (g_7 + g_6 \lll 8 + g_5) \approx m' \cdot (g_7 \oplus g_6 \lll 8 \oplus g_5) \tag{37}$$

Let Group One contain (32), (33), (36) and Group Two contain (34), (37). Following Sect. 3, we can demonstrate that the linear approximations in Group One are independent from those in Group Two. Nonetheless, given (35), the two groups are *not* independent. We let A denote the corresponding[4] Boolean function of (35), and let B, C denote the corresponding[5] Boolean function for Group One, Group Two respectively. Obviously, this is a conditional dependent bias problem as we proposed in Sect. 4. Using our first solution in Sect. 4, we compute the bias for $A \oplus B, C$ respectively and get $2^{-11.4}, 2^{-6}$. We estimate the combined bias for above six linear approximations by

$$2^{-11.4} \times 2^{-6} = 2^{-17.4}. \tag{38}$$

Now, we want to apply our black-box solution (31) in Sect. 4. In our case, we have $u_1 = g_7^{[31..16]}$. For $A \oplus B$, we compute with 2^{26} random samples for each randomly chosen u_1 and we run it 2^{14} times. We get in hexadecimal form: $q = 0x24a6/0x4000$, $p_+ = 0x1e0e6fc1/(0x24a6 * 2^{25})$, $p_- = 0x123210ab/(0x1b5a * 2^{25})$. They correspond to the percentage of positive bias 57.3%, the average of positive bias $+2^{-9.3}$, the average of negative bias $-2^{-9.6}$, and the average bias $+2^{-11.43}$ of all. For the function C, we compute with 2^{22} random samples for each randomly chosen u_1 and we run it 2^{16} times. We get $q' = 0xafed/0x10000$, $p'_+ = 0xd4698c87/(0xafed * 2^{21})$, $p'_- = 0x4ea00ceb/(0x5013 * 2^{21})$. They correspond to the percentage of positive bias 68.7%, the average of positive bias $+2^{-4.73}$, the average of negative bias $-2^{-5.03}$, and the average bias $+2^{-5.94}$ of all[6] . By (31), we estimate the bias $2^{-17.5}$ for $A \oplus B \oplus C$. This result agrees with our first estimation (38). Note that based on the naive independence assumption, this combined bias is estimated to be smaller, i.e., 2^{-20}, according to [5]. Consequently, we have an improved attack on Rabbit with complexity 2^{136}, based on [5].

7   Conclusion

In this paper, we take a first step towards the synthetic approach on bias analysis. We apply the "Divide-and-Conquer" method to our synthetic bias analysis.

---

[4] We obtain it by replacing '$\approx$' with '$\oplus$' in (35).

[5] We obtain it by replacing '$\approx$' with '$\oplus$' in all the linear approximations in the group and XORing them together.

[6] The computations were run several times and we always got these same statistics.

Our synthetic approach helps make the task of bias analysis easier when multiple Boolean functions are involved. We also propose a conditional dependent bias problem. Based on naive heuristics and certain ideal assumptions, we give the synthetic bias analysis to estimate the bias. Our synthetic approach is successfully applied to improve the best linear attacks [1,5] on CubeHash and Rabbit respectively. We present an improved attack on 11-round CubeHash with complexity $2^{414.2}$. Based on our 11-round attack, we give a new linear attack for 12-round CubeHash with complexity 2^{513}, which is sharply close to the security parameter 2^{512} of CubeHash. We also give an improved attack on Rabbit with complexity 2^{136}. Moreover, it seems that our results might be further improved, from our ongoing computations.

Acknowledgments. This work is supported by the National Science and Technology Major Project No. 2010ZX01036-001-002 and the Knowledge Innovation Key Directional Program of Chinese Academy of Sciences under Grant No. KGCX2-YW-125, and the National Natural Science Foundation of China under Grant No. 90818012.

References

1. Ashur, T., Dunkelman, O.: Linear Analysis of Reduced-Round CubeHash. In: Lopez, J., Tsudik, G. (eds.) ACNS 2011. LNCS, vol. 6715, pp. 462–478. Springer, Heidelberg (2011)
2. Bernstein, D.J.: CubeHash specification (2.B.1) Submission to NIST (2009)
3. Boesgaard, M., Vesterager, M., Christensen, T., Zenner, E.: The stream cipher Rabbit. In: The ECRYPT Stream Cipher Project, http://www.ecrypt.eu.org/stream/
4. Cho, J.Y., Pieprzyk, J.: Multiple Modular Additions and Crossword Puzzle Attack on NLSv2. In: Garay, J.A., Lenstra, A.K., Mambo, M., Peralta, R. (eds.) ISC 2007. LNCS, vol. 4779, pp. 230–248. Springer, Heidelberg (2007)
5. Lu, Y., Desmedt, Y.: Improved Distinguishing Attack on Rabbit. In: Burmester, M., Tsudik, G., Magliveras, S., Ilić, I. (eds.) ISC 2010. LNCS, vol. 6531, pp. 17–23. Springer, Heidelberg (2011)
6. Matsui, M.: Linear Cryptanalysis Method for DES Cipher. In: Helleseth, T. (ed.) EUROCRYPT 1993. LNCS, vol. 765, pp. 386–397. Springer, Heidelberg (1994)
7. Menezes, A.J., van Oorschot, P.C., Vanstone, S.A.: Handbook of Applied Cryptography. CRC (1996)
8. N.I.S.T., Cryptographic hash algorithm competition, http://www.nist.gov/hash-competition
9. Sepehrdad, P., Vaudenay, S., Vuagnoux, M.: Discovery and Exploitation of New Biases in RC4. In: Biryukov, A., Gong, G., Stinson, D.R. (eds.) SAC 2010. LNCS, vol. 6544, pp. 74–91. Springer, Heidelberg (2011)

Appendix A: New Linear Approximations on 12-Round CubeHash

The five new linear approximations on 12-round CubeHash, which we used in Section 5, are given below (x, x' denote the inputs, outputs respectively):

$$0x18199800 \cdot x_{00000} \oplus 0x18199800 \cdot x_{00010} \oplus 0xe7999f81 \cdot x_{01101}$$
$$\oplus \; 0xe7999f81 \cdot x_{01111} \oplus 0x18199800 \cdot x_{10001} \oplus 0x18199800 \cdot x_{10011}$$
$$\oplus \; 0x30333 \cdot x_{10101} \oplus 0x30333 \cdot x_{10111} \oplus 0x1819980 \cdot x_{11101}$$
$$\oplus \; 0x1819980 \cdot x_{11111} \approx 0x99800181 \cdot x'_{00000} \oplus 0x99800181 \cdot x'_{00010}$$
$$\oplus \; 0x18006018 \cdot x'_{01101} \oplus 0x18006018 \cdot x'_{01111} \oplus 0x99800181 \cdot x'_{10001}$$
$$\oplus \; 0x99800181 \cdot x'_{10011} \oplus 0x30333000 \cdot x'_{10101} \oplus 0x30333000 \cdot x'_{10111}$$
$$\oplus \; 0x19980018 \cdot x'_{11101} \oplus 0x19980018 \cdot x'_{11111}$$

$$0x6660006 \cdot x_{00000} \oplus 0x6660006 \cdot x_{00010} \oplus 0xe667e079 \cdot x_{01101}$$
$$\oplus \; 0xe667e079 \cdot x_{01111} \oplus 0x6660006 \cdot x_{10001} \oplus 0x6660006 \cdot x_{10011}$$
$$\oplus \; 0xc0ccc0 \cdot x_{10101} \oplus 0xc0ccc0 \cdot x_{10111} \oplus 0x60666000 \cdot x_{11101}$$
$$\oplus \; 0x60666000 \cdot x_{11111} \approx 0x60006066 \cdot x'_{00000} \oplus 0x60006066 \cdot x'_{00010}$$
$$\oplus \; 0x180606 \cdot x'_{01101} \oplus 0x180606 \cdot x'_{01111} \oplus 0x60006066 \cdot x'_{10001}$$
$$\oplus \; 0x60006066 \cdot x'_{10011} \oplus 0xccc000c \cdot x'_{10101} \oplus 0xccc000c \cdot x'_{10111}$$
$$\oplus \; 0x66000606 \cdot x'_{11101} \oplus 0x66000606 \cdot x'_{11111}$$

$$0x66600060 \cdot x_{00000} \oplus 0x66600060 \cdot x_{00010} \oplus 0x667e079e \cdot x_{01101}$$
$$\oplus \; 0x667e079e \cdot x_{01111} \oplus 0x66600060 \cdot x_{10001} \oplus 0x66600060 \cdot x_{10011}$$
$$\oplus \; 0xc0ccc00 \cdot x_{10101} \oplus 0xc0ccc00 \cdot x_{10111} \oplus 0x6660006 \cdot x_{11101}$$
$$\oplus \; 0x6660006 \cdot x_{11111} \approx 0x60666 \cdot x'_{00000} \oplus 0x60666 \cdot x'_{00010}$$
$$\oplus \; 0x1806060 \cdot x'_{01101} \oplus 0x1806060 \cdot x'_{01111} \oplus 0x60666 \cdot x'_{10001}$$
$$\oplus \; 0x60666 \cdot x'_{10011} \oplus 0xccc000c0 \cdot x'_{10101} \oplus 0xccc000c0 \cdot x'_{10111}$$
$$\oplus \; 0x60006066 \cdot x'_{11101} \oplus 0x60006066 \cdot x'_{11111}$$

$$0x30003033 \cdot x_{00000} \oplus 0x30003033 \cdot x_{00010} \oplus 0x3f03cf33 \cdot x_{01101}$$
$$\oplus \; 0x3f03cf33 \cdot x_{01111} \oplus 0x30003033 \cdot x_{10001} \oplus 0x30003033 \cdot x_{10011}$$
$$\oplus \; 0x6660006 \cdot x_{10101} \oplus 0x6660006 \cdot x_{10111} \oplus 0x33000303 \cdot x_{11101}$$
$$\oplus \; 0x33000303 \cdot x_{11111} \approx 0x3033300 \cdot x'_{00000} \oplus 0x3033300 \cdot x'_{00010}$$
$$\oplus \; 0xc0303000 \cdot x'_{01101} \oplus 0xc0303000 \cdot x'_{01111} \oplus 0x3033300 \cdot x'_{10001}$$
$$\oplus \; 0x3033300 \cdot x'_{10011} \oplus 0x60006066 \cdot x'_{10101} \oplus 0x60006066 \cdot x'_{10111}$$
$$\oplus \; 0x303330 \cdot x'_{11101} \oplus 0x303330 \cdot x'_{11111}$$

$$0x30333 \cdot x_{00000} \oplus 0x30333 \cdot x_{00010} \oplus 0xf03cf333 \cdot x_{01101}$$
$$\oplus \; 0xf03cf333 \cdot x_{01111} \oplus 0x30333 \cdot x_{10001} \oplus 0x30333 \cdot x_{10011}$$
$$\oplus \; 0x66600060 \cdot x_{10101} \oplus 0x66600060 \cdot x_{10111} \oplus 0x30003033 \cdot x_{11101}$$
$$\oplus \; 0x30003033 \cdot x_{11111} \approx 0x30333000 \cdot x'_{00000} \oplus 0x30333000 \cdot x'_{00010}$$
$$\oplus \; 0x303000c \cdot x'_{01101} \oplus 0x303000c \cdot x'_{01111} \oplus 0x30333000 \cdot x'_{10001}$$
$$\oplus \; 0x30333000 \cdot x'_{10011} \oplus 0x60666 \cdot x'_{10101} \oplus 0x60666 \cdot x'_{10111}$$
$$\oplus \; 0x3033300 \cdot x'_{11101} \oplus 0x3033300 \cdot x'_{11111}$$

Appendix B: Short Description on Stream Cipher Rabbit

We give a short description on Rabbit here. We refer to [3,5] for full description. Rabbit outputs the 128-bit keystream block s_i from the eight state variables x's of 32 bits at each iteration i,

$$s_i^{[15..0]} = x_{0,i}^{[15..0]} \oplus x_{5,i}^{[31..16]} \qquad s_i^{[31..16]} = x_{0,i}^{[31..16]} \oplus x_{3,i}^{[15..0]}$$
$$s_i^{[47..32]} = x_{2,i}^{[15..0]} \oplus x_{7,i}^{[31..16]} \qquad s_i^{[63..48]} = x_{2,i}^{[31..16]} \oplus x_{5,i}^{[15..0]}$$
$$s_i^{[79..64]} = x_{4,i}^{[15..0]} \oplus x_{1,i}^{[31..16]} \qquad s_i^{[95..80]} = x_{4,i}^{[31..16]} \oplus x_{7,i}^{[15..0]}$$
$$s_i^{[111..96]} = x_{6,i}^{[15..0]} \oplus x_{3,i}^{[31..16]} \qquad s_i^{[127..112]} = x_{6,i}^{[31..16]} \oplus x_{1,i}^{[15..0]}$$

The state variables x's are computed from intermediate variables g's of 32 bits,

$$x_{0,i+1} = g_{0,i} + (g_{7,i} \lll 16) + (g_{6,i} \lll 16) \tag{39}$$
$$x_{1,i+1} = g_{1,i} + (g_{0,i} \lll 8) + g_{7,i} \tag{40}$$
$$x_{2,i+1} = g_{2,i} + (g_{1,i} \lll 16) + (g_{0,i} \lll 16) \tag{41}$$
$$x_{3,i+1} = g_{3,i} + (g_{2,i} \lll 8) + g_{1,i} \tag{42}$$
$$x_{4,i+1} = g_{4,i} + (g_{3,i} \lll 16) + (g_{2,i} \lll 16) \tag{43}$$
$$x_{5,i+1} = g_{5,i} + (g_{4,i} \lll 8) + g_{3,i} \tag{44}$$
$$x_{6,i+1} = g_{6,i} + (g_{5,i} \lll 16) + (g_{4,i} \lll 16) \tag{45}$$
$$x_{7,i+1} = g_{7,i} + (g_{6,i} \lll 8) + g_{5,i} \tag{46}$$

where $\lll$ denotes left bit-wise rotation and all additions are computed modulo 2^{32}. The description of computing g's (see [3,5]) is not relevant for us and we omit it here.

On the Resistance of Boolean Functions against Fast Algebraic Attacks[*]

Yusong Du[1,2], Fangguo Zhang[1,3], and Meicheng Liu[3]

[1] School of Information Science and Technology,
Sun Yat-sen University, Guangzhou 510006, P.R. China
[2] Key Lab. of Network Security and Cryptology,
Fujian Normal University, Fuzhou 350007, P.R. China
[3] State Key Laboratory of Information Security, Institute of Software,
Chinese Academy of Sciences, Beijing 100190, P.R. China
yusongdu@hotmail.com, isszhfg@mail.sysu.edu.cn

Abstract. Boolean functions with large algebraic immunity resist algebraic attacks to a certain degree, but they may not resist fast algebraic attacks (FAA's). It is necessary to study the resistance of Boolean functions against FAA's. In this paper, we localize the optimal resistance of Boolean functions against FAA's and introduce the concept of e-fast algebraic immunity (e-FAI) for n-variable Boolean functions against FAA's, where e is a positive integer and $1 \leq e < \lceil \frac{n}{2} \rceil$. We give the sufficient and necessary condition of e-FAI. With e-FAI the problem of deciding the resistance of an n-variable Boolean function against FAA's can be converted into the problem of observing the properties of one given matrix. An algorithm for deciding e-FAI and the optimal resistance against FAA's is also described.

Keywords: stream ciphers, algebraic attacks, fast algebraic attacks, Boolean functions, algebraic immunity.

1 Introduction

Algebraic immunity (AI) has been an important cryptographic property for Boolean functions used in stream ciphers. The algebraic immunity of n-variable Boolean functions is upper bounded by $\lceil \frac{n}{2} \rceil$ [1,2]. Studying and constructing Boolean functions with the maximum AI (MAI Boolean functions) have received attention for years [3,4,5,6,7].

The existence of low degree multiples (or low degree annihilators) of Boolean functions is very necessary for an efficient algebraic attack. Boolean functions with large AI can resist algebraic attacks since large AI guarantees the nonexistence of low degree multiples. However, Boolean functions with large AI (even the maximum AI) may not resist fast algebraic attacks (FAA's) [8,9].

---

[*] This work is supported by Funds of Key Lab of Fujian Province University Network Security and Cryptology (2011008) and National Natural Science Foundations of China (Grant No. 61070168, 10971246, 61003244, 60803135).

H. Kim (Ed): ICISC 2011, LNCS 7259, pp. 261–274, 2012.
© Springer-Verlag Berlin Heidelberg 2012

This is because the existence of low degree multiples of Boolean functions is not necessary any more for FAA's. Therefore the resistance of Boolean functions against FAA's should be considered as another necessary cryptographic property for Boolean functions.

Some studies show that an n-variable Boolean function f has *optimal resistance against FAA's* if there does not exist any nonzero n-variable Boolean function g of degree lower than $\frac{n}{2}$ such that $fg = h$ and $\deg(g) + \deg(h) < n$ [8,9,5]. The concept of the optimal resistance of Boolean functions against FAA's can be implied from [8,9], but was firstly pointed out informally by Carlet *et al.* in [5] as far as we know.

In resent years several efforts have been made to find Boolean functions with good resistance against FAA's, but none of them gave a class of Boolean functions which can be proven to have optimal resistance against FAA's. F.Armknecht *et al.* introduced an effective algorithm with the purpose of observing the resistance of Boolean functions against FAA's [10]. In [5] Carlet *et al.* observed through computer experiments by Armknecht's algorithm that the class of MAI Boolean functions constructed by them may have good behavior against FAA's. E.Pasalic recursively constructed a class of Boolean functions with very good resistance against FAA's (called 'almostly' optimal resistance) [11]. M.Liu *et al.* proved that there does not exist a symmetric Boolean function with optimal resistance against FAA's [12]. P.Rizomiliotis studied the resistance against FAA's of a class of Boolean functions based on univariate polynomial representation [13].

In this paper, we further consider the optimal resistance of Boolean functions against FAA's. We note that the optimal resistance against FAA's is a global concept and it is not convenient for us to observe the optimal resistance of Boolean functions against FAA's. So our motivation is to find an alternative method of describing the resistance against FAA's so that some local properties of the optimal resistance can be manifested. This results the introduction to e-fast algebraic immunity (e-FAI) for n-variable Boolean functions where e is a positive integer and $1 \leq e < \lceil \frac{n}{2} \rceil$.

We give the sufficient and necessary condition of e-FAI. With e-FAI we can more conveniently describe Boolean functions with good resistance against FAA's. With e-FAI the problem of deciding the resistance of an n-variable Boolean function against FAA's can be converted into the problem of observing the properties of a given square matrix of order $\sum_{i=0}^{\lceil \frac{n}{2} \rceil - 1} \binom{n}{i}$. We also describe an algorithm for deciding e-FAI and the optimal resistance against FAA's. This algorithm can bring us more experimental information about Boolean functions against FAA's. Although we do not find more Boolean functions with good resistance against FAA's, we believe that our results can help better to understand the resistance of Boolean functions against FAA's.

The rest of the paper is organized as follows. Section 2 provides some preliminaries. Section 3 recalls the concept of the optimal resistance of Boolean functions against FAA's. Section 4 introduces the concept of e-FAI of Boolean functions. Section 5 gives the sufficient and necessary condition of e-FAI. Section 6 describes the algorithm for deciding e-FAI.

2   Preliminaries

Let n be a positive integer in this paper. We denote by $\mathbb{B}_n$ the set of all the n-variable Boolean functions. Any n-variable Boolean function has a unique representation as a multivariate polynomial over $\mathbb{F}_2$, called the *algebraic normal form* (ANF)

$$f(x_1, x_2, \cdots, x_n) = \sum_{I \in \mathcal{P}(N)} a_I \left(\prod_{i \in I} x_i \right),$$

where $\mathcal{P}(N)$ denotes the power set of $N = \{1, 2, \cdots, n\}$, $a_I \in \mathbb{F}_2$ is the coefficient of monomial $\prod_{i \in I} x_i$ and every coordinate x_i appears in this polynomial with exponents at most 1. The algebraic degree of Boolean function f, denoted by $\deg(f)$, is the degree of this polynomial.

A Boolean function $g \in \mathbb{B}_n$ is called an *annihilator* of $f \in \mathbb{B}_n$ if $fg = 0$. The lowest algebraic degree of all the nonzero annihilators of f and $1 + f$ is called *algebraic immunity* of f, denoted by $\mathcal{AI}_n(f)$ and it has been proved that $\mathcal{AI}_n(f) \leq \lceil \frac{n}{2} \rceil$ for a given $f \in \mathbb{B}_n$ [1,2,3]. A Boolean function $f \in \mathbb{B}_n$ has the *maximum algebraic immunity* (MAI) if $\mathcal{AI}_n(f) = \lceil \frac{n}{2} \rceil$.

Another representation of an n-variable Boolean Function $f(x_1, x_2, \cdots, x_n)$ is by the output column of its *truth table*, i.e., a column vector of dimension 2^n:

$$(f(0,0,\cdots,0), f(1,0,\cdots,0), f(0,1,\cdots,0), f(1,1,\cdots,0), \cdots, f(1,1,\cdots,1))^{tr}.$$

For $f \in \mathbb{B}_n$, the set of $x = (x_1, x_2, \cdots, x_n) \in \mathbb{F}_2^n$ for which $f(x) = 1$ (resp. $f(x) = 0$) is called the on-set (resp. off-set) of f, denoted by 1_f (resp. 0_f). The Hamming weight of f is the cardinality of 1_f, denoted by $\text{wt}(f)$. f is called balanced if $\text{wt}(f) = 2^{n-1}$.

Let $x = (x_1, x_2, \cdots, x_n) \in \mathbb{F}_2^n$. The Hamming weight of x is the number of its nonzero coordinates. We define $\text{supp}(x) = \{i \mid x_i = 1, 1 \leq i \leq n\}$. For $x, y \in \mathbb{F}_2^n$, we say that x is covered by y if $\text{supp}(x) \subseteq \text{supp}(y)$. For the sake of simplicity, $\text{supp}(x) \subseteq \text{supp}(y)$ is written as $x \subseteq y$. For any n-variable Boolean function, there is a relation between its ANF and its truth table, i.e.,

$$a_I = \sum_{\text{supp}(x) \subseteq I} f(x), \tag{1}$$

where $I \in \mathcal{P}(N)$. This equation means that the coefficient of every monomial in the ANF of f can be linearly expressed by the components of its truth table.

3   Optimal Resistance of Boolean Functions against FAA's

In this section, we recall the concept of the optimal resistance of Boolean functions against FAA's and discuss a note given by Gong in [14] about Theorem 7.2.1 in [8].

Suppose $f \in \mathbb{B}_n$ is the nonlinear filtering function in an LFSR-based binary nonlinear filter generator. If f has a low degree multiple $h \neq 0$ (or a nonzero annihilator of low degree), then the attacker can launch a standard algebraic attack and the attack may be converted into solving an over-defined system with multivariate equations of degree not more than the degree of h (or the degree of the annihilator) [1]. It has been proved that the lowest degree of all the nonzero multiples of f is equal to the lowest degree of all the nonzero annihilators of f [9]. Thus Boolean functions with large AI can resist standard algebraic attacks to a certain degree.

However, Boolean functions with large AI may not resist fast algebraic attacks (FAA's). If f has no low degree multiples or annihilators, but there exists a nonzero n-variable Boolean function g of low degree such that $fg \neq 0$ has not high degree, the attacker may launch a fast algebraic attack. The attack may be converted into solving an over-defined system with multivariate equations of degree not more than the degree of g and the complexity of establishing the over-defined system is mainly determined by the degree of fg [8,9].

Compared algebraic attacks with FAA's, excluding the precomputation for obtaining the over-defined system, the computation complexity of FAA's can be lower than that of algebraic attacks, since the algebraic degree of the over-defined system in a fast algebraic attack can be lower than that in an algebraic attack. This also means that the attacker will not launch a fast algebraic attack with a Boolean function g of degree not less than $\mathcal{AI}(f)$. About FAA's, there is a well-known observation given by N.Courtois [8].

Proposition 1. *[8] Let $f \in \mathbb{B}_n$, d_1 and d_2 be two positive integers not more than n. If $d_1 + d_2 \geq n$, then there exists $g \in \mathbb{B}_n$ with $\deg(g) \leq d_1$ such that $\deg(fg) \leq d_2$.*

In order to resist FAA's, we hope that $\deg(fg)$ can be as large as possible for any nonzero n-variable Boolean function g of degree less than $\mathcal{AI}(f) \leq \lceil \frac{n}{2} \rceil$. However, Proposition 1 reveals that there always exists a nonzero n-variable Boolean function g with $\deg(g) \leq e$ such that $\deg(fg) \leq n - e$. Therefore the best case for us against FAA's is that there does not exist any nonzero n-variable Boolean function g with $\deg(g) \leq e < \lceil \frac{n}{2} \rceil$ such that $deg(fg) \leq n - e - 1$. This means that $f \in \mathbb{B}_n$ has *optimal resistance against FAA's* if there does not exist any nonzero n-variable Boolean function g of degree lower than $\frac{n}{2}$ such that $\deg(g) + \deg(fg) < n$.

The concept of the optimal resistance against FAA's was firstly pointed out informally by Carlet *et al.* in [5] according to Proposition 1. However, it was noted by Gong in [14] that the Proposition 1 observed by Courtois (Theorem 7.2.1 in [8]) is not correct, since $d_1 + d_2 \geq n$ can not guarantee the existence of g with $\deg(g) \leq d_1$ such that $fg \neq 0$ and $fg \neq 0$ is necessary for FAA's.

We point out that the note on Proposition 1 given by Gong does not essentially affect the concept of the optimal resistance against FAA's. If there exists an n-variable Boolean function g with $\deg(g) < \frac{n}{2}$ such that $fg = 0$, then the attacker can not launch a fast algebraic attack, but she may launch a standard algebraic attack since g may be a annihilator of low degree. Our aim is to resist FAA's with Boolean function, but we can not neglect the resistance of Boolean function against standard algebraic attacks at the same time. Therefore $fg \neq 0$ must be involved in the optimal resistance of Boolean functions against FAA's.

By convention, $\deg(fg)$ can be any value when $fg = 0$. If we let $\deg(fg) = -\infty$ when $fg = 0$, then Proposition 1 is corrected. Thus, the description of the optimal resistance against FAA's above does not need a change and the requirement that $fg \neq 0$ is involved naturally.

4   Fast Algebraic Immunity for Boolean Functions

In this section, we discuss the local resistance of Boolean functions against FAA's and introduce the concept of e-fast algebraic immunity for n-variable Boolean functions.

The optimal resistance of Boolean functions against FAA's is a global concept. We usually need to consider the local resistance of Boolean functions against FAA's. We note that the (non-)existence of (d_1, d_2)-pairs considered by Rizomiliotis in [13] is one of typical methods of describing the local resistance of Boolean functions against FAA's. According to [13], $f \in \mathbb{B}_n$ has a (d_1, d_2)-pair if there is a nonzero $g \in \mathbb{B}_n$ with $\deg(g) \leq d_1$ such that $\deg(fg) \leq d_2$ where $1 \leq d_1 < \deg(fg) \leq d_2 < n$. According to the concept of the optimal resistance of Boolean functions against FAA's, in fact, we only need to consider the (non-)existence of (d_1, d_2)-pairs with $1 \leq d_1 < \frac{n}{2}$ and $d_2 < n - d_1$.

It is clear that some pairs, like the $(1, \lceil \frac{n}{2} \rceil)$-pair for n-variable Boolean functions, are favorable for FAA's. We can say n-variable Boolean functions with such pairs have poor resistance against FAA's, but we can not say an n-variable Boolean function without such pairs has good resistance against FAA's.

For examples, it was shown in [10] that a class of (non-symmetric) n-variable MAI Boolean functions presented in [3] may have poor resistance against FAA's since it has been tested for $n \leq 10$ that every of these functions has a $(1, \lceil \frac{n}{2} \rceil)$-pair, which is favorable for FAA's. On the contrary, it was shown in [13] that a class of MAI Boolean function in even n variables based on univariate polynomial representation may have some resistance against FAA's since it has been tested for $n \leq 20$ that every of these functions has no $(\frac{n}{2} - 3, \frac{n}{2})$-pairs. However, we can not say this class of function have good resistance against FAA's only based on these facts, since the non-existence of the $(1, \frac{n}{2} + 1)$-pair, which may also be favorable for FAA's, is still not sure.

It is clear that $f \in \mathbb{B}_n$ has optimal resistance against FAA's if f has no any (d_1, d_2)-pair such that $1 \leq d_1 < \frac{n}{2}$ and $d_1 + d_2 < n$. Let e be a fixed integer and $1 \leq e < \lceil \frac{n}{2} \rceil$. We note that the non-existence of the $(e, n - e - 1)$-pair implies the non-existence of all the (d_1, d_2)-pairs such that $1 \leq d_1 \leq e$ and $d_1 < d_2 \leq n - e - 1$. This means that an n-variable which has no $(e, n - e - 1)$-pair for every $e = 1, 2, \cdots, \lceil \frac{n}{2} \rceil - 1$ must have optimal resistance against FAA's and deciding an n-variable Boolean function to have optimal resistance can be divided into deciding these $\lceil \frac{n}{2} \rceil - 1$ pairs. Therefore it is interesting to study theses pairs independently. Based on this, we introduce the concept of e-fast algebraic immunity (e-FAI) for n-variable Boolean functions. We say an n-variable Boolean function has e-FAI if it has no $(e, n - e - 1)$-pair.

Definition 1. *Let $f \in \mathbb{B}_n$ and $1 \leq e < \lceil \frac{n}{2} \rceil$. f is called a Boolean function with e-fast algebraic immunity (e-FAI) if $\deg(fg) \geq n - e$ holds for any nonzero n-variable Boolean function g such that $\deg(g) \leq e$.*

The optimal resistance against FAA's can be divided into $\lceil \frac{n}{2} \rceil - 1$ parts according to e-FAI and e-FAI represents one of these parts. All these parts together are the optimal resistance. Thus Boolean functions in n variables with e-FAI have locally optimal resistance against FAA's.

Unfortunately, e-FAI can neither imply $(e-1)$-FAI nor $(e+1)$-FAI. We cannot say the resistance against FAA's of Boolean functions with e-FAI is better than that of Boolean functions with $(e - 1)$-FAI or $(e + 1)$-FAI. So with e-FAI how can we describe Boolean functions having good resistance against FAA's?

We firstly note that an n-variable Boolean function having 1-FAI, 2-FAI, $\cdots$ e-FAI at the same time possess good resistance against FAA's under a assumption.

Corollary 1. *Let $1 \leq e < \lceil \frac{n}{2} \rceil$. An n-variable Boolean function which has 1-FAI, 2-FAI, $\cdots$ e-FAI at the same time possesses good resistance against FAA's if the attacker only has the ability to solve systems with equations of degree not more than e.*

Proof. Only when (d_1, d_2)-pairs with $d_1 \leq e$ exist can the attacker launch a fast algebraic attack. The function has 1-FAI, 2-FAI, $\cdots$ e-FAI at the same time, so every (d_1, d_2)-pair with $d_1 \leq e$ which possibly exists must satisfy $d_2 \geq n - d_1 \geq n - e$. Thus the attacker can not find a (d_1, d_2)-pair with $d_1 + d_2 < n$ and $d_1 \leq e$. Although some (d_1, d_2)-pairs with $d_1 + d_2 < n$ and $d_1 > e$ can be found it is infeasible to launch a fast algebraic attack because of attacker's inability to solve systems with equations of degree more than e. $\qquad \square$

In order to launch a fast algebraic attack, the attacker expects the existence of (d_1, d_2)-pairs with $d_1 < \lceil \frac{n}{2} \rceil$ and small d_2. The exact value of d_2 for which a fast algebraic attack is feasible in practice depends on several parameters, like the size of the memory and the key size of the stream cipher [15]. Therefore for some (d_1, d_2)-pairs with d_2 not small it may be infeasible in practice for the attacker to launch a fast algebraic attack. We can assume that only when (d_1, d_2)-pairs with $d_2 \leq k$ exist does the attacker have the ability to launch a fast algebraic attack, then we have the following result.

Corollary 2. *Let k be a positive integer and $\lceil \frac{n}{2} \rceil < k < n$. Suppose only when (d_1, d_2)-pairs with $d_2 \leq k$ exist does the attacker have the ability to launch a fast algebraic attack. An n-variable Boolean function which has $(\lceil \frac{n}{2} \rceil - 1)$-FAI, $(\lceil \frac{n}{2} \rceil - 2)$-FAI, $\cdots$, $(n-k)$-FAI at the same time possesses good resistance against FAA's.*

Proof. The function has $(\lceil \frac{n}{2} \rceil - 1)$-FAI, $(\lceil \frac{n}{2} \rceil - 2)$-FAI, $\cdots$, $(n - k)$-FAI at the same time, so every (d_1, d_2)-pair with $d_2 \leq k$ which possibly exists must satisfy $d_1 + d_2 \geq n$. Thus the attacker can not find a (d_1, d_2)-pair with $d_1 + d_2 < n$ and $d_2 \leq k$. Although some (d_1, d_2)-pairs with $d_1 + d_2 < n$ and $d_2 > k$ can be found it is infeasible for the attacker to launch a fast algebraic attack. $\qquad \square$

Combining the two assumptions in Corollary 1 and Corollary 2, it is easy to see the following fact.

Corollary 3. *Let $1 \leq e < \lceil \frac{n}{2} \rceil$. Suppose only when (d_1, d_2)-pairs with $d_1 \leq e$ $d_2 \leq n-e-1$ exist does the attacker have the ability to launch a fast algebraic attack. An n-variable Boolean function with e-FAI possesses good resistance against FAA's.*

Boolean functions in n variables with e-FAI have also other interesting properties. In Definition 1, if f is a Boolean function with e-FAI then $\deg((1+f)g) = \deg(fg+g) \geq n-e$ since $\deg(g) \leq e < n-e \leq \deg(fg)$ for $1 \leq e < \lceil \frac{n}{2} \rceil$, thus $1+f$ is also a Boolean function with e-FAI if f has e-FAI. This means that we do not need to consider $1+f$ when discussing e-FAI of f.

Corollary 4. *Let $f \in \mathbb{B}_n$, $1 \leq e < \lceil \frac{n}{2} \rceil$. f is a Boolean function with e-FAI only if $\mathcal{AI}_n(f) > e$. Particularly, f is a Boolean function with $(\lceil \frac{n}{2} \rceil - 1)$-FAI only if $\mathcal{AI}_n(f) = \lceil \frac{n}{2} \rceil$.*

Proof. If $f \in \mathbb{B}_n$ is a Boolean function with e-FAI then f has no annihilators of degree less than e. Furthermore, $1+f$ is also a Boolean function with e-FAI and $1+f$ also has no annihilators of degree less than e. Therefore $\mathcal{AI}_n(f) > e$.   □

Corollary 5. *Let n be odd, $f \in \mathbb{B}_n$. f is a Boolean function with $(\lceil \frac{n}{2} \rceil - 1)$-FAI if and only if $\mathcal{AI}_n(f) = \lceil \frac{n}{2} \rceil$.*

Proof. By Corollary 4, f is a Boolean function with $(\lceil \frac{n}{2} \rceil - 1)$-FAI only if $\mathcal{AI}_n(f) = \lceil \frac{n}{2} \rceil$. If $\mathcal{AI}_n(f) = \lceil \frac{n}{2} \rceil$ then f has no annihilators of degree less than $\lceil \frac{n}{2} \rceil$, which implies that f has no multiples of degree less than $\lceil \frac{n}{2} \rceil$. Therefore for any nonzero n-variable Boolean function g with $\deg(g) \leq \lceil \frac{n}{2} \rceil - 1$ we have $\deg(fg) \geq \lceil \frac{n}{2} \rceil = n - (\lceil \frac{n}{2} \rceil - 1)$.   □

Corollary 6. *Let $f \in \mathbb{B}_n$, $1 \leq e < \lceil \frac{n}{2} \rceil$. f is a Boolean function with e-FAI, then $\deg(f) \geq n - e$.*

Proof. From the definition of e-FAI, $\deg(fg) \geq n - e$ holds for any nonzero n-variable Boolean function g such that $\deg(g) \leq e$. We let $\deg(g) = 0$, i.e., $g = 1$. Then we have $\deg(f) = \deg(fg) \geq n - e$.   □

5   Sufficient and Necessary Condition of e-Fast Algebraic Immunity

In this section we give the sufficient and necessary condition of e-FAI. Before this we need some prepared work including several definitions and lemmas.

Definition 2. *Let $\alpha = (a_1, a_2, \cdots, a_n)$, $\beta = (b_1, b_2, \cdots, b_n) \in \mathbb{F}_2^n$. $\alpha \prec \beta$ if and only if $\mathrm{wt}(\alpha) < \mathrm{wt}(\beta)$, or when $\mathrm{wt}(\alpha) = \mathrm{wt}(\beta)$ there exists $1 \leq i < n$ such that $a_i = 1$, $b_i = 0$ and $a_j = b_j$ for $1 \leq j < i$.*

We give an example to help understanding Definition 2. We consider three vectors in $\mathbb{F}_2^5$: $\alpha = (11000)$, $\beta = (01101)$ and $\gamma = (01011)$. According to Definition 2, $\alpha \prec \beta$ since $\mathrm{wt}(\alpha) < \mathrm{wt}(\beta)$, while $\mathrm{wt}(\beta) = \mathrm{wt}(\gamma)$ but there exists $i = 3$ satisfying the definition, thus $\beta \prec \gamma$. Similarly, we can write out all the vectors ordered by $\prec$ in $\mathbb{F}_2^5$.

$$(00000) \prec (10000) \prec (01000) \prec (00100) \prec (00010) \prec (00001) \prec$$
$$(11000) \prec (10100) \prec (10010) \prec (10001) \prec (01100) \prec (01010) \prec$$
$$(01001) \prec (00110) \prec (00101) \prec (00011) \prec \quad \cdots \quad \prec (11111) \ .$$

Let

$$\mathbb{F}_2^n = \{\gamma_i \,|\, 0 \le i \le 2^n - 1\}$$

where $\gamma_0 \prec \gamma_1 \prec \gamma_2 \prec \cdots \prec \gamma_{2^n-1}$ are ordered by $\prec$ according to Definition 2. We say that γ_i is the ith vector in $\mathbb{F}_2^n$.

Let $x = (x_1, x_2, \cdots, x_n)$ be a set of binary variables. For $\gamma = (c_1, c_2, \cdots, c_n) \in \mathbb{F}_2^n$, x^γ is defined to be the Boolean monomial $x_1^{c_1} x_2^{c_2} \cdots x_n^{c_n}$ where $x_i^0 = 1$ and $x_i^1 = x_i$ for $i = 1, 2, \cdots n$. Then all the Boolean monomials of n variables are

$$x^{\gamma_0}, x^{\gamma_1}, x^{\gamma_2}, \cdots, x^{\gamma_{2^n-1}}$$

where γ_i is the ith vector in $\mathbb{F}_2^n$. If x is taken as a vetoer in $\mathbb{F}_2^n$, it is clear that $x^{\gamma_i} = 1$ if and only if $\mathrm{supp}(\gamma_i) \subseteq \mathrm{supp}(x)$.

According to Definition 2, the truth table of an n-variable Boolean function $f(x_1, x_2, \cdots, x_n)$ can be written as:

$$T(f) = (f(\gamma_0), f(\gamma_1), \cdots, f(\gamma_{2^n-1}))^{tr}.$$

In the following content, we denote by $T(f)$ the truth table of $f \in \mathbb{B}_n$, in which the components are ordered by $\prec$ according to Definition 2.

The truth table of f can be also represented as a $2^n \times 2^n$ matrix, denoted by $R_{T(f)}$, whose entries on the main diagonal are the components of $T(f)$ respectively and the rest of entries are all zero. With these notations, for $g \in \mathbb{B}_n$ we have $T(h) = R_{T(f)}T(g)$ if $h = fg$. The multiplication of two n-variable Boolean functions can be represented as the product of a matrix and a column vector.

Definition 3. *Let $A \subseteq \mathbb{F}_2^n$, $|A|$ be the number of the elements in A and $k\, (\le n)$ be a positive integer. When x is taken as a vetoer in $\mathbb{F}_2^n$, $v_k(x)$ is defined to be a binary row vector of dimension $\sum_{i=0}^k \binom{n}{i}$, i.e.,*

$$v_k(x) = (x^{\gamma_0}, x^{\gamma_1}, \cdots, x^{\gamma_{\sum_{i=0}^k \binom{n}{i}-1}}).$$

Moreover, $V_k(A)$ is defined to be a matrix with $|A|$ row vectors $\{v_k(x) \,|\, x \in A\}$ and the order of its row vectors corresponds to the order of elements in A according to Definition 2.

Lemma 1. *The column vectors of $V_e(\mathbb{F}_2^n)$ are exactly the truth tables of all the n-variable monomials of degree not more than e.*

Proof. From Definition 3, the $(i+1)$th column vector is $(\gamma_0^{\gamma_i}, \gamma_1^{\gamma_i}, \cdots, \gamma_{2^n-1}^{\gamma_i})^{tr}$ and is also the truth table of Boolean monomial x^{γ_i} where $0 \le i \le \sum_{i=0}^{e} \binom{n}{i} - 1$. From Definition 2, $x^{\gamma_0}, x^{\gamma_1}, \cdots, x^{\gamma_E - 1}$ are all the n-variable Boolean monomials of degree not more than e where $E = \sum_{i=0}^{e} \binom{n}{i}$. The result follows.

Lemma 2. *Let* $\Delta_d^n = \{x \mid x \in \mathbb{F}_2^n, \mathrm{wt}(x) > d\}$. *Every component of the column vector* $V_n(\Delta_d^n)T(h)$ *uniquely corresponds to the coefficient of a monomial of degree more than* d *of* $h \in \mathbb{B}_n$.

Proof. From Equation (1) in section 2 we have

$$a_{\mathrm{supp}(\gamma_i)} = \sum_{x \subseteq \gamma_i} h(x) \quad i = 0, 1, \cdots, 2^n - 1. \tag{2}$$

By Definition 3 Equation (2) can be written as the form of matrix product of two vectors:

$$a_{\mathrm{supp}(\gamma_i)} = v_n(\gamma_i)T(h) \quad i = 0, 1, \cdots, 2^n - 1. \tag{3}$$

From Definition 3, $V_n(\Delta_d^n)$ consists of all the row vectors $v_n(\gamma)$ with $\gamma > \mathrm{wt}(d)$. Therefore every component of $V_n(\Delta_d^n)T(h)$ uniquely corresponds to $a_{\mathrm{supp}(\gamma)}$ for some γ with $\mathrm{wt}(\gamma) > d$. $\square$

We denote by $U_d(1_f)$ the matrix obtained by taking column $j_1, j_2, \cdots, j_{\mathrm{wt}(f)}$ in $V_n(\Delta_d^n)$ such that $f(\gamma_{j_i}) = 1$ $(i = 1, 2, \cdots, \mathrm{wt}(f))$. Then we can give the sufficient and necessary condition of e-FAI.

Theorem 1. *Let* $f \in \mathbb{B}_n$, $1 \le e < \lceil \frac{n}{2} \rceil$ *and* $d = n - e - 1$. f *is a Boolean function with* e-FAI *if and only if* $U_d(1_f)V_e(1_f)$ *is an invertible matrix.*

Proof. Let $g \in \mathbb{B}_n$, $g \ne 0$, $\deg(g) \le e$ and $h = fg \in \mathbb{B}_n$. From Lemma 2, every component of $V_n(\Delta_d^n)T(h)$ corresponds to the coefficient of a monomial of degree more than d of h. According to Definition 1, we let the coefficients of the monomials of degree more than d being zero in the ANF of h, i.e.,

$$V_n(\Delta_d^n)T(h) = \mathbf{0}.$$

In other words, $deg(h) \le d$ or $h = 0$ if and only if the equation above holds. Note that $T(h) = R_{T(f)}T(g)$. By Lemma 1 a column vector of $V_e(\mathbb{F}_2^n)$ is exactly the truth table of an n-variable monomial of degree not more than e. Then $T(g) = \sum_{i=0}^{E-1} k_i \alpha_i$, where $E = \sum_{i=0}^{e} \binom{n}{i}$, $k_i \in \mathbb{F}_2$, α_i is the $(i+1)$th column of $V_e(\mathbb{F}_2^n)$ and the truth table of monomial x^{γ_i}. Thus, we have

$$V_n(\Delta_d^n)R_{T(f)}\left(\sum_{i=0}^{E-1} k_i \alpha_i\right) = \mathbf{0}.$$

Viewing $\mathbf{K} = (k_0, k_1, \cdots, k_{E-1})^{tr}$ as the unknown, we can get a homogenous linear system with E unknowns and E equations:

$$V_n(\Delta_d^n)R_{T(f)}(\alpha_0, \alpha_1, \cdots, \alpha_{E-1})\mathbf{K} = \mathbf{0}, \tag{4}$$

where $V_n(\Delta_d^n)R_{T(f)}\alpha_i$ can be seen as the $(i+1)$th column in the coefficients matrix. It is clear that that nonzero g with $\deg(g) \le e$ does not exist if and only if system (4) only has zero solution, then f is a Boolean function with e-FAI if and only if the coefficient matrix of system (4) is invertible. Since $(\alpha_0, \alpha_1, \cdots, \alpha_{E-1}) = V_e(\mathbb{F}_2^n)$, the coefficient matrix can be written as

$$V_n(\Delta_d^n) R_{T(f)} V_e(\mathbb{F}_2^n).$$

Note that the zero columns in $V_n(\Delta_d^n) R_{T(f)}$ and their corresponding rows in $V_e(\mathbb{F}_2^n)$ do not give any contribution for the computation of the matrix. After omitting theses zero columns and corresponding rows, the coefficient matrix can be simplified into $U_d(1_f) V_e(1_f)$. Therefore, f is a Boolean function with e-FAI if and only if $U_d(1_f) V_e(1_f)$ is invertible.   □

For $1 \le e < \lceil \frac{n}{2} \rceil$ and $\mathcal{A} \subseteq \mathbb{F}_2^n$, in the following content we let

$$W_e(\mathcal{A}) = U_d(\mathcal{A}) V_e(\mathcal{A})$$

where $d = n - e - 1$. We can give the sufficient and necessary condition of Boolean functions to have optimal resistance against FAA's.

Theorem 2. Let $f \in \mathbb{B}_n$ and $1 \le e < \lceil \frac{n}{2} \rceil$. f has optimal resistance against FAA's if and only if $W_{\lceil \frac{n}{2} \rceil - 1}(1_f)_e$ is invertible for every e $(1 \le e < \lceil \frac{n}{2} \rceil)$, where $W_{\lceil \frac{n}{2} \rceil - 1}(1_f)_e$ is the submatrix which consists of the entries on the last E rows and the first E columns in $W_{\lceil \frac{n}{2} \rceil - 1}(1_f) = U_{\lfloor \frac{n}{2} \rfloor}(1_f) V_{\lceil \frac{n}{2} \rceil - 1}(1_f)$.

Proof. $U_d(1_f)$ is the submatrix of $U_{\lfloor \frac{n}{2} \rfloor}(1_f)$ obtained by taking the last E rows in $U_{\lfloor \frac{n}{2} \rfloor}(1_f)$, and $V_e(1_f)$ is the submatrix of $V_{\lceil \frac{n}{2} \rceil - 1}(1_f)$ obtained by taking the first E columns in $V_{\lceil \frac{n}{2} \rceil - 1}(1_f)$. Thus, $W_e(1_f) = U_d(1_f) V_e(1_f) = W_{\lceil \frac{n}{2} \rceil - 1}(1_f)_e$. From Theorem 1, the result follows.   □

From Theorem 2 we can see that all the $W_e(1_f)$ with $1 \le e \le \lceil \frac{n}{2} \rceil - 1$ are included in one square matrix of order $\sum_{i=0}^{\lceil \frac{n}{2} \rceil - 1} \binom{n}{i}$. Theorem 2 tells us studying the resistance of n-variable Boolean functions against FAA's can be converted into studying the properties of matrix $W_{\lceil \frac{n}{2} \rceil - 1}(1_f) = U_{\lfloor \frac{n}{2} \rfloor}(1_f) V_{\lceil \frac{n}{2} \rceil - 1}(1_f)$.

6   An Algorithm for Deciding e-FAI

For $x, y \in \mathbb{F}_2^n$, we let $x \cup y = (x_1 + y_1 + x_1 y_1, x_2 + y_2 + x_2 y_2, \cdots, x_n + y_n + x_n y_n) \in \mathbb{F}_2^n$ and $y \setminus x = (y_1 - x_1, y_2 - x_2, \cdots, y_n - x_n) = (y_1 + x_1, y_2 + x_2, \cdots, y_n + x_n) \in \mathbb{F}_2^n$ when $x \subseteq y$.

Lemma 3. Let $f \in \mathbb{B}_n$, $e = \lceil \frac{n}{2} \rceil - 1$, $E = \sum_{i=0}^{e} \binom{n}{i}$, $1 \le r, s \le E$. Denote by $w_{yz} = 1$ the entry on row r and column s of $W_{\lceil \frac{n}{2} \rceil - 1}(1_f)$ where $y = \gamma_{2^n - 1 - E + r}$ and $z = \gamma_{s-1}$. $w_{yz} = 1$ if and only if $z \subseteq y$ and $\sum_{u \subseteq z} a_{y \setminus u} = 1$ where $a_{y \setminus u} \in \mathbb{F}_2$ is the coefficient of the monomial $x^{y \setminus u}$ in the ANF of f.

Proof. w_{yz} is equal to the rth row of $U_{\lfloor \frac{n}{2} \rfloor}(1_f)$ multiplies (matrix multiplication) the sth column of $V_{\lceil \frac{n}{2} \rceil - 1}(1_f)$. For $1 \le k \le |1_f|$ we denote by $U_{\lfloor \frac{n}{2} \rfloor}(1_f)_{(r,k)}$ the kth component of the rth row of $U_{\lfloor \frac{n}{2} \rfloor}(1_f)$ and denote by $V_{\lceil \frac{n}{2} \rceil - 1}(1_f)_{(k,s)}$ kth component of the sth column of $V_{\lceil \frac{n}{2} \rceil -1}(1_f)$. From the definition of $U_{\lfloor \frac{n}{2} \rfloor}(1_f)$, $U_{\lfloor \frac{n}{2} \rfloor}(1_f)_{(r,k)} = 1$ if and only if $\gamma_{k-1} \in 1_f$ and $\gamma_{k-1} \subseteq y$, and $V_{\lceil \frac{n}{2} \rceil -1}(1_f)_{(k,s)} = 1$ if and only if $\gamma_{k-1} \in 1_f$ and $z \subseteq \gamma_{k-1}$. Note that $x \subseteq y$ if and only if $y^x = 1$ where $x, y \in \mathbb{F}_2^n$, thus,

$$w_{yz} = \sum_{x \in 1_f} y^x x^z = \sum_{x \in \mathbb{F}_2^n} y^x x^z f(x) = \sum_{x \subseteq y \setminus z} f(x \cup z).$$

Therefore, $w_{yz} = 1$ if and only if $z \subseteq y$ and $\sum_{x \subseteq y \setminus z} f(x \cup z) = 1$. When $z \subseteq y$, for $\sum_{x \subseteq y \setminus z} f(x \cup z)$ we have

$$\sum_{x \subseteq y \setminus z} f(x \cup z) = \sum_{x \subseteq y \setminus z} \sum_{u \subseteq z} f(x \cup u) \sum_{u \subseteq v \subseteq z} 1$$

$$= \sum_{x \subseteq y \setminus z} \sum_{u \subseteq z} \sum_{u \subseteq v \subseteq z} f(x \cup u)$$

$$= \sum_{x \subseteq y \setminus z} \sum_{v \subseteq z} \sum_{u \subseteq v} f(x \cup u)$$

$$= \sum_{v \subseteq z} \sum_{x \subseteq y \setminus z} \sum_{u \subseteq v} f(x \cup u)$$

$$= \sum_{v \subseteq z} \sum_{x \subseteq (y \setminus z) \cup v} f(x)$$

$$= \sum_{u \subseteq z} \sum_{x \subseteq y \setminus u} f(x)$$

$$= \sum_{u \subseteq z} a_{y \setminus u}$$

The Lemma is proved. □

Lemma 3 tells us that for a given $f \in \mathbb{B}_n$ and $\mathcal{A} = \{x \,|\, x \in \mathbb{F}_2^n, \mathrm{wt}(x) > \lfloor \frac{n}{2} \rfloor\}$, $W_{\lceil \frac{n}{2} \rceil -1}(1_f)$ is the Hadamard product (entrywise product) of two of matrices, $V_{\lceil \frac{n}{2} \rceil -1}(\mathcal{A})$ and a matrix defined by 1_f. On the ground of this, we can give an effective algorithm (Algorithm 1) for deciding whether a given $f \in \mathbb{B}_n$ is a function with e-FAI for $1 \le e < \lceil \frac{n}{2} \rceil$. We can also use Algorithm 1 for every integer e with $1 \le e < \lceil \frac{n}{2} \rceil$ to decide whether f is a Boolean function with optimal resistance against FAA's.

In Algorithm 1, the complexity of initializing matrix W is $\mathcal{O}(E^2)$. For every $W_{(i,j)} = 1$, we have a choice of y and z. The number of choices is not more than E^2. Given one choice of y and z, we have $|\mathcal{S}| = 2^{wt(z)}$, which corresponds to the number of operations in step 06. After three layers loop, a modified W is obtained and the complexity is $\mathcal{O}(E(\sum_{k=0}^{e} \binom{n}{k} 2^k))$. The complexity of deciding the invertibility of W is $\mathcal{O}(E^3)$. Therefore, the overall computation complexity of Algorithm 1 is $\mathcal{O}(E^3)$.

Algorithm 1. deciding whether $f \in \mathbb{B}_n$ is a Boolean function with e-FAI

Input: a_{γ^i} $(0 \leq i \leq 2^n - 1)$ (all the coefficients of monomials of $f \in \mathbb{B}_n$), positive integer e $(1 \leq e < \lceil \frac{n}{2} \rceil)$
Output: $True$ (f is a Boolean function with e-FAI) or $False$ (f is not a Boolean function with e-FAI)
Initialize: $d = n - e - 1$, $\mathcal{A} = \{x \mid x \in \mathbb{F}_2^n, \mathrm{wt}(x) > d\}$, $E \times E$ matrix $W = V_e(\mathcal{A})$ with $E = \sum_{i=0}^{e} \binom{n}{i}$, denote by $W_{(i,j)}$ its entry on row i and column j
01: **for** i from 1 to E **do**
02:    **for** j from 1 to E **do**
03:       **if** $W_{(i,j)} = 1$ **then**
04:          $W_{(i,j)} \leftarrow 0$, $y \leftarrow \gamma_{2^n - 1 - E + i}$, $z \leftarrow \gamma_{j-1}$
05:          Determine the set $\mathcal{S} \leftarrow \{u \mid u \subseteq z\}$
06:          $W_{(i,j)} \leftarrow W_{(i,j)} + \sum_{u \in \mathcal{S}} a_{y \setminus u}$
07:       **end if**
08:    **end for**
09: **end for**
10: **if** W invertible **then** return $True$ **else** return $False$

Another algorithm was introduced by Armknecht *et al.* (Algorithm 2 in [10], say Algorithm 2 simply). For a given $f \in \mathbb{B}_n$ and two positive integers e, d with $e + d < n$, it aimed at deciding the existence of a nonzero $g \in \mathbb{B}_n$ with $\deg(g) \leq e$ such that $\deg(fg) \leq d$. An $E \times E$ matrix is also established for determining its invertibility in Algorithm 2, and the complexity of obtaining it is $\mathcal{O}(E(\sum_{b=0}^{e} \binom{d+1}{b} 2^{d+1-b}))$.

We have to note that Algorithm 2 is valid only for pairs of (e, d) such that $E < \binom{n}{d+1}$ and it can be only used for denying the existence of g but determining e-FAI directly since $E > \binom{n}{d+1}$ always holds for $d = n - e - 1$.

Compared with Algorithm 2, Algorithm 1 may have better computation complexity if n is large and e is small, for instance $n \geq 11$ and $e \leq 2$, since in these cases, $\sum_{k=0}^{e} \binom{n}{k} 2^k < \sum_{b=0}^{e} \binom{d+1}{b} 2^{d+1-b}$ where $d = \lceil \frac{n}{2} \rceil - 1$ in order to guarantee $E < \binom{n}{d+1}$. This means that if n is large and e is small, Algorithm 1 may be better when one wants to deny the existence of g with $\deg(g) \leq e$ such that $\deg(fg) \leq d$.

It is not hard to describe a modified Algorithm 2 that can be used for determining e-FAI if we consider all the vectors of Hamming weight more than d rather than only the vectors of Hamming weight equal $d + 1$ (denoted by $\{\gamma : |\gamma| = d + 1\}$ in [10]) to construct the $E \times E$ matrix in Algorithm 2. In this case, the complexity of obtaining the matrix in Algorithm 2 increases to $\mathcal{O}(E \sum_{k=d+1}^{n} (\sum_{b=0}^{e} \binom{k}{b} 2^{k-b}))$. When two algorithm are used for deciding e-FAI, the computation complexity of Algorithm 1 is always better than that of Algorithm 2 since $\sum_{k=0}^{e} \binom{n}{k} 2^k < \sum_{k=d+1}^{n} (\sum_{b=0}^{e} \binom{k}{b} 2^{k-b})$ when $d = n - e - 1$.

We let $n = 5, 6, 7, 8, 9, 10$ respectively and select randomly 10^5 balanced n-variable Boolean functions. The results are listed in the following table.

Num. Var	$n=5$	$n=6$	$n=7$	$n=8$	$n=9$	$n=10$
Num. 1-FAI	4242	0	4184	0	4213	0
Num. 2-FAI	3330	2222	0	0	3959	4299
Num. 3-FAI	-	-	1576	0	4087	4123
Num. 4-FAI	-	-	-	-	2998	4261
Num. Optimal	1419	0	0	0	217	0

As shown in the table, the number of balanced Boolean functions with optimal resistance against FAA's seems to be not large especially for some numbers of variables.

It has been shown in [5] by algorithm 2 in [10] that the Carlet-Feng function of variable less than 13 have good behavior against FAA's. By our algorithm we can get more information about their resistance against FAA's.

Num. Var	$n=5$	$n=6$	$n=7$	$n=8$	$n=9$	$n=10$	$n=11$	$n=12$	$n=13$
1-FAI	✓	✗	✓	✗	✓	✗	✓	✗	✓
2-FAI	✓	✓	✗	✗	✓	✓	✗	✗	✓
3-FAI	-	-	✓	✗	✓	✓	✓	✗	✓
4-FAI	-	-	-	-	✓	✓	✓	✓	✗
5-FAI	-	-	-	-	-	-	✓	✓	✓
6-FAI	-	-	-	-	-	-	-	-	✓
Optimal	✓	✗	✗	✗	✓	✗	✗	✗	✗

As shown in the table, the Carlet-Feng function in 5 and 9 variables have optimal resistance. The Carlet-Feng function in 6, 7 10, 11, and 13 variables have almostly optimal resistance since there is only one (d_1, d_2)-pair with $d_1 < \frac{n}{2}$ and $d_1 + d_2 < n$ for each of them. The Carlet-Feng function in 12 variables behave not badly. But we can not say the Carlet-Feng function in 8 variables has very good resistance.

7 Conclusion

e-FAI is an alternative cryptographic property for measuring the resistance of Boolean functions against FAA's. e-FAI describes locally optimal resistance of Boolean functions against FAA's. A sufficient and necessary condition of Boolean functions to have e-FAI is provided. Thanks to this condition, the problem of deciding the resistance against FAA's of an n-variable Boolean function can be converted into the problem of observing the property of a given square matrix of order $\sum_{i=0}^{\lceil \frac{n}{2} \rceil -1} \binom{n}{i}$. Besides the algorithm given by Armknecht *et al.* there is an alternative algorithm for deciding the resistance against FAA's of Boolean functions. The computation complexity of this algorithm is better than that of the algorithm given by Armknecht *et al.* when deciding e-FAI or the optimal resistance against FAA's of Boolean functions.

References

1. Courtois, N., Meier, W.: Algebraic Attacks on Stream Ciphers with Linear Feedback. In: Biham, E. (ed.) EUROCRYPT 2003. LNCS, vol. 2656, pp. 345–359. Springer, Heidelberg (2003)
2. Meier, W., Pasalic, E., Carlet, C.: Algebraic Attacks and Decomposition of Boolean Functions. In: Cachin, C., Camenisch, J.L. (eds.) EUROCRYPT 2004. LNCS, vol. 3027, pp. 474–491. Springer, Heidelberg (2004)
3. Carlet, C., Dalai, D.K., Gupta, K.C., Maitra, S.: Algebraic Immunity for Cryptographically Significant Boolean Functions: Analysis and Construction. IEEE Trans. Inform.Theory 52(7), 3105–3121 (2006)
4. Li, N., Qi, W.: Boolean functions of an odd number of variables with maximum algebraic immunity. Sci China Ser F-Information Sciences 50(3), 307–317 (2007)
5. Carlet, C., Feng, K.: An Infinite Class of Balanced Functions with Optimal Algebraic Immunity, Good Immunity to Fast Algebraic Attacks and Good Nonlinearity. In: Pieprzyk, J. (ed.) ASIACRYPT 2008. LNCS, vol. 5350, pp. 425–440. Springer, Heidelberg (2008)
6. Liu, M., Pei, D., Du, Y.: Identification and construction of Boolean functions with maximum algebraic immunity. Sci China Ser F-Information Sciences 53(7), 1379–1396 (2010)
7. Tu, Z., Deng, Y.: A conjecture about binary strings and its applications on constructing Boolean functions with optimal algebraic immunity. Designs, Codes and Cryptography 60(1), 1–14 (2011)
8. Courtois, N.: Fast Algebraic Attacks on Stream Ciphers with Linear Feedback. In: Boneh, D. (ed.) CRYPTO 2003. LNCS, vol. 2729, pp. 176–194. Springer, Heidelberg (2003)
9. Armknecht, F.: Improving Fast Algebraic Attacks. In: Roy, B., Meier, W. (eds.) FSE 2004. LNCS, vol. 3017, pp. 65–82. Springer, Heidelberg (2004)
10. Armknecht, F., Carlet, C., Gaborit, P., Künzli, S., Meier, W., Ruatta, O.: Efficient Computation of Algebraic Immunity for Algebraic and Fast Algebraic Attacks. In: Vaudenay, S. (ed.) EUROCRYPT 2006. LNCS, vol. 4004, pp. 147–164. Springer, Heidelberg (2006)
11. Pasalic, E.: Almost Fully Optimized Infinite Classes of Boolean Functions Resistant to (Fast) Algebraic Cryptanalysis. In: Lee, P.J., Cheon, J.H. (eds.) ICISC 2008. LNCS, vol. 5461, pp. 399–414. Springer, Heidelberg (2009)
12. Liu, M., Lin, D.: Fast Algebraic Attacks and Decomposition of Symmetric Boolean Functions. IEEE Trans. Inform.Theory 57(7), 4817–4821 (2011)
13. Rizomiliotis, P.: On the Resistance of Boolean Functions Against Algebraic Attacks Using Univariate Polynomial Representation. IEEE Trans. Inform. Theory 56(8), 4014–4024 (2010)
14. Gong, G.: Sequences, DFT and Resistance against Fast Algebraic Attacks. In: Golomb, S.W., Parker, M.G., Pott, A., Winterhof, A. (eds.) SETA 2008. LNCS, vol. 5203, pp. 197–218. Springer, Heidelberg (2008)
15. Canteaut, A.: Open Problems Related to Algebraic Attacks on Stream Ciphers. In: Ytrehus, Ø. (ed.) WCC 2005. LNCS, vol. 3969, pp. 120–134. Springer, Heidelberg (2006)

CCA Secure IB-KEM from the Computational Bilinear Diffie-Hellman Assumption in the Standard Model

Yu Chen[1,2], Liqun Chen[3], and Zongyang Zhang[4,*]

[1] Institute of Information Engineering, Chinese Academy of Sciences
[2] State Key Laboratory of Information Security, Beijing, China
chenyu@is.iscas.ac.cn
[3] Hewlett-Packard Laboratories, Bristol, United Kingdom
liqun.chen@hp.com
[4] Department of Computer Science and Engineering,
Shanghai Jiao Tong University, Shanghai, China
zongyang.zhang@gmail.com

Abstract. In this paper, we propose several selective-identity chosen-ciphertext attack (IND-sID-CCA) secure identity based key encapsulation (IB-KEM) schemes that are provably secure under the computational bilinear Diffie-Hellman (CBDH) assumption in the standard model. Our schemes compare favorably to previous results in efficiency. With delicate modification, our schemes can be strengthened to be full-identity CCA secure easily.

Keywords: identity based encryption, standard model, CCA security, CBDH assumption.

1 Introduction

1.1 Background

Security against adaptive chosen ciphertext attack (CCA security for short) is nowadays considered the commonly accepted security notion for public key encryption (PKE)/identity based encryption (IBE). One of the most important research direction in this field is to design CCA-secure PKE/IBE schemes based on weak security assumptions in the standard model.

Cramer and Shoup [7] proposed the first practical CCA-secure PKE scheme without random oracles. Their construction was later generalized to hash proof systems [9]. However, all its variants [3, 14–17, 20] inherently rely on decisional assumption, e.g., the decisional Diffie-Hellman (DDH) assumption, the decisional bilinear Diffie-Hellman (DBDH) assumption or the decisional quadratic residuosity assumption. CCA security from computational assumptions was considered to be hard to obtain. Canetti, Halevi and Katz [4] made the breakthrough in 2004.

---

* Corresponding author.

H. Kim (Ed): ICISC 2011, LNCS 7259, pp. 275–301, 2012.
© Springer-Verlag Berlin Heidelberg 2012

They proposed the first practical CCA-secure PKE scheme based on CBDH assumption. Later, Cash et al. [5] presented a variant of Cramer-Shoup scheme [7] which is CCA-secure based on the strong twin CDH assumption, and in turn based on the standard CDH assumption. However, n group elements (where the value n is the bit-length of keys) have to be added into the ciphertext in order to prove CCA security. Hanaoka and Kurosawa [12] presented a CCA-secure PKE scheme enjoying the constant size ciphertext based on the CDH assumption from broadcast encryption. Hofheinz and Kitz [15] presented a construction assuming the hardness of factoring. Cramer, Hofheinz and Kitz [6] refined the well-known Naor-Yung paradigm [22] and constructed practical CCA-secure PKE schemes based on hard search problems, which includes the CDH and RSA type assumptions. Wee [24] gave more efficient and general transformations to CCA secure PKE schemes from extractable hash proof system, which again can be based on the hardness of CDH, RSA and factoring. Haralambiev, Jager, Kiltz and Shoup [13] then proposed a number of new PKE schemes that are provably secure under the CDH/CBDH assumption in the standard model, which improved efficiency of prior schemes from [5, 12].

For the time being, although there are several practical CCA-secure PKE schemes based on computational assumptions, CCA-secure IBE schemes based on weak assumptions are rare. This forms the main motivation of our work.

1.2 Our Contributions

In this paper we propose a number of new IB-KEM schemes that are CCA-secure under the CBDH assumption in the standard model. Our main idea is to extend the technique of constructing CCA-secure PKE schemes [13] to the IB-KEM version of Boneh-Boyen "commutative-blinding" IBE scheme (known as BB_1-IBE) [2]. We begin from a basic 1-bit IB-KEM, then extend it to n-bits IB-KEMs using different methods. As shown in Table 1 at the end of this section, our schemes improve efficiency of prior scheme [10].

A 1-BIT IB-KEM SCHEME. We first construct a 1-bit IB-KEM scheme. We denote it by Scheme 0 and briefly describe it as follows.

$$
\begin{aligned}
&\text{Setup}: \quad mpk = (g, h, X = g^a, X', Y), msk = a \\
&\text{KeyGen}: sk = (Y^a F(I)^s, g^s), \text{where } F(I) = X^I h \\
&\text{Encap}: \quad C = (g^r, (X^t X')^r, F(I)^r), \text{where } t = \mathsf{TCR}(g^r) \\
&\qquad\qquad K = f_{\mathsf{gl}}(e(X, Y)^r, R)
\end{aligned}
$$

Decryption only returns K if the ciphertext $C = (C_1, C_2, C_3)$ is consistent, i.e., $e(C_1, X^t X') = e(g, C_2) \wedge e(C_1, F(I)) = e(g, C_3)$. In all other cases it rejects and returns $\perp$. We defer the detailed construction and security proof to Section 3.

In what follows, we give a brief explanation of our strategy to achieve indistinguishability of ciphertext under selective-identity CCA attack (IND-sID-CCA) from two aspects, one is how to obtain selective-identity CCA security, and the other is how to reduce it to the CBDH assumption.

We first give the intuition behind the CCA security. From the attacker's view, the second part of the ciphertext $C_2 = (X^t X')^r$ prohibits an adversary from modifying a valid ciphertext in a meaningful way. From the challenger's view, the consistency of ciphertext is publicly verifiable, i.e., anyone could check the consistency of ciphertext with the help of bilinear map. Therefore any inconsistent ciphertext will be rejected. On the other hand, in the simulation all consistent ciphertexts can be classified into the following three types. Type-1 ciphertext is the one whose t value differs to t^* of the challenge ciphertext. Type-2 ciphertext is the one encrypted under an identity different from the challenge identity I^*. Type-3 ciphertext is exactly the challenge ciphertext. The reduction algorithm is able to decrypt all the consistent ciphertexts correctly by implementing dual all-but-one technique: set $X' := X^{-t^*} g^d$ to implement the all-but-one technique (with respect to $t \neq t^*$) to decrypt Type-1 ciphertexts $(t \neq t^*)$; set $F(I) := X^{I-I^*} g^z$ to implement the all-but-one technique (with respect to $I \neq I^*$) to extract a private key for all identities but the challenge identity I^*, thus to be able to decrypt Type-2 ciphertexts $(I \neq I^*)$. Type-3 ciphertext $(I = I^* \wedge t = t^*)$ is not allowed to be queried according to the definition of selective-identity chosen ciphertext security model. To summarize, the reduction algorithm can handle all the decryption queries correctly.

We then give our basic idea about how to reduce the IND-sID-CCA security to the CBDH assumption. Note that the indistinguishable type security notion is essentially defined as a decisional problem. Considering the gap between decisional problems and computational problems, it would be difficult to directly reduce the IND-sID-CCA security to the CBDH assumption. A natural approach is to find a stepping stone. More specifically, we first reduce the IND-sID-CCA security to some decisional assumption related to the CBDH assumption, then reduce the decisional assumption to the CBDH assumption. In this way the IND-sID-CCA security can be finally reduced to the CBDH assumption. We provide more details as follows. We select the Goldreich-Levin version decisional BDH (GL-DBDH) assumption [13] as the stepping stone, which states that there is no PPT algorithm that can distinguish the two distributions $\Delta_{\mathrm{bdh}} = (g, A, B, C, K, R)$ and $\Delta_{\mathrm{rand}} = (g, A, B, C, U, R)$. Here (g, A, B, C) are the inputs of a BDH problem, K is the output of a Goldreich-Levin hardcore predicate with $\mathrm{bdh}(A, B, C)$ and randomness R as input while U is a bit sampled from $\{0, 1\}$ uniformly random. Suppose a reduction algorithm $\mathcal{B}$ is asked to solve the GL-DBDH problem. $\mathcal{B}$ simulates a real attack game of Scheme 0 by embedding A into X, embedding B into Y, and embedding C into one part of the challenge ciphertext. We demonstrate that if there exists an IND-sID-CCA adversary $\mathcal{A}$ that can break the CCA security of Scheme 0, then $\mathcal{B}$ can break the GL-DBDH assumption. The GL-DBDH assumption can be thus reduced to the CBDH assumption according to the Goldreich-Levin theorem. Therefore, the IND-sID-CCA security of Scheme 0 is finally reduced to the CBDH assumption.

We note that Scheme 0 bears a close resemblance to the IB-KEM scheme [18]. The key difference between the two schemes is the derivation of the symmetric key. In [18] the Encap algorithm directly uses a BDH seed as a symmetric key, while in Scheme 0 the Encap algorithm uses the Goldreich-Levin hardcore predicate to derive a 1-bit symmetric key from a BDH seed.

Note that the element $(X^t X')^r$ and $F(I)^r$ in the ciphertext share the same randomness r, thus it is possible to further shrink the public parameters size and the ciphertext size. By using a technique similar to [19], the ciphertext can be reduced to two group elements at the cost of adding one group element in the private key and resorting to a stronger assumption, named the modified CBDH assumption. We denote the resulting scheme by Scheme $0'$. The concrete construction and security proof are included in Appendix A.

A SCHEME WITH CONSTANT SIZE PUBLIC PARAMETERS. To encapsulate a n-bits symmetric key, we can follow the standard multiple encapsulations method: perform the 1-bit IB-KEM n times using independent random coins. We denote the resulting scheme by Scheme 1 and describe it as follows.

Setup : $mpk = (g, h, X = g^a, X', Y), msk = a$
KeyGen : $sk = (Y^a F(I)^s, g^s)$
Encap : $C = (C_1, \ldots, C_n)$, where $C_i = (g^{r_i}, (X^t X')^{r_i}, F(I)^{r_i})$
           with $t = \mathsf{TCR}(C_{i,1}, \ldots, C_{n,1})$.
           $K = (K_1, \ldots, K_n)$, where $K_i = f_{gl}(e(X, Y)^{r_i}, R)$

We defer the detailed construction and security proof to Section 4.

A SCHEME WITH CONSTANT SIZE CIPHERTEXT. In contrast to the multiple encapsulations method used in Scheme 1, we may also adopt the randomness-reusing technique: include n group elements $(Y_1, \ldots, Y_n)$ into mpk (instead of a solo group element Y in previous schemes), then generate n BDH seeds using a single randomness r with respect to n different bases $e(X, Y_i)$. We denote the resulting scheme by Scheme 2 and describe it as follows.

Setup : $mpk = (g, h, X = g^a, X', Y_1, \ldots, Y_n), msk = a$
KeyGen : $sk = (sk_i, \ldots, sk_n)$, where $sk_i = (Y_i^a F(I)^{s_i}, g^{s_i})$
Encap : $C = (g^r, (X^t X')^r, F(I)^r)$, where $t = \mathsf{TCR}(g^r)$
           $K = (K_1, \ldots, K_n)$, where $K_i = f_{gl}(e(X, Y_i)^r, R)$

We defer the detailed construction and security proof to Section 5.

GENERALIZED SCHEME 1. Scheme 1 enjoys the constant-size mpk but its ciphertext size is linear in n, while Scheme 2 enjoys the constant-size ciphertext but its mpk size is linear in n. It is interesting to know if there exists a trade-off between mpk size and ciphertext size. From the above two schemes, it is easy to see that when generating n pair-wise independent BDH seeds, the roles of Y_i and the randomness r_j are exchangeable. With this observation, we propose the following generalized scheme that offers a trade-off between mpk and ciphertext. We denote it by Scheme 3 and described it as follows. The detailed construction and security proof are deferred to Section 6.

Setup : $\quad mpk = (g, h, X = g^a, X', Y_1, \ldots, Y_{n_1}), msk = a$
KeyGen : $sk = (sk_i, \ldots, sk_{n_1})$, where $sk_i = (Y_i^a F(I)^{s_i}, g^{s_i})$
Encap : $\quad C = (C_1, \ldots, C_{n_2})$, where $C_i = (g^{r_i}, (X^t X')^{r_i}, F(I)^{r_i})$
$\qquad$ with $t = \mathsf{TCR}(C_{i,1}, \ldots, C_{n_2,1})$
$\qquad K = (K_{i,j})$ for $1 \leq i \leq n_1, 1 \leq j \leq n_2$, where $K_{i,j} = f_{\mathrm{gl}}(e(X, Y_i)^{r_j}, R)$

In the above generalized scheme, $(Y_1, \ldots, Y_{n_1})$ are n_1 independent elements from $\mathbb{G}$. When performing encapsulation, the Encap algorithm picks $n_2 = n/n_1$ independent random integers $(r_1, \ldots, r_{n_2})$ from $\mathbb{Z}_p$, then mix-and-match them to generate n pair-wise independent BDH seeds of the form $e(X, Y_i)^{r_j}$. If we set $n_1 = n_2 = \sqrt{n}$, the yielding scheme has mpk of $O(\sqrt{n})$ group elements and ciphertext of $O(\sqrt{n})$ group elements. Scheme 1 and Scheme 2 can be viewed as special cases of the generalized scheme with the parameter choice $(n_1 = 1, n_2 = n)$ and $(n_1 = n, n_2 = 1)$, respectively. Interestingly, we find that the above trade-off method can naturally apply to the KEM schemes proposed in [12,13,24] and the IB-KEM scheme presented in [10]. Particularly, when implementing the trade-off method to the KEM scheme presented in [13, Section 3], the resulting scheme is exactly the one constructed by Liu et al. [21].

GENERALIZED SCHEME 2. Observe that one BDH seed bdh(A, B, C) is determined by three inputs, then mpk and ciphertext can be further shrunk to $O(\sqrt[3]{n})$ group elements by using the mix-and-match method twice. More precisely, instead of generating the BDH seed like $e(X, Y_i)^{r_j}$ as the above generalized scheme, we can generate the BDH seeds of the form $e(Y_i, Y_j)^{r_k}$. That is, first self mix-and-match the set $(Y_1, \ldots, Y_{n_1})$, then mix-and-match the resulting $n_1(n_1 - 1)/2$ bases $e(Y_i, Y_j)$ $(i \neq j)$ with n_2 random integers $(r_1, \ldots, r_{n_2})$. The self mix-and-match method is better than the "implicitly defining" method used in [13, Section 5.3] since it travels all the binary combinatorial pairs (Y_i, Y_j) over the set $(Y_1, \ldots, Y_{n_1})$, thus it can generate the same number of bases with smaller mpk. Based on this observation, we propose another generalized scheme called Scheme 4 as follows. The detailed construction and security proof are deferred to Section 7.

Setup : $\quad mpk = (g, h, X, X', Y_1 = g^{y_1}, \ldots, Y_{n_1} = g^{y_{n_1}}), msk = (y_1, \ldots, y_{n_1})$
KeyGen : $sk = (sk_{ij})$ for $1 \leq i < j \leq n_1$ where $sk_{ij} = (g^{y_i y_j} F(I)^{s_{ij}}, g^{s_{ij}})$
Encap : $\quad C = (C_1, \ldots, C_{n_2})$, where $C_k = (g^{r_k}, (X^t X')^{r_k}, F(I)^{r_k})$
$\qquad$ with $t = \mathsf{TCR}(C_{i,1}, \ldots, C_{n_2,1})$
$\qquad K = (K_{i,j,k} := f_{\mathrm{gl}}(e(Y_i, Y_j)^{r_k}, R))$ for $1 \leq i < j \leq n_1, 1 \leq k \leq n_2$

To generate n pair-wise independent BDH seeds we require that $n = n_1(n_1 - 1)n_2/2$. Let $n_1 = n_2$, then the public parameters and the ciphertext are both of $O(\sqrt[3]{n})$ groups elements. Not surprisingly, this trade-off technique can also apply to the KEM scheme [13, Section 5.3] and the IB-KEM scheme [10].

1.3   Related Work

Recently, Galindo [10] gave an IND-sID-CCA secure IB-KEM based on the CBDH assumption in the standard model by integrating the KEM scheme [12]

Table 1. Efficiency comparison of the proposed schemes

Scheme	Ciphertext Overhead	Efficiency [# exp, # pairing]		Key Sizes	
		Encap	Decap	$|mpk|$	$|msk|$
Galindo [10]	$4 \times |\mathbb{G}_T|$	$[4, 0]$	$[2, 2n + 2]$	$(2n + 9) \times |\mathbb{G}|$	$(n + 4) \times |\mathbb{Z}_p|$
Scheme 1 (§4)	$3n \times |\mathbb{G}_T|$	$[3n + 1, 0]$	$[1, 4n]$	$5 \times |\mathbb{G}|$	$1 \times |\mathbb{Z}_p|$
Scheme 2 (§5)	$3 \times |\mathbb{G}_T|$	$[4, 0]$	$[1, 2n + 2]$	$(n + 4) \times |\mathbb{G}|$	$1 \times |\mathbb{Z}_p|$
Scheme 3 (§6)	$3n_2 \times |\mathbb{G}_T|$	$[3n_2 + 1, 0]$	$[1, 2n + 2n_2]$	$(n_1 + 4) \times |\mathbb{G}|$	$1 \times |\mathbb{Z}_p|$
Scheme 4 (§7)	$3n_2 \times |\mathbb{G}_T|$	$[3n_2 + 1, 0]$	$[1, 2n + 2n_2]$	$(n_1 + 4) \times |\mathbb{G}|$	$n_1 \times |\mathbb{Z}_p|$

In Scheme 3 we have $n = n_1 n_2$, then n_1 and n_2 can be set to integers around $O(\sqrt{n})$. In Scheme 4 we have $n_1(n_1 - 1)n_2/2$, then n_1 and n_2 can be set to integers around $O(\sqrt[3]{n})$.

with the BB_1-IBE scheme [2]. Galindo's scheme is not conceptually simple due to the underlying KEM scheme [12], and its master secret consists of $O(n)$ group elements that might be impractical for some applications. Haralambiev et al. [13] mentioned that their KEM scheme with public key of size $O(\sqrt{n})$ can extend to selective-identity secure BB_1-IBE scheme [2]. They sketched their ideas as follows: the IBE scheme has the same parameters as their KEM scheme [13, Section 5.3], and a private key for identity I contains $2n$ group elements of the form $(g^{z_i z_j'} \cdot (X^I X')^{s_{ij}}, g^{s_{ij}}) \in \mathbb{G}^2$. However, we remark that a private key for identity I should be $(g^{z_i z_j'} \cdot F(I)^{s_{ij}}, g^{s_{ij}})$, where $F(I)$ is the Boneh-Boyen hash. Besides, the master secret key of their scheme is still a bit large ($2\sqrt{n}$ elements from $\mathbb{Z}_p$), which may render it less practical in use. Regarding to this, it would be very interesting to construct IBE schemes with short master secret key while provably secure under weak assumptions in the standard model.

2   Preliminaries

2.1   Notation

We use standard asymptotic notation O and o to denote the growth of functions. We denote with $\mathsf{poly}(\kappa)$ an unspecified function $f(\kappa) = O(\kappa^c)$ for some constant c. We denote with $\mathsf{negl}(\kappa)$ an unspecified function $f(\kappa)$ such that $f(\kappa) = o(\kappa^{-c})$ for every constant c. Throughout the paper, a *probabilistic polynomial-time* (PPT) algorithm is a randomized algorithm that runs in time $\mathsf{poly}(\kappa)$. For a positive integer n, we denote with $[n]$ the set $[n] = \{1, \ldots, n\}$. For a finite set S, we use $s \xleftarrow{R} S$ to denote that s is sampled from the set S uniformly at random.

2.2   Identity Based Key Encapsulation Mechanisms

An identity-based key encapsulation mechanism (IB-KEM) [1] consists of four PPT algorithms as follows:

Setup: takes the security parameter 1^κ as input and outputs the public parameter mpk and the master secret msk. Intuitively, mpk is the system parameters which will be public known, while the msk will be known only to the thrusted third party, called Private Key Generator.

KeyGen: takes mpk, msk, an identity I as input and outputs the associated private key sk.

Encap: takes mpk and an identity as input and outputs a pair (C, K) where C is the ciphertext and $K \in \mathcal{K}$ is a data encryption key.

Decap: takes mpk, private key sk, and a ciphertext C as input and outputs the data encryption key $K \in \mathcal{K}$.

We require that if $(mpk, msk) \xleftarrow{R} \mathsf{Setup}(1^\kappa)$, $sk \leftarrow \mathsf{KeyGen}(mpk, msk, I)$, and $(C, K) \leftarrow \mathsf{Encap}(mpk, I)$ then we have $\mathsf{Decap}(mpk, sk, C) = K$.

2.3 Chosen Ciphertext Security

CCA-security of an IB-KEM is defined by the following game playing between an adversary $\mathcal{A}$ and a challenger $\mathcal{CH}$.

Setup. $\mathcal{CH}$ takes a security parameter 1^κ and runs the KeyGen algorithm. It gives the adversary the resulting system parameters. It keeps the master key to itself.

Phase 1. $\mathcal{A}$ may make polynomially-many private key queries and decapsulation queries. $\mathcal{CH}$ answers these queries by running the algorithm KeyGen to extract the associated private keys.

Challenge. Once the adversary decides that Phase 1 is over it outputs an identity I^* on which it wishes to be challenged. The only constraint is that I^* did not appear in any private key extraction query in Phase 1. $\mathcal{CH}$ computes $(C^*, K_0^*) = \mathsf{Encap}(mpk, I^*)$, samples K_1^* uniform randomly from $\mathcal{K}$. Finally, $\mathcal{CH}$ picks a random bit $\beta \in \{0, 1\}$ and sends (C^*, K_β^*) as the challenge to the adversary.

Phase 2. $\mathcal{A}$ issues more private key queries with the restriction that $\langle I \rangle \neq \langle I^* \rangle$ and the decapsulation queries with the restriction that $\langle I, C \rangle \neq \langle I^*, C^* \rangle$.

Guess. Finally, $\mathcal{A}$ outputs a guess $\beta' \in \{0, 1\}$ and wins the game if $\beta = \beta'$.

We refer to such an adversary $\mathcal{A}$ as an IND-ID-CCA adversary. We define adversary $\mathcal{A}$'s advantage over the IB-KEM scheme $\mathcal{E}$ by $\mathrm{Adv}_{\mathcal{E}, \mathcal{A}}^{\mathsf{CCA}}(\kappa) = \left| \Pr[\beta = \beta'] - \frac{1}{2} \right|$, where κ is the security parameter. The probability is over the random bits used by the challenger and the adversary.

Definition 2.1. *We say that an IB-KEM scheme $\mathcal{E}$ is* IND-ID-CCA *secure if for any PPT* IND-ID-CCA *adversary $\mathcal{A}$ the advantage* $\mathrm{Adv}_{\mathcal{E}, \mathcal{A}}^{\mathsf{CCA}}(\kappa)$ *is negligible.*

Selective-identity CCA-security [4] can be defined in a similar game as the above game of full-identity CCA-security, except that the adversary needs to output a target identity at the very beginning of the game. We refer to such an adversary $\mathcal{A}$ as an IND-sID-CCA adversary. We define adversary $\mathcal{A}$'s advantage over the IB-KEM scheme $\mathcal{E}$ by $\mathrm{Adv}_{\mathcal{E},\mathcal{A}}^{\mathsf{CCA}}(\kappa) = \left|\Pr[\beta = \beta'] - \frac{1}{2}\right|$, where κ is the security parameter. The probability is over the random bits used by the challenger and the adversary.

Definition 2.2. *We say that an IB-KEM scheme $\mathcal{E}$ is* IND-sID-CCA *secure if for any PPT* IND-sID-CCA *adversary $\mathcal{A}$ the advantage* $\mathrm{Adv}_{\mathcal{E},\mathcal{A}}^{\mathsf{CCA}}(\kappa)$ *is negligible.*

2.4   Target Collision Resistant Hash Function

$\mathsf{TCR} = (\mathsf{TCR}_k)_{k \in \mathbb{N}}$ is a family of keyed hash function $\mathsf{TCR}_k^s : \mathbb{G} \to \mathbb{Z}_p$ for each k-bit key s. For an adversary $\mathcal{H}$, its tcr-advantage $\mathrm{Adv}_{\mathcal{H}}^{\mathsf{TCR}}(k)$ is defined as:

$$\Pr[\mathsf{TCR}^s(c^*) = \mathsf{TCR}^s(c) \ \wedge \ c \neq c^* : s \xleftarrow{R} \{0,1\}^k; c^* \xleftarrow{R} \mathbb{G}; c \leftarrow \mathcal{H}(s, c^*)]$$

Note that TCR is a weaker requirement than collision-resistance, so any practical collision-resistant function can be used. To simplify notation we will drop the superscript s and simply use TCR hereafter. Additionally, we can define multi-inputs TCR function in a natural way, that is $\mathsf{TCR}_k^s : (\mathbb{G})^n \to \mathbb{Z}_p$. The corresponding tcr-advantage of an adversary $\mathcal{H}$ is defined in a similar way except substituting c with $(c_1, \ldots, c_n)$ and c^* with $(c_1^*, \ldots, c_n^*)$.

2.5   Computational Bilinear Diffie-Hellman Assumption

Let $\mathbb{G}$ be a cyclic group generated by g and equipped with a bilinear map $e : \mathbb{G} \times \mathbb{G} \to \mathbb{G}_T$. Define

$$\mathrm{bdh}(A, B, C) := T, \text{ where } A = g^a, B = g^b, C = g^c, \text{ and } T = e(g, g)^{abc}$$

The computational bilinear Diffie-Hellman (CBDH) problem is computing the value $\mathrm{bdh}(A, B, C)$ given random $A, B, C \in \mathbb{G}$. The CBDH assumption asserts that the CBDH problem is hard, that is, $\Pr[\mathcal{A}(A, B, C) = \mathrm{bdh}(A, B, C)] \leq \mathrm{negl}(\kappa)$ for all PPT algorithms $\mathcal{A}$.

In the bilinear setting, the Goldreich-Levin theorem [11] gives us the following lemma for a Goldreich-Levin hardcore predicate $f_{\mathrm{gl}} : \mathbb{G}_T \times \{0,1\}^u \to \{0,1\}$.

Lemma 2.3. *Let $\mathbb{G}$ be a prime order group generated by g equipped with a pairing $e : \mathbb{G} \times \mathbb{G} \to \mathbb{G}_T$. Let $A, B, C \xleftarrow{R} \mathbb{G}$ be random group elements, $R \xleftarrow{R} \{0,1\}^u$, and let $K = f_{\mathrm{gl}}(\mathrm{bdh}(A, B, C), R)$. Let $U \xleftarrow{R} \{0,1\}$ be uniformly random. Suppose there exists a PPT algorithm $\mathcal{B}$ distinguishing the distributions $\Delta_{\mathrm{bdh}} = (g, A, B, C, K, R)$ and $\Delta_{\mathrm{rand}} = (g, A, B, C, U, R)$ with non-negligible advantage. Then there exists a PPT algorithm computing $\mathrm{bdh}(A, B, C)$ on input (A, B, C) with non-negligible success probability, hence breaking the CBDH problem.*

We assume that the global public parameters known to all the parties consist of the pairing parameters $(e, \mathbb{G}, \mathbb{G}_T, g, p) \leftarrow \mathsf{GroupGen}(1^\kappa)$, the descriptions of a target collision resistant hash function TCR and a suitable Goldreich-Levin hardcore predicate $f_{\mathrm{gl}}(\cdot, R)$ with randomness R to extract one pseudorandom bit from a BDH seed. It is well known that an IB-KEM scheme compares favorably to an IBE scheme in many ways [1, 8], and IB-KEM schemes can be readily bootstrapped to full functional IBE schemes by coupling with a DEM having appropriate properties. Therefore in this paper, we focus on the constructions of IB-KEM.

3   A 1-Bit IB-KEM Scheme

In this section we describe an 1-bit IB-KEM which is obtained by extending the techniques of [13] to the Boneh-Boyen IBE scheme [2]. The resulting IB-KEM scheme is IND-sID-CCA secure based on the CBDH assumption. It is defined as follows.

Setup. Pick $a \stackrel{R}{\leftarrow} \mathbb{Z}_p$, $h, X', Y \stackrel{R}{\leftarrow} \mathbb{G}$, set $X = g^a$, and define the function $F : \mathbb{Z}_p \to \mathbb{G}$ as $I \mapsto X^I h$. The public parameters and the master secret key are given by

$$mpk = (g, h, X, X', Y) \text{ and } msk = a$$

KeyGen. To generate a private key for an identity $I \in \mathbb{Z}_p$, pick $s \stackrel{R}{\leftarrow} \mathbb{Z}_p$ and output

$$sk = (Y^a F(I)^s, g^s)$$

Encap. Pick $r \stackrel{R}{\leftarrow} \mathbb{Z}_p$, then generate the ciphertext $C = (C_1, C_2, C_3)$ as $C_1 = g^r$, $C_2 = (X^t X')^r$ with $t = \mathsf{TCR}(C_1)$, and $C_3 = F(I)^r$. Compute

$$K = f_{\mathrm{gl}}(e(X, Y)^r, R)$$

Decap. To decapsulate ciphertext (C_1, C_2, C_3) under identity I, first compute $t = \mathsf{TCR}(C_1)$. If $e(C_1, X^t X') \neq e(g, C_2)$ or $e(C_1, F(I)) \neq e(g, C_3)$ then return $\perp$. Take the private key sk and the ciphertext $C = (C_1, C_2, C_3)$ as input and outputs $K = f_{\mathrm{gl}}\left(\frac{e(C_1, sk_1)}{e(C_3, sk_2)}, R\right)$. Indeed, for a valid ciphertext, we have

$$\frac{e(C_1, sk_1)}{e(C_3, sk_2)} = \frac{e(g^r, Y^a F(I)^s)}{e(F(I)^r, g^s)} = e(X, Y)^r.$$

Notice that the consistency of the ciphertext is publicly verifiable, i.e., anyone can verify a ciphertext being consistent or not.

Theorem 3.1. *Let* TCR *be a target collision-resistant hash function and suppose that the CBDH assumption holds in* $\mathbb{G}$. *Then the above scheme is an* IND-sID-CCA *secure IB-KEM.*

Proof. We proceed in a sequence of games. We write (C_1^*, C_2^*, C_3^*) to denote the challenge ciphertext with the corresponding key K^* of identity I^*, denote with U^* the random key chosen by the IND-sID-CCA experiment, and set $t^* = \mathsf{TCR}(C_1^*)$. Let W_i denote the event that $\mathcal{A}$ outputs β' such that $\beta' = \beta$ in Game i.

Game 0. This is the standard IND-sID-CCA game. By definition we have

$$\Pr[W_0] = \frac{1}{2} + \mathsf{AdvCCA}_{\mathsf{KEM}}^{\mathcal{A}}(\kappa) \tag{1}$$

Game 1. Let E_{01} be the event that the adversary issues a decapsulation query $\langle I^*, C_1', C_2', C_3' \rangle$ with $C_1' = C_1^*$ in Phase 1. Note that the probability that the adversary submits a decapsulation query such that $C_1' = C_1^*$ before seeing the challenge ciphertext is bounded by Q_d/p, where Q_d is the number of decapsulation queries issued by $\mathcal{A}$. Since $Q_d = \mathsf{poly}(\kappa)$, we have $\Pr[E_{01}] \le Q_d/p \le \mathsf{negl}(\kappa)$. We define Game 1 exactly the same as Game 0 except assuming that E_{01} never occurs in Game 1. It follows that

$$|\Pr[W_1] - \Pr[W_0]| \le \mathsf{negl}(\kappa) \tag{2}$$

Moreover, we remark that in Phase 2 a decapsulation query $\langle I^*, C_1', C_2', C_3' \rangle$ will be rejected if $C_1' = C_1^*$. Since if $C_2' \ne C_2^*$ or $C_3' \ne C_3^*$, the decapsulation query will be rejected for the inconsistency of the ciphertext. If $C_2' = C_2^*$ and $C_3' = C_3^*$, it will be rejected by definition of IND-sID-CCA game.

Game 2. Let E_{12} be the event that the adversary issues a decapsulation query $\langle I^*, C_1', C_2', C_3' \rangle$ with $C_1' \ne C_1^*$ and $\mathsf{TCR}(C_1') = \mathsf{TCR}(C_1^*)$. By the target collision resistance of TCR, we have $\Pr[E_{12}] \le \mathsf{negl}(\kappa)$. We define Game 2 exactly the same as Game 1 except assuming that E_{12} never occurs in Game 2. It follows that

$$|\Pr[W_2] - \Pr[W_1]| \le \mathsf{negl}(\kappa) \tag{3}$$

We claim that

$$\Pr[W_2] = \frac{1}{2} + \mathsf{negl}(\kappa) \tag{4}$$

We prove this statement by letting an adversary against the GL-DBDH assumption simulate the challenger in Game 2. $\mathcal{B}$ is given a challenge instance (g, A, B, C, L, R), where L is either $f_{\mathrm{gl}}(\mathrm{bdh}(A, B, C), R)$ or randomly sampled from $\{0, 1\}$. $\mathcal{B}$ plays the game with an adversary $\mathcal{A}$ against the IND-sID-CCA security of the 1-bit IB-KEM scheme.

Initialization. $\mathcal{A}$ first outputs an identity $I^* \in \mathbb{Z}_p$ that it intends to attack.

Setup. $\mathcal{B}$ picks $d \xleftarrow{R} \mathbb{Z}_p$, and then sets $X = A = g^a$, $X' = X^{-t^*} g^d$, $Y = B = g^b$, where $t^* = \mathsf{TCR}(C)$. $\mathcal{B}$ picks $z \xleftarrow{R} \mathbb{Z}_p$ and defines $h = X^{-I^*} g^z$. It gives $\mathcal{A}$ the public parameters $mpk = (g, h, X, X', Y)$. The corresponding msk, which is unknown to $\mathcal{B}$ is a. The function F is essentially of the form

$$F(I) = X^I h = X^{I-I^*} g^z$$

Phase 1 - Private Key Queries. $\mathcal{A}$ issues up to Q_e private key queries with the only restriction that $\langle I \rangle \neq \langle I^* \rangle$. To respond to a private query for identity $I \in \mathbb{Z}_p$, $\mathcal{B}$ generates sk as follows: for sk_ℓ algorithm $\mathcal{B}$ picks $s \xleftarrow{R} \mathbb{Z}_p$ and sets

$$sk_1 = Y^{\frac{-z}{I-I^*}} F(I)^s, \quad sk_2 = g^s Y^{\frac{-1}{I-I^*}}$$

Let $\widetilde{s} = s - b/(I - I^*)$. It is easy to see that sk is a valid random private key for I since

$$sk_1 = Y^{\frac{-z}{I-I^*}} (X^{I-I^*} g^z)^s = Y^a (X^{I-I^*} g^z)^{s - \frac{b}{I-I^*}} = Y^a F(I)^{\widetilde{s}_\ell}$$
$$sk_2 = g^s Y^{\frac{1}{I-I^*}} = g^{\widetilde{s}}$$

where s, $\widetilde{s}$ are uniform in $\mathbb{Z}_p$. This matches the definition for a private key for I. Hence, sk is a valid private key for I.

Phase 1 - Decapsulation Queries. Upon $\mathcal{A}$ issuing a decaptulation query $\langle I, C_1, C_2, C_3 \rangle$, $\mathcal{B}$ responds as follows. If $I \neq I^*$, $\mathcal{B}$ uses the corresponding private key to handle it. Otherwise, $\mathcal{B}$ computes $t = \mathsf{TCR}(C_1)$ and tests the consistency of the ciphertext by checking

$$e(C_1, X^t X') \overset{?}{=} e(g, C_2) \ \wedge \ e(C_1, F(I)) \overset{?}{=} e(g, C_3)$$

If the equality holds, $\mathcal{B}$ sets $K := f_{\mathrm{gl}}(e(\widetilde{X}, Y), R)$. The decapsulation is correct by observing that $\widetilde{X} = (C_2/C_1^d)^{1/(t-t^*)} = (X^{r(t-t^*)} g^{rd}/g^{rd})^{1/(t-t^*)} = X^r = \mathrm{dh}(X, C_1)$. By the definition of Game 2 we know that when $I = I^*$, if $C_1 \neq C_1^*$ then we have $t \neq t^*$. Therefore $\mathcal{B}$ can answer all decapsulation queries issued by $\mathcal{A}$ correctly.

Challenge. $\mathcal{B}$ sets $C_1^* = C$ (which implicitly assigns $r = c$), $C_2^* = C^d$, and $C_3^* = C^z$. The challenge ciphertext is $C^* = (C_1^*, C_2^*, C_3^*)$. Note that this is a consistent ciphertext since we have $(X^{t^*} X')^r = (g^d)^r = C^d$ and $F(I^*)^r = (g^z)^r = C^z$. Then $\mathcal{B}$ sets $K^* = L$ and gives $\mathcal{A}$ the challenge (C^*, K^*).

Phase 2. In Phase 2, all the queries are responded in the same way as in Phase 1 except the decapsulation query $\langle I^*, C^* \rangle$ will be rejected.

This finishes the description of simulation. It is easy to see that $\mathcal{B}$ simulates the challenger perfectly. If $\mathcal{A}$'s advantage is not negligible, then $\mathcal{B}$ has non-negligible advantage against the GL-DBDH problem. According to Lemma 2.3, $\mathcal{B}$ further implies an algorithm with non-negligible advantage against the CBDH problem, which contradicts to the CBDH assumption. Therefore, we prove the statement. The theorem follows by combining (1)-(4). □

4 CCA Secure IB-KEM with Constant Size Public Parameters

In this section we present a n-bit IB-KEM scheme based on the 1-bit IB-KEM scheme using multiple encapsulations method.

Setup. The same as Scheme 0.

KeyGen. The same as Scheme 0.

Encap. Pick $r_1, \ldots, r_n \stackrel{R}{\leftarrow} \mathbb{Z}_p$, then compute $C_{i,1} = g^{r_i}$, $t = \mathsf{TCR}(C_{1,1}, \ldots, C_{n,1})$, $C_{i,2} = (X^t X')^{r_i}$, $C_{i,3} = F(I)^{r_i}$. The final ciphertext is $C = (C_1, \ldots, C_n)$, where $C_i = (C_{i,1}, C_{i,2}, C_{i,3})$. Compute $K = (K_1, \ldots, K_n)$, where

$$K_i = f_{\mathrm{gl}}(e(X, Y)^{r_i}, R) \text{ for } 1 \leq i \leq n.$$

Decap. To decapsulate ciphertext $C = (C_1, \ldots, C_n)$ under identity I, first compute $t = \mathsf{TCR}(C_{1,1}, \ldots, C_{n,1})$. If $e(C_{i,1}, X^t X') \neq e(g, C_{i,2})$ or $e(C_{i,1}, F(I)) \neq e(g, C_{i,3})$ for any $i \in [n]$ then return $\perp$. Take the private key $sk = (sk_1, sk_2)$ and the ciphertext $C = (C_1, \ldots, C_n)$ as input and output

$$K_i = f_{\mathrm{gl}}\left(\frac{e(C_{i,1}, sk_1)}{e(C_{i,3}, sk_2)}, R\right) \text{ for } 1 \leq i \leq n.$$

Indeed, for a valid ciphertext, we have

$$\frac{e(C_{i,1}, sk_1)}{e(C_{i,3}, sk_2)} = \frac{e(g^{r_i}, Y^a F(I)^s)}{e(F(I)^{r_i}, g^s)} = e(X, Y)^{r_i} \text{ for } 1 \leq i \leq n.$$

Theorem 4.1. *Let* TCR *be a target collision-resistant hash function and suppose that the CBDH assumption holds in* $\mathbb{G}$. *Then the above scheme is an* IND-sID-CCA *secure IB-KEM.*

The security is somewhat straightforward by conducting the hybrid argument with the proof of Scheme 0. For completeness we put the proof in Appendix B.

5   CCA Secure IB-KEM with Constant Size Ciphertext

In this section we present a n-bit IB-KEM scheme based on the 1-bit IB-KEM scheme using the randomness-reuse technique.

Setup. Pick $a \stackrel{R}{\leftarrow} \mathbb{Z}_p$, $h, X', Y_1, \ldots, Y_n \stackrel{R}{\leftarrow} \mathbb{G}$, set $X = g^a$, and define the function $F : \mathbb{Z}_p \to \mathbb{G}$ as $I \mapsto X^I h$. The mpk and the msk are given by

$$mpk = (g, h, X, X', Y_1, \ldots, Y_n) \text{ and } msk = a$$

KeyGen. To generate a private key for an identity $I \in \mathbb{Z}_p$, pick $s_1, \ldots, s_n \stackrel{R}{\leftarrow} \mathbb{Z}_p$ and output $sk = (sk_1, \ldots, sk_n)$, where

$$sk_i = (Y_i^a F(I)^{s_i}, g^{s_i}) \text{ for } 1 \leq i \leq n.$$

Encap. Pick $r \stackrel{R}{\leftarrow} \mathbb{Z}_p$, then generate the ciphertext $C = (C_1, C_2, C_3)$ as $C_1 = g^r$, $C_2 = (X^t X')^r$ with $t = \mathsf{TCR}(C_1)$, and $C_3 = F(I)^r$. Compute $K = (K_1, \ldots, K_n)$, where

$$K_i = f_{\mathrm{gl}}(e(X, Y_i)^r, R) \text{ for } 1 \leq i \leq n.$$

Decap. To decapsulate ciphertext (C_1, C_2, C_3) under identity I, first compute $t = \mathsf{TCR}(C_1)$. If $e(C_1, X^t X') \neq e(g, C_2)$ or $e(C_1, F(I)) \neq e(g, C_3)$ then return $\perp$. Take the private key $sk = (sk_1, \ldots, sk_n)$ and the ciphertext $C = (C_1, C_2, C_3)$ as input and output

$$K_i = f_{\mathrm{gl}}\left(\frac{e(C_1, sk_{i,1})}{e(C_3, sk_{i,2})}, R\right) \text{ for } 1 \leq i \leq n.$$

Indeed, for a valid ciphertext, we have

$$\frac{e(C_1, sk_{i,1})}{e(C_3, sk_{i,2})} = \frac{e(g^r, Y_i^a F(I)^{s_i})}{e(F(I)^r, g^{s_i})} = e(X, Y_i)^r \text{ for } 1 \leq i \leq n.$$

Notice that the consistency of the ciphertext is publicly verifiable, i.e., anyone could verify a ciphertext being consistent or not.

Theorem 5.1. *Let* TCR *be a target collision-resistant hash function and suppose that the CBDH assumption holds in* $\mathbb{G}$. *Then the above scheme is an* IND-sID-CCA *secure IB-KEM.*

The security is somewhat straightforward by conducting the hybrid argument with the proof of Scheme 0. For completeness we put the proof in Appendix C.

6   Generalized Scheme 1

In this section we present the first generalized scheme which shows that there exists a trade-off between the ciphertext size and the public parameters size. We assume that n is the product of n_1 and n_2. The generalized scheme is defined as follows.

Setup. The same as Scheme 2 except we substitute n with n_1.

KeyGen. The same as Scheme 2 except that we substitute n with n_1.

Encap. Pick $r_1, \ldots, r_{n_2} \xleftarrow{R} \mathbb{Z}_p$, and set $C_{j,1} = g^{r_j}$ for $1 \leq j \leq n_2$. Set $t = \mathsf{TCR}(C_{1,1}, \ldots, C_{n_2,1})$, $C_{j,2} = (X^t X')^{r_j}$, $C_{j,3} = F(I)^{r_j}$ for $1 \leq j \leq n_2$. The ciphertext is $C = (C_1, \ldots, C_{n_2})$ where $C_j = (C_{j,1}, C_{j,2}, C_{j,3})$. Compute the symmetric key $K = (K_1, \ldots, K_n)$, where

$$K_{(i-1) \times n_1 + j} = f_{\mathrm{gl}}(e(X, Y_i)^{r_j}, R) \text{ for } 1 \leq i \leq n_1 \text{ and } 1 \leq j \leq n_2.$$

Decap. To decapsulate ciphertext $C = (C_1, \ldots, C_{n_2})$ encrypted under identity I, first compute $t = \mathsf{TCR}(C_{1,1}, \ldots, C_{n_2,1})$. If $e(C_{j,1}, X^t X') \neq e(g, C_{j,2})$ or $e(C_{j,1}, F(I)) \neq e(g, C_{j,3})$ for some $j \in [n_2]$ then return $\perp$. Take the private key $sk = (sk_1, \ldots, sk_{n_1})$ and $C = (C_1, \ldots, C_{n_2})$ as input and output

$$K_{(i-1) \times n_1 + j} = f_{\mathrm{gl}}\left(\frac{e(C_{j,1}, sk_{i,1})}{e(C_{j,3}, sk_{i,2})}, R\right) \text{ where } 1 \leq i \leq n_1 \text{ and } 1 \leq j \leq n_2.$$

Indeed, for a valid ciphertext, we have

$$\frac{e(C_{j,1}, sk_{i,1})}{e(C_{j,3}, sk_{i,2})} = \frac{e(g^{r_j}, Y_i^a F(I)^{s_i})}{e(F(I)^{r_j}, g^{s_i})} = e(X, Y_i)^{r_j} \text{ for } 1 \leq i \leq n_1 \text{ and } 1 \leq j \leq n_2.$$

Particularly, let n be a perfect square and $n_1 = n_2 = \sqrt{n}$, we obtain an IB-KEM scheme with $O(\sqrt{n})$ public parameters size and $O(\sqrt{n})$ ciphertext size.

Theorem 6.1. *Let* TCR *be a target collision-resistant hash function and suppose that the CBDH assumption holds in* $\mathbb{G}$*. Then the above scheme is an* IND-sID-CCA *secure IB-KEM.*

The security is somewhat straightforward by conducting the hybrid argument with the proof of Scheme 0. For completeness we put the proof in Appendix D.

7   Generalized Scheme 2

In this section we present the second generalized scheme. We assume that $n = n_1(n_1 - 1)n_2/2$.

Setup. mpk and msk are given by

$$mpk = (g, h, X, X', Y_1 = g^{y_1}, \ldots, Y_{n_1} = g^{y_{n_1}}, F) \text{ and } msk = (y_1, \ldots, y_{n_1})$$

KeyGen. To generate a private key $sk = (sk_{ij})$ for an identity $I \in \mathbb{Z}_p$, pick $s_{ij} \xleftarrow{R} \mathbb{Z}_p$ and set $sk_{ij} = (g^{y_i y_j} F(I)^{s_{ij}}, g^{s_{ij}})$ for $1 \le i < j \le n_1$.

Encap. Pick $r_1, \ldots, r_{n_2} \xleftarrow{R} \mathbb{Z}_p$, and set $C_{k,1} = g^{r_k}$ for $1 \le k \le n_2$. Set $t = \text{TCR}(C_{1,1}, \ldots, C_{n_2,1})$, $C_{j,2} = (X^t X')^{r_j}$, $C_{j,3} = F(I)^{r_j}$ for $1 \le j \le n_2$. The ciphertext is $C = (C_1, \ldots, C_{n_2})$ where $C_k = (C_{k,1}, C_{k,2}, C_{k,3})$. Compute the symmetric key $K = (K_{i,j,k})$, where

$$K_{i,j,k} = f_{\text{gl}}(e(Y_i, Y_j)^{r_k}, R) \text{ for } 1 \le i < j \le n_1 \text{ and } 1 \le k \le n_2.$$

Decap. To decapsulate ciphertext $C = (C_1, \ldots, C_{n_2})$ encrypted under identity I, first compute $t = \text{TCR}(C_{1,1}, \ldots, C_{n_2,1})$. If $e(C_{k,1}, X^t X') \ne e(g, C_{k,2})$ or $e(C_{k,1}, F(I)) \ne e(g, C_{k,3})$ for some $k \in [n_2]$ then return $\bot$. Take the private key $sk = (sk_{ij})$ and $C = (C_1, \ldots, C_{n_2})$ as input and output

$$K_{i,j,k} = f_{\text{gl}}\left(\frac{e(C_{k,1}, sk_{ij,1})}{e(C_{k,3}, sk_{ij,2})}, R\right) \text{ where } 1 \le i < j \le n_1 \text{ and } 1 \le k \le n_2.$$

Indeed, for a valid ciphertext, we have

$$\frac{e(C_{k,1}, sk_{ij,1})}{e(C_{k,3}, sk_{ij,2})} = \frac{e(g^{r_k}, g^{y_i y_j} F(I)^{s_{ij}})}{e(F(I)^{r_k}, g^{s_{ij}})} = e(Y_i, Y_j)^{r_k} \text{ for } 1 \le i < j \le n_1, 1 \le k \le n_2.$$

Theorem 7.1. *Let* TCR *be a target collision-resistant hash function and suppose that the CBDH assumption holds in* $\mathbb{G}$*. Then the above scheme is an* IND-sID-CCA *secure IB-KEM.*

The proof is similar to that of Scheme 1 in Section 4, Scheme 2 in Section 5, and Generalized Scheme 1 in Section 6, except that for a given CBDH challenge instance (A, B, C) the reduction algorithm first sets $Y_i = A$ for some $i \in [n_1]$ then sets $X = A^h$ for a random chosen exponent h instead of directly setting $X = A$ as before. For the limit of space, we omit the details here.

8   Extensions

Since BB_1-IBE [2] and Waters-IBE [23] share the same commutative-blinding framework, thus we can enhance our IB-KEM schemes with only selective-identity security to IB-KEM schemes with full-identity security by using the Waters-IBE as the underlying IBE scheme. The security proofs are somewhat straightforward by composing the proofs for IB-KEM schemes in Section 4, 5, and 6 based on BB_1-IBE and the proofs for Waters-IBE [18,23]. For a concrete example, we sketch the proof of Scheme 1*, which is the resulting scheme of replacing the underlying IBE scheme of Scheme 1 with Waters-IBE, as follows. The proof is conducted by a sequence of games. Game 0 is the standard IND-ID-CCA game. Game 1 is defined like Game 1 except that the reduction algorithm will terminate the simulation due to regular abort or artificial abort. Game 2, Game 3, and Game 4 are defined like Game 1, Game 2, and Game 3 in the proof for Scheme 1, respectively. The argument of the indistinguishability between Game 3 and Game 4 is similar to that between Game 2 and Game 3 in the proof for Scheme 1. Then the security result immediately follows.

Acknowledgments. We would like to thank Jiang Zhang, Cheng Chen, and Qiong Huang for helpful discussions. The work of the third author is supported in part by the National Natural Science Foundation of China under grant Nos. 60970110, 61033014, 61021004, 61170227, 61172085, 61103221, 11061130539 and 61161140320 and Science Foundation Project of Jiang Su Province under grant No. BM20101014.

References

1. Bentahar, K., Farshim, P., Malone-Lee, J., Smart, N.P.: Generic constructions of identity-based and certificateless kems. Journal of Cryptology 21(2), 178–199 (2008)
2. Boneh, D., Boyen, X.: Efficient Selective-ID Secure Identity-Based Encryption Without Random Oracles. In: Cachin, C., Camenisch, J.L. (eds.) EUROCRYPT 2004. LNCS, vol. 3027, pp. 223–238. Springer, Heidelberg (2004)
3. Boyen, X., Mei, Q., Waters, B.: Direct chosen ciphertext security from identity-based techniques. In: ACM CCS 2005, pp. 320–329 (2005)
4. Canetti, R., Halevi, S., Katz, J.: Chosen-Ciphertext Security from Identity-Based Encryption. In: Cachin, C., Camenisch, J.L. (eds.) EUROCRYPT 2004. LNCS, vol. 3027, pp. 207–222. Springer, Heidelberg (2004)
5. Cash, D., Kiltz, E., Shoup, V.: The Twin Diffie-Hellman Problem and Applications. In: Smart, N.P. (ed.) EUROCRYPT 2008. LNCS, vol. 4965, pp. 127–145. Springer, Heidelberg (2008)
6. Cramer, R., Hofheinz, D., Kiltz, E.: A Twist on the Naor-Yung Paradigm and Its Application to Efficient CCA-Secure Encryption from Hard Search Problems. In: Micciancio, D. (ed.) TCC 2010. LNCS, vol. 5978, pp. 146–164. Springer, Heidelberg (2010)

7. Cramer, R., Shoup, V.: A Practical Public Key Cryptosystem Provably Secure against Adaptive Chosen Ciphertext Attack. In: Krawczyk, H. (ed.) CRYPTO 1998. LNCS, vol. 1462, pp. 13–25. Springer, Heidelberg (1998)

8. Cramer, R., Shoup, V.: Design and analysis of practical public-key encryption schemes secure against adaptive chosen ciphertext attack. SIAM Journal on Computing 33, 167–226 (2001)

9. Cramer, R., Shoup, V.: Universal Hash Proofs and a Paradigm for Adaptive Chosen Ciphertext Secure Public-Key Encryption. In: Knudsen, L.R. (ed.) EUROCRYPT 2002. LNCS, vol. 2332, pp. 45–64. Springer, Heidelberg (2002)

10. Galindo, D.: Chosen-Ciphertext Secure Identity-Based Encryption from Computational Bilinear Diffie-Hellman. In: Joye, M., Miyaji, A., Otsuka, A. (eds.) Pairing 2010. LNCS, vol. 6487, pp. 367–376. Springer, Heidelberg (2010)

11. Goldreich, O., Levin, L.A.: A hard-core predicate for all one-way functions. In: Proceedings of the Twenty-First Annual ACM Symposium on Theory of Computing, STOC, pp. 25–32. ACM (1989)

12. Hanaoka, G., Kurosawa, K.: Efficient Chosen Ciphertext Secure Public Key Encryption under the Computational Diffie-Hellman Assumption. In: Pieprzyk, J. (ed.) ASIACRYPT 2008. LNCS, vol. 5350, pp. 308–325. Springer, Heidelberg (2008)

13. Haralambiev, K., Jager, T., Kiltz, E., Shoup, V.: Simple and Efficient Public-Key Encryption from Computational Diffie-Hellman in the Standard Model. In: Nguyen, P.Q., Pointcheval, D. (eds.) PKC 2010. LNCS, vol. 6056, pp. 1–18. Springer, Heidelberg (2010)

14. Hofheinz, D., Kiltz, E.: Secure Hybrid Encryption from Weakened Key Encapsulation. In: Menezes, A. (ed.) CRYPTO 2007. LNCS, vol. 4622, pp. 553–571. Springer, Heidelberg (2007)

15. Hofheinz, D., Kiltz, E.: Practical Chosen Ciphertext Secure Encryption from Factoring. In: Joux, A. (ed.) EUROCRYPT 2009. LNCS, vol. 5479, pp. 313–332. Springer, Heidelberg (2009)

16. Kiltz, E.: Chosen-Ciphertext Security from Tag-Based Encryption. In: Halevi, S., Rabin, T. (eds.) TCC 2006. LNCS, vol. 3876, pp. 581–600. Springer, Heidelberg (2006)

17. Kiltz, E.: Chosen-Ciphertext Secure Key-Encapsulation Based on Gap Hashed Diffie-Hellman. In: Okamoto, T., Wang, X. (eds.) PKC 2007. LNCS, vol. 4450, pp. 282–297. Springer, Heidelberg (2007)

18. Kiltz, E., Galindo, D.: Direct Chosen-Ciphertext Secure Identity-Based Key Encapsulation Without Random Oracles. In: Batten, L.M., Safavi-Naini, R. (eds.) ACISP 2006. LNCS, vol. 4058, pp. 336–347. Springer, Heidelberg (2006)

19. Kiltz, E., Vahlis, Y.: CCA2 Secure IBE: Standard Model Efficiency through Authenticated Symmetric Encryption. In: Malkin, T. (ed.) CT-RSA 2008. LNCS, vol. 4964, pp. 221–238. Springer, Heidelberg (2008)

20. Kurosawa, K., Desmedt, Y.: A New Paradigm of Hybrid Encryption Scheme. In: Franklin, M. (ed.) CRYPTO 2004. LNCS, vol. 3152, pp. 426–442. Springer, Heidelberg (2004)

21. Liu, Y., Li, B., Lu, X., Jia, D.: Efficient CCA-Secure CDH Based KEM Balanced between Ciphertext and Key. In: Parampalli, U., Hawkes, P. (eds.) ACISP 2011. LNCS, vol. 6812, pp. 310–318. Springer, Heidelberg (2011)

22. Naor, M., Yung, M.: Public-key cryptosystems provably secure against chosen ciphertext attacks. In: Proceedings of the Twenty Second Annual ACM Symposium on Theory of Computing - STOC, pp. 427–437. ACM (1990)

23. Waters, B.: Efficient Identity-Based Encryption Without Random Oracles. In: Cramer, R. (ed.) EUROCRYPT 2005. LNCS, vol. 3494, pp. 114–127. Springer, Heidelberg (2005)
24. Wee, H.: Efficient Chosen-Ciphertext Security via Extractable Hash Proofs. In: Rabin, T. (ed.) CRYPTO 2010. LNCS, vol. 6223, pp. 314–332. Springer, Heidelberg (2010)

A   A Variant of Scheme 0

In this section we describe a variant of Scheme 0 with shorter mpk and ciphertext at the cost of relying on a slightly strong assumption, named the modified computational bilinear Diffie-Hellman assumption.

A.1   The Modified Computational Bilinear Diffie-Hellman Assumption

Let $\mathbb{G}$ be a cyclic group generated by g and equipped with a bilinear map $e : \mathbb{G} \times \mathbb{G} \to \mathbb{G}_T$. Define

$$\text{mbdh}(A, B, B', C) := T, \text{ where } A = g^a, B = g^b, B' = g^{b^2}, C = g^c, T = e(g, g)^{abc}$$

The modified computational BDH (mCBDH) problem is computing the value $\text{mbdh}(A, B, B', C)$ given $A, B, B', C \in \mathbb{G}$ where $a, b, c \xleftarrow{R} \mathbb{Z}_p$. Compared to the BDH problem, the mBDH problem furthermore provide the adversary with the element g^{b^2}. The mCBDH assumption asserts that the mCBDH problem is hard, that is, $\Pr[\mathcal{A}(A, B, B', C) = \text{mbdh}(A, B, B', C)] \le \text{negl}(\kappa)$ for all PPT algorithms $\mathcal{A}$.

Lemma 1.1. *Let $\mathbb{G}$ be a prime order group generated by g equipped with a pairing $e : \mathbb{G} \times \mathbb{G} \to \mathbb{G}_T$. Let $a, b, c \xleftarrow{R} \mathbb{Z}_p$ be random integers, $R \xleftarrow{R} \{0,1\}^u$, and let $K = f_{gl}(\text{bdh}(A, B, C), R)$. Let $U \xleftarrow{R} \{0,1\}$ be uniformly random. Suppose there exists a PPT algorithm $\mathcal{B}$ distinguishing the distributions*

$$\Delta_{\text{mbdh}} = (g, A, B, B', C, K, R) \text{ and } \Delta_{\text{rand}} = (g, A, B, B', C, U, R)$$

with non-negligible advantage. Then there exists a PPT algorithm computing $\text{bdh}(A, B, C)$ on input (g, A, B, B', C) with non-negligible success probability, hence breaking the mCBDH assumption.

Setup. Pick $a \xleftarrow{R} \mathbb{Z}_p$, and then set $X = g^a$. Pick $h, Y \xleftarrow{R} \mathbb{G}$. Define the function $F : \mathbb{Z}_p \to \mathbb{G}$ as $I \mapsto X^I h$. The public parameters and the master secret key are given by

$$mpk = (g, h, X, Y) \text{ and } msk = a$$

KeyGen. To generate a private key for an identity $I \in \mathbb{Z}_p$, pick $s \xleftarrow{R} \mathbb{Z}_p$ and output $sk = (Y^a F(I)^s, g^{-s}, Y^s)$.

Encap. Pick $r \xleftarrow{R} \mathbb{Z}_p$, then compute $C_1 = g^r$, $C_2 = (F(I)Y^t)^r$ with $t = \text{TCR}(C_1)$. Compute $K = f_{gl}(e(X, Y)^r, R)$.

Decap. To decapsulate ciphertext (C_1, C_2) under identity I, first compute $t = \mathsf{TCR}(C_1)$. If $e(C_1, F(I)Y^t) \neq e(g, C_2)$ then return $\bot$. Otherwise, take the private key sk and $C = (C_1, C_2)$ as input, compute $K = f_{\mathsf{gl}}\left(e(C_1, sk_1 sk_3^t) e(C_2, sk_2), R\right)$. Indeed, for a valid ciphertext $C = (C_1, C_2)$, we have

$$e(C_1, sk_1 sk_3^t) e(C_2, sk_2) = e(g^r, Y^a F(I)^s Y^{st}) e(F(I)^r Y^{rt}, g^{-s}) = e(X, Y)^r$$

Notice that the consistency of the ciphertext is publicly verifiable, i.e., anyone could verify a ciphertext being consistent or not.

Theorem 1.2. *Let* TCR *be a target collision-resistant hash function and suppose that the mCBDH assumption holds in* $\mathbb{G}$. *Then the above scheme is an* IND-sID-CCA *secure IB-KEM.*

Proof. We proceed in a sequence of games. We write (C_1^*, C_2^*) to denote the challenge ciphertext with the corresponding key K^* of identity I^*, denote with U^* the random key chosen by the IND-sID-CCA experiment, and set $t^* = \mathsf{TCR}(C_1^*)$. Let W_i denote the event that $\mathcal{A}$ outputs β' such that $\beta' = \beta$ in Game i.

Game 0. This is the standard IND-sID-CCA game. By definition we have

$$\Pr[W_0] = \frac{1}{2} + \mathsf{AdvCCA}_{\mathsf{KEM}}^{\mathcal{A}}(\kappa) \tag{5}$$

Game 1. Let E_{01} be the event that the adversary issues a decapsulation query $\langle I^*, C_1', C_2' \rangle$ with $C_1' = C_1^*$ in Phase 1. Note that the probability that the adversary submits a decapsulation query such that $C_1' = C_1^*$ before seeing the challenge ciphertext is bounded by Q_d/p, where Q_d is the number of decapsulation queries issued by $\mathcal{A}$. Since $Q_d = \mathsf{poly}(\kappa)$, we have $\Pr[E_{01}] \leq Q_d/p \leq \mathsf{negl}(\kappa)$. We define Game 1 exactly the same as Game 0 except assuming that E_{01} never occurs in Game 1. It follows that

$$|\Pr[W_1] - \Pr[W_0]| \leq \mathsf{negl}(\kappa) \tag{6}$$

Moreover, we remark that in Phase 2 a decapsulation query $\langle I^*, C_1', C_2' \rangle$ will be rejected if $C_1' = C_1^*$. Since if $C_2' \neq C_2^*$, the decapsulation query will be rejected for the inconsistency of the ciphertext. If $C_2' = C_2^*$, it will be rejected by definition of IND-sID-CCA game.

Game 2. Let E_{12} be the event that the adversary issues a decapsulation query $\langle I^*, C_1', C_2' \rangle$ with $C_1' \neq C_1^*$ and $\mathsf{TCR}(C_1') = \mathsf{TCR}(C_1^*)$. By the target collision resistance of TCR, we have $\Pr[E_{12}] \leq \mathsf{negl}(\kappa)$. We define Game 2 exactly the same as Game 1 except assuming that E_{12} never occurs in Game 2. It follows that

$$|\Pr[W_2] - \Pr[W_1]| \leq \mathsf{negl}(\kappa) \tag{7}$$

We claim that

$$\Pr[W_2] = \frac{1}{2} + \mathsf{negl}(\kappa) \tag{8}$$

We prove this statement by letting an algorithm $\mathcal{B}$ against the GL-mDBDH assumption simulate the challenger in Game 2. Suppose $\mathcal{B}$ is given a challenge instance (g, A, B, B', C, L, R), where L is either uniform randomly sampled from $\{0, 1\}$ or $f_{gl}(\text{mbdh}(A, B, B', C), R)$. $\mathcal{B}$ plays Game 2 with an adversary $\mathcal{A}$ against the IB-KEM scheme as follows.

Initialization. $\mathcal{A}$ first outputs an identity $I^* \in \mathbb{Z}_p$ that it intends to attack.

Setup. $\mathcal{B}$ picks $d \xleftarrow{R} \mathbb{Z}_p$, and then sets $X = A = g^a$, $Y = B = g^b$, compute $t^* = \mathsf{TCR}(C)$. $\mathcal{B}$ picks $d \xleftarrow{R} \mathbb{Z}_p$ and defines $h = X^{-I^*} Y^{-t^*} g^d$. It gives $\mathcal{A}$ the public parameters $mpk = (g, h, X, Y)$. The corresponding msk, which is unknown to $\mathcal{B}$ is a. The function F is essentially of the form

$$F(I) = X^I h = X^{I-I^*} Y^{-t^*} g^d$$

Phase 1 - Private Key Queries. $\mathcal{A}$ issues up to Q_e private key queries with the only restriction that $\langle I \rangle \neq \langle I^* \rangle$. To respond to a private query for identity $I \in \mathbb{Z}_p$, $\mathcal{B}$ generates sk as follows: pick a random integer $s \in \mathbb{Z}_p$ and sets

$$sk_1 = Y^{\frac{-d}{I-I^*}} B'^{\frac{t^*}{I-I^*}} (X^{I-I^*} Y^{-t^*} g^d)^s, sk_2 = g^{-s} Y^{\frac{1}{I-I^*}}, sk_3 = Y^s B'^{\frac{-1}{I-I^*}}$$

Let $\widetilde{s} = s - b/(I - I^*)$. It is easy to see that sk is a valid private key for I since

$$sk_1 = Y^{\frac{-d}{I-I^*}} B'^{\frac{t^*}{I-I^*}} (X^{I-I^*} Y^{-t^*} g^d)^s = Y^a (X^{I-I^*} Y^{-t^*} g^d)^{s - \frac{b}{I-I^*}} = Y^a F(I)^{\widetilde{s}}$$
$$sk_2 = g^{-s} Y^{\frac{1}{I-I^*}} = g^{-s + \frac{b}{I-I^*}} = g^{-\widetilde{s}}$$
$$sk_3 = Y^s B'^{\frac{-1}{I-I^*}} = Y^{s - \frac{b}{I-I^*}} = Y^{\widetilde{s}}$$

where $s, \widetilde{s}$ are uniform in $\mathbb{Z}_p$. This matches the definition for a private key for I. Hence, sk is a valid private key for I.

Phase 1 - Decapsulation Queries. Upon $\mathcal{A}$ issuing a decaptulation query $\langle I, C_1, C_2 \rangle$, $\mathcal{B}$ responds as follows. If $I \neq I^*$, $\mathcal{B}$ uses the corresponding private key to handle it. Otherwise, $\mathcal{B}$ computes $t = \mathsf{TCR}(C_1)$ and tests the consistency of the ciphertext by checking

$$e(C_1, F(I)Y^t) \stackrel{?}{=} e(g, C_2)$$

If the above equality holds, $\mathcal{B}$ sets $K := f_{gl}(e(\widetilde{Y}, X), R)$. The answer is correct by observing that $\widetilde{Y} = (C_2/(C_1^d))^{\frac{1}{a(t-t^*)}} = (Y^{(t-t^*)r} g^{dr}/g^{rd})^{\frac{1}{a(t-t^*)}} = Y^r = \text{dh}(Y, g^r)$. By Game 2 we know that when $I = I^*$, if $C_1 \neq C_1^*$ then $t \neq t^*$. Therefore $\mathcal{B}$ can answer all decapsulation queries issued by $\mathcal{A}$ correctly.

Challenge. $\mathcal{B}$ sets $C_1^* = C$ (which implicitly assigns $r = c$), and $C_2^* = C^d$. The challenge ciphertext is $C^* = (C_1^*, C_2^*)$. Note that this is a consistent ciphertext since we have $(F(I^*)Y^{t^*})^r = (g^d)^r = C^d$. Then $\mathcal{B}$ sets $K^* = L$ and gives $\mathcal{A}$ the challenge (C^*, K^*).

Phase 2. In Phase 2, all the queries are responded in the same way as in Phase 1 except the decapsulation query $\langle I^*, C^* \rangle$ will be rejected.

This finishes the description of simulation. It is easy to see that $\mathcal{B}$ simulates the challenger perfectly. If $\mathcal{A}$'s advantage is not negligible, then $\mathcal{B}$ has non-negligible advantage against the GL-mDBDH problem. According to Lemma 2.3, $\mathcal{B}$ further implies an algorithm with non-negligible advantage against the mCBDH problem, which contradicts to the mCBDH assumption. Therefore, we prove the statement. The theorem follows by combining (5)-(8).   □

We compare Scheme 0 and Scheme 0' in Table 2. Scheme 0' can be extended to n-bits IB-KEMs in an analogous way as we did to Scheme 0.

Table 2. Comparison of Scheme 0 and Scheme 0'

Scheme	Assumption	Ciphertext Overhead	Efficiency [# exp, # pairing] Encap	Decap	Key Sizes $\|mpk\|$	$\|msk\|$
Scheme 0 (§3)	CBDH	$3 \times \|\mathbb{G}_T\|$	$[4, 0]$	$[1, 4]$	$5 \times \|\mathbb{G}\|$	$1 \times \|\mathbb{Z}_p\|$
Scheme 0' (§A)	mCBDH	$2 \times \|\mathbb{G}_T\|$	$[3, 0]$	$[2, 4]$	$4 \times \|\mathbb{G}\|$	$1 \times \|\mathbb{Z}_p\|$

B   The Proof of Scheme 1

Proof. We proceed in a sequence of games. Let $(C_1^*, \ldots, C_n^*)$ be the challenge ciphertext of the corresponding key K^* under I^*, denote with U^* the random key chosen by the IND-sID-CCA experiment, and set $t^* = \mathsf{TCR}(C_{1,1}^*, \ldots, C_{n,1}^*)$. We start with a game where the challenger proceeds like the standard IND-sID-CCA game (i.e., K^* is a real key and U^* is a random key), and end up with a game where both K^* and U^* are chosen uniformly random. Then we show that all games are computationally indistinguishable under the CBDH assumption. Let W_i denote the event that $\mathcal{A}$ outputs β' such that $\beta' = \beta$ in Game i.

Game 0. This is the standard IND-sID-CCA game. By definition we have

$$\Pr[W_0] = \frac{1}{2} + \mathsf{AdvCCA}_{\mathsf{KEM}}^{\mathcal{A}}(\kappa)$$

Game 1. Let E_{01} be the event that the adversary issues a decapsulation query $\langle I^*, C_1', \ldots, C_n' \rangle$ with $C_{i,1}' = C_{i,1}^*$ for all $1 \leq i \leq n$ in Phase 1. Note that the probability that the adversary submits a ciphertext such that $C_{i,1}' = C_{i,1}^*$ for all $1 \leq i \leq n$ before seeing the challenge ciphertext is bounded by Q_d/p^n, where Q_d is the number of decapsulation queries issued by $\mathcal{A}$. Since $Q_d = \mathsf{poly}(\kappa)$, we have $\Pr[E_{0,1}] \leq Q_d/p^n \leq \mathsf{negl}(\kappa)$. We define Game 1 like Game 0 except assuming that E_{01} never occurs in Game 1. It follows that

$$|\Pr[W_1] - \Pr[W_0]| \leq \mathsf{negl}(\kappa)$$

Moreover, we remark that in Phase 2 a decapsulation query $\langle I^*, C_1', \ldots, C_n' \rangle$ will be rejected if $C_{i,1}' = C_{i,1}^*$ for all $1 \leq i \leq n$. Since if $C_{i,2}' \neq C_{i,2}^*$ or $C_{i,3}' \neq C_{i,3}^*$ for some $i \in [n]$, the decapsulation query will be rejected for the inconsistency of the ciphertext. If $C'_{i,2} = C^*_{i,2}$ and $C'_{i,3} = C^*_{i,3}$ for all $1 \leq i \leq n$, it will be rejected by definition of IND-sID-CCA game.

Game 2. Let E_{12} be the event that the adversary issues a decapsulation query $\langle I^*, C'_1, \ldots, C'_n \rangle$ with $C'_{i,1} \neq C^*_{i,1}$ for some $i \in [n]$ and $\mathsf{TCR}(C'_{1,1}, \ldots, C'_{n,1}) = \mathsf{TCR}(C^*_{1,1}, \ldots, C^*_{n,1})$. By the target collision resistance of TCR we have $\Pr[E_{12}] \leq \mathsf{negl}(\kappa)$. We define Game 2 like Game 1 except assuming that E_{12} never occurs in Game 2. It follows that

$$| \Pr[W_2] - \Pr[W_1]| \leq \mathsf{negl}(\kappa)$$

Game 3. We define Game 3 like Game 2, except that we sample $K^* \xleftarrow{R} \{0,1\}^{n\nu}$ uniformly at random. Note that both K^* and U^* are chosen uniformly random, thus we have

$$\Pr[W_3] = \frac{1}{2}$$

We claim that $| \Pr[W_3] - \Pr[W_2]| \leq \mathsf{negl}(\kappa)$ under the CBDH assumption. We prove this by a hybrid argument. To this end, we define a sequence of hybrid games $H_0, \ldots, H_n$, such that H_0 equals Game 2 and H_n equals Game 3. Then we argue that hybrid H_i is indistinguishable from hybrid H_{i-1} for $i \in \{1, \ldots, n\}$ under the CBDH assumption. The claim follows, since $n = n(\kappa)$ is a polynomial. We define H_0 exactly like Game 2. Then, for i from 1 to n, in hybrid H_i we set the first $i\nu$ bits of K^* to independent random bits, and proceed otherwise exactly like in hybrid H_{i-1}. Thus, hybrid H_n proceeds exactly like Game 3. Let E_i denote the event that $\mathcal{A}$ outputs β' such that $\beta' = \beta$ in H_i. Suppose that

$$| \Pr[E_0] - \Pr[E_n]| = 1/\mathsf{poly}'(\kappa) \tag{9}$$

that is, the success probability of $\mathcal{A}$ in H_0 is not negligible close to the success probability in H_n. Note that then there must exist an index i such that $| \Pr[E_{i-1}] - \Pr[E_i]| = 1/\mathsf{poly}(\kappa)$ (since if $| \Pr[E_{i-1}] - \Pr[E_i]| \leq \mathsf{negl}(\kappa)$ for all i, then we should have $| \Pr[E_0] - \Pr[E_n]| \leq \mathsf{negl}(\kappa)$).

Suppose that there exists an algorithm $\mathcal{A}$ for which Equation (9) holds. Then we can construct an adversary $\mathcal{B}$ distinguishing the distributions Δ_{bdh} and Δ_{rand}, which by Lemma 2.3 is sufficient to prove security under the CBDH assumption in $\mathbb{G}$. Adversary $\mathcal{B}$ receives a challenge $D = (g, A, B, C, L, R)$ as input, guesses an index $\ell \in [n]$, which with probability at least $1/n$ such that $| \Pr[E_{\ell-1}] - \Pr[E_\ell]| = 1/\mathsf{poly}(\kappa)$, and proceeds as follows:

Initialization. $\mathcal{A}$ first outputs an identity $I^* \in \mathbb{Z}_p$ that it intends to attack.

Setup. For $i = [n]\backslash\ell$, $\mathcal{B}$ picks $r_i \xleftarrow{R} \mathbb{Z}_p$, then picks $d \xleftarrow{R} \mathbb{Z}_p$, and sets $X = A = g^a$, $Y = B = g^b$, and $X' = X^{-t^*}g^d$, where $t^* = \mathsf{TCR}(g^{r_1}, \ldots, g^{r_{\ell-1}}, C, g^{r_{\ell+1}}, \ldots, g^{r_n})$. Pick $z \xleftarrow{R} \mathbb{Z}_p$ and defines $h = X^{-I^*}g^z$. It gives $\mathcal{A}$ the system parameters $mpk = (g, h, X, X', Y, F)$. Note that the corresponding msk, which is unknown to $\mathcal{B}$ is a.

Phase 1 - Private Key Queries. $\mathcal{A}$ issues up to Q_e private key queries with the only restriction that $\langle I \rangle \neq \langle I^* \rangle$. To respond to a private query of $I \in \mathbb{Z}_p$, $\mathcal{B}$ picks $s \xleftarrow{R} \mathbb{Z}_p$ and sets

$$sk_1 = Y^{\frac{-z}{I-I^*}} F(I)^s, \quad sk_2 = g^s Y^{\frac{-1}{I-I^*}}$$

We claimed that sk is a valid private key for I. To see this, let $\tilde{s} = s - b/(I - I^*)$. Then we have

$$sk_1 = Y^{\frac{-z}{I-I^*}} (X^{I-I^*} g^z)^s = Y^a (X^{I-I^*} g^z)^{s - \frac{b}{I-I^*}} = Y^a F(I)^{\tilde{s}}$$
$$sk_2 = g^s Y^{\frac{-1}{I-I^*}} = g^{\tilde{s}}$$

where s, $\tilde{s}$ are uniform distributed in $\mathbb{Z}_p$. This matches the definition for a private key for I. Hence, sk is a valid private key for I.

Phase 1 - Decapsulation Queries. Upon $\mathcal{A}$ issuing a decapsulation query $\langle I, C_1, \ldots, C_n \rangle$, $\mathcal{B}$ responds as follows. If $I \neq I^*$, $\mathcal{B}$ uses the corresponding private key to handle it. Otherwise, $\mathcal{B}$ computes $t = \mathsf{TCR}(C_{1,1}, \ldots, C_{n,1})$ and tests the consistency of the ciphertext by checking

$$e(C_{i,1}, X^t X') \stackrel{?}{=} e(g, C_{i,2}) \ \wedge \ e(C_{i,1}, F(I)) \stackrel{?}{=} e(g, C_{i,3})$$

If the equality holds for all $1 \leq i \leq n$, $\mathcal{B}$ sets $K = (K_1, \ldots, K_n)$ as $K_i = f_{\mathsf{gl}}(e(X, Y)^{r_i}, R)$ for $i \in [n] \backslash \{\ell\}$ and $K_\ell = f_{\mathsf{gl}}(e(\widetilde{X}_\ell, Y), R)$. Here we compute $\widetilde{X}_\ell := (C_{\ell,2}/C_{\ell,1}^d)^{1/(t-t^*)} = (X^{r_\ell(t-t^*)} g^{r_\ell d}/g^{r_\ell d})^{1/(t-t^*)} = X^{r_\ell} = \mathsf{dh}(X, C_{\ell,1})$. By the definition of Game 2 we know that when $I = I^*$, if $C_{i,1} \neq C_{i,1}^*$ for some $i \in [n]$ then $t \neq t^*$. Therefore $\mathcal{B}$ can answer all decapsulation queries issued by $\mathcal{A}$ correctly.

Challenge. To generate the challenge ciphertext $C^* = (C_1^*, \ldots, C_n^*)$, for $i = [n] \backslash \{\ell\}$, $\mathcal{B}$ generates C_i^* normally. For $C_\ell^* = (C_{\ell,1}^*, C_{\ell,2}^*, C_{\ell,3}^*)$, $\mathcal{B}$ sets $C_{\ell,1}^* = C$ (which implicitly assigns $r_\ell = c$), $C_{i,2}^* = C^d$, and $C_{i,3}^* = C^z$. Note that C^* is a consistent ciphertext since we have $(X^{t^*} X')^{r_\ell} = (g^d)^{r_\ell} = C^d$ and $F(I^*)^c = (g^z)^c = C^z$. Then $\mathcal{B}$ samples $\ell - 1$ uniformly random groups of ν bits $K_1^*, \ldots, K_{\ell-1}^*$, sets $K_\ell^* = L$, $K_i^* = f_{\mathsf{gl}}(e(X, Y)^{r_j}, R)$ for i from $\ell + 1$ to n. $\mathcal{B}$ samples uniform randomly bits $U^* \in \{0, 1\}^{n\nu}$, picks a random bit $\beta \in \{0, 1\}$. If $\beta = 1$, it gives $\mathcal{A}$ the challenge (C^*, K^*). Otherwise it gives $\mathcal{A}$ the challenge (C^*, U^*).

Phase 2. In Phase 2, all the queries are responded the same way as in Phase 1 except the decapsulation query $\langle I^*, C^* \rangle$ will be rejected.

This completes the description of simulation. If $D \in \Delta_{\mathsf{bdh}}$ we have $K_\ell^* = f_{\mathsf{gl}}(\mathsf{bdh}(A, B, C), R)$. Thus $\mathcal{A}$'s view when interacting with $\mathcal{B}$ is identical to $H_{\ell-1}$. If $D \in \Delta_{\mathsf{rand}}$, then $\mathcal{A}$'s view is identical to H_ℓ. Thus $\mathcal{B}$ can use $\mathcal{A}$ to distinguish $D \in \Delta_{\mathsf{bdh}}$ from $D \in \Delta_{\mathsf{rand}}$. According to Lemma 2.3, $\mathcal{B}$ further implies a PPT algorithm which can break the CBDH problem, which contradicts to the CBDH assumption. □

C   The proof of Scheme 2

Proof. We proceed in a sequence of games. We write (C_1^*, C_2^*, C_3^*) to denote the challenge ciphertext with the corresponding key K^* of identity I^*, denote with U^* the random key chosen by the IND-sID-CCA experiment, and set $t^* = \mathsf{TCR}(C_1^*)$. We start with a game where the challenger proceeds like the standard IND-sID-CCA game (i.e., K^* is a real key and U^* is a random key), and end up with a game where both K^* and U^* are chosen uniformly random. Then we show that all games are computationally indistinguishable under the CBDH assumption. Let W_i denote the event that $\mathcal{A}$ outputs β' such that $\beta' = \beta$ in Game i.

Game 0. This is the standard IND-sID-CCA game. By definition we have

$$\Pr[W_0] = \frac{1}{2} + \mathsf{AdvCCA}_{\mathsf{KEM}}^{\mathcal{A}}(\kappa)$$

Game 1. Let E_{01} be the event that the adversary issues a decapsulation query $\langle I^*, C_1', C_2', C_3' \rangle$ with $C_1' = C_1^*$ in Phase 1. Note that the probability that the adversary submits a decapsulation query such that $C_1' = C_1^*$ before seeing the challenge ciphertext is bounded by Q_d/p, where Q_d is the number of decapsulation queries issued by $\mathcal{A}$. Since $Q_d = \mathsf{poly}(\kappa)$, we have $\Pr[E_{01}] \leq Q_d/p \leq \mathsf{negl}(\kappa)$. We define Game 1 exactly the same as Game 0 except assuming that E_{01} never occurs in Game 1. It follows that

$$|\Pr[W_1] - \Pr[W_0]| \leq \mathsf{negl}(\kappa)$$

Moreover, we remark that in Phase 2 a decapsulation query $\langle I^*, C_1', C_2', C_3' \rangle$ will be rejected if $C_1' = C_1^*$. Since if $C_2' \neq C_2^*$ or $C_3' \neq C_3^*$, the decapsulation query will be rejected for the inconsistency of the ciphertext. If $C_2' = C_2^*$ and $C_3' = C_3^*$, it will be rejected by definition of IND-sID-CCA game.

Game 2. Let E_{12} be the event that the adversary issues a decapsulation query $\langle I^*, C_1', C_2', C_3' \rangle$ with $C_1' \neq C_1^*$ and $\mathsf{TCR}(C_1') = \mathsf{TCR}(C_1^*)$. By the target collision resistance of TCR, we have $\Pr[E_{12}] \leq \mathsf{negl}(\kappa)$. We define Game 2 exactly the same as Game 1 except assuming that E_{12} never occurs in Game 2. It follows that

$$|\Pr[W_2] - \Pr[W_1]| \leq \mathsf{negl}(\kappa)$$

Game 3. We define Game 3 like Game 2, except that we sample $K_0^* \xleftarrow{R} \{0,1\}^{n\nu}$ uniformly random. Note that both K_0^* and K_1^* are chosen uniformly random, thus we have

$$\Pr[W_3] = \frac{1}{2}$$

We claim that $|\Pr[W_3] - \Pr[W_2]| \leq \mathsf{negl}(\kappa)$ under the CBDH assumption. We prove this by a hybrid argument. To this end, we define a sequence of hybrid games $H_0, \ldots, H_n$, such that H_0 equals Game 2 and H_n equals Game 3. Then we argue that hybrid H_i is indistinguishable from hybrid H_{i-1} for $i \in \{1, \ldots, n\}$ under the CBDH assumption. The claim follows, since $n = n(\kappa)$ is a polynomial.

We define H_0 exactly like Game 2. Then, for i from 1 to n, in hybrid H_i we set the first $i\nu$ bits of K^* to independent random bits, and proceed otherwise exactly like in hybrid H_{i-1}. Thus, hybrid H_n proceeds exactly like Game 3. Let E_i denote the event that $\mathcal{A}$ outputs β' such that $\beta' = \beta$ in H_i. Suppose that

$$|\Pr[E_0] - \Pr[E_n]| = 1/\mathsf{poly}'(\kappa) \tag{10}$$

that is, the success probability of $\mathcal{A}$ in H_0 is not negligible close to the success probability in H_n. Note that then there must exist an index i such that $|\Pr[E_{i-1}] - \Pr[E_i]| = 1/\mathsf{poly}(\kappa)$ (since if $|\Pr[E_{i-1}] - \Pr[E_i]| \le \mathsf{negl}(\kappa)$ for all i, then we should have $|\Pr[E_0] - \Pr[E_n]| \le \mathsf{negl}(\kappa)$).

Suppose that there exists an algorithm $\mathcal{A}$ for which Equation (10) holds. Then we can construct an adversary $\mathcal{B}$ distinguishing the distributions Δ_{bdh} and Δ_{rand}, which by Lemma 2.3 is sufficient to prove security under the CBDH assumption in $\mathbb{G}$. Adversary $\mathcal{B}$ receives a challenge $D = (g, A, B, C, L, R)$ as input, guesses an index $\ell \in [n]$, which with probability at least $1/n$ that $|\Pr[E_{\ell-1}] - \Pr[E_\ell]| = 1/\mathsf{poly}(\kappa)$, and proceeds as follows:

Initialization. $\mathcal{A}$ first outputs an identity $I^* \in \mathbb{Z}_p$ that it intends to attack.

Setup. $\mathcal{B}$ picks $d \xleftarrow{R} \mathbb{Z}_p$, and then sets $X = A = g^a$, $X' = X^{-t^*}g^d$, $Y_\ell = B = g^b$, where $t^* = \mathsf{TCR}(C)$. For $i \in [n]\backslash\{\ell\}$, $\mathcal{B}$ picks $y_j \xleftarrow{R} \mathbb{Z}_p$ and sets $Y_i = g^{y_j}$; picks $z \xleftarrow{R} \mathbb{Z}_p$ and defines $h = X^{-I^*}g^z$. It gives $\mathcal{A}$ the public parameters $mpk = (g, h, X, X', Y_1, \ldots, Y_n, F)$. The corresponding msk, which is unknown to $\mathcal{B}$ is a. The function F is essentially of the form

$$F(x) = X^x h = X^{x-I^*}g^z$$

Phase 1 - Private Key Queries. $\mathcal{A}$ issues up to Q_e private key queries with the only restriction that $\langle I \rangle \ne \langle I^* \rangle$. To respond to a private query for identity $I \in \mathbb{Z}_p$, $\mathcal{B}$ generates $sk = (sk_1, \ldots, sk_n)$ as follows: for sk_ℓ algorithm $\mathcal{B}$ picks $s_\ell \xleftarrow{R} \mathbb{Z}_p$ and sets

$$sk_{\ell,1} = Y_\ell^{\frac{-z}{I-I^*}} F(I)^{s_\ell}, \quad sk_{\ell,2} = g^{s_\ell} Y_\ell^{\frac{-1}{I-I^*}}$$

for sk_i where $i \in [n]\backslash\{\ell\}$, $\mathcal{B}$ picks a random integer $s_i \in \mathbb{Z}_p$ and sets

$$sk_{i,1} = X^{y_i} F(I)^{s_i} = Y_i^a F(I)^{s_i}, \quad sk_{i,2} = g^{s_i}$$

Let $\tilde{s}_\ell = s_\ell - b/(I - I^*)$. It is easy to see that sk is a valid random private key for I since

$$sk_{\ell,1} = Y_\ell^{\frac{-z}{I-I^*}}(X^{I-I^*}g^z)^{s_\ell} = Y_\ell^a(X^{I-I^*}g^z)^{s_\ell - \frac{b}{I-I^*}} = Y_\ell^a F(I)^{\tilde{s}_\ell}$$

$$sk_{\ell,2} = g^{s_\ell} Y_\ell^{\frac{1}{I-I^*}} = g^{\tilde{s}_\ell}$$

where s_ℓ, $\tilde{s}_\ell$ are uniform in $\mathbb{Z}_p$. This matches the definition for a private key for I. Hence, sk is a valid private key for I.

Phase 1 - Decapsulation Queries. Upon $\mathcal{A}$ issuing a decaptulation query $\langle I, C_1, C_2, C_3 \rangle$, $\mathcal{B}$ responds as follows. If $I \neq I^*$, $\mathcal{B}$ uses the corresponding private key to handle it. Otherwise, $\mathcal{B}$ computes $t = \mathsf{TCR}(C_1)$ and tests the consistency of the ciphertext by checking

$$e(C_1, X^t X') \stackrel{?}{=} e(g, C_2) \ \wedge \ e(C_1, F(I)) \stackrel{?}{=} e(g, C_3)$$

If the equality holds, $\mathcal{B}$ sets $K = (K_1, \ldots, K_n)$ as $K_i = f_{\mathrm{gl}}(e(X, C_1)^{y_i}, R)$ for $i \in [n] \backslash \{\ell\}$ and $K_\ell = f_{\mathrm{gl}}(e(\widetilde{X}, Y_\ell), R)$, where $\widetilde{X} := (C_2/C_1^d)^{1/(t-t^*)} = (X^{r(t-t^*)} g^{rd}/g^{rd})^{1/(t-t^*)} = X^r = \mathrm{dh}(X, C_1)$. By Game 2 we know that when $I = I^*$, if $C_1 \neq C_1^*$ then we have $t \neq t^*$. Therefore $\mathcal{B}$ can answer all decapsulation queries issued by $\mathcal{A}$ correctly.

Challenge. $\mathcal{B}$ sets $C_1^* = C$ (which implicitly assigns $r = c$), $C_2^* = C^d$, and $C_3^* = C^z$. The challenge ciphertext is $C^* = (C_1^*, C_2^*, C_3^*)$. Note that this is a consistent ciphertext since we have $(X^{t^*} X')^r = (g^d)^r = C^d$ and $F(I^*)^r = (g^z)^r = C^z$. Then $\mathcal{B}$ samples $i - 1$ uniformly random groups of ν bits $K_1^*, \ldots, K_{\ell-1}^*$, sets $K_\ell^* = L$, $K_i^* = f_{\mathrm{gl}}(e(X, C_1^*)^{y_i}, R)$ for i from $\ell + 1$ to n. $\mathcal{B}$ samples $U^* \in \{0, 1\}^{n\nu}$ uniformly at random, and picks a random bit $\beta \in \{0, 1\}$. If $\beta = 1$, it gives $\mathcal{A}$ the challenge (C^*, K^*). Otherwise it gives $\mathcal{A}$ the challenge (C^*, U^*).

Phase 2. In Phase 2, all the queries are responded in the same way as in Phase 1 except the decapsulation query $\langle I^*, C^* \rangle$ will be rejected.

This finishes the description of simulation. If $D \in \Delta_{\mathrm{bdh}}$ we have $K_\ell^* = f_{\mathrm{gl}}(\mathrm{bdh}(A, B, C), R)$, $\mathcal{A}$'s view is identical to $H_{\ell-1}$. If $D \in \Delta_{\mathrm{rand}}$, $\mathcal{A}$'s view is identical to H_ℓ. Thus $\mathcal{B}$ can use $\mathcal{A}$ to distinguish $D \in \Delta_{\mathrm{bdh}}$ from $D \in \Delta_{\mathrm{rand}}$. According to Lemma 2.3, $\mathcal{B}$ further implies a PPT algorithm which can break the CBDH problem, which contradicts to the CBDH assumption. $\qquad \square$

D   The proof of Generalized Scheme 1

Proof. We proceed in a sequence of games. We write $C^* = (C_1^*, \ldots, C_{n_2}^*)$ to denote the challenge ciphertext with the corresponding key K^* of I^*, denote with U^* the random key chosen by the IND-sID-CCA experiment, and set $t^* = \mathsf{TCR}(C_{1,1}^*, \ldots, C_{n_2,1}^*)$. We start with a game where the challenger proceeds as the standard IND-sID-CCA game (i.e., K^* is a real key and U^* is a random key), and end up with a game where both K^* and U^* are chosen uniformly random. Then we show that all games are computationally indistinguishable under the CBDH assumption. Let W_i denote the event that $\mathcal{A}$ outputs β' such that $\beta' = \beta$ in Game i.

Game 0. This is the standard IND-sID-CCA game. By definition we have

$$\Pr[W_0] = \frac{1}{2} + \mathsf{AdvCCA}^{\mathcal{A}}_{\mathsf{KEM}}(\kappa)$$

Game 1, Game 2, and Game 3 are defined in the same way as in the proof of Scheme 2. It is easy to verify that $|\Pr[W_1] - \Pr[W_0]| \leq \mathsf{negl}(\kappa)$, $|\Pr[W_2] - \Pr[W_1]| \leq \mathsf{negl}(\kappa)$, and $\Pr[W_3] = 1/2$. We claim that $|\Pr[W_3] - \Pr[W_2]| \leq \mathsf{negl}(\kappa)$ under the CBDH assumption. We prove this by a hybrid argument. To this end, we define a sequence of hybrid games $H_0, \ldots, H_n$, such that H_0 equals Game 2 and H_n equals Game 3. Then we argue that hybrid H_i is indistinguishable from hybrid H_{i-1} for $i \in \{1, \ldots, n\}$ under the CBDH assumption. The claim follows, since $n = n(\kappa)$ is a polynomial. We define H_0 exactly like Game 2. Then, for i from 1 to n, in hybrid H_i we set the first $i\nu$ bits of K^* to independent random bits, and proceed otherwise exactly like in hybrid H_{i-1}. Thus, hybrid H_n proceeds exactly like Game 3. Let E_i denote the event that $\mathcal{A}$ outputs β' such that $\beta' = \beta$ in H_i. Suppose that

$$|\Pr[E_0] - \Pr[E_n]| = 1/\mathsf{poly}'(\kappa) \tag{11}$$

that is, the success probability of $\mathcal{A}$ in H_0 is not negligible close to the success probability in H_n. Note that then there must exist an index ℓ such that $|\Pr[E_{i-1}] - \Pr[E_i]| = 1/\mathsf{poly}(\kappa)$ (since if $|\Pr[E_{i-1}] - \Pr[E_i]| \leq \mathsf{negl}(\kappa)$ for all i, then we should have $|\Pr[E_0] - \Pr[E_n]| \leq \mathsf{negl}(\kappa)$).

Suppose that there exists an algorithm $\mathcal{A}$ for which Equation (11) holds. Then we can construct an adversary $\mathcal{B}$ distinguishing the distributions Δ_{bdh} and Δ_{rand}, which by Lemma 2.3 is sufficient to prove security under the CBDH assumption in $\mathbb{G}$. Adversary $\mathcal{B}$ receives a challenge $D = (g, A, B, C, L, R)$ as input, guesses an index $\ell \in [n]$, which with probability at least $1/n$ such that $|\Pr[E_{\ell-1}] - \Pr[E_\ell]| = 1/\mathsf{poly}(\kappa)$. Let $(\bar{i}, \bar{j})$ be the unique tuple that satisfies $(\bar{i} - 1) \times n_1 + \bar{j} = \ell$, $\mathcal{B}$ proceeds as follows:

Initialization. $\mathcal{A}$ first outputs an identity $I^* \in \mathbb{Z}_p$ that it intends to attack.

Setup. $\mathcal{B}$ first picks $r_j \xleftarrow{R} \mathbb{Z}_p$ for $j \in [n_2] \backslash \bar{j}$, then sets the value t^* to be $\mathrm{TCR}(g^{r_1}, \ldots, g^{r_{\bar{j}-1}}, C, g^{r_{\bar{j}+1}}, \ldots, g^{r_{n_2}})$. $\mathcal{B}$ then picks $d \xleftarrow{R} \mathbb{Z}_p$, and sets $X = A = g^a$, $X' = X^{-t^*} g^d$, $Y_{\bar{i}} = B = g^b$; picks $y_i \xleftarrow{R} \mathbb{Z}_p$ and sets $Y_i = g^{y_i}$ for $i \in [n_1] \backslash \bar{i}$. It gives $\mathcal{A}$ the system public parameters $mpk = (g, h, X, X', Y_1, \ldots, Y_{n_2}, F)$. Note that the corresponding msk, which is unknown to $\mathcal{B}$ is a.

Phase 1 - Private Key Queries. $\mathcal{A}$ issues up to Q_e private key queries with the only restriction that $\langle I \rangle \neq \langle I^* \rangle$. To respond to the query of $I \in \mathbb{Z}_p$, for $sk_{\bar{i}}$ algorithm $\mathcal{B}$ picks $s_{\bar{i}} \xleftarrow{R} \mathbb{Z}_p$ and sets

$$sk_{\bar{i},1} = Y_{\bar{i}}^{\frac{-z}{I-I^*}} F(I)^{s_{\bar{i}}}, \quad sk_{\bar{i},2} = g^{s_{\bar{i}}} Y_{\bar{i}}^{\frac{-1}{I-I^*}}$$

for sk_i where $i \in [n_1] \backslash \{\bar{i}\}$, $\mathcal{B}$ picks a random $s_i \in \mathbb{Z}_p$ and sets

$$sk_{i,1} = X^{y_i} F(I)^{s_i} = Y_i^a F(I)^{s_i} \quad sk_{i,2} = g^{s_i}$$

Let $\tilde{s}_{\bar{i}} = s_{\bar{i}} - b/(I - I^*)$. It is easy to see that sk is a valid random private key for I since

$$sk_{\bar{i},1} = Y_{\bar{i}}^{\frac{-z}{I-I^*}}(X^{I-I^*}g^z)^{s_{\bar{i}}} = Y_{\bar{i}}^a(X^{I-I^*}g^z)^{s_{\bar{i}}-\frac{b}{I-I^*}} = Y_{\bar{i}}^a F(I)^{\tilde{s}_{\bar{i}}}$$

$$sk_{\bar{i},2} = g^{s_{\bar{i}}}Y_{\bar{i}}^{\frac{1}{I-I^*}} = g^{\tilde{s}_{\bar{i}}}$$

where $s_{\bar{i}}$ and $\tilde{s}_{\bar{i}}$ are uniform distributed in $\mathbb{Z}_p$. This matches the definition of a private key for I. Hence, sk is a valid private key for I.

Phase 1 - Decapsulation Queries. Upon $\mathcal{A}$ issuing a decaptulation query $\langle I, C_1, \ldots, C_{n_2} \rangle$, $\mathcal{B}$ responds as follows. If $I \neq I^*$, $\mathcal{B}$ uses the corresponding private key to handle it. Otherwise, $\mathcal{B}$ computes $t = \mathsf{TCR}(C_{1,1}, \ldots, C_{n_2,1})$ and then tests the consistency of the ciphertext by checking

$$e(C_{j,1}, X^t X') \overset{?}{=} e(g, C_{j,2}) \ \wedge \ e(C_{j,1}, F(I)) \overset{?}{=} e(g, C_{j,3})$$

If the equality holds for all $1 \leq j \leq n_2$, then $\mathcal{B}$ computes $K = (K_1, \ldots, K_n)$ as follows. Suppose $e = (i - 1) \times n_1 + j$,

1. If $i \neq \bar{i}$, set $K_e = f_{\mathrm{gl}}(e(X, C_j)^{y_i}, R)$.
2. If $i = \bar{i}$, compute $\widetilde{X}_j := (C_{j,2}/C_{j,1}^d)^{1/(t-t^*)} = (X^{r_j(t-t^*)}g^{r_j d}/g^{r_j d})^{1/(t-t^*)} = X^{r_j} = \mathrm{dh}(X, C_{j,1})$, set $K_e = f_{\mathrm{gl}}(e(\widetilde{X}_j, Y_{\bar{i}}), R)$.

By the definition of Game 2 we know that when $I = I^*$, if $C_{j,1} \neq C_{j,1}^*$ for some $j \in [n_2]$ then $t \neq t^*$. Therefore $\mathcal{B}$ can answer all decapsulation queries issued by $\mathcal{A}$ correctly.

Challenge. To generate the challenge ciphertext $C^* = (C_1^*, \ldots, C_{n_2}^*)$, for $j = [n_2]\backslash\bar{j}$, $\mathcal{B}$ sets $C_j^* = (C_{j,1}^*, C_{j,2}^*, C_{j,3}^*) = (g^{r_j}, (X^{t^*}X')^{r_j}, F(I^*)^{r_j})$; for $C_{\bar{j}}^* = (C_{\bar{j},1}^*, C_{\bar{j},2}^*, C_{\bar{j},3}^*)$, $\mathcal{B}$ sets $C_{\bar{j},1}^* = C$ (which implicitly assigns $r_{\bar{j}} = c$), $C_{\bar{j},2}^* = C^d$, and $C_{\bar{j},3}^* = C^z$. Note that this is a consistent ciphertext since we have $(X^{t^*}X')^{r_{\bar{j}}} = (g^d)^{r_{\bar{j}}} = C^d$ and $F(I^*)^{r_{\bar{j}}} = (g^z)^{r_{\bar{j}}} = C^z$. Then $\mathcal{B}$ samples $\ell - 1$ uniformly random groups of ν bits $K_1^*, \ldots, K_{\ell-1}^*$, sets $K_\ell^* = L$. For $\ell \leq e \leq n$, $\mathcal{B}$ generates K_e in a similar way as it did when answering decapsulation queries, that is, suppose $e = (i-1) \times n_1 + j$, if $i \neq \bar{i}$, set $K_e = f_{\mathrm{gl}}(e(X, C_j)^{y_i}, R)$; if $j \neq \bar{j}$, set $K_e = f_{\mathrm{gl}}(e(X, Y_i)^{r_j}, R)$. $\mathcal{B}$ samples $U^* \in \{0, 1\}^{n\nu}$ uniformly at random, then picks a random bit $\beta \in \{0, 1\}$. If $\beta = 1$, it gives $\mathcal{A}$ the challenge (C^*, K^*). Otherwise it gives $\mathcal{A}$ the challenge (C^*, U^*).

Phase 2. In Phase 2, all the queries are responded in the same way as in Phase 1 except the decapsulation query $\langle I^*, C^* \rangle$ will be rejected.

This completes the description of simulation. If $D \in \Delta_{\mathrm{bdh}}$ we have $K_\ell^* = f_{\mathrm{gl}}(\mathrm{bdh}(A, B, C), R)$, $\mathcal{A}$'s view when interacting with $\mathcal{B}$ is identical to $H_{\ell-1}$. If $D \in \Delta_{\mathrm{rand}}$, $\mathcal{A}$'s view is identical to H_ℓ. Thus $\mathcal{B}$ can use $\mathcal{A}$ to distinguish $D \in \Delta_{\mathrm{bdh}}$ from $D \in \Delta_{\mathrm{rand}}$. According to Lemma 2.3, $\mathcal{B}$ further implies a PPT algorithm which can break the CBDH problem, which contradicts to the CBDH assumption. $\square$

Design, Implementation, and Evaluation of a Vehicular Hardware Security Module

Marko Wolf and Timo Gendrullis*

ESCRYPT GmbH, Embedded Security, Munich, Germany
{marko.wolf,timo.gendrullis}@escrypt.com

Abstract. Todays in-vehicle IT architectures are dominated by a large network of interactive, software driven digital microprocessors called electronic control units (ECU). However, ECUs relying on information received from open communication channels created by other ECUs or even other vehicles that are not under its control leaves the doors wide open for manipulations or misuse. Thus, especially safety-relevant ECUs need effective, automotive-capable security measures that protect the ECU and its communications efficiently and dependably. Based on a requirements engineering approach that incorporates all security-relevant automotive use cases and all distinctive automotive needs and constraints, we present an vehicular hardware security module (HSM) that enables a holistic protection of in-vehicle ECUs and their communications. We describe the hardware design, give technical details on the prototypical implementation, and provide a first evaluation on the performance and security while comparing our approach with HSMs already existing.

Keywords: hardware security module, automotive, in-vehicle, on-board.

1 Introduction and Motivation

Over the last two decades vehicles have silently but dramatically changed from rather "dumb" electro-mechanical devices into interactive, mobile information and communication systems already carrying dozens of digital microprocessors, various external radio interfaces, and several hundred megabytes of embedded software. In fact, information and communication technology is *the* driving force behind most innovations in the automotive industry, with perhaps 90% of all innovations in vehicles based on digital IT systems [18]. This "digital revolution" enables very sophisticated solutions considerably increasing flexibility, safety and efficiency of modern vehicles and vehicle traffic [28]. It further helps saving fuel, weight, and costs. Whereas in-vehicle IT safety (i.e., protection against [random] technical failures) is already a relatively well-established (if not necessarily well-understood) field, the protection of vehicular IT systems against systematic manipulations has only very recently started to emerge. In fact, automotive

---

* The authors thank Mirko Lange for his extensive help, Oliver Mischke for his valuable comments, the EC, and all partners involved in the EVITA project [15]. The work was done in scope of the European FP7 research project EVITA (FP7-ICT-224275).

H. Kim (Ed): ICISC 2011, LNCS 7259, pp. 302–318, 2012.
© Springer-Verlag Berlin Heidelberg 2012

IT systems were never designed with security in mind. But with the increasing application of digital software and various radio interfaces to the outside world (including the Internet), modern vehicles are becoming even more vulnerable to all kinds of malicious encroachments like hackers or malware [2]. This is especially noteworthy, since in contrast to most other IT systems, a successful malicious encroachment on a vehicle will not only endanger critical services or business models, but can also endanger human lives [26]. Thus strong security measures should be mandatory when developing vehicular IT systems. Today most vehicle manufacturer (hopefully) incorporate security as a design requirement. However, realizing dependable IT security solutions in a vehicular environment considerably differs from realizing IT security for typical desktop or server environments. In a typical vehicular attack scenario an attacker, for instance, has extended attack possibilities (i.e., insider attacks, offline attacks, physical attacks) and could have many different attack incentives and attack points (e.g., tachometer manipulations by the vehicle owner vs. theft of the vehicle components vs. industrial espionage). Thus, just porting "standard" security solutions to the, moreover, very heterogeneous IT environment usually will not work. However, there already exist some first automotive-capable (software) security solutions [29,32]. But, especially with regard to potential internal and physical attackers, these software solutions have to be protected against manipulations as well. In order to reliably enforce the security of software security mechanisms, the application of hardware security modules (HSM) is one effective countermeasure as HSMs:

– protect software security measures by acting as trusted security anchor,
– securely generate, store, and process security-critical material shielded from any potentially malicious software,
– restrict the possibilities of hardware tampering attacks by applying effective tamper-protection measures,
– accelerate security measures by applying specialized cryptographic hardware,
– reduce security costs on high volumes by applying highly optimized special circuitry instead of costly general purpose hardware.

Unfortunately, there are currently no automotive-capable HSMs available (cf. Section 2.2). Thus, the objective of this research was to design and prototype a standardized automotive-capable HSM for automotive on-board networks where security-relevant components are protected against tampering and sensitive data are protected against compromise. The HSM was especially designed for protecting e-safety applications such as emergency break based on communications between vehicles (V2V) or emergency call based on communications between vehicles and (traffic) infrastructures (V2I).

Our Contributions and Paper Outline. After this motivation, Section 2 gives a short introduction into our design rationale for a vehicular HSM including a short state-of-the-art review of the *related work*. Section 3 then presents the main objectives of this article, the design of a vehicular HSM concretely its system architecture, communication interface, security building blocks and security functionalities. Section 4 gives a detailed, technical overview of the HSM prototype implementation followed by a performance and security evaluation in Section 5 that compares our HSM approach with other HSMs currently available.

2 The Need For Efficient Hardware Security

In this section we provide an introduction into our design rationale for realizing a vehicular HSM. However, as this article focuses on the design, implementation, and evaluation of a vehicular HSM, this section mainly summarizes the work done before in [15] where the authors were involved, too. There, with a requirements engineering approach we identified all automotive use cases with a security impact [6], that means, all use cases involving a security-critical asset or allow for potential misuses. The security impacts identified are then transformed into high-level security objectives that could thwart these impacts accordingly. For fulfilling these security objectives, we then derived concrete (technical) security requirements [7] including all related functional (security) requirements using an appropriate security and trust model and reasonable "attack scenarios" [8].

2.1 Security and Functional Requirements Engineering

The security and functional requirements engineering is described in more detail in [6,7,9]. It has yielded to the following HSM security requisites (SR) and functional requisites (FR) as outlined below.

SR.1 Autonomous, strongly isolated security processing environment
SR.2 Minimal immutable trusted code to be executed prior to ECU processor
SR.3 Internal non-volatile memory for storing root security artifacts
SR.4 Non-detachable (tamper-protected) connection with ECU hardware
SR.5 Authentic, confidential, fresh comm. channel between HSM and ECU
SR.6 Autonomously controlled alert functionality (e.g., log entry, ECU halt)
SR.7 Only standardized, established security algorithms (e.g., NIST[1], BSI[2])

FR.1 Physical stress resistance to endure an automotive life-cycle of $\geq$20 years
FR.2 Bandwidth and latency performance that meets at least ISO 11898 [24]
FR.3 Compatibility with existing ECU security modules, i.e. with HIS-SHE [21]
FR.4 Compatibility with existing ECU microprocessor architectures
FR.5 Open, patent free specifications for cost-efficient OEM-wide application

2.2 Related Work

The vehicular HSM presented here is not the first security processor applied in the automotive domain. Hence, there already exist some proprietary and single-purpose HSM realizations used, for instance, by vehicle immobilizers, digital tachographs [19] or tolling solutions [34]. However, these are no general-purpose,

---

[1] US National Institute of Standards and Technology (www.nist.gov).
[2] German Federal Office for Information Security (www.bsi.bund.de).

private HSMs and hence cannot be reused by other vehicular security solutions. On the other hand, general-purpose HSMs that are currently available, for instance, the IBM 4758 cryptographic co-processor [4], the TCG Mobile/Trusted Platform Module [35], or typical cryptographic smartcards are not applicable for use within an automotive security context. They, for instance, lack of cost efficiency, performance, physical robustness, or security functionality. Solely, the secure hardware extension (SHE) as proposed by the HIS consortium [21] takes an exceptional position as it was explicitly designed for application in a automotive security context. However, the SHE module is mainly built for securing cryptographic key material against software attacks, but cannot be used, for instance, to protect V2X communications. An overview comparison of the HSM proposed in this work with the general-purpose HSMs is given later in Table 5.

3   Design

This section describes the hardware architecture and security functionality of our HSM. The corresponding full detailed HSM specification can be found in [9].

3.1   System Architecture

As shown in Figure 1, the hardware architecture design is based on a closed on-chip realization with a standard ECU application core and the HSM together on the same chip connected by an internal communication link. Hence, HSM commands are not explicitly and individually protected at hardware level. That means they are communicated in plain and without any replay and authenticity protection between HSM and application core. However, a TCG like command protection approach based on session keys and rotating nonces [35] is possible. The interface to the internal communication bus and external communication peripherals (if existing) are managed by the application core. The HSM consists of an internal processor (*secure processor*), some internal volatile and non-volatile memories (*secure memory*), a set of hardware security building blocks, hardware security functionality, and the interface to the application core. In order to enable a holistic but cost-efficient in-vehicle security architecture, there exist at least three different HSM variants – *full*, *medium*, and *light* – each focusing on

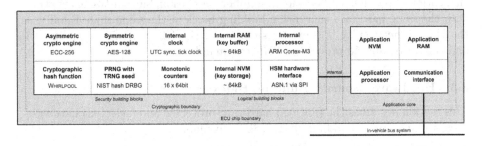

Fig. 1. Overall vehicular hardware security module architecture (full module)

different security use cases with different cost, functional and security requirements. However, these variants are no isolated developments, in fact, the light and medium module are proper subsets of the full module. The **full module** focuses on securing V2X communications and securely connecting the external vehicle interface(s) with the in-vehicle IT infrastructure. Due to its central importance and due to the strong V2X requirements, the full HSM provides the maximum level of functionality, security, and performance applying amongst others a very powerful hardware-accelerated asymmetric cryptographic building block, as shown in Figure 1. The **medium module** focuses on securing the in-vehicle communication. Hence the medium module has no dedicated hardware-accelerated asymmetric cryptographic building block, no dedicated hardware-accelerated hashing function, and a somewhat less performing internal processor. Even though, the medium HSM has no asymmetric cryptography in hardware, it is nonetheless able to perform some non-time-critical asymmetric cryptography operations (e.g., key exchange protocols) using the internal processor and firmware. As for efficiency and cost reasons virtually all internal communication protection is based on symmetric cryptographic algorithms, omitting the asymmetric cryptography and hashing hardware blocks is reasonable to save costs, size and power consumption. The **light module** focuses on securing the interaction of ECUs with sensors and actuators and is able to fulfill the strict cost and efficiency requirements typical there. In comparison with the medium module, the light module is again minimized and contains only a hardware-accelerated symmetric cryptography engine, a hardware random number generator, a UTC clock together with some very small optional volatile and non-volatile memories.

3.2   Hardware Interface in General

The HSM applies an asynchronous (i.e., non-blocking) hardware interface. The interface is multi-session capable (i.e., interruptible) for all security building blocks (cf. Section 3.3) that use a session identifier (e.g., encryptions, signatures, hash functions). Single session (i.e., non-interruptible) security building blocks, in contrast, are the random number generator, all counter functionality and all higher-level security functionality (cf. Section 3.4). Internally, the HSM is single-threaded in general, but provides limited multi-threading if different hardware functionality is accessed in parallel. Thus, for instance, one can invoke a hardware encryption in parallel with a random number generation, but not two encryptions with the same hardware encryption building block at the same time. Finally, as the HSM functionality is a proper superset of the HIS SHE functionality [21] and should be able to process all SHE commands.

3.3   Hardware Security Building Blocks

This section describes the (always) hardware-protected and (selected) hardware-accelerated security building blocks (SBB) available.

Asymmetric crypto engine enables creation and verification of digital signatures with the asymmetric signature algorithm specified on invocation using different hashing functions, different padding schemes and optional time-stamping. The full and medium prototype modules provide an ECC-256 (i.e., NIST P-256 [16] based) signature function that is hardware-accelerated at the full module. Full and medium prototype modules also provide the elliptic curve integrated asymmetric encryption scheme (ECIES) [1].

Symmetric crypto engine enables symmetric encryption and decryption with the cipher specified on invocation including different modes of operation (e.g., ECB, CBC, GCM) and different padding schemes (e.g., bit padding, PKCSx) – if available. As shown in Figure 1, all prototype modules provide at least the AES-128 block cipher [17].The symmetric crypto engine further enables generation and verification of message authentication codes (MACs) that optionally can get time-stamped using the internal tick or UTC clock (if synchronized). Thus, all prototype modules provide at least AES-128 CMAC [33] functionality.

Cryptographic hash function enables generation and verification of plain hash fingerprints and additionally HMACs (hash-based message authentication code) calculated with a secret key with the hashing algorithm specified on invocation. This SBB also provides optional time-stamping of the hashes/HMACs generated using the internal UTC clock (if synchronized). The full and medium prototype modules provide an ISO 10118 WHIRLPOOL [23] hashing function that is hardware-accelerated at the full module.

Pseudo random number generator (PRNG) creates pseudo random numbers with a PRNG algorithm specified on invocation that can be seeded internally from a physical true random number generator (TRNG) or from an external TRNG during production in a controlled environment of the chip manufacturer. The latter case additionally requires a proper seed update protocol. All prototype modules provide at least an officially evaluated PRNG according to E.4 [31] (e.g., AES- or hash-based).

Internal clock serves as hardware-protected time reference that can be synchronized with UTC time. For further details see Section 3.4.

Monotonic counters servs as a simple secure clock alternative while providing at least 16 monotonically increasing 64-bit counters together with corresponding access control similar to TCG's monotonic counters [35].

A distinctive feature of the HSM is the possibility for very fine-grained application specific authorizations for the processing and the migration of internal security assets. Concretely, a key can have several individual authorizations that allow or forbid processing it in different SBBs specified by so-called use_flags. As shown in the example in Table 1, a symmetric key, for instance, can have a use_flag for using it for MAC verifications but has a different use_flag (or even *no* use_flag) for using it for the creation of MACs. Our HSM prototype supports key use_flags each with individual authorizations for processing and migration at least for signing, signature/MAC verification, data encryption and decryption, time stamping, secure boot, secure storage, clock synchronization, key creations and key transports. Moreover, these use_flags can have individual migration authorizations that specify their transport restrictions to locations outside the respective HSM (cf. Section 3.4). Thus, a use_flag can be restricted to be moved (i) to no other location (*internal*), (ii) only between HSMs identical in construction (*migratable*), (iii) only between HSMs and trusted OEM locations (*oem*), or freely to any other location (*external*). For example, the use_flag *verify* for signature verification of a certain key can be allowed to become migrated to another HSM (*migratable*), while the use_flag *sign* for signature creation of the same key cannot be moved to a location outside its local HSM (*internal*). Lastly, each use_flags can have also its individual authorizations required for each invocation that can be simple passwords, can be based on the individual ECU platform configuration as measured at ECU bootstrap (cf. Section 3.4), or could be even a combination of both (i.e., configuration and password).

Table 1. Internal key structure including # of instances per field

key identifier [1]			algorithm identifier [1..n]
expiration date [1]			key signatures [0..n]
public key data [0..1]			private key data [0..1]
use flag [1..i..n]	transport flag [i]	auth. flag [i]	auth. value [0..i]
sign	*internal*	*password*	*Hash("abc")*
verify	*migratable*	*ecr*	*ECR(0;1;7) = 0x123*

3.4   Hardware Security Logic and Functionality

This section gives a short description of some central HSM keys and the underlying key hierarchy used by our HSMs. The *manufacturer verification key* (MVK) is a key from the module manufacturer to verify the authenticity of other HSMs or to authenticate HSM firmware updates. The *device identity key* (IDK) enables global HSM identification and authentication. The IDK is unique per module, fixed for HSM lifetime, and signed by MVK. The *OEM verification key* (OVK) is a key from an OEM to have an additional OEM managed trust domain similar to the manufacturer trust domain controlled via MVK. The OVK is unique per OEM and is also fixed for HSM lifetime. The *clock synchronization key(s)* (CSK) are a verification keys from a trusted time reference (e.g., a certain GPS module trusted by the HSM manufacturer) accepted for synchronizing the internal tick counter to absolute UTC time. The CSK is signed by MVK. The *storage root key* (SRK) is the (symmetric) master parent key for securely swapping internally created keys to external storages similar to the SRK as introduced by the TCG [35]. The *stakeholder key(s)* (SxK) finally are all externally created symmetric (SSK) or asymmetric (SAK) keys for stakeholder individual usage such as authentication, secure feature activation, or content protection. To increase (external) trust into SxK and OVK, they *can* be signed by MVK as well.

Key management provides functionalities for internal key creation (using the internal RNG), key import and export. The HSM further provides hardware-protected functionality for Diffie-Hellman key agreements [3] and for symmetric key derivations, for instance, according to ISO 18033 [25]. The creator of a key has the possibility to set individual usage authorizations (use_flags) and transport authorizations (trnsp_flags) for each key usage as introduced in Section 3.3 and Table 1. Note that some key flags cannot always be set freely but are inherently set by the HSM (e.g., the *internal* transport flag). For moving keys between different HSMs, between HSMs and external (trusted) locations (if permitted), the HSM provides key import and export functionality that ensures confidentiality of private key internals via (symmetric or asymmetric) transport encryption as well as authenticity of all key data structures via (symmetric or asymmetric) so-called transport authenticity codes (i.e., a digital signature or a MAC). Strictly speaking, not the whole key itself is moved, but only individual key use_flags if they have proper transport authorizations (trnsp_flags). The trnsp_flags inherently define also the keys that can be used for transport encryption and authenticity enforcement. Hence, use flags of keys marked *internal* are only permitted to become swapped out to offline storage and imported again to the same HSM via the SRK. Use flags of keys marked *migratable* are additionally permitted to become moved between HSMs identical in construction. This is enforced by accepting only target IDKs for transport encryption that are signed by a trusted MVK (e.g., MVKs from the same manufacturer used for HSMs of at least equal physical security). A similar approach is foreseen for use flags of keys marked *oem* that accept for transport encryption only keys that are signed by a trusted OVK. This was introduced to support an OEM managed trusted domain that can be differentiated (but not enforced!) by the HSM in contrast to use flags of keys marked *external* that are fully out of the control of the HSM for enforcing any trust assumptions.

Secure boot and authenticated boot is realized using HSM-internal so-called *ECU configuration registers* (ECR) that are similar to TCG's platform configuration registers (PCR) [35]. In contrast to the TCG approach, our HSM is also acting as the – by all involved parties a priori trusted[3] – core root of trust (CRT) that initializes the chain of trust for ECU integrity and authenticity verification. Assuming a multi-stage bootstrap with the CRT executed first, the trust chain is built by a hierarchical step by step *measuring* of the program code of all upper layers i such as the boot ROM, the boot routine, the boot loader, and all consecutive layers that are part of the Trusted Computing Base(TCB)[4]. For the HSM, the corresponding measurement routine $m()$ that creates a small unique fingerprint f_i for each program code measured, is the WHIRLPOOL one-way hash function for *full* and *medium* modules and the AES-MAC for *light* modules[5]. The fingerprint f_i in turn is saved to an individual ECR_n protected against manipulations inside the HSM before the next stage becomes executed. In order to prevent the overwriting of a previously saved ECR by layers executed later, ECRs in fact can only become *extended* via $ECR_n[t] = m(ECR_n[t-1], f_i)$ that connects the new fingerprint f_i cryptographically with the old $ECR_n[t-1]$ value and hence prevents overwriting. To detect and counteract possible vali-

---

[3] It is per definition impossible to self-verify the integrity of the *core root* of trust.

[4] The TCB means all code of the ECU that can be critical for ECU security.

[5] The bootstrap security functionality is optional for light HSMs.

dation failures, there exist at least two different approaches, usually known as *secure boot* and *authenticated boot*. In case the HSM can be deployed as active module (e.g., as ECU master) and hence having autonomous control to the corresponding ECU or an alarm function, the HSM can realize the active secure boot approach. Then the procedure executed at each step of the bootstrap uses the HSM to compare the actual value of an $ECR_n[t]$ with the corresponding reference value $ECR_{n,ref}$ that can securely preset for each ECR (cf. [9] for further details). In case of a mismatch between $ECR_n[t]$ and $ECR_{n,ref}$, secure boot automatically yields to an immediate response (e.g., ECU halt or alarm). Authenticated boot, in contrast, remains rather passive while only measuring the bootstrap without any direct interventions. It can be used if the HSM can be deployed as passive add-on module only having no autonomous control to the ECU or no alarm function. By using the authenticated boot approach, some essential key use_flags) (e.g., decryption) can become inaccessible afterwards in case the actual bootstrap measurements do not match the ECR reference individually linked for this particular use_flag) by the key owner. Our HSM therefore can enforce individual ECR references $ECR_{n,ref}$ as additional use_flag invocation authorization (cf. Section 3.3) that makes the corresponding key use_flag inaccessible (and only this) in case of an ECR mismatch.

Secure clock is realized by a so-called *tick counter* t_c that is monotonically increasing during HSM runtime using a fixed `tick()` interval of at least 1 Hz. As shown in Table 2, the initial value of t_c on HSM's first initialization in life is $t_c = 0$ or $t_c = UTC(t_0)$ that represents a UNIX/POSIX encoded default time value, for instance, the HSM release date. After any reset, the HSM continues increasing t_c unsynchronized (i.e., $t_c < UTC(t)$), but starting with the last internally saved[6] value of t_c. This provides a simple relative secure clock that – even if seldom or never synchronized – never "runs slow". However, optionally t_c can be synchronized to absolute UTC time received from an external UTC source (e.g., in-vehicle GPS sensor) trusted by the HSM manufacturer (e.g., in-vehicle GPS sensor). Therefore, the HSM can be invoked with the actual UTC time and the HSM synchronization challenge (as requested from the HSM before) both signed with a trusted CSK (cf. Section 3.4).

Table 2. HSM clock with relative *tick time* optionally synchronized to absolute UTC

On clock event	t_c	t_c vs. $UTC(t)$
First initialization in HSM life	0 or $UTC(t_0)$	$t_c < UTC(t)$
Internal `tick()` after reset	t_c++	$t_c < UTC(t)$
External UTC synchronization	$UTC(t)$	$t_c = UTC(t)$
Internal `tick()` after sync	t_c++	$t_c = UTC(t)$

---

[6] The internal saving interval for t_c depends on the rewrite capabilities of the HSM internal non-volatile memory and can vary from seconds up to hours or even days, but is always inherently invoked after a successful UTC synchronization.

Administration and audit functionality provides HSM status information, self tests, internal state backup, migration, and updates. Auditing the HSM operations can enhance the security by enabling an (external) auditor to retrace for instance critical operations (e.g., key migrations) or critical incidents that have occurred (e.g., authorization failures, resource exhaustions).

3.5   Driver Software and Software Security Framework

The HSM provides a set of basic hardware security functionalities on which a larger set of more sophisticated software security mechanisms can be built on. However, these higher-level software security mechanisms are not part of this work. But the corresponding HSM low-level software driver [12], an appropriate high-level software security framework [13] and proposals for implementing secure on-board protocols using the HSM as basis [10] are already available.

4   Implementation

The HSM architecture has been prototypically implemented [11] on a *Xilinx Virtex-5* FPGA (*XC5VFX70T-1FF1136C*). This FPGA has a total amount of 11,200 slices (i.e., 44,800 both flip-flops (FF) and 6-input look-up tables (LUT)), 128 dedicated digital signal processing (DSP) blocks, and 148 Block-RAMs (BRAMs) with 36 Kb of memory each. Additionally, the FPGA comprises an embedded hard-coded microprocessor block, that is, a PowerPC 440 (PPC440) with a maximum frequency of 550MHz. The FPGA comes on the *Virtex-5 FXT FPGA ML507 Evaluation Platform* with all necessary peripherals such as various interfaces, both volatile memory and non-volatile memory (NVM), and power supply. The *application processor* was implemented on an *Infineon TriCore TC1797* with all necessary peripherals and resources (i.e., application NVM, application RAM, communication interfaces) available on a corresponding development board. Both development boards are mechanically and electrically attached to each other via a *custom-made connection board* that provides access from the application core to the HSM. For this purpose the connection board wires the TC1797's external SPI bus with general purpose in-/output pins (GPIOs) of the FPGA. Figure 2 gives an overview of the layered approach of the HSM implementation. In the prototypical design, the HSM's *secure processor* is mapped to the embedded PPC440 of the FPGA that runs at 400 MHz with a standard Linux 2.6.34 kernel as operating system. This is the basis to host software implementations of the HSM firmware, the cryptographic library, and the Linux kernel drivers to access the hardware cores. The hardware cores implemented in the configurable logic of the FPGA are connected to the PPC440 via the 32 bit wide Xilinx specific processor local bus (PLB). Thus, the interface between software and hardware implementation has a theoretical throughput of 3 Gbit/s at the given PLB bus speed of 100 MHz.

First, all SBBs of Section 3.3 (also cf. Figure 1) and all HSM security functionalities of Section 3.4 were implemented in software (i.e., *HSM library*) as a reference for verification and performance analysis. The underlying cryptographic primitives of the SBBs were made available in a cryptographic software

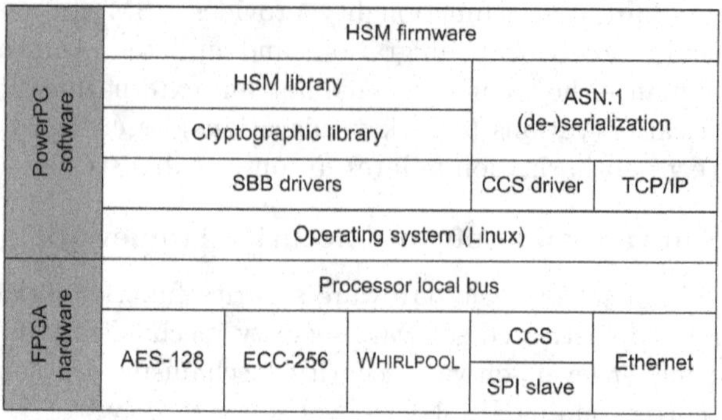

	HSM firmware		
PowerPC software	HSM library	ASN.1 (de-)serialization	
	Cryptographic library		
	SBB drivers	CCS driver	TCP/IP
	Operating system (Linux)		
FPGA hardware	Processor local bus		
	AES-128	ECC-256	WHIRLPOOL	CCS	Ethernet
				SPI slave	

Fig. 2. Architectural overview of the prototypical HSM implementation

library (i.e., CycurLIB [5]). Afterwards, those cryptographic primitives involved in the vast majority of all SBBs were additionally implemented in hardware (i.e., AES-128, ECC-256, and WHIRLPOOL). In a straight-forward approach, they were wrapped with the PLB interface and cryptographic algorithms in software were enhanced by non-blocking hardware calls via custom SBB Linux kernel drivers. Different modes and protocols based on one of the three cryptographic primitives in hardware are also part of the cryptographic library using the hardware calls. SBBs complementary to these three algorithms remain available as pure software. The HSM firmware is the top level entity that handles all communication requests, that means, all requests from internal hardware cores and also all external requests (i.e., from the application processor). Access from the application processor to the secure processor of the HSM is provided via the configuration, control and status module (CCS). The CCS is also part of the HSM and its basic task is to make the communication interface independent (i.e., to abstract it) from the underlying physical communication medium. The CCS includes a shared message RAM to buffer issued commands and their answers. The instantiation of the CCS module in the HSM prototype utilizes an SPI slave module with a data rate of 22.5 Mbit/s to connect the HSM to the *TC1797*. However, because of the separate CCS abstraction layer, the design can be easily enhanced to physical communication mediums other than SPI (e.g., to a socket based communication using TCP/IP over Ethernet for demonstration purposes with standard PCs). Since a serial bus (i.e., SPI) connects the HSM with the application processor all data has to be serialized before transmitting and de-serialized after receiving. This is done by a separate (de-)serialization layer using abstract syntax notation number one (ASN.1 [22]). At the application processor side, commands are serialized accordingly into a binary encoding and return values are deserialized with the common ASN.1 representation of the HSM API. The SBBs implemented in hardware are briefly described in the following. Afterwards, Table 3 shows a comparison of the entire HSM system design and the single SBB hardware implementation results (time and area) after synthesis. The synthesis results for the SBBs are mentioned as well for the sole hardware core as for the core with hardware overhead for PLB integration.

AES-128 was implemented in ECB mode with a focus on a very low area, still providing a reasonable performance. To keep the design small, the implementation is comprised of four 8 bit S-boxes that are shared between all round functions and the key scheduling. With this architecture, encrypting one 128 bit block in ECB mode takes 53 clock-cycles. As the AES operates with a clock of 100 MHz, it has a theoretical throughput of 242 Mbit/s. For decryption, the AES needs another 53 clock-cycles to perform the key scheduling each time a new decryption key is used. Thus, in the worst case, the decryption data rate decreases to 50% compared to encryption. Due to the small hardware footprint, several instances could be mapped to the design in parallel to increase the performance. Since all remaining operation modes of AES (e.g., CBC) share the ECB mode as their common atomic operation, they are realized in software finally using hardware calls of the AES core in ECB mode.

WHIRLPOOL was implemented with a moderate time-area trade off. Both, the underlying 512 bit block cipher W (which is very similar to AES) and the enclosing *Miyaguchi-Preneel* compression function were implemented in hardware. In our implementation, the 512 bit state of W consists of eight 8 Byte blocks operating on eight 8 bit S-boxes. Both the round function and the key scheduling share all computational modules (i.e., S-boxes, ShiftColumns, MixRows). With this architecture, one 512 bit block operation takes 308 clock-cycles. To keep the number of different clock domains in the HSM design at a minimum, WHIRLPOOL operates at the same clock of 100 MHz as the AES-128 resulting in a theoretical throughput of 166 Mbit/s.

ECC-256 was implemented based on the work of Güneysu and Paar [20] with a strong emphasis on performance optimization. Only the point multiplication of the NIST P-256 [16] ECC operation is implemented in hardware while all remaining parts of the algorithm are still performed in software. Even though this is only a small portion of the algorithm it covers over 90% of the entire ECC computation time. The design is highly optimized for the Xilinx DSP architecture and uses a pipelined architecture with input and output queues. Operated at a clock of 200 MHz, one ECC point multiplication (for signature generation) with 303,450 clocks takes 1.52 ms on average (i.e., 659 OP/s) and two point multiplications plus one addition with 366,905 clocks takes 1.83 ms (i.e., 545 OP/s).

Table 3. Synthesis results for the prototypical HSM implementation

Module	FF	LUT	BRAM	DSP	Critical Path
AES standalone	279	1,137	0	0	9.878 ns
AES with PLB	744	1,399	0	0	9.860 ns
WHIRLPOOL standalone	2,656	2,932	0	0	6.364 ns
WHIRLPOOL with PLB	3,445	3,826	0	0	6.386 ns
ECC standalone	1,854	1,964	13	32	3.539 ns
ECC with PLB	2,102	2,095	13	32	4.426 ns
HSM system	16,273	18,664	37	45	—

5 Evaluation

This section provides an evaluation on HSM performance and HSM security based on the security requirements stated in Section 2 together with a comparison of our HSM approach with other relevant HSMs already existing.

5.1 Performance Analysis

After fully implementing and extensively testing the HSM hardware design, the implementation was finally deployed on the target platform. Table 4 shows a comparison of the hardware cores' theoretical throughput from the previous section and the SBB performances measured in both pure software and hardware accelerated versions. All measurements were performed on the internal PPC440 based on the cryptographic library CycurLIB [5] enhanced by drivers for hardware core access. On first sight, the AES-128 hardware accelerated implementation gains only between $5 - 40\%$ compared to the pure software solution. Moreover, it reaches only $20 - 41\%$ of the theoretical throughput of the hardware core depending on the mode of operation. This seemingly small yield of performance is mainly caused by the very high bidirectional data traffic over the PLB bus. Since the AES core operates only on single 128 bit blocks in ECB mode, each en-/decryption requires first sending and afterwards receiving a small block of 128 bit data. Additionally, all remaining modes (including MAC generation) have to perform their protocol overhead in software based on the ECB block operation in hardware. Together with (bus) latencies this explains the relatively small performance gain. With DMA (direct memory access) usage and slight modifications to the hardware implementation, e.g., enhancing wrappers for different operation modes, buffers for in-/output data or using pipelining, the throughput could be increased vastly. On the other hand, the WHIRLPOOL implementation reaches 77% of its theoretical performance in hardware. Similar to the AES, it operates on single blocks (of 512 bit for WHIRLPOOL) but the communication remains unidirectional except for the last block of the hash computation when the result is returned. This saves a vast amount of bus traffic (and latencies). While the hardware performance benefits from the larger block size of W (i.e., 512 bit), it makes computation in software much harder. Actually, the hardware accelerated WHIRLPOOL outperforms the software version by a factor of 25. Optimizing the software implementation might reduce this gap slightly but will never close it entirely. From all SBBs, ECC-256 benefits the most from the hardware acceleration. The more complex ECDSA signature verification (two ECC point multiplications plus one addition vs. one ECC point multiplication only) is 29 times faster in hardware than in software. Since one ECC operation requires only sending 1024 bit and receiving 512 bit data, the requirements on communication bandwidth are very low. This allows the ECC core to reach up to 80% of its theoretical throughput although the complete ECDSA protocol (except for the point multiplication) is performed in software.

Table 4. Comparison of SBB Performance-Measurements in Software and Hardware

SBB Mode	Throughput		
	SW (measured)	HW (measured)	HW (theoretical)
AES-128, ECB encrypt	53 Mbit/s	76 Mbit/s	242 Mbit/s
AES-128, ECB decrypt	53 Mbit/s	58 Mbit/s	121 Mbit/s
AES-128, CBC encrypt	46 Mbit/s	68 Mbit/s	242 Mbit/s
AES-128, CBC decrypt	44 Mbit/s	46 Mbit/s	242 Mbit/s
AES-128, CMAC generate	44 Mbit/s	60 Mbit/s	242 Mbit/s
WHIRLPOOL, hash generate	5 Mbit/s	128 Mbit/s	166 Mbit/s
ECC-256, ECDSA generate	30 sig/s	480 sig/s	659 OP/s
ECC-256, ECDSA verify	15 sig/s	436 sig/s	545 OP/s

5.2   Security Analysis

This section shortly analyses the fulfillment of the security requirements (SR) stated in Section 2 by our HSM approach. We fulfill *SR.1* (*isolated security processing environment*) as our HSM is realized as an autonomous microprocessor with its own independent memories (physically) isolated from the ECU main processor (cf. Section 3.1). The shielded execution environment can also be strengthened further against physical manipulations by applying appropriate physical tamper-protection measures [27]. We fulfill *SR.2* (*immutable core root of trust code*) by having foreseen an HSM deployment architecture that either assumes an active HSM (part) executed prior to the ECU main processor or an additional, small, immutable component that initializes the hierarchical bootstrap measurements (cf. Section 3.1). We fulfill *SR.3* (*internal non-volatile memory*) by having our HSM foreseen at least about 64kB non-volatile memory for shielded storing of security root artifacts (cf. Section 3.1 and Figure 1). We fulfill *SR.4* (*non-detachable connection with ECU hardware*) by assuming a system-on-chip (SoC) design, for instance, that ensures at least tamper evidence if someone tries to detach the HSM from the corresponding ECU. However, in the end, this requirement has to be fulfilled by the ASIC system designer, who can apply different powerful tamper protection measures (e.g., cf. [27]). We fulfill *SR.5* (*secure HSM-ECU communication channel*) by assuming a secure communication channel between HSM and ECU realized by appropriate physical tamper protection, for example, by assuming an SoC communication line. In case this is not possible, we propose to realize a secure communication channel by cryptographic means, for instance, as proposed by TCG's command transport encryption [35]. However, the additional cryptographic protection of HSM commands is not part of our current design. We fulfill *SR.6* (*autonomously controlled alert functionality*) by moving this alert functionality inside the HSM's cryptographic boundary (cf. Figure 1). The specification of the actual alert response (e.g., from autonomous log entry up to ECU full stop), however, should be defined by the respective ECU application (engineer). We finally also fulfill *SR.7* (*standardized, well-established security algorithms*) for all security building blocks, concretely by applying the NIST advance encryption standard [17] for symmetric encryption, the NIST digital signature standard [16] for asymmetric encryption, the ISO hash function standard [23] for cryptographic hashing, a BSI evaluated PRNG according to E.4 [31] for random number generation, and the ISO standardized ECDH [25] for key agreements.

5.3   Comparison with Other Hardware Security Modules

This section provides an overview comparison of our HSM with other hardware security modules currently available. Table 5 compares our three different HSM variants – full, medium, and light – with the Secure Hardware Extension (SHE) as proposed by HIS [21], with the Trusted Platform Module (TPM) as proposed by the TCG [35], and with a typical cryptographic smartcard (SmC). While the TPM and the smartcard were neither designed nor applicable for use within an automotive security context (as they lack, for instance, in cost efficiency, performance, physical robustness, or security functionality), SHE was explicitly designed for application in an automotive security context. It can be seen that the HSM is quite successful in merging the relevant security functionalities from SHE, from the TPM and from a typical smartcard while transferring them into the automotive domain with its strict cost efficiency and performance requirements. Beyond this, Table 5 clearly indicates the distinctive security features, for instance, the possibility for very fine-grained application-specific authorizations

Table 5. Comparison with other hardware security modules

	Full	Medium	Light	SHE	TPM	SmC
Cryptographic algorithms						
ECC/RSA	■/■	■/■	□/□	□/□	□/■	⊞/⊞
AES/DES	■/⊞	■/⊞	■/□	■/□	□/□	⊞/⊞
WHIRLPOOL/SHA	■/■	■/■	□/□	□/□	□/■	⊞/⊞
Hardware acceleration						
ECC/RSA	■/□	□/□	□/□	□/□	□/□	□/□
AES/DES	■/□	■/□	■/□	■/□	□/□	□/□
WHIRLPOOL/SHA	■/□	□/□	□/□	□/□	□/□	□/□
Security features						
Secure/authenticated boot	■/■	■/■	⊞/⊞	■/□	□/■	□/□
Key AC per use/bootstrap	■/■	■/■	■/⊞	□/■	⊞/■	□/□
PRNG with TRNG seed	■	■	■	■	■	■
Monotonic counters 32/64 bit	■/■	■/■	□/□	□/□	■/□	□/□
Tick/UTC-synced clock	■/■	■/■	■/■	□/□	□/□	□/□
Internal processing						
Programmable/preset CPU	■/⊞	■/⊞	□/⊞	□/■	□/■	⊞/⊞
Internal V/NV (key) memory	■/■	■/■	⊞/⊞	■/■	■/□	■/□
Asynchronous/parallel IF	■/⊞	■/□	■/□	■/□	□/□	□/□

Annotation: ■ = available, □ = not available, ⊞ = partly or optionally available for the processing and the migration of internal security assets or the possibility for a secure (UTC) time reference, that are exclusively available with this HSM.

6 Conclusion and Outlook

Based on a tight automotive-driven security requirements engineering, we have designed, implemented, and evaluated a vehicular hardware security module that enables a holistic protection of all relevant in-vehicle ECUs and their communications. Practical feasibility will be proven with first HSM-equipped passenger vehicle demonstrators [14] by end of 2011. Further large-scale field operational tests with hundreds of HSMs and HSM-equipped vehicles that amongst others address performance, scalability, and deployment of our HSM approach are already scheduled in a subsequent research project [30]. We are convinced that future interactive vehicles can be realized in dependable manner only based on an effective, efficient holistic in-vehicle security architecture, which in turn is based upon effective and efficient vehicular HSM as presented by this work.

References

1. Abdalla, M., Bellare, M., Rogaway, P.: DHAES: An encryption scheme based on the Diffie-Hellman problem. Submission to P1363a: Standard Specifications for Public-Key Cryptography, Additional Techniques 5 (2000)
2. Checkoway, S., et al.: Comprehensive Experimental Analyses of Automotive Attack Surfaces. National Academy of Sciences Committee on Electronic Vehicle Controls and Unintended Acceleration (2011)
3. Diffie, W., Hellman, M.: New Directions in Cryptography. IEEE Transactions on Information Theory 22(6) (1976)
4. Dyer, J., Lindemann, M., Perez, R., Sailer, R., Van Doorn, L., Smith, S., Weingart, S.: Building the IBM 4758 Secure Coprocessor. IEEE Computer 34(10) (2001)
5. escrypt GmbH – Embedded Security: CycurLIB - Cryptographic Software Library (2011), http://www.escrypt.com/products/cycurlib/overview/
6. EVITA: Deliverable 2.1: Specification and Evaluation of E-Security Relevant Use Cases (2008)
7. EVITA: Deliverable 2.3: Security Requirements for Automotive On-Board Networks Based on Dark-Side Scenarios (2009)
8. EVITA: Deliverable 3.1.2: Security and Trust Model (2009)
9. EVITA: Deliverable 3.2: Secure On-board Architecture Specification (2010)
10. EVITA: Deliverable 3.3: Secure On-Board Protocols Specification (2010)
11. EVITA: Deliverable 4.1.3: Security Hardware FPGA Prototype (2011)
12. EVITA: Deliverable 4.2.2: Basic Software (2011)
13. EVITA: Deliverable 4.3.2: Implementation of Software Framework (2011)
14. EVITA: Deliverable 5.1.2: On-board Communication Demonstrator (2011)
15. EVITA Project: E-safety Vehicle Intrusion proTected Applications, European Commission research grant FP7-ICT-224275 (2008), http://www.evita-project.org
16. FIPS-186-3: Digital Signature Standard (DSS). NIST (1994, 2006)
17. FIPS-197: Advanced Encryption Standard (AES). NIST (2001)

18. Frischkorn, H.G.: Automotive Software – The Silent Revolution. In: Workshop on Future Generation Software Architectures in the Automotive Domain, San Diego, CA, USA, January 10- 12 (2004)
19. Furgel, I., Lemke, K.: A Review of the Digital Tachograph System. In: Embedded Security in Cars: Securing Current and Future Automotive IT Applications. Springer (2006)
20. Güneysu, T., Paar, C.: Ultra High Performance ECC over NIST Primes on Commercial FPGAs. In: Oswald, E., Rohatgi, P. (eds.) CHES 2008. LNCS, vol. 5154, pp. 62–78. Springer, Heidelberg (2008)
21. Herstellerinitiative Software (HIS): SHE Secure Hardware Extension Version 1.1 (2009), http://portal.automotive-his.de
22. International Telecommunication Union – ITU-T Study Group 7: Abstract Syntax Notation number One – ASN.1 (1995), http://www.itu.int/ITU-T/asn1/
23. ISO/IEC 10118-3:2004: Information technology – Security techniques – Hash-functions – Part 3: Dedicated hash-functions. ISO/IEC (2004)
24. ISO/IEC 11898:2003-2007: Information technology – Road vehicles Controller area network. ISO/IEC (2007)
25. ISO/IEC 18033-2:2006: Information technology - Security techniques - Encryption algorithms - Part 2: Asymmetric ciphers. ISO/IEC (2006)
26. Koscher, K., et al.: Experimental Security Analysis of a Modern Automobile. In: IEEE Symposium on Security and Privacy (2010)
27. Lemke, K.: Physical Protection against Tampering Attacks. In: Embedded Security in Cars: Securing Current and Future Automotive IT Applications. Springer (2006)
28. Luo, J., Hubaux, J.: A Survey of Inter-Vehicle Communication. EPFL, Lausanne, Switzerland, Tech. Rep (2004)
29. PRECIOSA Project: Privacy Enabled Capability in Co-operative Systems and Safety Applications (2008), http://www.preciosa-project.org
30. PRESERVE Project: Preparing Secure Vehicle-to-X Communication Systems (2011), http://www.preserve-project.eu
31. Schindler, W.: AIS 20 – Functionality classes and evaluation methodology for deterministic random number generators. German Federal Office for Information Security (BSI) (1999)
32. SeVeCom Project: Secure Vehicular Communication (2006), http://www.sevecom.org
33. Song, J., Poovendran, R., Lee, J., Iwata, T.: The AES-CMAC Algorithm. RFC4493, IETF (June 2006)
34. Toll Collect GmbH (2011), http://www.toll-collect.com
35. Trusted Computing Group (TCG): TPM Specification 1.2 Revision 116 (2011), http://www.trustedcomputinggroup.org

Efficient Modular Exponentiation-Based Puzzles for Denial-of-Service Protection

Jothi Rangasamy, Douglas Stebila, Lakshmi Kuppusamy,
Colin Boyd, and Juan Gonzalez Nieto

Information Security Institute, Queensland University of Technology,
GPO Box 2434, Brisbane, Queensland 4001, Australia
{j.rangasamy,stebila,l.kuppusamy,c.boyd,j.gonzaleznieto}@qut.edu.au

Abstract. Client puzzles are moderately-hard cryptographic problems — neither easy nor impossible to solve — that can be used as a counter-measure against denial of service attacks on network protocols. Puzzles based on modular exponentiation are attractive as they provide important properties such as non-parallelisability, deterministic solving time, and linear granularity. We propose an efficient client puzzle based on modular exponentiation. Our puzzle requires only a few modular multi-plications for puzzle generation and verification. For a server under de-nial of service attack, this is a significant improvement as the best known non-parallelisable puzzle proposed by Karame and Čapkun (ESORICS 2010) requires at least $2k$-bit modular exponentiation, where k is a secu-rity parameter. We show that our puzzle satisfies the unforgeability and difficulty properties defined by Chen et al. (Asiacrypt 2009). We present experimental results which show that, for 1024-bit moduli, our proposed puzzle can be up to 30× faster to verify than the Karame-Čapkun puzzle and 99× faster than the Rivest et al.'s time-lock puzzle.

Keywords: client puzzles, time-lock puzzles, denial of service resistance, RSA, puzzle difficulty.

1 Introduction

DENIAL-OF-SERVICE (DoS) attacks are a growing concern due to the advance-ment in information technology and its application to electronic commerce. The main goal of DoS attacks is to make a service offered by a service provider unavailable by exhausting the service provider resources. In recent years, DoS attacks disabled several Internet e-commerce sites including eBay, Yahoo!, Ama-zon and Microsoft's name server [16].

Since millions of computers are connected through the Internet, DoS attacks on any of these systems would lead to a large scale impact on the whole network. Many essential services such as communications, defense, health systems, bank-ing and financial systems have become Internet-based applications. There is an immense need for keeping these services alive and available on request. However mounting a DoS attack is very easy for the sophisticated attackers while defend-ing them is very hard for the victim servers. A promising way to deal with this

H. Kim (Ed): ICISC 2011, LNCS 7259, pp. 319–331, 2012.
© Springer-Verlag Berlin Heidelberg 2012

problem is for a defending server to identify and segregate the malicious requests as early as possible.

CLIENT PUZZLES, also known as *proofs of work*, can guard against resource exhaustion attacks such as DoS attacks and spam [2,7,10]. When client puzzles are employed, a defending server will process a client's request only after the client has provided the correct solution to its puzzle challenge. In this way, a client can prove to the server its legitimate intentions in getting a connection. Although employing client puzzles adds an additional cost for legitimate clients, a big cost will be imposed on an attacker who is trying to make multiple connections. In this case, the attacker would need to invest its own resources in solving a large number of puzzles before exhausting the server resources.

1.1 Puzzle Properties

The essential property of a client puzzle is that it be *difficult* to solve: not impossible, but not too easy, either. Many cryptographic puzzles are based on inverting a hash function [3,8,10].

Puzzles based on modular exponentiation have the potential to provide additional properties:

- NON-PARALLELISABILITY. A client puzzle is called non-parellelisable if the time to solve the puzzle cannot be reduced by using many computers in parallel. This property ensures that a DoS attacker cannot divide a puzzle into multiple small tasks and therefore gains no advantage in puzzle solving with the large number of machines it may have in its control.
- DETERMINISTIC SOLVING-TIME. If the puzzle issuing server has specified a value as a difficulty parameter, then a client needs to do at least the specified number of operations to solve a puzzle. This will help the server decide the minimum work each client must do before getting a connection. Many puzzles in contrast, only determine the average work required to obtain a solution.
- FINE GRANULARITY. A puzzle construction achieves finer granularity if the server is able to set the difficulty level accurately. That is, the gap between two adjacent difficulty levels should be small. This property helps servers switch between different difficulty levels easily. If there is a large gap between two difficulty levels, then increasing the difficulty to the next level might have an impact on computationally-poor legitimate clients.

It is imperative that both the puzzle generation and solution verification algorithms add only minimal computation and memory overhead to the server. Otherwise, this puzzle mechanism itself may become a target for resource exhaustion DoS attacks when a malicious client sends a large number of fake requests for puzzle generation or a large number of puzzle solutions for verification.

Rivest *et al.* [17] described a concrete modular exponentiation puzzle based on the RSA modulus factorisation problem. This was the first puzzle to provide the three properties listed above: non-parallelisability, deterministic solving time, and finer granularity. However, the main disadvantage of these puzzles is that

Table 1. Verification costs and timings (in microseconds) for modular exponentiation-based puzzles; n is an RSA modulus, k is a security parameter. Timings are for 1024-bit modulus n with $k = 80$ and for 512-bit modulus n with $k = 56$, both with puzzle difficulty 1 million.

Puzzle	Verification Cost	Verification Time (μs) 512-bit n	1024-bit n
Rivest *et al.* [17]	$|n|$-bit mod. exp.	474.68	2903.99
Karame-Čapkun [11]	$2k$-bit mod. exp.	263.35	895.17
This paper	**3 mod. mul.**	**14.75**	**29.24**

they require a busy server to perform computationally intensive exponentiation to verify solutions. Recently, Karame and Čapkun [11] improved the verification efficiency of Rivest *et al.*'s puzzle by a factor of $\frac{|n|}{2k}$ for a given RSA modulus n, where k is the security parameter. More details on these two puzzles will be provided in Section 2.

Although Karame and Čapkun's performance gain in verification cost is impressive, it is still sufficiently expensive that it could be burdensome on a defending server, as verification still requires modular exponentiations. This is the main reason preventing modular exponentiation-based puzzles being deployed widely, despite having some attractive characteristics. Construction of modular exponentiation-based puzzles which avoid a big modular exponentiation for puzzle generation and solution verification has not been attained until now.

1.2 Contributions

1. We propose an efficient modular exponentiation-based puzzle which achieves non-parallelisability, deterministic solving time, and finer granularity. Our puzzle can be seen as an efficient alternative to Rivest *et al.*'s time-lock puzzle [17]. Our puzzle requires only a few modular multiplications to generate and verify puzzle solutions. The verification costs and timings for our puzzle and other puzzles of the same type are presented in Table 1.
2. We analyse the security properties of our puzzle in the puzzle security model of Chen *et al.* [6] and show that our puzzle is unforgeable and difficult.
3. In order to validate the performance of our puzzle, we give experimental results and compare them with the performances of Rivest *et al.*'s time-lock puzzle [17] and Karame and Čapkun's puzzle [11], which is the most efficient non-parallelisable puzzle in the literature. Our results suggest that our puzzle reduces the solution verification time by approximately 99 times when compared to Rivest *et al.*'s time-lock puzzle and 30 times when compared to Karame and Čapkun, for 1024-bit moduli.

Organization of paper: The rest of the paper is organized as follows. Section 2 presents the background and motivation for our work. Section 3 describes our proposed puzzle and Section 4 analyses the security properties of the proposed puzzle in the Chen *et al.* model. Section 5 presents our experimental results validating the efficiency of the proposed puzzle scheme and we conclude the paper in Section 6.

2   Background: Modular Exponentiation-Based Puzzles

In this section, we review known modular exponentiation-based puzzles and follow the definition of a client puzzle proposed by Chen *et al.* [6].

Notation. If n is an integer, then we use $|n|$ to denote the length in bits of n, and $\phi(n)$ is the Euler phi function of n, which is equivalent to the size of the multiplicative group $\mathbb{Z}_n^*$. We denote the set $\{a, \ldots, b\}$ of integers by $[a, b]$. We use $x \leftarrow_r S$ to denote choosing x uniformly at random from S. If A is an algorithm, then $x \leftarrow A(y)$ denotes assigning to x the output of A when run with the input y. If k is a security parameter, then negl(k) denotes a function that is negligible in k (asymptotically smaller than the inverse of any polynomial in k). By p.p.t. algorithm, we mean probabilistic polynomial time algorithm.

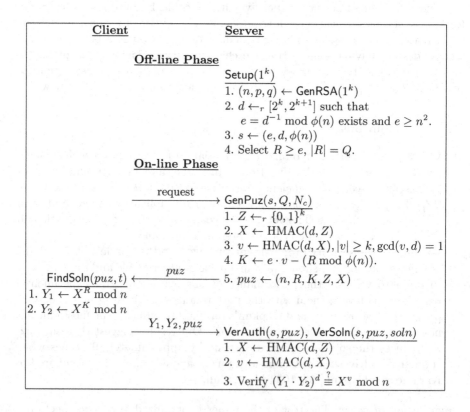

Fig. 1. KCPuz: Karame-Čapkun's Variable-Exponent Client Puzzle [11]

2.1   Rivest *et al.*'s Puzzle

Given a RSA modulus n, Rivest *et al.*'s puzzle [17] requires $|n|$-bit exponentiation to verify puzzle solutions. In detail, the server generates two RSA primes p and q, and computes the associated RSA modulus $n = pq$ and the Euler totient function $\phi(n) = (p-1) \cdot (q-1)$. Then sets the difficulty level Q or the amount of work a client needs to do. Now, the server picks an integer $a \leftarrow_r \mathbb{Z}_n^*$ and sends the client the tuple (a, Q, n). The client's task is to compute and return $b \leftarrow a^{2^Q} \bmod n$. The server first computes $c = 2^Q \bmod \phi(n)$ and then checks if $a^c \overset{?}{\equiv} b \bmod n$. Here the server can compute c once and use it for all the solution verifications unless it changes the difficulty level Q. With the trapdoor information $\phi(n)$, the server is able to verify the solution in one $|n|$-bit exponentiation whereas the client should perform Q repeated squarings and typically $Q \gg |n|$.

2.2   Karame-Čapkun Puzzle

Recently, Karame and Čapkun [11] reduced $|n|$-bit exponentiation in time-lock puzzle verification to $2k$-bit exponentiation modulo n, thereby significantly reducing the computational burden of the server by a factor of $\frac{|n|}{2k}$, where k is a security parameter. The Karame and Čapkun puzzle KCPuz is depicted in Figure 1. Karame and Čapkun showed that their puzzle is unforgeable and difficult in the puzzle security model of Chen *et al.* [6]. In this paper, we are considering the variable-exponent version of Karame and Čapkun's puzzle.

Although the verification cost is improved significantly in KCPuz, the server still needs to engage in at least $2k$-bit exponentiation for each puzzle solution it receives. Since it is expected that the defending server may receive a large number of fake requests/solutions, puzzle generation and solution verification should be as efficient as possible. Otherwise this mitigation mechanism itself opens door for resource exhaustion DoS attacks when a malicious client sends a number of fake requests/solutions for puzzles triggering the server engage in those expensive operations.

Parallelisability. Puzzle solving in KCPuz can be partially parallelised by decomposing the exponent R into multiple parts. For example, consider a malicious client C with two compromised machines, namely M_1 and M_2, under its control. In order to parallelise the computation of $x^R \bmod n$, C first decomposes R into two parts R_1 and R_2 such that $R = R_1 \| R_2$, where $\|$ denotes the concatenation. Then C gives $R_1 \| 0^{\frac{\ell}{2}}$ to M_1 and R_2 to M_2, along with the public values (X, n). Now, using the square and multiply algorithm, M_1 computes $X^{R_1 \| 0^{\frac{\ell}{2}}} \bmod n$ in $\frac{5\ell}{4}$ modular multiplications and M_2 computes $X^{R_2} \bmod n$ in $\frac{3\ell}{4}$ modular multiplications. Note that, without decomposition, $\frac{3\ell}{2}$ modular multiplications would have been required if the malicious client chose to compute $X^R \bmod n$ itself. Since M_1 and M_2 could work in parallel, the time taken by C to compute $X^R \bmod n$ is the time taken by M_1 to compute $X^{R_1 \| 0^{\frac{\ell}{2}}} \bmod n$, which requires to do more operations than M_2. Therefore, this decomposition saves the malicious client $\frac{1}{6}$ of the total time needed to solve the puzzle. This parallelisation via exponent decomposition is gainful only if R is not a power of 2. Rivest et $al.$ set R to be power of 2 to achieve non-parallelisability. Moreover when $|R| \gg 2^{20}$ bits and $R = 2^Q$ for some $Q \in \mathbb{N}^+$, sending Q for each puzzle instead of R will save communication cost as well.

Granularity. Unlike Rivest et $al.$'s modular exponentiation-based puzzle, the Karame-Čapkun puzzle does not provide fine control over granularity of difficulty levels. In KCPuz, a client is given the pair (K, R) where $K \leftarrow e \cdot v - (R \bmod \phi(n))$ and therefore for security reasons, R must be large enough so that $R > n$. This condition rules out difficulty levels between 0 and n. Also, if R is the current difficulty level, then the next difficulty level R' must satisfy the following: $\frac{R'}{R} \geq n^2$. This implies that there will be a large gap between the two successive difficulty levels. Hence, KCPuz does not support fine granularity.

Example Parameter Sizes. In a DoS scenario, a client is given a puzzle whose hardness is typically set between 0 to 2^{25} operations. Since a client needs to perform at most 2^{25} for each puzzle, the 40-bit security level is enough for the puzzle scheme and is higher than the work needed to solve a puzzle. Lenstra and Verheul [13] suggest using a 512-bit RSA modulus n which is widely believed to match the 56-bit security of Data Encryption Standard (DES). Since $|n| = 512$, $|R| \geq 512$. Suppose $R = 2^{512}$. From $\frac{R'}{R} \geq n^2$, the possible values for the next two difficulty levels are $R' = 2^{1536}$ and $R'' = 2^{2560}$.

In this work, we give an efficient modular exponentiation-based puzzle which achieves both non-parallelism and finer granularity.

3   Our Client Puzzle Protocol

In this section, we present a non-parallelisable client puzzle scheme that requires $only$ a few modular multiplications for puzzle generation and solution verification. First we review the cryptographic ingredients required and then present our puzzle construction.

3.1   Tools

Our puzzle construction makes use of algorithm GenRSA that generates an RSA-style modulus $n = pq$ as follows:

Definition 1 (Modulus Generation Algorithm). *Let k be a security parameter. A modulus generation algorithm is a probabilistic polynomial time algorithm GenRSA that, on input 1^k, outputs (n, p, q) such that $n = pq$ and p and q are k-bit primes.*

In our puzzle generation algorithm, the server needs to produce a pair (x, x^u) for each puzzle. Since the generation of these pairs are expensive, we utilise a technique due to Boyko et $al.$ [5] for efficient generation of many pairs $(x_i, x_i^u \bmod n)$ for a fixed u using a relatively small amount of pre-computation.

Definition 2 (BPV Generator). *Let k, ℓ, and N, with $N \geq \ell \geq 1$, be parameters. Let $n \leftarrow \mathsf{GenRSA}(1^k)$ be an RSA modulus. Let u be an element in $\mathbb{Z}_{\phi(n)}$ of length m. A BPV generator consists of the following two algorithms:*

- $\mathsf{BPVPre}(u, n, N)$: *This is a pre-processing algorithm that is run once. The algorithm generates N random integers $\alpha_1, \alpha_2, \ldots, \alpha_N \leftarrow_r \mathbb{Z}_n^*$ and computes $\beta_i \leftarrow \alpha_i^u \bmod n$ for each i. Finally, it returns a table $\tau \leftarrow ((\alpha_i, \beta_i))_{i=1}^N$.*
- $\mathsf{BPVGen}(n, \ell, \tau)$: *This is run whenever a pair $(x, x^u \bmod n)$ is needed. Choose a random set $S \subseteq_r \{1, \ldots, N\}$ of size ℓ. Compute $x \leftarrow \prod_{j \in S} \alpha_j \bmod n$. If $x = 0$, then stop and generate S again. Otherwise, compute $X \leftarrow \prod_{j \in S} \beta_j \bmod n$ and return (x, X). In particular, the indices S and the corresponding pairs $((\alpha_j, \beta_j))_{j \in S}$ are not revealed.*

Indistinguishability of the BPV Generator. Boyko and Goldwasser [4] and Shparlinski [18] showed that the values x_i generated by the BPV generator are statistically close to the uniform distribution. To analyse the security properties of the proposed puzzle, we use the following results by Boyko and Goldwasser [4, Chapter 2]. Let N be the number of pre-computed pairs (α_i, β_i) such that α_i's are chosen independently and uniformly from $[1, n]$ and $\beta_i = \alpha_i^u \bmod n$. Each time a random set $S \subseteq \{1, \ldots N\}$ of ℓ elements is chosen and a new pair (x, X) is computed such that $x = \prod_{j \in S} \alpha_j \bmod n$ and $X = \prod_{j \in S} \beta_j \bmod n$. Then, with overwhelming probability on the choice of α_i's, the distribution of x is statistically close to the uniform distribution of a randomly chosen $x' \in \mathbb{Z}_n^*$. Here we also note that although BPV outputs a pair (x, x^u), only x is made available to clients and x^u is kept secret by the server. That is, each time clients are given the pair $(x, 1)$, not (x, x^u).

Theorem 1. *[4, Chapter 2] If $\alpha_1, \ldots, \alpha_N$ are chosen independently and uniformly from $\mathbb{Z}_n^*$ and if $x = \prod_{j \in S} \alpha_j \bmod n$ is computed from a random set $S \subseteq \{1, \ldots N\}$ of ℓ elements, then the statistical distance between the computed x and a randomly chosen $x' \in \mathbb{Z}_n^*$ is bounded by $2^{-\frac{1}{2}\left(\log \binom{N}{\ell} + 1\right)}$. That is,*

$$\left| \Pr\left(\prod_{j \in S} \alpha_j = x \bmod n \right) - \frac{1}{\phi(n)} \right| \leq 2^{-\frac{1}{2}\left(\log \binom{N}{\ell} + 1\right)} .$$

Parameters for BPV. As discussed in Section 2, in a DoS scenario, the difficulty level Q for a puzzle is typically set between 0 and 2^{25} operations. Therefore, it can be anticipated that factoring of n and hence computing $\phi(n)$ for solving puzzles easily, will be much more difficult than performing Q squarings as $Q \ll n$ when $Q \leq 2^{25}$ and $|n| \geq 512$. Lenstra and Verheul [13] suggest using a 512-bit RSA modulus n to match the 56-bit security of Data Encryption Standard (DES). Since a client needs to perform at most 2^{25} for each puzzle, the 40-bit security level could be enough for the puzzle scheme and hence breaking the scheme is much harder than solving a puzzle.

Boyko *et al.*[4,5] suggest to set N and ℓ so that subset product problem is intractable and birthday attacks becomes infeasible. To achieve the above security level, we can select N and ℓ such that $\binom{N}{\ell} > 2^{40}$. Boyko *et al.* [4,5] suggest setting $N = 512$ and $\ell = 6$ for the BPV generator. Alternatively, we could achieve this with $N = 2500$ and $\ell = 4$; this increases the amount of precomputation required in BPVPre but reduces the number of modular multiplications performed online in BPVGen from 12 to 8.

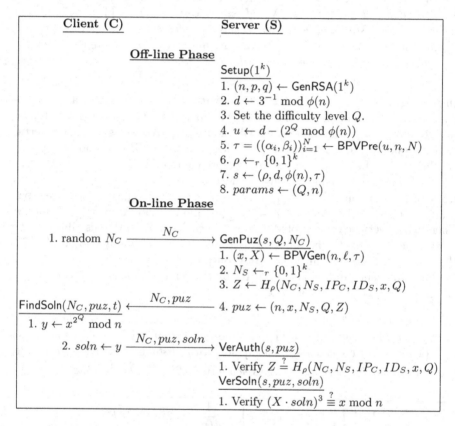

Fig. 2. RSAPuz: A new client puzzle based on modular exponentiation

3.2   The Proposed Puzzle: **RSAPuz**

The main idea behind our puzzle construction is: given a RSA modulus n, an integer Q and $X \in \mathbb{Z}_n^*$, the task of a client is to compute $X^{2^Q} \bmod n$.

Our client puzzle RSAPuz is presented in Fig 2 as an interaction between a server issuing puzzles and a client solving them. RSAPuz is parameterized by a security parameter k and a difficulty parameter Q. In practice, a server using puzzles as a DoS countermeasure can vary Q based on the severity of the attack it is experiencing. However once a difficulty level is set, it is increased only if the server still receives a large number of requests with correct puzzle solutions.

In RSAPuz, the server does the following:

- PUZZLE PRE-COMPUTATION. Generating (n, p, q) and computing d is a one time process. Whenever a server is required to change the difficulty parameter Q, it selects an integer R such that $|R| = Q$ and computes u. Then it runs the BPV pre-processing step with inputs (u, n, N) and obtains N pairs of (α_i, β_i). Since all the required pre-computations are done off-line, the defending server can be more effective on-line against DoS attacks.

In a DoS setting, an attacker could mount a resource depletion attack by asking the server to generate many puzzles and to verify many fake puzzle solutions. Hence the following algorithms run online by the server many times should be very efficient to resist such flooding attacks.

- PUZZLE GENERATION. The dominant cost in puzzle generation is the BPV pair generation BPVGen, which requires $2(\ell - 1)$ modular multiplications: $\ell - 1$ to compute x and $\ell - 1$ to compute X. There is also a single call to the pseudo-random function H_ρ to compute the authentication tag Z. As suggested by Boyko et al., ℓ could be set between 4 and 16 so that our puzzle requires only 8 modular multiplications in the best case.
- PUZZLE AUTHENTICITY VERIFICATION. Puzzle authenticity verification is quite cheap, requiring just a single call to the pseudo-random function H to verify the authentication tag Z.
- PUZZLE VERIFICATION. To verify correctness of a solution, the server has to perform only 3 modular multiplications.

Our puzzle construction dramatically reduces the puzzle verification cost incurred by the server and is the only modular exponentiation-based puzzle that does not require a big exponentiation to be performed by the server on-line. The efficiency of our puzzle is compared with the efficiency of Karame-Čapkun and Rivest et al.'s puzzles in Section 5.

After receiving the puzzle, the client finds the solution to the puzzle as follows:

- PUZZLE SOLVING. One typical method for a legitimate client to implement the FindSoln algorithm is to use square-and-multiply algorithm, which is the most commonly used algorithm for computing modular exponentiations. Upon receiving a puzzle puz from the server with an integer Q, the client computes y as $x^{2^Q} \bmod n$. We note however that a client could also choose to factor n first and then can solve the puzzle efficiently.

The best known method to solve our puzzle is to perform Q repeated squarings and this is an inherently sequential process [9,11,17]. Therefore a client needs to do exactly Q sequential modular multiplications to correctly solve the given puzzle, thereby achieving deterministic solving time property and non-parallelisability. We also get finer granularity as Q can be set to any positive integer regardless of the previously used difficulty values.

Table 2. Puzzle properties and operation counts for puzzle solving, generation and verification. Q is the difficulty level, n is an RSA modulus, k is a security parameter, and ℓ is a small integer.

PUZZLE	NON-PARALLELISABLE & DETERMINISTIC SOLVING TIME	FINER GRANULARITY	PUZZLE SOLVING COST	GENERATION COST	VERIFICATION COST
TLPuz [17]	Yes	Yes	Q mod. mul.	1 hash	1 hash $|n|$-bit mod. exp.
KCPuz [11]	Yes	No	$O(Q)$ mod. mul. $O(|n^2|)$ mod. mul.	2 HMAC (4 hash) 1 gcd	2 HMAC (4 hash) $2k$-bit mod. exp.
RSAPuz	Yes	Yes	Q mod. mul.	1 hash $2(\ell-1)$ mod. mul.	1 hash 3 mod. mul.

In Table 2, we compare the puzzle properties and asymptotic costs for FindSoln, GenPuz and VerSoln algorithms for the non-parallelisable puzzles examined in this paper. In particular, we compare the performance of the proposed puzzle RSAPuz with that of Rivest *et al.*'s time-lock puzzle (TLPuz) and Karame-Čapkun's variable exponent puzzle (KCPuz).

Remark 1. In RSAPuz as illustrated in Figure 2, the server requires a short-term secret X for verifying the puzzle solution. Storing X for each puzzle may introduce a memory-based DoS attack on the server. Fortunately, the server may avoid this type of attack by employing *stateless connections* [1] to offload storage of X to the client. That is, the server can use a long-term symmetric key s_k to encrypt X and send it along with each puzzle. Then the client has to send back this encrypted value while returning the solution to the puzzle. In this way, the server remains stateless and obtains X by decrypting the encrypted value using the same key s_k. With an efficient symmetric encryption algorithm, the server will not experience any significant computational burden.

Table 3. Timings for modular exponentiation-based puzzles. For RSAPuz, $N = 2500$ and $\ell = 4$.

Puzzle	512-bit modulus, $k = 56$				1024-bit modulus, $k = 80$			
	Setup (ms)	GenPuz (μs)	FindSoln (s)	VerSoln (μs)	Setup (ms)	GenPuz (μs)	FindSoln (s)	VerSoln (μs)
Difficulty: 1 million								
TLPuz [17]	13.92	4.80	1.54	474.68	56.10	4.86	4.13	2903.99
KCPuz [11]	11.52	8.37	1.59	263.35	42.30	8.66	4.27	895.17
RSAPuz	140.11	16.66	1.54	14.75	8510.92	35.15	4.29	29.24
Difficulty: 10 million								
TLPuz [17]	49.99	4.80	15.17	474.83	103.95	4.87	42.62	2917.25
KCPuz [11]	28.95	8.37	15.18	265.28	85.09	8.62	43.31	907.03
RSAPuz	1419.78	16.66	15.34	14.53	8669.75	34.72	43.08	28.97
Difficulty: 100 million								
TLPuz [17]	416.29	4.81	157.10	470.61	607.87	4.84	429.31	2924.01
KCPuz [11]	218.76	8.35	160.97	259.39	327.46	8.70	426.04	899.00
RSAPuz	1609.83	16.76	158.22	14.88	8966.74	34.76	422.58	29.18

4   Security Analysis of **RSAPuz**

We analyse the security properties of RSAPuz using the security model of Chen *et al.* [6].In particular, we show that RSAPuz satisfies the unforgeability and difficulty properties introduced by Chen *et al.* Since we use a secure pseudo-random function H in puzzle generation, proof of unforgeability for RSAPuz is quite straightforward. Due to lack of space, we omit the proof for unforgeability and give only the intuition behind the proof for difficulty.

4.1   Difficulty of **RSAPuz**

The time-lock puzzle was first proposed in 1996 and to date the best known method of solving the puzzle is sequential modular squaring, provided that factoring the modulus is more expensive. Indeed, it has been widely accepted that given a large RSA modulus n, the computation of a^{2^Q} mod n can be obtained by Q repeated squarings and no algorithm with better complexity than Q squarings is known [9,11,14,17].

Karame and Čapkun proved that their puzzle KCPuz is $\epsilon_{k,R}(t)$- difficult in the Chen *et al.* model, where

$$\epsilon_{k,R}(t) = \min\left\{\left\lfloor \frac{t}{\log R}\right\rfloor + O\left(\frac{1}{2^k}\right), 1\right\}$$

for all probabilistic algorithms $\mathcal{A}$ running in time at most t.

If a solver knows a multiple of $\phi(n)$, then it can compute $\phi(n)$ and the factors of n very efficiently [15]. Then the solver can efficiently compute x^R mod n by computing $c \leftarrow R$ mod $\phi(n)$ first and then computing x^c mod n. However, it is computationally infeasible for a client to compute a multiple of $\phi(n)$ from the transcripts of the puzzle scheme, so computing x^R mod n requires at least $O(\log R)$ modular multiplications. Hence, the success probability of solving the puzzle is bounded by $\epsilon_{k,R}(t)$ for any algorithm running in time at most t.

Detailed examination of Karame and Čapkun's proof reveals that they are essential making the assumption that the best approach for solving the time-lock puzzle is sequential modular squaring and multiplication. Moreover, their proof further makes the assumption that the time-lock puzzle is difficult in the Chen *et al.* model, in other words, when the adversary is allowed to see valid puzzle-solution pairs returned from the CreatePuzSoln query.

We show in the following theorem that our puzzle RSAPuz is difficult in the Chen *et al.* model [6] as long as Rivest *et al.*'s time-lock puzzle is difficult. Due to lack of space the proof of the theorem will appear in full version.

Theorem 2 (Difficulty of RSAPuz**).** *Let k be a security parameter and let Q be a difficulty parameter. Let GenRSA be a modulus generation algorithm. If TLPuz with GenRSA is $\epsilon_{k,Q}(t)$-difficult, then RSAPuz is $\epsilon'_{k,Q}(t)$-difficult for all probabilistic algorithms $\mathcal{A}$ running in time at most t, where*

$$\epsilon'_{k,Q}(t) = 2 \cdot \epsilon_{k,Q}\left(t + (q_\mathsf{C} + 1)\left(2(\ell - 1)T_\mathrm{Mul}\right) + c\right).$$

Here, q_C is the number of CreatePuzSoln *queries and T_{Mul} is the time complexity for computing a multiplication modulo n, and c is a constant.*

5   Performance Comparison

Table 3 presents timings from an implementation of these puzzle variants for both 512-bit and 1024-bit RSA moduli with $k = 56$ and $k = 80$, respectively, for puzzle difficulty levels 1 million, 10 million, and 100 million. The software was implemented using big integer arithmetic from OpenSSL 0.9.8ℓ and run on a single core of a 3.06 GHz Intel Core i3 with 4GB RAM, compiled using gcc -O2 with architecture x86_64.

In the 512-bit case, our puzzle reduces the solution verification time by approximately 32 times when compared to TLPuz and 17 times when compared to KCPuz. For the 1024-bit case, the gain in the verification time is approximately 99 times when compared to TLPuz and 30 times when compared to KCPuz.

Since VerSoln cost in RSAPuz is independent of k the security parameter, the verification gain increases as the size of RSA moduli increases. Note that, for both the moduli, the puzzle generation algorithm GenPuz is 2 to 7 times slower than the GenPuz in TLPuz and KCPuz. However, the cumulative puzzle generation and puzzle verification time in RSAPuz is still substantially less than in TLPuz or KCPuz. Furthermore, GenPuz cost in RSAPuz can still be improved by reducing ℓ from 4 to 2 and increasing the number N of pairs precomputed by BPVPre in the puzzle setup algorithm.

6   Conclusion

In this paper, we presented the most efficient non-parallelisable puzzle based on RSA. A DoS defending server needs to perform only $2(\ell - 1)$ modular multiplications online, where ℓ could be as low as 2, for a given RSA modulus. For the comparable difficulty level, the best known non-parallelisable puzzle requires a busy server perform online at least $2k$-bit modular exponentiation, where k is a security parameter.

We have also proved that the proposed puzzle satisfies the two security notions proposed by Chen et al. In particular, we have reduced the difficulty of solving our puzzle to the difficulty of solving Rivest et al.'s time-lock puzzle.

Experimental results show that our puzzle reduces the solution verification time by a factor of 99 when compared to Rivest et al.'s time-lock puzzle and a factor of 30 when compared to Karame and Čapkun puzzle, for 1024-bit moduli.

Acknowledgements. The authors are grateful to anonymous referees for their comments. This work is supported by Australia-India Strategic Research Fund project TA020002.

References

1. Aura, T., Nikander, P.: Stateless Connections. In: Han, Y., Okamoto, T., Qing, S. (eds.) ICICS 1997. LNCS, vol. 1334, pp. 87–97. Springer, Heidelberg (1997)
2. Aura, T., Nikander, P., Leiwo, J.: DOS-Resistant Authentication with Client Puzzles. In: Christianson, B., Crispo, B., Malcolm, J.A., Roe, M. (eds.) Security Protocols 2000. LNCS, vol. 2133, pp. 170–177. Springer, Heidelberg (2001)
3. Back, A.: Hashcash: A denial-of-service countermeasure (2002), http://www.hashcash.org/papers/hashcash.pdf
4. Boyko, V.: A pre-computation scheme for speeding up public-key cryptosystems. Master's thesis, Massachusetts Institute of Technology (1998), http://hdl.handle.net/1721.1/47493
5. Boyko, V., Peinado, M., Venkatesan, R.: Speeding up Discrete Log and Factoring Based Schemes via Precomputations. In: Nyberg, K. (ed.) EUROCRYPT 1998. LNCS, vol. 1403, pp. 221–235. Springer, Heidelberg (1998)
6. Chen, L., Morrissey, P., Smart, N.P., Warinschi, B.: Security Notions and Generic Constructions for Client Puzzles. In: Matsui, M. (ed.) ASIACRYPT 2009. LNCS, vol. 5912, pp. 505–523. Springer, Heidelberg (2009)
7. Dwork, C., Naor, M.: Pricing via Processing or Combatting Junk Mail. In: Brickell, E.F. (ed.) CRYPTO 1992. LNCS, vol. 740, pp. 139–147. Springer, Heidelberg (1993)
8. Feng, W., Kaiser, E., Luu, A.: Design and implementation of network puzzles. In: INFOCOM 2005, vol. 4, pp. 2372–2382. IEEE (2005)
9. Hofheinz, D., Unruh, D.: Comparing two notions of simulatability. In: Kilian [2], pp. 86–103
10. Juels, A., Brainard, J.: Client puzzles: A cryptographic countermeasure against connection depletion attacks. In: NDSS 1999, pp. 151–165. Internet Society (1999)
11. Karame, G., Čapkun, S.: Low-Cost Client Puzzles Based on Modular Exponentiation. In: Gritzalis, D., Preneel, B., Theoharidou, M. (eds.) ESORICS 2010. LNCS, vol. 6345, pp. 679–697. Springer, Heidelberg (2010)
12. Kilian, J. (ed.): TCC 2005. LNCS, vol. 3378. Springer, Heidelberg (2005)
13. Lenstra, A., Verheul, E.: Selecting cryptographic key sizes. J. Cryptology 14(4), 255–293 (2001)
14. Mao, W.: Timed-Release Cryptography. In: Vaudenay, S., Youssef, A.M. (eds.) SAC 2001. LNCS, vol. 2259, pp. 342–358. Springer, Heidelberg (2001)
15. Miller, G.L.: Riemann's hypothesis and tests for primality. In: STOC, pp. 234–239. ACM (1975)
16. Moore, D., Shannon, C., Brown, D.J., Voelker, G.M., Savage, S.: Inferring internet denial-of-service activity. ACM Transactions on Computer Systems (TOCS) 24(2), 115–139 (2006)
17. Rivest, R.L., Shamir, A., Wagner, D.A.: Time-lock puzzles and timed-release crypto. Technical Report TR-684, MIT Laboratory for Computer Science (March 1996)
18. Shparlinski, I.: On the uniformity of distribution of the RSA pairs. Mathematics of Computation 70(234), 801–808 (2001)

Implementing Information-Theoretically Secure Oblivious Transfer from Packet Reordering

Paolo Palmieri and Olivier Pereira

Université catholique de Louvain,
UCL Crypto Group,
Place du Levant 3, B-1348 Louvain-la-Neuve, Belgium
{paolo.palmieri,olivier.pereira}@uclouvain.be

Abstract. If we assume that adversaries have unlimited computational capabilities, secure computation between mutually distrusting players can not be achieved using an error-free communication medium. However, secure multi-party computation becomes possible when a noisy channel is available to the parties. For instance, the Binary Symmetric Channel (BSC) has been used to implement Oblivious Transfer (OT), a fundamental primitive in secure multi-party computation. Current research is aimed at designing protocols based on real-world noise sources, in order to make the actual use of information-theoretically secure computation a more realistic prospect for the future.

In this paper, we introduce a modified version of the recently proposed Binary Discrete-time Delaying Channel (BDDC), a noisy channel based on communication delays. We call our variant Reordering Channel (RC), and we show that it successfully models *packet reordering*, the common behavior of packet switching networks that results in the reordering of the packets in a stream during their transit over the network. We also show that the protocol implementing oblivious transfer on the BDDC can be adapted to the new channel by using a different sending strategy, and we provide a functioning implementation of this modified protocol. Finally, we present strong experimental evidence that reordering occurrences between two remote Internet hosts are enough for our construction to achieve statistical security against honest-but-curious adversaries.

Keywords: Oblivious transfer, secure multi-party computation, noisy channels, packet reordering, delay.

1 Introduction

When a source transmits information over a packet-switching network, it produces an in-order sequence of packets. However, depending on the network properties and the communication protocol used, the sequence received at the destination might be a different one. In this paper we show how this noise introduced by the network can be used in practice to achieve oblivious transfer and, more generally, secure computation.

H. Kim (Ed): ICISC 2011, LNCS 7259, pp. 332–345, 2012.
© Springer-Verlag Berlin Heidelberg 2012

At the network level of the ISO/OSI model, the Internet Protocol (IP) offers no guarantee that packets are received at destination in the same order in which they were sent at the source. This task is taken on by some of the protocols at the transport layer, most notably the Transmission Control Protocol (TCP), while others leave the problem unaddressed, as in the case of the User Datagram Protocol (UDP). The phenomenon for which the ordering of a sequence of packets in a stream is modified during its transit on a network is commonly known as *packet reordering*. Common causes of packet reordering are packet striping at the data-link and network layers [2,11], priority scheduling and route fluttering [16,3]. Reordering is a common behavior over the Internet. For instance, tests conducted in [1] for 50 hosts, 35 of which chosen randomly, show an occurrence rate of over 40%, with a mean reordering rate roughly fluctuating between 10 and 20% per occurrence. Current Internet trends, like increasing link speeds and increased parallelism within routers, wireless ad hoc routing, and the widespread use of quality of service (QoS) mechanisms and overlay routing, all indicate an expected increase in packet reordering occurrences.

For its ability to deteriorate the responsiveness and quality of data transmission, especially in applications featuring real time communication or streaming of multimedia content, packet reordering is generally treated as any other form of noise: a problem that needs to be solved. However, cryptographers have a history in transforming noise into something useful and desirable. It is the case of secure multi-party computation, that can be achieved only through the use of noisy channels when the adversaries are computationally unbounded. Multi-party computation deals with the problem of performing a shared task between two or more players who do not trust each other. Security is achieved when the privacy of each player's input and the correctness of the result are guaranteed [4]. A basic primitive and a fundamental building block for any secure computation is Oblivious Transfer (OT), introduced by Rabin in 1981 [19]. In fact, when oblivious transfer is available, any two-party computation can be implemented in a secure way [13]. A commonly used variant of the primitive is the 1-out-of-2 oblivious transfer, proposed by Even, Goldreich and Lempel [10], and later proved to be equivalent to the original OT by Crépeau [5]. In this protocol, a sender Sam knows two secrets, and is interested in transmitting one to a receiver Rachel without disclosing the other. Rachel wants to choose which secret to receive, but does not want to reveal her choice. Privacy of the inputs and correctness of the result are achieved without implying any degree of mutual trust between the players.

The first protocol to implement oblivious transfer over a noisy channel used the well-known Binary Symmetric Channel (BSC) [6]. The BSC is a simple binary channel that flips with probability p each bit passing trough it. While being a common reference in information theory, the BSC proved not to satisfy cryptographers, more interested in modeling advantages a potential adversary might have. Many modifications of the channel were consequently proposed, in the direction of allowing dishonest players an edge over honest ones. An Unfair Noisy Channel (UNC) let the adversary choose the crossover probability within a specific (narrow) range [9,8], while the Weak Binary Symmetric Channel (WBSC) introduces the possibility for the dishonest player to know with a certain probability if a bit was received correctly [20]. The aim of these constructions is to gain generality by easing the security assumptions, but, while we know that OT can be built on any non-trivial noisy channel [7], none of these proved to be suitable for actual implementation in a real world communication scenario, due to the lack of flexibility and strong requirements imposed by the channel model.

A different approach was taken in [15], where the proposed oblivious transfer protocol uses a different source of noise: communication delays. The protocol is built over a new noisy channel model, called Binary Discrete-time Delaying Channel (BDDC), that accepts binary string inputs at discrete times, and output each string at the following discrete time. Strings passing through the channel have a probability p of being delayed, and therefore being kept in the channel until the following output time.

1.1   Contribution

In this paper we present a new channel model, the Reordering Channel (RC). The reordering channel is a modified version of the BDDC, that modifies the concept of delay from a temporal one to that of shifting positions in a sequence. We observe that the RC provides enough ambiguity (noise) for building an oblivious transfer protocol, and we show that using a different strategy for sending packets at the sender's end we are able to build oblivious transfer on the channel using a modified version of the same protocol used on the BDDC. Since the reordering channel models the behavior of packet reordering over the Internet, we provide an actual, functioning implementation of the protocol based on the transmission of UDP packets. The source code of the application is provided.[1] Finally, we present strong experimental evidence supporting the effectiveness and security of the construction and we show how different specific packet reordering behaviors can be used to improve the efficiency of the protocol.

To the best of our knowledge, this is the first actual implementation of oblivious transfer over the Internet that provides security against computationally unbounded adversaries, that is, adversaries with unlimited computational capabilities.

1.2   Outline of the Paper

In section 2 we introduce some preliminary notions and definitions relative to oblivious transfer that will be useful in the following. The definition of BDDC and the protocol implementing OT over it are also presented. In section 3 we discuss the implementation of oblivious transfer from packet reordering. We initially introduce the reordering channel and we show how the OT protocol for the BDDC can be modified to work on the new channel. In Section 3.1 we discuss

---

[1] The source code of the latest version of the program is available for download over the internet at the address: http://www.uclouvain.be/crypto/ifyd-latest.tar.gz.

common reordering behaviors that influence the design of our implementation of the protocol, which is presented in Section 3.2. In Section 3.3 we introduce some metrics used to analyze the data gathered during the testing of our application, of which we relate in Section 3.4, where we present statistical evidence of the security of our construction.

2   Preliminaries

Our construction is based on the chosen 1-out-of-2 binary oblivious transfer (in the following simply called oblivious transfer). A protocol implements oblivious transfer in a secure fashion when three conditions are satisfied after a successful execution: the receiver party learns the value of the selected bit b_s (correctness); the receiver party gains no further information about the value of the other bit b_{1-s} (security for Sam); the sender party learns nothing about the value of the selection bit s (security for Rachel) [6].

The behavior of the players defines the level of security that a protocol can achieve. The oblivious transfer protocol for the BDDC is secure against *honest-but-curious* players. In practice, the player strictly follows the protocol but tries to gain extra information from her inputs and output, in order to gain an advantage in guessing the other player's secret.

2.1   Binary Discrete-Time Delaying Channel

A protocol for achieving oblivious transfer from communication delays in the information-theoretic model has been proposed in [15]. The noisy channel used to model delay is the Binary Discrete-time Delaying Channel (Figure 1).

Definition 1. [15] *A Binary Discrete-time Delaying Channel with delaying probability p consists of: an input alphabet $\{0,1\}^n$, an output alphabet $\{0,1\}^n$, a set of consecutive input times $T = \{t_0, t_1, \ldots\} \subseteq \mathbb{N}$, a set of consecutive output times $U = \{u_0, u_1, \ldots\} \subseteq \mathbb{N}$ where $\forall u_i \in U, t_i \in T, u_i \geq t_i$. Each input admitted into the channel at input time $t_i \in T$ is output once by the channel, with probability of being output at time $u_j \in U$*

$$\Pr[u_j] = p^{(j-i)} - p^{(j-i+1)} . \tag{1}$$

2.2   Oblivious Transfer over a BDDC

The following protocol, also proposed in [15], implements oblivious transfer over a BDDC with error probability p. The sender party, Sam, inputs two secret bits b_0, b_1 and gets no output; the receiver Rachel inputs the selection bit s and receives output b_s. In the following, we introduce a modified version of this protocol, which serves as the base for our construction over packet reordering.

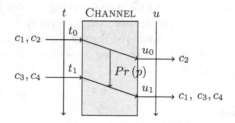

Fig. 1. A schematization representing a Binary Discrete-time Delaying Channel accepting two strings at time t_0, one of which gets delayed once, and two at time t_1, none of which gets delayed. This results in the channel emitting one string at time u_0 and three at u_1.

Protocol 1. [15] *Before starting the communication between the parties, Sam selects two disjoints sets E, E' each composed of n distinct binary strings of length l: $e_1, \ldots, e_n$ and $e'_1, \ldots, e'_n$. Then, Sam builds the sets $C = \{c_1, \ldots, c_n\}$ and $C' = \{c'_1, \ldots, c'_n\}$, where $c_i := e_i \| i$ and $c'_i := e'_i \| i$.*

1. *Sam sends to Rachel the set C at instant t_0, and C' at t_1, using a p-BDDC.*
2. *Rachel receives over the BDDC the strings in $\{C \cup C'\}$, in the order produced by the channel. She keeps listening on the channel at instants $u_2, u_3, \ldots$ until all the delayed strings have been received.* [2]
3. *Rachel selects the set I_s, where $s \in \{0, 1\}$ is her selection bit, such that $|I_s| = \frac{n}{2}$ and so that $i \in I_s$ only if $c_i \in C$ has been received at u_0. Then she selects $I_{1-s} := \{1, \ldots, n\} \setminus I_s$ and sends I_0 and I_1 to Sam over a clear channel.* [3]
4. *Sam receives I_0, I_1 and chooses two universal hash functions f^0 and f^1, whose output is 1-bit long for any input. Let $E_j \subset E$ be the set containing every $e_i \in E$ corresponding to an $i \in I_j$, such that*

$$e_i \in E_j \Leftrightarrow i \in I_j \ . \tag{2}$$

For each set I_j, Sam computes the string g_j by concatenating each $e^j_k \in E_j$, ordering them for increasing binary value, so that

$$g_j = \left(e^j_1 \| \ldots \| e^j_{\frac{n}{2}} \right) \quad \text{with } e^j_1, \ldots, e^j_{\frac{n}{2}} \in E_j \ . \tag{3}$$

Sam computes $h_0 = f^0(g_0)$, $h_1 = f^1(g_1)$ and sends to Rachel the functions f^0, f^1 and the two values

$$i_0 = (h_0 \oplus b_0) \ , \qquad i_1 = (h_1 \oplus b_1) \ . \tag{4}$$

---

[2] If less than $\frac{n}{2}$ strings are received at u_0 Rachel instructs Sam to abort the communication.

[3] Or Rachel can just send one of these two sets in order to save bandwidth as Sam can easily reconstruct the other.

5. *Rachel computes her guess for* b_s

$$b_s = f^s(g_s) \oplus i_s \; . \tag{5}$$

The internal working of the protocol is easily explained: when listening on the BDDC, Rachel receives at instant u_0 all the strings in C that have not been delayed by the channel. This subset will constitute the correct information she needs to decode the selected bit. Any string received at u_1 or at a later time is instead ambiguous, and guarantees that Rachel can not decode both b_0 and b_1. At u_1, the strings from C delayed once and the strings of set C' that have not been delayed can not be distinguished, and so on and so forth for $u_2, u_3, \ldots$.

3   Packet Reordering as a Noisy Channel

While the BDDC is able to model discrete delays in a communication, it is not suitable to simulate the delaying behavior of packet switching networks, which is usually visible in the form of packet reordering. Therefore, we introduce a new channel model, called Reordering Channel (RC), that redefines the concept of delay using the relative position of a packet in a stream.

Definition 2. *A* Reordering Channel *consists of: an input sequence of binary strings* $T = (t_1, \ldots, t_n)$, *an output sequence of binary strings* $U = (u_1, \ldots, u_n)$, *a sequence of identically distributed discrete random variables* $\{X_n\}$ *over* $\mathbb{N}$ *and its probability distribution* $P_X : \mathbb{N} \to [0,1]$, *with* $\sum_{m \in \mathbb{N}} P_X(n) = 1$. *Each string in* T *is output once by the channel in* U, *i.e.* $\{t_1, \ldots, t_n\} \equiv \{u_1, \ldots, u_n\}$. *The ordering of the output sequence is determined by the channel, that selects for the next available position in the output sequence the* $t_i \in T$ *not already selected with the smallest value* $v = i + X_i$. *In case more than one string shares the same value* v, *the channel selects among them the one with the smallest value* i.

In practice, the channel takes a stream of packets as input, and outputs the same packets in a reordered fashion. For an appropriate distribution function, where the probability of a packet not to be delayed ($X = 0$) is high enough, this channel simulates accurately the reordering behavior of standard Internet connections. We discuss experimental results regarding the amount of reordering that is observed on the Internet in Section 3.4.

With an appropriate discrete probability distribution, that is, a distribution that respects (1) for a value p, the delaying behavior of the reordering channel follows that of the corresponding BDDC with probability p. However, the reordering channel, due to its continuous nature, opposed to the discrete one of the BDDC, lacks the reference points in time that are needed by the receiver to make sure that a string has not been delayed. Therefore, the protocol that implements oblivious transfer over the BDDC will not work on the RC. To adapt the protocol to the new channel, we need to use a different sending strategy at the sender's end. Instead of sending the two sets C and C' from step 1 of the protocol into the reordering channel sequentially, we can send them as a stream,

by interleaving the strings in the two sets. We observe, in fact, that we obtain
an ambiguity similar to the one of two strings output at the same time by the
BDDC, if two strings with the same value happen to be received consecutively as
a result of reordering by the RC. As illustrated in Figure 2, we can start sending
a first batch of i strings form C, where i is the arbitrarily selected interleaving
value. After c_i, we interleave the strings from C with the ones from C'. Using
this sending sequence, the receiver is unable to distinguish between two strings
c_i and c'_i when c_i is reordered, and received at least $i - 1$ positions far from its
original place. Adopting this sending strategy, we can implement the oblivious
transfer protocol for the BDDC on the reordering channel and, therefore, on the
Internet.

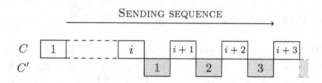

Fig. 2. Package interleaving

3.1   Reordering Dynamics

For an implementation of the protocol to be effective, the selected value of i must
be consistent with the actual amount of reordering observed over the Internet.
However, it is not the only parameter that will affect reordering.

The probability of occurrence of packet reordering depends on a number of
factors, such as physical distance and number of intermediate hops between the
hosts, transmission medium, quality and speed of hop-to-hop links, traffic on
the network and so on. Packet reordering also frequently displays a consistent
behavior over time between two given hosts.

As already experimentally observed by Bellardo and Savage, the inter-spacing
of packets effectively reduces the reordering probability [1]. In the test they
conducted, the probability is significantly reduced when adding an inter-packet
gap of 100 microseconds (μs), while a longer spacing of 500 μs brings the number
of reorderings close to 0. An increase in the size of the packets has the same effect,
since the longer serialization delay increases the delay between the leading edge of
each subsequent packet. This, in turn, decreases the possibility that two packets
will be reordered if assigned to different queues, when subject to parallelization
during routing. We can actively use this property by adding an inter-packet gap
to stabilize a path affected by a high reordering probability.

3.2   Protocol Implementation

In our implementation of the protocol, the receiver acts as a server, waiting online for a client (the sender) to connect. The protocol used to transmit packets
is the User Datagram Protocol (UDP). UDP provides no guarantees of message

delivery to the upper layer, offers no reordering detection or correction mechanism, and retains no state of the messages once sent. The simple structure of a packet, called datagram and defined in RFC 768, minimizes the size and does not include any sequence number [18]. The structure of the packets sent follows the structure of the strings in C and C' as defined in Protocol 1. Each packet p_j is composed of the sequence number j and an unique identification value e.

To select the receiver mode of operation, the option -r {S} must be specified, where S is the selection bit. The receiver algorithm is structured as follows. After network initialization, the program waits for incoming connections on port 9930. Once packets are received, they are put in the arrival order in a buffer. The number of packets to be received is determined in advance with the sender. The buffer is then read packet by packet, and each pair of packets sharing the same sequence number j and satisfying any of the following conditions is marked as ambiguous: the first packet with sequence number j is found in a position higher than $(j + i - 1)$; the two packets sharing sequence number j are less than $\frac{i}{2}$ position apart. Sets I_s and I_{1-s} are created by putting all the sequence numbers corresponding to an ambiguous pair of packets in I_{1-s}, plus enough non-ambiguous sequence numbers to reach half the total, and putting the remaining values in I_s. The two sets are then sent back to the sender's address. The software then waits for the encoded bits, and the chosen bit b_s is decoded using (5).

The sender mode of operation is selected by using the option -s {B0:B1}. The two secret bits are passed in the argument, separated by a colon character. The mandatory option -a {IP_ADDRESS} is used to specify the receiver's IP address. -w USEC can be used to add a USEC microseconds long waiting gap between packets. The algorithms is structured as follows. Two sets P^0 and P^1 of n packets each are created, with each packet p composed of an increasing sequence number j and a randomly selected unique identifier e. The two sets are then sent to the receiver's address, using the sending sequence described in Section 3 for a predetermined i. Once all the packets have been sent, the software starts waiting for the sets I_0, I_1 from the receiver. Using the information received, the secret bits are encoded according to (4), using an hash function, and sent to the receiver.

The reference platform for our implementation of the protocol is Linux. The programming language used is C++. Only standard POSIX libraries have been used. The compiler of choice is the GNU Compiler Collection (gcc), version 4. Full logging capabilities are implemented, including on-screen and file logging.

3.3   Metrics

In order to measure the incidence and relevance of the reorderings observed in our tests, we use the metrics proposed in RFC 5236 [12] and in [21], adapting them to the needs of our specific application when necessary.

When packets are received at the destination they are assigned a *receive index* (RI), according to the order of arrival. *Displacement* (D) of a packet is defined as the difference between RI and the sequence number of the packet. For example, the displacement of packet p_j^0 from the first set is $RI\left(p_j^0\right) - j$, while $D\left(p_j^1\right) = RI\left(p_j^1\right) + i - j$. Therefore, a negative displacement value indicates the earliness of a packet and a positive value the lateness. We call *absolute displacement* the modulus of D. The *displacement frequency* $FD(k)$ is the number of received packets having a displacement of k. The *reorder density* (RD) is the distribution of the displacement frequencies, normalized with respect to the number of received packets, ignoring lost and duplicate packets. The *mean displacement of packets* (M_D) is defined as

$$M_D = \left|\sum_{i=-D_r}^{i=+D_r} (|i| \times RD\,[i])\right| / \left|\sum_{i=-D_r}^{i=+D_r} RD\,[i]\right| , \tag{6}$$

while the *mean displacement of late packets* (M_L) is

$$M_L = \left|\sum_{i=1}^{i=+D_r} (|i| \times RD\,[i])\right| / \left[\sum_{i=1}^{i=+D_r} RD\,[i]\right] \tag{7}$$

in the case of packets with positive displacement. The *reorder entropy* (E_R) is an indicator of the reorder density (a discrete probability distribution) to be concentrated or dispersed. It is defined as $E_R = (-1) \times \sum_{i=-D_r}^{i=+D_r} (RD\,[i] \times \ln RD\,[i])$.

For simplicity, and without loss of generality, in the following we study only absolute displacements. In fact, for our purposes, displacements values of $\pm i$ are equally ambiguous.

3.4 Experiment

Contrary to what is common in the study of packet reordering, our experiment uses a an active approach, instead of passively monitoring traffic. We do so by using the testing capabilities included in our protocol implementation, which let us produce a stream of UDP datagrams from the sender to the receiver. For testing the software we use two hosts:

- merlin.dice.dice.ac.ucl.be, IP 130.104.205.236, located in Belgium. Debian GNU/Linux (kernel 2.6.32-5, i686);
- ec2-50-18-108-9.us-west-1.compute.amazonaws.com, IP 50.18.108.9, located in Northern California (USA). Ubuntu GNU/Linux (kernel 2.6.35, x86 64-bit).

Both hosts are connected to the Internet through wired, high-speed links. A sample tracert shows 18 hops between the two. The mean round trip time (RTT) is 149.6 milliseconds, with a standard deviation of 0.6 ms.

In the following, we base our analysis on two sample traffic datasets, produced by observing the behavior of our protocol implementation in two different settings.

The dataset corresponding to the first experiment is the result of a single prolonged execution of the protocol test routine. In total, 60167 datagrams are recorded. The test session took place in May, 2011. The aim of the test is to observe behavior under ideal traffic conditions: during the execution, both hosts had no network activity beside the traffic generated by our protocol implementation itself. A first analysis of the data shows a total of 7009 reordering occurrences, equal to 11.65% of all packets received. The maximum displacement value observed is 59. Detailed figures for low displacement values are displayed in Fig. 3. The mean displacement is $M_D = 0.44$, while the mean displacement of late packets is $M_L = 3.75$. The reorder entropy of the set is $E_R = 0.60$, which shows a good variance and therefore a uniformity of displacement frequency and probability. These values appear to be consistent with those observed in [21].

D	FD	$\%$	D	FD	$\%$
0	53157	88.35	6	246	0.41
1	1876	3.12	7	137	0.23
2	1697	2.82	8	79	0.13
3	1240	2.06	9	59	0.10
4	860	1.43	10+	347	0.58
5	468	0.78			

Fig. 3. Number of occurrences and percentage for each absolute displacement value

The security of our construction is based on the assumption that the sender can not accurately predict reorderings that will happen during the datagrams transit over the network. Predictions of future reorderings are based on the observation of past occurrences, and the research of patterns in the frequency and magnitude of displacements. In order to show the independence between past reordering occurrences and future ones, we evaluate the *autocorrelation* function. Autocorrelation is the cross correlation of a vector of random variables with itself, and is a useful tool frequently used in signal processing to find repeating patterns. In Fig. 4 the number of reordering occurrences for three different subset into which the first dataset has been divided for a more detailed analysis are provided, along with the relative autocorrelation. It is evident that none of the functions shows any recurring pattern.

The second experiment aims to reproduce reordering behavior under intense traffic over the network. In order to generate traffic we use the software suite composed of RUDE (Real-time UDP Data Emitter) and CRUDE (Collector for RUDE), developed by Laine and Saaristo [14]. This client-server tool allows us to produce UDP traffic from the sender to the receiver with a great deal of precision. In particular, we use the following routine:

- for the first 5 seconds, RUDE sends 1000 packets per second, with a packet size of size 200 bytes, generating a traffic of 200 KB/s;
- for the following 5 seconds, RUDE sends 10000 packets/second with 20 bytes/packet (200 KB/s);
- for the last 5 seconds, RUDE sends 500 packets/second with 2000 bytes/packet (1 MB/s);

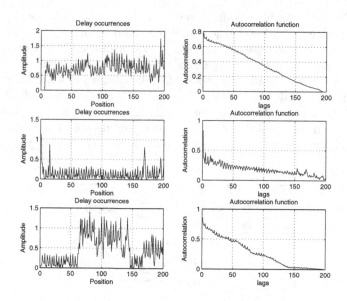

Fig. 4. The number of occurrences (left) and the autocorrelation function (right) for the three subsets composing the first dataset

- for all the duration of the test, RUDE also sends 2500 packets/second with 100 bytes/packet (250 KB/s). These last datagrams have the Type of Service (ToS) priority flag of the IPv4 header set to LOW_DELAY (0x10).

In addition to the traffic generated by RUDE, the receiver host also performs a large file download from a remote host (mirror.garr.it, 131.175.1.35), using the HTTP protocol. The file chosen is large enough for the download to last over the entire execution of the experiment. The average download speed observed is 96.5 KB/s.

During the experiment, a total of 4293 datagrams generated by our protocol implementation are received by the receiver host. About 4 seconds after the testing routine start, a peak of reorderings is recorded (Fig. 5): this is due to a set of packets being lost due to the traffic congestion. The particular routine of external traffic generated with RUDE allows us to analyze how the frequency of reordering occurrences is affected by manipulation of the traffic reaching the receiver host. Fig. 5 clearly shows that the reordering behavior is only marginally affected by external traffic, even when the capacity of the network is almost fully used. In fact, no significant differences both in the number of occurrences and mean displacement value can be seen at the change of bandwidth used at second 10, where the external traffic goes from 200 KB/s to 1 MB/s. The lower values recorded for the first 4 seconds can be instead explained by the progressive increase in TCP traffic generated by the HTTP download, that take full advantage of the available bandwidth only after a few seconds, thanks to the mechanisms regulating TCP traffic.

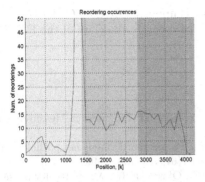 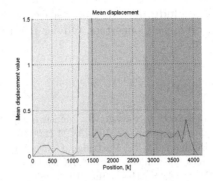

Fig. 5. The number of occurrences (left) and the mean displacement value (right) for the second dataset, calculated over subsets of 100 packets each. The peak of both functions is due to the loss of a set of packets. The three areas colored with different shades of gray picture the three time lapses into which the set is divided (5 seconds each)

4   Conclusion

In this paper we propose an implementation of oblivious transfer over the Internet. The construction we propose is secure against adversaries with unlimited computational capabilities in the honest-but-curious model, and uses packet reordering, a common phenomenon present in any packet-switching network. Reordering of packets in a stream are due to a number of different causes, among which intrinsic network characteristics, parallelism both at the node and the link level, and traffic control and congestion, all of which are increasingly present in today's Internet.

Our construction is based on the protocol proposed for the Binary Discrete-time Delaying Channel. In order to adapt the protocol for practical use over the Internet, we introduce a new channel, the Reordering Channel, that models packet reordering. We then build a modified version of the protocol, adapted to work on the RC, and we present a practical implementation of this modified protocol based on the transmission of UPD packets over the Internet. We also present extensive experimental evidence of the security of the implementation: we show that reordering occurrences are found consistently under both intense traffic load and minimal network usage, statistical analysis of the reorderings shows no sign of recurring patterns in frequency or magnitude and no strong statistical correlation is found over reordering occurrences over time.

To the best of our knowledge, our implementation is the first oblivious transfer protocol secure against computationally unbounded capabilities being implemented over the Internet. The novelty of our construction, based on network behavior, opens the way to new security constructions entirely based on channel characteristics.

Acknowledgments. This research work was supported by the SCOOP Action de Recherche Concertées. Olivier Pereira is a Research Associate of the F.R.S.-FNRS.

References

1. Bellardo, J., Savage, S.: Measuring packet reordering. In: Internet Measurement Workshop, pp. 97–105. ACM (2002)
2. Bennett, J.C.R., Partridge, C., Shectman, N.: Packet reordering is not pathological network behavior. IEEE/ACM Trans. Netw. 7(6), 789–798 (1999)
3. Bohacek, S., Hespanha, J.P., Lee, J., Lim, C., Obraczka, K.: A new tcp for persistent packet reordering. IEEE/ACM Trans. Netw. 14(2), 369–382 (2006)
4. Chaum, D., Damgård, I., van de Graaf, J.: Multiparty computations ensuring privacy of each party's input and correctness of the result. In: Pomerance [17], pp. 87–119
5. Crépeau, C.: Equivalence between two flavours of oblivious transfers. In: Pomerance [17], pp. 350–354
6. Crépeau, C., Kilian, J.: Achieving oblivious transfer using weakened security assumptions (extended abstract). In: FOCS, pp. 42–52. IEEE (1988)
7. Crépeau, C., Morozov, K., Wolf, S.: Efficient Unconditional Oblivious Transfer from Almost Any Noisy Channel. In: Blundo, C., Cimato, S. (eds.) SCN 2004. LNCS, vol. 3352, pp. 47–59. Springer, Heidelberg (2005)
8. Damgård, I., Fehr, S., Morozov, K., Salvail, L.: Unfair Noisy Channels and Oblivious Transfer. In: Naor, M. (ed.) TCC 2004. LNCS, vol. 2951, pp. 355–373. Springer, Heidelberg (2004)
9. Damgård, I.B., Kilian, J., Salvail, L.: On the (Im)possibility of Basing Oblivious Transfer and Bit Commitment on Weakened Security Assumptions. In: Stern, J. (ed.) EUROCRYPT 1999. LNCS, vol. 1592, pp. 56–73. Springer, Heidelberg (1999)
10. Even, S., Goldreich, O., Lempel, A.: A randomized protocol for signing contracts. Commun. ACM 28(6), 637–647 (1985)
11. Jaiswal, S., Iannaccone, G., Diot, C., Kurose, J., Towsley, D.: Measurement and classification of out-of-sequence packets in a tier-1 ip backbone. IEEE/ACM Trans. Netw. 15, 54–66 (2007), http://dx.doi.org/10.1109/TNET.2006.890117
12. Jayasumana, A., Piratla, N., Banka, T., Bare, A., Whitner, R.: Improved packet reordering metrics. RFC 5236 (Informational) (June 2008), http://www.ietf.org/rfc/rfc5236.txt
13. Kilian, J.: Founding cryptography on oblivious transfer. In: STOC, pp. 20–31. ACM (1988)
14. Laine, J., Saaristo, S.: RUDE: Real-time UDP data emitter (1999–2002), http://rude.sourceforge.net/
15. Palmieri, P., Pereira, O.: Building Oblivious Transfer on Channel Delays. In: Lai, X., Yung, M., Lin, D. (eds.) Inscrypt 2010. LNCS, vol. 6584, pp. 125–138. Springer, Heidelberg (2011)
16. Paxson, V.E.: Measurements and Analysis of End-to-End Internet Dynamics. Ph.D. thesis, EECS Department, University of California, Berkeley (June 1997), http://www.eecs.berkeley.edu/Pubs/TechRpts/1997/5498.html
17. Pomerance, C. (ed.): CRYPTO 1987. LNCS, vol. 293. Springer, Heidelberg (1988)
18. Postel, J.: User datagram protocol. RFC 768 (Standard) (August 1980), http://www.ietf.org/rfc/rfc768.txt

19. Rabin, M.O.: How to exchange secrets by oblivious transfer. Technical Report TR-81, Aiken Computation Laboratory, Harvard University (1981) (manuscript)
20. Wullschleger, J.: Oblivious Transfer from Weak Noisy Channels. In: Reingold, O. (ed.) TCC 2009. LNCS, vol. 5444, pp. 332–349. Springer, Heidelberg (2009)
21. Ye, B., Jayasumana, A.P., Piratla, N.M.: On monitoring of end-to-end packet reordering over the internet. In: International Conference on Networking and Services (2006)

Compression Functions Using a Dedicated Blockcipher for Lightweight Hashing

Shoichi Hirose[1], Hidenori Kuwakado[2], and Hirotaka Yoshida[3,4]

[1] Graduate School of Engineering, University of Fukui
[2] Graduate School of Engineering, Kobe University
[3] Yokohama Research Laboratory, Hitachi, Ltd.
[4] Department of Electrical Engineering ESAT/SCD-COSIC,
Katholieke Universiteit Leuven

Abstract. This article presents a model of compression functions using a blockcipher for lightweight hashing on memory-constrained devices. The novelty of the proposed model is that the key length of the underlying blockcipher is half of its block length, which enables the reduction of the size of the internal state without sacrificing the security. Security of iterated hash functions composed of compression functions in the model is also discussed. First, their collision resistance and preimage resistance are quantified in the ideal cipher model. Then, a keyed hashing mode is defined, and its security as a pseudorandom function is reduced to the security of the underlying blockcipher as a pseudorandom permutation. The analysis supports the security of Lesamnta-LW, which is a lightweight hash function proposed in ICISC 2010. Finally, preimage resistance is quantified assuming a computationally secure blockcipher.

Keywords: hash function, blockcipher, collision resistance, preimage resistance, pseudorandom function, pseudorandom permutation.

1 Introduction

Background and Motivation. Secure communication in resource-constrained environments such as RFID tags and sensor networks is getting one of the important research topics. To achieve secure communication, both of confidentiality and authentication are important. A hash function is a useful primitive for them. This article considers hash functions suitable for implementation on memory-constrained devices. The construction of a hash function is roughly classified into blockcipher-based, sponge-function-based and stream-cipher-based. Among them, the blockcipher-based construction has been extensively studied. However, this article reconsiders this construction in terms of lightweight implementation on memory-constrained devices.

The most well-known blockcipher-based compression functions are Davies-Meyer (DM), Matyas-Meyer-Oseas (MMO), and Miyaguchi-Preneel (MP) (Fig. 1). All of them have feedforwards of inputs which make them one-way. On the other hand, since feedforwards require extra memory and make their internal

H. Kim (Ed): ICISC 2011, LNCS 7259, pp. 346–364, 2012.
© Springer-Verlag Berlin Heidelberg 2012

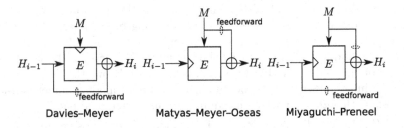

Fig. 1. Compression functions including the feedforward

states larger, they are not preferable to memory-constrained devices. Hence, this article focuses on blockcipher-based compression functions without feedforwards. Actually, we already know some blockcipher-based compression functions without feedforwards such as Rabin's scheme in the PGV model [15]. In this model, it is assumed that the key length of the underlying blockcipher is equal to its block length. On the other hand, if we shorten the key, we can further reduce the size of the internal state. We only consider the case where the key length is half of the plaintext length because this case gives a very good tradeoff between the internal state size and the query complexity of collision attacks. In fact, the lightweight hash function using such a compression function, *Lesamnta-LW* [9], was proposed in ICISC 2010. The hardware implementation of Lesamnta-LW requires only 8.24 Kgates. Its software implementation requires only 50 bytes of RAM and runs fast on short messages on 8-bit CPUs.

Our Contribution. This article first presents a model of compression functions using a blockcipher for lightweight hashing. It is similar to the PGV model [15]. The novelty of the proposed model is that it uses a dedicated blockcipher whose key length is half of its block length, which helps the reduction of the size of the internal state without sacrificing the security. We will explain that this key length is good for achieving collision resistance. This article focuses on showing the validity of the new blockcipher-based approach for hashing on memory-constrained devices and how to design an underlying blockcipher itself is out of scope. Security of iterated hash functions using compression functions in the proposed model is also discussed. First, their collision resistance and preimage resistance are quantified in the ideal cipher model. Their preimage resistance is as large as their collision resistance since the compression functions in the model are invertible and collision attacks also work for finding a preimage. We will see that some of them have optimal collision resistance and others have nearly optimal collision resistance. As for the near optimality, the degradation is at most an inverse factor of the output length, and it is reasonably small. This analysis includes the security of Lesamnta-LW.

Second, a keyed hashing mode (a keyed-via-IV mode) is defined, and its security as a pseudorandom function (PRF) is reduced to the security of the underlying blockcipher as a pseudorandom permutation (PRP). Some of the compression functions require their blockcipher to be secure against related-key differential attacks. Notice that, for Rabin's scheme, such reduction is impossible since only a message block is fed into the key of the underlying blockcipher.

Finally, their preimage resistance is also analyzed in a computationally secure blockcipher model. It is reduced to the pseudorandomness or the key-finding hardness of the underlying blockcipher. As far as we know, this is the first result on preimage resistance of blockcipher-based hash functions under non-ideal assumptions though the reduction requires an unconventional padding.

Related Work. The PGV model [15] is a model of compression functions using a blockcipher whose key length equals its block length. In the same paper, their security against several generic attacks are intensively discussed. Collision resistance and preimage resistance of the PGV schemes are formally analyzed in the ideal cipher model [4]. Stam [16] recently presented a model of compression functions, which includes the PGV model. He also discussed the collision resistance and preimage resistance of iterated hash functions using compression functions in the model. Actually, his model also includes our proposed model. However, we treat compression functions not treated in [16], and we further discuss the security of a keyed hashing mode as a PRF as well as collision resistance and preimage resistance. The security as a PRF of the keyed hashing mode of a hash function using the MMO compression function and the MDP domain extension [11] is almost reduced to the security of the underlying blockcipher as a PRP [10].

There are few proposals for lightweight hash functions and their building blocks compared to block/stream ciphers. MAME [17] is a dedicated lightweight compression function. Bogdanov et al. [7] discussed hardware implementations of hash functions using AES or a lightweight blockcipher PRESENT [6]. The wide pipe construction [12] is quite popular among candidates of the NIST SHA-3 competition, but, it does not seem suitable for lightweight hashing. QUARK [1], PHOTON [8] and SPONGENT [5] are recently proposed algorithms for lightweight hashing. They adopt the sponge construction [3]. To achieve the same levels of security, a blockcipher-based construction generally needs a larger internal state than a sponge construction. As far as the authors know, however, an advantage of the former approach seems that the security of keyed hashing modes as a PRF can be reduced to a weaker and standard security assumption on the underlying blockcipher (PRP).

Organization. Section 2 gives definitions of security properties discussed in this article, and presents the model of compression functions. Collision resistance and preimage resistance are discussed in the ideal cipher model in Section 3. The security of the keyed hashing mode is discussed in Sections 4. Preimage resistance is discussed assuming a computationally secure blockcipher in Section 5.

2   Preliminaries

2.1   Definitions

For a set S, let $S^{\leq \ell} = \bigcup_{i=0}^{\ell} S^i$ and $S^+ = \bigcup_{i=1}^{\infty} S^i$. The number of elements in S is denoted by $\#S$. Let $s \xleftarrow{\$} S$ represent that an element s is selected from S under the uniform distribution. For sequences x and y, let $x\|y$ be their concatenation. Let $\mathcal{F}(\mathcal{X},\mathcal{Y})$ be the set of all functions from $\mathcal{X}$ to $\mathcal{Y}$. Let $\mathcal{P}(\mathcal{X})$ be the set of all permutations on $\mathcal{X}$. Let $\mathcal{BC}(\kappa,n)$ be the set of all (κ,n) blockciphers, where κ and n represent their key size and block size, respectively. Let H^E be a hash function using a blockcipher E.

Collision Resistance. The collision resistance of a blockcipher-based hash function is often discussed in the ideal cipher model. We follow this convention. In the ideal cipher model, the underlying blockcipher E is assumed to be uniformly distributed over $\mathcal{BC}(\kappa,n)$. An encryption/decryption operation is an encryption/decryption query to the oracle E. Let A be an adversary trying to find a collision for H^E, that is, a pair of distinct inputs mapped to the same output by H^E. The col-advantage of A against H^E is given by

$$\mathbf{Adv}^{\mathrm{col}}_{H^E}(A) = \Pr[A^E = (M,M') \wedge H^E(M) = H^E(M') \wedge M \neq M'] \ ,$$

where E is uniformly distributed over $\mathcal{BC}(\kappa,n)$. It is assumed that A makes all the queries necessary to compute $H^E(M)$ and $H^E(M')$. Notice that A^E can be regarded as a random variable.

Preimage Resistance. The preimage resistance is also discussed in the ideal cipher model. Let A be an adversary trying to find a preimage of a given output for H^E. The pre-advantage of A against H^E is given by

$$\mathbf{Adv}^{\mathrm{pre}}_{H^E}(A) = \Pr[A^E(v) = M \wedge H^E(M) = v] \ ,$$

where v is uniformly distributed over $\mathcal{Y}$ and E is uniformly distributed over $\mathcal{BC}(\kappa,n)$. It is assumed that A makes all the queries necessary to compute $H^E(M)$.

Pseudorandom Function and Permutation (PRF & PRP). Let $f : \mathcal{K} \times \mathcal{X} \to \mathcal{Y}$ be a keyed function from $\mathcal{X}$ to $\mathcal{Y}$, where $\mathcal{K}$ is its key space. $f(K,\cdot)$ is often denoted by f_K. Let A be an adversary which has oracle access to a function from $\mathcal{X}$ to $\mathcal{Y}$ and outputs 0 or 1. The prf-advantage of A against f is given by

$$\mathbf{Adv}^{\mathrm{prf}}_{f}(A) = \left| \Pr[A^{f_K} = 1] - \Pr[A^{\rho} = 1] \right| \ ,$$

where K is uniformly distributed over $\mathcal{K}$ and ρ is uniformly distributed over $\mathcal{F}(\mathcal{X},\mathcal{Y})$. We sometimes consider the case where A has access to multiple oracles. Each query by A is directed to just one of them. Let us define the following notation: $\langle u_j \rangle_{j=1}^m = u_1, u_2, \ldots, u_m$. The m-prf-advantage of A against f is given by

$$\mathbf{Adv}^{m\text{-}\mathrm{prf}}_{f}(A) = \left| \Pr[A^{\langle f_{K_j} \rangle_{j=1}^m} = 1] - \Pr[A^{\langle \rho_j \rangle_{j=1}^m} = 1] \right| \ ,$$

where K_j's are independent random variables uniformly distributed over $\mathcal{K}$, and ρ_j's are independent random functions uniformly distributed over $\mathcal{F}(\mathcal{X},\mathcal{Y})$.

The prp-advantage of A against f is given by

$$\mathbf{Adv}^{\mathrm{prp}}_{f}(A) = \left| \Pr[A^{f_K} = 1] - \Pr[A^{\rho} = 1] \right| \ ,$$

where K is uniformly distributed over $\mathcal{K}$ and ρ is uniformly distributed over $\mathcal{P}(\mathcal{X})$. The m-prp-advantage is defined similarly to the m-prf-advantage.

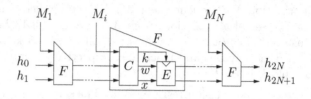

Fig. 2. Hash function $(E \circ C)^+$

PRF & PRP under Related-Key Attack. Let $\varPhi \subset \mathcal{F}(\mathcal{K}, \mathcal{K})$. Let A be an adversary which has oracle access to a function $u \in \mathcal{F}(\mathcal{K} \times \mathcal{X}, \mathcal{Y})$ with a key $K \in \mathcal{K}$, and outputs 0 or 1. A can make queries of the form $(\phi, x) \in \varPhi \times \mathcal{X}$, and obtain $u(\phi(K), x)$. The prf-rka-advantage of A against f restricted by $\varPhi$ is given by

$$\mathbf{Adv}_{\varPhi,f}^{\mathrm{prf\text{-}rka}}(A) = \left| \Pr[A^{(f,K)} = 1] - \Pr[A^{(\rho,K)} = 1] \right| ,$$

where K is uniformly distributed over $\mathcal{K}$ and ρ is uniformly distributed over $\mathcal{F}(\mathcal{K} \times \mathcal{X}, \mathcal{Y})$.

Let $\mathcal{P}(\mathcal{K} \times \mathcal{X}, \mathcal{X})$ be a set of all keyed permutations on $\mathcal{X}$, where $\mathcal{K}$ is their key space. The prp-rka-advantage of A against f restricted by $\varPhi$ is given by

$$\mathbf{Adv}_{\varPhi,f}^{\mathrm{prp\text{-}rka}}(A) = \left| \Pr[A^{(f,K)} = 1] - \Pr[A^{(\rho,K)} = 1] \right| ,$$

where K is uniformly distributed over $\mathcal{K}$ and ρ is uniformly distributed over $\mathcal{P}(\mathcal{K} \times \mathcal{X}, \mathcal{X})$. The m-prf(prp)-rka-advantage is defined similarly to the m-prf(prp)-advantage.

2.2   Model

In the remaining parts of the paper, let $n > 0$ be an even integer and $\check{n} = n/2$. We consider constructions of an iterated hash function H^E with a compression function based on $E \in \mathcal{BC}(\check{n}, n)$ (Fig. 2). The compression function $F : \{0,1\}^n \times \{0,1\}^{\check{n}} \to \{0,1\}^n$ is specified as follows: $h_{2i} \| h_{2i+1} = F(h_{2i-2} \| h_{2i-1}, M_i) = E_k(w \| x)$, where $h_{2i-2}, h_{2i-1}, h_{2i}, h_{2i+1}, M_i \in \{0,1\}^{\check{n}}$, $k, w, x \in \{0,1\}^{\check{n}}$,

$$\begin{pmatrix} k \\ w \\ x \end{pmatrix} = C \begin{pmatrix} h_{2i-2} \\ h_{2i-1} \\ M_i \end{pmatrix} = \begin{pmatrix} c_{1,1} & c_{1,2} & c_{1,3} \\ c_{2,1} & c_{2,2} & c_{2,3} \\ c_{3,1} & c_{3,2} & c_{3,3} \end{pmatrix} \begin{pmatrix} h_{2i-2} \\ h_{2i-1} \\ M_i \end{pmatrix} ,$$

and C is a non-singular 3×3 binary matrix. F is denoted by $E \circ C$.

For $1 \le i \le N$, let $M_i \in \{0,1\}^{\check{n}}$. $F^+ : \{0,1\}^n \times (\{0,1\}^{\check{n}})^+ \to \{0,1\}^n$ is an iterated hash function such that $F^+(IV, M_1 \| \cdots \| M_N) = h_{2N} \| h_{2N+1}$, where $h_0 \| h_1 = IV$ is an initial value and $h_{2i} \| h_{2i+1} = F(h_{2i-2} \| h_{2i-1}, M_i)$ for $1 \le i \le N$. For $M \in \{0,1\}^*$, an unambiguous padding function $\mathsf{pad} : \{0,1\}^* \to (\{0,1\}^{\check{n}})^+$ is necessary to apply F^+ to M. Any unambiguous padding suffices for most cases. Thus, unless explicitly stated otherwise, an input M is assumed to be in $(\{0,1\}^{\check{n}})^+$.

Remark 1. For Lesamnta-LW,

$$C = \begin{pmatrix} 1 & 0 & 0 \\ 0 & 0 & 1 \\ 0 & 1 & 0 \end{pmatrix} .$$

3   Collision Resistance and Preimage Resistance

In this section, the collision resistance of $(E \circ C)^+$ is analyzed in the ideal cipher model. We will see that it depends on the value of $c_{1,3}$.

Lemma 1. *The upper right 2×2 submatrix of C^{-1} is* .

1. *singular but non-zero if $c_{1,3} = 0$.*
2. *non-singular if $c_{1,3} = 1$.*

The proof is given in Appendix A.1. The case where $c_{1,3} = 0$ is first discussed, which is not discussed in [16]. Theorem 1 implies that the complexity of collision attacks on $(E \circ C)^+$ with $c_{1,3} = 0$ is $\Omega((\log n)2^{\tilde{n}}/n)$ if they do not explore the internal structure of the underlying blockcipher.

Theorem 1. *Let A be a col-adversary against $(E \circ C)^+$ asking at most q queries to E. Suppose that $c_{1,3} = 0$. Then, in the ideal cipher model,*

$$\mathbf{Adv}^{col}_{(E \circ C)^+}(A) \leq \frac{(\mu(n) + 3)q}{2^{\tilde{n}} - 1} ,$$

where $\mu(n) = e \ln 2^{\tilde{n}} / \ln \ln 2^{\tilde{n}} = (e/2)n/(\log_2 n - \log_2 \log_2 e - 1)$.

Proof. For $1 \leq i \leq q$, let $(t_i, k_i, w_i \| x_i, y_i \| z_i)$ be a tuple such that $E(k_i, w_i \| x_i) = y_i \| z_i$ and $t_i \in \{e, d\}$ obtained by the i-th query. t_i represents the type of the i-th query: encryption (e) or decryption (d). Moreover, let $(g_i^0, g_i^1, m_i)^{\mathrm{T}} = C^{-1}(k_i, w_i, x_i)^{\mathrm{T}}$. Notice that $y_i \| z_i = (E \circ C)(g_i^0 \| g_i^1, m_i)$. Let $G_0, G_1, \ldots, G_q$ be a sequence of directed graphs such that $G_i = (V_i, L_i)$, where

- $V_0 = L_0 = \emptyset$, and
- $V_i = V_{i-1} \cup \{g_i^0 \| g_i^1, y_i \| z_i\}$, $L_i = L_{i-1} \cup \{(g_i^0 \| g_i^1, y_i \| z_i)\}$ for $1 \leq i \leq q$.

Each edge $(g_i^0 \| g_i^1, y_i \| z_i)$ is labeled by (t_i, m_i).

Suppose that the adversary A first finds a collision of $(E \circ C)^+$ with the i-th query. Then, there must be a path in G_i from IV to some colliding output, which does not exist in $G_1, \ldots, G_{i-1}$. This path also contains the nodes $g_i^0 \| g_i^1$ and $y_i \| z_i$, and the edge (t_i, m_i).

If $t_i = e$, that is, the i-th query is an encryption query, then there must be an event such that $y_i \| z_i \in \{y_j \| z_j \mid 1 \leq j \leq i - 1\} \cup \{g_j^0 \| g_j^1 \mid 1 \leq j \leq i - 1\} \cup \{IV\}$. If $t_i = d$, then there must be an event such that $g_i^0 \| g_i^1 \in \{y_j \| z_j \mid 1 \leq j \leq i - 1\} \cup \{IV\}$. For the case where $t_i = d$ and $g_i^0 \| g_i^1 \in \{y_j \| z_j \mid 1 \leq j \leq i - 1\}$, let us look into the new path in G_i mentioned above. Let $IV \xrightarrow{(t_{j_1}, M_{j_1})} v_{j_1} \xrightarrow{(t_{j_2}, M_{j_2})}$

$\cdots \xrightarrow{(t_{j_{l-1}}, M_{j_{l-1}})} v_{j_{l-1}} \xrightarrow{(t_{j_l}, M_{j_l})} v_{j_l}$ be the prefix of the path, where $v_{j_{l-1}} = g_i^0 \| g_i^1$, $(t_{j_l}, M_{j_l}) = (\mathsf{d}, m_i)$ and $v_{j_l} = y_i \| z_i$. We start from v_{j_l} and go back toward IV until we first find an edge (e, M_{j_k}) or reach the node IV without finding such an edge. Suppose that we reach IV. Then, it implies that there is an event such that $t_{i'} = \mathsf{d}$ and $g_{i'}^0 \| g_{i'}^1 = IV$ for some i' such that $1 \leq i' < i$. On the other hand, suppose that we find an edge (e, M_{j_k}). Then, it implies that there is an event such that $t_{i'} = \mathsf{e}$ and $y_{i'} \| z_{i'} \in \{g_j^0 \| g_j^1 \,|\, 1 \leq j < i'\}$ for some i' such that $1 < i' < i$, or an event such that $t_{i'} = \mathsf{d}$ and $g_{i'}^0 \| g_{i'}^1 \in \{y_j \| z_j \,|\, 1 \leq j < i' \wedge t_j = \mathsf{e}\}$ for some i' such that $1 < i' \leq i$.

From the discussions above, if A finds a collision with at most q queries, then there must be at least one of the following events for some i such that $1 \leq i \leq q$:

Ea_i $t_i = \mathsf{e}$ and $y_i \| z_i = IV$,
Eb_i $t_i = \mathsf{e}$ and $y_i \| z_i \in \{y_j \| z_j \,|\, 1 \leq j \leq i - 1\} \cup \{g_j^0 \| g_j^1 \,|\, 1 \leq j \leq i - 1\}$,
Ec_i $t_i = \mathsf{d}$ and $g_i^0 \| g_i^1 = IV$,
Ed_i $t_i = \mathsf{d}$ and $g_i^0 \| g_i^1 \in \{y_j \| z_j \,|\, 1 \leq j < i \wedge t_j = \mathsf{e}\}$.

It is easy to see that

$$\Pr[\mathrm{Ea}_i] \leq \frac{1}{2^n - (i - 1)} \quad \text{and} \quad \Pr[\mathrm{Eb}_i] \leq \frac{2(i - 1)}{2^n - (i - 1)} .$$

For Ec_i, since k_i is fixed by the query, and the upper right 2×2 submatrix of C^{-1} is non-zero and singular from Lemma 1,

$$\Pr[\mathrm{Ec}_i] \leq \frac{2^{\tilde{n}}}{2^n - (i - 1)} .$$

For Ed_i, since k_i, which equals g_i^0, g_i^1 or $g_i^0 \oplus g_i^1$, is fixed, the multicollision on y_j, z_j or $y_j \oplus z_j$ should be taken into consideration. From Theorem 3.1 of [13], for $1 \leq q \leq 2^n$,

$$\Pr[\mathrm{Ed}_i] \leq \frac{\mu(n) 2^{\tilde{n}}}{2^n - (i - 1)} + \frac{1}{2^{\tilde{n}}} .$$

Precisely speaking, the distribution of $y_j \| z_j$ is not uniform on $\{0,1\}^n$ since E is a keyed permutation. However, the probability of multicollision is smaller in this case since $\Pr[y_j \in \{y_1, \ldots, y_{j-1}\}] \leq \Pr[y_j \notin \{y_1, \ldots, y_{j-1}\}]$, $\Pr[z_j \in \{z_1, \ldots, z_{j-1}\}] \leq \Pr[z_j \notin \{z_1, \ldots, z_{j-1}\}]$, and $\Pr[y_j \oplus z_j \in \{y_1 \oplus z_1, \ldots, y_{j-1} \oplus z_{j-1}\}] \leq \Pr[y_j \oplus z_j \notin \{y_1 \oplus z_1, \ldots, y_{j-1} \oplus z_{j-1}\}]$. Thus, for $1 \leq q \leq 2^{\tilde{n}}$,

$$\mathbf{Adv}_{(E \circ C)+}^{\mathrm{col}}(A) \leq \sum_{i=1}^{q} (\Pr[\mathrm{Ea}_i] + \Pr[\mathrm{Eb}_i] + \Pr[\mathrm{Ec}_i] + \Pr[\mathrm{Ed}_i])$$

$$\leq \frac{(\mu(n) + 2)q}{2^{\tilde{n}} - q/2^{\tilde{n}}} + \frac{q}{2^{\tilde{n}}} \leq \frac{(\mu(n) + 3)q}{2^{\tilde{n}} - 1} .$$

The upper bound exceeds 1 for $q > 2^{\tilde{n}}$.   □

If $c_{1,3} = 1$, then the compression function is of Type-II defined by Stam [16]. Thus, the following theorem follows from Theorem 9 in [16]. It implies that any collision attack is at most as effective as the birthday attack if it does not explore the internal structure of the underlying blockcipher. The proof is given in Appendix A.2.

Theorem 2. *Let A be a col-adversary against $(E \circ C)^+$ asking at most q queries to E. Suppose that $c_{1,3} = 1$. Then, in the ideal cipher model,*

$$\mathbf{Adv}^{\mathrm{col}}_{(E \circ C)^+}(A) \leq \frac{q^2 + q}{2\,(2^n - q)} \; .$$

For the current model of compression functions, shorter keys would reduce the size of the internal state. However, if the key length is αn for some constant $\alpha \leq 1/2$, then the query complexity of collision attacks is $o(2^{\alpha n})$ for the case where the key of the underlying blockcipher does not depend on the message blocks. Thus, we only consider the case of $\alpha = 1/2$.

Let us compare the upper bounds of Theorems 1 and 2. For Theorem 1, Table 1 presents some values of q which satisfy $(\mu(n) + 3)q/(2^n - 1) = 1/2$. For Theorem 2, q almost equals $2^{n/2}$ if $(q^2 + q)/(2\,(2^n - q)) = 1/2$.

Table 1. Some values of q such that $(\mu(n) + 3)q/(2^{\tilde{n}} - 1) = 1/2$ for Theorem 1

n	128	160	192	224	256
q	57.87	73.65	89.47	105.3	121.1

Since the compression function $E \circ C$ is invertible, the preimage resistance of the hash function $(E \circ C)^+$ is as large as its collision resistance.

Theorem 3. *Let A be a pre-adversary against $(E \circ C)^+$ asking at most q queries to E. Then, in the ideal cipher model,*

$$\mathbf{Adv}^{\mathrm{pre}}_{(E \circ C)^+}(A) \leq \begin{cases} \dfrac{(\mu(n) + 3)q}{2^{\tilde{n}} - 1} & \text{if } c_{1,3} = 0 \\ \dfrac{q^2 + q}{2\,(2^n - q)} & \text{if } c_{1,3} = 1 \end{cases},$$

where $\mu(n) = \mathrm{e}\ln 2^{\tilde{n}} / \ln \ln 2^{\tilde{n}}$.

4 Keyed Hashing Mode

We consider a keyed hashing mode of $(E \circ C)^+$: keyed-via-IV (KIV) mode, which is accompanied by an output function $\omega : \{0,1\}^n \to \{0,1\}^{\tilde{n}}$. The KIV mode of $(E \circ C)^+$ is denoted by kiv-$(E \circ C)^+$. For a secret key $K \in \{0,1\}^n$ and a message $M \in (\{0,1\}^{\tilde{n}})^+$, kiv-$(E \circ C)^+(K, M) = \omega((E \circ C)^+(K, M))$.

In the remaining part of this section, the security of the keyed hashing mode as a PRF will be reduced to the security of the underlying blockcipher E as a PRP. First, we discuss the conditions on C.

The reduction is impossible if adversaries know the value fed to the key of E. Thus, we will assume that the key depends on the chaining value, that is, $(c_{1,1}, c_{1,2}) \neq (0,0)$.

For some PRP E, its non-random behaviour may be observed if a query related to the key value is allowed. This fact should be taken into account when C is chosen.

Example 1. Let $\tilde{E}$ be a blockcipher in $\mathcal{BC}(\check{n}, n)$ such that $\tilde{E}_k(w\|x) = w\|x$ if $w \oplus k$ is a constant c. $\tilde{E}$ can be a PRP. However, it is easy to distinguish it from a random permutation if a query $(k \oplus c)\|x$ is allowed. Thus, for example, it does not seem a good idea to use the compression function $E_{h_{2i-2}}(h_{2i-2} \oplus M_i \| h_{2i-1})$.

For the general case, suppose that $M_i = 0^{\check{n}}$ since it is fully controlled by adversaries. If the lower left 2×2 submatrix of C is non-singular, then k is represented by a linear combination of w and x. Thus, we will assume that the rank of the submatrix is 1.

Let $(c_{a,1}\ c_{a,2})$ be a non-zero row of the lower left 2×2 submatrix of C. The output function ω is defined as follows:

$$\omega(h_{2N}\|h_{2N+1}) = c_{a,1}h_{2N} \oplus c_{a,2}h_{2N+1} .$$

4.1   If the Message Blocks Are Not Fed into the Key of E

Let us first consider the case where the message blocks are not fed into the key of the underlying blockcipher, that is, $c_{1,3} = 0$ for C. The following theorem implies that the KIV mode of $(E \circ C)^+$ is a PRF if E is a PRP.

Theorem 4. *Let A be a prf-adversary against the KIV mode of $(E \circ C)^+$. Suppose that $c_{1,3} = 0$. Suppose that A runs in time at most t, and makes at most q queries, and each query has at most ℓ message blocks. Then, there exists a prp-adversary B against E such that*

$$\mathbf{Adv}^{\mathrm{prf}}_{\mathrm{kiv}\text{-}(E\circ C)^+}(A) \leq \ell q \cdot \mathbf{Adv}^{\mathrm{prp}}_E(B) + \ell q(q-1)/2^{n+1} .$$

B makes at most q queries and runs in time at most $t + O(\ell q T_E)$, where T_E represents the time required to compute E.

Theorem 4 directly follows from the two lemmas given below. Their proofs are given in Appendices B.1 and B.2, respectively.

Lemma 2. *Let A be a prf-adversary against $\mathrm{kiv}\text{-}(E\circ C)^+$. Suppose that $c_{1,3} = 0$. Suppose that A runs in time at most t, and makes at most q queries, and each query has at most ℓ message blocks. Then, there exists a prf-adversary B against E with access to q oracles such that*

$$\mathbf{Adv}^{\mathrm{prf}}_{\mathrm{kiv}\text{-}(E\circ C)^+}(A) \leq \ell \cdot \mathbf{Adv}^{q\text{-}\mathrm{prf}}_E(B) .$$

B makes at most q queries and runs in time at most $t + O(\ell q T_E)$, where T_E represents the time required to compute E.

Lemma 3. *Let A be a prf-adversary against E with m oracles. Suppose that A runs in time at most t, and makes at most q queries. Then, there exists a prp-adversary B against E such that*

$$\mathbf{Adv}_E^{m\text{-prf}}(A) \leq m \cdot \mathbf{Adv}_E^{\text{prp}}(B) + q(q-1)/2^{n+1} .$$

B makes at most q queries and runs in time at most $t + O(qT_E)$, where T_E represents the time required to compute E.

4.2   If the Message Blocks Are Fed into the Key of E

Let us now consider the case where the message blocks are fed into the key of the underlying blockcipher, that is, $c_{1,3} = 1$. In this case, adversaries are more powerful than in the previous case because they can directly affect key inputs of the underlying blockcipher through message blocks. Thus, the underlying blockcipher should be secure against the related-key differential attacks where the attacker can control only the difference between the two related keys. The proof is omitted since it is similar to that in Sect. 4.1.

Theorem 5. *Let $\Phi = \{\phi \mid \phi : \{0,1\}^{\tilde{n}} \to \{0,1\}^{\tilde{n}}, \ \phi(x) = x \oplus c, \ c \in \{0,1\}^{\tilde{n}}\}$. Let A be a prf-adversary against the KIV mode of $(E \circ C)^+$. Suppose that A runs in time at most t, and makes at most q queries, and each query has at most ℓ message blocks. Then, there exists a prp-rka-adversary B against E such that*

$$\mathbf{Adv}_{\text{kiv-}(E \circ C)^+}^{\text{prf}}(A) \leq \ell q \cdot \mathbf{Adv}_E^{\text{prp-rka}}(B) + \ell q(q-1)/2^{n+1} .$$

B makes at most q queries restricted by Φ and runs in time at most $t + O(\ell q T_E)$, where T_E represents the time required to compute E.

5   Preimage-Resistance in the Computational Model

This section discusses the preimage resistance of $(E \circ C)^+$ when an underlying blockcipher E is not ideal, that is, an efficient algorithm of E is known to an adversary. This section assumes that the matrix C satisfies the conditions given in Sect. 4: C is non-singular, the upper-left 1×2 submatrix of C is non-zero, and the rank of the lower-left 2×2 submatrix of C is 1. When the distributions of h_{2i-2} and h_{2i-1} are uniform and M_i is 0^n, k is distributed on $\{0,1\}^{\tilde{n}}$ uniformly and independently from the distributions of w, x. This property plays an important role in proving theorems.

This section also assumes the following unambiguous padding. Let M be an l-bit message to be hashed. A single '1' bit is appended to the head of M, and then '0' bits are appended to the head so that the length in bits of the padded message becomes a multiple of $\tilde{n}$. Finally, all zero block, $0^{\tilde{n}}$, is appended to the tail of M. Unlike usual padding, some padding bits are located in the front of the original message. The number of padded message blocks is given by $N = \lfloor l/\tilde{n} \rfloor + 2$. As shown in the following, this padding allows us to reduce the preimage resistance to computational security of E. The disadvantage of this padding is that the length of M has to be known before hashing M.

We define two types of preimage resistance that differ in how to produce a given digest. The first assumes that a digest is uniformly chosen from the digest space. This definition is the computational version of the one given in Sect. 2. The second assumes that a digest is given by the same computation as the last compression function. The second is necessary only to discuss the relation between the preimage resistance and the security of an underlying blockcipher.

Definition 1. *Let A be an adversary finding a preimage for a given digest. The pre-advantage of A is defined as*

$$\mathbf{Adv}^{\mathrm{pre}}_{(E \circ C)^+}(A) = \Pr\left[A(h) = M \wedge (E \circ C)^+(M) = h\right] \quad (1)$$

where h is chosen uniformly at random from $\{0,1\}^n$. This adversary is called a pre-finder. The pseudo-pre-advantage of A is defined as

$$\mathbf{Adv}^{\mathrm{ppre}}_{(E \circ C)^+}(A) = \Pr\left[A(h) = M \wedge (E \circ C)^+(M) = h\right] \quad (2)$$

where h is produced in the following manner:

$$u \| d \xleftarrow{\$} \{0,1\}^n, \ (k,w,x)^T = C(u,d,0^{\tilde{n}})^T, \ h = E_k(w \| x), \quad (3)$$

where T is a transposition.

We show that the pre-advantage and the pseudo-pre-advantage are characterized with the prp-advantage for E. The choice of h in Eq. (1) is equivalent to

$$u \| d \xleftarrow{\$} \{0,1\}^n, \ (k,w,x)^T = C(u,d,0^{\tilde{n}})^T, \ h = \rho(w \| x) \quad (4)$$

where ρ is a random permutation. In Eq. (2), k is chosen uniformly at random from $\{0,1\}^{\tilde{n}}$ and is independent of $w \| x$ because of the conditions on C. Comparing Eq. (3) with Eq. (4), we see that the difference between Eq. (1) and Eq. (2) is the difference between E_k and ρ. Recall the prp-advantage $\mathbf{Adv}^{\mathrm{prp}}_E$ that measures the difference between E_k and ρ. Let $\mathbf{Adv}^{\mathrm{prp}}_E(q,t)$ be the best prp-advantage among adversaries B with queries q and time t, that is,

$$\mathbf{Adv}^{\mathrm{prp}}_E(q,t) = \max_B \mathbf{Adv}^{\mathrm{prp}}_E(B) \ .$$

We then have the following lemma, which is proved in Appendix C.1. This lemma means that if E is a pseudorandom permutation, then the pre-advantage is nearly equal to the ppre-advantage.

Lemma 4. *Let $F = E \circ C$. Let A be an adversary that finds a preimage for a given digest. Suppose that the running time of A is at most t and the maximum length of a preimage is l. Then, we have*

$$\left|\mathbf{Adv}^{\mathrm{ppre}}_{F^+}(A) - \mathbf{Adv}^{\mathrm{pre}}_{F^+}(A)\right| \le \mathbf{Adv}^{\mathrm{prp}}_E(1, t + Nt_F), \quad (5)$$

where t_F is the time for computing F and $N = \lfloor l/\tilde{n} \rfloor + 2$.

Let $\tilde{F}(v) = (E \circ C)(v, 0^{\tilde{n}})$. The *non-injectivity* of E is defined as

$$\nu_E = \max_{v \in \{0,1\}^n} \#\tilde{F}^{-1}(v) \ .$$

The non-injectivity is a property of the mapping. Since the block length of E is twice the key length, the value of ν_E is expected to be small if E is well-designed. The following theorem implies that $(E \circ C)^+$ is secure against any efficient pre-finder if the non-injectivity of E is small and E is a pseudorandom permutation. The proof is given in Appendix C.2. The last zero message block, which is appended in the padding method, plays an important role in the proof.

Theorem 6. *Let* $F = E \circ C$. *Let* A *be a pre-finder for the hash function* F^+. *Suppose that the running time of* A *is at most* t *and the length of a preimage produced by* A *is at most* l. *Then, we have*

$$\left(\frac{1}{\nu_E} - \frac{1}{2^n - 1}\right) \mathbf{Adv}^{\mathrm{pre}}_{(E \circ C)^+}(A) \leq 2\mathbf{Adv}^{\mathrm{prp}}_E(2, t + (N+1)t_F),$$

where t_F *is the time for computing* F *and* $N = \lfloor l/\tilde{n} \rfloor + 2$.

6   Concluding Remarks

This article has first presented a model of compression functions using a dedicated blockcipher for lightweight hashing. Then, it has discussed the security of iterated hash functions composed of compression functions in the model. We can find the following trade-off between collision resistance and the PRF property of keyed hashing modes: if $c_{1,3} = 0$, then collision resistance is nearly optimal but the KIV mode only requires a blockcipher secure as a PRP; if $c_{1,3} = 1$, then collision resistance is optimal but the KIV mode requires a blockcipher secure as a PRP against related-key attacks.

Acknowledgement. The first author was partially supported by KAKENHI 21240001. The second author was partially supported by KAKENHI 22560376.

References

1. Aumasson, J.-P., Henzen, L., Meier, W., Naya-Plasencia, M.: QUARK: A Lightweight Hash. In: Mangard, S., Standaert, F.-X. (eds.) CHES 2010. LNCS, vol. 6225, pp. 1–15. Springer, Heidelberg (2010)
2. Bellare, M., Canetti, R., Krawczyk, H.: Pseudorandom functions revisited: The cascade construction and its concrete security. In: Proceedings of the 37th IEEE Symposium on Foundations of Computer Science, pp. 514–523 (1996)
3. Bertoni, G., Daemen, J., Peeters, M., Van Assche, G.: Sponge functions. In: ECRYPT Hash Workshop (2007)

4. Black, J., Rogaway, P., Shrimpton, T.: Black-Box Analysis of the Block-Cipher-Based Hash-Function Constructions from PGV. In: Yung, M. (ed.) CRYPTO 2002. LNCS, vol. 2442, pp. 320–335. Springer, Heidelberg (2002)
5. Bogdanov, A., Knežević, M., Leander, G., Toz, D., Varıcı, K., Verbauwhede, I.: SPONGENT: A Lightweight Hash Function. In: Preneel, B., Takagi, T. (eds.) CHES 2011. LNCS, vol. 6917, pp. 312–325. Springer, Heidelberg (2011)
6. Bogdanov, A., Knudsen, L.R., Leander, G., Paar, C., Poschmann, A., Robshaw, M.J.B., Seurin, Y., Vikkelsoe, C.: PRESENT: An ultra-lightweight block cipher. In: Paillier, Verbauwhede [14], pp. 450–466
7. Bogdanov, A., Leander, G., Paar, C., Poschmann, A., Robshaw, M.J.B., Seurin, Y.: Hash Functions and RFID Tags: Mind the Gap. In: Oswald, E., Rohatgi, P. (eds.) CHES 2008. LNCS, vol. 5154, pp. 283–299. Springer, Heidelberg (2008)
8. Guo, J., Peyrin, T., Poschmann, A.: The PHOTON Family of Lightweight Hash Functions. In: Rogaway, P. (ed.) CRYPTO 2011. LNCS, vol. 6841, pp. 222–239. Springer, Heidelberg (2011)
9. Hirose, S., Ideguchi, K., Kuwakado, H., Owada, T., Preneel, B., Yoshida, H.: A Lightweight 256-Bit Hash Function for Hardware and Low-End Devices: Lesamnta-LW. In: Rhee, K.-H., Nyang, D. (eds.) ICISC 2010. LNCS, vol. 6829, pp. 151–168. Springer, Heidelberg (2011)
10. Hirose, S., Kuwakado, H.: Efficient pseudorandom-function modes of a block-cipher-based hash function. IEICE Transactions on Fundamentals E92-A(10), 2447–2453 (2009)
11. Hirose, S., Park, J.H., Yun, A.: A Simple Variant of the Merkle-Damgård Scheme with a Permutation. In: Kurosawa, K. (ed.) ASIACRYPT 2007. LNCS, vol. 4833, pp. 113–129. Springer, Heidelberg (2007)
12. Lucks, S.: A Failure-Friendly Design Principle for Hash Functions. In: Roy, B. (ed.) ASIACRYPT 2005. LNCS, vol. 3788, pp. 474–494. Springer, Heidelberg (2005)
13. Motwani, R., Raghavan, P.: Randomized Algorithms. Cambridge University Press (1995)
14. Paillier, P., Verbauwhede, I. (eds.): CHES 2007. LNCS, vol. 4727. Springer, Heidelberg (2007)
15. Preneel, B., Govaerts, R., Vandewalle, J.: Hash Functions Based on Block Ciphers: A Synthetic Approach. In: Stinson, D.R. (ed.) CRYPTO 1993. LNCS, vol. 773, pp. 368–378. Springer, Heidelberg (1994)
16. Stam, M.: Blockcipher-Based Hashing Revisited. In: Dunkelman, O. (ed.) FSE 2009. LNCS, vol. 5665, pp. 67–83. Springer, Heidelberg (2009)
17. Yoshida, H., Watanabe, D., Okeya, K., Kitahara, J., Wu, H., Küçük, Ö., Preneel, B.: MAME: A compression function with reduced hardware requirements. In: Paillier, Verbauwhede [14], pp. 148–165

A   Proofs of Lemma and Theorem in Sect. 3

A.1   Proof of Lemma 1

It is clear that the upper right 2×2 submatrix of C^{-1} is not a zero matrix. Let

$$C^{-1} = \begin{pmatrix} d_{1,1} & d_{1,2} & d_{1,3} \\ d_{2,1} & d_{2,2} & d_{2,3} \\ d_{3,1} & d_{3,2} & d_{3,3} \end{pmatrix} .$$

For $1 \le i \le 3$, let c_{ri} be the i-th row of C and d_{ci} be the i-th column of C^{-1}. Suppose that $c_{1,3} = 0$. Then,

$$c_{r1} \cdot d_{c2} = c_{1,1}d_{1,2} \oplus c_{1,2}d_{2,2} = 0 \quad \text{and} \quad c_{r1} \cdot d_{c3} = c_{1,1}d_{1,3} \oplus c_{1,2}d_{2,3} = 0 .$$

They imply that the rows of the upper right 2×2 submatrix of C^{-1} are linearly dependent since $(c_{1,1}, c_{1,2}) \neq (0,0)$.

Suppose that $c_{1,3} = 1$. Then, the following matrix is the inverse of the upper right 2×2 submatrix of C^{-1}:

$$\begin{pmatrix} c_{2,1} & c_{2,2} \\ c_{3,1} & c_{3,2} \end{pmatrix} \oplus c_{2,3} \begin{pmatrix} c_{1,1} & c_{1,2} \\ 0 & 0 \end{pmatrix} \oplus c_{3,3} \begin{pmatrix} 0 & 0 \\ c_{1,1} & c_{1,2} \end{pmatrix} .$$

A.2   Proof of Theorem 2

Following the formalization in [16], a compression function $F : \{0,1\}^n \times \{0,1\}^{\tilde{n}} \to \{0,1\}^n$ is specified as follows: $h_{2i} \| h_{2i+1} = F(h_{2i-2} \| h_{2i-1}, M_i)$, where

1. $(k, w \| x) \leftarrow C^{\mathrm{PRE}}(M_i, h_{2i-2} \| h_{2i-1})$,
2. $y \| z \leftarrow E_k(w \| x)$,
3. $h_{2i} \| h_{2i+1} \leftarrow C^{\mathrm{POST}}(M_i, h_{2i-2} \| h_{2i-1}, y \| z)$.

$C^{\mathrm{PRE}} : \{0,1\}^{\tilde{n}} \times \{0,1\}^n \to \{0,1\}^{\tilde{n}} \times \{0,1\}^n$ is called preprocessing, and $C^{\mathrm{POST}} : \{0,1\}^{\tilde{n}} \times \{0,1\}^n \times \{0,1\}^n \to \{0,1\}^n$ is called postprocessing.
A compression function F is of Type-II if the following three conditions hold:

1. C^{PRE} is bijective.
2. For all M_i and $h_{2i-2} \| h_{2i-1}$, $C^{\mathrm{POST}}(M_i, h_{2i-2} \| h_{2i-1}, \cdot)$ is bijective.
3. For all k, $C^{-\mathrm{PRE}}(k, \cdot)$ restricted to its second output is bijective.

The compression function $E \circ C$ satisfies the first condition since its C^{PRE} is the multiplication of the non-singular C. It also satisfies the second condition since $C^{\mathrm{POST}}(M_i, h_{2i-2} \| h_{2i-1}, y \| z) = y \| z$. For the third condition, notice that

$$\begin{pmatrix} h_{2i-2} \\ h_{2i-1} \\ M_i \end{pmatrix} = C^{-1} \begin{pmatrix} k \\ w \\ x \end{pmatrix}$$

and that the upper right 2×2 submatrix of C^{-1} is non-singular from Lemma 1. Thus, there is one-to-one correspondence between (w, x) and (h_{2i-2}, h_{2i-1}) for any fixed k.

B   Proofs of Lemmas in Sect. 4

B.1   Proof of Lemma 2

The proof basically follows the hybrid argument given in [2]. Let $M_{[1,l]} = M_1\|M_2\|\cdots\|M_l$. For $i \in \{0,1,\ldots,\ell\}$ ($\ell \geq 1$), let $I_i : (\{0,1\}^{\tilde{n}})^{\leq \ell} \to \{0,1\}^{\tilde{n}}$ be a random function such that

$$I_i(M_{[1,l]}) = \begin{cases} \alpha_1(M_{[1,l]}) & \text{if } 1 \leq l \leq i, \\ \text{kiv-}(E \circ C)^+(K_{i,0}\|K_{i,1}, M_{[i+1,l]}) & \text{if } i+1 \leq l \leq \ell, \end{cases}$$

where

$$\begin{pmatrix} K_{i,0} \\ K_{i,1} \end{pmatrix} = \begin{pmatrix} c_{1,1} & c_{1,2} \\ c_{a,1} & c_{a,2} \end{pmatrix}^{-1} \begin{pmatrix} \alpha_0(M_{[1,i]}) \\ \alpha_1(M_{[1,i]}) \end{pmatrix},$$

and α_0 and α_1 are random functions uniformly distributed over $\mathcal{F}((\{0,1\}^{\tilde{n}})^i, \{0,1\}^{\tilde{n}})$ and $\mathcal{F}((\{0,1\}^{\tilde{n}})^{\leq i}, \{0,1\}^{\tilde{n}})$, respectively. If $i = 0$, then both α_0 and α_1 are just random elements uniformly and independently distributed over $\{0,1\}^{\tilde{n}}$. Then,

$$\mathbf{Adv}^{\mathrm{prf}}_{\mathrm{kiv-}(E \circ C)^+}(A) = \left| \Pr[A^{I_0} = 1] - \Pr[A^{I_\ell} = 1] \right|.$$

A q-prf-adversary B with q oracles $\langle u_j \rangle_{j=1}^q$ is constructed using A as a subroutine. B first selects $i \in \{1,\ldots,\ell\}$ uniformly at random. Then, B runs B_i described below.

B_i runs A. B_i simulates a random function β uniformly distributed over $\mathcal{F}((\{0,1\}^{\tilde{n}})^{\leq i-1}, \{0,1\}^{\tilde{n}})$ via lazy sampling. When B_i receives the p-th query $M^{(p)} = M_{[1,l]}^{(p)}$ of A, B_i returns

$$\begin{cases} \beta(M_{[1,l]}^{(p)}) & \text{if } 1 \leq l \leq i-1, \\ \omega(u_{\mathrm{idx}(M_{[1,i-1]}^{(p)})}(w^{(p)}\|x^{(p)})) & \text{if } l = i, \\ \text{kiv-}(E \circ C)^+(u_{\mathrm{idx}(M_{[1,i-1]}^{(p)})}(w^{(p)}\|x^{(p)}), M_{[i+1,l]}^{(p)}) & \text{if } i+1 \leq l \leq \ell, \end{cases}$$

where

$$\begin{pmatrix} w^{(p)} \\ x^{(p)} \end{pmatrix} = \begin{pmatrix} \beta_2 \\ \beta_3 \end{pmatrix} \oplus M_i^{(p)} \begin{pmatrix} c_{2,3} \\ c_{3,3} \end{pmatrix},$$

and, for $j \in \{2,3\}$,

$$\beta_j = \begin{cases} \beta(M_{[1,i-1]}^{(p)}) & \text{if } (c_{j,1}, c_{j,2}) = (c_{a,1}, c_{a,2}), \\ 0 & \text{if } (c_{j,1}, c_{j,2}) = (0,0). \end{cases}$$

In the above, for each p, $\mathrm{idx}(M_{[1,i-1]}^{(p)})$ is a unique integer in $\{1,\ldots,q\}$, which depends on $M_{[1,i-1]}^{(p)}$. If there is a previous query $M^{(p')}$ ($p' < p$) such that $M_{[1,i-1]}^{(p')} = M_{[1,i-1]}^{(p)}$, then $\mathrm{idx}(M_{[1,i-1]}^{(p)}) = \mathrm{idx}(M_{[1,i-1]}^{(p')})$. Otherwise, $\mathrm{idx}(M_{[1,i-1]}^{(p)}) = p$.

Now, suppose that B_i is given oracles $E_{K_1}, E_{K_2}, \ldots, E_{K_q}$, where K_j's are independent random variables uniformly distributed over $\{0,1\}^{\tilde{n}}$. Then, in response to $M_{[1,l]}^{(p)}$, B_i returns

$$\begin{cases} \beta(M_{[1,l]}^{(p)}) & \text{if } 1 \leq l \leq i-1, \\ \text{kiv-}(E \circ C)^+(K_{i-1,0} \| K_{i-1,1}, M_{[i,l]}^{(p)}) & \text{if } i \leq l \leq \ell, \end{cases}$$

where

$$\begin{pmatrix} K_{i-1,0} \\ K_{i-1,1} \end{pmatrix} = \begin{pmatrix} c_{1,1} & c_{1,2} \\ c_{a,1} & c_{a,2} \end{pmatrix}^{-1} \begin{pmatrix} K_{\text{idx}(M_{[1,i-1]}^{(p)})} \\ \beta(M_{[1,i-1]}^{(p)}) \end{pmatrix} .$$

Since $K_{\text{idx}(M_{[1,i-1]}^{(p)})}$ can be regarded as a random function of $M_{[1,i-1]}^{(p)}$, we can say that A has oracle access to I_{i-1}. Therefore,

$$\Pr[B_i^{\langle E_{K_j} \rangle_{j=1}^q} = 1] = \Pr[A^{I_{i-1}} = 1] .$$

Next, suppose that B_i has oracle access to $\rho_1, \ldots, \rho_q$, where ρ_j's are independent random functions uniformly distributed over $\mathcal{F}(\{0,1\}^n, \{0,1\}^n)$. Since $(c_{2,3}, c_{3,3}) \neq (0,0)$, the first half and the second half of $\rho_{\text{idx}(M_{[1,i-1]}^{(p)})}(w^{(p)} \| x^{(p)})$ are independent random functions of $M_{[1,i]}^{(p)}$. Thus, we can say that A has oracle access to I_i and

$$\Pr[B_i^{\langle \rho_j \rangle_{j=1}^q} = 1] = \Pr[A^{I_i} = 1] .$$

From the discussions above,

$$\begin{aligned} \mathbf{Adv}_E^{q\text{-prf}}(B) &= \left| \Pr[B^{\langle E_{K_j} \rangle_{j=1}^q} = 1] - \Pr[B^{\langle \rho_j \rangle_{j=1}^q} = 1] \right| \\ &= \frac{1}{\ell} \left| \sum_{i=1}^{\ell} \Pr[A^{I_{i-1}} = 1] - \sum_{i=1}^{\ell} \Pr[A^{I_i} = 1] \right| \\ &= \frac{1}{\ell} \left| \Pr[A^{I_0} = 1] - \Pr[A^{I_\ell} = 1] \right| = \frac{1}{\ell} \mathbf{Adv}_{\text{kiv-}(E \circ C)^+}^{\text{prf}}(A) . \end{aligned}$$

B makes at most q queries and runs in time at most $t + O(\ell q T_E)$. There may exist an algorithm with the same resources and larger advantage. Let us also call it B. Then,

$$\mathbf{Adv}_{\text{kiv-}(E \circ C)^+}^{\text{prf}}(A) \leq \ell \cdot \mathbf{Adv}_E^{q\text{-prf}}(B) .$$

B.2   Proof of Lemma 3

The proof is also based on the hybrid argument. Let $K_1, \ldots, K_m$ be independent random variables uniformly distributed over $\{0,1\}^{\tilde{n}}$. Let $\rho_1, \ldots, \rho_m$ be independent random functions uniformly distributed over $\mathcal{F}(\{0,1\}^n, \{0,1\}^n)$. Let

$\varpi_1, \ldots, \varpi_m$ be independent random permutations uniformly distributed over $\mathcal{P}(\{0,1\}^n)$. Then,

$$\mathbf{Adv}_E^{m\text{-}prf}(A) = \left| \Pr[A^{\langle E_{K_j} \rangle_{j=1}^m} = 1] - \Pr[A^{\langle \rho_j \rangle_{j=1}^m} = 1] \right|$$

$$\leq \left| \Pr[A^{\langle E_{K_j} \rangle_{j=1}^m} = 1] - \Pr[A^{\langle \varpi_j \rangle_{j=1}^m} = 1] \right| +$$

$$\left| \Pr[A^{\langle \varpi_j \rangle_{j=1}^m} = 1] - \Pr[A^{\langle \rho_j \rangle_{j=1}^m} = 1] \right| .$$

For $0 \leq i \leq m$, let $\mathcal{O}_i$ be m oracles such that $E_{K_1}, \ldots, E_{K_i}, \varpi_{i+1}, \ldots, \varpi_m$. Notice that $\mathcal{O}_0 = \langle \varpi_j \rangle_{j=1}^m$ and $\mathcal{O}_m = \langle E_{K_j} \rangle_{j=1}^m$.

A prp-adversary B with an oracle u is constructed using A as a subroutine. u is either E_K or ϖ, where K is a random variable uniformly distributed over $\{0,1\}^{\tilde{n}}$ and ϖ is a random permutation uniformly distributed over $\mathcal{P}(\{0,1\}^n)$.

B first selects i from $\{1, 2, \ldots, m\}$ uniformly at random. Then, B runs A with oracles $E_{K_1}, \ldots, E_{K_{i-1}}, u, \varpi_{i+1}, \ldots, \varpi_m$ by simulating $E_{K_1}, \ldots, E_{K_{i-1}}$, and $\varpi_{i+1}, \ldots, \varpi_m$. Finally, B outputs A's output.

Then,

$$\Pr[B^{E_K} = 1] = \frac{1}{m} \sum_{i=1}^m \Pr[A^{\mathcal{O}_i} = 1] \quad \text{and} \quad \Pr[B^{\varpi} = 1] = \frac{1}{m} \sum_{i=0}^{m-1} \Pr[A^{\mathcal{O}_i} = 1] .$$

Thus,

$$\mathbf{Adv}_E^{prp}(B) = \left| \Pr[B^{E_K} = 1] - \Pr[B^{\varpi} = 1] \right| = \frac{1}{m} \left| \Pr[A^{\mathcal{O}_m} = 1] - \Pr[A^{\mathcal{O}_0} = 1] \right| .$$

B makes at most q queries and runs in time at most $t + O(q T_E)$. There may exist an algorithm with the same resources and larger advantage. Let us also call it B. Then,

$$\left| \Pr[A^{\mathcal{O}_m} = 1] - \Pr[A^{\mathcal{O}_0} = 1] \right| \leq m \cdot \mathbf{Adv}_E^{prp}(B) .$$

It is possible to distinguish $\varpi_1, \ldots, \varpi_m$ and $\rho_1, \ldots, \rho_m$ only by the fact that there may be a collision for ρ_i. Thus, since A makes at most q queries,

$$\left| \Pr[A^{\langle \varpi_j \rangle_{j=1}^m} = 1] - \Pr[A^{\langle \rho_j \rangle_{j=1}^m} = 1] \right| \leq \frac{q(q-1)}{2^{n+1}} .$$

C   Proofs of Lemma and Theorem in Sect. 5

C.1   Proof of Lemma 4

Consider the following prp-adversary B. The aim of B is to determine whether its oracle $\mathcal{O}$ is E_k or ρ by using A as a subroutine.

1. Compute (v, w, x) as $u \| d \xleftarrow{\$} \{0,1\}^n$, $(v, w, x)^T = C(u, d, 0^{\tilde{n}})^T$.
2. Make a query $w \| x$ to $\mathcal{O}$ and receive its response $h = y \| z$.

3. Input a digest h to $A_{t,l}$ and let M be output of A.
4. If $(E \circ C)^+(M) = h$, then output 1, otherwise output 0.

The number of queries made by B is one and the running time of B, which is dominated by the running time of A and the computation of $(E \circ C)^+(M)$ in step 4, is $t + (\lfloor \lambda/\check{n} \rfloor + 2)t_F$ where λ is the length of M. Since $\lambda \le l$, the running time of B is at most $t + Nt_F$. The probabilities that B outputs 1 are

$$\Pr\left[B^{E_k} = 1\right] = \Pr\left[A_{t,l}(h) = M \wedge (E \circ C)^+(M) = h\right] = \mathbf{Adv}^{\mathrm{ppre}}_{(E \circ C)^+}(A),$$

$$\Pr\left[B^\rho = 1\right] = \Pr\left[A_{t,l}(h) = M \wedge (E \circ C)^+(M) = h\right] = \mathbf{Adv}^{\mathrm{pre}}_{(E \circ C)^+}(A).$$

We consequently obtain

$$\begin{aligned}
\mathbf{Adv}^{\mathrm{prp}}_E(B_{1,t+(\lfloor \lambda/\check{n} \rfloor + 2)t_F}) &= \left|\Pr\left[B^{E_k} = 1\right] - \Pr\left[B^\rho = 1\right]\right| \\
&= \left|\mathbf{Adv}^{\mathrm{pre}}_{(E \circ C)^+}(A) - \mathbf{Adv}^{\mathrm{ppre}}_{(E \circ C)^+}(A)\right| \\
&\le \mathbf{Adv}^{\mathrm{prp}}_E(1, t + Nt_F).
\end{aligned}$$

C.2 Proof of Theorem 6

Suppose that A is a pre-finder running in time at most t and producing a preimage of length at most l. Consider the following adversary B which uses A as a subroutine. The aim of B is to determine whether its oracle $\mathcal{O}$ is E_k or a random permutation ρ by finding k if $\mathcal{O} = E_k$.

1. Compute (v, w, x) as $u \| d \xleftarrow{\$} \{0,1\}^n$, $(v, w, x)^T = C(u, d, 0^{\check{n}})^T$.
2. Make a query $w \| x$ to $\mathcal{O}$ and receive its response $h = y \| z$.
3. Run A with a digest h, and receive its output M.
4. If $(E \circ C)^+(M) \ne h$, then output 0 and terminate the algorithm.
5. Suppose that $(E \circ C)^+(M) = h$. Let ξ be the number of (padded) message blocks. Compute $(\tilde{k}, \tilde{w}, \tilde{x})^T = C(h_{2\xi-2}, h_{2\xi-1}, 0^{\check{n}})^T$. Note that $(h_{2\xi-2} \| h_{2\xi-1}, 0^{\check{n}})$ is the input to the last compression function of $(E \circ C)^+(M)$.
6. Choose $w' \| x'$ from $\{0,1\}^n \setminus \{w \| x\}$ uniformly. If

$$\mathcal{O}(w' \| x') = E(\tilde{k}, w' \| x'), \tag{6}$$

then output 1, otherwise output 0 and terminate the algorithm.

Since the length of M is at most l, the running time of B is at most $t + (N+1)t_F$, where $N = \lfloor l/\check{n} \rfloor + 2$. The number of queries made by B is at most two.

We next evaluate the probability that B outputs 1. If $\mathcal{O} = E_k$, then the probability that $(\tilde{k}, \tilde{w}, \tilde{x}) = (k, w, x)$ is at least $1/\nu_E$. In this case, Eq. (6) always holds because $\tilde{k} = k$. Note that even if $(\tilde{k}, \tilde{w}, \tilde{x}) \ne (k, w, x)$, Eq. (6) may hold. Hence,

$$\Pr\left[\mathcal{O}(w' \| x') = E(\tilde{k}, w' \| x')\right] \ge \frac{1}{\nu_E}.$$

If $\mathcal{O} = \rho$, then the probability that $\mathcal{O}(w'\|x') = E(\tilde{k}, w'\|x')$ is $1/(2^n - 1)$ because ρ is a random permutation. Noting that two cases differ in how to produce $y\|z$, we have

$$\Pr\left[B^{E_k} = 1\right] = \Pr\left[A(h) = M \wedge (E \circ C)^+(M) = h \wedge \mathcal{O}(w'\|x') = E(\tilde{k}, w'\|x')\right]$$

$$\geq \frac{1}{\nu_E}\mathbf{Adv}^{\mathrm{ppre}}_{(E \circ C)^+}(A),$$

$$\Pr\left[B^{\rho} = 1\right] = \Pr\left[A(h) = M \wedge (E \circ C)^+(M) = h \wedge \mathcal{O}(w'\|x') = E(\tilde{k}, w'\|x')\right]$$

$$= \frac{1}{2^n - 1}\mathbf{Adv}^{\mathrm{pre}}_{(E \circ C)^+}(A).$$

The above equations yield

$$\mathbf{Adv}^{\mathrm{prp}}_E(B) = \left|\Pr\left[B^{E_k} = 1\right] - \Pr\left[B^{\rho} = 1\right]\right|$$

$$\geq \frac{1}{\nu_E}\mathbf{Adv}^{\mathrm{ppre}}_{(E \circ C)^+}(A) - \frac{1}{2^n - 1}\mathbf{Adv}^{\mathrm{pre}}_{(E \circ C)^+}(A). \qquad (7)$$

From Lemma 4,

$$\frac{1}{\nu_E}\mathbf{Adv}^{\mathrm{prp}}_E(1, t + Nt_F) \geq \frac{1}{\nu_E}\left|\mathbf{Adv}^{\mathrm{pre}}_{(E \circ C)^+}(A) - \mathbf{Adv}^{\mathrm{ppre}}_{(E \circ C)^+}(A)\right|. \qquad (8)$$

Combining Eq. (7) and Eq. (8) gives

$$\mathbf{Adv}^{\mathrm{prp}}_E(B) + \frac{1}{\nu_E}\mathbf{Adv}^{\mathrm{prp}}_E(1, t + Nt_F) \geq \left(\frac{1}{\nu_E} - \frac{1}{2^n - 1}\right)\mathbf{Adv}^{\mathrm{pre}}_{(E \circ C)^+}(A).$$

Since $1 \leq \nu_E \leq 2^{\tilde{n}}$, simplifying this inequality gives

$$2\mathbf{Adv}^{\mathrm{prp}}_E(2, t + (N + 1)t_F) \geq \left(\frac{1}{\nu_E} - \frac{1}{2^n - 1}\right)\mathbf{Adv}^{\mathrm{pre}}_{(E \circ C)^+}(A).$$

Since the right-hand term of the above inequality is non-negative, this inequality is meaningful.

Biclique Attack on the Full HIGHT

Deukjo Hong, Bonwook Koo, and Daesung Kwon

ETRI
{hongdj,bwkoo,ds_kwon}@ensec.re.kr

Abstract. HIGHT is a lightweight block cipher proposed at CHES 2006 and included in ISO/IEC 18033-3. In this paper, we apply recently proposed biclique cryptanalysis to attack HIGHT. We show that bicliques can be constructed for 8 rounds in HIGHT, and those are used to recover the 128-bit key for the full rounds of HIGHT with the computational complexity of $2^{126.4}$, faster than exhaustive search. This is the first single-key attack result for the full HIGHT.

Keywords: HIGHT, Biclique, Cryptanalysis.

1 Introduction

HIGHT is a lightweight block cipher which was proposed at CHES 2006 for low-resource device such as radio frequency identifications (RFID) [4]. It is approved by Telecommunications Technology Association (TTA) of Korea and included in ISO/IEC 18033-3 [5]. It consists of 32 rounds in 8-branch type-2 generalized Feistel network with 64-bit block and 128-bit key. The round function is designed with ARX structure using addition modulo 2^8, XOR, and bitwise rotation.

In its proposal, it was claimed that at least 20 rounds of HIGHT was secure. However, in 2007, Lu presented an impossible differential attack on 25 rounds, an related-key rectangle attack on 26 rounds, and an related-key impossible differential attack on 28 rounds [7]. In 2009, Özen et al. presented an impossible differential attack on 26 rounds and an related-key impossible differential attack on 31 rounds [8], and Zhang et al. presented an saturation attack on 22 rounds [10]. In 2010, Koo et al. presented the first key recovery attack on the full HIGHT using a related-key rectangle distinguisher for weak key space with the fraction of 1/4 [6].

Some block ciphers such as AES and HIGHT adopt simple key schedule for efficiency in hardware implementation. However, the simplicity of the key schedule has been exploited for various related- and weak-key attacks. On the other hand, those kinds of attacks have been regarded as relatively impractical, while recently proposed biclique cryptanalysis [2] yields a single-key attack on such block ciphers.

Biclique cryptanalysis introduced by Bogdanov, Khovratovich, and Rechberger [2] is a kind of meet-in-the-middle attack such that bicliques improve the efficiency. They provided two approaches for key recovery using bicliques,

H. Kim (Ed): ICISC 2011, LNCS 7259, pp. 365–374, 2012.
© Springer-Verlag Berlin Heidelberg 2012

and gave several cryptanalytic results for AES including the first single-key attack results for full rounds.

The notion of biclique is as follows. Let f be a subcipher that maps an internal state S to the ciphertext C: $f_K(S) = C$. We consider 2^d internal states $\{S_0, ..., S_{2^d-1}\}$, 2^d ciphertexts $\{C_0, ..., C_{2^d-1}\}$, and 2^{2d} keys $\{K_{\langle i,j\rangle}\}$:

$$\{K_{\langle i,j\rangle}\} = \begin{bmatrix} K_{\langle 0,0\rangle} & K_{\langle 0,1\rangle} & \cdots & K_{\langle 0,2^d-1\rangle} \\ \vdots & & & \\ K_{\langle 2^d-1,0\rangle} & K_{\langle 2^d-1,1\rangle} & \cdots & K_{\langle 2^d-1,2^d-1\rangle} \end{bmatrix}. \tag{1}$$

The 3-tuple $[\{C_i\}, \{S_j\}, \{K_{\langle i,j\rangle}\}]$ is called a d-dimensional biclique, if

$$C_i = f_{K_{\langle i,j\rangle}}(S_j) \text{ for all } i, j \in \{0, ..., 2^d - 1\}. \tag{2}$$

In this paper, we study biclique cryptanalysis of the full HIGHT. We find that the slow and limited diffusion of the key schedule and encryption in HIGHT leads to relatively long biclqiues with high dimension and the efficient matching check with precomputations. Our attack recovers the 128-bit key of HIGHT with the computational complexity of $2^{126.4}$, faster than exhaustive search. This is the first single-key attack result for the full HIGHT.

This paper is organized as follows. Section 2 introduces biclique cryptanalysis and techniques which we use for our attack. In Section 3, the block cipher HIGHT is briefly described. In Section 4, we explain how to construct bicliques for 8 rounds of HIGHT. In Section 5, the key recovery procedure for the full HIGHT are explained and the complexity is evaluated. In Section 6, the complexities of our attack are evaluated. In Section 7, we conclude our work.

Table 1. Summary of the attacks on HIGHT(Imp.:Impossible, Diff.:Differential, Rel.:Related, and Rec.:Rectangle). Time complexities marked by '*' should be compared with 2^{127}, the complexity of the brute force attack with two distinct keys. Time complexities marked by '**' should be compared with 2^{126}, the complexity of the brute force attack with four distinct keys.

Rounds	Attack	Complexities		References
		Data	Time	
18	Imp. Diff.	$2^{46.8}$	$2^{109.2}$	[4]
22	Saturation	$2^{62.04}$	$2^{118.7}$	[10]
25	Imp. Diff.	2^{60}	$2^{126.8}$	[7]
26	Imp. Diff.	2^{61}	$2^{119.5}$	[8]
26	Rel.-Key Rec.	$2^{51.2}$	$2^{120.4}$ **	[7]
28	Rel.-Key Imp.	2^{60}	$2^{125.5}$ *	[7]
31	Rel.-Key Imp.	2^{63}	$2^{127.3}$ *	[8]
32(Full)	Rel.-Key Rec.	$2^{57.84}$	$2^{125.8}$ **	[6]
32(Full)	Biclique	2^{48}	$2^{126.4}$	This paper

2   Biclique Cryptanalysis

We introduce Bogdanov et al.'s biclique cryptanalysis [2].

2.1   Attack Procedure

The biclique attack procedure consists of the following phases.

Preparation. The adversary partitions the key space into sets of 2^{2d} keys each. The block cipher is considered as the composition of two subciphers: $e = f \circ g$. The key in a set is indexed as an element of a $2^d \times 2^d$ matrix like (1): $\{K_{\langle i,j \rangle}\}$.

Constructing Bicliques. For each set of keys, the adversary build the structure consisting of $\{C_0, ..., C_{2^d-1}\}$, $\{S_0, ..., S_{2^d-1}\}$, and $\{K_{\langle i,j \rangle}\}$ satisfying (2).

Collecting Data. The adversary obtains the plaintexts $\{P_i\}$ from the ciphertexts $\{C_i\}$ through the decryption oracle.

Testing Keys. The right key K maps the plaintext P_i to the intermediate S_j. So, the adversary checks

$$\exists i,j \; : \; P_i \xrightarrow[g]{K_{\langle i,j \rangle}} S_j, \tag{3}$$

which proposes a key candidate. If the right key is not found in the key set, he chooses another key set and repeats the above phases.

2.2   Constructing Bicliques from Independent Related-Key Differentials

We introduce one of two methods to construct bicliques, described in [2]. Firstly, we consider two sets of 2^d keys $\mathcal{A} = \{K_{\langle i,0 \rangle} | 0 \le i \le 2^d - 1\}$ and $\mathcal{B} = \{K_{\langle 0,j \rangle} | 0 \le j \le 2^d - 1\}$ such that $\mathcal{A} \cap \mathcal{B} = \{K_{\langle 0,0 \rangle}\}$. We assume that n is the bit-length of the block and S_0 is an intermediate string which is a randomly chosen n-bit string. Let the key $K_{\langle 0,0 \rangle}$ map intermediate state S_0 to ciphertext C_0 with f:

$$S_0 \xrightarrow[f]{K_{\langle 0,0 \rangle}} C_0. \tag{4}$$

$\{C_i\}$ and $\{S_j\}$ are obtained through the following computations:

$$S_0 \xrightarrow[f]{K_{\langle i,0 \rangle}} C_i, \tag{5}$$

$$S_j \xleftarrow[f^{-1}]{K_{\langle 0,j \rangle}} C_0. \tag{6}$$

Let $\Delta_i = C_0 \oplus C_i$, $\Delta_i^K = K_{\langle 0,0 \rangle} \oplus K_{\langle i,0 \rangle}$, $\nabla_j = S_0 \oplus S_j$, and $\nabla_j^K = K_{\langle 0,0 \rangle} \oplus K_{\langle 0,j \rangle}$. The Δ_i-differential is the related-key differential trail between (4) and (5) which is denoted by $0 \xrightarrow{\Delta_i^K} \Delta_i$. The ∇_j-differential is the related-key differential trail between (4) and (6) which is denoted by $\nabla_j \overset{\nabla_j^K}{\leadsto} 0$. If the two related-key differential trails $0 \overset{\Delta_i^K}{\leadsto} \Delta_i$ and $\nabla_j \overset{\nabla_j^K}{\leadsto} 0$ do not share active nonlinear components (such as modular additions with nonzero input differences in HIGHT) for all i and j, the following relation is satisfied.

$$S_0 \oplus \nabla_j \xrightarrow[f]{K_{\langle 0,0 \rangle} \oplus \Delta_i^K \oplus \nabla_j^K} C_0 \oplus \Delta_i \text{ for } i, j \in \{0, ..., 2^d - 1\}. \qquad (7)$$

This is proved by the concept of S-box switch [1] and a sandwich attack [3] in the theory of boomerang attacks [9]. Let $K_{\langle i,j \rangle} = K_{\langle 0,0 \rangle} \oplus \Delta_i^K \oplus \nabla_j^K$. Then, we get bicliques satisfying the definition of (2). The construction of a biclique requires less than $2 \cdot 2^d$ computations of f.

Note that we could not construct any good bicliques suitable for the full-round attack on HIGHT, with the other approach for biclique construction in [2].

2.3   Matching with Precomputations

The matching with precomputations is an efficient way to check equation (3) in the attack procedure. Let v be a part of an internal state between $\{P_i\}$ and $\{S_j\}$. v is called the matching variable. We denote v computed from P_i by $\overrightarrow{v}_i$ and v computed from S_j by $\overleftarrow{v}_j$. First, the adversary computes and store in memory the followings:

$$\text{for } i = 0, 1, ..., 2^d - 1, \quad P_i \xrightarrow{K_{\langle i,0 \rangle}} \overrightarrow{v}_i \text{ and}$$
$$\text{for } j = 0, 1, ..., 2^d - 1, \quad \overleftarrow{v}_j \xleftarrow{K_{\langle 0,j \rangle}} S_j.$$

Then, for particular i and j, the adversary checks the matching at v by recomputing only those parts of the cipher which differ from the stored ones.

The cost of recomputation depends on the diffusion properties of both internal rounds and the key schedule of the cipher. Since the HIGHT key schedule has very slow and limited diffusion, the adversary can skip most recomputations of the key schedule operations.

3   Description of HIGHT

We use the following notations for describing HIGHT.

- $\oplus$: bitwise exclusive OR(XOR)
- $\boxplus$: addition modulo 2^8

HIGHT takes a 64-bit plaintext P and a 128-bit key K, and its 32-round encryption procedure produces a 64-bit ciphertext C. From now on, we present any 64-bit variable A and any 128-bit variable B as a tuple of eight bytes $(A[7], ..., A[1], A[0])$ and a tuple of sixteen bytes $(B[15], ..., B[1], B[0])$.

The key schedule produces 128 8-bit subkeys $SK[0], ..., SK[127]$ from a 128-bit key $K = (K[15], ..., K[0])$: for $0 \leq i \leq 7$ and $0 \leq j \leq 7$,

$$\begin{cases} SK[16i + j] \leftarrow K[j - i \bmod 8] \boxplus \delta[16i + j], \\ SK[16i + j + 8] \leftarrow K[(j - i \bmod 8) + 8] \boxplus \delta[16i + j + 8], \end{cases}$$

where $\delta[0], ..., \delta[127]$ are public constants.

Let $X_{i-1} = (X_{i-1}[7], ..., X_{i-1}[0])$ and $X_i = (X_i[7], ..., X_i[0])$ be the input and output of the Round $i - 1$ for $1 \leq i \leq 32$, respectively, where 'Round i' denotes the $(i + 1)$-th round(i.e. Round 0 implies the first round).

The encryption procedure of HIGHT is as follows.

1. Initial Transformation:

$$X_0[0] \leftarrow P[0] \boxplus K[12]; X_0[2] \leftarrow P[2] \oplus K[13];$$
$$X_0[4] \leftarrow P[4] \boxplus K[14]; X_0[6] \leftarrow P[6] \oplus K[15];$$
$$X_0[1] \leftarrow P[1]; X_0[3] \leftarrow P[3]; X_0[5] \leftarrow P[5]; X_0[7] \leftarrow P[7].$$

2. Round Iteration for $1 \leq i \leq 32$:

$$X_i[0] \leftarrow X_{i-1}[7] \oplus (F_0(X_{i-1}[6], SK[4i - 1]));$$
$$X_i[2] \leftarrow X_{i-1}[1] \boxplus (F_1(X_{i-1}[0], SK[4i - 2]));$$
$$X_i[4] \leftarrow X_{i-1}[3] \oplus (F_0(X_{i-1}[2], SK[4i - 3]));$$
$$X_i[6] \leftarrow X_{i-1}[5] \boxplus (F_1(X_{i-1}[4], SK[4i - 4]));$$
$$X_i[1] \leftarrow X_{i-1}[0]; X_i[3] \leftarrow X_{i-1}[2]; X_i[5] \leftarrow X_{i-1}[4]; X_i[7] \leftarrow X_{i-1}[6],$$

where for two 8-bit inputs x and sk, the functions F_0 and F_1 are defined by

$$\begin{cases} F_0(x, sk) = (x^{\lll 1} \oplus x^{\lll 2} \oplus x^{\lll 7}) \boxplus sk, \\ F_1(x, sk) = (x^{\lll 3} \oplus x^{\lll 4} \oplus x^{\lll 6}) \oplus sk. \end{cases}$$

3. Final Transformation:

$$C[0] \leftarrow X_{32}[1] \boxplus K[0]; C[2] \leftarrow X_{32}[3] \oplus K[1];$$
$$C[4] \leftarrow X_{32}[5] \boxplus K[2]; C[6] \leftarrow X_{32}[7] \oplus K[3];$$
$$C[1] \leftarrow X_{32}[2]; C[3] \leftarrow X_{32}[4]; C[5] \leftarrow X_{32}[6]; C[7] \leftarrow X_{32}[0].$$

4   Constructing Bicliques for 8 Rounds

In this section, we explain how to construct bicliques for 8 rounds of HIGHT. Table 2 lists key bytes used at every round. For example, the subkeys $SK[43]$, $SK[42]$, $SK[41]$, and $SK[40]$ are added at Round 10(=8+2), and you can see four bytes $K[9], K[8], K[15]$, and $K[14]$ of the key K are used for generating them from Table 2. Considering the table, we found varying $K[7]$ and $K[9]$ give bicliques for attack on the full rounds of HIGHT.

Table 2. Key bytes used for generating subkeys

Round i	Round i	Round $8+i$	Round $16+i$	Round $24+i$
0	3, 2, 1, 0	1, 0, 7, 6	7, 6, 5, 4	5, 4, 3, 2
1	7, 6, 5, 4	5, 4, 3, 2	3, 2, 1, 0	1, 0, 7, 6
2	11, 10, 9, 8	9, 8, 15, 14	15, 14, 13, 12	13, 12, 11, 10
3	15, 14, 13, 12	13, 12, 11, 10	11, 10, 9, 8	9, 8, 15, 14
4	2, 1, 0, 7	0, 7, 6, 5	6, 5, 4, 3	4, 3, 2, 1
5	6, 5, 4, 3	4, 3, 2, 1	2, 1, 0, 7	0, 7, 6, 5
6	10, 9, 8, 15	8, 15, 14, 13	14, 13, 12, 11	12, 11, 10, 9
7	14, 13, 12, 11	12, 11, 10, 9	10, 9, 8, 15	8, 15, 14, 13

The base keys $K_{\langle 0,0 \rangle}$ are all possible 2^{112} 16-byte values with $K[7]$ and $K[9]$ fixed to 0 whereas the remaining 14 bytes run over all values. The keys $\{K_{\langle i,j \rangle}\}$ in a set are enumerated by all possible byte differences i and j with respect to the base key $K_{\langle 0,0 \rangle}$. This partitions the HIGHT key space into the 2^{112} sets of 2^{16} keys each. Note that $K[7]$ and $K[9]$ are not used for whitening keys. So, for simplicity, we do not bother to consider whitening key additions and do not depict them in figures because they have no effect on our attack at all.

Let f be the subcipher from Round 24 to Round 31. The adversary fixes $C_0 = 0$ and derives $S_0 = f_{K_{\langle 0,0 \rangle}}^{-1}(C_0)$. The Δ_i-differentials are based on the difference Δ_i^K where the difference of $K[9]$ is i and the other bytes have zero difference, and ∇_j-differentials are based on the difference ∇_j^K where the difference of $K[7]$ is j and the other bytes have zero difference. Both of differentials are depicted in Fig. 1. Note that the right F_1 in Round 26 and the right F_0 in Round 30 are affected by the key byte $K[9]$, and the left F_1 in Round 24 and the left F_0 in Round 28 are affected by the key byte $K[7]$. They are colored blue (in Δ_i-differential) and red (in ∇_j-differential) in Fig. 1. As two differentials share no active nonlinear elements (modular additions), the resulting combined differentials yield a biclique of dimension 8.

Since the Δ_i-differential affects only 6 bytes of the ciphertext, all the ciphertexts can be forced to share the same values in two bytes $C[0]$ and $C[1]$. As a result, the data complexity does not exceed 2^{48}.

5   Key Recovery for the Full HIGHT

We describe the key recovery procedure using bicliques for the full HIGHT. We rewrite the decomposition of the cipher:

$$E: \quad P \xrightarrow{g_1} V \xrightarrow{g_2} S \xrightarrow{f} C,$$

where g_1 is the subcipher from Round 0 to Round 11, g_2 from Round 12 to from Round 23, f from Round 24 to Round 31. In Section 4, 8-round bicliques are constructed for Round 24 to Round 31. We assume the plaintexts set $\{P_i\}$ corresponding to an 8-round biclique are obtained through the decryption oracle.

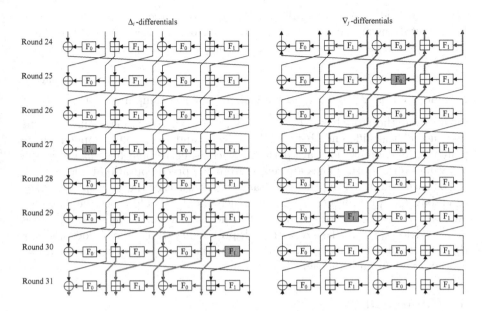

Fig. 1. Δ_i- and ∇_j-differentials for 8-round bicliques

Applying (3) to $g_2 \circ g_1$, the adversary detects the right key by computing an intermediate variable v in both directions:

$$P_i \xrightarrow[g_1]{K_{\langle i,j \rangle}} \vec{v} \stackrel{?}{=} \overleftarrow{v} \xleftarrow{K_{\langle i,j \rangle}}{g_2^{-1}} S_j. \tag{8}$$

This is performed through the followings. Please recall that we do not bother to consider whitening key additions and do not depict them in figures because they have no effect on our attack at all.

Precomputation. For the efficient meet-in-the-middle attack on the $g_2 \circ g_1$, we use the matching with precomputations. According to 2.3, we prepare precomputations with $2^{d+1}(= 2^9)$ computations.

For all $i = 0, ..., 2^8 - 1$, the adversary computes the 4-th byte of the output of the Round 11, $X_{11}[4]$ from P_i and $K_{\langle i,0 \rangle}$ in forward direction, and store it as $\vec{v}_i$, together with the intermediate states and subkeys in memory. For all $j = 0, ..., 2^8 - 1$, the adversary computes $X_{11}[4]$ from S_i and $K_{\langle 0,j \rangle}$ in backward direction, and store it as $\overleftarrow{v}_j$, together with the intermediate states and subkeys in memory.

Computation in backward direction. In backward direction, the adversary should compute $\overleftarrow{v}(= X_{11}[4])$ from S_j and $K_{\langle i,j \rangle}$ for all i and j, and store them in memory. We look at how the computation $\overleftarrow{v} \xleftarrow{K_{\langle i,j \rangle}} S_j$ differs from the stored one $\overleftarrow{v}_j \xleftarrow{K_{\langle 0,j \rangle}} S_j$. It is determined by the influence of the difference between keys $K_{\langle i,j \rangle}$ and $K_{\langle 0,j \rangle}$. The full area to be recomputed is depicted in Fig. 2. Recomputations are performed according to bold lines in Fig. 2, and the values on the other lines are reused from the precomputation table.

Computation in forward direction. In forward direction, the adversary should compute $\vec{v}$ from P_i and $K_{\langle i,j \rangle}$. We look at how the computation $P_i \xrightarrow{K_{\langle i,j \rangle}} \vec{v}$ differs from the stored one $P_i \xrightarrow{K_{\langle i,0 \rangle}} \vec{v}_i$. Similarly, it is determined by the influence of the difference between keys $K_{\langle i,j \rangle}$ and $K_{\langle i,0 \rangle}$, now applied to the plaintext. Recomputations are performed according to bold lines in Fig.2, and the values on the other lines are reused from the precomputation table.

For each computed $\vec{v}$, the adversary checks whether the corresponding key candidate $K_{\langle i,j \rangle}$ satisfies (8). If he finds such one, he should check the matching on whole bytes at X_{11} for the $K_{\langle i,j \rangle}$, P_i, and S_j. The matching on whole bytes at X_{11} yields the right key K with high probability. If a biclique does not give the right key, the adversary should choose another biclique and repeat the above procedure again until the right key is found.

6   Complexities

The total computational complexity of the biclique attack on the full HIGHT is evaluated as follows:

$$C_{\text{total}} = 2^{n-2d}[C_{\text{biclique}} + C_{\text{precomp}} + C_{\text{recomp}} + C_{\text{falsepos}}], \qquad (9)$$

where

- $n = 128$ and $d = 8$.
- C_{biclique} is the complexity of constructing a single biclique. In our attack, it is at most $2^{d+1}(= 2^9)$ 8-round computations.
- C_{precomp} is the complexity of preparing the precomputation for the matching check in (8). In our attack, it is less than $2^d(= 2^8)$ 23-round computations.
- C_{recomp} is the complexity of the recomputation of the internal variable v $2^{2d}(= 2^{16})$ times. In one backward recomputation, F_0 and F_1 are computed 9 and 8 times, respectively, and 11 XOR and 10 addition modulo 2^8 operations are required. Furthermore, 2 newly generated subkey bytes are needed for a different key. In one forward recomputation, F_0 and F_1 are computed 10 and 10 times, respectively, and 11 XOR and 11 addition modulo 2^8 operations are required. Furthermore, 2 newly generated subkey bytes are needed for a different key.
  Altogether, C_{recomp} is less than 2^{16} 10-round computations plus $2^{16}\alpha$ where α means the computational cost for generating 4 subkey bytes.
- C_{falsepos} is the complexity caused by false positives, which have to be matched on other byte positions. Since the matching check is performed on a single byte in our attack, $C_{\text{false-pos}}$ is less than $2^{2d-8}(= 2^8)$ computations. The unit of the computation is less than 23-round if the adversary checks partial matching for another byte, e.g. $X_{11}[0]$, $X_{11}[2]$, or $X_{11}[6]$.

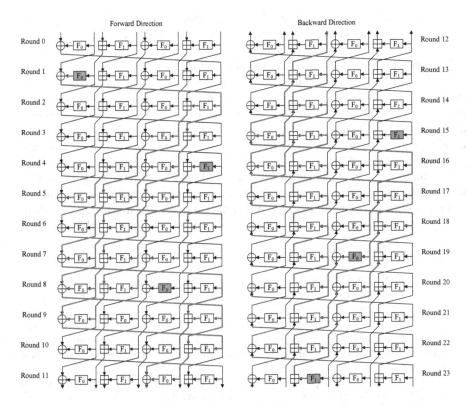

Fig. 2. Recomputations in forward and backward directions

Consequently, the most dominant complexity is C_{recomp}, and $C_{total} \simeq 2^{112} \cdot 2^{16}(10R+\alpha)$ where R means one-round computation. We consider the unit of the computation for the brute-force attack as $32R$ and the key schedule operation. For a clever exhaustive search, one can use the fact that only 8 modular additions are required for key scheduling if he makes the sequence of the tested keys such that neighbored keys differ only on a single byte. However, even if we compare our attack with this smart exhaustive search, C_{total} is upper bounded by $2^{128} \cdot \frac{1}{3} \simeq 2^{126.4}$. The memory requirement is upper bounded by the storage of 2^8 computations of 23 rounds from Round 0 to Round 11 and from Round 13 to Round 23.

7   Conclusion and Discussion

Biclique cryptanalysis provides advanced techniques of the meet-in-the-middle attack. We found the slow and limited diffusion of the key schedule and encryption in HIGHT leads to relatively long bicliques with high dimension and the efficient matching check with precomputations. Our attack recovers the 128-bit key of the full HIGHT with the computational complexity of $2^{126.4}$. This is the first single-key attack result for it.

Since our attack causes a marginal improvement of time complexity from 2^{128}, we need to examine carefully time complexities of generic attacks. A naive implementation of exhaustive search with a single plaintext-ciphertext pair costs 2^{128} encryptions, but more decent implementations will not.

If we divide the key space into 2^{120} subspaces where only the key byte $K[11]$ changes, the exhaustive search considering such partition of the key space can take the advantage of partial computations in the first several rounds. Moreover if we add a meet-in-the-middle approach, we can apply partial computations for middle rounds and last rounds. Totally, 18 rounds are saved per trial. Its time complexity is around 2^{127}.

The time complexity of our attack is computed as $2^{126.4}$ with the same estimation way. In fact, the exhaustive with a single plaintext-ciphertext pair is more simple and parallelizable than our attack, but a high-cost dedicated machine is out of our consideration. We address our work reports a weakness of an existing block cipher.

References

1. Biryukov, A., Khovratovich, D.: Related-Key Cryptanalysis of the Full AES-192 and AES-256. In: Matsui, M. (ed.) ASIACRYPT 2009. LNCS, vol. 5912, pp. 1–18. Springer, Heidelberg (2009)
2. Bogdanov, A., Khovratovich, D., Rechberger, C.: Biclique Cryptanalysis of the Full AES. In: Lee, D.H. (ed.) ASIACRYPT 2011. LNCS, vol. 7073, pp. 344–371. Springer, Heidelberg (2011); Full paper is availavle at Cryptology ePrint Archive 2011/449
3. Dunkelman, O., Keller, N., Shamir, A.: A Practical-Time Related-Key Attack on the KASUMI Cryptosystem Used in GSM and 3G Telephony. In: Rabin, T. (ed.) CRYPTO 2010. LNCS, vol. 6223, pp. 393–410. Springer, Heidelberg (2010)
4. Hong, D., Sung, J., Hong, S., Lim, J., Lee, S., Koo, B.-S., Lee, C., Chang, D., Lee, J., Jeong, K., Kim, H., Kim, J., Chee, S.: HIGHT: A New Block Cipher Suitable for Low-Resource Device. In: Goubin, L., Matsui, M. (eds.) CHES 2006. LNCS, vol. 4249, pp. 46–59. Springer, Heidelberg (2006)
5. International Organization for Standardization, "Information technology – Security techniques – Encryption algorithms – Part 3: Block ciphers," ISO/IEC 18033-3:2005 (2005)
6. Koo, B., Hong, D., Kwon, D.: Related-Key Attack on the Full HIGHT. In: Rhee, K.-H., Nyang, D. (eds.) ICISC 2010. LNCS, vol. 6829, pp. 49–67. Springer, Heidelberg (2011)
7. Lu, J.: Cryptanalysis of Reduced Versions of the HIGHT Block Cipher from CHES 2006. In: Nam, K.-H., Rhee, G. (eds.) ICISC 2007. LNCS, vol. 4817, pp. 11–26. Springer, Heidelberg (2007)
8. Özen, O., Varıcı, K., Tezcan, C., Kocair, Ç.: Lightweight Block Ciphers Revisited: Cryptanalysis of Reduced Round PRESENT and HIGHT. In: Boyd, C., González Nieto, J. (eds.) ACISP 2009. LNCS, vol. 5594, pp. 90–107. Springer, Heidelberg (2009)
9. Wagner, D.: The Boomerang Attack. In: Knudsen, L.R. (ed.) FSE 1999. LNCS, vol. 1636, pp. 156–170. Springer, Heidelberg (1999)
10. Zhang, P., Sun, B., Li, C.: Saturation Attack on the Block Cipher HIGHT. In: Garay, J.A., Miyaji, A., Otsuka, A. (eds.) CANS 2009. LNCS, vol. 5888, pp. 76–86. Springer, Heidelberg (2009)

Preimage Attacks on Step-Reduced SM3 Hash Function

Jian Zou[1,2], Wenling Wu[1], Shuang Wu[1],
Bozhan Su[1], and Le Dong[1]

[1] State Key Laboratory of Information Security,
Institute of Software, Chinese Academy of Sciences, Beijing 100190, P.R. China
{zoujian,wwl,wushuang,subozhan,dongle}@is.iscas.ac.cn
[2] Graduate University of Chinese Academy of Sciences, Beijing 100049, P.R. China

Abstract. This paper proposes a preimage attack on SM3 hash function reduced to 30 steps. SM3 is an iterated hash function based on the Merkle-Damgård design. It is a hash function used in applications such as the electronic certification service system in China. Our cryptanalysis is based on the Meet-in-the-Middle (MITM) attack. We utilize several techniques such as initial structure, partial matching and message compensation to improve the standard MITM preimage attack. Moreover, we use some observations on the SM3 hash function to optimize the computation complexity. Overall, a preimage of 30 steps SM3 hash function can be computed with a complexity of 2^{249} SM3 compression function computation, and requires a memory of 2^{16}. As far as we know, this is yet the first preimage result on the SM3 hash function.

Keywords: SM3, hash function, preimage attack, Meet-in-the-Middle.

1 Introduction

Cryptographic hash functions play a very important role in cryptology. In general, hash function must fulfill three requirements made up of preimage resistance, second preimage resistance and collision resistance. In other word, it should be impossible for an adversary to find a collision faster than birthday attack or a (second) preimage faster than brute force attack. In recent years, after the pioneering work of Wang[1,2,3], many cryptanalysis tools and new ideas have been applied to attack hash function such as MD4, MD5, RIPEMD, SHA-0 and SHA-1.

At FSE 2008, Leurent proposed a way to construct the preimage of full MD4 hash function[4]. From then on, many techniques are proposed to improve the preimage attacks. In this paper, we adopt the technique of MITM preimage attack proposed by Aoki and Sasaki[5]. The basic idea of the MITM technique is to divide hash function into two chunks (sub-functions are called chunks) with regard to the neutral words. Using the independence of neutral words, we can calculate the two chunks to the matching point respectively. Due to the independence, we compute a preimage faster than the brute force attack.

H. Kim (Ed): ICISC 2011, LNCS 7259, pp. 375–390, 2012.
© Springer-Verlag Berlin Heidelberg 2012

The MITM technique has been widely used to search for the preimage of many famous hash functions, such as MD4[5,6,7], HAVAL[8,9], Tiger[10,6,11], GOST[12], MD5[13,14,9], SHA-2[10,15,6], reduced SHA-0 and SHA-1[16].

SM3[17] is a hash function used in applications such as the electronic certification service system. It has the similar structure as SHA-2. However, as explained below, SM3 has a more complex step function and stronger message dependency than SHA-256. Choosing neutral words from SM3 hash function is more difficult than SHA-256. In other word, finding preimage of SM3 based on MITM method seems to be more difficult than SHA-256.

Our Contributions. In this paper, we utilize several techniques to present a preimage attack on reduced version of SM3 hash function to 30 steps. Firstly We use the MITM technique to build up the general framework of the preimage attack on SM3 compression function. Secondly we utilize the property of message expansion of SM3 hash function and several techniques to achieve better results. As far as we know, this is yet the first preimage result on the SM3 hash function. The preimage of the reduced SM3 hash function can be computed in the computational complexity of 2^{249}, and the memory requirement is 2^{16}.

Outline of the Paper. This paper is organized as follows. We describe SM3 hash function in Section 2. In Section 3, we make a brief introduction to the Meet-in-the-Middle preimage attack. Then we explain the details of the our techniques and show how to use these techniques to attack more steps of compression function of SM3 hash function in Section 4. Finally, we summarize the attack process and the complexity of the attack in Section 5.

2   Description of SM3

SM3 is a hash function which compresses a message not more than $(2^{64}-1)$ bits into 256 bits. SM3 is an iterated hash function based on the Merkle-Damgård design. Its compression function maps 256-bit state and 512-bit message block to a new 256-bit state, that is, the hash value is computed as follows:

$$\begin{cases} V_0 \leftarrow IV \\ V_{i+1} \leftarrow CF(V_i, M_i) \text{ for } i = 0, 1, \ldots, n-1, \end{cases}$$

where IV = 7380166f 4914b2b9 172442d7 da8a0600 a96f30bc 163138aa e38dee4d b0fb0e4e is the initial value defined in the specification[17] and V_0 is a 256-bit value. At last, V_n is output as a hash value of M.

Before applying the compression function, the input message M is processed to be a multiple of 512 bits by the padding procedure. According to the padding procedure, a single bit '1' and len_0 '0's are put at the end of the message M. Here len_0 satisfies the following equation $len_M + 1 + len_0 \equiv 448 \mod 512$ (len_M and len_0 are short for the length of M and the number of '0' respectively). After the above step, we put another 64 bits including the length of the message at the end of the padding. Then the padded message M^* is divided into 512-bit blocks M_i ($i = 0, 1, \ldots, n-1$).

The compression function $V_{i+1} \leftarrow CF(V_i, M_i)$ is computed as follows:

1. Divide M_i into 32-bit message words W_j ($j = 0, 1, \ldots, 15$) and expand them into vectors of 68 32-bit words W_j and 64 32-bit words W'_j by the following formula:

   a FOR $j = 16$ TO 67

   $$W_j \leftarrow P_1(W_{j-16} \oplus W_{j-9} \oplus (W_{j-3} \lll 15)) \oplus (W_{j-13} \lll 7) \oplus W_{j-6} \quad (1)$$

   ENDFOR

   Here we generate W_{67} for calculating W'_{63}. $P_1()$ is a permutation that will be defined later.

   b FOR $j = 0$ TO 63
   $$W'_j = W_j \oplus W_{j+4}$$

   ENDFOR

2. $S_0 \leftarrow V_i$, where $S_j = (A_j, \ldots, H_j)$.
3. S_j is updated by step function f which is defined in Fig.1, and should be operated 64 identical steps.
4. Output $V_{i+1} \leftarrow S_0 \oplus S_{64}$, i.e. $(A_0 \oplus A_{64}, \ldots, H_0 \oplus H_{64})$.

Three bitwise Boolean functions and two diffusion functions are employed by the step transformation. The Boolean functions

$$FF_j(X, Y, Z) = \begin{cases} X \oplus Y \oplus Z & 0 \le j \le 15 \\ (X \wedge Y) \vee (X \wedge Z) \vee (Y \wedge Z) & 16 \le j \le 63 \end{cases}$$

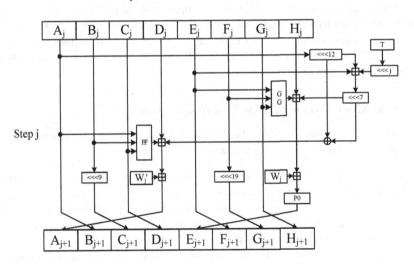

Fig. 1. j-th step function of SM3 uses words W_j and W'_j of the expanded message to update the the state of eight chaining variables $A_j, \ldots, H_j$

$$GG_j(X,Y,Z) = \begin{cases} X \oplus Y \oplus Z & 0 \le j \le 15 \\ (X \wedge Y) \vee (\neg X \wedge Z) & 16 \le j \le 63, \end{cases}$$

and diffusion permutations

$$P_0(X) = X \oplus (X \lll 9) \oplus (X \lll 17),$$

$$P_1(X) = X \oplus (X \lll 15) \oplus (X \lll 23),$$

and the inverse function of $P_0(X)$ and $P_1(X)$ are as follows,

$$P_0^{-1}(X) = X \oplus (X \lll 2) \oplus (X \lll 3) \oplus (X \lll 9) \oplus (X \lll 11)$$
$$\oplus (X \lll 17) \oplus (X \lll 18) \oplus (X \lll 19) \oplus (X \lll 27),$$

$$P_1^{-1}(X) = X \oplus (X \lll 5) \oplus (X \lll 13) \oplus (X \lll 14) \oplus (X \lll 15)$$
$$\oplus (X \lll 21) \oplus (X \lll 23) \oplus (X \lll 29) \oplus (X \lll 30).$$

3   Previous Works: Techniques for Preimage Attacks

3.1   Converting Pseudo-preimages to a Preimage

A pseudo-preimage is a pair of (x, M) satisfying $CF(x, M) = y$, where y is a given hash value and CF is the compression function. The difference between a pseudo-preimage and a preimage is that we don't need x to be equal to the initial value. There is a generic algorithm that converts pseudo-preimages to a preimage. Assume we can find a pseudo-preimage with computational complexity 2^k, then the computational complexity of the conversion algorithm will be $2^{\frac{n+k}{2}+1}$. The detail of the algorithm is described in [18, Fact 9.99]. At FSE 2008, a new way called unbalanced-tree multi-target pseudo-preimage (MTPP) method to convert pseudo-preimages to a preimage was provided by Leurent[4]. The MTPP method can convert pseudo-preimages to a preimage more efficiently than the generic approach. Since our attack cannot satisfy the MTPP requirements, we omit the details of the new method.

3.2   The Meet-in-the-Middle Preimage Attack

The general idea of the Meet-in-the-Middle Preimage Attack is shown in Fig.2. We explain the idea in detail as follows:

1. Choose neutral words W_a and W_b respectively, and split the compression function into two chunks (sub-functions are called chunks) according to the neutral words, where one chunk is independent from neutral word W_a and the other chunk is independent from neutral word W_b. Here we call the two chunk the forward chunk and the backward chunk respectively.
2. Assign random value to the chaining registers at the splitting point and fix all other message words except W_a and W_b. For all possible values of W_a, we compute backward from the splitting point and obtain the value at the matching point. Store the values in a list L_a.

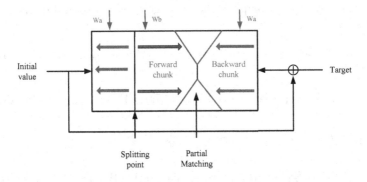

Fig. 2. Pseudo-Preimage Attack on Davies-Meyer Hash Functions

3. For all possible values of W_b, we compute forward from the splitting point and obtain the result at the matching point. Check if there exists an entry in L_a that matches the result (all the state bits or only some bits of the state) at the matching point.
4. Repeat the above two steps with different initial assignments until we find a full match.
5. The above four steps offer a way to return a pseudo-preimage of the given hash function because the initial value is determined during the attack. Furthermore, we can convert pseudo-preimages to a preimage by the method introduced in section 3.1.

4   Our Strategies of the Pseudo-preimage on the Compression Function of Reduced SM3 Hash Function

As shown in Section 2, W'_j is determined by W_j and W_{j+4}. In other word, we need to process W_j and W_{j+4} in the j-th step, which means $step_j$ is affected by W_j and W_{j+4} both. Compared to SM3, SHA-256 only processes W_j in the j-th step. As a result, SM3 has stronger message dependency than SHA-256, and it is more difficult to select neutral words for SM3 than SHA-256. In this section, we will show several techniques that allow us to attack more steps on SM3.

4.1   The Message Expansion of SM3 Hash Function

Note that we want to choose the neutral words from message M_i. According to (1), we can select the neutral words from $\{W_0, \ldots, W_{15}\}$ instead. Note that the message expansion (1) of SM3 hash function is invertible; as a result, all the expanded message words can be uniquely determined by any consecutive 16 words $\{W_z, \ldots, W_{z+15}\}$, $0 \le z \le 52$. Once we choose any 16 consecutive words, we can determine the other ones recursively in both directions. Suppose we start with $\{W_z, \ldots, W_{z+15}\}$. To determine the optimal choice of the splitting point

and the neutral words, we expand the message words in both directions.
For backward direction:

$$W_{z-1} = P_1^{-1}(W_{z+15} \oplus W_{z+9} \oplus (W_{z+2} \lll 7)) \oplus W_{z+6} \oplus (W_{z+12} \lll 15),$$

$$W_{z-2} = P_1^{-1}(W_{z+14} \oplus W_{z+8} \oplus (W_{z+1} \lll 7)) \oplus W_{z+5} \oplus (W_{z+11} \lll 15), \quad (2)$$

$$W_{z-3} = P_1^{-1}(W_{z+13} \oplus W_{z+7} \oplus (W_z \lll 7)) \oplus W_{z+4} \oplus (W_{z+10} \lll 15),$$

$$W_{z-4} = P_1^{-1}(W_{z+12} \oplus W_{z+6} \oplus (W_{z-1} \lll 7)) \oplus W_{z+3} \oplus (W_{z+9} \lll 15),$$

$$W_{z-5} = P_1^{-1}(W_{z+11} \oplus W_{z+5} \oplus (W_{z-2} \lll 7)) \oplus W_{z+2} \oplus (W_{z+8} \lll 15),$$

$$W_{z-6} = P_1^{-1}(W_{z+10} \oplus W_{z+4} \oplus (W_{z-3} \lll 7)) \oplus W_{z+1} \oplus (W_{z+7} \lll 15),$$

$$W_{z-7} = P_1^{-1}(W_{z+9} \oplus W_{z+3} \oplus (W_{z-4} \lll 7)) \oplus W_z \oplus (W_{z+6} \lll 15),$$

$$W_{z-8} = P_1^{-1}(W_{z+8} \oplus W_{z+2} \oplus (W_{z-5} \lll 7)) \oplus W_{z-1} \oplus (W_{z+5} \lll 15).$$

For forward direction:

$$W_{z+16} = P_1(W_z \oplus W_{z+7} \oplus (W_{z+13} \lll 15)) \oplus (W_{z+3} \lll 7) \oplus W_{z+10}, \quad (3)$$

$$W_{z+17} = P_1(W_{z+1} \oplus W_{z+8} \oplus (W_{z+14} \lll 15)) \oplus (W_{z+4} \lll 7) \oplus W_{z+11},$$

$$W_{z+18} = P_1(W_{z+2} \oplus W_{z+9} \oplus (W_{z+15} \lll 15)) \oplus (W_{z+5} \lll 7) \oplus W_{z+12}, \quad (4)$$

$$W_{z+19} = P_1(W_{z+3} \oplus W_{z+10} \oplus (W_{z+16} \lll 15)) \oplus (W_{z+6} \lll 7) \oplus W_{z+13}, \quad (5)$$

$$W_{z+20} = P_1(W_{z+4} \oplus W_{z+11} \oplus (W_{z+17} \lll 15)) \oplus (W_{z+7} \lll 7) \oplus W_{z+14},$$

$$W_{z+21} = P_1(W_{z+5} \oplus W_{z+12} \oplus (W_{z+18} \lll 15)) \oplus (W_{z+8} \lll 7) \oplus W_{z+15}, \quad (6)$$

$$W_{z+22} = P_1(W_{z+6} \oplus W_{z+13} \oplus (W_{z+19} \lll 15)) \oplus (W_{z+9} \lll 7) \oplus W_{z+16}, \quad (7)$$

$$W_{z+23} = P_1(W_{z+7} \oplus W_{z+14} \oplus (W_{z+20} \lll 15)) \oplus (W_{z+10} \lll 7) \oplus W_{z+17},$$

$$W_{z+24} = P_1(W_{z+8} \oplus W_{z+15} \oplus (W_{z+21} \lll 15)) \oplus (W_{z+11} \lll 7) \oplus W_{z+18},$$

$$W_{z+25} = P_1(W_{z+9} \oplus W_{z+16} \oplus (W_{z+22} \lll 15)) \oplus (W_{z+12} \lll 7) \oplus W_{z+19},$$

$$W_{z+26} = P_1(W_{z+10} \oplus W_{z+17} \oplus (W_{z+23} \lll 15)) \oplus (W_{z+13} \lll 7) \oplus W_{z+20},$$

$$W_{z+27} = P_1(W_{z+11} \oplus W_{z+18} \oplus (W_{z+24} \lll 15)) \oplus (W_{z+14} \lll 7) \oplus W_{z+21},$$

$$W_{z+28} = P_1(W_{z+12} \oplus W_{z+19} \oplus (W_{z+25} \lll 15)) \oplus (W_{z+15} \lll 7) \oplus W_{z+22},$$

$$W_{z+29} = P_1(W_{z+13} \oplus W_{z+20} \oplus (W_{z+26} \lll 15)) \oplus (W_{z+16} \lll 7) \oplus W_{z+23}.$$

After many tests, we find out that choosing W_{z+9} and W_{z+11} as neutral words enables us to attack more steps. After that, the forward chunk from $step_{z+10}$ to $step_{z+17}$ is independent from W_{z+9}, while the backward chunk from $step_{z+8}$ to $step_{z+9}$ is independent from W_{z+11}. Although the length of backward is short now, we can increase it with the below techniques.

4.2 Initial Structure

An Initial structure is a method that can swap the order of the neutral message words near the splitting point, as shown in Fig.3. Note that such shifting is allowed only when the swap dose not change the behavior of the step function.

In this paper, initial structure utilizes the absorption property of the Boolean functions $MAJ(X, Y, Z) = (X \wedge Y) \vee (X \wedge Z) \vee (Y \wedge Z)$ and $IF(X, Y, Z) = (X \wedge Y) \vee (\neg X \wedge Z)$. For MAJ, if $X = Y$, then $MAJ(X, Y, Z) = X = Y$, which means Z does not affect the output of MAJ function. Similarly, when $Y = Z$ or $X = Z$, X or Y dose not affect the output. For IF, if X is 1 (all bits of X are 1), then $IF(1, Y, Z) = Y$ which means Z does not affect the output of IF function. Similarly when X is 0 (all bits of X are 0), Y does not affect the output. The result is also right when we want to control some bits of the output. The only thing we need to do is to fix the corresponding bits of input of the Boolean function instead of all bits of input.

In Fig.3, we consider 3 consecutive step functions, i.e. from $step_{z+7}$ to $step_{z+9}$. We show a way to move messages W_{z+9} and W'_{z+9} to $step_{z+7}$ and move message W'_{z+7} (which contains W_{z+11}) to $step_{z+8}$ without changing the behavior of the step function. Note that message words W'_{z+9} and W_{z+9} can be added to D_{z+9} and H_{z+9} respectively with no constraint.

As shown in Fig.3, l most significant bits (MSB) of W_{z+9} and W'_{z+9} are arbitrary (neutral bits, denoted by right oblique dashed box) and the rest bits (gray) are set to 0, while l MSB of W'_{z+7} (gray) are set to 0 and the rest bits are arbitrary (neutral bits, denoted by left oblique dashed box). Meanwhile, we fix A_{z+8} to 0 (to avoid interference with addition on least significant bits) and set l MSB of B_{z+8} to 0 (to utilize the absorption property of MAJ). Due to the absorption property of MAJ, we ensure that the output of the FF_{z+8} is not

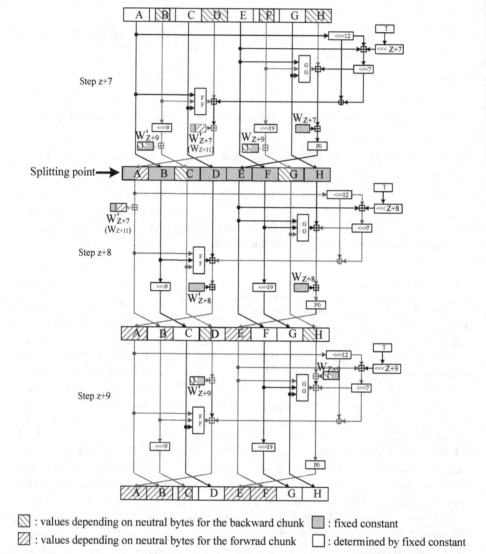

: values depending on neutral bytes for the backward chunk   : fixed constant

: values depending on neutral bytes for the forwrad chunk   : determined by fixed constant

Fig. 3. Initial structure allows us to move W_{z+9}(right oblique dashed box) two steps upwards and W_{z+11}(left oblique dashed box) one step downward. The right oblique dashed boxes and left oblique dashed boxes are used to denote messages that are influenced by messages W_{z+9} and W_{z+11} respectively, the gray color means we can set them freely at the beginning, and the white color means messages are determined by the gray color messages.

affected by the neutral bits of W_{z+9}. Therefore, we can move W'_{z+9} to $step_{z+7}$ (adding W'_{z+9} to C_{z+8}), and transfer W'_{z+7} to $step_{z+8}$ (adding W'_{z+7} to B_{z+9}). Similarly, due to the absorption property of IF, if we set l MSB of E_{z+8} to 1, we can move message W_{z+9} to $step_{z+7}$ too (adding W_{z+9} to G_{z+8}). If we don't utilize the absorption property of MAJ or IF, A_{z+9} or E_{z+9} is affected by W_{z+9} and W_{z+11} both, which means we fail to separate the forward chunk from W_{z+9}. We choose the optimal value $l = 16$ to reduce the complexity.

As we must utilize the absorption property of MAJ and IF, we should make sure $FF = MAJ$ and $GG = IF$, which means $z + 8 \geq 16$. If we just want to find a pseudo-preimage of the compression function of SM3, we don't need to consider the message padding.

4.3   Message Compensation

Using the initial structure technique explained in Section 4.2, then, the forward chunk is from $step_{z+10}$ to $step_{z+17}$, and the backward chunk is from $step_{z-1}$ to $step_{z+6}$, and the initial structure is from $step_{z+7}$ to $step_{z+9}$.

In the forward chunk, W_{z+9} is used in equation (4), (6), and (7), because W_{z+18} contains W_{z+9} and appears in (6). In order to extend the forward chunk for several more steps, we compensate W_{z+9} by W_{z+2}, W_{z+3}, W_{z+6} and W_z. Here the compensation means making the equation values of (4), (7), (5), and (3) independent from W_{z+9} by forcing

$$W_{z+2} \oplus W_{z+9} = C,$$

$$P_1(W_{z+6}) \oplus (W_{z+9} \lll 7) = C,$$

$$P_1(W_{z+3}) \oplus (W_{z+6} \lll 7) = C,$$

$$P_1(W_z) \oplus (W_{z+3} \lll 7) = C,$$

(C is some constant here, and we set $C = 0$ for simplicity). Because these messages are independent from each other, we can satisfy the above four equations with W_{z+2}, W_{z+3}, W_{z+6} and W_z. Similarly, in the backward chunk, W_{z+11} is used in equation (2), we compensate it by forcing

$$P_1^{-1}(W_{z+14}) = (W_{z+11} \lll 15).$$

Note that the state of $step_{z+21}$ is affected by W_{z+25} (which contains W_{z+9}). As a result, the forward chunk from $step_{z+10}$ to $step_{z+20}$ is independent from W_{z+9}. Similarly, the backward chunk from $step_{z-4}$ to $step_{z+6}$ is independent from W_{z+11}. Besides, the initial structure is from $step_{z+7}$ to $step_{z+9}$. There are 25 steps totally, regardless of the choice of z.

We give a brief description of the messages W_j and W'_j influenced by the neutral words in Fig.4.

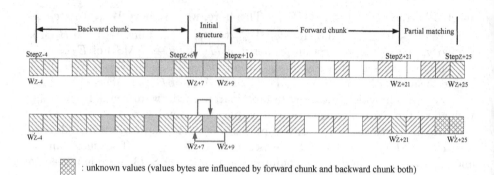

: unknown values (values bytes are influenced by forward chunk and backward chunk both)

Fig. 4. Messages that are influenced by the neutral words

4.4 Partial Matching

SM3 step function can be thought to be composed of two MD5 step functions. The basic partial matching for MD5 would offer us 3 more steps. However, using the property of SM3 we can get 2 more steps.

Partial matching is from $step_{z+21}$ to $step_{z+25}$. As shown in Fig.5, it is clear that message W_{z+25} is independent from W_{z+11}, so that registers from B_{z+26} to H_{z+26} are still known to us at $step_{z+25}$. In other word, we only lose register A_{z+26} at $step_{z+25}$. Note that we need at least one register to perform the matching, and the computation backward loses one register per step. Utilizing the property of the Feistel structure, register E_{z+22} is independent from message word W_{z+11} at $step_{z+21}$, and registers from A_{z+21} to H_{z+21} are independent from message word W_{z+9}. We use the point E_{z+22} shown in Fig.5 as the matching point.

Aoki proposed partial fixing technique to improve partial matching in [5]. Owing to the diffusion effect of the permutation P_0 and P_1, we cannot utilize partial fixing technique to improve our attack.

Jian Guo also extended the partial matching by using the indirect partial matching technique in [15]. However, as we know, some errors occur when we modify modular equations by exchanging the order of an addition and a bit rotation. Due to the rotation operation of the step function of SM3, we can't find such simple relation between states as Jian Guo did.

5   Preimage of the 30 Steps SM3 Hash Function

We should preset the message words W_{13}, W_{14} and W_{15} of the last block to satisfy the message padding of SM3, if we try to find preimage of the SM3 hash function.

We fix $z = 10$, then the neutral words are W_{19} and W_{21}, and the message words W_{13}, W_{14} and W_{15} are original W_{z+3}, W_{z+4} and W_{z+5} respectively. According to our previous cryptanalysis in Section 4, we preset the lower 16 bits of W_{19} in initial structure, so we can utilize the inverse function of $P_1(X)$, as well as 1

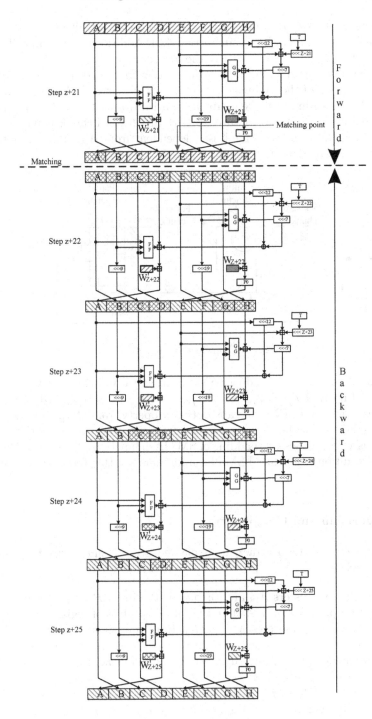

Fig. 5. Partial Matching

bit freedom of the lower 16 bits of W_{19} to ensure the last bit of W_{13} is '1'(the relation between W_{19} and W_{13} is in Section 4.3). Additionally we can utilize the freedom of W_{z+4} and W_{z+5} to satisfy the message padding.

Here we attack the reduced SM3 hash function from $step_6$ to $step_{35}$. We give a brief description of the preimage cryptanalysis of 30-step SM3 Hash Function in Fig.6. As shown in Fig.6, we swap the neutral words W_{19} and W_{21} near the splitting point with initial structure, so that the forward chunk is independent from W_{19} and the backward chunk is independent from W_{21}. Partial matching gives us 5 more steps from $step_{31}$ to $step_{35}$.

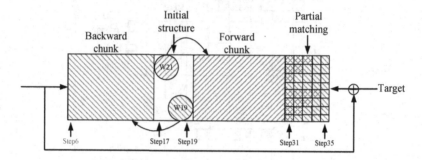

Fig. 6. Overview of the attack

If we want to attack the reduced SM3 from $step_0$, we will meet with several difficulties. Firstly, $FF_j(X, Y, Z) = GG_j(X, Y, Z) = X \oplus Y \oplus Z$ $(0 \le j \le 15)$, which means we can't utilize the absorption property from $step_0$ to $step_{15}$. Secondly, it is difficult to find three successive messages (gray, we can set them value freely) satisfying the message padding of SM3 hash function. We present a 28 step preimage result on the reduced SM3 hash function starting from $step_0$(shown in Appendix A).

5.1   Algorithm and Complexity

As all the elements of our attack are explained, we can summarize them in one algorithm below.

1. Set all bits of registers A_{18} and the upper 16 bits of B_{18} to 0 (to utilize the absorption property of MAJ) and the upper 16 bits of register E_{18} to 1 (to utilize the absorption property of IF). Randomly choose the registers C_{18}, D_{18}, F_{18}, G_{18}, H_{18}, the lower 16 bits of B_{18} and E_{18}. Set message words that are independent from neutral words randomly, i.e. W_{11}, W_{14}, W_{15}, W_{17}, W_{18}, W_{20}, W_{22}, W_{23}, W_{25}.
2. Set the lower 16 bits of W_{19} to special values to ensure the last bit of W_{13} is '1'. For all upper 16 bits of W_{19}, compute the corresponding W_{10}, W_{12}, W_{13} and W_{16}. For all possible values of W_{19}, we compute backward and obtain the value of E_{32} shown in Fig.5. Store the results in a list L_a.

3. Set the upper 16 bits of W_{21} to 0. For all lower 16 bits of W_{21}, compute the corresponding W_{24}. For all possible values of W_{21} and W_{24}, we compute forward and obtain the value of E_{32}. Here we use E_{32} as the matching point. Using partial matching technique, we check if there exists an entry in L_a that matches the result.
4. If a match is found, compute other registers A_{32}, B_{32}, C_{32}, D_{32}, F_{32}, G_{32}, and H_{32} to check whether they match from both directions. If they do, a pseudo-preimage is found.
5. Repeat the above four steps with different initial values until a pseudo-preimage is found.
6. Repeat step 5 to find sufficiently many pseudo-preimages, then use the conversion algorithm in [18, Fact 9.99] to find a preimage.

The length of preimage of SM3 is at least two blocks, the last block is used to find the pseudo-preimage while the second last block is used to find the link point of the initial value to the last block. If we consider the two blocks situation, we should preset W_{14} and W_{15} special values. The corresponding computational complexity is computed as follows. The computational complexity of step 2 and 3 is 2^{16} and 2^{16} respectively, and it generates 2^{32} pairs. After that, we examine 32-bit matching for 2^{32} pairs, and we obtain $2^{32} \times 2^{-32} = 1$ pair whose 32 bits are matched. Finally, by repeating step 1 to 3 of the above procedure $2^{256-32} = 2^{224}$ times, we obtain a pair, where all 256 bit are matched (with the degree of freedom of messages and states we can do this). As a result, the final complexity of the pseudo-preimage attack is $2^{256-32+16} = 2^{240}$. According to [18, Fact 9.99] the overall complexity of the preimage attack is $2^{(256+240)/2+1} = 2^{249}$, and the memory requirement is 2^{16}.

5.2   A Brief Comparison between SM3 and SHA-256

We show the compression function of SHA-256 in Fig.8 in Appendix B. Although SM3 is very similar in structure to SHA-256, it turns out to be difficult to achieve the same preimage result as SHA-256 due to the following three facts. Firstly, SM3 has stronger message dependency than SHA-256, since we need to process W_j and W_{j+4} in the j-th step while SHA-256 only processes W_j in the j-th step. Secondly, it is impossible to utilize the indirect partial matching technique as the step function contains the rotation operation. Finally, SM3 employs Boolean functions $X \oplus Y \oplus Z$, MAJ and IF in its step function, while SHA-256 only employs Boolean functions MAJ and IF in its step function. In conclusion, it is more difficult to find preimage of SM3 than SHA-256 based on MITM method.

6   Conclusion

In this paper, we propose a preimage attack on SM3 hash function reduced to 30 steps out of 64. Here we use several techniques to increase steps that we can attack. These techniques include two-way message expansion, initial structure, message compensation and partial matching. As SM3 adopts a more complex step function and stronger message dependency, it is difficult to achieve the same preimage result as SHA-256.

Acknowledgments. We would like to thank anonymous referees for their helpful comments and suggestions. This work is supported by the National Natural Science Foundation of China (No.60873259), and the Knowledge Innovation Project of The Chinese Academy of Sciences.

References

1. Wang, X., Yin, Y.L., Yu, H.: Finding Collisions in the Full SHA-1. In: Shoup, V. (ed.) CRYPTO 2005. LNCS, vol. 3621, pp. 17–36. Springer, Heidelberg (2005)
2. Wang, X., Yu, H.: How to Break MD5 and Other Hash Functions. In: Cramer, R. (ed.) EUROCRYPT 2005. LNCS, vol. 3494, pp. 19–35. Springer, Heidelberg (2005)
3. Wang, X., Yu, H., Yin, Y.L.: Efficient Collision Search Attacks on SHA-0. In: Shoup, V. (ed.) CRYPTO 2005. LNCS, vol. 3621, pp. 1–16. Springer, Heidelberg (2005)
4. Leurent, G.: MD4 is Not One-Way. In: Nyberg, K. (ed.) FSE 2008. LNCS, vol. 5086, pp. 412–428. Springer, Heidelberg (2008)
5. Aoki, K., Sasaki, Y.: Preimage Attacks on One-Block MD4, 63-Step MD5 and More. In: Avanzi, R.M., Keliher, L., Sica, F. (eds.) SAC 2008. LNCS, vol. 5381, pp. 103–119. Springer, Heidelberg (2009)
6. Guo, J., Ling, S., Rechberger, C., Wang, H.: Advanced Meet-in-the-Middle Preimage Attacks: First Results on Full Tiger, and Improved Results on MD4 and SHA-2. In: Abe, M. (ed.) ASIACRYPT 2010. LNCS, vol. 6477, pp. 56–75. Springer, Heidelberg (2010)
7. Dobbertin, H.: The First Two Rounds of MD4 are Not One-Way. In: Vaudenay, S. (ed.) FSE 1998. LNCS, vol. 1372, pp. 284–292. Springer, Heidelberg (1998)
8. Sasaki, Y., Aoki, K.: Preimage Attacks on 3, 4, and 5-Pass HAVAL. In: Pieprzyk, J. (ed.) ASIACRYPT 2008. LNCS, vol. 5350, pp. 253–271. Springer, Heidelberg (2008)
9. Aumasson, J.-P., Meier, W., Mendel, F.: Preimage Attacks on 3-Pass HAVAL and Step-Reduced MD5. In: Avanzi, R.M., Keliher, L., Sica, F. (eds.) SAC 2008. LNCS, vol. 5381, pp. 120–135. Springer, Heidelberg (2009)
10. Isobe, T., Shibutani, K.: Preimage Attacks on Reduced Tiger and SHA-2. In: Dunkelman, O. (ed.) FSE 2009. LNCS, vol. 5665, pp. 139–155. Springer, Heidelberg (2009)
11. Indesteege, S., Preneel, B.: Preimages for Reduced-Round Tiger. In: Lucks, S., Sadeghi, A.-R., Wolf, C. (eds.) WEWoRC 2007. LNCS, vol. 4945, pp. 90–99. Springer, Heidelberg (2008)
12. Mendel, F., Pramstaller, N., Rechberger, C.: A (Second) Preimage Attack on the GOST Hash Function. In: Nyberg, K. (ed.) FSE 2008. LNCS, vol. 5086, pp. 224–234. Springer, Heidelberg (2008)
13. Sasaki, Y., Aoki, K.: Preimage Attacks on Step-Reduced MD5. In: Mu, Y., Susilo, W., Seberry, J. (eds.) ACISP 2008. LNCS, vol. 5107, pp. 282–296. Springer, Heidelberg (2008)

14. Sasaki, Y., Aoki, K.: Finding Preimages in Full MD5 Faster Than Exhaustive Search. In: Joux, A. (ed.) EUROCRYPT 2009. LNCS, vol. 5479, pp. 134–152. Springer, Heidelberg (2009)
15. Aoki, K., Guo, J., Matusiewicz, K., Sasaki, Y., Wang, L.: Preimages for Step-Reduced SHA-2. In: Matsui, M. (ed.) ASIACRYPT 2009. LNCS, vol. 5912, pp. 578–597. Springer, Heidelberg (2009)
16. Cannière, C.D., Rechberger, C.: Preimages for Reduced SHA-0 and SHA-1. In: Wagner, D. (ed.) CRYPTO 2008. LNCS, vol. 5157, pp. 179–202. Springer, Heidelberg (2008)
17. Sepecification of SM3 cryptographic hash function (in Chinese), http://www.oscca.gov.cn/UpFile/20101222141857786.pdf/
18. Menezes, A.J., van Oorschot, P.C., Vanstone, S.A.: Handbook of Applied Cryptography. CRC Press (1996)

A   28 Steps Preimage of the Reduced SM3 Hash Function (from Step 0 to Step 27)

As mentioned above, we will meet with a lot of difficulties if we want to attack from the first step of the reduced SM3 hash function. In this case, we choose W_{z+8} and W_{z+11} as the neutral words respectively. According to the message expansion algorithm, we can swap W_{z+8} with W_{z+11} without any condition from $step_{z+7}$ to $step_{z+8}$. As explained above, we use message compensation to attack more steps. In the forward chunk, W_{z+8} is used in equation (11), (12) and (15), we compensate it by forcing

$$W_{z+1} = W_{z+8},$$

$$P_1(W_{z+2}) = (W_{z+5} \lll 7),$$

$$P_1(W_{z+5}) = (W_{z+8} \lll 7).$$

In the backward chunk, W_{z+11} is used in equation (3), we compensate it by forcing

$$P_1^{-1}(W_{z+14}) = (W_{z+11} \lll 15).$$

Here we set $z = 5$, the neutral words are W_{13} and W_{16}, and the message words W_{13}, W_{14} and W_{15} are original W_{z+8}, W_{z+9} and W_{z+10} respectively. Since W_{z+9} and W_{z+10} are set random values in our algorithm, we can preset them special values. We can preset message words W_{14} and W_{15} some special values. Using 1 bit freedom of W_{z+8} to ensure the last bit of W_{13} is '1'. Due to the neutral words, we can only increase 4 more steps by partial matching technique. We will show the attack in Fig.7. The preimage search needs time complexity of $2^{241.5}$, and requires a memory of about 2^{31}.

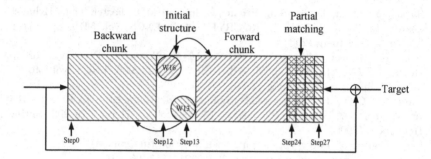

Fig. 7. 28 steps preimage of the reduced SM3 hash function

B Graph of SHA-256 Compression Function

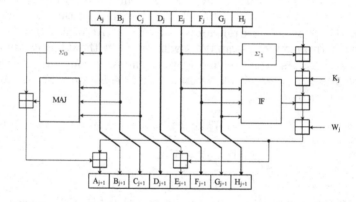

Fig. 8. the compression function of SHA-256

Breaking a 3D-Based CAPTCHA Scheme

Vu Duc Nguyen[1], Yang-Wai Chow[2], and Willy Susilo[1,*]

[1] Centre for Computer and Information Security Research,
[2] Advanced Multimedia Research Laboratory,
School of Computer Science and Software Engineering,
University of Wollongong, Australia
{dvn108,caseyc,wsusilo}@uow.edu.au

Abstract. CAPTCHA is a standard defence mechanism against bots, or automated programs, that attempt to use web-based services meant for human users. While there are many different types of CAPTCHA schemes that have emerged over the years, to date, the most widely used type is 2D text-based CAPTCHAs. Unfortunately, a large number of 2D CAPTCHA schemes have been successfully broken. Thus, 3D-based CAPTCHAs are seen as an alternative paradigm which has been explored by a number of CAPTCHA designers. 3D CAPTCHAs are meant to overcome the limitations of 2D CAPTCHAs and are supposed to be more robust and secure against automated attacks. To investigate the robustness of 3D text-based CAPTCHAs, this paper presents an approach to breaking a representative 3D CAPTCHA scheme called Teabag 3D. In particular, this paper describes the techniques that were used to break this CAPTCHA, and as such highlights various security issues that have to be considered in order to design better 3D CAPTCHA schemes.

Keywords: 3D CAPTCHA, character extraction, segmentation, optical character recognition.

1 Introduction

CAPTCHA, an acronym which stands for 'Completely Automated Public Turing test to tell Computers and Humans Apart', is an automated challenge and response test to ensure that a human is making an online transaction rather than a computer [18]. Typically, CAPTCHAs can generate and grade tests that a human being should be able to solve easily, but are infeasible for an automated program. At present, CAPTCHAs are used on many web-based services as a standard security mechanism against spam bots or other malicious automated programs. For example, it is used to prevent bots from sending out thousands of comment spams every minute or signing up for thousands of email accounts from free email services [19]. CAPTCHAs also offer potential solutions against email worms or even to prevent cheating in online multiplayer games via the use of bot programs [5].

---

* This work is supported by ARC Future Fellowship FT0991397.

H. Kim (Ed): ICISC 2011, LNCS 7259, pp. 391–405, 2012.
© Springer-Verlag Berlin Heidelberg 2012

A CAPTCHA can be classified based on its content type, and there are three main categories in existence; namely, text-based, image-based and sound-based [21]. To date, 2D text-based CAPTCHAs are the most pervasive type of CAPTCHA employed in real online applications. Among other reasons, this is due to its human friendliness, intuitiveness, ease of use and low implementation cost [3]. In general, a text-based CAPTCHA challenge typically takes the form of a word or a sequence of characters, or digits, embedded within an image, which contains distortion and noise to deter automated computer attacks.

Unfortunately, a large number of 2D text-based CAPTCHAs have successfully been broken. It has been shown that various design flaws can be exploited by automated programs to break these CAPTCHAs. Well known examples include a Microsoft CAPTCHA that was broken by a low-cost attack [20] or the EZ-Gimpy CAPTCHA, previously used by Yahoo, that could be solved automatically by a computer 92% of the time [12]. To increase the security strength and to confuse automated Optical Character Recognition (OCR) programs, 2D text-based CAPTCHAs rely on techniques like the warping of text and the overlaying of visual noise. However, this often makes the resulting CAPTCHA difficult for humans to recognize.

This has driven CAPTCHA developers to explore alternate paradigms in order to design more secure and usable text-based CAPTCHAs. In recent years, 3D text-based CAPTCHAs have been proposed to overcome the limitations of 2D text-based CAPTCHAs. For instance, a 3D CAPTCHA was proposed by Ince et al. [7] which presents randomly selected characters and numbers on individual faces of a 3D cube or STE3D-CAP, a text-based CAPTCHA that is built from stereoscopic 3D images, was proposed by Susilo et al. [17]. Others have also proposed 3D text-based CAPTCHA challenges that present the user with 3D text objects [2,6,10]. These proposals are based on the assumption that a human can recognize 3D images of text characters better than computer vision systems [2,16], and some have even suggested that 3D text-based CAPTCHAs is the next generation in CAPTCHA design [6].

This paper addresses the question of whether 3D text-based CAPTCHAs are really more secure. In particular, this paper presents a method of breaking a representative 3D CAPTCHA scheme called Teabag 3D [14], which at the time of writing was implemented on rediff.com when registering for Rediffmail [15]. The results of this research show that although this 3D CAPTCHA initially appears to be secure, as it is effectively resistant against one of the best OCR programs on the market, this paper shows that by performing some automated processing on the CAPTCHA prior to the character recognition stage, the OCR can correctly solve the CAPTCHA challenge at a high success rate.

Our Contribution. While there is much research on breaking 2D CAPTCHAs, to our knowledge, this is the first time in literature that describes a method of breaking a 3D-based CAPTCHA. We demonstrate that 3D text objects contain additional side surface information which can be exploited to break 3D CAPTCHAs. In particular, this paper introduces a novel method of extracting side surface information, and we show that this information can be used to extract characters from the background as well as to segment the character string into individual characters. This highlights a certain flaw in the design of 3D CAPTCHAs that are rendered with an additional dimension and emphasizes the need to hide or distort this exploitable information.

2   Related Work

2.1   Breaking CAPTCHAs

A number of researchers have documented techniques that they have used to break a variety of diverse CAPTCHA schemes. Mori and Malik [12] developed an approach to break the Gimpy and EZ-Gimpy CAPTCHAs using object recognition techniques to identify words amidst background clutter. In their work, they presented a holistic approach of recognizing entire words at once, rather than attempting to identify individual characters in severe clutter. Moy et al. [13] described a method of breaking the EZ-Gimpy and Gimpy-r by estimating the distortion of the text in the CAPTCHA image. After implementing the distortion estimation techniques, their approach then proceeded to undistort the text prior to object recognition.

Machine learning algorithms have also been used to break CAPTCHAs. In work by Chellapilla and Simard [4], they demonstrated that machine learning algorithms could successfully be used to break a variety of CAPTCHA schemes. In addition, unlike sophisticated computer vision or machine learning algorithms, Yan and Ahmad [19] showed that simple pattern recognition algorithms could be used to exploit flaws and design errors in CAPTCHA schemes, making them susceptible to simple attacks like counting the number of pixels to identify individual characters. Li et al. [9] have also shown that image processing and pattern recognition algorithms, such as k-means clustering, digital image inpainting, character recognition based on cross-correlation, etc. have been successful in breaking a variety of e-Banking CAPTCHAs.

2.2   Segmentation Resistant

It is widely accepted that the design of a secure text-based CAPTCHA must adhere to the segmentation-resistant principle. This principle is based on the work by Chellapilla et al. [3], where they established that computers could perform better at character recognition tasks compared to humans. As such, if a CAPTCHA can be segmented into its constituting characters, it is essentially broken. Techniques of segmenting text into individual characters and then performing recognition using OCR programs have also been used in areas like developing text readers for the visually impaired [11].

The segmentation-resistant principle required for robust CAPTCHA design, led a research team at Microsoft to develop a CAPTCHA scheme that was meant to be segmentation-resistant. Unfortunately, it was shown that the Microsoft CAPTCHA could in fact be segmented, and thus broken, by a low-cost attack [20]. In addition, a number of researchers have demonstrated using novel segmentation techniques to break various CAPTCHAs [19]. Nevertheless, the success of these attacks do not negate the segmentation-resistant principle in the design of robust CAPTCHAs.

3   The Targeted CAPTCHA

To investigate the robustness of a 3D text-based CAPTCHA scheme, the Teabag 3D CAPTCHA was selected. This CAPTCHA scheme was designed by the OCR Research Team [14], whose aim is to break known CAPTCHAs to identify weaknesses and to create new secure CAPTCHAs. Their website gives a good overview of a number of 2D text-based CAPTCHAs and their corresponding weaknesses. Additionally, some of the team's valuable experiences are presented in [8]. While there are a number of versions of Teabag 3D, the research in this paper deals with the commercial version as implemented on rediff.com [15], shown in Figure 1, as well as version 1.2 [14].

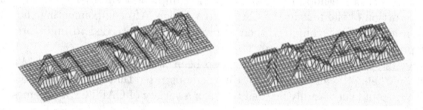

Fig. 1. Examples showing the version of Teabag 3D as implemented on rediff.com [15]

A number of characteristics regarding the CAPTCHA in question were identified as follows:

- The CAPTCHA challenge appears on a grid in 3D space as depicted in Figure 2.
- Four characters are used in each challenge.
- Only a selection of upper case letters and digits are used.
- Characters in close proximity may touch (i.e. are connected together).
- Each challenge appears to be generated from slightly different viewpoints.
- There are small variations in the grid direction and the shape of background cells between challenges.

4   Our Approach

Our framework to break the 3D-based CAPTCHA can be divided into a number of different phases. An overview of these stages is depicted in Figure 3.

4.1   Pre-processing

The first challenge faced in breaking Teabag 3D was in how to ascertain regions containing characters. This involved effectively separating characters from the

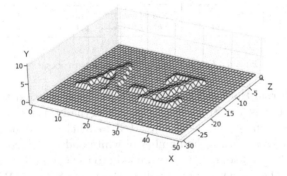

Fig. 2. The text appears on a grid in 3D space

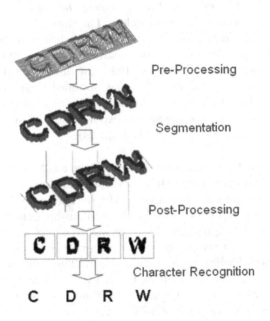

Fig. 3. Overview of the stages

background grid by identifying and extracting key features from the CAPTCHA image. Our method to achieve this is described as follows:

Adaptive Binarization. Teabag 3D challenges are presented as colored .png image files. Therefore, to process the image it was initially binarized into a black-and-white image. In order to find an appropriate binarization threshold, an adaptive binarization approach was employed for this. The steps for this are depicted in Figure 4.

To find the initial threshold, the image was first converted to greyscale. From this, a histogram representing the distribution of greyscale pixel intensities within the CAPTCHA was constructed. The initial binarization threshold

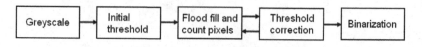

Fig. 4. The adaptive binarization process

was determined from the histogram, as the value that bisected the area of the histogram by half. Pixels with an intensity greater than this threshold were converted to white, while others were converted to black. This specific threshold value was selected to balance the number of white and black pixels.

To fine tune the threshold, an algorithm akin to the flood fill algorithm is used to count the number of white pixels within the cells of the CAPTCHA. If the white pixel count was more than half the total number of CAPTCHA pixels, this meant that the cell's borders were broken during the binarization process and that the selected threshold value was too low. The threshold was then increased and the process was repeated until all the cells had clear borders.

To facilitate the identification of front character surfaces (described later in the 'front surface identification' section), the value of the lowest greyscale pixel intensity, i.e. the darkest greyscale pixel, was obtained from the histogram. In other words, this value would give us the lowest binarization threshold and if it were to be used to binarize the image, this would result in the least number of black pixels (excluding the case of no black pixels).

Side Surface Identification. After binarizing the challenge into a black-and-white image, it was observed that the side surfaces of the characters were often represented in larger cells as compared to the background cells. As such, this information could be exploited to facilitate distinguishing the sides of text objects from the surrounding regions. A simple way to detect possible side surface cells of the characters in the image was to compute the average cell size (i.e. the average number of white pixels within a cell). Then, for each cell the total number of white pixels contained within the cell was compared with the average cell size. Any cell with a size that was greater than the average cell size, would be identified as a side surface cell of a character.

In some situations, the number of white pixels in the side surface cells was smaller than the average size. This was because the shape of these cells were often longer and/or narrower than the other cells. To identify and extract these cells, the average horizontal and vertical distances of the cell boundaries were calculated. Consequently when processing each cell, their maximum horizontal and vertical cell boundary distances were computed and compared with the average values. Cells that had maximum distances that were much less than the norm would also be identified as cells representing the side surface of a character. An example illustrating the result of side surface identification is shown in Figure 5(a).

Front Surface Identification. From the CAPTCHA's image, one can see that some borders between the characters and the background can be distinguished because certain borders are clearly darker than the rest of the image. Hence, these

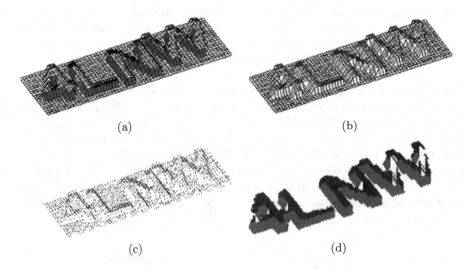

(a)   (b)

(c)   (d)

Fig. 5. Extracting characters from the background. (a) Side surface identification; (b) Front surface identification from clusters of black pixels; (c) Front surface identified by using the lowest binarization threshold; (d) Final results showing the front and side surfaces of the characters.

borders were obtained by simply identifying black pixels that were surrounded by 4 neighboring black pixels in the black and white image. An example of this is shown in Figure 5(b).

Furthermore, to get additional front surface pixels, the lowest binarization threshold value was used in conjunction with the greyscale image to identify the set of the darkest pixels (as previously described in the section on 'adaptive binarization'). Figure 5(c) shows an example resulting from the implementation of this approach and it can be seen that additional information was obtained for identifying the front surfaces of the characters.

Extracting Characters. The pixels identified as belonging to the front surface were often only the borders of the front character surface. Therefore, to obtain the front surface pixels within the borders of the characters, we scanned the pixels column by column from top to bottom. In each column, whenever we encountered a short section that started with a front surface pixel and that ended with either a front or side surface pixel, we would fill in all the pixels between these to represent front surface pixels. Figure 5(d) shows an example of results obtained using this approach.

4.2   Segmentation

This stage involves the decomposition of the image into sub-images which only contain single characters. This is a challenging task, as it can be seen from the results of the extraction process (Figure 5(d)) that the extracted characters may have missing pixels and some characters may be connected together. As such, a number of segmentation techniques were used to obtain a set of possible splitters for the characters. The best splitters were then selected from this candidate set. The segmentation and splitter selection methods are described as follows.

Segmentation Using Vertical Projection. This approach involves creating a histogram representing the number of character pixels per column in the image (note that this is a different histogram from the one used for binarization), then separating the image into chunks by identifying columns containing no character pixels [20]. In the case of Teabag 3D, this was done by projecting pixels in two directions respectively: vertically based on the image's vertical axis, and diagonally based on the projected vertical axis (since the CAPTCHA was rendered in 3D, this corresponds to the characters' vertical axis in 3D space), as shown in Figure 6(a) and 6(b) respectively.

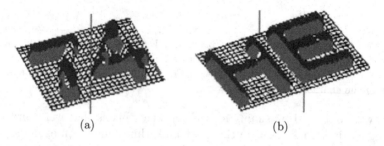

(a)   (b)

Fig. 6. Segmentation using vertical projection. (a) Segmentation based on the image's vertical axis; (b) Segmentation in the direction of the characters' vertical axis in 3D space.

Side Surface Segmentation. Vertical segmentation [20] and other existing methods such as Caliper distance [11] and Snake segmentation [19] have previously been used in character segmentation. However, in the case of Teabag 3D, these are not the best choice as they are less effective when it comes to dealing with connected characters. Furthermore, these methods are too sensitive when it comes to character regions that are broken due to lack of pixel information, as shown on Figure 7. To improve segmentation results, we implemented a novel method that we named 'Side Surface Scan' (SSS) which is segmentation based on the side surfaces of 3D characters. Our method is more effective for touching or broken characters in Teabag 3D and can possibly be extended to other 3D text-based CAPTCHAs.

The basic idea behind the Side Surface Scan method is depicted in Figure 8(a). An SSS can be done by performing a line wise scan from the left to the right side of the image and counting the number of continuous side surface pixels along each line. By lines, we mean the lines of pixels that are traced along the contour of the character's left side surface boundary, in the downward direction. In the straight forward case, these lines will be parallel to the left border of the grid. If the number of continuous side surface pixels in a line is greater than

Fig. 7. Example of connected characters and a character lacking pixel information

three quarters of the total number of pixels per line, this will be treated as a side surface line. In view of the fact that the side surfaces of the characters have a certain width, if there are more than six consecutive side surface lines, the left most side surface line will be used as a splitter (i.e. segmentation line).

Other than the straight forward case, in other situations, such as the letter 'Y' in Figure 8(b), the lines are not completely parallel to the left grid border. For example, in Figure 8(b) it can be seen that the splitter was determined to be at the left of the letter 'Y', as lines were traced along the contour of the character's left side surface boundary (note that there is no splitter inside the letter 'N', because the number of continuous side surface pixels was less than three quarters of the total number of pixels per line). Figure 8(c) in turn shows a splitter to the left of the letter 'W'.

In addition, SSS is flexible and can be used in different areas with different parameters. Thus, we also performed SSS on the lower half of the CAPTCHA. When performed on the lower half, the parameters were changed and all lines containing continuous side surface pixels were considered to be side surface lines. However, a splitter was only set if the number of consecutive side surface lines was greater than half the average character width. This allowed us to set splitters for characters like the number '4', as can be seen in Figure 8(d).

Splitter Selection. After the above segmentation methods were performed, the result is a set of splitters (i.e. segmentation line), as can be seen in the example shown in Figure 9(a). Noise or the lack of character pixels, may cause some splitters to be in the wrong position and other splitters may not be ideal. To get the 'best splitters' from the set of splitters, we assigned weights to the splitters with the values of 1, 2, 3 and 4 (shown in Figure 9(a) as blue, pink, orange and red respectively).

Initially, the lowest weight of 1 is assigned to the splitters. The initial weights will then be changed based on certain conditions. In essence, a splitter obtained from the vertical projection method or the SSS method will have its weight changed to 2. For three consecutive splitters obtained from the same segmentation method, their weight will be changed to 3, and a weight of 4 is given for four or more consecutive splitters obtained from the same segmentation method.

The splitter with the highest weight will be chosen as the best splitter. In cases where splitters in close proximity have the same weight, priority is given to the splitter on the right. The result of splitter selection can be seen from the example in Figure 9(b).

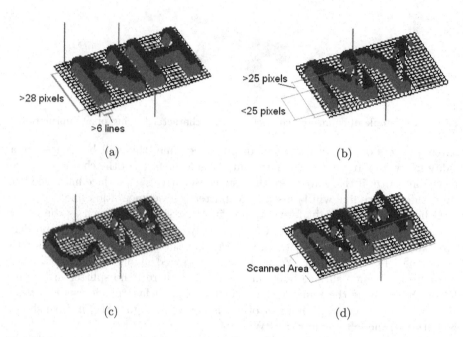

Fig. 8. Side Surface Scan (SSS) segmentation. (a) Lines parallel to left grid border; (b) Lines not completely parallel to the left border, but instead traced along the contour of a character's left side surface boundary; (c) Another example of lines traced along a character's contour; (d) SSS performed on lower half of the CAPTCHA with different parameters.

Fig. 9. Example of splitter selection results. (a) Splitters resulting of the segmentation methods; (b) Results after selecting the best splitters.

4.3 Post-processing and Character Recognition

After the segmentation stage, the result will be four individual characters. The ABBYY OCR[1] [1] was used for character recognition, as this is one of the best OCR programs currently available on the market. However, before the characters were passed to the OCR program, some post-processing steps had

---

[1] The ABBYY FineReader 10 Professional Edition was used in this work.

to be performed to ensure optimal character recognition accuracy. These post-processing steps are listed as follows:

- Combining the side and front surfaces: As can be seen in Figure 10(a), the image quality from the front surface alone is rather poor. As such, the side and front surface pixels were combined and converted to black to get a higher quality image of the character. An example of this is shown in Figure 10(b).
- Character de-skewing: Characters are originally skewed, or slanted, as they were rendered in 3D space. This step straightens the characters, using a de-skewing angle that was calculated based on the four extreme corners of CAPTCHA's grid. Figure 10(c) shows an example of a character after the de-skewing step.
- Character resizing: This step was necessary, otherwise the OCR would mis-interpret the image as containing a 'word' rather than an individual charac-ter. The OCR would then proceed by performing segmentation on the word and using a dictionary during recognition, hence leading to incorrect results. Simply shrinking the character's width by 50% was enough for the OCR to interpret the image as containing a single character.
- Character refining: To increase the accuracy of character recognition, this step removed noise, filled holes and smoothed the character borders. During the automated hole filling process, it was important to retain the original holes at the center of the characters (e.g. for letters like O, P, B, etc.), as these were key features required by the OCR program to recognize the characters. As such, holes in the center of the characters were left untouched. Figure 10(d) shows an example after character resizing and refining.

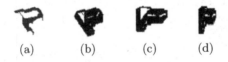

(a)   (b)   (c)   (d)

Fig. 10. Post-processing steps. (a) Front surface only; (b) After combining side and front surfaces; (c) After de-skewing; (d) After resizing and refining.

The four individual character images resulting from the post-processing stage were then passed, in order, as input into the OCR program. The ABBYY OCR uses machine learning to get more accurate results and needed to be trained using a training set of character samples. In addition to the ABBYY FineReader's embedded training database, we added 100 of our own character samples to the training set. Furthermore, since Teabag 3D only used a selection of uppercase letters and digits, these were defined as the input language for the OCR.

5   Results

An experiment was set up to test the accuracy of our method. We used a total of
1,000 CAPTCHA samples collected from rediff.com [15]. To compare the results
of our approach, we tested the samples using three different approaches. First,
the samples were input into the OCR program in their original form. Then the
pre-processing and post-processing stages were applied to the samples, without
segmentation, and input into the OCR. This was done to test the effectiveness of
the segmentation stage. Finally, the samples were processed using all the stages
in our method and passed into the OCR.

Table 1 shows the experimental results for the different approaches. This includes the individual character recognition accuracy (i.e. each challenge contains
four individual characters) and the accuracy of the entire challenge (i.e. correct
recognition of all four characters). It can be observed from the results that without any processing, Teabag 3D is robust against the OCR program. However,
with our automated extraction and segmentation approaches, the CAPTCHA
can be broken with a high success rate of 76%. The effectiveness of the segmentation process can also be seen, as the accuracy of 29% without segmentation
is much lower than with segmentation. Experiments were conducted on an Intel
Core 2 Duo 3.33GHz PC, and the average attack speed was around 7 seconds
per challenge.

Table 1. Experimental results

Different Approaches	Accuracy	
	Individual Characters	*Entire Challenge*
Unprocessed CAPTCHA image	0%	0%
Pre-processing and post-processing, without segmentation	68%	29%
All stages applied	92%	76%

5.1   Discussion

Our attack exploited a number of weaknesses in the 3D CAPTCHA. For one
thing, the 3D characters could be separated from the background because of the
regular grid. This meant that the side surfaces could be identified based on the
fact that the size and shape of the background cells were somewhat constant.

Furthermore, the front surfaces could be identified based on the density of pixels at certain locations.

Another design flaw was that the borders of the grid could be used to estimate the orientation of the characters. This would have been more difficult if the borders were irregularly cut off or distorted. Between CAPTCHA challenges, there was little variation in the size and orientation of the characters. This made it easier for us to determine appropriate factors like the average cell size, character width and height. The absence of any character distortion made the CAPTCHA easy for humans to read, however, it also made it easier for the OCR to correctly recognize the characters.

In essence, one of the key factors governing the success our attack, lies in the fact that the CAPTCHA was rendered in 3D with no attempt to hide the additional dimension. As such, the addition of the third dimension, while visually attractive, also increased the amount of information that could be used to break the CAPTCHA. In our approach, we used this information to identify the side surfaces of the characters and also to facilitate segmentation. In addition, the side surfaces were combined with the front surfaces to increase the accuracy of character recognition.

Version 1.2. The OCR Research Team have released a new commercial version on their website [14], i.e. version 1.2, which at the time of writing was not yet implemented on rediff.com. As can be seen in Figure 11(a), this new version appears to be much more robust compared to the previous version. In particular, the grid and the characters are distorted using waves. This causes significant variation in the size and shape of the cells.

We were still able to apply our approach to breaking this new version, albeit at a lower success rate. To do this, we had make some changes to the pre-processing and segmentation stages. In particular, for the pre-processing stage, instead of using a global average cell size for side surface identification, a local average of cell sizes was used instead. These local averages were calculated dynamically for each cell based on the size its neighboring cells. In this manner, any large change in the size of adjacent cells would be detected. Examples of the results after front and side surface identification can be seen in Figure 11(b).

It can be seen from the figure that the resulting image typically contains some noise. These are removed based on their distances to the side surface cells, and the results after character extraction are shown in Figure 11(c). In the segmentation stage, the SSS method was improved by approximating the size, width and height of each side surface region, rather than using a pixel by pixel approach, and applying this information for locating splitters. Figure 11(d) shows examples of the segmentation results. For a sample size of 100, our approach achieved a 51% accuracy for individual character recognition and an accuracy of 24% in terms of correctly recognizing entire challenges.

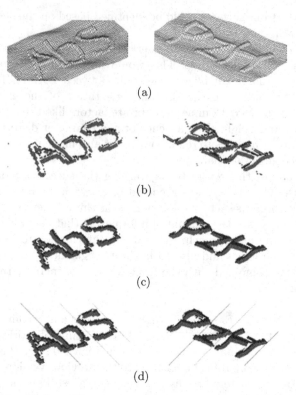

(a)

(b)

(c)

(d)

Fig. 11. Processing Teabag 3D version 1.2 using our approach. (a) Teabag 3D version 1.2; (b) Extracting side and front surfaces of the characters; (c) Noise removal and refinement; (d) Segmentation.

6   Conclusion

In this paper, we explored an attack on Teabag 3D which is a 3D text-based CAPTCHA scheme. While this CAPTCHA effectively resists direct OCR attacks, we show that after some processing our attack can successfully break this CAPTCHA at a high success rate. Our attack exploits certain characteristics of the 3D CAPTCHA. In particular, the 3D text characters were rendered with additional information in the form of side surfaces which we used to separate the text from the background and also in character segmentation. As such, this additional third dimension made the CAPTCHA scheme less secure. It is highly likely that our approach can be extended and be applied to break other 3D CAPTCHA schemes.

References

1. ABBYY. ABBYY FineReader, http://finereader.abbyy.com
2. Chaudhari, S.K., Deshpande, A.R., Bendale, S.B., Kotian, R.V.: 3D Drag-n-drop CAPTCHA Enhanced Security through CAPTCHA. In: Mishra, B.K. (ed.) ICWET, pp. 598–601. ACM (2011)

3. Chellapilla, K., Larson, K., Simard, P.Y., Czerwinski, M.: Designing Human Friendly Human Interaction Proofs (HIPs). In: van der Veer, G.C., Gale, C. (eds.) CHI, pp. 711–720. ACM (2005)

4. Chellapilla, K., Simard, P.Y.: Using Machine Learning to Break Visual Human Interaction Proofs (HIPs). In: NIPS (2004)

5. Chow, Y.-W., Susilo, W., Zhou, H.-Y.: CAPTCHA Challenges for Massively Multiplayer Online Games: Mini-game CAPTCHAs. In: Proceedings of the 2010 International Conference on Cyberworlds, CW 2010, pp. 254–261. IEEE Computer Society, Washington, DC (2010)

6. Imsamai, M., Phimoltares, S.: 3D CAPTCHA: A Next Generation of the CAPTCHA. In: Proceedings of the International Conference on Information Science and Applications (ICISA 2010), Seoul, South Korea, April 21-23, pp. 1–8. IEEE Computer Society (2010)

7. Ince, I.F., Salman, Y.B., Yildirim, M.E., Yang, T.-C.: Execution Time Prediction for 3D Interactive CAPTCHA by Keystroke Level Model. In: Proceedings of the 2009 Fourth International Conference on Computer Sciences and Convergence Information Technology, ICCIT 2009, pp. 1057–1061. IEEE Computer Society, Washington, DC (2009)

8. Kolupaev, A., Ogijenko, J.: CAPTCHAs: Humans vs. Bots. IEEE Security & Privacy 6(1), 68–70 (2008)

9. Li, S., Shah, S.A.H., Khan, M.A.U., Khayam, S.A., Sadeghi, A.-R., Schmitz, R.: Breaking e-Banking CAPTCHAs. In: Gates, C., Franz, M., McDermott, J.P. (eds.) ACSAC, pp. 171–180. ACM (2010)

10. Macias, C., Izquierdo, E.: Visual Word-based CAPTCHA using 3D Characters. IET Seminar Digests 2009(2), P41–P41 (2009)

11. Mancas-Thillou, C., Ferreira, S., Demeyer, J., Minetti, C., Gosselin, B.: A Multifunctional Reading Assistant for the Visually Impaired. J. Image Video Process. 2007, 5:1–5:11 (2007)

12. Mori, G., Malik, J.: Recognizing Objects in Adversarial Clutter: Breaking a Visual CAPTCHA. In: CVPR (1), pp. 134–144 (2003)

13. Moy, G., Jones, N., Harkless, C., Potter, R.: Distortion Estimation Techniques in Solving Visual CAPTCHAs. In: CVPR (2), pp. 23–28 (2004)

14. OCR Research Team. Teabag 3D CAPTCHA, http://ocr-research.org.ua

15. Rediff Inc. Rediffmail, http://register.rediff.com/register/register.php

16. Ross, S.A., Halderman, J.A., Finkelstein, A.: Sketcha: a CAPTCHA based on Line Drawings of 3D Models. In: Rappa, M., Jones, P., Freire, J., Chakrabarti, S. (eds.) WWW, pp. 821–830. ACM (2010)

17. Susilo, W., Chow, Y.-W., Zhou, H.-Y.: STE3D-CAP: Stereoscopic 3D CAPTCHA. In: Heng, S.-H., Wright, R.N., Goi, B.-M. (eds.) CANS 2010. LNCS, vol. 6467, pp. 221–240. Springer, Heidelberg (2010)

18. von Ahn, L., Blum, M., Hopper, N.J., Langford, J.: CAPTCHA: Using Hard AI Problems for Security. In: Biham, E. (ed.) EUROCRYPT 2003. LNCS, vol. 2656, pp. 294–311. Springer, Heidelberg (2003)

19. Yan, J., Ahmad, A.S.E.: Breaking Visual CAPTCHAs with Naive Pattern Recognition Algorithms. In: ACSAC, pp. 279–291. IEEE Computer Society (2007)

20. Yan, J., Ahmad, A.S.E.: A Low-Cost Attack on a Microsoft CAPTCHA. In: ACM Conference on Computer and Communications Security, pp. 543–554 (2008)

21. Yan, J., Ahmad, A.S.E.: Usability of CAPTCHAs or Usability Issues in CAPTCHA Design. In: Cranor, L.F. (ed.) SOUPS. ACM International Conference Proceeding Series, pp. 44–52. ACM (2008)

Multi-User Keyword Search Scheme for Secure Data Sharing with Fine-Grained Access Control

Fangming Zhao[1,2], Takashi Nishide[1], and Kouichi Sakurai[1]

[1] Department of Computer Science and Communication Engineering,
Kyushu University, 744 Motooka, Nishi-ku, Fukuoka, 819-0395, Japan
[2] Corporate Research & Development Center, TOSHIBA Corporation,
1 Komukai-Toshiba-cho, Saiwai-ku, Kawasaki, 212-8582, Japan
fangming.zhao@toshiba.co.jp, {nishide,sakurai}@inf.kyushu-u.ac.jp

Abstract. We consider the problem of searchable encryption scheme for the cryptographic cloud storage in such a way that it can be efficiently and privately executed under the multi-user setting. Searchable encryption schemes allow users to perform keyword searches on encrypted files to retrieve their interested data without decryption. All existing such schemes only consider the straightforward search approach where for searching one encrypted keyword, the cloud server must look round all encrypted files on the storage to compare that encrypted keyword to each keyword index. Since the file number can be very huge and the user may be unable to decrypt all files, that approach is not efficient and secure enough. In this paper, we first propose a keyword search scheme for the cryptographic cloud storage based on attribute-based cryptosystems. Our scheme presents a new keyword search notion: fine-grained access control aware keyword search. By narrowing the search scope to the user's decryptable files' group before executing the keyword search, our approach can both decrease information leakage from the query process and be more efficient than other existing schemes.

Keywords: keyword search, multi-user, fine-grained and flexible access control, data sharing.

1 Introduction

1.1 Background and Motivation

For reasons of management cost and convenience, users often store their data not on their own machine, but on remote servers, i.e., cloud storage. To address users' concerns of data confidentiality on the cloud storage, a common approach is using cryptography. Encryption at the server's side is not appropriate when the server is not fully trusted. Data owners encrypt all data before sending to the cloud servers and later the encrypted data can be retrieved and decrypted by users who have a decryption key. This kind of cloud storage is often called *cryptographic cloud storage* [6]. Even if this will ease user's concerns of data leakage, it also introduces a new problem: because the encryption of data is not meaningful to

H. Kim (Ed): ICISC 2011, LNCS 7259, pp. 406–418, 2012.
© Springer-Verlag Berlin Heidelberg 2012

the cloud servers, many useful data processing operations performed by cloud servers become infeasible. One of the most important operations for efficient data retrieval and sharing in the cloud is the keyword search.

Many protocols have been proposed to partially solve the above problems. However, most of existing schemes are limited to the single-user setting where the owner who generates the encrypted data on the cloud is also the single user to perform encrypted keyword searches on it. They can not satisfy the characteristics of cryptographic cloud storage: sharing encrypted data with multiple users who have the appropriate access rights. In the *multi-user cryptographic cloud storage* setting, the data owner shares his encrypted data with multiple users and also allow users who have the access permission to perform encrypted keyword searches over the owner's shared data on the cloud server side.

In an existing work Zhao et al [12], an attribute-based cryptosystems based multi-user cryptographic cloud storage was proposed for secure data sharing. Users with different attributes can access the cryptographic cloud storage for secure data sharing, and the fine-grained and flexible access control is realized by an access tree [2]. However, no specified keyword search method is given in that work. In this paper, for efficient use of cryptographic cloud storage, we are focusing on the keyword search toward the cryptographic cloud storage under the multi-user setting.

1.2   Related Work

Several schemes have been developed to encrypt data on the client-side and enable server-side searches on encrypted data. Song et al. [10] proposed the first practical scheme for searching on encrypted data. The scheme enables clients to perform searches on encrypted text without disclosing any information about the plaintext and the keyword to the untrusted server. The untrusted server cannot learn the plaintext given only the ciphertext, and it cannot search without the user's authorisation, and it learns nothing more than the encrypted search results. The basic idea is to generate a keyed hash for the keywords and store this information inside the ciphertext, then the server can search the keywords by recalculating and matching the hash value. In [4], Goh presents a scheme for keyword search on encrypted data using Bloom Filters. Golle et al. [5] first considers keyword conjunctions which is based on pairings on elliptic curves. The first public key schemes for keyword search over encrypted data are presented in [3]. In that work, authors consider a setting in which the sender of an email encrypts keywords under the public key of the recipient in such a way that the recipient is able to give capabilities for any particular keyword to their mail gateway for routing purposes. Their scheme allows multiple users to encrypt data using the public key, but only the user who has the private key can search and decrypt the data. Bao et al. [1,11] consider the multi-user query over encrypted data. Their scheme allows each user to possess a distinct secret key for generating the encrypted keyword (or called a trapdoor) respectively.

1.3 Challenging Issues

For existing keyword search works [3,4,5,10], if we want to apply their protocols to the multi-user cryptographic cloud storage setting, a naive approach is sharing the secret key with all valid users. However, sharing keys is generally not a good idea since it increases the risk of key exposure. The keys must be changed if a user is no longer qualified to access the data. Moreover, changing keys may result in decrypting all the data with the old key and re-encrypting it using the new keys. For the cloud storage with a large number of users and files, this is not practical.

All existing schemes have not considered the user's access right while designing the keyword search process on the cloud server's side. (i): In the multi-user's cryptographic cloud storage environment, since not all users can read(decrypt) all data, each user is not able to search through data that is not decryptable to that user for the cloud data confidentiality. For example, a curious user of the cloud storage who possesses a common secret key for searching, can perform an investigation over all encrypted data to know whether some data on the cloud storage contains a specified keyword w, by generating an encrypted query based on that keyword. Especially in the cryptographic cloud storage using the latest cryptographic technique: attribute-based encryption(ABE) [2], like [12], such a result of the curious user's investigation also leaks confidential information: encrypted data using CP-ABE shows what kind of user attributes are required to decrypt that encrypted data. That is, the relationship between encrypted data and its access requirement can be known to the curious user from his keyword search. (ii): We consider the search efficiency problem. Assuming n and m ($m \subseteq n$) are numbers of total files and decryptable files on the cryptograhic cloud storage for a specific user. Let r be the average number of keywords associated in a file. Traditionally, the cloud storage's computation cost for executing a keyword search shall be $O(n \times r)$. Thus the cloud server wastes $O((n - m) \times r)$ computation cost in each query process. Allowing the cloud server to narrow the search scope from n to m(the user's decryptable scope) without leaking any secret keys and plaintexts is a new challenge for the multi-user cryptographic cloud storage.

1.4 Our Contributions

We study the encrypted keyword search problem in the multi-user cryptographic cloud storage setting, and we present a first encrypted keyword search concept: fine-grained access control aware keyword search, which requires the user's keyword search is performed by the cloud server over the user's decryptable data scope considering differential access right of the user. This new characteristic is never considered by any other existing works. The formal definition of our concept will be given in Section 4.1.

Our construction uses as building blocks some of the schemes mentioned above. We stress, however, that it is not sufficient to use the schemes *as-is*. We show an approach of access structure (also called access tree) computation that evolves from the attribute-based cryptosystems, and then we apply it to the query process for specifying the user's access permission from the encrypted keyword in order to narrow the search scope to the user's decryptable file group. Advantages of our approach can be summarized as follows:

- Decreasing information leakage from the keyword search process executed between users and cloud servers.
- Being more efficient than existing works since our method does not search unrelated files which can not be decrypted by that user.

Since its new characteristics on the security and the protocol convenience, attribute-based encryption(ABE) technique is widely used on the cryptographic cloud storage. However, no keyword search scheme is designed for the ABE based cryptographic cloud storage because of the complex composition of the ciphertext. Our work gives a simple but more effective scheme that first proposes an encrypted keyword search approach for the attribute-based encryption. Moreover, our scheme innovates the application of the attribute-based signature(ABS)[7] protocol, which is not applied widely yet.

From the viewpoint of scheme functionalities, under the multi-user cloud setting, by providing fine-grained and flexible access control to the data on the cloud, not only the data owner, but authorized users can also update the encrypted data and encrypted keyword list. Since each user has a distinct key for keyword search, the key management becomes simpler. For example, key update and user revocation can be easily achieved without complicated process of decrypt and re-encrypt.

2   System Models and Definitions

2.1   System Models

We consider a multi-user cryptographic cloud storage which is described in [12]. In this system, a group of authorized users(E.g. readers and writers in [12]) can share encrypted data and perform keyword search on the encrypted data without decrypting them.

Cloud Server: The main responsibility of the cloud storage server is to store and retrieve encrypted data according to authorized users' requests. Moreover, an new external functionality is provided by the cloud server: before executing the keyword search for each user, the cloud server first needs to sort out those files which can be decrypted by that user, and then the cloud server searches the keyword only from his decryptable file group;

Trusted authority(TA): Being similar to the assumption in [12], TA is a trusted third party in our system. Firstly, it is responsible for managing all attributes and their related keys used in ABE and ABS. Secondly, about users' keyword search, it is also responsible for user enrollment and revocation, i.e. managing keys for user' query generation.

Users: Being similar to the assumption in [12], we consider a multi-user setting. Not only the owner can perform the keyword search, but other users, e.g.

readers(who has the decrypt right, can read data) and writers(who has both the read right and the update right), can also perform the keyword search corresponding to their access right. Only writers can update the encrypted file and its associated encrypted keyword list.

Data: As described in the previous work [12], all data is encrypted with CP-ABE on the user's side before sending to the cloud storage. Here, the data file can be documents, videos, images, etc. Each file can be associated with a list of keywords which is also encrypted by the user.

2.2 Definitions

In this subsection, we define our scheme: decryptable keyword search for cryptographic cloud storage. Our proposed scheme consists of a tuple of algorithms (Setup, BuildIndex, Write, Query, Search) such that:

- Setup(1^k): The initialization algorithm *Setup* is run by the TA which takes as input the security parameter k and outputs the unique master secret key K_{msk} and the key pair $\langle K_{U_{id}}, CK_{U_{id}} \rangle$ for each valid user whose user ID is U_{id}. The TA respectively distributes the $K_{U_{id}}$ and the $\langle U_{id}, CK_{U_{id}} \rangle$ to the user and the cloud server.
- BuildIndex($w, CK_{U_{id}}$): The *BuildIndex* algorithm is run by the data owner and the cloud server interactively. This algorithm outputs an index I(w) for all keywords $w = \{w_1, w_2, ...\}$.
- Write(I(w), CT, $T_{decrypt}$)): This *Write* algorithm is run by the data owner. After the owner generates encrypted data CT, $T_{decrypt}$ and the I(w), he writes(or, uploads) them to the cloud server. CT is the ciphertext of the original data. We consider CP-ABE [2] for the data encryption as described in [12].
- Query(U_{id}, Q(w), Sig(Q(w))): The *Query* algorithm is run by the user to generate a trapdoor for the keyword w and query to the cloud server. The user first computes the trapdoor Q(w) by his keys' material, and then generates a signature Sig(Q(w)) of the trapdoor. We consider the ABS [7] for the digital signature. Finally, the user will send the query data: $\langle U_{id}, Q(w), Sig(Q(w)) \rangle$ to the cloud server for a keyword w.
- Search(U_{id}, CK_{Uid}, CT, I(w), Q(w), Sig(Q(w)), T_{sign}): The *Search* algorithm is run by the cloud server. For each user, cloud server will only search for the keyword Q(w) on the data's group which can be decrypted by that user.
- Revoke(U_{id}): The user search revocation algorithm is run by the TA and the cloud server. Given user ID U_{id}, they revoke the user by updating its user's key list $L = L \setminus \langle U_{id}, CK_{U_{id}} \rangle$, then the user is no longer able to search the cloud storage.

3   Technical Preliminaries

We build on the work by Bethencourt et al. [2] and Maji et al. [7] respectively (We do not describe their computation process here, and please refer to their works for details). We also review some notions about efficiently computable bilinear maps.

3.1 Ciphertext-Policy Attribute-Based Encryption

CP-ABE [2] is one of the latest public key cryptography primitives for secure data sharing. More precisely, a user's private key will be associated with an arbitrary number of attributes expressed as strings. When a party encrypts a message, they first specify an associated access structure over attributes. A user will only be able to decrypt a ciphertext if that user's attributes satisfy the ciphertext's access structure. At a mathematical level, access structures in our system are described by a monotone access structure (or access tree) $T_{decrypt}$ [2], where nodes of the access structure are composed of threshold gates and the leaves describe attributes. Usually, AND gates can be constructed as n-of-n threshold gates and OR gates as 1-of-n threshold gates. If a set of attributes U satisfies the access structure $T_{decrypt}$, we denote it as $T_{decrypt}(U) = 1$.

Setup is probabilistic and run by the TA: on input the security parameter and a universe of attributes, the master key MK and public key PK are generated.

Encryption$(PK, m, T_{decrypt})$ is probabilistic and run by a user who wants to encrypt a plaintext message m for a user with a set of attributes in the access structure $T_{decrypt}$, this algorithm generates a ciphertext CT.

Key-Generation(MK, U) is probabilistic and run by the TA: on input the master key MK and a set of attributes U belonging to a user, a secret key SK for these attributes is generated.

Decryption(CT, SK) is deterministic and run by a user with a set of attributes U. On input CT and SK, this algorithm outputs the underlying plaintext m, if CT is a valid encryption of m and U is contained in the access structure $T_{decrypt}$ specified in the computation of CT. Otherwise an error will be returned.

3.2 Attribute-Based Signature

Like the CP-ABE, there are two entities in ABS: a central trust authority(TA) and users. The authority is in charge of the issue of attribute private key to users requesting them. Denote the universe of attributes as U, as the access structure in the CP-ABE, there is a a monotone boolean claim-predicate(access structure) T_{verify} over U whose inputs are associated with attributes of U. We say that an attribute set U satisfies a predicate T_{verify} if $T_{verify}(U) = 1$. The algorithms are defined as follows.

Setup The authority obtains a key pair (PK, MK) and outputs public parameters PK and keeps a private master key MK.

Key-Generation(MK, U) To assign a set of attributes U to a user, the authority computes a signing key SK_U and gives it to the user.

Sign$(PK, SK_U, m, T_{verify})$ To sign a message m with a claim-predicate T_{verify}, and a set of attributes U such that $T_{verify}(U) = 1$, the user computes a signature σ by $(PK, SK_U, m, T_{verify})$.

Verify$(PK, m, T_{sign}, \sigma)$ To verify a signature σ on a message m with a claim-predicate T_{sign}, a user runs $Verify(PK, m, T_{sign}, \sigma)$, which outputs a boolean value, accept or reject.

3.3 Bilinear Map

Let $\mathbb{G}_0$ and $\mathbb{G}_1$ be two bilinear groups of prime order p. Let $\hat{e} : \mathbb{G}_0 \times \mathbb{G}_0 \to \mathbb{G}_1$ denote the bilinear map. Let g be a generator of $\mathbb{G}_0$. The bilinear map $\hat{e}$ has the following properties:

- Bilinearity: for all $u, v \in G_0$ and $a, b \in \mathbb{Z}_p$, we have $\hat{e}(u^a, v^b) = \hat{e}(u, v)^{ab}$
- Non-degeneracy: $\hat{e}(g, g) \neq 1$.
- Computable: $\hat{e}(u, v)$ can be efficiently computed for any $u, v \in G_0$.

4 Concrete Constructions

We describe details of our fine-grained access control aware multi-user keyword search scheme in this section. As mentioned, access control needs to be enforced before the cloud server searches a keyword and a user is not allowed to search through data which is not decryptable for him. We make our new proposal by modifying two existing schemes: an attribute-based cryptographic cloud storage as described in [12] and a query protocol by Bao et al. [1,11].

4.1 Access Tree Based Fine-Grained Access Control Verification Mechanism

Attribute based cryptosystems [2,7,12] use the access tree (also called access structure) to provide fine-grained access control. In CP-ABE [2], a user will be able to decrypt a ciphertext with a given key if and only if there is an assignment of attributes from the ciphertexts to nodes of the tree such that the tree is satisfied. In ABS [7], a signer, who possesses a set of attributes from the authority, can sign a message with a predicate that is satisfied by his attributes. The signature reveals no more than the fact that a single user certainly with some set of attributes satisfying the predicate has attested to the message.

Access Tree: T Let T be a tree representing an access structure. Each non-leaf node of the tree represents a threshold gate, described by its children and a threshold value. If num_x is the number of children of a node x and k_x is its threshold value, then $0 < K_x \leq num_x$. When $k_x = 1$, the threshold gate is an *OR* gate and when $k_x = num_x$, it is an *AND* gate. Each leaf node of the tree simply represents an attribute.

Satisfying an Access Tree: Let T be an access tree with root r. Denote by T_x the subtree of T rooted at the node x. Hence T is the same as T_r. If a set of attributes r satisfies the access tree T_x, we denote it as $T_x(r) = 1$. We compute $T_x(r)$ recursively as follows. If x is a non-leaf node, evaluate $T_{x'}(r)$ for all children x' of node x. $T_x(r)$ returns 1 if and only if at least k_x children return 1. If x is a leaf node, then $T_x(r)$ returns 1 if and only if $att(x) \in r$.

In our proposed scheme, we present a first encrypted keyword search concept: fine-grained access control aware keyword search, which is formally defined as:

Definition 1. *Access control aware keyword search: Let n be the number of encrypted files on the storage server, and a user u wants the server to search encrypted files that contain an encrypted w. Let m be the file number on the storage server which can be decrypted by u, and ($m \subseteq n$). The user u's access right is unknown to the storage server and the server knows nothing about the plaintext of both the encrypted files and keywords. The access control aware keyword search requires that the server execute the keyword search scheme after narrowing the search scope from n to m aware of u's access right.*

To realize our idea, we take advantage of the access tree to (i) allow the cloud server to focus on the user's decryptable file group by his attributes; (ii) make a user show the cloud server that he really holds those attributes which is needed to decrypt some files before his keyword query is executed. In our scheme, (ii) is achieved by generating an attribute based signature(ABS) using the *AND* of all his attributes: $T_{sign} = \{Att_1 \wedge Att_2 \wedge Att_3...\}$. After verifying the signature, the result proves whether the user holds those attributes as he claims. Two examples of access trees T_{sign} and $T_{decrypt}$ are shown in Figure 1. In this example, a user can prove the possession of his attributes $\{Professor \wedge Dean \wedge Trustee\}$ in T_{sign} by generating an attribute based signature(ABS). If the verification succeeds, the cloud server can judge whether that user can decrypt a file by checking: $T_{sign} \models T_{decrypt} = 1$ or 0, $\models$ is formally defined in Definition 2:

Definition 2. *Let T_1 and T_2 be two access trees (also called access structure) in attribute-based cryptosystems. $T_1 \models T_2$ is an access tree computation that outputs 1 or 0, where 1 means that there is an attribute set x in T_1 that satisfies T_2. 0 means no such attribute set exists in T_1 and T_2.*

As shown in Figure 1, since the attribute *"Trustee"* that exists in T_{sign} also satisfies $T_{decrypt}$, the computation result of $T_{sign} \models T_{decrypt} = 1$, and the cloud server verifys that file is decryptable to that user. Next, we will describe our scheme in details.

4.2   Proposed Scheme

Setup(1^k): The initialization algorithm Setup(1^k) is run by the TA which takes as input the security parameter 1^k and outputs the TA's unique master secret key $K_{msk} \in \mathbb{Z}_p$ and the key pair $\langle K_{U_{id}} \in \mathbb{Z}_p, CK_{U_{id}} \rangle$ for each valid user whose user ID is U_{id}, where $CK_{U_{id}} = g^{K_{msk}/K_{U_{id}}}$ is a complementary key for a user.

The TA respectively distributes the $K_{U_{id}}$ and the $\langle U_{id}, CK_{U_{id}} \rangle$ to the user and the cloud server.

BuildIndex($w, CK_{U_{id}}$): The BuildIndex algorithm is run by the data owner and the cloud server interactively. This algorithm outputs an index $I(w)$ for the keywords set $w = \{w_1, w_2, ...\}$. Data owner first uploads the $\langle U_{id}, h(w)^r \rangle$ to the cloud server. $h(): \{0, 1\}^* \rightarrow \mathbb{G}_0$ is the hash function and $r \in \mathbb{Z}_p$ is a random number. After receiving the request, the cloud server calculates the $Cap_w = \hat{e}(h(w)^r, CK_{U_{id}})$ for each w and then sends it back to the data owner. The data owner can build the index for w as $I(w) = [R, HMAC_k(R)]$, where the key for the $HMAC$ calculation is $k = h(Cap_w^{K_{U_{id}}/r})$, $R \in \mathbb{Z}_p$ is also a random number.

Write($I(w)$, CT, $T_{decrypt}$): After the owner generated the encrypted data CT and the $I(w)$, he writes(or, uploads) them to the cloud server. CT is the ciphertext of CP-ABE [2] as described in [12]. $CT = Enc(PK_{enc}, M, T_{decrypt})$, PK_{enc} is the public key for encryption, M is the data's plaintext, $T_{decrypt}$ is the access tree for the CP-ABE generated by the owner. Finally, $\langle CT, I(w), T_{decrypt} \rangle$ is writen to the cloud server.

Query(U_{id}, $Q(w)$, $Sig(Q(w))$): For a specific keyword w, the user first generates a trapdoor $Q(w) = h(w)^{K_{U_{id}}}$. Then he generates an attribute-based signature(ABS) for that trapdoor: $Sig(Q(w)) = \{PK, SK, Q(w), T_{sign}\}$. Note that the T_{sign} is made by all of the user's attribute: $T_{sign} = \{Att_1 \wedge Att_2 \wedge Att_3...\}$. Note, the signature ABS shows that the user certainly possesses a set of attributes from the authority as he/she declared in the access tree T_{sign}. The cloud server can verify the user's identification (attribute) by public keys from the TA. In this step, the user sends $\langle U_{id}, Q(w), Sig(Q(w)) \rangle$ to the cloud server.

Search(U_{id}, $CK_{U_{id}}$, CT, $I(w)$, $Q(w)$, $Sig(Q(w))$, T_{sign}): After receiving the query from a user, the server first checks the complementary key $CK_{U_{id}}$ by the user ID U_{id}. If the U_{id} is valid, the server shall confirms the user's decryptable

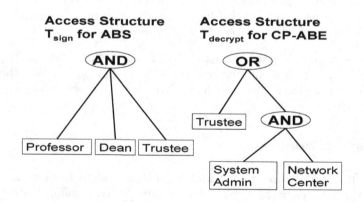

Fig. 1. Access Trees

file group by: (i). Verify user's attribute set $\{Att_1 \wedge Att_2 \wedge Att_3...\}$ as described in T_{sign} by the ABS-verification process,: Verify$(PK, Q(w), T_{sign}, Sig(Q(w))) =$ $True$ or $False$. The verification key PK are published by the TA. If the ABS verification result is true, the user's attributes as he/she declared in the T_{sign} are confirmed. (ii). Using $T_{decrypt}$ of each CT stored on the server, the cloud server can confirm the search scope S as the following procedure:

```
{
S = Null;
for(i = 0; i < n; i + +)
//i is the file index number; n is the total file number.
{
if((T_sign |= T_decrypt[i])! = 0)
S = S ∪ i;
}
return S; }
```

Then the cloud server performs the the keyword search only in the scope S. It first computes $k' = \hat{e}(Q(w), CK_{U_{id}})$, and then checks each index of the data CT in the scope S as: $HMAC_k(R) \overset{?}{=} HMAC_{k'}(R)$. Finally, the server sends the search result to the user.

Revoke(U_{id}): Since TA and the cloud server manage all users' pair $\langle U_{id}, CK_{U_{id}} \rangle$. For a comprised user, TA just instructs the cloud server to delete the entry from the user list L: $L = L \setminus \langle U_{id}, CK_{U_{id}} \rangle$, then the user is no longer able to search the cloud storage.

5   Discussions on Security and Performance

5.1   Security Analysis

In this subsection, we discuss our access control aware keyword search scheme from three security requirements: Data confidentiality, Query privacy, Query unforgeability.

Data Confidentiality. The notion of data confidentiality here requires that user's data must be protected from both unauthorized users and the service provider. In our proposed system, since both the data and the index of keyword are encrypted before uploading to the server, the cloud server does not know the plaintext of the data. Also, the server can not get any information about the secret keys for decryption. In the keyword search process, since the query (encrypted keyword) is also encrypted at the user's side before sending to the cloud server, the cloud server can not know what is queried from a user.

A characteristic of our keyword search scheme is that the index of all keywords is generated by the data owner and the cloud server interactively. As described in Section 4.2, the user first sends $\langle U_{id}, h(w)^r \rangle$ to the cloud server, then the cloud server calculates the $Cap_w = \hat{e}(h(w)^r, CK_{U_{id}})$ for each keyword w and then sends it back to the user. Finally, the index for a keyword w is generated as $I(w)$

$= [R, HMAC_k(R)]$. In this index generation process, both the cloud server and any attacker who can get the $\langle U_{id}, h(w)^r \rangle$ can not deduce any information about w because the one-way hash function is used. Replay attack can also be prevented by the random number r. In the whole process, no plaintext information is leaked to the cloud server or unauthorized users.

Query Privacy. In traditional attribute-based cryptosystems [2,12], access structure (also called access tree) which is associated with ciphertexts indicates the access policies. This information leaks what kind of policies (attributes) are needed to decrypt the ciphertext. Recently, some research has contributed to this privacy related problem, such as [8,9]. Actually, the problem of query privacy also exists in our keyword search scheme.

The notion of query privacy in this paper means that the privacy information of users or data which may leak to the unauthorized party from the query process. In the *Query* phase, the user sends an attribute-based signature *Sig(Q(w))* with the encrypted keyword *Q(w)*. The signature is generated with all of his attributes, $T_{sign} = \{Att_1 \wedge Att_2 \wedge Att_3...\}$. Thus the user's attributes information is inevitably leaked to the cloud server. The cloud server can also deduce deduce which encrypted file can be decrypted by that user by comparing T_{sign} and $T_{decrypt}$. This privacy information leakage to the cloud server is tolerable, because no data confidentiality (such as plaintexts or secret keys) is disclosed to the cloud server or other users.

Query Unforgeability. In our scheme, an individual secret query key is used by each user for encrypting keywords, so a query issued by a user is distinct to his query key. In the multi-user cloud storage setting, this property is very important. Neither another user nor the cloud server can generate a fake query on behalf the valid user. Our scheme offers query unforgeability towards the cloud server and a dishonest user unless a user's secret query key is compromised. And for compromised keys, our scheme allows the cloud administrator to dynamically and efficiently revoke users.

5.2 Performance Analysis

Main processes of our scheme are performed by the user and the cloud server interactively. In the *BuildIndex* phase, the main computation at the user side is simply an exponentiation computation and an paring computation is run by the server for each keyword w. In the *Query* phase, an exponentiation computation and an ABS generation which is mainly based on the paring computation is run by the user for each query. The *Search* process is run by the cloud server. In this process, main overhead is composed of a paring computation and an ABS verification for each query. Consequently, comparing to the overhead of Bao et al. [1,11] our scheme increase the computation cost of ABS generation and ABS verification to the user's side for each query.

However, when considering the viewpoint of cloud server's keyword search computation complexity, our scheme is much efficient than the scheme [1,11] of Bao et al.. Assuming n and m ($m \subseteq n$) are numbers of total files and decryptable

Table 1. A comparison of cloud server's keyword search computation complexity between our scheme and existing works

	Keyword Search Type	Computation Complexity
Bao *et al.* [1,11]	single-user	$O(n \times r)$
Boneh *et al.* [3]	multi-user	$O(n \times r)$
Goh *et al.* [4]	single-user	$O(n \times r)$
Golle *et al.* [5]	single-user	$O(n \times r)$
Song *et al.* [10]	multi-user	$O(n \times r)$
Our Scheme	multi-user	$O(m \times r)$, where $m \subseteq n$

files on the cryptograhic cloud storage for a specific user. Let r be the average number of keywords associated in a file. In the scheme of Bao et al., the cloud storage's computation complexity for executing a keyword search is $O(n \times r)$. In our approach, by allowing the cloud server to narrow the search scope from n to m (to the user's decryptable data group), the computation complexity of our scheme is optimized from $O(n \times r)$ to $O(m \times r)$ for each query, $m \subseteq n$ (m is a subset of n), see Table 1. Especially for the large scale cloud storage environment which has numerous users with different access right and numerous files, the computation complexity advantage of our approach is significant.

6   Concluding Remarks and Future Works

In this paper, we considered the keyword search of a multi-user cryptographic cloud storage(such as [12]) using attribute-based encryption. Our scheme proposes a new keyword search concept for the first time: fine-grained access control aware keyword search. An access structure computation taking advantage of access structures from CP-ABE and ABS is defined in our scheme to achieve fine-grained access control, which facilitate granting differential access rights to a set of users and allow flexibility in specifying the access rights of individual users. Thus our approach also puts forward a new application of ABS protocol, which can not be substituted by traditional signature schemes. Finally, by narrowing the search scope to the user's decryptable files' group, our keyword search scheme decreases information leakage from the query process and is more efficient than other existing schemes.

In our complimentary work, we plan to construct an entire trust model of our proposed multi-user cryptographic cloud storage with flexible, fine-grained access control and secure keyword search. Then, we hope to publish the full result to accelerate the research of cryptographic cloud storage.

Acknowledgments. This work is partially supported by Grant-in-Aid for Young Scientists (B) (23700021), Japan Society for the Promotion of Science (JSPS).

References

1. Bao, F., Deng, R.H., Ding, X., Yang, Y.: Private Query on Encrypted Data in Multi-user Settings. In: Chen, L., Mu, Y., Susilo, W. (eds.) ISPEC 2008. LNCS, vol. 4991, pp. 71–85. Springer, Heidelberg (2008)
2. Bethencourt, J., Sahai, A., Waters, B.: Ciphertext-policy attribute-based encryption. In: IEEE Symposium on Security and Privacy (2007)
3. Boneh, D., Di Crescenzo, G., Ostrovsky, R., Persiano, G.: Public Key Encryption with Keyword Search. In: Cachin, C., Camenisch, J.L. (eds.) EUROCRYPT 2004. LNCS, vol. 3027, pp. 506–522. Springer, Heidelberg (2004)
4. Goh, E.-J.: Secure indexes. Cryptology ePrint Archive, Report 2003/216 (2003), http://eprint.iacr.org/2003/216/
5. Golle, P., Staddon, J., Waters, B.: Secure Conjunctive Keyword Search over Encrypted Data. In: Jakobsson, M., Yung, M., Zhou, J. (eds.) ACNS 2004. LNCS, vol. 3089, pp. 31–45. Springer, Heidelberg (2004)
6. Kamara, S., Lauter, K.: Cryptographic Cloud Storage. In: Sion, R., Curtmola, R., Dietrich, S., Kiayias, A., Miret, J.M., Sako, K., Sebé, F. (eds.) FC 2010 Workshops. LNCS, vol. 6054, pp. 136–149. Springer, Heidelberg (2010)
7. Maji, H.K., Prabhakaran, M., Rosulek, M.: Attribute-Based Signatures. In: Kiayias, A. (ed.) CT-RSA 2011. LNCS, vol. 6558, pp. 376–392. Springer, Heidelberg (2011); An full version on Cryptology ePrint Archive, http://eprint.iacr.org/2010/595
8. Nishide, T., Yoneyama, K., Ohta, K.: Attribute-Based Encryption with Partially Hidden Encryptor-Specified Access Structures. In: Bellovin, S.M., Gennaro, R., Keromytis, A.D., Yung, M. (eds.) ACNS 2008. LNCS, vol. 5037, pp. 111–129. Springer, Heidelberg (2008)
9. Nishide, T., Yoneyama, K., Ohta, K.: Attribute-based encryption with partially hidden ciphertext policies. IEICE Transactions on Fundamentals of Electronics, Communications and Computer Sciences 92-A(1), 22–32 (2009)
10. Song, D.X., Wagner, D., Perrig, A.: Practical techniques for searches on encrypted data. In: IEEE Symposium on Security and Privacy (2000)
11. Yang, Y., Ding, X., Deng, R.H., Bao, F.: Multi-User Private Queries over Encrypted Databases. International Journal of Applied Cryptography Archive 1(4) (August 2009)
12. Zhao, F., Nishide, T., Sakurai, K.: Realizing Fine-Grained and Flexible Access Control to Outsourced Data with Attribute-Based Cryptosystems. In: Bao, F., Weng, J. (eds.) ISPEC 2011. LNCS, vol. 6672, pp. 83–97. Springer, Heidelberg (2011)

Reaction Attack on Outsourced Computing with Fully Homomorphic Encryption Schemes*

Zhenfei Zhang, Thomas Plantard, and Willy Susilo

Centre for Computer and Information Security Research
School of Computer Science & Software Engineering (SCSSE)
University of Wollongong, Australia
{zz920,thomaspl,wsusilo}@uow.edu.au

Abstract. Outsourced computations enable more efficient solutions towards practical problems that require major computations. Nevertheless, users' privacy remains as a major challenge, as the service provider can access users' data freely. It has been shown that fully homomorphic encryption schemes might be the perfect solution, as it allows one party to process users' data homomorphically, without the necessity of knowing the corresponding secret keys. In this paper, we show a reaction attack against full homomorphic schemes, when they are used for securing outsourced computation. Essentially, our attack is based on the users' reaction towards the output generated by the cloud. Our attack enables us to retrieve the associated secret key of the system. This secret key attack takes $O(\lambda \log \lambda)$ time for both Gentry's original scheme and the fully homomorphic encryption scheme over integers, and $O(\lambda)$ for the implementation of Gentry's fully homomorphic encryption scheme.

Keywords: Cloud Computing, Fully Homomorphic Encryption, Reaction Attack, CCA security, Secured Outsource Computation.

1 Introduction

Cloud computing has changed the phenomena in the Information Technology (IT) industry completely. It allows access to highly scalable, inexpensive, on-demand computing resources that can execute the code and store the data that are provided to them. This aspect, known as data outsourced computation, is very attractive, as it alleviates most of the burden on IT services from the consumer (or data owner). Nevertheless, the adoption of data outsourced computation by business has a major obstacle, since the data owner does not want to allow the untrusted cloud provider to have access to the data being outsourced. Merely encrypting the data prior to storing it on the cloud is not a viable solution, since encrypted data cannot be further manipulated. This means that if the data owner would like to search for particular information, then the data would need to be retrieved and decrypted - a very costly operation, which limits the usability of the cloud to merely be used as a data storage centre.

---

* This work is supported by ARC Future Fellowship FT0991397.

H. Kim (Ed): ICISC 2011, LNCS 7259, pp. 419–436, 2012.
© Springer-Verlag Berlin Heidelberg 2012

In [1], van Dijk and Juels argued that, cryptography tools alone is not sufficient for providing privacy for cloud computing. Yet, they found that in a single-client scenario, fully homomorphic encryption schemes deliver the required security.

Indeed, a fully homomorphic encryption is a solution for enabling operations on the encrypted data. Essentially, fully homomorphic encryption schemes enable one to apply homomorphic operations over an arbitrary number (n) of given ciphertexts $c_1, c_2, \ldots, c_n$ without the need to know the corresponding plaintexts $m_1, m_2, \ldots, m_n$.

This feature is useful in the outsourced computation scenario, where one can upload encrypted data to the cloud and enable the cloud to process the data *without* the need for decryption. Nevertheless, whether a fully homomorphic encryption by itself is sufficient enough to secure the outsourced computation in practice remains unclear. This will require some further research.

Our Contribution

In this paper, we show a negative result to the above question. Before addressing our contribution, we should highlight some important factors to motivate our work. In fact, one can categorize an outsourced computation into the following models:

1. The user possesses the data and the computation circuit, and the service provider provides the computation power;
   Example: a stock share holder buys/sells his/her stocks via the cloud, and then retrieve the receipt from the cloud to obtain his/her updated financial status.
2. The user possesses the data, and the service provider provides its computational power, while the computation circuit can be made available publicly to both of the entities;
   Example: a hospital outsources its patients' information to a research institute for acquiring further analysis from the institute (such as the result of the prostate cancer), as the institute has more computational power compared to the hospital.
3. The user possesses the data, and the service provider provides its computational power, only the cloud has access to the computation circuit;
   Example: a company outsources its financial status to an auditing company, however, the auditing algorithm is auditing the company's private property.

In all of the above models, the users' data privacy has to be ensured. The difference among them lies on the privacy of the computational circuit.

In this paper, we present a practical reaction attack that can be applied to all of the above models, in which every time a user interacts with the cloud, he/she is under the risk of leaking some information. As a result, known approaches, i.e., secured outsourced computation (SOC) [2], becomes essential in the first two models, while for the third model, even SOC technique will not be sufficient.

Further, using our attack, one can construct a probabilistic decryption oracle. Consequently, we argue that for any fully homomorphic encryption schemes, the CCA-1 security is essential. When applying our attack to Gentry's framework [3,4], our attack recovers the secret key for *all* published fully homomorphic encryption schemes. This secret key attack take $O(\lambda \log \lambda)$ time for both Gentry's original scheme and the fully homomorphic encryption scheme over integers. Furthermore, for the implementation of Gentry's fully homomorphic encryption scheme, this secret key attack requires $O(\lambda)$ time.

Related Work

In [5], Hall et al. presented a reaction attack against several public key cryptosystems, mainly on lattice based cryptosystems and coding based cryptosystems. By observing the reaction of the decryption procedure, one obtains information about the secret key and/or message. Compared to the result of [5], where one does not observe the reaction of the decryption procedure, our attack relies on the difference of users' reactions on receiving different results.

In [2], Gennaro et al. presented a non-interactive verifiable computing protocol, that allows one to outsource its computation to untrusted workers. This method can be applied to the first two models, where the users can check if the cloud has modified the demanded computation circuit. However, it is impossible for this method to be applied to the third model. In addition, the main obstacle of using this technique is its computation efficiency. Initially, this protocol requires a one time pre-processing to generate a minimum garbled circuit, which takes $O(C) \times poly(\lambda)$ time, where C is in function of the computation circuit, and λ is the security parameter. Then, for each new computation circuit, one is required to modify the minimum garbled circuit, using a fully homomorphic encryption scheme. Thus, we argue that this method is impractical.

Other related work of this paper can be found in [6].

2 Background

2.1 Fully Homomorphic Encryptions

The idea of fully homomorphic encryption was raised by Rivest, Adleman and Dertouzos [7], shortly after the invention of RSA [8]. A fully homomorphic encryption scheme consists of following four algorithms:

- KEYGEN(λ): Input a security parameter λ, it outputs public key **pk**, secret key **sk**.
- ENCRYPT(m, **pk**): Input a message m and the public key **pk**, it outputs a corresponding ciphertext c.
- DECRYPT(c, **sk**): Input a ciphertext c and the secret key **sk**, it outputs a corresponding message m.
- EVAL(**pk**, $c_1, c_2, \ldots, c_n, C^n$): Input a public key **pk**, n ciphertext $c_1, c_2, \ldots, c_n$ and a permitted circuit C^n, it outputs $C^n(c_1, c_2, \ldots, c_n)$.

Following this notion, schemes that support partial homomorphism have been proposed. Recently, Gentry [3,4] successfully provided a framework for constructing homomorphic encryption schemes (referred to as the GENTRY scheme) and, further, he provided a concrete construction. In addition, subsequent works based on his framework have been proposed recently (such as [9,10,11]). For instance, in [12] (referred to as GENTRY-HALEVI scheme), the author optimized the performance of GENTRY scheme, while in [11] (referred to as vDGHV scheme), the author proposed an integer variant of GENTRY scheme. In the following, for clarity, we will review Gentry's framework.

2.2   Gentry's Framework

Gentry's framework for constructing fully homomorphic encryption schemes is based on creating a function to perform two atomic operations which will allow the user to build any kind of circuit. Effectively, any circuit can be built with two atomic functions, namely addition + and multiplication × over $\mathbb{F}_2$ (see [3,4]). Therefore, to evaluate any circuit, we are only required to be able to add and multiply over $\mathbb{F}_2$ two encrypted bits.

We note that, to ensure security, such an encryption function is required to be indistinguishable, namely $Enc(m_0) \neq Enc(m_1) \not\Rightarrow m_0 \neq m_1$. To build such a function, $\oplus$ and $\otimes$, Gentry used a simple model. Gentry defined the two functions f_+ and $f_\times$ which are equivalent to decrypting both encrypted bits, adding or multiplying such decrypted bits and then encrypting the resulting bits (See Figure 1).

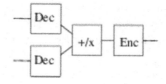

Fig. 1. $f_{+,\times}$

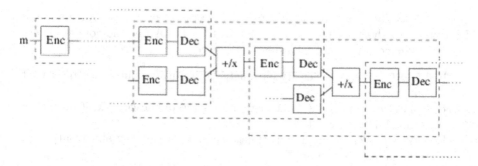

Fig. 2. Gentry's fully homomorphic encryption model

However, if f_+ and $f_\times$ return the desired result for $\oplus$ and $\otimes$, the bits are clearly readable and therefore they do not maintain the intended security requirement.

To achieve this required property, Gentry used an encryption scheme which allows evaluation of short circuits. Therefore, it encrypts the ciphertext with a second cryptosystem. Hence, it can remove the first encryption securely to perform the addition or the multiplication (See Figure 2).

Using such a technique, Gentry simplified the quest of constructing a fully homomorphic encryption that can evaluate any circuit on encrypted data by finding an encryption system that can evaluate only some short circuits, namely f_+ and $f_\times$. In [4,3], Gentry built such an encryption scheme using ideal lattices. This work was followed by other fully homomorphic encryption schemes based also on ideal lattices [10,9]. Another type of encryption scheme respecting Gentry's model requirement was also proposed in [11] using integers.

2.3 The vDGHV Fully Homomorphic Encryption Scheme

In this subsection, we describe the fully homomorphic encryption scheme over integers (vDGHV scheme), instead of GENTRY scheme, since this scheme uses integers rather than ideal lattice, and therefore, it is easier to demonstrate and explain, and later incorporate our idea into.

vDGHV scheme consists of a somewhat homomorphic encryption scheme (SHE) that supports limited additions and multiplications, and the bootstrapping technique to break such limitation.

The somewhat homomorphic encryption scheme consists of four algorithms:

- KEYGEN(λ): Input a security parameter λ, it firstly generates parameters $\{\alpha, \beta, \gamma, t, n\}$ in function of λ. It then generates a secret *odd* integer $p \in (2^\beta, 2^{\beta+1})$, n different integers $\{r_i \in [-2^\alpha, 2^\alpha)\}$ and another n different integers $\{g_i \in [0, 2^{\gamma-\beta})\}$, respectively. It finally outputs the public key $\mathbf{pk} = \{x_i = g_i p + 2r_i\}$ and secret key $\mathbf{sk} = \{p\}$.
- ENCRYPT(m, $\mathbf{pk}$): Input the public key $\mathbf{pk}$ and a message $m \in \{0,1\}$, it chooses a random subset $\mathbf{s} \subseteq \mathbf{pk}$ and output $c = m + 2r + \sum_{x_i \in \mathbf{s}} x_i \bmod x_0$, where x_0 is the smallest in $\{x_i\}$, $r \in [-2^\alpha, 2^\alpha)$ is a random noise.
- DECRYPT(c, $\mathbf{sk}$): Input the secret key $\mathbf{sk} = \{p\}$ and a ciphertext c, it outputs $m = (c \bmod 2) \oplus (\lfloor c/p \rceil \bmod 2)$, where $\lfloor c/p \rceil$ returns the closest integer of c/p.
- EVALUATE($c_1, c_2, ..., c_k, \mathcal{P}, \mathbf{pk}$). It outputs $\mathcal{P}(c_1, c_2, ..., c_k)$, where $\mathcal{P}$ is a k-inputs evaluation polynomial whose circuit depth is lower than the maximum circuit depth allowed by this SHE.

This SHE supports homomorphic additions and multiplications, when $\alpha \ll \beta$. For instance, suppose $c_1 = m_1 + g_1 p + 2r_1$ and $c_2 = m_2 + g_2 p + 2r_2$ for certain g_1, r_1, g_2, r_2, the product of two ciphertext $c_1 c_2 = m_1 m_2 + 2(r_1 m_2 + r_2 m_1 + 2r_1 r_2) + p(g_1 m_2 + 2g_1 r_2 + g_2 m_1 + 2g_2 r_1 + g_1 g_2 p)$. One can observe that the decryption of $c_1 c_2$ is $m_1 m_2$, as long as $2(r_1 m_2 + r_2 m_1 + 2r_1 r_2) \in (-p/2, p/2]$. Therefore, the above SHE is somewhat homomorphic.

However, the homomorphic circuit depth is limited, *i.e.*, the noise grows after each operation, and eventually it is possible that the absolute value of the noise will be greater than $p/2$ and a decryption error is then being generated.

Suppose we want to evaluate a circuit whose depth is greater than this SHE permits, we break the circuit into several sub-circuits. For each sub-circuit, the absolute value of resulted noise is less than the threshold $(p/2)$. Then we refresh the resulted ciphertext using the bootstrapping technique. We describe the bootstrapping technique in general. We refer the readers to their original scheme[11] for more details.

To bootstrap, firstly, they modify the decryption circuit. As we have shown earlier, the noise grows significantly faster in a multiplication than in an addition. Therefore, vDGHV scheme used a squashing method that breaks the decryption circuit from one multiplication into several additions. The squashing technique is as follows:

- Generate $x = \lfloor 2^\kappa/p \rfloor$, where κ is a parameter in λ that is greater than $\beta + 1$.
- Build a bit sequence $S =< s_1, s_2, \ldots, s_\eta >$, $s_i \in \{0, 1\}$, with $\sum s_i = \theta$. S becomes the new secret key.
- Choose n random integers u_i between 0 and $2^{\kappa+1}$, such that $\sum_i^n s_i u_i = x \bmod 2^{\kappa+1}$.
- Set $y_i = u_i/2^\kappa$. Then $\sum s_i y_i = 1/p + \epsilon$, where ϵ is negligible compared with $1/p$.
- New ciphertext is a vector $z =< z_1, z_2, \ldots, z_\eta >$, generated by $z_i = [c \times y_i]_2$.
- New decryption circuit becomes $m = [c - \lfloor \sum s_i \times z_i \rfloor]_2$

As a result, the decryption circuit now consists only additions, while the growth of noise in additions is extremely slow. Then, because the modified decryption circuit depth is relatively low, now it is possible to carry out the decryption circuit homomorphically, through the proposed SHE.

In practice, they encrypt ciphertexts, denoted by $\{Enc(z_i)\}$ and the secret keys, denoted by $\{Enc(s_i)\}$. Denote $\mathcal{C}_D$ the decryption circuit, then

$$\text{DECRYPT}(\mathcal{C}_D, \{Enc(z_i)\}, \{Enc(s_i)\}) = Enc(m).$$

This is because firstly $\mathcal{C}_D(\{z_i\}, \{s_i\}) = m$ and secondly, $\mathcal{C}_D$ can be carried out homomorphically. Therefore, we obtain a new ciphertext $Enc(m)$.

The new ciphertext, $Enc(m)$ has a refreshed noise level (less than 2^α), which means $Enc(m)$ can be evaluated again. By doing this repeatedly, we can evaluate circuit with any depth homomorphically. Therefore, a fully homomorphic encryption scheme is achieved.

2.4   Security Models

In the following, we describe briefly both Chosen-Plaintext Attack (CPA) [13] and Chosen-Ciphertext Attack (CCA) [14] attacks for completeness.

The IND-CPA security game is defined as follows:

1. The challenger runs KEYGEN algorithm and outputs a secret key **sk** and a public key **pk**;
2. The attacker is given an encryption oracle that computes the functionality ENCRYPT(m, **pk**);
3. The attacker then generates two ciphertexts m_0 and m_1;

4. The challenger generates $c = \text{ENCRYPT}(m_b, \textbf{sk})$, where $b \in \{0, 1\}$;
5. The attacker outputs b'.

We say that an encryption scheme is CPA secure if the advantage of the attacker to win the game $(Pr[b = b'] - 1/2)$ is negligible.

The IND-CCA-1/2 security game is defined as follows:

1. The challenger runs KEYGEN algorithm and outputs a secret key $\textbf{sk}$ and a public key $\textbf{pk}$;
2. The attacker is given two oracles, an encryption oracle and a decryption oracle;
3. The attacker then generates two ciphertexts m_0 and m_1;
4. The challenger generates $c = \text{ENCRYPT}(m_b, \textbf{sk})$, where $b \in \{0, 1\}$;
5. (Only for CCA-2) The attacker is given the two oracles again, but it can not query on c;
6. The attacker outputs b'.

We say that an encryption scheme is CCA-1/2 secure if the advantage of the attacker to win the game $(Pr[b = b'] - 1/2)$ is negligible.

3   Our Reaction Attack

In this section, we will firstly introduce our message attack that recovers a message of any kind of fully homomorphic systems used in the outsourced computation, in probabilistic manner. Then, we show that by adapting our attack in Gentry's framework, we can recover the secret key. We note that our attack described in this section is applicable to all three models in the first section. For simplicity, we uses the first model.

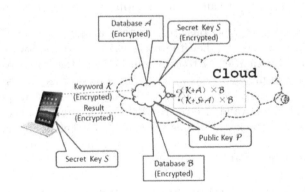

Fig. 3. Using Fully Homomorphic Encryption in a Cloud Search Scenario

3.1   A Message Attack

The Idea. In this subsection, we describe our message attack. For any given ciphertext, our attack recovers the message with provability ε.

We will first illustrate our high level construction for clarity. To use fully homomorphic encryption schemes in outsourced computation scenarios, the users firstly upload their encrypted data to the cloud. Then, they submit their demanded circuits to the cloud, in an on-demand fashion. The demanded circuit consists either some data and an evaluation function, or merely an evaluation function only. The cloud processes users' data through the requested circuits, and returns the result.

Ideally, all the data, including the results, are encrypted, and hence, a malicious cloud provider cannot gain information from the users, $i.e.$, let ε_1 be the possibility of $m = 1$, then for any ciphertext, $|\varepsilon_1 - 1/2|$ is negligible from the cloud provider's point of view.

Nevertheless, we notice that the attacker can modify the encrypted circuits/results by adding some random ciphertext c that encrypts a message m. Because homomorphism is enabled, modifying the demanded circuits/results will affect the plaintext eventually. To be more precise, if the added random ciphertext encrypts a 0, the returned result remains the same; while if the added random ciphertext encrypts a 1, the returned result is modified. By observing the users' reactions, the service provider can increase or decrease ε_1 accordingly, and eventually recover m.

Generally speaking, the cloud provider can compare users' reaction with their former reaction, if the users are acting "unexpectedly", then the cloud can expect $m = 1$. For completeness, we list some (*but not* all) possible reactions that can be defined as "unexpected" behaviors.

- The users set up a new task much sooner than usual, after they acquire the result sent by the cloud;
- The circuit of a new task is identical from a former one;
- The number of tasks is significantly higher than average - this occurs when the cloud provider feeds same faulty information for a certain period.

We note that these users' reactions are very natural and practical. To anticipate the reactions is even easier, when the users use a certain software, instead of expecting the results themselves, to communicate with the cloud. We argue that this is very common in practice as nobody will conduct this process manually.

However, the success of our attack relies highly on the actions performed by the users after receiving valid or error results. Hence, if the users act completely randomly, then our attack will be unsuccessful. Nevertheless, we argue that the latter usually will not happen in practice, as it is the users' interest to acquire the results that they would like to obtain.

An Example. In this subsection, we demonstrate an example of our attack. Suppose we have a cloud search engine (see Figure 3), which looks up keywords from database A, and outputs the corresponding results in database B. Database

A consists of names of stocks, while database B shows corresponding price for each stock.

To distinguish from a traditional search engine, the databases of the cloud search engine are all encrypted, and the search circuit is a homomorphic circuit. Without losing generality, we use vDGHV scheme to demonstrate our attack.

Table 1. Databases in plaintext

Entry	A	B
1	AAPL	335
2	GOOG	494
3	MSFT	027
4	SPRD	013
5	NDAQ	024
...	...	...

Table 1 shows the databases in plaintext for our example. We note that this is not exactly the databases stored in the cloud. The cloud maintains multiple copies of the databases for different users, each copy is encrypted under different users' FHE secret key.

Let K, A and B denote the binary form of the keyword, database A and database B, respectively. Let k_i, a_i and b_i be the i-th digit of K, A and B. Then, the database in the cloud consists of $\{Enc(a_i)\}$ and $\{Enc(b_i)\}$. Also, let $\bigotimes_1^n c_i$ be $c_1 \otimes c_2 \otimes \cdots \otimes c_n$. A basic search algorithm is defined in Algorithm 1, and we achieve a fully homomorphic searching algorithm in Algorithm 2 using the fully homomorphic encryption scheme over integers.

Algorithm 1. Basic Search (K, A, B)

$l_a \leftarrow \text{LEN\_OF\_WORD\_A} \; //l_a = 32$
$l_b \leftarrow \text{LEN\_OF\_WORD\_B} \; //l_b = 24$
for $j = 0 \rightarrow \text{END\_OF\_ENTRY} - 1$ **do**
  for $i = 1 \rightarrow l_b$ **do**
    $r_i \leftarrow b_{i+jl_b} \otimes (\bigotimes_{t=1}^{l_a} (a_{t+jl_a} \oplus k_{t+jl_a})) \oplus r_i$
  end for
end for

As shown in Table 2, suppose we want to look for the price of GOOG, with the basic search algorithm, we obtain $52, 57, 52$ in ASCII code, which is 494. While with the homomorphic search algorithm, we obtain the third column of Table 2. The cloud cannot decrypt $Enc(52)/Enc(57)/Enc(52)$, hence, the users' privacy is guaranteed. The user holds the secret key, therefore, he/she is the only one who knows the searching results, while the cloud cannot even distinguish the difference between the two $Enc(52)$.

Algorithm 2. Homomorphical Search $(\{Enc(k_i)\}, \{Enc(a_i)\}, \{Enc(b_i)\})$

$l_a \leftarrow$ LEN_OF_WORD_A $//l_a = 24$
$l_b \leftarrow$ LEN_OF_WORD_B $//l_b = 32$
for $j = 0 \rightarrow$ END_OF_ENTRY-1 **do**
    for $i = 1 \rightarrow l_a$ **do**
        $r_i \leftarrow Enc(b_{i+jl_b}) \times \prod_{t=1}^{l_a}(Enc(a_{t+jl_a}) + Enc(k_{t+jl_a})) + r_i$
    end for
end for

Table 2. Searching Results in ASC II

Database A	Basic Search	Homomorphic Search	Homo Search + Faulty Info
AAPL	0,0,0	Enc(0), Enc(0), Enc(0)	Enc(0),Enc(0),Enc(0)
GOOG	52,57,52	Enc(52), Enc(57), Enc(52)	Enc(0),Enc(0),Enc(0)
MSFT	52,57,52	Enc(52), Enc(57), Enc(52)	Enc(0),Enc(0),Enc(0)
SPRD	52,57,52	Enc(52), Enc(57), Enc(52)	Enc(0),Enc(0),Enc(0)
NDAQ	52,57,52	Enc(52), Enc(57), Enc(52)	Enc(0),Enc(0),Enc(0)
...	...	...	...
result	494	Enc("494")	error

However, if the cloud acts maliciously, it can recover one bit of message from c through our attack model, since it adds a value to the keyword. Hence, if $Dec(c) = 0$, the algorithm will search for $Enc($GOOG$)$ as before, and therefore, no error will occur, and the user will most likely do nothing. However, if $Dec(c) = 1$, instead of searching for $Enc($GOOG$)$, the input of the algorithm is actually $Enc($GOOH$)$. Therefore, no match will be found. It is reasonable to believe that the user will start to execute another search, in which case the cloud increases ε_1.

As we have stated, a malicious cloud can also modify the circuit/result accordingly. However, we notice that in the above example, modifying the result merely helps our attack, as if the cloud induces an $Enc(1)$, the user will receive 495, which will be recognized as a valid result.

Formal Construction. In this section we show a formal construction of evaluating users' reactions in terms of timing. We note that our formal construction is only a function of the time the user responses. In practice, with the aid of other parameters, the malicious service provider can further increase the successful rate of this attack.

Let t_{res} be the time period when the user starts a new task after receiving the result from the cloud. Let t_n be the average time when the user sets up his/her tasks when no error is induced, and t_e be the minimum time when the user starts a new task if some error occurs. Assume that t_{res} follows a certain distribution $\mathcal{D}_0 = f_0(t)$ depending on the average time t_n when no error occurs, and another distribution $\mathcal{D}_1 = f_1(t)$ depending on the t_e when there exists an error. As a service provider, we assume the cloud is aware of above information.

We note that the success of the message attack relies on the difference between $\mathcal{D}_0$ and $\mathcal{D}_1$. For users whose reactions are completely random, i.e., $\mathcal{D}_0 = \mathcal{D}_1$, our attack will not be successful. However, we argue that, in practice, a reasonable user will have different distributions for different results.

For any time frame t, $\mathcal{D}_0$ has a $f_0(t)$ probability to distribute a new task, while $\mathcal{D}_1$ has a $f_1(t)$ probability to distribute a new task. As a result, if the cloud receives a new task at time t, the confidence of $m = 0$ and $m = 1$ can be determined by $g_0(x) = \frac{f_0(t)}{f_0(t)+f_1(t)}$ and $g_1(x) = \frac{f_1(t)}{f_0(t)+f_1(t)}$, respectively.

Later, we can anticipate the successful rate of our attack as follows. Let $\psi = g_1(t_1) = g_2(t_2)$ $(t_2 > t_1)$, when the cloud wants to build a minimum ψ confidence. Therefore, any task delivered prior to t_1 will give the cloud at least ψ confidence of $m = 1$, while any task delivered after t_2 will give the cloud at least ψ confidence of $m = 0$. Then we can evaluate the successful rate ε. Generally, it can be expressed by the volume of $f_1(x)$ in $[0, t_1]$ and $[t_2, \infty)$ over the total volume. More formally,

$$\varepsilon = \frac{\displaystyle\int_0^{t_1} f_1(x)\,dx + \int_{t_2}^{\infty} f_1(x)\,dx}{\displaystyle\int_0^{\infty} f_1(x)\,dx}.$$

In the case the cloud has not developed sufficient confidence for a certain message (i.e., it receives a new task at time between t_1 and t_2), it will induce the same faulty information to the next several tasks. Suppose the cloud takes n tasks to develop the confidence, and the corresponding response time are $t_1, t_2, \ldots, t_n$, then the confidence can be determined by $\psi' = \max(\psi_0', \psi_1')$, where

$$\psi_0' = \frac{\prod_i^n g_0(t_i)}{\prod_i^n g_0(t_i) + \prod_i^n g_1(t_i)}, \qquad \psi_1' = \frac{\prod_i^n g_1(t_i)}{\prod_i^n g_0(t_i) + \prod_i^n g_1(t_i)}.$$

For instance, if the adversary sends an identical message twice, with response time t_1 and t_2, then the possibility of two continuous 0-s and 1-s are $g_0(t_1)g_0(t_2)$ and $g_1(t_1)g_0(t_2)$, respectively. We note that it is not possible to have 10 or 01, as we are using the same ciphertext. As a result, the possibility of 10 and 01 are eliminated. Therefore, we have

$$\psi_0' = \frac{g_0(t_1)g_0(t_2)}{g_0(t_1)g_0(t_2) + g_1(t_1)g_1(t_2)}, \qquad \psi_1' = \frac{g_1(t_1)g_1(t_2)}{g_0(t_1)g_0(t_2) + g_1(t_1)g_1(t_2)}.$$

Meanwhile, the new successful rate can be determined by $\varepsilon' = \frac{\varepsilon^n}{\varepsilon^n + (1-\varepsilon)^n}$. This guarantees that we can achieve a very high confidence/successful rate by attacking the same ciphertext repeatedly.

To exemplify our construction, without losing generality, we show a example where users' reaction follows a normal distribution (also known as Gaussian distribution) [15] with the distribution factor $\sigma = 2$ (see Figure 4). We note that for simplicity we uses smooth curves to illustrate the distributions, while in practise, the actual distribution will be more discrete.

In this example, $t_n = 10$ minute. Meanwhile, we assume the transmission time/evaluation time is negligible, which implies $t_e = 0$. From this figure, we obtain a figure that illustrates the confidence of messages in Figure 5. Hence, if the adversary requires 96% confidence in a single round, the corresponding t_1 and t_2 are 4 and 6.5, respectively. Therefore, the successful rate by one single round can be determined by $\frac{V_1+V_0}{V_1+V_0+V_e}$, which is essentially $\frac{\int_0^4 f_1(x)\,dx + \int_{6.5}^\infty f_1(x)\,dx}{\int_0^\infty f_1(x)\,dx}$.

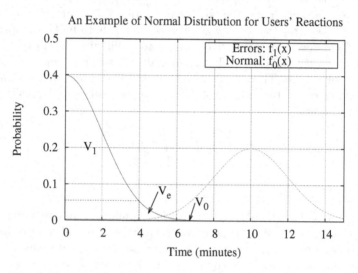

Fig. 4. An example of Users' Reactions Using Normal Distribution

On CCA Security. It is well known that for any public key encryption schemes, the CPA security is essential. Meanwhile, constructing a CCA-2 secure fully homomorphic encryption scheme is impossible, since homomorphic operations on ciphertexts are enabled (and also it is due to the "malleability" of the ciphertext). Moreover, unfortunately, fully homomorphic encryption schemes that follow Gentry's framework cannot be CCA-1 secure due to the bootstrapping technique. In fact, if a somewhat homomorphic encryption scheme is CCA-1 secure is questionable. Indeed, in [16], the authors presented a CCA-1 attack against GENTRY-HALEVI SHE scheme.

We note that the consequence of such an attack might be severe. A CCA-1 attack is an attack model that assumes there exists a decryption oracle. People may argue that this is just merely an attack model, since in practice, having such an oracle (or an honest user who is helping the attacker during the learning phase) is impractical. Further, constructing such an oracle in general is not really feasible. Nevertheless, with our message attack, it is possible to construct a probabilistic decryption oracle in practice. Consequently, if for a certain fully homomorphic encryption scheme, a CCA-1 attack is successful with a non-negligible advantage

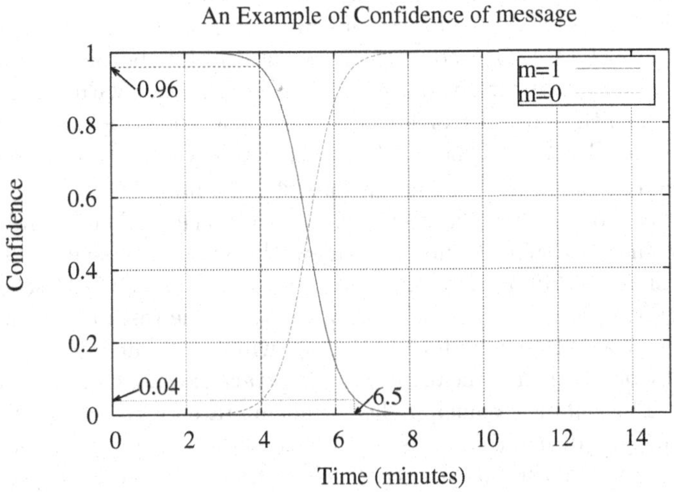

An Example of Confidence of message

Fig. 5. An example of Confidence Curve

of ϕ using a hypothetical deterministic oracle, then one can achieve this attack with an advantage of $\phi\varepsilon^n$ using our message attack, where n is the number of ciphertext required for the CCA-1 attack. Hence, the CCA-1 attack becomes really practical.

To sum up, we argue that in practice, if a fully homomorphic encryption scheme is not CCA-1 secure, then it alone cannot deliver secured outsourced computation.

3.2 The Secret Key Attack

In this subsection, we propose our secret key attack. We show that for *any* FHE scheme, even its SHE schem is CCA-1 secure, as long as it uses Gentry's bootstrapping technique, it is vulnerable to our attack.

Recall that in Gentry's framework, binary operations (*i.e.*, $\oplus, \otimes$) of messages are eventually + and × of ciphertext in a FHE scheme; also, to enable "bootstrappability", one has to publish an encryption of its secret key. Therefore, for any FHE schemes that follows Gentry's framework, no matter what kind of circuit the user demands, with our message attack, a malicious cloud recovers one bit of the secret key through each attack. Eventually, it will recover the whole secret key.

For instance, as in vDGHV scheme, one publishes $\{Enc(s_i)\}$. The cloud can recover s_i through each message attack, and recover S as a result. As another example, the secret key in GENTRY-HALEVI scheme [17] is a bit sequence of 1024, while the number of encrypted 1-s is 15. Hence, it takes a maximum 1024 message attacks to recover the secret key.

One may argue that the secret key may contain too many bits, and therefore, it is impractical to recover them all. For instance, S in vDGHV scheme contains λ^5 digits, *i.e.*, $\eta \sim O(\lambda^5)$. Consequently, to recover all bits becomes impractical.

However, we note that, firstly, our attack can be launched with other attacks. For instance, with the existence of a decryption oracle, the attack in [16] recovers the secret key of GENTRY-HALEVI scheme within $O(\lambda^2)$ queries, consequently, a CCA attack can break the system with an ε^{λ^2} advantage.

Secondly, we notice that the secret keys of GENTRY scheme, and that of all other schemes following Gentry's framework, are sparse sequences with a significantly smaller Hamming weight. As an example, in vDGHV scheme, the secret key $S = \langle s_1, s_2, \ldots, s_\eta \rangle$ only has θ 1-s, while the rest are 0-s and $\theta \ll \eta$ (*i.e.*, $\theta = \lambda$, $\eta = \lambda^5$). Hence, inducing $Enc(s_i)$ will not result into any error, in most cases. In practice, this means using our attack merely increases the error rate by $1/\lambda^4$. This enables a much higher attacking rate for the cloud. The cloud can induce encrypted secret key bits at all time without being detected.

Finally, because of the special structure of the secret key, we propose an optimized secret key attack that improves the performance of our secret key attack significantly. As a result, it only requires $O(\lambda \log \lambda)$ operations to recover all λ^5 bits. We will describe this optimized attack in the following section.

Optimized Secret Key Attack. In Gentry's framework, the secret key $S = \langle s_1, s_2, \ldots, s_\eta \rangle$ contains η bits, while the Hamming weight $\theta = \sum_i^\eta s_i$ is significantly smaller than η. Therefore, we propose two optimizations, using fully Dichotomy search and block Dichotomy search algorithm.

For the block Dichotomy search algorithm, if we cut the whole secret key into k blocks, with each block $l = \eta/k$ bits, with a very high probability the block will only consist of 0-s. Now, instead of testing the parity of each bit, we test if a block contains only 0-s. To perform such a task, one generates

$$b = Enc(1) + \prod_i^l (Enc(1) + Enc(s_i)),$$

and test the parity of b. Because homomorphic operations are enabled, if all the bits are 0-s, then decrypting b will give us 0, and vice versa. As a result, the adversary recovers l bits in one message attack.

There is one exception, where there are at least one 1 in the block. Hence, a Dichotomy search algorithm is required. The average case takes place when there are no more than one encrypted 1 in a block. Therefore, for each encrypted 1, it requires additional $\log(l)$ message attacks.

To sum up, this optimization reduces the number of message attacks from η to $k + \theta \log(\eta/k)$. With the configuration of vDGHV scheme/GENTRY scheme ($\theta \sim \lambda$ and $\eta \gg \theta$), the minimum number for message attacks is $O(\lambda \log \lambda)$. While with the configuration of GENTRY-HALEVI scheme ($\theta = \lambda/\log \lambda$ and $\eta = 2^{\lceil 2 \log \lambda \rceil}$), the minimum number for message attacks is $O(\lambda)$.

While for the fully Dichotomy search algorithm, one cuts the secret key into two halves for each round. For each half, one generates $b' = Enc(1) +$

$\prod_i(Enc(1) + Enc(s_i))$. If one or more 1-s is anticipated in any piece, the adversary feeds the user with the inverse of b', and vice versa. This method is to ensure that the induced ciphertexts always have a higher probability of encrypting 0-s, and consequently, the error rate will be minimized. For instance, for the first round, any half has $1 - \binom{\eta/2}{\theta}/\binom{\eta}{\theta}$ possibility of having at least a 1. Therefore, the cloud sends $b' + 1$.

In the worst case, the fully Dichotomy search algorithm requires $\theta \log(\eta/\theta) + \theta$ enquiries. Applying over the FHE schemes, we observe a similar result with the previous optimization, i.e., $O(\lambda \log \lambda)$ for VDGHV scheme/GENTRY scheme and $O(\lambda)$ for GENTRY-HALEVI scheme. However, we note that in practice, this optimization might work better, because we will have a higher probability to eliminate a big block of 0-s.

We use VDGHV scheme to exemplify our secret key attack. The recommended configuration for this scheme is $\theta = 80$ and $\eta = 80^5$. Our optimization reduces the number of messages attacks from 80^5 to approximately 2100. Meanwhile, the actual successful rate for the secret key attack depends on the confidence from the message attack. Suppose through the message attack the cloud develops 0.999 confidence on the message (this ratio can be achieved by launching the message attack repeatedly on a same message), then the cloud can recover the secret key with a probability of $0.999^{2100} = 0.122$. That is to say, in average case, the cloud needs to use the secret key attack for 8 times to recover the secret key, which is unacceptable by the users.

4 Discussion

4.1 Practicality of Our Attack

In practice, errors cannot be eliminated in the outsourced computation scenario. The users cannot distinguish if it is caused by a malicious service provider, or by some connection/transmission error, or even by the algorithm itself. As we have shown before, most of our message attacks in the (optimized) secret key attack will not cause errors. Even with our optimization, the error rate is still significantly small, compared with other causes. If $\lambda = 80$, the error rate can be as small as 2.44×10^{-8} for the original secret key attack. Meanwhile, the cloud can manipulate the attacking rate, i.e., it attacks only when the current error rate of the network is lower than normal. Therefore, the user is incapable of detecting this attack.

As a result, most of the decryption errors is generated by other reasons. The user cannot afford to transfer its data to another cloud every time an error occurs, since transferring encrypted data is too costly.

One may also argue that when the attacker induces $Enc(0)$ and it might also be so coincidence that a user is proceeding another search subsequently. Consequently, the attacker would recognize the corresponding bit to be 1. Will this mislead the attacker? Possibly. This is the reason why we stated that our attack is probabilistic, and we have shown that our attack can minimize the impact of this event to occur. The attacker performs one attack during a period, instead of one interaction. During this period, ε_1 may have already been decreased a lot times for other reasons. Therefore, one increase will *not* mislead the cloud after all.

4.2 Protecting FHE with Verifiable Computation

In theory, using FHE with verifiable computation will stop our attack in model 1 and 2. Since the user is able to verify the computation, the cloud will not be able to modify the computation circuit without be detected. As a result, our attack will be unsuccessful. However, we argue that, in order to use FHE in those models, one *must* use verifiable computation all the time.

Moreover, we also observe that, although the cost of verification is low, to generate the minimum circuit and to homomorphically modify it is costly. Thus, whether this technique can be used in practise is doubtful.

Finally, as far as model 3 is concerned, the computation circuit is private to the cloud, hence, verifiable computation protocols are not applicable in this scenario.

4.3 Other Possible Protections

A possible solution to the secret key attack could be removing the necessity of bootstrapping technique. If there is such a FHE scheme that supports arbitrary circuit depth without the bootstrapping technique, then the users can avoid publishing his/her encrypted secret keys. Unfortunately, so far all FHE schemes except [18] follow Gentry's framework, and to bootstrap is essential for them to achieve fully homomorphic. Meanwhile, even if there exists a FHE scheme without bootstrapping, by incorporating our attack, the attacker is still capable of recovering data (*but not* the secret key) from the users.

To stop the message attack is much more difficult, as one cannot determine where an error is actually from. A possible solution could be setting up certain protocols between the service provider and the users. This protocol has to minimize the error rate. Further, whenever an error occurs, the provider has to show the users' full details of the evaluation circuit, in order to convince the users.

Another possible but expensive solution would be letting the users to generate random "meaningless" tasks (or "stubs") periodically. Whenever the user needs to use the service, he/she replaces a stub with the one he/she really requires. Even though the user receives some errors and needs to set up the same task again, he/she does not process immediately. Instead, he/she will wait till the next period of sending tasks. As a result, the attacker will not be able to distinguish if a task is a valid one, or a repeated one (due to message attack), or merely a random one. Also, the overall average task rate remains the same.

However, we note that this solution is very expensive, as it requires periodic communications between the users and the cloud. Meanwhile, the user sometimes has to wait for several periods when he/she requires multiple tasks. As a result, the availability of the cloud is not always ensured. Further, in practice, the service providers charge users based on their tasks, and additional random meaningless tasks will significantly increase the cost of using cloud computing.

5   Conclusion

It is widely believed that cloud computing has become the next stage of the Internet, as it enables outsourced computation. However, how to ensure information security and users' privacy remains a challenging open problem. In this paper, we showed that, fully homomorphic encryption schemes, although seem to be a promising candidate, have some problems when they are used in the context of cloud computing.

Subsequently, we presented a practical message attack against *all* fully homomorphic encryption schemes, in that a malicious cloud can recover the messages by observing users reactions. With several examples, we showed that in practice, our message attack has a very high probability to be successful.

In addition, this message attack can be extended to construct a probabilistic decryption oracle. This brings CCA-1 security as an essential requirement for constructing a secure fully homomorphic encryption scheme.

Further, because of the bootstrapping technique that is used in Gentry's framework, we obtain a secret key attack against *all* fully homomorphic encryption schemes that follow this framework [11,4,3,9,17,19,10,12,16], and this secret key attack is very practical that only takes a maximum $O(\lambda \log \lambda)$ time.

Finally, we argued that CCA-1 security and *no* bootstrappability are the two essential requirements for fully homomorphic encryption schemes that can be used to secure cloud computing scenarios.

References

1. van Dijk, M., Juels, A.: On the impossibility of cryptography alone for privacy-preserving cloud computing. Cryptology ePrint Archive, Report 2010/305 (2010), http://eprint.iacr.org/
2. Gennaro, R., Gentry, C., Parno, B.: Non-interactive Verifiable Computing: Outsourcing Computation to Untrusted Workers. In: Rabin, T. (ed.) CRYPTO 2010. LNCS, vol. 6223, pp. 465–482. Springer, Heidelberg (2010)
3. Gentry, C.: A Fully Homomorphic Encyrption Scheme. PhD thesis, Stanford University (2009)
4. Gentry, C.: Fully homomorphic encryption using ideal lattices. In: Mitzenmacher, M. (ed.) STOC, pp. 169–178. ACM (2009)
5. Hall, C., Goldberg, I., Schneier, B.: Reaction Attacks against Several Public-Key Cryptosystem. In: Varadharajan, V., Mu, Y. (eds.) ICICS 1999. LNCS, vol. 1726, pp. 2–12. Springer, Heidelberg (1999)
6. Myers, S., Shelat, A.: Bit encryption is complete. In: FOCS, pp. 607–616. IEEE Computer Society (2009)
7. Rivest, R., Adleman, L., Dertouzos, M.: On data banks and privacy homomorphisms. In: Foundations of Secure Computation, pp. 169–177. Academic Press (1978)
8. Rivest, R.L., Shamir, A., Adleman, L.M.: A method for obtaining digital signatures and public-key cryptosystems. Commun. ACM 21(2), 120–126 (1978)
9. Smart, N.P., Vercauteren, F.: Fully Homomorphic Encryption with Relatively Small Key and Ciphertext Sizes. In: Nguyen, P.Q., Pointcheval, D. (eds.) PKC 2010. LNCS, vol. 6056, pp. 420–443. Springer, Heidelberg (2010)

10. Stehlé, D., Steinfeld, R.: Faster Fully Homomorphic Encryption. In: Abe, M. (ed.) ASIACRYPT 2010. LNCS, vol. 6477, pp. 377–394. Springer, Heidelberg (2010)

11. van Dijk, M., Gentry, C., Halevi, S., Vaikuntanathan, V.: Fully Homomorphic Encryption over the Integers. In: Gilbert, H. (ed.) EUROCRYPT 2010. LNCS, vol. 6110, pp. 24–43. Springer, Heidelberg (2010)

12. Gentry, C., Halevi, S.: Implementing Gentry's Fully-Homomorphic Encryption Scheme. In: Paterson, K.G. (ed.) EUROCRYPT 2011. LNCS, vol. 6632, pp. 129–148. Springer, Heidelberg (2011)

13. Goldwasser, S., Micali, S.: Probabilistic encryption. J. Comput. Syst. Sci. 28(2), 270–299 (1984)

14. Bellare, M., Desai, A., Pointcheval, D., Rogaway, P.: Relations among Notions of Security for Public-Key Encryption Schemes. In: Krawczyk, H. (ed.) CRYPTO 1998. LNCS, vol. 1462, pp. 26–45. Springer, Heidelberg (1998)

15. Georgii, H.O.: Stochastics: Introduction to Probability and Statistics (de Gruyter Textbook), 1st edn. Walter de Gruyter (2008)

16. Loftus, J., May, A., Smart, N.P., Vercauteren, F.: On CCA-Secure Somewhat Homomorphic Encryption. In: Miri, A., Vaudenay, S. (eds.) SAC 2011. LNCS, vol. 7118, pp. 55–72. Springer, Heidelberg (2012)

17. Gentry, C., Halevi, S., Vaikuntanathan, V.: i-Hop Homomorphic Encryption and Rerandomizable Yao Circuits. In: Rabin, T. (ed.) CRYPTO 2010. LNCS, vol. 6223, pp. 155–172. Springer, Heidelberg (2010)

18. Brakerski, Z., Gentry, C., Vaikuntanathan, V.: Fully homomorphic encryption without bootstrapping. Electronic Colloquium on Computational Complexity (ECCC) 18, 111 (2011)

19. Gentry, C.: Computing arbitrary functions of encrypted data. Commun. ACM 53(3), 97–105 (2010)

A Blind Digital Image Watermarking Method Based on the Dual-Tree Complex Discrete Wavelet Transform and Interval Arithmetic

Teruya Minamoto and Ryuji Ohura

Department of Information Science, Saga University, Saga, Japan
{minamoto,ohura}@ma.is.saga-u.ac.jp

Abstract. We propose a new digital image watermarking method based on the dual-tree complex discrete wavelet transform (DT-CDWT) and interval arithmetic (IA). Both the DT-CDWT and IA produce redundancy from the original data. This implies that there is a possibility of developing a new watermarking method based on the DT-CDWT and IA. We describe our watermarking procedure in detail and show experimental results demonstrating that our method gives watermarked images that have better quality and that are robust against attacks such as marking, clipping, JPEG and JPEG2000 compressions, median filtering, addition of Gaussian white noise, addition of salt & pepper noise, rotation and resizing.

Keywords: Digital watermarking, Dual-tree complex discrete wavelet transform, Interval arithmetic.

1 Introduction

Many digital image watermarking methods have been proposed over the last decade. According to whether or not the original signal is available during the watermark detection process, digital watermarking methods can be roughly categorized into two types: non-blind and blind. Non-blind methods require the original image at the detection end, whereas blind methods do not. Blind methods are more useful than non-blind ones because the original image may not be available in actual scenarios. The majority of watermarking schemes can be categorized as algorithms operating either in the spatial domain or in the transform domain. The spatial domain schemes embed a watermark by modifying the pixel values directly so that common image processing operations can eliminate the watermark. In contrast, the transform domain schemes are more robust against signal processing attacks. Popular transforms used in digital watermarking are based on the frequency domain, such as the discrete cosine transform (DCT) and the discrete wavelet transform (DWT). These discussions are summarized in Ref. [2].

Up to now, since the growing adoption of the JPEG2000 standard and the shift from DCT-based to DWT-based image compression methods, many DWT-based

H. Kim (Ed): ICISC 2011, LNCS 7259, pp. 437–449, 2012.
© Springer-Verlag Berlin Heidelberg 2012

watermarking schemes have been proposed. Most of them use the downsampling-type wavelet transforms described in Ref. [3]. One drawback with this approach, however, is the lack of translation invariance. Against this background, several watermarking methods based on the dual-tree complex discrete wavelet transform (DT-CDWT) have been proposed recently [1,5,6]. The DT-CDWT approximates translation invariance [4] and has a redundancy ratio of $2^m : 1$ in m dimensions. The watermarking scheme in Ref. [5] is a non-blind type that embeds the watermark into the low-frequency components. It seems that the quality of watermarked images is relatively low, though the authors do not mention the quality of watermarked images and do not describe how to determine the parameters used in their scheme. The watermarking scheme in Ref. [1] is a blind type that embeds the watermark into the six high-frequency components or the low-frequency components. This schemes resists at most 50% JPEG compression ratio, and is not robust to image rotation. In Ref. [6], the subimage extracted from the original image is transformed using the DT-CDWT, and the watermark is embedded into each low-frequency component using a pseudo random matrix. This scheme is the blind type and resists some attacks; however, for rotation attacks, this scheme requires the synchronization of orthogonal axes to reconstruct the watermark.

This paper proposes a new blind digital image watermarking method based on DT-CDWT. Instead of Refs. [1,5,6], we use a new DT-CDWT proposed in Ref. [10]. This DT-CDWT allows reconstruction by introducing redundancy into the transform and achieves perfect translation invariance (PTI). Our method is an extension of the watermarking method proposed in Ref. [7] where the downsampling-type DWT [3] and interval arithmetic (IA) [8] are used. According to Ref. [7], a combination of the DWT and IA produces high-frequency components containing low-frequency components, and both the DT-CDWT and IA introduce redundancy produced by the original data. Thus we expect that a new robust image watermarking method based on the DT-CDWT and IA may be developed. Our proposed method is not only robust to spatial and frequency attacks, such as marking, clipping, JPEG and JPEG2000 compressions, median filtering, addition of Gaussian white noise, and addition of salt & pepper noise, but also rotation and resizing. Moreover, we do not require any synchronization of orthogonal axes to extract the watermark designed for a rotation attack.

The remainder of this paper is organized as follows: in Section 2, we briefly describe the basics of the DT-CDWT and IA. In Section 3, we introduce the DT-CDWT based on IA, and in Section 4, we propose a new digital watermarking method. Experimental results are presented in Section 5, and Section 6 concludes the paper.

2   Preliminaries

Since many readers are probably not familiar with the DT-CDWT proposed in Refs. [10] and IA [8], an outline is presented here for the sake of convenience.

2.1   Dual-Tree Complex Discrete Wavelet Transform (DT-CDWT)

According to Ref. [10], the interpolation of the target digital signal $\{f_l\}$ with the real and imaginary scaling functions $\phi^R(t-k)$, $\phi^I(t-k)$, and $k \in \mathbb{Z}$ is represented as

$$f(t) = \sum_k \{c_{0,k}^R \phi^R(t-k) + c_{0,k}^I \phi^I(t-k)\},$$

$$c_{0,k}^R = \frac{1}{2} \sum_l f_l \overline{\phi^R(l-k)}, \quad c_{0,k}^I = \frac{1}{2} \sum_l f_l \overline{\phi^I(l-k)}, \tag{1}$$

and the function $f(t)$ interpolates the target digital signal $\{f_l\}$, that is, $f_n = f(n), n \in \mathbb{Z}$. Here $\overline{\phi(t)}$ is the complex conjugate of $\phi(t)$, and $\mathbb{Z}$ denotes the set of integer numbers. Then, the DT-CDWT is calculated by the following decomposition algorithm:

$$c_{j-1,n}^R = \sum_k a_{2n-k}^R c_{j,k}^R, \quad d_{j-1,n}^R = \sum_k b_{2n-k}^R c_{j,k}^R,$$

$$c_{j-1,n}^I = \sum_k a_{2n-k}^I c_{j,k}^I, \quad d_{j-1,n}^I = \sum_k b_{2n-k}^I c_{j,k}^I, \quad j = 0, -1, -2, \ldots, \tag{2}$$

where $\{a_n^R, b_n^R\}$ are the real decomposition sequences, and $\{a_n^I, b_n^I\}$ are the imaginary decomposition sequences. Here, $\{a_n^R, a_n^I\}$ are low-pass filters and $\{b_n^R, b_n^I\}$ are high-pass filters.

The inverse DT-CDWT is calculated by the following reconstruction algorithm:

$$c_{j,n}^R = \sum_k \{g_{n-2k}^R c_{j-1,k}^R + h_{n-2k}^R d_{j-1,k}^R\},$$

$$c_{j,n}^I = \sum_k \{g_{n-2k}^I c_{j-1,k}^I + h_{n-2k}^I d_{j-1,k}^I\}. \tag{3}$$

Here, $\{g_n^R, h_n^R\}$ are the real reconstruction sequences, and $\{g_n^I, h_n^I\}$ are the imaginary reconstruction sequences.

Using the relations (3) and (1), the original discrete signal $\{f_n\}$ is obtained.

2.2   Interval Arithmetic (IA)

An interval is a set of the form $A = [a_1, a_2] = \{t | a_1 \le t \le a_2, a_1, a_2 \in \mathbb{R}\}$, where $\mathbb{R}$ denotes the set of real numbers. We denote the lower and upper bounds of an interval A by $\inf(A) = a_1$ and $\sup(A) = a_2$, respectively, and the width of any non-empty interval A is defined by $w(A) = a_2 - a_1$. The four basic operations, namely, addition $(+)$, subtraction $(-)$, multiplication $(*)$ and division $(/)$, on two intervals $A = [a_1, a_2]$ and $B = [b_1, b_2]$ are defined as follows:

$$A + B = [a_1 + b_1, a_2 + b_2], \quad A - B = [a_1 - b_2, a_2 - b_1],$$

$$A * B = [\min\{a_1 b_1, a_1 b_2, a_2 b_1, a_2 b_2\}, \max\{a_1 b_1, a_1 b_2, a_2 b_1, a_2 b_2\}], \tag{4}$$

$$A/B = [a_1, a_2] * [1/b_2, 1/b_1], \quad 0 \notin B.$$

For interval vectors and matrices whose elements consist of intervals, these operations are executed at each element.

From the basic operations (4), in general, the width of the interval expands in proportion to the number of computations. This phenomena is sometimes called "interval expansion". We regard interval expansion as a useful tool to produce the redundant part from the original signal.

3   DT-CDWT Based on IA

We define the DT-CDWT based on IA by

$$
I(c^R_{j-1,n}) = \sum_k I(\Delta_k) a^R_{2n-k} c^R_{j,k}, \quad I(d^R_{j-1,n}) = \sum_k I(\Delta_k) b^R_{2n-k} c^R_{j,k},
$$
$$
I(c^I_{j-1,n}) = \sum_k I(\Delta_k) a^I_{2n-k} c^I_{j,k}, \quad I(d^I_{j-1,n}) = \sum_k I(\Delta_k) b^I_{2n-k} c^I_{j,k},
$$

(5)

where $I(\Delta_k) = [1 - \Delta_k, 1 + \Delta_k]$ and Δ_k are positive real numbers. All operations are executed using IA, and then $c^R_{j-1,n} \subset I(c^R_{j-1,n})$, $d^R_{j-1,n} \subset I(d^R_{j-1,n})$, $c^I_{j-1,n} \subset I(c^I_{j-1,n})$, and $d^I_{j-1,n} \subset I(d^I_{j-1,n})$ hold. Thus, the DT-CDWT based on IA contains the original DT-CDWT. In this sense, the DT-CDWT based on IA is the redundant set from the original signal. In actual computation, the decompositions (5) are computed by using the decomposition sequences listed in Ref. [10] and interval arithmetic software.

In the case of images, the formulas (2) or (5) are applied in each direction, that is, the vertical and horizontal directions. This procedure is shown in Fig. 1. Here C, D, E and F indicate the low-frequency components and high-frequency components in the vertical, horizontal, and diagonal directions, respectively. The superscripts R and I stand for real and imaginary, respectively.

4   Watermarking Algorithm

We assume that the binary-valued watermark W consists of -1 and 1. To simplify the remainder of the discussion, we discuss only the DT-CDWT from level 0 to -1.

The embedding procedure is as follows:

1. Apply the DT-CDWT based on IA and the usual DT-CDWT without IA to the horizontal direction and to the vertical direction, respectively, and obtain the 16 interval components corresponding to C^{RR}_{-1}, D^{RR}_{-1}, ..., F^{II}_{-1} in Fig. 1 and represent these interval components as $I(C^{RR}_{-1})$, $I(D^{RR}_{-1})$,..., $I(F^{II}_{-1})$.
2. Choose several components of the 16 interval components, and represent the chosen components as $I(S_i)$ ($i = 1, 2, ..., N, 1 < N < 16$).
3. Set $S'_i = \sup(I(S_i))$.
4. Replace other components with floating point ones.

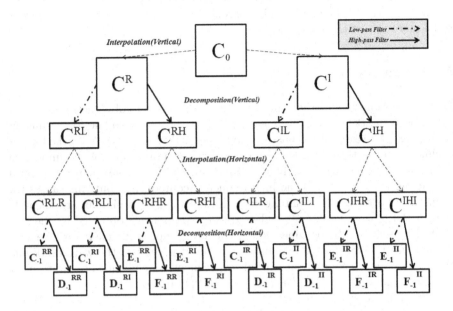

Fig. 1. Two dimensional DT-CDWT

5. Compute the following sequence using a block-type sliding window of $(2k+1) \times (2l+1)$:

$$\underline{S_i}(i,j) = \text{sgn}(S_i'(i,j)) \cdot \frac{1}{2k+1} \frac{1}{2l+1} \sum_{K=-k}^{k} \sum_{L=-l}^{l} |S_i'(i+K, j+L)|, \quad (6)$$

where k and l are fixed natural numbers and $\text{sgn}(a)$ is the usual signum function of a real number a.

6. Embed the watermark W by computing

$$\widetilde{S}_i(i,j) = \underline{S_i}(i,j)(1 + \alpha W(i,j)), \quad (7)$$

where $0 < \alpha < 1$ is a given hiding factor which adjusts the robustness.

7. Reconstruct the image using $\widetilde{S}_i$, the components that were replaced with floating point ones, and the inverse DT-CDWT. Then, the watermarked image $\widetilde{C}_0$ is obtained.

In the extraction procedure, we only need information about the components where the watermark is embedded. This information is critical in extracting the watermark. The extracting procedure is as follows:

1. Decompose $\widetilde{C}_0$ into the 16 components $\widetilde{C}_{-1}^{RR}, \widetilde{D}_{-1}^{RR}, ..., \widetilde{F}_{-1}^{II}$ by using the DT-CDWT and choose the watermarked components $\widetilde{S}_i$.
2. Compute $\widetilde{W}_i = \text{sgn}(|\widetilde{S}_i| - |\underline{\widetilde{S}_i}|)$ and $\widetilde{W}_e = \sum_{i=1}^{N} \widetilde{W}_i$.
3. Extract the binary-valued watermark $\widetilde{W}$ by thresholding $\widetilde{W}_e$.

In this extraction procedure, we sum up several watermarked pixels W_i to extract the watermark $\widetilde{W}$. Thanks to this approach, the proposed method is expected to be more robust than the methods in Ref. [7].

5   Experimental Results

Digital watermarking methods involve a trade-off between robustness and quality of the watermarked image. Embedding the watermark into the low-frequency components increases the robustness, but the quality of the watermarked image declines because the low-frequency components contain most of the important information representing an image. To maintain the image quality, we must embed the watermark into the high-frequency components, but this causes the robustness to decrease simultaneously. However, using a similar argument to that in Ref. [7], we know that high-frequency components obtained by the DT-CDWT based on IA contain a low-frequency component. Therefore, we may expect that the robustness is maintained even if we embed the watermark into the high-frequency components. In this experiment, we embedded the watermark into four components: $I(D_{-1}^{RR})$, $I(D_{-1}^{RI})$, $I(D_{-1}^{IR})$, and $I(D_{-1}^{II})$.

Moreover, we set the parameters $\Delta_k = 0.02$ in (5), $k = l = 10$ in (6) and $\alpha = 0.9$ in (7) so as not to considerably decrease the value of the peak signal to noise ratio (PSNR), expressed in decibels, which is computed by $PSNR = 20 \log_{10} \left(\dfrac{255}{\sqrt{\frac{1}{N_x N_y} \sum_{i=1}^{N_x} \sum_{j=1}^{N_y} (C_0(i,j) - \widetilde{C_0}(i,j))^2}} \right)$, where N_x and N_y are the sizes of the image in the horizontal and vertical directions, respectively.

To evaluate the performance of the proposed method, we adopted 256-grayscale images of size 256×256 pixels, namely Lenna, Woman, Boat, and Pepper images, and a binary watermark of size 128×128 pixels, as shown in Fig. 2. We implemented our method using INTLAB[9], which is a MATLAB toolbox that supports interval arithmetic.

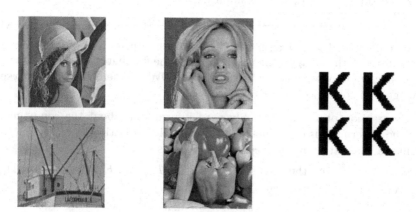

Fig. 2. Original images and watermark

Fig. 3 shows the watermarked images, together with PSNRs, obtained by the proposed method and the watermarks extracted from the watermarked images without any attack. If k and l in (6) are getting smaller, $\underline{S_i} \approx S_i'$ holds, and the quality of the extracted watermark will be declining. Since there are many points on which $\underline{S_i} \approx S_i'$ holds even if we set $k = l = 10$, the watermarks are typically removed. However, these watermarked images have better quality, and the inserted watermark is invisible to the naked eye.

Fig. 3. Watermarked images and extracted watermarks without any attack

5.1   Non-geometric Attacks

Figs. 4–7 illustrate watermarked images and the extracted watermarks under attacks such as marking, median filtering, and addition of Gaussian white noise and salt & pepper noise. Although the extracted images are degraded, we are able to identify the existence of the watermark at a single glance.

Fig. 4. Watermarked images with marked areas and extracted watermarks

Fig. 5. Watermarked images median-filtered in the 3 by 3 neighborhood and extracted watermarks

Fig. 6. Watermarked images with added Gaussian white noise (mean: 0, variance: 0.001) and extracted watermarks

Fig. 7. Watermarked images with added salt & pepper noise (noise density: 0.005) and extracted watermarks

Figs. 8 and 9 illustrate the watermarks extracted from the watermarked images under JPEG and JPEG2000 attacks. We are able to barely identify the existence of the watermark.

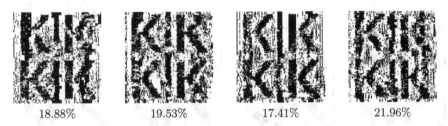

| 18.88% | 19.53% | 17.41% | 21.96% |

Fig. 8. Watermarks extracted from watermarked JPEG images compressed with different compression ratios (from the left, Lenna, Woman, Boat, and Pepper)

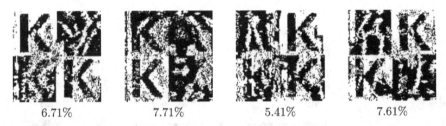

| 6.71% | 7.71% | 5.41% | 7.61% |

Fig. 9. Watermarks extracted from watermarked JPEG2000 images compressed with different compression ratios (from the left, Lenna, Woman, Boat, and Pepper)

5.2 Geometric Attacks

We examine for the robustness against geometric attacks such as clipping, rotation and resizing. Figs. 10–12 illustrate the watermarked images and the extracted watermarks under clipping, rotation, and resizing attacks, respectively. We are able to identify the existence of the watermark at a single glance.

5.3 Multi-attacks

We also examine for the robustness against several multi-attacks. Figs. 13–14 demonstrate the watermarked images and the extracted watermarks under several multi-attacks. We are able to identify the existence of the watermark.

5.4 Comparison with Existing Methods

For comparison of the proposed method with existing methods, we pick up several methods described in Refs. [1,5,6,7]. The method in Ref. [7] uses the IA and DWT, the other methods are based on the DT-CDWT.

Fig. 10. 190×190 fragments of watermarked images and extracted watermarks

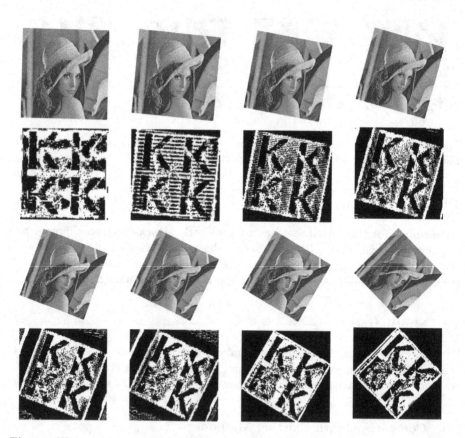

Fig. 11. Watermarks extracted from watermarked images rotated by angles 1, 5, 10, and 15 degrees (from left to right at top), and by angles 20, 25, 30, and 45 degrees (from left to right at bottom)

Fig. 12. Watermarks extracted from watermarked image scaled by multiplying ratios 0.8, 0.9, 1.1 and 1.2 (from left to right)

Fig. 13. Watermarked images and watermarks extracted from watermarked images subjected to JPEG 2000 compression with compression ratio 10.81% and several attacks (from the left, clipping, addition of Gaussian white noise, addition of salt & pepper noise, resizing, and rotation)

Fig. 14. Watermarked images and watermarks extracted from watermarked images subjected to several attacks (from the left, resizing & rotation & clipping, JPEG compression with compression ratio 25.25%& marking, and JPEG 2000 compression with compression ratio 10.81%& marking

Non-geometric attacks

The results shown in Figs. 8 and 9 are superior to those of the methods described in Ref. [7] (JPEG compression ratio : 32.3%, JPEG2000 compression ratio : 24.4%), in Ref. [6](JPEG compression ratio : 35%) and in Ref. [1] (JPEG compression ratio : 50%). In Ref. [5], the authors do not touch on the robustness against compression attacks. For other attacks, the performance of our method is almost the same as the other methods.

Geometric attacks

The methods described in Refs. [1] can not resist geometric attacks. Unlike the scheme in Ref. [6], we do not require any synchronization of orthogonal axes to reconstruct the watermark from the rotated watermarked image. Thanks to the translation invariance property of the DT-CDWT, we can extract the watermark, whereas the method based on the DWT in Ref. [7] can not.

We would like to remark that the authors in Ref. [5] did not describe how to choose the parameters which are needed to carry out their method, so we can not implement their method. Moreover, we would like to point out that the extraction process in Ref. [6] does not depend on their embedding rule, and it uses the pixel coordinates where the watermark pixels are embedded, if we understand correctly. Therefore, we can extract (or make) the watermark depending on these coordinates from any image in our simulations based on Ref. [6].

6   Conclusion

We proposed a new watermarking method using the DT-CDWT and IA. To the best of our knowledge, this work is the first application of the DT-CDWT proposed in Ref. [10] based on IA. Experimental results demonstrated that our method gives better-quality watermarked images and is robust to marking, clipping, JPEG and JPEG2000 compressions, median filtering, addition of Gaussian white noise, addition of salt & pepper noise, rotation and resizing. We have compared our results with the ones described in [1,5,6,7]. From the comparison with these methods, we can conclude that our proposed method is superior to these. Moreover, we would like to emphasize that this work opens up new possibilities for interval arithmetic, because interval arithmetic has rarely been used for image processing except computer graphics [8]. In this sense, we believe that our approach makes meaningful contributions to the fields of not only digital watermarking but also interval arithmetic.

References

1. Baolong, G., Leida, L., Jeng-Shyang, P., Liu, Y., Xiaoyue, W.: Robust Image Watermarking Using Mean Quantization in DTCWT Domain. In: Eighth International Conference on Intelligent Systems Design and Applications, pp. 19–22 (2008)

2. Cox, I.J., Miller, M.L., Bloom, J.A., Fridrich, J., Kalker, T.: Digital Watermarking and Steganography. Morgan Kaufmann Publishers (2008)
3. Daubechies, I.: Ten Lectures on Wavelets. SIAM (1992)
4. Kingsbury, N.G.: Complex wavelets for shift invariant analysis and filtering of signals. Applied Computational Harmonic Analysis 10(3), 234–253 (2001)
5. Lee, J.J., Kim, W., Lee, N.Y., Kim, G.Y.: A new incremental watermarking based on dual-tree complex wavelet transform. The Journal of Supercomputing 33, 133–140 (2005)
6. Mabtoul, S., Ibn-Elhaj, E., Aboutajdine, D.: A Blind Chaos-Based Complex Wavelet-Domain Image Watermarking Technique. IJCSNS International Journal of Computer Science and Network Security 6(3), 134–139 (2006)
7. Minamoto, T., Aoki, K.: A blind digital image watermarking method using interval wavelet decomposition. International Journal of Signal Processing, Image Processing and Pattern Recognition 3(2), 59–72 (2010)
8. Moore, R.E., Kearfott, R.B., Cloud, M.J.: Introduction to Interval Analysis. SIAM (2009)
9. Rump, S.M.: INTLAB - Interval Laboratory, http://www.ti3.tu-harburg.de/rump/intlab/
10. Toda, H., Zhang, Z.: Perfect translation invariance with a wide range of shapes of Hilbert transform pairs of wavelet bases. International Journal of Wavelets, Multiresolution and Information Processing 8(4), 501–520 (2010)

On the Communication Complexity of Reliable and Secure Message Transmission in Asynchronous Networks

Ashish Choudhury[1],[*] and Arpita Patra[2]

[1] Department of Computer Science
University of Bristol
Ashish.Choudhary@bristol.ac.uk, partho31@gmail.com
[2] Department of Computer Science
ETH Zuruch
arpitapatra10@gmail.com, arpita.patra@inf.ethhz.ch

Abstract. In this paper, we study the communication complexity of *Reliable Message Transmission* (RMT) and *Secure Message Transmission* (SMT) protocols in *asynchronous* settings. We consider two variants of the problem, namely *perfect* (where no error is allowed in the protocol outcome) and statistical (where the protocol may output a wrong outcome with negligible probability). RMT and SMT protocols have been investigated rigorously in synchronous settings. But not too much attention has been paid to the asynchronous version of the problem. In a significant work, Choudhury et al. (ICDCN 2009 and JPDC 2011) have studied the network connectivity requirement for asynchronous perfect and statistical SMT protocols. Their investigation reveals the following two important facts:

1. *perfect* SMT protocols require *more* network connectivity in asynchronous network than synchronous network.
2. Connectivity requirement of *statistical* SMT protocols is *same* for both synchronous and asynchronous network.

Unfortunately, nothing is known about the communication complexity of RMT and SMT protocols in asynchronous settings. In this paper, we derive tight bounds on the communication complexity of the above problems and compare our results with the existing bounds for synchronous protocols. The interesting conclusions derived from our results are:

1. **RMT:** Asynchrony *increases* the communication complexity of *perfect* RMT protocols. However, asynchrony has *no impact* on the communication complexity of *statistical* RMT protocols.
2. **SMT:** Communication complexity of SMT protocols is *more* in asynchronous network, for *both* perfect as well as statistical case.

* Part of this work was done when the author was working at Center of Excellence in Cryptology, Indian Statistical Institute Kolkata India. The work in this paper was partially supported by EPSRC via grant EP/I03126X/1, and by the European Commission through the ICT Programme under Contract ICT2007216676 ECRYPT II.

H. Kim (Ed): ICISC 2011, LNCS 7259, pp. 450–466, 2012.
© Springer-Verlag Berlin Heidelberg 2012

1   Introduction

Reliable Message Transmission (RMT) and *Secure Message Transmission* (SMT) [4] are fundamental problems in secure distributed computing as well as in cryptography. In the problem of RMT, there are n disjoint channels (also called as *wires*) between a sender **S** and a receiver **R**. **S** and **R** shares no information in advance. There is a *computationally unbounded active* adversary, denoted as $\mathcal{A}_t$, who can listen and forge the communication over t out of the n wires, where $t < n$. **S** has a message $m^{\mathbf{S}}$, which is a sequence of ℓ elements, chosen from a finite field $\mathbb{F}$, where $\ell \geq 1$ and $|\mathbb{F}| > n$. The challenge is to design a protocol, such that at the end of the protocol, **R** correctly outputs $m^{\mathbf{R}} = m^{\mathbf{S}}$. Now there are two flavors of RMT:

1. Perfect RMT (**PRMT**): Here $m^{\mathbf{R}} = m^{\mathbf{S}}$, without any error.
2. Statistical RMT (**SRMT**): Here $m^{\mathbf{R}} = m^{\mathbf{S}}$ with probability at least $1 - \delta$, where $0 < \delta < 1/2$ and is called the error probability.

Notice that there is no issue of privacy in RMT protocols; i.e., the adversary can also know $m^{\mathbf{S}}$ during the protocol execution. If we add the issue of privacy to RMT protocols, then we arrive at the notion of SMT protocols. In SMT protocols, we require that not only **R** outputs $m^{\mathbf{R}} = m^{\mathbf{S}}$, but also $\mathcal{A}_t$ should not learn any information about $m^{\mathbf{S}}$ in *information theoretic* sense. We can have two types of SMT protocols:

1. Perfect SMT (**PSMT**): Here $m^{\mathbf{R}} = m^{\mathbf{S}}$, without any error.
2. Statistical SMT (**SSMT**): Here $m^{\mathbf{R}} = m^{\mathbf{S}}$ with probability at least $1 - \delta$, where $0 < \delta < 1/2$ and is called error probability. However, there is no compromise in the privacy which should be error free and information theoretic.

RMT and SMT problem were first formulated by Dolev et al. [4]. Any RMT or SMT protocol has the following important parameters:

1. *Connectivity*: It is the total number of wires n (expressed as a function of t) required in the protocol. We consider following two types of wires:
   (a) *Uni-directional wires*: where all the n wires are uni-directional, directed from **S** to **R**, allowing only one way communication (i.e., no interaction) from **S** to **R**;
   (b) *Bi-directional wires*: where all the n wires are bi-directional, allowing bi-directional communication (i.e., interaction) between **S** and **R**.
2. *Communication Complexity*: It is the number of field elements communicated by **S** and **R** (expressed as a function of n, ℓ) in the protocol.

RMT and SMT problem have been studied rigorously by several researchers (see for example [11,13,1,5,7,9,3]) and tight bounds have been established on connectivity and communication complexity of PRMT, SRMT, PSMT and SSMT protocols. These bounds are summarized in Table 1.

Table 1. Existing bounds for RMT and SMT protocols

Type of Protocol	Type of Channels	n	Bound on the Communication Complexity
PRMT	Uni-directional	$n \geq 2t + 1$ [4]	$\Theta\left(\frac{n\ell}{n-2t}\right)$ [13,15]
PRMT	Bi-directional	$n \geq 2t + 1$ [4]	$\Theta(\ell)$ [15,8]
SRMT	Uni-directional	$n \geq 2t + 1$ [6]	$\Theta(\ell)$ [9]
SRMT	Bi-directional	$n \geq 2t + 1$ [6]	$\Theta(\ell)$ [9]
PSMT	Uni-directional	$n \geq 3t + 1$ [4]	$\Theta\left(\frac{n\ell}{n-3t}\right)$ [5]
PSMT	Bi-directional	$n \geq 2t + 1$ [4]	$\Theta\left(\frac{n\ell}{n-2t}\right)$ [13,15,8,7]
SSMT	Uni-directional	$n \geq 2t + 1$ [6]	$\Theta\left(\frac{n\ell}{n-2t}\right)$ [9]
SSMT	Bi-directional	$n \geq 2t + 1$ [6]	$\Theta(\ell)$ [9]

Remark 1. (**Note on the Communication Complexity of SRMT and SSMT Protocols**) In any SRMT/SSMT, the field size $|\mathbb{F}|$ is selected as a function of the error probability δ. So though δ is not figuring out explicitly in the communication complexity expressions of SRMT/SSMT protocols in Table 1, it is implicitly present. More specifically, each element of $\mathbb{F}$ can be represented by $\log |\mathbb{F}|$ bits, which will be a function of δ. So if we consider the total number of bits communicated during any SRMT/SSMT protocol, δ will be present in the communication complexity expression. This point will be made more clear, when we discuss our protocols. □

Motivation of Our Work. The results given in Table 1 assume that the underlying network is *synchronous*, where there is a global clock and the transmission delay over each wire is bounded by an upper bound. Though theoretically interesting, this does not model the real life scenario (like the Internet) appropriately, as the delay in the transmission of even a single message will affect the overall properties of the protocol. In a typical large network like the Internet, every message can have arbitrary delay and this can be modeled more appropriately by *asynchronous* networks, where no timing assumptions are made. Unfortunately, unlike synchronous networks, not much attention has been paid to RMT and SMT protocols in asynchronous settings. In this paper, we improve this situation by deriving tight bounds on the communication complexity of asynchronous RMT and SMT protocols.

Asynchronous Network Model. In an asynchronous network, every wire can have *arbitrary*, yet finite delay. That is, the messages are assumed to be delivered *eventually*. To model the worst case scenario, it is assumed that $\mathcal{A}_t$ can schedule the messages over each wire and hence can control the transmission delay over each wire. *However, note that $\mathcal{A}_t$ can only schedule the messages sent over an honest wire, without having any access to them.* The inherent difficulty that arises in designing a protocol in asynchronous settings is that we cannot distinguish between a *slow wire* and a *corrupted wire*. That is, if in the protocol, some information is supposed to arrive over a wire and if no information arrives, then it cannot be distinguished whether the wire is honest and the information is simply delayed (due to the malicious scheduling by $\mathcal{A}_t$) or whether $\mathcal{A}_t$ has simply blocked the transmission over the wire by taking its control. Due to this, neither **S** nor **R** can afford to wait for all the n wires to transmit their information, as waiting for all of them may turn out to be endless. So they have to start the computation, as soon as they receive information over at least $n - t$ wires and they may have to ignore the transmission over t (potentially honest) wires. Due to this limitation, the techniques from the synchronous world cannot be adapted straight forwardly to the asynchronous settings.

We call the asynchronous PRMT, SRMT, PSMT and SSMT protocols as APRMT, ASRMT, APSMT and ASSMT respectively. Now in addition to the reliability and secrecy condition (as in the synchronous protocols), these asynchronous protocols also have to explicitly satisfy **termination** condition, according to which both **S** and **R** should eventually terminate the protocol.

Existing Results for Asynchronous Protocols. The first asynchronous SMT protocol was proposed in [10], where the authors have designed an APSMT protocol with $n = 2t + 1$ uni-directional wires from **S** to **R**. However, in [2], Choudhury et al. have shown that the protocol of [10] is insecure. Moreover, they have also studied the connectivity requirement for APSMT and ASSMT protocols. More specifically, they have shown the following two *surprising* results:

1. Any APSMT protocol requires $n \geq 3t + 1$ wires, irrespective of whether the wires are uni-directional or bi-directional. This is quiet surprising, since we can design PSMT protocols in synchronous settings with $n \geq 2t + 1$ bi-directional wires (see sixth row of Table 1). This shows that *asynchrony affects the connectivity of PSMT protocols.*
2. Any ASSMT protocol requires $n \geq 2t + 1$ wires, irrespective of whether the wires are uni-directional or bi-directional. The same connectivity is required even for SSMT protocols (see the last two rows of Table 1). This implies *asynchrony has no affect on the connectivity of SSMT protocols.*

Our Results and Their Significance. So far nothing is known about the communication complexity of APRMT, ASRMT, APSMT and ASSMT protocols. We derive tight bounds on the communication complexity of the above problems. These bounds are summarized in Table 2.

Comparing Table 1 and Table 2, we find the following surprising facts:

1. **PRMT:** *Asynchrony increases the communication complexity of PRMT protocols.* With $n = 2t + 1$ bi-directional wires, PRMT protocol can be designed with a communication complexity of $\Theta(\ell)$ (second row of Table 1), where as APRMT protocol must have a communication complexity of $\Theta(n\ell)$ (second row of Table 2).
2. **SRMT:** *Asynchrony does not affect the communication complexity of SRMT protocols.* In this case, the communication complexity is *same* for both

Table 2. Our Bounds for Asynchronous RMT and SMT Protocols

Type of Protocol	Type of Channels	n	Bound on the Communication Complexity
APRMT	Uni-directional	$n \geq 2t + 1$	$\Theta\left(\frac{n\ell}{n-2t}\right)$
APRMT	Bi-directional	$n \geq 2t + 1$	$\Theta\left(\frac{n\ell}{n-2t}\right)$
ASRMT	Uni-directional	$n \geq 2t + 1$	$\Theta(\ell)$
ASRMT	Bi-directional	$n \geq 2t + 1$	$\Theta(\ell)$
APSMT	Uni-directional	$n \geq 3t + 1$	$\Theta\left(\frac{n\ell}{n-3t}\right)$
APSMT	Bi-directional	$n \geq 3t + 1$	$\Theta\left(\frac{n\ell}{n-3t}\right)$
ASSMT	Uni-directional	$n \geq 2t + 1$	$\Theta\left(\frac{n\ell}{n-2t}\right)$
ASSMT	Bi-directional	$n \geq 2t + 1$	$\Theta\left(\frac{n\ell}{n-2t}\right)$

synchronous as well as asynchronous protocols (third and fourth row of Table 1 and Table 2 respectively).

3. **PSMT:** *Asynchrony increases the communication complexity of PSMT protocols.* From [2], asynchrony increases the connectivity requirement of PSMT protocols. Our results show that the same holds even for the communication complexity (see the sixth row of Table 1 and Table 2).

4. **SSMT:** Interestingly, we find that *asynchrony even increases the communication complexity of SSMT protocols.* Specifically, for $n = 2t+1$ bi-directional wires, SSMT protocol can be designed with a communication complexity of $\Theta(\ell)$ (last row of Table 1), where as ASSMT scheme must have a communication complexity of $\Theta(n\ell)$ (last row of Table 2). However, [2] shows that asynchrony does not increase the connectivity of SSMT protocols.

The Road-Map. We present our results on APRMT, ASRMT, APSMT and ASSMT in Section 2, 3, 4 and 5 respectively. We conclude the paper and discuss few open problems in Section 6.

2   Bound on the Communication Complexity of APRMT

Throughout Section 2, we assume $n \geq 2t+1$, as $n \geq 2t+1$ wires (uni-directional or bi-directional) are required for any PRMT protocol (Table 1), we require the same for APRMT protocols as well.

2.1   Bounds for Uni-directional Wires

Theorem 1. *Any APRMT protocol, executed over n ($n \geq 2t+1$) uni-directional wires has a communication complexity of $\Omega\left(\frac{n\ell}{n-2t}\right)$.*

PROOF: Easy, as the bound holds for PRMT protocols [13].   □

Theorem 2. *Let there* n *($n = 2t + 1$) uni-directional wires from* **S** *to* **R***. Then there exists an APRMT protocol with communication complexity of* $\mathcal{O}(n\ell) = \mathcal{O}\left(\frac{n\ell}{n-2t}\right)$.

PROOF: Consider the following protocol: To reliably send a message of size ℓ, **S** sends the message over all the n wires. **R** waits for a message received identically over $t + 1$ wires and output the message. The output is correct, since at least one wire out of these $t + 1$ wires is honest, which will deliver the original message. Moreover, termination is guaranteed since there are at least $t + 1$ honest wires, which will eventually deliver correct message. It is easy to verify that the communication complexity is $\mathcal{O}(n\ell)$. □

2.2   Bounds for Bi-directional Wires

Let **S** and **R** be connected by n bi-directional wires, denoted by $\mathcal{W} = \{w_1, \ldots, w_n\}$, where $n \geq 2t + 1$. Then we show that any APRMT protocol has a communication complexity of $\Omega\left(\frac{n\ell}{n-2t}\right)$. For this, we prove the following:

1. We first show that the information exchanged over *any* $n - 2t$ wires should completely determine the message in any APRMT protocol executed over n bi-directional wires (Lemma 1).
2. Next, we show that any APRMT protocol where the information exchanged over *any* $n - 2t$ wires completely determine the message, must communicate $\Omega\left(\frac{n\ell}{n-2t}\right)$ (Lemma 2).

Lemma 1. *In any APRMT protocol executed over* $n \geq 2t + 1$ *bi-directional wires, the information exchanged over any* $n - 2t$ *wires completely determines the message.*

PROOF: On contrary, let Π^{APRMT} be an APRMT protocol where the information exchanged over any $n-2t$ wires is independent of $m^{\mathbf{S}}$. We divide the set of n wires into *three* groups, namely G_1, G_2 and G_3. The group G_1 consists of the first $n-2t$ wires $w_1, \ldots, w_{n-2t}$, group G_2 consists of the next t wires $w_{n-2t+1}, \ldots, w_{n-t}$ and group G_3 consists of the last t wires $w_{n-t+1}, \ldots, w_n$. *Now according to our assumption, the information exchanged over the wires in group* G_1 *(consisting of* $n - 2t$ *wires) in any execution of* Π^{APRMT} *will be independent of message.* That is, there exist a pair of messages, say $m_1^{\mathbf{S}}$ and $m_2^{\mathbf{S}}$ such that the information communicated over G_1 while sending $m_1^{\mathbf{S}}$ and respectively $m_2^{\mathbf{S}}$ are same. We define the following variables with respect to any execution E of Π^{APRMT}:

1. $time(E, \mathbf{R}, w_i)$: lists the arrival time-stamps of different messages (with respect to the local clock of **R**) received by **R** along wire w_i, for $i = 1, \ldots, n$ in execution E.
2. $time(E, \mathbf{S}, w_i)$: lists the arrival time-stamps of the different messages (with respect to the local clock of **S**) received by **S** along wire w_i, for $i = 1, \ldots, n$ in execution E.
3. E^{time}: denotes the total time taken (with respect to **R**) by execution E; i.e., the time at which **R** terminates the protocol in execution E.

Since Π^{APRMT} is an APRMT protocol, any execution E of Π^{APRMT} must terminate. Now consider the following two possible executions of Π^{APRMT}, E_1 and E_2. Let $\mathbf{R}$ terminates E_1 (E_2) at time E_1^{time} (E_2^{time}), correctly outputting $m_1^{\mathbf{S}}$ $(m_2^{\mathbf{S}})$.

1. Execution E_1: The random coins of $\mathbf{S}$ and $\mathbf{R}$ are r_1 and r_2 respectively. $\mathbf{S}$ wants to reliably send the message $m_1^{\mathbf{S}}$. The adversary strategy is to passively listen (without modifying them) the communication over the wires in group G_2 and arbitrarily delaying the communication over the wires in group G_3, till the time $E_1^{time} + E_2^{time} + 1$. Let α and β_1 denote the messages that are exchanged between $\mathbf{S}$ and $\mathbf{R}$, along the wires in group G_1 and G_2 respectively.

2. Execution E_2: The random coins of $\mathbf{S}$ and $\mathbf{R}$ are r_3 and r_2 respectively. $\mathbf{S}$ wants to reliably send the message $m_2^{\mathbf{S}} \neq m_1^{\mathbf{S}}$. The adversary strategy is to passively listen (without modifying them) the communication over the wires in group G_2 and arbitrarily delaying the communication over the wires in group G_3, till the time $E_1^{time} + E_2^{time} + 1$. Let α and β_2 denote the messages that are exchanged between $\mathbf{S}$ and $\mathbf{R}$, along the wires in group G_1 and G_2 respectively. *Notice that α is same as in execution E_1 due to our assumption about the distribution of information over the wires in G_1 in Π^{APRMT}.*

We now show another possible execution of Π^{APRMT} and an adversary strategy, where $\mathbf{R}$ outputs an incorrect message.

3. Execution E^{Cor}: The random coins of $\mathbf{S}$ and $\mathbf{R}$ are r_1 and r_2 respectively. $\mathbf{S}$ wants to reliably send the message $m_1^{\mathbf{S}}$. Let α denote the messages exchanged over the wires in G_1. *Notice that α is same as in execution E_1 and E_2.* Now the adversary strategy in E^{Cor} is as follows: adversary delay any information along the wires in group G_3 for time $E_1^{time} + E_2^{time} + 1$. In addition, the adversary controls the wires in group G_2 in Byzantine fashion and change the communication over these wires, such that $\mathbf{R}$ gets messages corresponding to β_2 along G_2, while $\mathbf{S}$ receives messages corresponding to β_1 along G_2. Moreover, adversary schedules the messages along the wires in G_1 and G_2 in such a way that $time(E^{\text{Cor}}, \mathbf{S}, w_i) = time(E_1, \mathbf{S}, w_i)$, for every $w_i \in G_1 \cup G_2$ and $time(E^{\text{Cor}}, \mathbf{R}, w_i) = time(E_2, \mathbf{R}, w_i)$, for every $w_i \in G_1 \cup G_2$. Thus the view of $\mathbf{S}$ is $\alpha \, \beta_1$, while view of $\mathbf{R}$ is $\alpha \, \beta_2$.

Thus the view of $\mathbf{S}$ in E_1 and E^{Cor} are same, so $\mathbf{S}$ will assume that $m_1^{\mathbf{S}}$ has been communicated reliably. However, the view of $\mathbf{R}$ in E^{Cor} is same as in E_2 and hence $\mathbf{R}$ will output $m_2^{\mathbf{S}}$. But this violates the perfect reliability property of Π^{APRMT}, which is a contradiction. Hence Π^{APRMT} does not exist.   □

Lemma 2. *Any APRMT protocol tolerating $\mathcal{A}_t$ executed over n $(n \geq 2t+1)$ bidirectional wires, in which the information exchanged over any $n - 2t$ wires completely determine the message, has a communication complexity of $\Omega\left(\frac{n\ell}{n-2t}\right)$.*

PROOF: Let Π^{APRMT} be an APRMT protocol, executed over n bi-directional wires (where $n \geq 2t + 1$), to reliably send a message of size ℓ, such that the information exchanged over *any* $n - 2t$ wires completely determine the message. We now define the following notations:

1. $\mathcal{M}$ denotes the message space from where $\mathbf{S}$ selects the message to be sent. So $\mathcal{M} = \mathbb{F}^{\ell}$.
2. $\mathbf{T}_i^m$ denotes the set of all possible transmissions that can occur on wire $w_i \in \{w_1, \ldots, w_n\}$, when $\mathbf{S}$ transmits message $m \in \mathcal{M}$ using Π^{APRMT}.
3. For $j \geq i$, $\mathbf{M}_{i,j}^m \subseteq \mathbf{T}_i^m \times \mathbf{T}_{i+1}^m \times \ldots \times \mathbf{T}_j^m$ denotes the set of all possible transmissions that can occur over the wires $\{w_i, w_{i+1}, \ldots, w_j\}$, when $\mathbf{S}$ transmits message $m \in \mathcal{M}$ using protocol Π^{APRMT}.
4. $\mathbf{M}_{i,j} = \bigcup_{m \in \mathcal{M}} \mathbf{M}_{i,j}^m$ and $\mathbf{T}_i = \bigcup_{m \in \mathcal{M}} \mathbf{T}_i^m$. We call $\mathbf{T}_i$ as the *capacity* of wire w_i and $\mathbf{M}_{i,j}$ as the *capacity* of the set of wires $\{w_i, w_{i+1}, \ldots, w_j\}$.

In protocol Π^{APRMT}, one element from the set $\mathbf{T}_i$ is transmitted over each wire w_i, for $i = 1, \ldots, n$. Moreover, each element of the set $\mathbf{T}_i$ can be represented by $\log |\mathbf{T}_i|$ bits. Thus, the lower bound on the communication complexity of Π^{APRMT} is $\Sigma_{i=1}^n \log |\mathbf{T}_i|$ bits. In the sequel, we try to estimate $\mathbf{T}_i$.

Since the transmission over any set of $n - 2t$ wires in Π^{APRMT} completely determines the message, it must hold that $|\mathbf{M}_{2t+1,n}| \geq |\mathcal{M}|$. Though the above relation must hold for any set of $n - 2t$ wires, for simplicity, we have focussed specifically on the last $n - 2t$ wires. From the definition of $\mathbf{T}_i$ and $\mathbf{M}_{i,j}$, we get

$$\prod_{i=2t+1}^n |\mathbf{T}_i| \geq |\mathbf{M}_{2t+1,n}| \geq |\mathcal{M}|.$$

Let $g = n - 2t$. The above inequality holds for any selection of g wires $\mathcal{D} \subset \{w_1, \ldots, w_n\}$, where $|\mathcal{D}| = g$; i.e., $\prod_{w_i \in \mathcal{D}} |\mathbf{T}_i| \geq |\mathcal{M}|$. In particular, it holds for every selection $\mathcal{D}_k = \{w_{kg+1 \bmod n}, w_{kg+2 \bmod n}, \ldots, w_{kg+g \bmod n}\}$, with $k \in \{0, \ldots, n-1\}$. If we consider all the $\mathcal{D}_k$ sets collectively, then each wire is counted exactly g times in the collection. Thus, the product of the capacities of all $\mathcal{D}_k$ yields the capacity of the full wire set to the gth power and also since each $\mathcal{D}_k$ has capacity at least $|\mathcal{M}|$, we get

$$\prod_{k=0}^{n-1} \prod_{w_j \in \mathcal{D}_k} |\mathbf{T}_j| = \left(\prod_{i=1}^n |\mathbf{T}_i| \right)^g, \text{ and } |\mathcal{M}|^n \leq \prod_{k=0}^{n-1} \prod_{w_j \in \mathcal{D}_k} |\mathbf{T}_j|$$

and therefore

$$n \log(|\mathcal{M}|) \leq g \sum_{i=1}^n \log(|\mathbf{T}_i|).$$

As $\log(|\mathcal{M}|) = \ell \log(|\mathbb{F}|)$, from the above inequality, we get

$$\sum_{i=1}^n \log(|\mathbf{T}_i|) \geq \left(\frac{n\ell \log(|\mathbb{F}|)}{g} \right) \geq \left(\frac{n\ell \log(|\mathbb{F}|)}{n - 2t} \right).$$

As mentioned earlier, $\sum_{i=1}^n \log(|\mathbf{T}_i|)$ denotes the lower bound on the communication complexity in bits. From the above inequality, we find that the lower bound is $\left(\frac{n\ell \log(|\mathbb{F}|)}{n-2t} \right)$ bits. Now each field element can be represented by $\log(|\mathbb{F}|)$ bits. Thus the lower bound is $\left(\frac{n\ell}{n-2t} \right)$ field elements. $\square$

From the previous two lemmas, we get the following theorem.

Theorem 3. *Any APRMT protocol executed over n (n ≥ 2t + 1) bi-directional wires has a communication complexity of $\Omega\left(\frac{n\ell}{n-2t}\right)$.*

Any protocol executed over n uni-directional wires can also be executed over n bi-directional wires. From Theorem 2, there exists an APRMT protocol which can be executed over $n = 2t + 1$ wires and requires a communication complexity of $\mathcal{O}\left(\frac{n\ell}{n-2t}\right)$. Thus the bound in Theorem 3 is tight.

3   Bounds on the Communication Complexity of ASRMT

Throughout Section 3, we assume $n \geq 2t + 1$. These many wires wires (uni-directional or bi-directional) are required for any SRMT protocol (Table 1). So it will also be required for ASRMT protocols.

3.1   Bounds for Uni-directional Wires

Theorem 4. *Any ASRMT protocol executed over the n (n ≥ 2t + 1) uni-directional wires has a communication complexity of $\Omega(\ell)$.*

PROOF: Easy, as any ASRMT protocol has to at least send the message.   □

We now show that the bound in Theorem 4 is *asymptotically tight*. That is, suppose there exists $n = 2t + 1$ uni-directional wires from **S** to **R** and consider a finite field $\mathbb{F}$, where $|\mathbb{F}| = \frac{t^2}{\delta}$. Then we design an ASRMT protocol tolerating $\mathcal{A}_t$ called ASRMT-Uni-Directional, which reliably sends a message $m^{\mathbf{S}}$ of size $(t + 1)^2 = \Theta(n^2)$ field elements and has a total communication complexity of $\mathcal{O}(n^2)$. The protocol has an error probability of δ. The high level idea of the protocol is as follows: let the n wires be denoted by $\mathcal{W} = \{w_1, \ldots, w_n\}$ and let $m^{\mathbf{S}} = \{m^{\mathbf{S}}_{i,j} : i, j = 0, \ldots, t\}$, consisting of $(t + 1)^2$ elements of $\mathbb{F}$. **S** selects a bi-variate polynomial $Q^{\mathbf{S}}(x, y)$ of degree-t in x and y, whose $(t + 1)^2$ coefficients are elements of $m^{\mathbf{S}}$. Now $Q^{\mathbf{S}}(x, y)$ is evaluated at $y = 1, \ldots, n$ to obtain the uni-variate polynomials $f^{\mathbf{S}}_i(x) = Q^{\mathbf{S}}(x, i)$ and $f^{\mathbf{S}}_i(x)$ is sent over wire w_i (by sending its coefficients). To recover $m^{\mathbf{S}}$, **R** should correctly recover $Q^{\mathbf{S}}(x, y)$ which requires **R** to know $t + 1$ correct $f^{\mathbf{S}}_i(x)$'s. In order to facilitate **R** to identify the correct $f^{\mathbf{S}}_i(x)$'s, **S** authenticates each $f^{\mathbf{S}}_i(x)$ using n different secret authentication keys and sends the authentication information and authentication key across the n wires. Now at the receiving end, **R** will consider an $f^{\mathbf{S}}_i(x)$ as *valid* only if it passes the authentication test with respect to the keys of $t + 1$ wires. Since at least one of these $t + 1$ wires is honest and the adversary will have no information about the authentication keys delivered over an honest wire, with very high probability a polynomial considered as valid by **R** will be indeed a correct polynomial. Moreover, there are at least $t + 1$ honest wires, who will eventually deliver $t + 1$ correct $f^{\mathbf{S}}_i(x)$'s. The complete details are in Fig. 1.

Computation and Communication by S:

1. Corresponding to the message $m^{\mathbf{S}} = \{m_{i,j}^{\mathbf{S}} : i,j = 0,\ldots,t\}$, **S** forms the bivariate polynomial $Q^{\mathbf{S}}(x,y) = \sum_{i=0,j=0}^{i=t,j=t} m_{i,j}^{\mathbf{S}} x^i y^j$.
2. For $i = 1,\ldots,n$, **S** computes $f_i^{\mathbf{S}}(x) = Q^{\mathbf{S}}(x,i)$ and the *authentication values* $Auth_{ij}^{\mathbf{S}} = f_i^{\mathbf{S}}(Key_{ij}^{\mathbf{S}})$, corresponding to random *authentication keys* $Key_{ij}^{\mathbf{S}}$, for $j = 1,\ldots,n$.
3. For $i = 1,\ldots,n$, **S** sends the following to **R** over wire w_i and terminates:
   (a) The degree-t polynomial $f_i^{\mathbf{S}}(x)$;
   (b) n authentication keys $Key_{ji}^{\mathbf{S}}$, for $j = 1,\ldots,n$;
   (c) n authentication values $Auth_{ji}^{\mathbf{S}}$, for $j = 1,\ldots,n$.

Message Recovery by R:

For $r = 0,\ldots,t$, **R** does the following in iteration r:

1. Let $\mathcal{W}^{\mathbf{R}}$ be the set of wires w_i over which **R** receives a complete set of values; i.e.,
   (a) A degree-t polynomial $f_i^{\mathbf{R}}(x)$;
   (b) n Authentication keys $Key_{ji}^{\mathbf{R}}$, for $j = 1,\ldots,n$;
   (c) n authentication values $Auth_{ji}^{\mathbf{R}}$, for $j = 1,\ldots,n$.
   Let $W_r^{\mathbf{R}}$ denote the contents of $\mathcal{W}^{\mathbf{R}}$, when $\mathcal{W}^{\mathbf{R}}$ contains exactly $t+1+r$ wires.
2. Wait until $|\mathcal{W}^{\mathbf{R}}| \geq t+1+r$. Now corresponding to every $w_i \in W_r^{\mathbf{R}}$, **R** computes

$$Support_i = \{w_j \in W_r^{\mathbf{R}} : Auth_{ij}^{\mathbf{R}} = f_i^{\mathbf{R}}(Key_{ij}^{\mathbf{R}})\}$$

3. If $Support_i \geq t+1$, then **R** concludes that $f_i^{\mathbf{R}}(x)$ is a *valid* polynomial.
4. If **R** finds $t+1$ valid polynomials, then using them **R** constructs the bi-variate polynomial $Q^{\mathbf{R}}(x,y) = \sum_{i=0,j=0}^{i=t,j=t} m_{i,j}^{\mathbf{R}} x^i y^j$, outputs $m^{\mathbf{R}} = \{m_{i,j}^{\mathbf{R}} : i,j = 0,\ldots,t\}$ and terminates the protocol. Otherwise **R** proceeds to the next iteration.

Fig. 1. Protocol ASRMT-Uni-Directional with $n = 2t+1$ unidirectional wires

Lemma 3. *In protocol ASRMT-Uni-Directional, if* **R** *concludes that* $f_i^{\mathbf{R}}(x)$ *is a valid polynomial, then* $f_i^{\mathbf{R}}(x) = f_i^{\mathbf{S}}(x)$ *except with probability* $\frac{t}{|\mathbb{F}|}$.

PROOF: The lemma trivially holds without any error if w_i is honest. So let w_i be a corrupted wire, which delivers $f_i^{\mathbf{R}}(x) \neq f_i^{\mathbf{S}}(x)$. In order that $f_i^{\mathbf{R}}(x)$ is considered as a valid polynomial by **R**, it must hold that $Support_i \geq t+1$. This further implies that there exists at least one honest wire, say w_j, such that $w_j \in Support_i$. This implies that $Auth_{ij}^{\mathbf{R}} = f_i^{\mathbf{R}}(Key_{ij}^{\mathbf{R}})$. Now notice that w_j is an honest wire and so $Auth_{ij}^{\mathbf{R}} = Auth_{ij}^{\mathbf{S}} = f_i^{\mathbf{S}}(Key_{ij}^{\mathbf{S}})$ and $Key_{ij}^{\mathbf{R}} = Key_{ij}^{\mathbf{S}}$. However $\mathcal{A}_t$ will have no information about $Auth_{ij}^{\mathbf{R}}$ and $Key_{ij}^{\mathbf{R}}$, as they are sent over w_j. So the probability that $f_i^{\mathbf{R}}(Key_{ij}^{\mathbf{S}}) = f_i^{\mathbf{S}}(Key_{ij}^{\mathbf{S}})$, even if $f_i^{\mathbf{R}}(x) \neq f_i^{\mathbf{S}}(x)$ is at most $\frac{t}{|\mathbb{F}|}$. This is because two different polynomials of degree-t can agree on at most t points and $Key_{ij}^{\mathbf{S}}$ is selected randomly by **S**. So except with error probability $\frac{t}{|\mathbb{F}|}$, $f_i^{\mathbf{R}}(x) = f_i^{\mathbf{S}}(x)$ for every valid polynomial $f_i^{\mathbf{R}}(x)$. $\square$

Lemma 4 (Termination). R *will eventually terminate ASRMT-Uni-Directional.*

PROOF: The proof follows from the fact that there always exists at least $t + 1$ honest wires, who will eventually deliver valid polynomials.   □

Lemma 5 (Communication Complexity). *ASRMT-Uni-Directional has a communication complexity of $\mathcal{O}(n^2)$ to send a message of size $(t+1)^2 = \Theta(n^2)$.*

PROOF: Easy and follows from the protocol description.   □

Lemma 6 (Reliability). *In protocol ASRMT-Uni-Directional, R will output the correct message, except with error probability δ.*

PROOF: From Lemma 3, $f_i^{\mathbf{R}}(x) = f_i^{\mathbf{S}}(x)$ for every valid polynomial $f_i^{\mathbf{R}}(x)$, except with probability $\frac{t}{|\mathbb{F}|}$. In the worst case, out of the $t+1$ wires which have delivered valid polynomials, t could be corrupted. So the probability that R outputs an incorrect message is at most $\frac{t^2}{|\mathbb{F}|} = \delta$ (since $|\mathbb{F}| = \frac{t^2}{\delta}$).   □

Theorem 5. *Assume that there are n ($n = 2t + 1$) uni-directional wires from S to R. Then there exists an ASRMT scheme which can reliably send a message containing $\Theta(n^2)$ elements from $\mathbb{F}$ by communicating $\mathcal{O}(n^2)$ elements from $\mathbb{F}$, where $|\mathbb{F}| = \frac{t^2}{\delta}$ and δ is the error probability.*

3.2   Bounds for Bi-directional Wires

It is obvious that $\Theta(\ell)$ can be the most tight bound on the communication complexity of any ASRMT protocol, irrespective of whether the wires are uni-directional or bi-directional. Now in the previous section, we have already shown that this bound is achieved if we consider only uni-directional wires. The same bound will also hold even if we consider bi-directional wires.

Till now, we have focussed only on RMT protocols, without worrying about the privacy. We next begin our discussion on SMT protocols, where we have to ensure privacy, in addition to reliability.

4   Bounds on the Communication Complexity of APSMT

Throughout Section 4, we assume that $n \geq 3t + 1$ since $n \geq 3t + 1$ wires (uni-directional or bi-directional) are required for any APSMT protocol [2].

4.1   Bounds for Uni-directional Wires

In [5], it is shown that any PSMT protocol has a communication complexity of $\Omega\left(\frac{n\ell}{n-3t}\right)$, when there exists $n \geq 3t + 1$ uni-directional wires from S to R. The lower bound will also hold for APSMT protocols. Moreover, in [2], the authors have designed an APSMT protocol, which requires a communication complexity of $\mathcal{O}(n\ell) = \mathcal{O}\left(\frac{n\ell}{n-3t}\right)$ to send a message of size ℓ, provided there are $n = 3t + 1$ uni-directional wires from S to R. From this discussion, we can state the following theorem.

Theorem 6. *Any APSMT scheme executed over n uni-directional wires from* **S** *to* **R***, where $n \geq 3t+1$ has a communication complexity of $\Theta\left(\frac{n\ell}{n-3t}\right)$.*

4.2   Bounds for Bi-directional Wires

Let **S** and **R** be connected by n bi-directional wires, denoted by $\mathcal{W} = \{w_1, \ldots, w_n\}$, where $n \geq 3t+1$. Then we show that any APSMT protocol has a communication complexity of $\Omega\left(\frac{n\ell}{n-3t}\right)$. To derive the lower bound, we use an approach similar to the one used Theorem 3. Specifically, we show the following:

1. We first show that in any APSMT protocol executed over n bi-directional wires, where $n \geq 3t+1$, the information exchanged over *any* $n-2t$ wires completely determine the message (Lemma 7).
2. Next, we show that any APSMT protocol where the information exchanged over $n-2t$ wires completely determine the message has a communication complexity of $\Omega\left(\frac{n\ell}{n-3t}\right)$ (Lemma 8).

Lemma 7. *In any APSMT protocol executed over n bi-directional wires, where $n \geq 3t+1$, the information exchanged over any $n-2t$ wires should completely determine the secret message $m^{\mathbf{S}}$.*

PROOF: The proof follows using same arguments as in Lemma 1.   □

Lemma 8. *Any APSMT protocol executed over n $(n \geq 3t+1)$ bi-directional wires, in which the information exchanged over any $n-2t$ wires completely determine the message, has a communication complexity of $\Omega\left(\frac{n\ell}{n-3t}\right)$.*

PROOF: Here we will use same arguments as used in Lemma 2. But we will also use an additional fact about APSMT protocols. Let Π^{APSMT} be an APSMT protocol, executed over n bi-directional wires (where $n \geq 3t+1$), to securely send a message of size ℓ, such that the information exchanged over *any* $n-2t$ wires completely determine the message. We now define the notations $\mathcal{M}$, $\mathbf{T}_i^m$, $\mathbf{M}_{i,j}^m$ and $\mathbf{M}_{i,j}$, which are exactly the same as in Lemma 2.

Since Π^{APSMT} is an APSMT protocol, it implies that in Π^{APSMT}, the transmission on any set of t wires is *independent* of the secret message. If it is not the case, then adversary will also know the secret message by passively listening the t wires. Thus, for any two messages $m_1, m_2 \in \mathcal{M}$, it must hold that

$$\mathbf{M}_{2t+1,3t}^{m_1} = \mathbf{M}_{2t+1,3t}^{m_2}.$$

Notice that the above relation must hold for any selection of t wires. We focussed on the set $\{w_{2t+1}, \ldots, w_{3t}\}$ just for simplicity. Now in Π^{APSMT}, the transmission over any set of $n-2t$ wires has *full* information about the secret message. Thus it must also hold that

$$\mathbf{M}_{2t+1,n}^{m_1} \cap \mathbf{M}_{2t+1,n}^{m_2} = \emptyset.$$

We again stress that the above relation must hold for any selection of $n-2t$ wires. We focussed on the set $\{w_{2t+1}, \ldots, w_n\}$ just for simplicity. As mentioned earlier, $\mathbf{M}_{2t+1,3t}^m$ will be same for all messages $m \in \mathcal{M}$. Thus, in order that the above relation holds, it must hold that $\mathbf{M}_{3t+1,n}^m$ is *unique* for every message $m \in \mathcal{M}$. This implies that

$$|\mathbf{M}_{3t+1,n}| = |\mathcal{M}|.$$

From the definition of $\mathbf{T}_i$ and $\mathbf{M}_{i,j}$, we get

$$\prod_{i=3t+1}^{n} |\mathbf{T}_i| \geq |\mathbf{M}_{3t+1,n}| \geq |\mathcal{M}|.$$

Let $g = n - 3t$. The above inequality holds for any selection of g wires $\mathcal{D} \subset \{w_1, \ldots, w_n\}$, where $|\mathcal{D}| = g$; i.e., $\prod_{w_i \in \mathcal{D}} |\mathbf{T}_i| \geq |\mathcal{M}|$. In particular, it holds for every selection $\mathcal{D}_k = \{w_{kg+1 \bmod n}, w_{kg+2 \bmod n}, \ldots, w_{kg+g \bmod n}\}$, with $k \in \{0, \ldots, n-1\}$. If we consider all the $\mathcal{D}_k$ sets collectively, then each wire is counted exactly g times in the collection. Thus, the product of the capacities of all $\mathcal{D}_k$ yields the capacity of the full wire set to the g-th power, and since each $\mathcal{D}_k$ has capacity at least $|\mathcal{M}|$, we get

$$|\mathcal{M}|^n \leq \prod_{k=0}^{n-1} \Pi_{w_j \in \mathcal{D}_k} |\mathbf{T}_j| = \left(\prod_{i=1}^{n} |\mathbf{T}_i| \right)^g,$$

and therefore

$$n \log(|\mathcal{M}|) \leq g \sum_{i=1}^{n} \log(|\mathbf{T}_i|).$$

As $\log(|\mathcal{M}|) = \ell \log(|\mathbb{F}|)$, from the above inequality, we get

$$\sum_{i=1}^{n} \log(|\mathbf{T}_i|) \geq \left(\frac{n\ell \log(|\mathbb{F}|)}{g} \right) \geq \left(\frac{n\ell \log(|\mathbb{F}|)}{n - 3t} \right).$$

Now $\sum_{i=1}^{n} \log(|\mathbf{T}_i|)$ denotes the lower bound on the communication complexity of protocol Π^{APSMT} in bits. From the above inequality, we find that the lower bound is $\left(\frac{n\ell \log(|\mathbb{F}|)}{n-3t} \right)$ bits. Now each field element can be represented by $\log(|\mathbb{F}|)$ bits. Thus the lower bound is $\left(\frac{n\ell}{n-3t} \right)$ field elements.   □

Theorem 7. *Any APSMT protocol executed over n ($n \geq 3t+1$) bi-directional wires has a communication complexity of $\Omega\left(\frac{n\ell}{n-3t} \right)$.*

Now any protocol executed over n uni-directional wires can also be executed over n bi-directional wires. From Theorem 6, there exists an APSMT protocol which can be executed over $n = 3t + 1$ uni-directional wires and requires a communication complexity of $\mathcal{O}\left(\frac{n\ell}{n-3t} \right)$. So the bound in Theorem 7 is tight.

5   Bounds on the Communication Complexity of ASSMT

Any ASSMT protocol requires $n \geq 2t+1$ wires, irrespective of whether the wires are uni-directional or bi-directional [2]. So we assume that $n \geq 2t+1$ throughout Section 5.

5.1   Bounds for Uni-directional Wires

Theorem 8. *Any ASSMT protocol executed over* n *($n \geq 2t + 1$) uni-directional wires has a communication complexity of* $\Omega\left(\frac{n\ell}{n-2t}\right)$.

PROOF: The theorem follows from the fact that any SSMT protocol with $n \geq 2t+1$ uni-directional wires [9] requires the same communication complexity.   □

We now show that the bound in Theorem 8 is asymptotically tight. Let there exists $n = 2t + 1$ uni-directional wires $\mathcal{W} = \{w_1, \ldots, w_n\}$ from **S** to **R**. Then we design a protocol called ASSMT-Uni-Directional, which securely sends a message $m^{\mathbf{S}} = \{m_k^{\mathbf{S}} : k = 1, \ldots, n\}$ containing $\ell = n$ elements from the field $\mathbb{F}$ and has a communication complexity of $\mathcal{O}(n^2) = \mathcal{O}\left(\frac{n\ell}{n-2t}\right)$, where $|\mathbb{F}| = \frac{nt}{\delta}$.

The high level idea of the protocol is as follows: for each $m_k^{\mathbf{S}}$, sender generates n Shamir shares [12]. Now the ith share of each $m_k^{\mathbf{S}}$ is sent over wire w_i. However, it is not enough to just send the shares, as the adversary can delay the communication over t honest wires and it can also change the shares over t corrupted wires. So **S** also sends some authentication information, which will enable **R** to identify the corrupted shares with very high probability. For performing the authentication, we use similar idea as used in our ASRMT protocol (see Fig. 1), with some *additional* steps. More specifically, we interpret the ith shares of n secrets as the coefficients of a polynomial $f_i^{\mathbf{S}}(x)$ of degree-n. This polynomial will be sent over w_i (this is same as sending the ith shares for the n messages). Now the polynomial $f_i^{\mathbf{S}}(x)$ can be authenticated by n random authentication keys $Key_{ij}^{\mathbf{S}}$ by computing $Auth_{ij}^{\mathbf{S}} = f_i^{\mathbf{S}}(Key_{ij}^{\mathbf{S}})$. However, the communication of $Key_{ij}^{\mathbf{S}}, Auth_{ij}^{\mathbf{S}}$ over wire w_j will breach the privacy of $f_i^{\mathbf{S}}(x)$, if w_i is honest and w_j is corrupted. To avoid this, we perform the authentication in the following way: corresponding to $Key_{ij}^{\mathbf{S}}$, we select a random masking key $Mask_{ij}^{\mathbf{S}}$ and define $Auth_{ij}^{\mathbf{S}} = f_i^{\mathbf{S}}(Key_{ij}^{\mathbf{S}}) + Mask_{ij}^{\mathbf{S}}$. Finally, $Key_{ij}^{\mathbf{S}}, Auth_{ij}^{\mathbf{S}}$ will be sent over w_j, while $Mask_{ij}^{\mathbf{S}}$ will be sent over w_i, along with $f_i^{\mathbf{S}}(x)$. As we will show later, this will help to maintain the perfect privacy and will also help to identify the corrupted shares with very high probability. Once **R** receives $t + 1$ correct shares (which he will receive eventually), **R** will correctly recover each $m_k^{\mathbf{S}}$ with very high probability. The details are in Fig. 2.

We now prove the properties of protocol ASSMT-Uni-Directional.

Lemma 9. *In protocol ASSMT-Uni-Directional, if* **R** *concludes that* $f_i^{\mathbf{R}}(x)$ *is a valid polynomial, then* $f_i^{\mathbf{R}}(x) = f_i^{\mathbf{S}}(x)$ *except with probability* $\frac{n}{|\mathbb{F}|}$.

PROOF (SKETCH): Follows using similar arguments as in Lemma 3 and the fact that two different polynomials of degree-(n) can agree on at most n points.   □

Lemma 10 (Termination). **R** *will eventually terminate ASSMT-Uni-Directional.*

PROOF: The proof follows from the fact that there always exists at least $t + 1$ honest wires, who will eventually deliver valid polynomials.   □

Lemma 11 (Communication Complexity). *ASSMT-Uni-Directional has a communication complexity of* $\mathcal{O}(n^2)$ *to send a message of size* n.

Computation and Communication by S:

1. For $k = 1, \ldots, n$, corresponding to $m_k^{\mathbf{S}}$, **S** selects a random degree-t polynomial $p_k^{\mathbf{S}}(x)$, where $p_k^{\mathbf{S}}(0) = m_k^{\mathbf{S}}$ and computes $Sh_{ki}^{\mathbf{S}} = p_k^{\mathbf{S}}(i)$, for $i = 1, \ldots, n$.
2. For $i = 1, \ldots, n$, **S** forms a polynomial $f_i^{\mathbf{S}}(x) = Sh_{1i}^{\mathbf{S}} \cdot x + Sh_{2i}^{\mathbf{S}} \cdot x^2 + \ldots + Sh_{ni}^{\mathbf{S}} \cdot x^n$.
3. For $i = 1, \ldots, n$, corresponding to the polynomial $f_i^{\mathbf{S}}(x)$, **S** selects n random authentication keys $Key_{ij}^{\mathbf{S}}$ and n random masking keys $Mask_{ij}^{\mathbf{S}}$, for $j = 1, \ldots, n$. **S** then computes $Auth_{ij}^{\mathbf{S}} = f_i^{\mathbf{S}}(Key_{ij}^{\mathbf{S}}) + Mask_{ij}^{\mathbf{S}}$.
4. For $i = 1, \ldots, n$, **S** sends the following to **R** over wire w_i and terminates:
   (a) The polynomial $f_i^{\mathbf{S}}(x)$ of degree-n by sending its coefficients;
   (b) n Masking keys $Mask_{ij}^{\mathbf{S}}$, for $j = 1, \ldots, n$;
   (c) n authentication keys $Key_{ji}^{\mathbf{S}}$ and n authentication values $Auth_{ji}^{\mathbf{S}}$, for $j = 1, \ldots, n$.

Message Recovery by R:

For $r = 0, \ldots, t$, **R** does the following in iteration r:

1. Let $\mathcal{W}^{\mathbf{R}}$ be the set of wires w_i over which **R** receives a complete set of values; i.e.,
   (a) A polynomial $f_i^{\mathbf{R}}(x)$ of degree-n;
   (b) n Masking keys $Mask_{ij}^{\mathbf{R}}$, for $j = 1, \ldots, n$;
   (c) n authentication keys $Key_{ji}^{\mathbf{R}}$ and n authentication values $Auth_{ji}^{\mathbf{R}}$, for $j = 1, \ldots, n$.
   Let $W_r^{\mathbf{R}}$ denote the contents of $\mathcal{W}^{\mathbf{R}}$, when $\mathcal{W}^{\mathbf{R}}$ contains exactly $t + 1 + r$ wires.
2. Wait until $|\mathcal{W}^{\mathbf{R}}| \geq t + 1 + r$. Now corresponding to every $w_i \in W_r^{\mathbf{R}}$, **R** computes
$$Support_i = \{w_j \in W_r^{\mathbf{R}} : Auth_{ij}^{\mathbf{R}} = f_i^{\mathbf{R}}(Key_{ij}^{\mathbf{R}}) + Mask_{ij}^{\mathbf{R}}\}$$
3. If $Support_i \geq t + 1$, then **R** concludes that $f_i^{\mathbf{R}}(x)$ is a *valid* polynomial. Let $f_i^{\mathbf{R}}(x) = Sh_{1i}^{\mathbf{R}} \cdot x + Sh_{2i}^{\mathbf{R}} \cdot x^2 + \ldots + Sh_{ni}^{\mathbf{R}} \cdot x^n$. Then $Sh_{ki}^{\mathbf{R}}$ is considered as a valid share for $m_k^{\mathbf{S}}$, for $k = 1, \ldots, n$.
4. If **R** finds $t + 1$ valid polynomials, then from their coefficients, **R** finds $t + 1$ valid shares for each $m_k^{\mathbf{S}}$, for $k = 1, \ldots, n$. Now using these valid shares, **R** reconstructs the degree-t polynomials $p_k^{\mathbf{R}}(x)$, outputs $m^{\mathbf{R}} = \{p_k^{\mathbf{R}}(0) : k = 1, \ldots, n\}$ and terminates the protocol. Otherwise **R** proceeds to the next iteration.

Fig. 2. Protocol ASSMT-Uni-Directional. Let $m^{\mathbf{S}} = \{m_k^{\mathbf{S}} : k = 1, \ldots, n\}$.

PROOF: Easy and follows from the protocol description. □

Lemma 12 (Reliability). *In protocol ASSMT-Uni-Directional,* **R** *will output the correct message, except with error probability δ.*

PROOF(SKETCH): The proof follows using similar arguments as used in Lemma 6 and the fact that $|\mathbb{F}| = \frac{nt}{\delta}$. □

Lemma 13 (Perfect Secrecy). *In protocol ASSMT-Uni-Directional, the message $m^{\mathbf{S}}$ will be perfectly secure.*

PROOF: Without loss of generality, let $w_1, \ldots, w_t$ be under the control of $\mathcal{A}_t$. So the adversary will know t shares for each m_k^S, for $k = 1, \ldots, n$ through the polynomials $f_1^S(x), \ldots, f_t^S(x)$. The adversary will also know $Auth_{ji}$, for $j = t+1, \ldots, n$ and $i = 1, \ldots, t$. But this will not reveal any new information about $f_j^S(x)$, for $j = t+1, \ldots, n$, as the adversary will not know the corresponding masking keys $Mask_{ji}$, for $j = t+1, \ldots, n$ and $i = 1, \ldots, t$ because they are sent over wires w_j, for $j = t+1, \ldots, n$, which are honest. Now the secrecy of each m_k^S follows from the properties of Shamir secret sharing [12]. □

Theorem 9. *Let there exists n $(n \geq 2t+1)$ uni-directional wires from $\mathbf{S}$ to $\mathbf{R}$. Then there exists an ASSMT protocol, which securely sends a message of size $\ell = n$ and requires a communication complexity of $\mathcal{O}(n^2) = \mathcal{O}\left(\frac{n\ell}{n-2t}\right)$.*

5.2 Bounds for Bi-directional Wires

Theorem 10. *Any ASSMT protocol executed over n $(n \geq 2t+1)$ bi-directional wires has a communication complexity of $\Omega\left(\frac{n\ell}{n-2t}\right)$.*

PROOF(SKETCH): We give the high level idea. We first claim that in any ASSMT protocol executed over n $(n \geq 2t+1)$ bi-directional wires, the communication over any set of $n-t$ wires should completely determine the secret message. This is obvious, since the adversary can arbitrarily delay the communication over t wires. So $\mathbf{R}$ should have the capacity to recover the message even from the communication done over $n-t$ wires. We next claim that in any ASSMT protocol, the communication over any set of t wires should be completely independent of the secret message. Now from these two facts, we can derive that the communication complexity will be $\Omega\left(\frac{n\ell}{n-2t}\right)$. □

From Theorem 9, there exists an ASSMT scheme, which can be executed over $n = 2t+1$ uni-directional wires and which has an asymptotic communication complexity of $\mathcal{O}\left(\frac{n\ell}{n-2t}\right)$. The same protocol can also be executed over $n = 2t+1$ bi-directional wires. Thus the bound in Theorem 10 is asymptotically tight.

6   Conclusion and Open Problems

In this paper, we have resolved the communication complexity of asynchronous RMT and SMT protocols. Our investigation reveals several insightful facts. We have considered settings where all the n wires are either uni-directional or bi-directional. It is interesting to consider a more general setting, where certain wires are directed from $\mathbf{S}$ to $\mathbf{R}$ and certain wires are directed from $\mathbf{R}$ to $\mathbf{S}$.

References

1. Agarwal, S., Cramer, R., de Haan, R.: Asymptotically Optimal Two-Round Perfectly Secure Message Transmission. In: Dwork, C. (ed.) CRYPTO 2006. LNCS, vol. 4117, pp. 394–408. Springer, Heidelberg (2006)

2. Choudhary, A., Patra, A., Ashwinkumar, B.V., Srinathan, K., Pandu Rangan, C.: On Minimal Connectivity Requirement for Secure Message Transmission in Asynchronous Networks. In: Garg, V., Wattenhofer, R., Kothapalli, K. (eds.) ICDCN 2009. LNCS, vol. 5408, pp. 148–162. Springer, Heidelberg (2008); Full version to appear in JPDC 2011

3. Choudhury, A.: Protocols for reliable and secure message transmission. Cryptology ePrint Archive, Report 2010/281 (2010)

4. Dolev, D., Dwork, C., Waarts, O., Yung, M.: Perfectly secure message transmission. JACM 40(1), 17–47 (1993)

5. Fitzi, M., Franklin, M.K., Garay, J.A., Vardhan, S.H.: Towards Optimal and Efficient Perfectly Secure Message Transmission. In: Vadhan, S.P. (ed.) TCC 2007. LNCS, vol. 4392, pp. 311–322. Springer, Heidelberg (2007)

6. Franklin, M., Wright, R.: Secure communication in minimal connectivity models. Journal of Cryptology 13(1), 9–30 (2000)

7. Kurosawa, K., Suzuki, K.: Truly Efficient 2-Round Perfectly Secure Message Transmission Scheme. In: Smart, N.P. (ed.) EUROCRYPT 2008. LNCS, vol. 4965, pp. 324–340. Springer, Heidelberg (2008)

8. Patra, A., Choudhary, A., Srinathan, K., Pandu Rangan, C.: Constant Phase Bit Optimal Protocols for Perfectly Reliable and Secure Message Transmission. In: Barua, R., Lange, T. (eds.) INDOCRYPT 2006. LNCS, vol. 4329, pp. 221–235. Springer, Heidelberg (2006)

9. Patra, A., Choudhary, A., Srinathan, K., Pandu Rangan, C.: Unconditionally reliable and secure message transmission in undirected synchronous networks: Possibility, feasibility and optimality. IJACT 2(2), 159–197 (2010)

10. Sayeed, H., Abu-Amara, H.: Perfectly secure message transmission in asynchronous networks. In: IEEE Symposium on Parallel and Distributed Processing, pp. 100–105 (1995)

11. Sayeed, H., Abu-Amara, H.: Efficient perfectly secure message transmission in synchronous networks. Information and Computation 126(1), 53–61 (1996)

12. Shamir, A.: How to share a secret. Communications of the ACM 22(11), 612–613 (1979)

13. Srinathan, K., Narayanan, A., Pandu Rangan, C.: Optimal Perfectly Secure Message Transmission. In: Franklin, M. (ed.) CRYPTO 2004. LNCS, vol. 3152, pp. 545–561. Springer, Heidelberg (2004)

14. Srinathan, K., Patra, A., Choudhary, A., Pandu Rangan, C.: Probabilistic Perfectly Reliable and Secure Message Transmission – Possibility, Feasibility and Optimality. In: Srinathan, K., Pandu Rangan, C., Yung, M. (eds.) INDOCRYPT 2007. LNCS, vol. 4859, pp. 101–122. Springer, Heidelberg (2007)

15. Srinathan, K., Prasad, N.R., Pandu Rangan, C.: On the optimal communication complexity of multiphase protocols for perfect communication. In: IEEE S&P, pp. 311–320 (2007)

Two-Party Round-Optimal Session-Policy Attribute-Based Authenticated Key Exchange without Random Oracles

Kazuki Yoneyama

NTT Information Sharing Platform Laboratories
yoneyama.kazuki@lab.ntt.co.jp

Abstract. In this paper, we propose a new one-round session-policy attribute-based (implicitly) authenticated key exchange (SP-ABAKE) scheme which allows expressive access controls and is secure in the standard model (StdM). Our scheme enjoys the best of both worlds: efficiency and security. The number of rounds is one (optimal) while the known secure scheme in the StdM is not one-round protocol. Our scheme is comparable in communication complexity with the most efficient known scheme whereas it cannot be proved in the StdM. Also, our scheme is proved to satisfy security against advanced attacks like key compromise impersonation under a non-interactive number-theoretic assumption. We construct our scheme based on Waters' ciphertext-policy attribute-based encryption with the generic conversion technique to the CCA-security from the CPA-security.

Keywords: authenticated key exchange, attribute-based authenticated key exchange, session-policy, CK model.

1 Introduction

1.1 Background

How to provide flexible access control mechanisms over encrypted data is getting important today because people ordinarily exchange sensitive information over the Internet in various web services. Attribute-based encryption (ABE) solves this issue cryptographically. There are two flavors of ABE; one is *key-policy* ABE (KP-ABE) and the other is *ciphertext-policy* ABE (CP-ABE). In a KP-ABE scheme, a party possesses a secret key corresponding to an access policy and can decrypt a ciphertext corresponding to a set of attributes that satisfies the access policy of the party. In a CP-ABE scheme, a party possesses a secret key corresponding to a set of attributes and can decrypt a ciphertext corresponding to an access policy that the set of attributes of the party satisfies. Thus, a typical application of KP-ABE is content delivery services and that of CP-ABE is data storage services.

On the other hand, how to provide flexible access control mechanisms for establishing a secure channel is also important because a kind of web services is unsuitable to the authentication based on identity. Attribute-based authenticated key exchange (ABAKE)

H. Kim (Ed): ICISC 2011, LNCS 7259, pp. 467–489, 2012.
© Springer-Verlag Berlin Heidelberg 2012

is recently studied to solve such an issue. For example, ABAKE is useful in the situation that some sensitive information (e.g., medical history) is sent with the secure channel established by an ABAKE scheme. Then, parties may hope to hide their identities from the peer of the session though the peer is needed to be a qualified registered person. By using ABAKE, parties can establish the secure channel with a qualified registered person without revealing their identities. ABAKE has also two flavors like ABE; one is *key-policy* ABAKE (KP-ABAKE) and the other is *session-policy* ABAKE (SP-ABAKE). KP-ABAKE is an analogy of KP-ABE; that is, a key generation center (KGC) issues the static secret key according to an access policy of a party by using the master secret key and the party specifies a set of attributes as an authentication condition in a session. If the set of attributes a party specifies satisfies the policy of the peer and vice versa, the common session key is established. SP-ABAKE is an analogy of CP-ABE; that is, the role of the set of attributes and the access policy is swapped from KP-ABAKE.

There are several previous studies about ABAKE. Wang et al. [1,2,3] proposed simple variants of ABAKE as prior works. In their schemes, attributes are regarded as identification strings and there is no mechanism for evaluating policy. Gorantla et al. [4] proposed a one-round (i.e., round-optimal) generic construction of group SP-ABAKE based on attribute-based KEM. Group ABAKE provides qualified parties to establish a common session key. If underlying attribute-based KEM is *expressive* (i.e., fine-grained access policies are allowed), their scheme is also expressive. In the setting of their scheme, parties do not specify access policies individually. Parties obtain the common session key if sets of attributes of all of them satisfy a *common* access policy that is set in advance. Thus, their scheme fits in a special scenario as each party cannot specify the access policy which the peer is expected to satisfy each other in the session. A drawback of their scheme is that the proposed instantiation is proved in the random oracle model (ROM) and the generic group model (GGM). Steinwandt and Suárez Corona [5] proposed another two-round generic construction of group SP-ABAKE based on attribute-based signcryption. Like the construction by Gorantla et al. [4], the access policy in a session is common for all parties in their scheme. Unfortunately, the proposed instantiation (using the encrypt-then-sign signcryption) is also proved in the ROM and in the selective security setting (i.e., the adversary must specify an attribute or an access policy of the attack target party before receiving public parameters). Fujioka et al. [6] and Yoneyama [7] proposed a KP-ABAKE scheme and a SP-ABAKE scheme which satisfy resistance to leakage of session-specific randomness (LSR). Though their schemes are round-optimal and expressive, security proofs are also given in the ROM and in the selective security setting.

Beyond the above schemes, Birkett and Stebila [8] proposed a generic construction (BS scheme) of ABAKE based on predicate-based signatures. The BS scheme can be used as both KP-ABAKE and SP-ABAKE according to underlying signatures. Compared with the constructions by Gorantla et al. [4] and Steinwandt and Suárez Corona [5], the BS scheme allows parties to specify access policies individually; thus, it has wider applicative scenarios. Also, an additional property that is called credential privacy (CP) is guaranteed as well as the standard security of the session key. CP implies that the adversary cannot distinguish between two parties satisfying the same access policy. However, the BS scheme has several drawbacks. First, it needs sequential protocol execution in a session (i.e., a party cannot compute the next message to a peer without receiving the previous message from the peer; thus, the party must wait the previous message from the peer). Three-move communication is necessary. Note that we must distinguish 'move' from 'round'. A message sent by an initiator in one-round protocols must be independent from a message sent by a responder; that is, parties can send messages simultaneously. On the other hand, in a kind of two-move protocols, a message sent by a responder may depend on a received message; that is, the responder must wait the message from the initiator before sending his message. Thus, such a protocol (including the BS scheme) is sensitive about the network condition because parties must send messages sequentially. Most of ABAKE schemes need only one-round in a session and allow simultaneous computation of each message. This drawback comes from adopting the signed-Diffie-Hellman (DH) paradigm [9]. Secondly, possible instantiations have several problems. There are some predicate-based (attribute-based) signature schemes [10,11,12,13]. Instantiations with [10,11] cannot achieve expressive ABAKE (i.e., these signatures only allow threshold access policies). Since three signature schemes are proposed in [12], three instantiations of SP-ABAKE with [12] are possible. One instantiation with [12] is very efficient and expressive, but the security proof is given in the GGM. Other instantiations with [12] need large communication complexity. The instantiation with [13] provides fully secure and expressive SP-ABAKE scheme in the standard model (StdM). However, communication complexity of it is larger than that of the efficient instantiation with [12], depending on the size to represent access policies.

1.2   Our Contribution

We introduce a two-party SP-ABAKE scheme which has several attractive points compared to existing schemes as follows:

Round complexity. The number of rounds of our scheme is optimal; that is, one-round. All of previous one-round ABAKE schemes are proved in the ROM or GGM. The only known ABAKE scheme in the StdM is the BS scheme that needs three-moves. Moreover, since messages depend on previous one, messages must be sent sequentially in the BS scheme.

Security in the standard model. The security of our scheme can be proved in the StdM under a non-interactive assumption (the decisional parallel bilinear DH exponent (DPBDHE) assumption [14]). Most of existing schemes [4,6,7,5] rely on the ROM or the GGM. Though one of instantiations of the BS scheme is secure in the StdM, our scheme is more efficient both in round complexity and communication complexity. Our scheme is proved in the selective security setting due to the security of the underlying Waters CP-ABE [14]. Though some of previous schemes achieve full security, one [4] relies on the ROM and the GGM, and the other [8] is less efficient than our scheme.

Communication complexity. In our scheme, a message sent by a party in a session contains $2\ell + 2$ group elements and a signature and a verification key of a one-time signature scheme, where ℓ is the size of rows of the matrix according to the access

Table 1. Comparison among expressive ABAKE schemes

	Access policy	Round complexity	Attack model	Resource model	Additional property	Communication complexity
[4]	SP	1 round	full	GGM & ROM	none	$6 + |ssig| + |svk| \rightarrow |aCT|$
[6]	KP	1 round	selective	ROM	wPFS, KCI, LSR	$k + 1$
[7]	SP	1 round	selective	ROM	wPFS, KCI, LSR	$\ell n + 1$
[5]	SP	2 rounds	selective	ROM	wPFS	$3.5 + |asig| + |aCT|$
[8] with [12]	SP	3 moves	full	GGM	wPFS, CP	$\ell + n + 3$
[8] with [13]	KP & SP	3 moves	full	StdM	wPFS, CP	$7\ell + 12$
Ours	SP	1 round	selective	StdM	wPFS, KCI	$2\ell + 2 + |sig| + |vk|$

k is the number of attributes specified in a session. $|asig|$ and $|aCT|$ are the sizes of a signature of an attribute-based signature scheme and a ciphertext of a CP-ABE scheme. $|sig|$ and $|vk|$ are the sizes of the signature and the verification key of a one-time signature scheme. For example, when the Mohassel signature [16] is used, $|sig| = 2$ and $|vk| = 4$.

policy. For example, when the Boneh-Boyen signature [15] is used, the signature and the verification key contain 2 and 4 group elements, respectively. It is comparable with the most efficient instantiation of the BS scheme secure in the GGM (a message sent by a party in a session contains $\ell + n + 3$ group elements where n is the size of columns of the matrix according to the access policy.). Also, our scheme is more efficient than the instantiation of [8] with [13] in the StdM (a message sent by a party in a session contains $7\ell + 12$ group elements), especially when an access policy is complicated (i.e., when ℓ becomes large).

Resistance to advanced attacks. Our scheme satisfies (weak) perfect forward secrecy (wPFS) and resistance to key compromise impersonation (KCI). wPFS in ABAKE is formulated as an adversary cannot obtain any information about the session key even if the adversary can obtain the master secret key of the KGC *after* the completion of the session. Resistance to KCI means that an adversary cannot impersonate an honest party A to another honest party B even if the adversary can obtain the static secret key of B. The generic construction in [4] does not satisfy wPFS. Also, the security models of previous ABAKE schemes other than [6] and [7] do not capture resistance to KCI; that is, this property is not guaranteed in these schemes. Thus, our scheme is the first ABAKE scheme proved to be resilient to KCI in the StdM.

Table. 1 shows a comparison among our scheme and existing schemes, which are expressive.

Our Technique. We construct our scheme based on the CP-ABE scheme by Waters [14]. The Waters CP-ABE is expressive (due to adopting linear secret sharing) and efficient in the ciphertext size and the computational cost. Also, security is proved in the StdM under the DPBDHE assumption. The intuitive strategy to construct our scheme is as follows: First, parties exchange (a part of) ciphertexts of the Waters CP-ABE with their respectively specifying access policies. If their attributes satisfy the specified access policy each other, then they can share common values. Next, shared values are converted to be uniformly distributed with a strong randomness extractor. Finally, they obtain the session key derived from the outputs of a pseudo-random function (PRF) where keys of the PRF are the outputs of the strong randomness extractor. This session key derivation technique is similar to the generic construction of AKE [17,18]. Though it seems to work correctly, there is a problem in the security proof. We must simulate responses to ciphertexts sent by the adversary. However, since the Waters CP-ABE is not CCA secure, we cannot decrypt such ciphertexts and the simulation is impossible. Thus, we use a one-time signature additionally in order to simulate such a case based on the conversion technique [19,20] to the CCA security from the CPA security.

To prove the security of our scheme, we extend the security model of [8] to capture wPFS and resilient to KCI. By allowing the adversary to obtain the master secret key for completed (matching) sessions, we represent wPFS. Also, by allowing the adversary to obtain the static secret key for the target session that has no matching session, we represent resilient to KCI.

1.3   Related Works

CP-ABE.   The first ABE scheme is proposed by Sahai and Waters [21], called the fuzzy ID-based encryption, which parties must match at least a certain threshold of attributes. Bethencourt et al. [22] proposed the first CP-ABE scheme which allows the ciphertext policies to be very expressive, but the security proof is in the generic group model. Cheung and Newport [23] proposed a provably secure CP-ABE scheme and their scheme deals with negative attributes explicitly and supports wildcards in the ciphertext policies. Kapadia et al. [24] and Nishide et al. [25] also proposed CP-ABE schemes which achieves hidden ciphertext policies in a limited way, respectively. Shi et al. [26] proposed a predicate encryption scheme that focuses on range queries over huge numbers, which can also achieve a CP-ABE scheme with range queries. Boneh and Waters [27] proposed a predicate encryption scheme based on the primitive called the hidden vector encryption, which needs bilinear groups whose order is a product of two large primes; thus, it needs to deal with large group elements and the number of attributes is fixed at the system setup. Katz et al. [28] proposed a novel predicate encryption scheme and their scheme is very general and can achieve both KP-ABE and CP-ABE schemes. Waters [14] proposed expressive and efficient CP-ABE schemes based on non-interactive assumptions. Lewko et al. [29] proposed the first fully secure CP-ABE scheme. Okamoto and Takashima [30] proposed a CP-ABE scheme which allows non-monotone access policies and is fully secure under non-interactive assumptions.

2   Preliminaries

2.1   Access Structure

We introduce the notion of the access structure to represent the access control by the policy. We show the definition given in [31].

Definition 1 (Access Structure [31]). *Let* $\{P_1, P_2, \ldots, P_n\}$ *be a set of parties. A collection* $\mathbb{A} \subseteq 2^{\{P_1, P_2, \ldots, P_n\}}$ *is monotone if* $\forall Att_1, Att_2 :$ *if* $Att_1 \in \mathbb{A}$ *and* $Att_1 \subseteq Att_2$ *then* $Att_2 \in \mathbb{A}$. *An access structure (resp. monotone access structure) is a collection (resp. monotone collection)* $\mathbb{A}$ *of non-empty subsets of* $\{P_1, P_2, \ldots, P_n\}$, *i.e.,* $\mathbb{A} \subseteq 2^{\{P_1, P_2, \ldots, P_n\}} \backslash \{\emptyset\}$. *The sets in* $\mathbb{A}$ *are called the authorized sets, and the sets not in* $\mathbb{A}$ *are called the unauthorized sets.*

Though this definition restricts monotone access structures, it is also possible to (inefficiently) realize general access structures by having the 'not' of an attribute as a separate attribute altogether. Thus, the number of attributes in the system will be doubled.

2.2 Linear Secret Sharing

We use linear secret sharing schemes (LSSSs) to obtain the fine-grained access control. The LSSS can provide arbitrary conditions for the reconstruction of the secret with monotone access structures. We show the definition given in [31].

Definition 2 (Linear Secret Sharing Schemes [31]). *A secret sharing scheme* Π *over a set of parties* $\mathbb{P}$ *is called linear (over* $\mathbb{Z}_p$*) if*

1. *The shares for each party form a vector over* $\mathbb{Z}_p$.
2. *There exists a matrix* M *with* ℓ *rows and* n *columns called the share-generating matrix for* Π. *For all* $i = 1, \ldots, \ell$, *the ith row of* M *we let a labeling function* ρ *defined the party labeling row* i *as* $\rho(i)$. *When we consider the column vector* $v = (s, r_2, \ldots, r_n)$, *where* $s \in \mathbb{Z}_p$ *is the secret to be shared, and* $r_2, \ldots, r_n \in \mathbb{Z}_p$ *are randomly chosen, then* Mv *is the vector of* ℓ *shares of the secret* s *according to* Π. *The share* $(Mv)_i$ *belongs to party* $\rho(i)$.

The important property of LSSSs is the linear reconstruction property, defined as follows: Suppose that Π is an LSSS for the access structure $\mathbb{A}$. Let $S \in \mathbb{A}$ be any authorized set, and let $I \subset \{1, 2, \ldots \ell\}$ be defined as $I = \{i : \rho(i) \in S\}$. Then, there exist constants $\{w_i \in \mathbb{Z}_p\}_{i \in I}$ such that, if $\{\lambda_i\}$ are valid shares of any secret s according to Π, then $\sum_{i \in I} w_i \lambda_i = s$. In [31], it is shown that these constants $\{w_i\}$ can be found in time polynomial in the size of the share generating matrix M.

Note on Convention. We note that we use the convention that vector $(1, 0, 0, \ldots, 0)$ is the "target" vector for any linear secret sharing scheme. For any satisfying set of rows I in M, we will have that the target vector is in the span of I. For any unauthorized set of rows I the target vector is not in the span of the rows of the set I. Moreover, there will exist a vector w such that $w \cdot (1, 0, 0, \ldots, 0) = -1$ and $w \cdot M_i = 0$ for all $i \in I$.

2.3 Bilinear Maps

Definition 3 (Bilinear Maps). *Let* G *be a cyclic group of prime order* p *and* g *is a generator of* G. *We say that* $e : G \times G \to G_T$ *is a bilinear map if the following holds:*

– *For all* $X, Y \in G$ *and* $a, b \in \mathbb{Z}_p$, *we have* $e(X^a, Y^b) = e(X, Y)^{ab}$,
– $e(g, g) \neq 1$.

We say that G is a bilinear group if e and the group operation in G and G_T can be computed efficiently.

2.4 Decisional Parallel Bilinear Diffie-Hellman Exponent Assumption

Let κ be the security parameter and p be a κ-bit prime. Let G be a cyclic group of a prime order p with a generator g and G_T be a cyclic group of the prime order p with a generator g_T. Let $e : G \times G \to G_T$ be a bilinear map. We say that G, G_T are bilinear groups with the pairing e where $g_T = e(g, g)$.

The decisional parallel bilinear DH exponent problem is as follows. An adversary $\mathcal{A}$ is given inputs $\mathbf{m} =$

$$G, G_T, e, p, g, g_T, g^c, g^a, \ldots, g^{(a^q)}, g^{(a^{q+2})}, \ldots, g^{(a^{2q})},$$

$$\forall_{1 \le j \le q} \; g^{cb_j}, g^{a/b_j}, \ldots, g^{a^q/b_j}, g^{a^{q+2}/b_j}, \ldots, g^{a^{2q}/b_j},$$

$$\forall_{1 \le j,k \le q, k \ne j} \; g^{acb_k/b_j}, \ldots, g^{(a^q cb_k/b_j)}$$

where $c, a, b_1, \ldots, b_q \in \mathbb{Z}_p$ are randomly chosen. For adversary $\mathcal{A}$, we define advantage

$$\mathsf{Adv}^{\mathrm{DPBDHE}}(\mathcal{A}) = |\Pr[\mathcal{A}(\mathbf{m}, \mathcal{T} = g_T^{a^{q+1}c}) = 1] - \Pr[\mathcal{A}(\mathbf{m}, \mathcal{T} = R) = 1]|,$$

where $R \in G_T$ is randomly chosen and the probability is taken over the choices of $\mathbf{y}$ and the random tape of $\mathcal{A}$.

Definition 4 (Decisional Parallel Bilinear Diffie-Hellman Exponent Assumption). *We say that the DPBDHE assumption in G and G_T holds if for all polynomial-time adversary $\mathcal{A}$, the advantage $\mathsf{Adv}^{\mathrm{DPBDHE}}(\mathcal{A})$ is negligible in security parameter κ.*

The validity of the DPBDHE assumption is proved in the GGM in [14].

2.5 Strong Randomness Extractor

Let $Ext : S \times X \to Y$ be a function with finite seed space S, finite domain X, and finite range Y.

Definition 5 (Strong Randomness Extractor). *We say that the function Ext is the (k, ϵ)-strong randomness extractor if for any distribution D_X over X with $H_\infty(D_X) \ge k$, $\Delta((U_S, Ext(U_S, D_X)), (U_S, U_Y)) \le \epsilon$ holds, where two U_S in $(U_S, Ext(U_S, D_X))$ denotes the same random variable, H_∞ denotes min-entropy, Δ denotes statistical distance, and U_S, U_X, U_Y denotes uniform distribution over S, X, Y, respectively.*

2.6 Pseudo-random Function

Let κ be a security parameter and $\mathsf{F} = \{F_k : Dom_\kappa \to Rng_\kappa\}_\kappa$ be a function family with a family of domains $\{Dom_\kappa\}_\kappa$, a family of ranges $\{Rng_\kappa\}_\kappa$ and a key k.

Definition 6 (Pseudo-Random Function). *We say that function family $\mathsf{F} = \{F_s\}_\kappa$ is the (t, ϵ)-PRF family if for any distinguisher $\mathcal{D}$ with a time-complexity at most t, $\mathsf{Adv}^{\mathrm{prf}} = |\Pr[\mathcal{D}^{F_k(\cdot)} \to 1] - \Pr[\mathcal{D}^{RF(\cdot)} \to 1]| \le \epsilon$, where k is randomly chosen the key space and $RF : Dom_\kappa \to Rng_\kappa$ be a truly random function.*

3 Security Model

In this section, we introduce a security model for SP-ABAKE. Our attribute-based Canetti-Krawczyk (ABCK) model is an extension of the CK security model [9].

The proposed ABCK model is different from the original CK model in the following points: (1) the session is identified by a set of attributes $\mathbb{S}_P$ of party P, (2) resistance to KCI is captured by giving the static secret key of the target party to the adversary, and (3) wPFS is captured by giving the master secret key to the adversary same as in the ID-based AKE.

Syntax. An ABAKE scheme consists of the following algorithms. We denote a party by P and his associated set of attributes by $\mathbb{S}_P$. The party P and other parties are modeled as a probabilistic polynomial-time Turing machine.

Setup. The setup algorithm **Setup** takes a security parameter κ as input, and outputs a master secret key MSK and a master public key MPK, i.e.,

$$\textbf{Setup}(1^{\kappa}) \rightarrow (MSK, MPK).$$

Key Generation. The key generation algorithm **KeyGen** takes the master secret key MSK, the master public key MPK, and a set of attributes $\mathbb{S}_P$ given by a party P, and outputs a static secret key $SK_{\mathbb{S}_P}$ corresponding to $\mathbb{S}_P$, i.e.,

$$\textbf{KeyGen}(MSK, MPK, \mathbb{S}_P) \rightarrow SK_{\mathbb{S}_P}.$$

Key Exchange. The party A and the party B share a session key by performing the following N-move protocol. A (resp. B) selects a policy $\mathbb{A}_A$ (resp. $\mathbb{A}_B$) as an access structure, respectively.

A starts the protocol by computing the 1st message m_1 by the algorithm **Message**, that takes the master public key MPK, the set of attributes $\mathbb{S}_A$, the static secret key $SK_{\mathbb{S}_A}$ and the policy $\mathbb{A}_A$, and outputs 1st message m_1. A sends m_1 to the other party B.

For $i = 2, ..., N$, upon receiving the $(i-1)$th message m_{i-1}, the party P ($P = A$ or B) computes the ith message by algorithm **Message**, that takes the master public key MPK, the set of attributes $\mathbb{S}_P$, the static secret key $SK_{\mathbb{S}_P}$, the policy $\mathbb{A}_P$ and the sent and received messages $m_1, \ldots, m_{i-1}$, and outputs the ith message m_i, i.e.,

$$\textbf{Message}(MPK, \mathbb{S}_P, SK_{\mathbb{S}_P}, \mathbb{A}_P, m_1, \ldots, m_{i-1}) \rightarrow m_i.$$

The party P sends m_i to the other user $\bar{P}$ ($\bar{P} = B$ or A).

Upon receiving or after sending the final nth message m_n, P computes a session key by algorithm **SessionKey**, that takes the master public key MPK, the set of attributes $\mathbb{S}_P$, the static secret key $SK_{\mathbb{S}_P}$, the policy $\mathbb{A}_P$ and the sent and received messages $m_1, ..., m_N$, and outputs a session key K, i.e.,

$$\textbf{SessionKey}(MPK, \mathbb{S}_P, SK_{\mathbb{S}_P}, \mathbb{A}_P, m_1, \ldots, m_N) \rightarrow K.$$

Both parties A and B can compute the same session key if and only if $\mathbb{S}_A \in \mathbb{A}_B$ and $\mathbb{S}_B \in \mathbb{A}_A$.

Session. An invocation of a protocol is called a *session*. A session is activated with an incoming message of the forms $(\mathcal{I}, \mathbb{S}_A, \mathbb{S}_B)$ or $(\mathcal{R}, \mathbb{S}_B, \mathbb{S}_A, m_1)$, where $\mathcal{I}$ and $\mathcal{R}$ with role identifiers, and A and B with user identifiers. If A was activated with $(\mathcal{I}, \mathbb{S}_A, \mathbb{S}_B)$, then A is called the session *initiator*. If B was activated with $(\mathcal{R}, \mathbb{S}_B, \mathbb{S}_A, m_1)$, then B is called the session *responder*. After activated with an incoming message of the forms $(\mathcal{I}, \mathbb{S}_A, \mathbb{S}_B, m_1, \ldots, m_{k-1})$ from the responder B, the initiator A outputs m_k, then may be activated next by an incoming message of the forms $(\mathcal{I}, \mathbb{S}_A, \mathbb{S}_B, m_1, \ldots, m_{k+1})$ from the responder B. After activated by an incoming message of the forms $(\mathcal{R}, \mathbb{S}_B, \mathbb{S}_A, m_1, \ldots, m_k)$ from the initiator A, the responder B outputs m_{k+1}, then may be activated next by an incoming message of the forms $(\mathcal{R}, \mathbb{S}_B, \mathbb{S}_A, m_1, \ldots, m_{k+2})$ from the initiator A. Upon receiving or after sending the final nth message m_n, both parties A and B computes a session key K.

If A is the initiator of a session, the session is identified by $\mathsf{sid} = (\mathcal{I}, \mathbb{S}_A, \mathbb{S}_B, m_1)$, $(\mathcal{I}, \mathbb{S}_A, \mathbb{S}_B, m_1, m_2, m_3)$, $\ldots$, $(\mathcal{I}, \mathbb{S}_A, \mathbb{S}_B, m_1, \ldots, m_N)$. If B is the responder of a session, the session is identified by $\mathsf{sid} = (\mathcal{R}, \mathbb{S}_B, \mathbb{S}_A, m_1, m_2)$, $(\mathcal{R}, \mathbb{S}_B, \mathbb{S}_A, m_1, m_2, m_3, m_4)$, $\ldots$, $(\mathcal{R}, \mathbb{S}_B, \mathbb{S}_A, m_1, \ldots, m_N)$. We say that a session is *completed* if a session key is computed in the session. The *matching session* of a completed session $(\mathcal{I}, \mathbb{S}_A, \mathbb{S}_B, m_1, \ldots, m_N)$ is a completed session with identifier $(\mathcal{R}, \mathbb{S}_B, \mathbb{S}_A, m_1, \ldots, m_N)$ such that $\mathbb{S}_A \in \mathbb{A}_B$ and $\mathbb{S}_B \in \mathbb{A}_A$, and vice versa.

Adversary. The adversary $\mathcal{A}$ that is modeled as a probabilistic polynomial-time Turing machine controls all communications between parties including the session activation by performing the following queries.

- Send(message): The message has one of the following forms: $(\mathcal{I}, \mathbb{S}_A, \mathbb{S}_B, m_1, \ldots, m_k)$, or $(\mathcal{R}, \mathbb{S}_B, \mathbb{S}_A, m_1, \ldots, m_{k+1})$. The adversary obtains the response from the party.

Revealing secret information of parties is captured via the following queries.

- KeyReveal(sid): The adversary obtains the session key for the session sid if the session is completed.
- StateReveal(sid): The adversary obtains the ephemeral secret key associated with the session sid. The adversary obtains session state of owner of the session sid, if the session is not completed (the session key is not established yet). Session state includes all chosen randomness and intermediate computation results but not the static secret key.
- Corrupt(P): This query allows the adversary to obtain all information of the party P (including the static secret key corresponding to the set of attributes $\mathbb{S}_P$) and the adversary totally controls that party. Then, we call P dishonest. If a party P' possesses $\mathbb{S}_{P'}$ such that $\mathbb{S}_{P'} \subseteq \mathbb{S}_P$, then we also call P' dishonest. Otherwise, we call P' honest.

Freshness. For the security definition, we need the notion of freshness.

Definition 7 (Freshness). *Let* $\mathsf{sid}^* = (\mathcal{I}, \mathbb{S}_A, \mathbb{S}_B, m_1, \ldots, m_N)$ *or* $(\mathcal{R}, \mathbb{S}_B, \mathbb{S}_A, m_1, \ldots, m_N)$ *be a completed session between honest users A with the set of attributes $\mathbb{S}_A$ and B with*

$\mathbb{S}_B$. *If the matching session exists, then let* $\overline{\mathsf{sid}}^*$ *be the matching session of* sid*. *We say* sid* *to be* fresh *if none of the following conditions hold:*

1. *The adversary poses* KeyReveal(sid*) *or* KeyReveal($\overline{\mathsf{sid}}^*$) *query if* $\overline{\mathsf{sid}}^*$ *exists,*
2. $\overline{\mathsf{sid}}^*$ *exists and the adversary poses* StateReveal(sid*) *or* StateReveal($\overline{\mathsf{sid}}^*$),
3. $\overline{\mathsf{sid}}^*$ *does not exist and the adversary poses* StateReveal(sid*).

Security Experiment. For our security definition, we consider the following security experiment. Initially, the adversary $\mathcal{A}$ is given a set of honest users, and makes any sequence of the queries described above. During the experiment, $\mathcal{A}$ makes the following query.

- Test(sid*): Here, sid* must be a fresh session. Select random bit $b \in \{0, 1\}$, and return the session key held by sid* if $b = 0$, and return a random key if $b = 1$.

The experiment continues until $\mathcal{A}$ makes a guess b'. The adversary *wins* the game if the test session sid* is still fresh and if $\mathcal{A}$'s guess is correct, i.e., $b' = b$. The advantage of $\mathcal{A}$ in the experiment with the ABAKE scheme Π is defined as

$$\mathsf{Adv}_{\Pi}^{\mathrm{ABAKE}}(\mathcal{A}) = \Pr[\mathcal{A} \ wins] - \frac{1}{2}.$$

We define the security as follows.

Definition 8 (ABCK Security). *We say that an ABAKE scheme Π is secure in the ABCK model if the following conditions hold:*

1. *If two honest parties completing matching sessions and* $\mathbb{S}_A \in \mathbb{A}_B$ *and* $\mathbb{S}_B \in \mathbb{A}_A$ *hold, then, except with negligible probability, they both compute the same session key.*
2. *For any probabilistic polynomial-time adversary $\mathcal{A}$,* $\mathsf{Adv}_{\Pi}^{\mathrm{ABAKE}}(\mathcal{A})$ *is negligible if* $\overline{\mathsf{sid}}^*$ *exists and the master secret key is given to the adversary.*
3. *For any probabilistic polynomial-time adversary $\mathcal{A}$,* $\mathsf{Adv}_{\Pi}^{\mathrm{ABAKE}}(\mathcal{A})$ *is negligible if* $\overline{\mathsf{sid}}^*$ *does not exist and the static secret key of the owner of* sid* *is given to the adversary.*

Moreover, we say that the ABAKE scheme is selectively secure in the ABCK model, if $\mathcal{A}$ specifies $\mathbb{A}_A$ *in* sid* *(and* $\mathbb{A}_B$ *in* $\overline{\mathsf{sid}}^*$ *if* $\overline{\mathsf{sid}}^*$ *exists) at the beginning of the security experiment.*

We remark that the item 2 of Def. 8 corresponds to wPFS and the item 3 corresponds to resistance to KCI. Both cases imply the standard CK security for ABAKE [8].

4 Round-Optimal SP-ABAKE without Random Oracles

In this section, we provide our two-party SP-ABAKE scheme that allows fine-grained access structure. Expressiveness of access structures is due to the direct application of LSSSs for the access control same as the Waters CP-ABE [14]. Our construction is parameterized by *att* which specifies the number of attributes in the system.

Waters CP-ABE. First, we review the protocol of the Waters CP-ABE as a warm-up for our main result.

Setup : For input a security parameter κ, choose p, G, G_T, g and g_T such that G and G_T are bilinear groups with pairing $e : G \times G \rightarrow G_T$ of order κ-bit prime p with generators g and $g_T = e(g,g)$, respectively. Then, output a master public key $MPK := (g, g^r, g_T^z, h_1, \ldots, h_{att})$ and a master secret key $MSK := g^z$ such that $r, z \in \mathbb{Z}_p$ and $h_1, \ldots, h_{att} \in G$ are randomly chosen.

Encrypt : For input the master public key MPK, a plaintext m and an LSSS access structure (M, ρ) where the function ρ associates rows of $\ell \times n$ share-generating matrix M to attributes, randomly choose $\mathbf{u} = (u_1, \ldots, u_n) \in \mathbb{Z}_p^n$ and $(x_1, \ldots, x_\ell) \in \mathbb{Z}_p^\ell$. For $i = 1$ to ℓ, find $\lambda_i = \mathbf{u} \cdot M_i$ where M_i is the vector corresponding to the ith row of M. Then, output the ciphertext $CT := (U', U, \{U\}, \{X\})$ such that $U' = m \cdot (g_T^z)^{u_1}$, $U = g^{u_1}$, $U_i = g^{r\lambda_i} h_{\rho(i)}^{-x_i}$ and $X_i = g^{x_i}$ for $1 \le i \le \ell$ (let $\{U\}$ denote the set of U_i and $\{X\}$ denote the set of X_i for $1 \le i \le \ell$).

KeyGen : For input the master secret key MSK and a set of attributes $\mathbb{S}$, choose $t \in \mathbb{Z}_p$, and compute $S' = g^z g^{rt}$, $T = g^t$ and $S_i = h_i^t$ for $i \in \mathbb{S}$ (let $\{S\}$ denote the set of S_k for $i \in \mathbb{S}$). Then, output a secret key $SK := (S', T, \{S\})$.

Decrypt : For input a ciphertext CT for the access structure (M, ρ) and a secret key SK for a set $\mathbb{S}$, let $I \subset \{1, 2, \ldots, \ell\}$ be defined as $I = \{i : \rho(i) \in \mathbb{S}\}$. We suppose that $\mathbb{S}$ satisfies M and ρ. Then, find $\{w_i \in \mathbb{Z}_p\}_{i \in I}$ such that $\sum_{i \in I} w_i \lambda_i = \alpha$ for valid shares $\{\lambda_i\}$ of any secret α according to M and output the plaintext m as

$$m = U' \cdot \left(\prod_{i \in I} e(U_i, T) e(X_i, S_{\rho(i)}) \right) / e(U, S').$$

Design Principle. The Waters CP-ABE can be regarded as an attribute-based KEM scheme. Specifically, a ciphertext contains $U' = mg_T^{u_1 z}$ and the decryption algorithm derives m by computing $g_T^{u_1 z}$ where m is a plaintext. Thus, we can regard $g_T^{u_1 z}$ as a KEM key. In our construction, parties exchange ciphertexts except U' with their specified access policies and share two KEM keys $g_T^{u_1 z}$ and $g_T^{v_1 z}$. The shared secret information $g_T^{u_1 z}$ and $g_T^{v_1 z}$ are used to derive the session key. However, only $g_T^{u_1 z}$ and $g_T^{v_1 z}$ are not enough to achieve the security in the ABCK model in Section 3. The ABCK model allows the adversary to reveal the master secret key in the case that the test session has the matching session, in order to capture wPFS. We cannot prove the security in such a case because the simulator cannot embed the instance of the DPBDHE problem to the master secret key. Thus, we must share g_T^{rxy} as an additional shared secret to create the session key in order to simulate such a case. We can embed $g^r = g^c$, $g^x = g^a$ and $g^y = g^{(a^q)}$ because r is not contained in the master secret key. Also, we must consider the case that the adversary poses Send query and StateReveal query to a session when the simulator embeds the instance of the DPBDHE problem to the static secret key. In this case, though the simulator must respond the shared secret $(g_T^{u_1 z}, g_T^{v_1 z}, g_T^{rxy})$ of the session key, it cannot be computed because the Waters CP-ABE is only CPA secure. Thus, we apply the generic conversion technique [19,20] to the CCA security from the CPA security with a one-time signature scheme (Gen, Sign, Ver). Specifically, we use a set $\mathbb{W}$ of *dummy attributes* as well as the set $\mathbb{S}$ of real attributes. $\mathbb{W}$ is associated to a verification key of the one-time signature and a ciphertext is specified with a real access policy and the dummy access policy corresponding to the verification key.

If the simulator does not know the static secret key corresponding to $\mathbb{S}$, he can correctly respond queries of the adversary with the static secret key corresponding to $\mathbb{W}$.

Our Construction.   The protocol of our scheme is as follows.

Setup : We set $\mathbb{W} = \{W_{1,0}, W_{1,1}, \ldots, W_{k,0}, W_{k,1}\}$ and $att' = att + 2k$. For input att' and a security parameter κ, choose p, G, G_T, g and g_T such that G and G_T are bilinear groups with pairing $e : G \times G \rightarrow G_T$ of order κ-bit prime p with generators g and $g_T = e(g,g)$, respectively. $F : \{0,1\}^* \times Kspace \rightarrow \{0,1\}^\kappa$ be a pseudo-random function, $Ext : G \rightarrow Kspace$ be a strong randomness extractor where $Kspace$ is the key space for F, and (Gen, Sign, Ver) be a one-time signature scheme where the bit length of a verification key is k. Then, randomly choose $h_1, \ldots, h_{att'} \in G$ and $r, z \in \mathbb{Z}_p$, and output a master public key $MPK := (g, g^r, g_T^z, h_1, \ldots, h_{att'})$ and a master secret key $MSK := g^z$.

KeyGen : For input a set of attributes $\mathbb{S}_P$ from a party P and MSK, randomly choose $t_P \in \mathbb{Z}_p$, and compute $S'_P = g^z g^{rt_P}$, $T_P = g^{t_P}$ and $S_{P_i} = h_i^{t_P}$ for $i \in \mathbb{S}_P$ (let $\{S_P\}$ denote the set of S_{P_i} for $i \in \mathbb{S}_P$). Then, output a static secret key $SK_P := (S'_P, T_P, \{S_P\})$.

Exchange : We suppose that the party A is the session initiator and the party B is the session responder. A has the static secret key $SK_A = (S'_A, T_A, \{S_A\})$ corresponding to the set of his attributes $\mathbb{S}_A$ and B has the static secret key $SK_B = (S'_B, T_B, \{S_B\})$ corresponding to the set of his attributes $\mathbb{S}_B$. Then, A sends to B the message mem_A corresponding to the access structure $\mathbb{A}_A$, and B sends to A the message mem_B corresponding to the access structure $\mathbb{A}_B$. Finally, both parties A and B compute the shared session key SK if and only if the set of attributes $\mathbb{S}_A$ satisfies the access structure $\mathbb{A}_B$ and the set of attributes $\mathbb{S}_B$ satisfies the access structure $\mathbb{A}_A$.

1. A decides an access structure $\mathbb{A}_A$ which he requires that the set of attributes $\mathbb{S}_B$ of B satisfies $\mathbb{A}_A$. Next, A runs $(vk_A, sk_A) \leftarrow \text{Gen}(1^\kappa)$ and sets a dummy attributes set $\mathbb{W}_A = \{W_{1,vk_{A1}}, \ldots, W_{k,vk_{Ak}}\} \subset \mathbb{W}$ where vk_{Ai} is the ith bit of vk_A. A derives the $\ell_A \times n_A$ share-generating matrix M_A and the labeling function ρ_A in an LSSS for the access structure $\mathbb{A}_A \vee (\wedge_{W \in \mathbb{W}_A} W)$. A randomly chooses $\mathbf{u} = (u_1, \ldots, u_{n_A}) \in \mathbb{Z}_p^{n_A}$ and $(x, x_1, \ldots, x_{\ell_A}) \in \mathbb{Z}_p^{\ell_A+1}$. For $i = 1$ to ℓ_A, A finds $\lambda_i = \mathbf{u} \cdot M_{Ai}$ where M_{Ai} is the vector corresponding to the ith row of M_A. Then, A computes $U = g^{u_1}$, $U_i = g^{r\lambda_i} h_{\rho_A(i)}^{-x_i}$, $X = g^x$ and $X_i = g^{x_i}$ for $1 \le i \le \ell_A$ (let $\{U\}$ denote the set of U_i and $\{X\}$ denote the set of X_i for $1 \le i \le \ell_A$). Also, A runs $s_A \leftarrow \text{Sign}_{sk_A}(U, \{U\}, X, \{X\})$. A sends $mem_A := (U, \{U\}, X, \{X\}, M_A, \rho_A, vk_A, s_A)$ to B.

2. B decides an access structure $\mathbb{A}_B$ which he requires that the set of attributes $\mathbb{S}_A$ of A satisfies $\mathbb{A}_B$. Next, B runs $(vk_B, sk_B) \leftarrow \text{Gen}(1^\kappa)$ and sets a dummy attributes set $\mathbb{W}_B = \{W_{1,vk_{B1}}, \ldots, W_{k,vk_{Bk}}\} \subset \mathbb{W}$ where vk_{Bi} is the ith bit of vk_B. B derives the $\ell_B \times n_B$ share-generating matrix M_B and the labeling function ρ_B in an LSSS for the access structure $\mathbb{A}_B \vee (\wedge_{W \in \mathbb{W}_B} W)$. B randomly chooses $\mathbf{v} = (v_1, \ldots, v_{n_B}) \in \mathbb{Z}_p^{n_B}$ and $(y, y_1, \ldots, y_{\ell_B}) \in \mathbb{Z}_p^{\ell_B+1}$. For $i = 1$ to ℓ_B, B finds $\lambda_i = \mathbf{v} \cdot M_{Bi}$ where M_{Bi} is the vector corresponding to the ith row of M_B. Then, B computes $V = g^{v_1}$, $V_i = g^{r\lambda_i} h_{\rho_B(i)}^{-y_i}$, $Y = g^y$ and $Y_i = g^{y_i}$ for $1 \le i \le \ell_B$ (let $\{V\}$ denote the set of V_i and $\{Y\}$ denote the set of Y_i for $1 \le i \le \ell_B$). Also, B runs $s_B \leftarrow \text{Sign}_{sk_B}(V, \{V\}, Y, \{Y\})$. B sends $mem_B := (V, \{V\}, Y, \{Y\}, M_B, \rho_B, vk_B, s_B)$ to A.

3. Upon receiving mem_B, A checks whether the set of his attributes $\mathbb{S}_A$ satisfies the access structure M_B and ρ_B. We suppose that $\mathbb{S}_A$ satisfies M_B and ρ_B, and let $I_A, I'_A \subset \{1, 2, \ldots, \ell_B\}$ be defined as $I_A = \{i : \rho_B(i) \in \mathbb{S}_A\}$ and $I'_A = \{i : \rho_B(i) \in \mathbb{W}_B\}$. A can efficiently find $\{w_{Ai} \in \mathbb{Z}_p\}_{i \in I_A}$ and $\{w'_{Ai} \in \mathbb{Z}_p\}_{i \in I'_A}$ such that $\sum_{i \in I_A} w_{Ai}\lambda_i = \sum_{i \in I'_A} w'_{Ai}\lambda_i = \alpha$ for valid shares $\{\lambda_i\}$ of any secret α according to M_B. Note that, if $\mathbb{S}_A$ does not satisfy M_B and ρ_B, A cannot find all w_{Ai} for $i \in I_A$ from the property of LSSSs. A verifies whether the following condition holds:

$$(V, \{V\}, Y, \{Y\} \in G) \wedge (1 \leftarrow \mathsf{Ver}_{vk_B}(V, \{V\}, Y, \{Y\}, s_B)) \wedge$$

$$\left(\prod_{i \in I_A}(e(V_i, g)e(Y_i, h_{\rho_B(i)}))^{w_{Ai}} = \prod_{i \in I'_A}(e(V_i, g)e(Y_i, h_{\rho_B(i)}))^{w'_{Ai}} = e(g^r, V) \right).$$

If not, A aborts. Otherwise, A computes the session key as follows: First, A sets shared information

$$\sigma_2 = e(V, S'_A) / \left(\prod_{i \in I_A}(e(V_i, T_A)e(Y_i, S_{A\rho_B(i)}))^{w_{Ai}} \right),$$

$$\sigma_1 = (g_T^z)^{u_1}, \quad \sigma_3 = e(g^r, Y)^x.$$

A computes $\sigma'_1 \leftarrow Ext(\sigma_1)$, $\sigma'_2 \leftarrow Ext(\sigma_2)$ and $\sigma'_3 \leftarrow Ext(\sigma_3)$, sets the session identity $\mathsf{sid} = (mem_A, mem_B)$ and the session key $SK = F_{\sigma'_1}(\mathsf{sid}) \oplus F_{\sigma'_2}(\mathsf{sid}) \oplus F_{\sigma'_3}(\mathsf{sid})$, completes the session and erases all temporary information other than SK.

4. Upon receiving mem_A, B checks whether the set of his attributes $\mathbb{S}_B$ satisfies the access structure M_A and ρ_A. We suppose that $\mathbb{S}_B$ satisfies M_A and ρ_A, and let $I_B, I'_B \subset \{1, 2, \ldots, \ell_A\}$ be defined as $I_B = \{i : \rho_A(i) \in \mathbb{S}_B\}$ and $I'_B = \{i : \rho_A(i) \in \mathbb{W}_A\}$. B can efficiently find $\{w_{Bi} \in \mathbb{Z}_p\}_{i \in I_B}$ and $\{w'_{Bi} \in \mathbb{Z}_p\}_{i \in I'_B}$ such that $\sum_{i \in I_B} w_{Bi}\lambda_i = \sum_{i \in I'_B} w'_{Bi}\lambda_i = \alpha$ for valid shares $\{\lambda_i\}$ of any secret α according to M_A. Note that, if $\mathbb{S}_B$ does not satisfy M_A and ρ_A, B cannot find all w_{Bi} for $i \in I_B$ from the property of LSSSs. B verifies whether the following condition holds:

$$(U, \{U\}, X, \{X\} \in G) \wedge (1 \leftarrow \mathsf{Ver}_{vk_A}(U, \{U\}, X, \{X\}, s_A)) \wedge$$

$$\left(\prod_{i \in I_B}(e(U_i, g)e(X_i, h_{\rho_A(i)}))^{w_{Bi}} = \prod_{i \in I'_B}(e(U_i, g)e(X_i, h_{\rho_A(i)}))^{w'_{Bi}} = e(g^r, U) \right).$$

If not, B aborts. Otherwise, B computes the session key as follows: First, B sets shared information

$$\sigma_1 = e(U, S'_B) / \left(\prod_{i \in I_B}(e(U_i, T_B)e(X_i, S_{B\rho_A(i)}))^{w_{Bi}} \right),$$

$$\sigma_2 = (g_T^z)^{v_1}, \quad \sigma_3 = e(g^r, X)^y.$$

B computes $\sigma'_1 \leftarrow Ext(\sigma_1)$, $\sigma'_2 \leftarrow Ext(\sigma_2)$ and $\sigma'_3 \leftarrow Ext(\sigma_3)$, sets the session identity $\mathsf{sid} = (mem_A, mem_B)$ and the session key $SK = F_{\sigma'_1}(\mathsf{sid}) \oplus F_{\sigma'_2}(\mathsf{sid}) \oplus F_{\sigma'_3}(\mathsf{sid})$, completes the session and erases all temporary information other than SK.

Correctness. Seeds of the session key that both parties compute are

$$\sigma_1 = e(U, S'_B)/\left(\prod_{i\in I_B}(e(U_i, T_B)e(X_i, S_{B\rho_A(i)}))^{w_{Bi}}\right)$$

$$= e(g^{u_1}, g^z g^{rt_B})/\left(\prod_{i\in I_B}(e(g^{r\lambda_i}h_{\rho_A(i)}^{-x_i}, g^{t_B})e(g^{x_i}, h_{\rho_A(i)}^{t_B}))^{w_{Bi}}\right)$$

$$= g_T^{u_1(z+rt_B)}/\prod_{i\in I_B}g_T^{rt_B\lambda_i w_{Bi}}$$

$$= g_T^{u_1(z+rt_B)}/g_T^{rt_B u_1}$$

$$= g_T^{u_1 z}(= (g_T^z)^{u_1}),$$

$$\sigma_2 = e(V, S'_A)/\left(\prod_{i\in I_A}(e(V_i, T_A)e(Y_i, S_{A\rho_B(i)}))^{w_{Ai}}\right)$$

$$= e(g^{v_1}, g^z g^{rt_A})/\left(\prod_{i\in I_A}(e(g^{r\lambda_i}h_{\rho_B(i)}^{-y_i}, g^{t_A})e(g^{y_i}, h_{\rho_B(i)}^{t_A}))^{w_{Ai}}\right)$$

$$= g_T^{v_1(z+rt_A)}/\prod_{i\in I_A}g_T^{rt_A\lambda_i w_{Ai}}$$

$$= g_T^{v_1(z+rt_A)}/g_T^{rt_A v_1}$$

$$= g_T^{v_1 z}(= (g_T^z)^{v_1}),$$

$$\sigma_3 = e(g^r, X)^y = g_T^{rxy} = e(g^r, Y)^x,$$

and therefore they can compute the same session key SK.

5 Security

We prove that our SP-ABAKE scheme is secure in the ABCK model. Since the underlying ABE scheme just satisfies selective security, our scheme also satisfies selective security.

Theorem 1. *Suppose that the DPBDHE assumption holds, F is a pseudo-random function, Ext is a strong randomness extractor, and* (Gen, Sign, Ver) *is an existentially unforgeable one-time signature scheme against chosen message attacks. Then, our scheme is selectively secure in the ABCK model.*

The proof of Theorem 1 is shown in Appendix A.

6 Concluding Remark

Our SP-ABAKE scheme is superior to the BS scheme with [13] except security model (i.e., selective vs. full). The selective security of our scheme comes from the underlying Waters CP-ABE. We may be able to achieve full security if we construct a scheme based on a fully secure CP-ABE scheme. For example, the CP-ABE scheme by Okamoto-Takashima [30] achieves full security in the StdM. However, communication complexity will heavily increase. Thus, a remaining problem of future researches is to achieve full security in the StdM with small communication complexity.

References

1. Wang, H., Xu, Q., Ban, T.: A Provably Secure Two-Party Attribute-Based Key Agreement Protocol. In: IIH-MSP 2009, pp. 1042–1045 (2009)
2. Wang, H., Xu, Q., Fu, X.: Revocable Attribute-based Key Agreement Protocol without Random Oracles. JNW 4(8), 787–794 (2009)

3. Wang, H., Xu, Q., Fu, X.: Two-Party Attribute-based Key Agreement Protocol in the Standard Model. In: ISIP 2009, pp. 325–328 (2009)
4. Gorantla, M.C., Boyd, C., González Nieto, J.M.: Attribute-Based Authenticated Key Exchange. In: Steinfeld, R., Hawkes, P. (eds.) ACISP 2010. LNCS, vol. 6168, pp. 300–317. Springer, Heidelberg (2010)
5. Steinwandt, R., Suárez Corona, A.: Attribute-based group key establishment. Advances in Mathematics of Communications 4(3), 381–398 (2010)
6. Fujioka, A., Suzuki, K., Yoneyama, K.: Predicate-Based Authenticated Key Exchange Resilient to Ephemeral Key Leakage. In: Chung, Y., Yung, M. (eds.) WISA 2010. LNCS, vol. 6513, pp. 15–30. Springer, Heidelberg (2011)
7. Yoneyama, K.: Strongly Secure Two-Pass Attribute-Based Authenticated Key Exchange. In: Joye, M., Miyaji, A., Otsuka, A. (eds.) Pairing 2010. LNCS, vol. 6487, pp. 147–166. Springer, Heidelberg (2010)
8. Birkett, J., Stebila, D.: Predicate-Based Key Exchange. In: Steinfeld, R., Hawkes, P. (eds.) ACISP 2010. LNCS, vol. 6168, pp. 282–299. Springer, Heidelberg (2010)
9. Canetti, R., Krawczyk, H.: Analysis of Key-Exchange Protocols and Their Use for Building Secure Channels. In: Pfitzmann, B. (ed.) EUROCRYPT 2001. LNCS, vol. 2045, pp. 453–474. Springer, Heidelberg (2001)
10. Shahandashti, S.F., Safavi-Naini, R.: Threshold Attribute-Based Signatures and Their Application to Anonymous Credential Systems. In: Preneel, B. (ed.) AFRICACRYPT 2009. LNCS, vol. 5580, pp. 198–216. Springer, Heidelberg (2009)
11. Li, J., Au, M.H., Susilo, W., Xie, D., Ren, K.: Attribute-based signature and its applications. In: ASIACCS 2010, pp. 60–69 (2010)
12. Maji, H.K., Prabhakaran, M., Rosulek, M.: Attribute-Based Signatures. In: Kiayias, A. (ed.) CT-RSA 2011. LNCS, vol. 6558, pp. 376–392. Springer, Heidelberg (2011)
13. Okamoto, T., Takashima, K.: Efficient Attribute-Based Signatures for Non-monotone Predicates in the Standard Model. In: Catalano, D., Fazio, N., Gennaro, R., Nicolosi, A. (eds.) PKC 2011. LNCS, vol. 6571, pp. 35–52. Springer, Heidelberg (2011)
14. Waters, B.: Ciphertext-Policy Attribute-Based Encryption: An Expressive, Efficient, and Provably Secure Realization. In: Catalano, D., Fazio, N., Gennaro, R., Nicolosi, A. (eds.) PKC 2011. LNCS, vol. 6571, pp. 53–70. Springer, Heidelberg (2011)
15. Boneh, D., Boyen, X.: Short Signatures Without Random Oracles. In: Cachin, C., Camenisch, J.L. (eds.) EUROCRYPT 2004. LNCS, vol. 3027, pp. 56–73. Springer, Heidelberg (2004)
16. Mohassel, P.: One-Time Signatures and Chameleon Hash Functions. In: Biryukov, A., Gong, G., Stinson, D.R. (eds.) SAC 2010. LNCS, vol. 6544, pp. 302–319. Springer, Heidelberg (2011)
17. Boyd, C., Cliff, Y., Gonzalez Nieto, J.M., Paterson, K.G.: Efficient One-Round Key Exchange in the Standard Model. In: Mu, Y., Susilo, W., Seberry, J. (eds.) ACISP 2008. LNCS, vol. 5107, pp. 69–83. Springer, Heidelberg (2008)
18. Boyd, C., Cliff, Y., González Nieto, J.M., Paterson, K.G.: One-round key exchange in the standard model. IJACT 1(3), 181–199 (2009)
19. Boneh, D., Canetti, R., Halevi, S., Katz, J.: Chosen-Ciphertext Security from Identity-Based Encryption. SIAM J. Comput. 36(5), 1301–1328 (2007)
20. Yamada, S., Attrapadung, N., Hanaoka, G., Kunihiro, N.: Generic Constructions for Chosen-Ciphertext Secure Attribute Based Encryption. In: Catalano, D., Fazio, N., Gennaro, R., Nicolosi, A. (eds.) PKC 2011. LNCS, vol. 6571, pp. 71–89. Springer, Heidelberg (2011)
21. Sahai, A., Waters, B.: Fuzzy Identity-Based Encryption. In: Cramer, R. (ed.) EUROCRYPT 2005. LNCS, vol. 3494, pp. 457–473. Springer, Heidelberg (2005)
22. Bethencourt, J., Sahai, A., Waters, B.: Ciphertext-Policy Attribute-Based Encryption. In: IEEE Symposium on Security and Privacy 2007, pp. 321–334 (2007)

23. Cheung, L., Newport, C.C.: Provably secure ciphertext policy ABE. In: ACM Conference on Computer and Communications Security, pp. 456–465 (2007)
24. Kapadia, A., Tsang, P.P., Smith, S.W.: Attribute-Based Publishing with Hidden Credentials and Hidden Policies. In: NDSS 2007, pp. 179–192 (2007)
25. Nishide, T., Yoneyama, K., Ohta, K.: Attribute-Based Encryption with Partially Hidden Encryptor-Specified Access Structures. In: Bellovin, S.M., Gennaro, R., Keromytis, A.D., Yung, M. (eds.) ACNS 2008. LNCS, vol. 5037, pp. 111–129. Springer, Heidelberg (2008)
26. Shi, E., Bethencourt, J., Chan, H.T.-H., Song, D.X., Perrig, A.: Multi-Dimensional Range Query over Encrypted Data. In: IEEE Symposium on Security and Privacy 2007, pp. 350–364 (2007)
27. Boneh, D., Waters, B.: Conjunctive, Subset, and Range Queries on Encrypted Data. In: Vadhan, S.P. (ed.) TCC 2007. LNCS, vol. 4392, pp. 535–554. Springer, Heidelberg (2007)
28. Katz, J., Sahai, A., Waters, B.: Predicate Encryption Supporting Disjunctions, Polynomial Equations, and Inner Products. In: Smart, N.P. (ed.) EUROCRYPT 2008. LNCS, vol. 4965, pp. 146–162. Springer, Heidelberg (2008)
29. Lewko, A., Okamoto, T., Sahai, A., Takashima, K., Waters, B.: Fully Secure Functional Encryption: Attribute-Based Encryption and (Hierarchical) Inner Product Encryption. In: Gilbert, H. (ed.) EUROCRYPT 2010. LNCS, vol. 6110, pp. 62–91. Springer, Heidelberg (2010)
30. Okamoto, T., Takashima, K.: Fully Secure Functional Encryption with General Relations from the Decisional Linear Assumption. In: Rabin, T. (ed.) CRYPTO 2010. LNCS, vol. 6223, pp. 191–208. Springer, Heidelberg (2010)
31. Beimel, A.: Secure Schemes for Secret Sharing and Key Distribution. PhD thesis, Israel Institute of Technology, Technion (1996)

A   Proof of Theorem 1

In the experiment of the ABCK model, we suppose that sid^* is the session identity for the test session, and that there are N users and at most L sessions are activated. Let κ be the security parameter, and let $\mathcal{A}$ be a PPT (in κ) bounded adversary. Suc denotes the event that $\mathcal{A}$ wins. We consider the following events that cover all cases of the behavior of $\mathcal{A}$.

- Let E_1 be the event that the test session sid^* has matching session $\overline{sid}^*$, and the master secret key is given to $\mathcal{A}$.
- Let E_2 be the event that the test session sid^* has no matching session $\overline{sid}^*$, and the static secret key of the owner of sid^* is given to $\mathcal{A}$.

To finish the proof, we investigate events $E_i \wedge Suc$ ($i = 1, 2$) that cover all cases of event Suc.

Proposition 1. $\Pr[E_1 \wedge Suc]$ *is negligible.*

Proof. We change the interface of oracle queries and the computation of the session key. These instances are gradually changed over six hybrid experiments, depending on specific sub-cases. In the last hybrid experiment, the session key in the test session does not contain information of the bit b. Thus, the adversary clearly only output a random guess. We denote these hybrid experiments by $\mathbf{H}_0, \ldots, \mathbf{H}_5$ and the advantage of the adversary $\mathcal{A}$ when participating in experiment $\mathbf{H}_i$ by $\mathsf{Adv}(\mathcal{A}, \mathbf{H}_i)$.

Hybrid Experiment H_0: This experiment denotes the real experiment for the ABCK model and in this experiment the environment for $\mathcal{A}$ is as defined in the protocol. Thus, $\mathsf{Adv}(\mathcal{A}, H_0)$ is the same as the advantage of the real experiment.

Hybrid Experiment H_1: In this experiment, if session identities in two sessions are identical, the experiment halts.

When $(U, \{U\}, X, \{X\}, vk_A, s_A, V, \{V\}, Y, \{Y\}, vk_B, s_B)$ in two sessions (i.e., randomness in two sessions are independently and uniformly chosen) are identical and two access policies are identical, session identities in two sessions are also identical. Such an event occurs with the probability that collision of randomness occur, that is negligible. Thus, $|\mathsf{Adv}(\mathcal{A}, H_1) - \mathsf{Adv}(\mathcal{A}, H_0)| \leq negl$.

Hybrid Experiment H_2: In this experiment, the experiment selects a party A intending that the peer is a party B and integer $i \in [1, L]$ randomly in advance. If $\mathcal{A}$ poses Test query to a session except i-th session of A with intended peer B, the experiment halts.

Since guess of the test session matches with $\mathcal{A}$'s choice with probability $1/N^2 L$, $\mathsf{Adv}(\mathcal{A}, H_2) \geq 1/N^2 L \cdot \mathsf{Adv}(\mathcal{A}, H_1)$.

Hybrid Experiment H_3: In this experiment, the computation of σ_3^* in the test session is changed. Instead of computing $\sigma_3^* = g_T^{x^* y^* r}$, it is changed as choosing $\sigma_3^* \leftarrow G_T$ randomly, where we suppose that B is the intended partner of A in the test session.

We construct a DPBDHE solver S from $\mathcal{A}$ in H_2 or H_3; that is, if $|\mathsf{Adv}(\mathcal{A}, H_3) - \mathsf{Adv}(\mathcal{A}, H_2)|$ is non-negligible, we can construct successful S. The simulation in $E_1 \wedge Suc$ is very simple; S can perfectly respond to all queries of $\mathcal{A}$ as the same way as the protocol (except the test session) because S can know the master secret key and all static secret keys. S performs the following steps.

Init. The DPBDHE solver S receives a DPBDHE tuple $(\mathsf{m}, \mathcal{T})$ as a challenge. Also, S receives $\mathbb{A}_A^*$ and $\mathbb{A}_B^*$ as a challenge access structure from $\mathcal{A}$.

Setup. S sets $\mathbb{W} = \{W_{1,0}, W_{1,1}, \ldots, W_{k,0}, W_{k,1}\}$ and $att' = att + 2k$, and embeds $g^r = g^c$. S randomly choose $h_1, \ldots, h_{att'} \in G$ and $z \in \mathbb{Z}_p$, and output a master public key $MPK := (g, g^r, g_T^z, h_1, \ldots, h_{att'})$ and a master secret key $MSK := g^z$. MSK is given to $\mathcal{A}$. S generates all static secret keys of N parties with g^z.

S sets the messages mem_A^* and mem_B^* of i_Ath session of A as follows: First, S embeds $X^* = g^a$ and $Y^* = g^{(a^q)}$, runs $(vk_A^*, sk_A^*) \leftarrow \mathsf{Gen}(1^k)$ and $(vk_B^*, sk_B^*) \leftarrow \mathsf{Gen}(1^k)$, and sets dummy attributes sets $\mathbb{W}_A^* = \{W_{1, vk_{A1}^*}, \ldots, W_{k, vk_{Ak}^*}\} \subset \mathbb{W}$ and $\mathbb{W}_B^* = \{W_{1, vk_{B1}^*}, \ldots, W_{k, vk_{Bk}^*}\} \subset \mathbb{W}$. S derives the $\ell_A^* \times n_A^*$ share-generating matrix M_A^* and the labeling function ρ_A^* in an LSSS for the access structure $\mathbb{A}_A^* \vee (\wedge_{W \in \mathbb{W}_A^*} W)$, and the $\ell_B^* \times n_B^*$ share-generating matrix M_B^* and the labeling function ρ_B^* in an LSSS for the access structure $\mathbb{A}_B^* \vee (\wedge_{W \in \mathbb{W}_B^*} W)$. S randomly chooses $\mathbf{u}^* = (u_1^*, \ldots, u_{n_A^*}^*) \in \mathbb{Z}_p^{n_A^*}$, $\mathbf{v}^* = (v_1^*, \ldots, v_{n_B^*}^*) \in \mathbb{Z}_p^{n_B^*}$, $(x_1^*, \ldots, x_{\ell_A^*}^*) \in \mathbb{Z}_p^{\ell_A^*}$ and $(y_1^*, \ldots, y_{\ell_B^*}^*) \in \mathbb{Z}_p^{\ell_B^*}$. For $i = 1$ to ℓ_A^*, S finds $\lambda_{Ai} = \mathbf{u}^* \cdot M_{Ai}^*$ where M_{Ai}^* is the vector corresponding to the ith row of M_A^*. For $i = 1$ to ℓ_B^*, S finds $\lambda_{Bi} = \mathbf{v}^* \cdot M_{Bi}^*$ where M_{Bi}^* is the vector corresponding to the ith row of M_B^*. Then, S computes $U^* = g^{u_1^*}$, $U_i^* = g^{r\lambda_{Ai}} h_{\rho_A^*(i)}^{-x_i^*}$ and $X_i^* = g^{x_i^*}$ for $1 \leq i \leq \ell_A^*$ (let $\{U^*\}$ denote the set of U_i^* and $\{X^*\}$

denote the set of X_i^* for $1 \le i \le \ell_A^*$). Also, S computes $V^* = g^{v_1^*}$, $V_i^* = g^{r\lambda_{Bi}} h_{\rho_B^*(i)}^{-y_i^*}$ and $Y_i^* = g^{y_i^*}$ for $1 \le i \le \ell_B^*$ (let $\{V^*\}$ denote the set of V_i^* and $\{Y^*\}$ denote the set of Y_i^* for $1 \le i \le \ell_B^*$). S runs $s_A^* \leftarrow \mathsf{Sign}_{sk_A^*}(U^*, \{U\}^*, X^*, \{X\}^*)$ and $s_B^* \leftarrow \mathsf{Sign}_{sk_B^*}(V^*, \{V\}^*, Y^*, \{Y\}^*)$. Finally, S sets the messages $mem_A^* = (U^*, \{U^*\}, X^*, \{X^*\}, M_A^*, \rho_A^*, vk_A^*, s_A^*)$ and $mem_B^* = (V^*, \{V^*\}, Y^*, \{Y^*\}, M_B^*, \rho_B^*, vk_B^*, s_B^*)$ for i_Ath session of A.

Simulation. S simulates oracle queries by $\mathcal{A}$ as follows. S maintains the list $\mathcal{L}_K$ that contains queries and answers of KeyReveal.

1. Send($I, \mathbb{S}_P, \mathbb{S}_{\bar{P}}$): If $P = A$ and the session is i_A-th session of A, S returns the message mem_A^* computed in the setup. Otherwise, S computes mem_P obeying the protocol, returns it and records $(\mathbb{S}_P, \mathbb{S}_{\bar{P}}, mem_P)$.
2. Send($R, \mathbb{S}_{\bar{P}}, \mathbb{S}_P, mem_P$) and Send($I, \mathbb{S}_{\bar{P}}, \mathbb{S}_P, mem_P$): S computes the message $mem_{\bar{P}}$ and the session key SK obeying the protocol, returns it, and records all session state and $(\mathbb{S}_P, \mathbb{S}_{\bar{P}}, mem_P, mem_{\bar{P}})$ as the completed session and SK in $\mathcal{L}_K$.
3. KeyReveal(sid):
   (a) If the session sid is not completed, S returns an error message.
   (b) Otherwise, S returns the recorded value $SK \in \mathcal{L}_K$ for sid.
4. StateReveal(sid):
   (a) If the session sid is completed, S returns an error message.
   (b) Otherwise, S returns all recorded state of sid.
5. Corrupt(P): S returns SK_P generated in the setup.
6. Test(sid*): S computes $\sigma_1^* = g_T^{u_1^* z}$ and $\sigma_2^* = g_T^{v_1^* z}$, and embeds $\sigma_3^* = \mathcal{T}$. S returns SK^* computed from σ_1^*, σ_2^* and σ_3^*.
7. If $\mathcal{A}$ outputs a guess b', S outputs b'.

Analysis. The simulation for S is same as $\mathbf{H}_2$ for $\mathcal{A}$ when $\mathcal{T} = g_T^{a^{q+1}c}$ and same as $\mathbf{H}_3$ for $\mathcal{A}$ when $\mathcal{T} = R$. Hence, if $|\mathsf{Adv}(\mathcal{A}, \mathbf{H}_3) - \mathsf{Adv}(\mathcal{A}, \mathbf{H}_2)|$ is non-negligible, S is successful with negligible probability.

Therefore, $|\mathsf{Adv}(\mathcal{A}, \mathbf{H}_3) - \mathsf{Adv}(\mathcal{A}, \mathbf{H}_2)| \le negl$ from the DPBDHE assumption.

Hybrid Experiment $\mathbf{H}_4$: In this experiment, the computation of $\sigma_3'^*$ in the test session is changed. Instead of computing $\sigma_3'^* \leftarrow Ext(\sigma_3^*)$, it is changed as choosing $\sigma_3'^* \in Kspace$ randomly.

Since σ_3^* is randomly chosen in $\mathbf{H}_3$, it has sufficient min-entropy. Thus, by the definition of the strong randomness extractor, $|\mathsf{Adv}(\mathcal{A}, \mathbf{H}_4) - \mathsf{Adv}(\mathcal{A}, \mathbf{H}_3)| \le negl$.

Hybrid Experiment $\mathbf{H}_5$: In this experiment, the computation of SK^* in the test session is changed. Instead of computing $SK^* = F_{\sigma_1'^*}(\mathsf{sid}^*) \oplus F_{\sigma_2'^*}(\mathsf{sid}^*) \oplus F_{\sigma_3'^*}(\mathsf{sid}^*)$, it is changed as $SK^* = F_{\sigma_1'^*}(\mathsf{sid}^*) \oplus F_{\sigma_2'^*}(\mathsf{sid}^*) \oplus \alpha$ where $\alpha \in \{0, 1\}^k$ is chosen randomly and we suppose that B is the intended partner of A in the test session.

We construct a distinguisher $\mathcal{D}$ between PRF $F : \{0, 1\}^* \times Kspace \to \{0, 1\}^k$ and a random function RF from $\mathcal{A}$ in $\mathbf{H}_4$ or $\mathbf{H}_5$; that is, if $|\mathsf{Adv}(\mathcal{A}, \mathbf{H}_5) - \mathsf{Adv}(\mathcal{A}, \mathbf{H}_4)|$ is non-negligible, we can construct successful $\mathcal{D}$. $\mathcal{D}$ performs the following steps.

Setup. $\mathcal{D}$ sets the master secret key and all N parties' static secret keys.

Simulation. $\mathcal{D}$ simulates oracle queries by $\mathcal{A}$ as follows. $\mathcal{D}$ maintains the list $\mathcal{L}_K$ that contains queries and answers of **KeyReveal**.

1. **Send**$(I, \mathbb{S}_P, \mathbb{S}_{\bar{P}})$: $\mathcal{D}$ computes mem_P obeying the protocol, returns it and records $(\mathbb{S}_P, \mathbb{S}_{\bar{P}}, mem_P)$.
2. **Send**$(\mathcal{R}, \mathbb{S}_{\bar{P}}, \mathbb{S}_P, mem_P)$ and **Send**$(I, \mathbb{S}_{\bar{P}}, \mathbb{S}_P, mem_P)$: $\mathcal{D}$ computes the message $mem_{\bar{P}}$ and the session key SK obeying the protocol, returns it, and records all session state and $(\mathbb{S}_P, \mathbb{S}_{\bar{P}}, mem_P, mem_{\bar{P}})$ as the completed session and SK in $\mathcal{L}_K$.
3. **KeyReveal**(sid):
   (a) If the session sid is not completed, $\mathcal{D}$ returns an error message.
   (b) Otherwise, $\mathcal{D}$ returns the recorded value $SK \in \mathcal{L}_K$ for sid.
4. **StateReveal**(sid):
   (a) If the session sid is completed, $\mathcal{D}$ returns an error message.
   (b) Otherwise, $\mathcal{D}$ returns all recorded state of sid.
5. **Corrupt**(P): $\mathcal{D}$ returns SK_P generated in the setup.
6. **Test**(sid*): $\mathcal{D}$ poses sid* to the function F^* (F^* is the PRF F or a random function RF) and obtains $\alpha = F^*(\text{sid}^*)$. $\mathcal{D}$ returns $SK^* = F_{\sigma_1''}(\text{sid}^*) \oplus F_{\sigma_2''}(\text{sid}^*) \oplus \alpha$.
7. If $\mathcal{A}$ outputs a guess b', $\mathcal{D}$ outputs b'.

Analysis. The simulation for $\mathcal{D}$ is same as $\mathbf{H}_4$ for $\mathcal{A}$ when F^* is F with a random key because $\sigma_3'^*$ is randomly chosen in $\mathbf{H}_4$. Also, the simulation for $\mathcal{D}$ is same as $\mathbf{H}_5$ for $\mathcal{A}$ when F^* is RF. Hence, if $|\mathsf{Adv}(\mathcal{A}, \mathbf{H}_5) - \mathsf{Adv}(\mathcal{A}, \mathbf{H}_4)|$ is non-negligible, $\mathcal{D}$ is successful with negligible probability.

Therefore, $|\mathsf{Adv}(\mathcal{A}, \mathbf{H}_5) - \mathsf{Adv}(\mathcal{A}, \mathbf{H}_4)| \leq negl$.

In $\mathbf{H}_5$, the session key in the test session is perfectly randomized. Thus, $\mathcal{A}$ cannot obtain any advantage from **Test** query.

Therefore, $\mathsf{Adv}(\mathcal{A}, \mathbf{H}_5) = 0$ and $\Pr[E_1 \wedge Suc]$ is negligible. $\square$

Proposition 2. $\Pr[E_2 \wedge Suc]$ *is negligible.*

Proof. We change the interface of oracle queries and the computation of the session key. These instances are gradually changed over six hybrid experiments, depending on specific sub-cases. In the last hybrid experiment, the session key in the test session does not contain information of the bit b. Thus, the adversary clearly only output a random guess. We denote these hybrid experiments by $\mathbf{H}_0, \ldots, \mathbf{H}_5$ and the advantage of the adversary $\mathcal{A}$ when participating in experiment $\mathbf{H}_i$ by $\mathsf{Adv}(\mathcal{A}, \mathbf{H}_i)$.

Hybrid Experiment $\mathbf{H}_0$: This experiment denotes the real experiment for the ABCK model and in this experiment the environment for $\mathcal{A}$ is as defined in the protocol. Thus, $\mathsf{Adv}(\mathcal{A}, \mathbf{H}_0)$ is the same as the advantage of the real experiment.

Hybrid Experiment $\mathbf{H}_1$: In this experiment, if session identities in two sessions are identical, the experiment halts.

When $(U, \{U\}, X, \{X\}, vk_A, s_A, V, \{V\}, Y, \{Y\}, vk_B, s_B)$ in two sessions (i.e., randomness in two sessions are independently and uniformly chosen) are identical and two access policies are identical, session identities in two sessions are also identical. Such an event occurs with the probability that collision of randomness occur, that is negligible. Thus, $|\mathsf{Adv}(\mathcal{A}, \mathbf{H}_1) - \mathsf{Adv}(\mathcal{A}, \mathbf{H}_0)| \leq negl$.

Hybrid Experiment H_2: In this experiment, the experiment selects a party A and integer $i \in [1, L]$ randomly in advance. If $\mathcal{A}$ poses Test query to a session except i-th session of A, the experiment halts.

Since guess of the test session matches with $\mathcal{A}$'s choice with probability $1/N^2L$, $\mathsf{Adv}(\mathcal{A}, \mathbf{H}_2) \geq 1/N^2L \cdot \mathsf{Adv}(\mathcal{A}, \mathbf{H}_1)$.

Hybrid Experiment H_3: In this experiment, the computation of σ_1^* in the test session is changed. Instead of computing $\sigma_1^* = g_T^{u_1^* z}$, it is changed as choosing $\sigma_1^* \leftarrow G_T$ randomly, where we suppose that B is the intended partner of A in the test session.

We construct a DPBDHE solver S from $\mathcal{A}$ in $\mathbf{H}_2$ or $\mathbf{H}_3$; that is, if $|\mathsf{Adv}(\mathcal{A}, \mathbf{H}_3) - \mathsf{Adv}(\mathcal{A}, \mathbf{H}_2)|$ is non-negligible, we can construct successful S. The simulation in $E_2 \wedge Suc$ is quite complicated; S should respond to all queries of $\mathcal{A}$ without knowing the master secret key. S performs the following steps.

Init. The DPBDHE solver S receives a DPBDHE tuple $(\mathsf{m}, \mathcal{T})$ as a challenge. Also, S receives $\mathbb{A}_A^*$ as a challenge access structure from $\mathcal{A}$.

Setup. First, S must determine the $\ell_A^* \times n_A^*$ share-generating matrix M_A^* and the labeling function ρ_A^* for i-th session of A where $\ell_A^*, n_A^* \leq q$. S sets $\mathbb{W} = \{W_{1,0}, W_{1,1}, \ldots, W_{k,0}, W_{k,1}\}$ and $att' = att + 2k$, runs $(vk_A^*, sk_A^*) \leftarrow \mathsf{Gen}(1^\kappa)$, and sets dummy attributes sets $\mathbb{W}_A^* = \{W_{1,vk_{A1}^*}, \ldots, W_{k,vk_{Ak}^*}\} \subset \mathbb{W}$. S derives the $\ell_A^* \times n_A^*$ share-generating matrix M_A^* and the labeling function ρ_A^* in an LSSS for the access structure $\mathbb{A}_A^* \vee (\wedge_{W \in \mathbb{W}_A^*} W)$. Then, S randomly chooses $(z', z_1, \ldots, z_{att'}) \in \mathbb{Z}_p^{att'+1}$ and embeds $g_T^z = e(g^a, g^{a^q})g_T^{z'}$ and $g^r = g^a$; that is, z is implicitly set as $a^{q+1} + z'$. Let I denote the set of indices i such that $\rho_A^*(i) = j$ for $1 \leq j \leq att'$. S programs $h_j = g^{z_j} \prod_{i \in I} g^{aM_{A(i,1)}^*/b_i} \cdots g^{a^{n_A^*}M_{A(i,n_A^*)}^*/b_i}$ for $1 \leq j \leq att'$. S outputs the master public key $MPK := (g, g^r, g_T^z, h_1, \ldots, h_{att'})$.

Next, S generates all static secret keys of N parties. We suppose that for a party P according to the set of attributes $\mathbb{S}_P$ that does not satisfy M_A^*,[1] S randomly chooses r_P. S finds a vector $\mathbf{w} = (w_1, \ldots, w_{n_A^*}) \in \mathbb{Z}_p^{n_A^*}$ such that $w_1 = -1$ and $\mathbf{w} \cdot M_{Ai}^* = 0$ for all i where $\rho_A^*(i) \in \mathbb{S}_P$. S embeds $T_P = g^{r_P} \prod_{i=1,\ldots,n_A^*} (g^{a^{q+1-i}})^{w_i}$; that is, t_P is implicitly set as $r_P + w_1 a^q + \cdots + w_{n_A^*} a^{q-n_A^*+1}$. Then, S can compute $S_P' = g^{z'} g^{ar_P} \prod_{i=2,\ldots,n_A^*} (g^{a^{q+2-i}})^{w_i}$ without knowing g^z because g^{at_P} cancels out the unknown term in g^z. Also, S can compute $S_{Pi} = T_P^{z_i} \prod_{j \in J} \prod_{l=1,\ldots,n_A^*} \left(g^{(a^l/b_j)r_P} \prod_{k=1,\ldots,n_A^*(k \neq l)} (g^{a^{q+1+l-k/b_j}})^{w_k}\right)^{M_{A(j,l)}^*}$ without knowing g^{a^{q+1}/b_j} because $\mathbf{w} \cdot M_{Ai}^* = 0$ and all these terms are canceled out. The static secret key $(S_A', T_A, \{S_A\})$ of A is given to $\mathcal{A}$.

Finally, S sets the message mem_A^* of i_Ath session of A as follows: S embeds $U^* = g^c$ and randomly chooses $u_2^*, \ldots, u_{n_A^*}^*, x^*, x_1^*, \ldots, x_{\ell_A^*}^* \in \mathbb{Z}_p$. S sets $\{U^*\}$ and $\{X^*\}$ such that $\mathbf{u}^*$ is implicitly defined as $(c, ca + u_2^*, \ldots, ca^{n_A^*-1} + u_{n_A^*}^*)$. For $i = 1, \ldots, n_A^*$, let J_i be a set of all $j(\neq i)$ such that $\rho_A^*(i) = \rho_A^*(j)$; that is, J_i is a set of all other row indices that have the same attributes as row i. S embeds $U_i^* = h_{\rho_A^*(i)}^{x_i^*} \left(\prod_{j=2,\ldots,n_A^*} (g^a)^{M_{A(i,j)}^* u_j^*}\right) (g^{b_i c})^{-z_{\rho_A^*(i)}}$

---

[1] S does not need to create static secret keys of parties according to sets of attributes that satisfy M_A^*. $\mathcal{A}$ cannot pose Corrupt queries for $\mathbb{S}_P$ that satisfies M_A^* because of the freshness in Def. 7.

$\cdot \left(\prod_{k \in J_i} \prod_{j=1,\dots,n_A^*} (g^{a^j c(b_i/b_k)})^{M_{A(k,j)}^*} \right)$ and $X_i^* = g^{-x_i^*} g^{-cb_i}$. Also, $\mathcal{S}$ computes $X^* = g^{x^*}$ and runs $s_A^* \leftarrow \mathsf{Sign}_{sk_A^*}(U^*, \{U\}^*, X^*, \{X\}^*)$. $\mathcal{S}$ sets the message $mem_A^* = (U^*, \{U^*\}, X^*, \{X^*\}, M_A^*, \rho_A^*, vk_A^*, s_A^*)$.

Simulation. $\mathcal{S}$ simulates oracle queries by $\mathcal{A}$ as follows. $\mathcal{S}$ maintains the list $\mathcal{L}_K$ that contains queries and answers of **KeyReveal**.

1. **Send**$(\mathcal{I}, \mathbb{S}_P, \mathbb{S}_{\bar{P}})$: If $P = A$ and the session is i_A-th session of A, $\mathcal{S}$ returns the message mem_A^* computed in the setup. Otherwise, $\mathcal{S}$ computes mem_P obeying the protocol, returns it, and records all session state and $(\mathbb{S}_P, \mathbb{S}_{\bar{P}}, mem_P)$.

2. **Send**$(\mathcal{R}, \mathbb{S}_{\bar{P}}, \mathbb{S}_P, mem_P)$ and **Send**$(\mathcal{I}, \mathbb{S}_{\bar{P}}, \mathbb{S}_P, mem_P)$: $\mathcal{S}$ parses mem_P into $(U, \{U\}, X, \{X\}, M, \rho, vk, s)$. Then, $\mathcal{S}$ checks that $1 \leftarrow \mathsf{Ver}_{vk}(U, \{U\}, X, \{X\}, s)$. If not, $\mathcal{S}$ returns nothing. If the verification holds and $vk = vk_A^*$ holds, $\mathcal{S}$ fails in the simulation. Otherwise, we suppose that $\mathbb{S}_{\bar{P}}$ satisfies M and ρ, and let $I, I' \subset \{1, 2, \dots, \ell\}$ be defined as $I = \{i : \rho(i) \in \mathbb{S}_{\bar{P}}\}$ and $I' = \{i : \rho(i) \in \mathbb{W}_P\}$. $\mathcal{S}$ can efficiently find $\{w_i \in \mathbb{Z}_p\}_{i \in I}$ and $\{w_i' \in \mathbb{Z}_p\}_{i \in I'}$ such that $\sum_{i \in I} w_i \lambda_i = \sum_{i \in I'} w_i' \lambda_i = \alpha$ for valid shares $\{\lambda_i\}$ of any secret α according to M. $\mathcal{S}$ verifies whether the following condition holds:

$$\left(\prod_{i \in I} (e(U_i, g) e(X_i, h_{\rho(i)}))^{w_i} = \prod_{i \in I'} (e(U_i, g) e(X_i, h_{\rho(i)}))^{w_i'} = e(g^r, U) \right).$$

If not, $\mathcal{S}$ returns nothing. Otherwise, $\mathcal{S}$ generates a static secret key according to the set of attributes $\mathbb{W}_P$ as in the setup and computes the session key obeying the protocol. Note that $\mathbb{W}_P$ does not satisfy M_A^* because $vk \neq vk_A^*$. $\mathcal{S}$ records all session state and $(\mathbb{S}_P, \mathbb{S}_{\bar{P}}, mem_P, mem_{\bar{P}})$ as the completed session and SK in $\mathcal{L}_K$.

3. **KeyReveal**(sid):
   (a) If the session sid is not completed, $\mathcal{S}$ returns an error message.
   (b) Otherwise, $\mathcal{S}$ returns the recorded value $SK \in \mathcal{L}_K$ for sid.

4. **StateReveal**(sid):
   (a) If the session sid is completed, $\mathcal{S}$ returns an error message.
   (b) Otherwise, $\mathcal{S}$ returns all recorded state of sid.

5. **Corrupt**(P): $\mathcal{S}$ returns SK_P generated in the setup.

6. **Test**(sid*): $\mathcal{S}$ embeds $\sigma_1^* = \mathcal{T}e(g^c, g^z)$, and computes $\sigma_2^* = e(V^*, S_A') / (\prod_{i \in I_A}(e(V_i^*, T_A) \cdot e(Y_i^*, S_{A\rho_B(i)^*}))^{w_{Ai}})$ and $\sigma_3^* = e(g^a, Y^*)^{x^*}$. $\mathcal{S}$ returns SK^* computed from σ_1^*, σ_2^* and σ_3^*.

7. If $\mathcal{A}$ outputs a guess b', $\mathcal{S}$ outputs b'.

Analysis. First, we have to estimate the probability that $\mathcal{S}$ aborts. $\mathcal{S}$ aborts when $1 \leftarrow \mathsf{Ver}_{vk}(U, \{U\}, X, \{X\}, s)$ and $vk = vk_A^*$ occur. This event corresponds to a forge of a signature in the unforgeability game of the one-time signature because sk_A^* is hidden from $\mathcal{A}$. Thus, such a probability is negligible.

Next, we show that if $\prod_{i \in I}(e(U_i, g)e(X_i, h_{\rho(i)}))^{w_i} = \prod_{i \in I'}(e(U_i, g)e(X_i, h_{\rho(i)}))^{w_i'} = e(g^r, U)$ holds, the session key is computable with the static secret key $(S', T, \{S\})$ according to the set of attributes $\mathbb{W}_P$. σ_1 is defined as

$$e(U, S'_{\bar{p}})/\left(\prod_{i \in I}(e(U_i, T_{\bar{p}})e(X_i, S_{\bar{p}\rho(i)}))^{w_i}\right) = e(U, S'_{\bar{p}})/\left(\prod_{i \in I}(e(U_i, g)^{t_p}e(X_i, h_{\rho(i)})^{t_p})^{w_i}\right)$$

$$= e(U, S'_{\bar{p}})/\left(\prod_{i \in I'}(e(U_i, g)e(X_i, h_{\rho(i)}))^{w'_i}\right)^{t_p}$$

$$= e(U, S'_{\bar{p}})/e(g^r, U)^{t_p}$$

$$= g_T^{u_1 z}.$$

σ_2 is also similarly computable. Thus, the simulation of Send queries is perfect.

The simulation for S is same as $\mathbf{H}_2$ for $\mathcal{A}$ when $T = g_T^{a^{q+1}c}$ and same as $\mathbf{H}_3$ for $\mathcal{A}$ when $T = R$ except negligible probability. Hence, if $|\mathsf{Adv}(\mathcal{A}, \mathbf{H}_3) - \mathsf{Adv}(\mathcal{A}, \mathbf{H}_2)|$ is non-negligible, S is successful with negligible probability.

Therefore, $|\mathsf{Adv}(\mathcal{A}, \mathbf{H}_3) - \mathsf{Adv}(\mathcal{A}, \mathbf{H}_2)| \leq negl$ from the DPBDHE assumption.

Hybrid Experiment $\mathbf{H}_4$: In this experiment, the computation of $\sigma_1'^*$ in the test session is changed. Instead of computing $\sigma_1'^* \leftarrow Ext(\sigma_1^*)$, it is changed as choosing $\sigma_1'^* \in Kspace$ randomly.

Since σ_1^* is randomly chosen in $\mathbf{H}_3$, it has sufficient min-entropy. Thus, by the definition of the strong randomness extractor, $|\mathsf{Adv}(\mathcal{A}, \mathbf{H}_4) - \mathsf{Adv}(\mathcal{A}, \mathbf{H}_3)| \leq negl$.

Hybrid Experiment $\mathbf{H}_5$: In this experiment, the computation of SK^* in the test session is changed. Instead of computing $SK^* = F_{\sigma_1'^*}(\mathsf{sid}^*) \oplus F_{\sigma_2'^*}(\mathsf{sid}^*) \oplus F_{\sigma_3'^*}(\mathsf{sid}^*)$, it is changed as $SK^* = \alpha \oplus F_{\sigma_2'^*}(\mathsf{sid}^*) \oplus F_{\sigma_3'^*}(\mathsf{sid}^*)$ where $\alpha \in \{0, 1\}^k$ is chosen randomly and we suppose that B is the intended partner of A in the test session.

We construct a distinguisher $\mathcal{D}$ between PRF $F : \{0, 1\}^* \times Kspace \rightarrow \{0, 1\}^k$ and a random function RF from $\mathcal{A}$ in $\mathbf{H}_4$ or $\mathbf{H}_5$; that is, if $|\mathsf{Adv}(\mathcal{A}, \mathbf{H}_5) - \mathsf{Adv}(\mathcal{A}, \mathbf{H}_4)|$ is non-negligible, we can construct successful $\mathcal{D}$. $\mathcal{D}$ performs the following steps.

Setup. $\mathcal{D}$ sets the master secret key and all N parties' static secret keys.

Simulation. $\mathcal{D}$ simulates oracle queries by $\mathcal{A}$ as follows. $\mathcal{D}$ maintains the list $\mathcal{L}_K$ that contains queries and answers of KeyReveal.

1. Send($\mathcal{I}, \mathbb{S}_P, \mathbb{S}_{\bar{p}}$): $\mathcal{D}$ computes mem_P obeying the protocol, returns it and records $(\mathbb{S}_P, \mathbb{S}_{\bar{p}}, mem_P)$.
2. Send($\mathcal{R}, \mathbb{S}_{\bar{p}}, \mathbb{S}_P, mem_P$) and Send($\mathcal{I}, \mathbb{S}_{\bar{p}}, \mathbb{S}_P, mem_P$): $\mathcal{D}$ computes the message $mem_{\bar{p}}$ and the session key SK obeying the protocol, returns it, and records all session state and $(\mathbb{S}_P, \mathbb{S}_{\bar{p}}, mem_P, mem_{\bar{p}})$ as the completed session and SK in $\mathcal{L}_K$.
3. KeyReveal(sid):
   (a) If the session sid is not completed, $\mathcal{D}$ returns an error message.
   (b) Otherwise, $\mathcal{D}$ returns the recorded value $SK \in \mathcal{L}_K$ for sid.
4. StateReveal(sid):
   (a) If the session sid is completed, $\mathcal{D}$ returns an error message.
   (b) Otherwise, $\mathcal{D}$ returns all recorded state of sid.

5. Corrupt(P): $\mathcal{D}$ returns SK_P generated in the setup.
6. Test(sid*): $\mathcal{D}$ poses sid* to the function F^* (F^* is the PRF F or a random function RF) and obtains $\alpha = F^*(\text{sid}^*)$. $\mathcal{D}$ returns $SK^* = \alpha \oplus F_{\sigma_2'^*}(\text{sid}^*) \oplus F_{\sigma_3'^*}(\text{sid}^*)$.
7. If $\mathcal{A}$ outputs a guess b', $\mathcal{D}$ outputs b'.

Analysis. The simulation for $\mathcal{D}$ is same as $\mathbf{H}_4$ for $\mathcal{A}$ when F^* is F with a random key because $\sigma_1'^*$ is randomly chosen in $\mathbf{H}_4$. Also, the simulation for $\mathcal{D}$ is same as $\mathbf{H}_5$ for $\mathcal{A}$ when F^* is RF. Hence, if $|\mathsf{Adv}(\mathcal{A}, \mathbf{H}_5) - \mathsf{Adv}(\mathcal{A}, \mathbf{H}_4)|$ is non-negligible, $\mathcal{D}$ is successful with negligible probability.

Therefore, $|\mathsf{Adv}(\mathcal{A}, \mathbf{H}_5) - \mathsf{Adv}(\mathcal{A}, \mathbf{H}_4)| \leq negl.$

In $\mathbf{H}_5$, the session key in the test session is perfectly randomized. Thus, $\mathcal{A}$ cannot obtain any advantage from Test query.

Therefore, $\mathsf{Adv}(\mathcal{A}, \mathbf{H}_5) = 0$ and $\Pr[E_2 \wedge Suc]$ is negligible.   □

From Proposition 1 and 2, we obtain that $\Pr[Suc]$ is negligible.   □

Sufficient Condition for Identity-Based Authenticated Key Exchange Resilient to Leakage of Secret Keys

Atsushi Fujioka and Koutarou Suzuki

NTT Information Sharing Platform Laboratories
3-9-11 Midori-cho Musashino-shi Tokyo 180-8585, Japan
{fujioka.atsushi,suzuki.koutarou}@lab.ntt.co.jp

Abstract. In this paper, we provide a sufficient condition, called *admissible polynomials*, to construct a two-pass identity-based authenticated key exchange (ID-AKE) protocol secure in the identity-based extended Canetti-Krawczyk (id-eCK) model. The proposed ID-AKE protocol is secure under the gap Bilinear Diffie-Hellman assumption in the random oracle model.

Keywords: identity-based authenticated key exchange, gap Bilinear Diffie-Hellman assumption, identity-based extended Canetti-Krawczyk model, random oracle model.

1 Introduction

Key exchange is one of the important cryptographic protocols since it can be used to establish secure channels. Recent progress in its research has reached a formulation, *authenticated key exchange* (AKE), where AKE enables two parties to share a key via a public communication channel, and both parties are assured that only their intended peers can derive the session key.

In ordinary AKE, public-key infrastructure (PKI) is necessary since each static public-key needs to be linked with the identity of each user to provide authenticity, and such AKE is called *PKI-based* AKE. It is natural to introduce an identity-based version of AKE, called *identity-based authenticated key exchange* (ID-AKE). In order to avoid such requirement like a PKI system, a *key generation center* (KGC) in an ID-AKE protocol generates a pair of master public and secret keys, and the KGC extracts each user's secret key corresponding to the user's identity. Every user has own secret key and can use the identity as public information. Then, a party, called *initiator*, who wants to share a key with another party, called *responder*, sends ephemeral public information to the responder, the responder sends back another ephemeral public information to the initiator. This type of ID-AKE is called *two-pass* ID-AKE. Each party generates the session key from the master public key, own secret key given by the KGC, own secret values of ephemeral information, the peer's identity, and the received ephemeral information.

H. Kim (Ed): ICISC 2011, LNCS 7259, pp. 490–509, 2012.
© Springer-Verlag Berlin Heidelberg 2012

The several security models of ID-AKE has been investigated, and they are influenced by the security models of PKI-based AKE. For PKI-based AKE, the Canetti-Krawczyk (CK) model [9] and the extended Canetti-Krawczyk (eCK) model [22] followed the Bellare-Rogaway (BR) model [4], which is the first formal security model for AKE. The security notions defined in the above models are given as an indistinguishability game, where an adversary is required to differentiate between a random key and a session key of a session. The session is called *target session*, and is chosen by the adversary. Based on these models, the id-BR[1] [6,11], id-CK [7], and id-eCK [21] models were defined, respectively. Note that in ID-AKE, although the KGC has much power than users, any session key between users should not be reveal even to the KGC. This property is called *forward secrecy against KGC* (KGC-FS). The security and the eCK security are stronger than the BR security [9,13,22], however, the CK security and the eCK security are incompatible [14,15]. These relations hold in the security definitions of ID-AKE, also.

In this paper, we adopt the id-eCK model since it ensures the security against an adversary who tries to distinguish the session key from a random value under the disclosure of any pair of static secret keys and ephemeral secret keys of the initiator and the responder in the session except both static and ephemeral secret key of the initiator or the responder. In addition, it is complicated to precisely define state information in the CK model.

1.1   Related Works

In a literature, many ID-AKE protocols secure in the id-BR model have been investigated. (See survey papers for them [6,11].)

However, the BR model does not consider exposure of secret information, and then, the CK model was formulated to treat such situations. The CK model considers leakage of information in sessions, and the adversary in the CK model is allowed to access the state information, named *session state*. The session state contains not only the ephemeral secret key but also internal values computed with the static secret key, however, the adversary is not allowed to access the static secret key itself. The same discussion can be applied to the id-BR and id-CK models, and several id-CK secure protocols [7,17,18] have followed the first attempt by Chow and Choo [12]. It is worthy to note that it is complicated to precisely define state information in the CK model.

For (ID-)AKE, several security notions have been proposed in addition to the (id-)CK security: The *weak Perfect Forward Secrecy* (wPFS) implies that an adversary cannot recover an already established session key before the compromise of static secret keys. The resilient to *Key Compromised Impersonation* (KCI) means that given a static secret key an adversary tries to impersonate some honest party against the owner of the leaked secret key. The resilient to *Reflection Attack* (REF) implies that the indistinguishability test must be passed even

---

[1] The BR model is defined in symmetric key setting, and the model in public key setting is defined by Blake-Wilson, Johnson, and Menezes [5].

when the adversary is allowed to make a session between the same party. The resilience to *Maximal EXposure attack* (MEX) means that an adversary tries to distinguish the session key from a random value under the disclosure of any pair of secret static keys and ephemeral secret keys of the initiator and the responder in the session except both static and ephemeral secret key of the initiator or the responder.

Regarding to the id-eCK security, Huang and Cao defined the model, and proposed an id-eCK secure ID-AKE protocol [21]. The extended Canetti-Krawczyk (eCK) model for AKE was defined by LaMacchia, Lauter, and Mityagin [22], and roughly speaking, the eCK security assures that any adversary cannot distinguish the session key of a target session from a random value even when the adversary is allowed to access either the static secret key or the ephemeral secret key of each party establishing the target session, and the similar assurance is achieved in the id-eCK model.

Fujioka, Suzuki, and Ustaoğlu [20] proposed a shared security model for ID-AKE, where two different protocols can be securely carried out even when the identity and the secret key are shared in these protocols. They adopted and modified the id-eCK model, and proved that their proposed protocols are secure in both id-eCK and shared security models.

For ID-AKE, it is required that the KGC-FS security should be ensured, also, and the id-eCK security guarantees the all previous security notions: wPFS, KGC-FS, KCI, REF, and MEX.

Fujioka and Suzuki [19] provide a sufficient condition for PKI-based AKE secure in the eCK model and introduce the notion of admissible polynomials. The notion of admissible polynomials closely related to the generalization of DDH assumption introduced by Bresson, Lakhnech, Mazaré, and Warinschi [8].

1.2   Our Contributions

This paper provides a sufficient condition, called *admissible polynomials*, to construct an id-eCK secure ID-AKE protocol under the gap Bilinear Diffie-Hellman (gap BDH) assumption [1]. Here, the gap BDH problem is to solve the Bilinear Diffie-Hellman (BDH) problem with the help of a decisional BDH (DBDH) oracle, the BDH problem is to compute g_T^{xyz} from g^x, g^y, and g^z, and the DBDH oracle returns a bit which means the input tuple (g^x, g^y, g^z, g_T^w) is a BDH instance or not, i.e., $g_T^w = g^{xyz}$ or not, where g_T is $e(g,g)$ and e is a polynomial-time computable bilinear non-degenerate map called a *pairing*. We adopt the id-eCK model, where the security ensures the ID-based version of the eCK security model [22]. We give the condition regarding the exponents of shared values computed as the intermediate value in the protocols.

As seen in the above, there are fewer id-CK or id-eCK secure protocols than the id-BR secure protocols. The reason is that we need complex analysis to prove that a protocol is id-CK or id-eCK secure. To prove it, the detail analysis on exposure of secret information is required.

In an original Diffie-Hellman (DH) protocol [16], a party uses a single key to compute a shared value, that is Y^x from x and Y, and the peer also computes

X^y from y and X, where $X = g^x$, $Y = g^y$, and g is a generator of a cyclic group. We extend this exponent of the shared value to weighted inner product of two-dimensional vectors related to the exponents of the static and ephemeral public keys. For two vectors $u = (u_0, u_1)$, $v = (v_0, v_1)$ and two-dimensional square matrix C, the shared value is computed as g^{uCv^T}, where T is a transposition operation. Then, the exponent of the shared value is given as a quadratic polynomial, $p = uCv^T$, of u_0, u_1, v_0, and v_1. When all polynomials p are admissive polynomial, the underlying protocol can be proved as id-eCK secure. In the considering protocols of this paper, the shared value of the ID-AKE protocol is computed as $g_T^{(uCv^T)z}$ $(= g_T^{p \cdot z})$, where g_T is $e(g, g)$, e is a pairing function, and z is a master secret key of the KGC.

When the exponents of the shared value in an ID-AKE protocol are expressed by admissible polynomials, we can construct a reduction algorithm, which interacts with the adversary and solves a BDH problem with the help of a decisional BDH (DBDH) oracle. The algorithm simulates all queries the adversary requires and extracts the answer of the BDH instance. The resulting ID-AKE protocols based on admissible polynomials contain not only the existing efficient protocols but also new id-eCK secure protocols. That is, our sufficient condition is useful for constructing two-pass ID-AKE protocols.

Once the exponents of the shared values in an ID-AKE protocol are expressed by admissible polynomials, the ID-AKE protocol is id-eCK secure. It is required only to confirm that the exponents are expressed by admissible polynomials, and this confirmation is an easier task than the proof of id-eCK security.

It can be viewed as an application of a sufficient condition for PKI-based AKE in the eCK model, investigated by Fujioka and Suzuki [19], to ID-AKE in the id-eCK model. In the security proof, the gap BDH assumption and the random oracle model (ROM) [3] are needed. Furthermore, the twin DH technique [10] makes the constructed protocols secure under the BDH assumption.

The resulting protocol which satisfies the proposed condition guarantees the wPFS, KGC-FS, KCI, REF, and MEX. Although the security of the protocols constructed under the proposed condition is proved in the random oracle model, its security proof is done without the Forking Lemma [23]. Notice that in the case of using the Forking Lemma, the security parameter in the protocols must be bigger than the expected one in the underlying problem since the security degrades according to the number of hash queries. Thus, the protocols need longer key-length to meet the security parameter and they may loose the advantage in efficiency. The resulting protocols have an advantage in efficiency as number of the static (secret) keys and the ephemeral keys are related to the sizes of storage and communication data in the system, respectively.

Organization

In Section 2, we provide the id-eCK security model for ID-AKE. In Section 3, we propose a sufficient condition for id-eCK secure ID-AKE protocol and the resultant two-pass ID-AKE protocols, and discuss security arguments. In Section 4, we conclude the paper. In Appendix, we provide the security proof.

2   Security Model for ID-Based AKE

We recall the id-eCK security model for ID-AKE by Huang and Cao [21] that is the ID-based version of the eCK security model by LaMacchia, Lauter and Mityagin [22].

We denote a party by U_i and the identifier of U_i by ID_i. We outline our model for two-pass ID-AKE protocol, where parties U_A and U_B exchange ephemeral public keys X_A and X_B, i.e., U_A sends X_A to U_B and U_B sends X_B to U_A, and thereafter compute a session key. The session key depends on the exchanged ephemeral keys, identities of the parties, the static keys corresponding to these identities and the protocol instance that is used.

In the model, each party is a probabilistic polynomial-time Turing machine in security parameter κ and obtains a static private key corresponding to its identity string from a key generation center (KGC) via a secure and authenticated channel. The KGC uses a master secret key to generate individual private keys.

Session. An invocation of a protocol is called a *session*. A session is activated via an incoming message of the forms $(\Pi, \mathcal{I}, ID_A, ID_B)$ or $(\Pi, \mathcal{R}, ID_A, ID_B, X_B)$, where Π is a protocol identifier. If U_A was activated with $(\Pi, \mathcal{I}, ID_A, ID_B)$, then U_A is the session *initiator*, otherwise the session *responder*. After activation, U_A appends an ephemeral public key X_A to the incoming message and sends it as an outgoing response. If U_A is the responder, U_A computes a session key. If U_A is the initiator, U_A that has been successfully activated via $(\Pi, \mathcal{I}, ID_A, ID_B)$ can be further activated via $(\Pi, \mathcal{R}, ID_A, ID_B, X_A, X_B)$ to compute a session key. We say that U_A is *owner* of session sid if the third coordinate of session sid is ID_A. We say that U_A is *peer* of session sid if the fourth coordinate of session sid is ID_A. We say that a session is *completed* if its owner computes a session key.

A session initiator U_A identifies the session via $(\Pi, \mathcal{I}, ID_A, ID_B, X_A, \times)$ or $(\Pi, \mathcal{I}, ID_A, ID_B, X_A, X_B)$. If U_A is the responder, the session is identified via $(\Pi, \mathcal{R}, ID_A, ID_B, X_B, X_A)$. For session $(\Pi, \mathcal{I}, ID_A, ID_B, X_A, X_B)$ the *matching session* has identifier $(\Pi, \mathcal{R}, ID_B, ID_A, X_A, X_B)$ and vice versa. From now on we omit $\mathcal{I}$ and $\mathcal{R}$ since these "role markers" are implicitly defined by the order of X_A and X_B.

Adversary. The adversary $\mathcal{A}$ is modeled as a probabilistic Turing machine that controls all communications between parties including session activation, performed via a Send(message) query. The message has one of the following forms: (Π, ID_A, ID_B), (Π, ID_A, ID_B, X_A), or $(\Pi, ID_A, ID_B, X_A, X_B)$. Each party submits its responses to the adversary, who decides the global delivery order. Note that the adversary does not control the communication between each party and the key generation center.

A party's private information is not accessible to the adversary, however, leakage of private information is captured via the following adversary queries.

– SessionKeyReveal(sid) The adversary obtains the session key for the session sid, provided that the session holds a session key.

- EphemeralKeyReveal(sid) The adversary obtains the ephemeral secret key associated with the session sid.
- StaticKeyReveal(ID_i) The adversary learns the static secret key of party U_i.
- MasterKeyReveal() The adversary learns the master secret key of the KGC.
- EstablishParty(ID_i) This query allows the adversary to register a static public key on behalf of a party U_i, the adversary totally controls that party. If a party is established by an EstablishParty(ID_i) query issued by the adversary, then we call the party *dishonest*, and if not, we call the party *honest*. This query models malicious insiders.

Freshness. Our security definition requires the notion of "freshness".

Definition 1 (Freshness). *Let* sid* *be the session identifier of a completed session, owned by an honest party* U_A *with peer* U_B, *who is also honest. If the matching session exists, then let* $\overline{\text{sid}^*}$ *be the session identifier of the matching session of* sid*. *Define* sid* *to be fresh if none of the following conditions hold:*

1. $\mathcal{A}$ *issues* SessionKeyReveal(sid*) *or* SessionKeyReveal($\overline{\text{sid}^*}$) *(if* $\overline{\text{sid}^*}$ *exists).*
2. $\overline{\text{sid}^*}$ *exists and* $\mathcal{A}$ *makes either of the following queries*
   - *both* StaticKeyReveal(ID_A) *and* EphemeralKeyReveal(sid*), *or*
   - *both* StaticKeyReveal(ID_B) *and* EphemeralKeyReveal($\overline{\text{sid}^*}$).
3. $\overline{\text{sid}^*}$ *does not exist and* $\mathcal{A}$ *makes either of the following queries*
   - *both* StaticKeyReveal(ID_A) *and* EphemeralKeyReveal(sid*), *or*
   - StaticKeyReveal(ID_B).

Note that if $\mathcal{A}$ *issues* MasterKeyReveal(), *we regard* $\mathcal{A}$ *as having issued both* StaticKeyReveal(ID_A) *and* StaticKeyReveal(ID_B).

Security Experiment. The adversary $\mathcal{A}$ starts with a set of honest parties, for whom $\mathcal{A}$ adaptively selects identifiers. The adversary makes an arbitrary sequence of the queries described above. During the experiment, $\mathcal{A}$ makes a special query Test(sid*) and is given with equal probability either the session key held by sid* or a random key. The experiment continues until $\mathcal{A}$ makes a guess whether the key is random or not. The adversary *wins* the game if the test session sid* is fresh at the end of $\mathcal{A}$'s execution and if $\mathcal{A}$'s guess was correct.

Definition 2 (security). *The advantage of the adversary* $\mathcal{A}$ *in the experiment with ID-AKE protocol* Π *is defined as*

$$\text{Adv}_{\Pi}^{\text{ID-AKE}}(\mathcal{A}) = \Pr[\mathcal{A} \ wins] - \frac{1}{2}.$$

We say that Π *is secure ID-AKE protocols in the id-eCK model if the following conditions hold.*

1. *If two honest parties complete matching session, then, except with negligible probability in security parameter* κ, *they both compute the same session key.*
2. *For any probabilistic polynomial-time bounded adversary* $\mathcal{A}$, $\text{Adv}_{\Pi}^{\text{ID-AKE}}(\mathcal{A})$ *is negligible in security parameter* κ.

3 Sufficient Condition for id-eCK Secure ID-AKE Protocol

In this section, we propose a sufficient condition for id-eCK secure ID-AKE protocol and resultant two-pass ID-AKE protocols. The sufficient condition can be seen as the ID-based version of the notion of *admissible polynomials* [19].

The proposed ID-AKE protocol is a natural extension of the DH key exchange, where shared value g^{xy} is computed w.r.t. the ephemeral public keys g^x of user U_A and g^y of user U_B. The proposed ID-AKE protocol is a two-dimensional and ID-based generalization of the DH key exchange, i.e., shared value $e(g,g)^{zp(a,x,b,y)}$ is computed w.r.t. the master public key g^z, the static and ephemeral public keys (g^a, g^x) of user U_A and (g^b, g^y) of user U_B, where e is a pairing and

$$p(a,x,b,y) = \begin{pmatrix} a & x \end{pmatrix} \begin{pmatrix} c_{0,0} & c_{0,1} \\ c_{1,0} & c_{1,1} \end{pmatrix} \begin{pmatrix} b \\ y \end{pmatrix}$$

is a weighted inner product of vectors (a, x) and (b, y) of secret keys. For (KGC) forward security, we additionally adopt shared value g^{xy} that cannot be computed even if master key or both static keys are leaked. We show that if $p(a, x, b, y)$ satisfies the condition of *admissible polynomials*, the resultant two-pass ID-AKE protocol is id-eCK secure, i.e., we propose a sufficient condition for id-eCK secure ID-AKE protocol.

3.1 Admissible Polynomials

We recall the notion of *admissible polynomials* introduced by [19]. We define the notion of admissible polynomials over $\mathbb{Z}_q$, where $\mathbb{Z}_q$ is the additive group with prime modulus q.

Definition 3 (Admissible Polynomials [19]). *We say m polynomials $p_i \in \mathbb{Z}_q[u_0, u_1, v_0, v_1]$ $(i = 1, ..., m)$ are admissible if the following conditions are satisfied.*

1. $p_i(u_0, u_1, v_0, v_1) = c_{i,0,0}u_0v_0 + c_{i,0,1}u_0v_1 + c_{i,1,0}u_1v_0 + c_{i,1,1}u_1v_1$.
2. *For any $f\ (= 0, 1)$, there exist i, j $(1 \le i, j \le m)$, s.t.*

$$(c_{i,f,0}, c_{i,f,1}) \text{ and } (c_{j,f,0}, c_{j,f,1})$$

   are linearly independent, and for any $f' = 0, 1$, there exist i, j $(1 \le i, j \le m)$, s.t.

$$(c_{i,0,f'}, c_{i,1,f'}) \text{ and } (c_{j,0,f'}, c_{j,1,f'})$$

   are linearly independent.
3. *For any $i\ (= 1, ..., m)$, either of the following conditions holds: a) $p_i(u_0, u_1, v_0, v_1)$ is expressed as a product of $\ell_i(u_0, u_1)$ and $\ell'_i(v_0, v_1)$, where $\ell_i(u_0, u_1)$ and $\ell'_i(v_0, v_1)$ are linear combinations of u_0, u_1 and v_0, v_1, respectively, s.t.*

$$p_i(u_0, u_1, v_0, v_1) = \ell_i(u_0, u_1)\ell'_i(v_0, v_1).$$

Or b) for any f $(= 0, 1)$, $c_{i,f,0}u_f v_0 + c_{i,f,1}u_f v_1$ is expressed as a product of $\ell_{i,f,*}(u_0, u_1)$ and $\ell'_{i,f,*}(v_0, v_1)$, where $\ell_{i,f,*}(u_0, u_1)$ and $\ell'_{i,f,*}(v_0, v_1)$ are linear combinations of u_0, u_1 and v_0, v_1, respectively, s.t.

$$c_{i,f,0}u_f v_0 + c_{i,f,1}u_f v_1 = \ell_{i,f,*}(u_0, u_1)\ell'_{i,f,*}(v_0, v_1),$$

and for any f' $(= 0, 1)$, $c_{i,0,f'}u_0 v_{f'} + c_{i,1,f'}u_1 v_{f'}$ is expressed as a product of $\ell_{i,*,f'}(u_0, u_1)$ and $\ell'_{i,*,f'}(v_0, v_1)$, where $\ell_{i,*,f'}(u_0, u_1)$ and $\ell'_{i,*,f'}(v_0, v_1)$ are linear combinations of u_0, u_1 and v_0, v_1, respectively, s.t.

$$c_{i,0,f'}u_0 v_{f'} + c_{i,1,f'}u_1 v_{f'} = \ell_{i,*,f'}(u_0, u_1)\ell'_{i,*,f'}(v_0, v_1).$$

We provide examples of admissible polynomials below.

Example 1. The first example of admissible polynomials is

$$m = 4, \; p_1(a, x, b, y) = ab, \; p_2(a, x, b, y) = ay, \; p_3(a, x, b, y) = xb, \; p_4(a, x, b, y) = xy.$$

Example 2. The second example of admissible polynomials is

$$m = 3, \; p_1(a, x, b, y) = ab, \; p_2(a, x, b, y) = (a + x)(b + y), \; p_3(a, x, b, y) = xy.$$

The three conditions of admissible polynomials relate to the security proof of the (ID-)AKE protocol constructed from admissible polynomials. From the first condition, both users can compute the shared values. From the second condition, the simulator can extract the answer of a (B)DH problem in the security proof. From the third condition, the simulator can check that the shared values are correctly formed in the security proof. See sketch of the proof of Theorem 1 for details.

3.2 Resultant ID-Based AKE Protocol

We propose the ID-AKE protocol $\Pi_{p_1,...,p_m}$ constructed from admissible polynomials p_i $(i = 1, ..., m)$. We then prove in Theorem 1 that if polynomials p_i $(i = 1, ..., m)$ satisfy the conditions of admissible polynomials, the proposed ID-AKE protocol $\Pi_{p_1,...,p_m}$ is id-eCK secure, i.e., we provide a sufficient condition for id-eCK secure ID-AKE protocols.

The proposed ID-AKE protocol $\Pi_{p_1,...,p_m}$ is described as follows. Let p_i $(i = 1, ..., m)$ be admissible polynomials s.t. $p_m(q_A, x_A, q_B, x_B) = x_A x_B$. Let κ be the security parameter. Let G and G_T be a cyclic group with generator g and $g_T = e(g, g)$ and of order κ-bit prime q, and $e : G \times G \to G_T$ be pairing. Let $H : \{0,1\}^* \to \{0,1\}^\kappa$ and $H_1 : \{0,1\}^* \longrightarrow G$ be cryptographic hash functions modeled as random oracles. Let Π be the protocol identifier of the protocol $\Pi_{p_1,...,p_m}$.

KGC randomly selects master secret key $z \in \mathbb{Z}_q$, and publishes master public key $Z = g^z \in G$. These are provided as part of the system parameters.

User U_i with identity ID_i is assigned static secret key $D_i = Q_i^z \in G$, where $Q_i = H_1(ID_i) = g^{q_i} \in G$.

Thus, U_A's identity and static secret key are ID_A and $D_A = Q_A^z = H_1(ID_A)^z = g^{zq_A} \in G$, and U_B's identity and static secret key are ID_B and $D_B = Q_B^z = H_1(ID_B)^z = g^{zq_B} \in G$

In the description, user U_A is the session initiator and user U_B is the session responder.

1. U_A selects a random ephemeral private key $x_A \in_U \mathbb{Z}_q$, computes the ephemeral public key $X_A = g^{x_A}$, and sends (Π, ID_B, ID_A, X_A) to U_B.
2. Upon receiving (Π, ID_B, ID_A, X_A), U_B selects a random ephemeral private key $x_B \in_U \mathbb{Z}_q$, computes the ephemeral public key $X_B = g^{x_B}$, and sends $(\Pi, ID_A, ID_B, X_A, X_B)$ to U_A.
   U_B computes m shared values

$$Z_i = e(Q_A, D_B^{c_{i,0,0}} Z^{c_{i,0,1} x_B}) \cdot e(X_A, D_B^{c_{i,1,0}} Z^{c_{i,1,1} x_B}) \quad (i = 1, ..., m-1),$$

$$Z_m = X_A^{x_B}$$

   computes the session key $K = H(Z_1, ..., Z_m, \Pi, ID_A, ID_B, X_A, X_B)$, and completes the session.
3. Upon receiving $(\Pi, ID_A, ID_B, X_A, X_B)$, U_A checks if U_A has sent (Π, ID_B, ID_A, X_A) to U_B or not, and aborts the session if not.
   U_A computes m shared values

$$Z_i = e(D_A^{c_{i,0,0}} Z^{c_{i,1,0} x_A}, Q_B) \cdot e(D_A^{c_{i,0,1}} Z^{c_{i,1,1} x_A}, X_B) \quad (i = 1, ..., m-1),$$

$$Z_m = X_B^{x_A}$$

   computes the session key $K = H(Z_1, ..., Z_m, \Pi, ID_A, ID_B, X_A, X_B)$, and completes the session.

Both parties compute the same shared values

$$Z_i = g_T^{z p_i(q_A, x_A, q_B, x_B)} \quad (i = 1, ..., m-1),$$

$$Z_m = g^{p_m(q_A, x_A, q_B, x_B)}$$

and compute the same session key K.

The proposed ID-AKE protocol $\Pi_{p_1, ..., p_m}$ requires $2(m-1)$ pairing operations, $4(m-1) + 2$ exponential operations (including the exponentiation for the ephemeral public key), and m shared values, at most.

Example 3. From the second example of admissible polynomials, i.e.,

$$m = 3, \ p_1(a, x, b, y) = ab, \ p_2(a, x, b, y) = (a+x)(b+y), \ p_3(a, x, b, y) = xy,$$

we have ID-AKE protocol $\Pi_{q_A q_B, (q_A + x_A)(q_B + x_B), x_A x_B}$, where user U_A computes

$$Z_1 = e(D_A, Q_B) = g_T^{z q_A q_B}, \ Z_2 = e(D_A Z^{x_A}, Q_B X_B) = g_T^{z(q_A + x_A)(q_B + x_B)},$$

$$Z_3 = X_B^{x_A} = g^{x_A x_B}.$$

This protocol requires 2 pairing operations, 3 exponential operations (including the exponentiation for the ephemeral public key), and 3 shared values. The protocol has been proposed as protocol Π_2 in [20], and is the most efficient id-eCK secure ID-AKE protocol based on pairing as far as our best knowledge.

3.3   Security

For the security of the proposed protocol, we need the gap Bilinear Diffie-Hellman (gap BDH) assumption described below. The computational BDH function $\text{BCDH} : G^3 \to G_T$ is $\text{BCDH}(U, V, W) = e(g, g)^{\log U \log V \log W}$, and the decisional BDH predicate $\text{BDDH} : G^4 \to \{0, 1\}$ is a function which takes an input $(g^u, g^v, g^w, e(g, g)^x)$ and returns the bit 1 if $uvw = x \bmod q$ and the bit 0 otherwise, where log denote discrete logarithm. An adversary $\mathcal{A}$ is given input $U, V, W \in_U G$ selected uniformly random and oracle access to $\text{BDDH}(\cdot, \cdot, \cdot, \cdot)$ oracle, and tries to compute $\text{BCDH}(U, V, W)$. For adversary $\mathcal{A}$, we define advantage

$$Adv^{\text{gapBDH}}(\mathcal{A}) = \Pr[U, V, W \in_R G, \mathcal{A}^{\text{BDDH}(\cdot,\cdot,\cdot,\cdot)}(U, V, W) = \text{BCDH}(U, V, W)],$$

where the probability is taken over the choices of U, V, W and $\mathcal{A}$'s random tape.

Definition 4 (gap BDH assumption). *We say that G, G_T satisfy the gap BDH assumption if, for all polynomial-time adversaries $\mathcal{A}$, advantage Adv^{gapBDH} $(\mathcal{A})$ is negligible in security parameter κ.*

The proposed ID-AKE protocol is secure in the id-eCK model [21] under the gap BDH assumption in the random oracle model.

Theorem 1. *If G, G_T are groups where the gap BDH assumption holds, H, H_1 are random oracles, and p_i $(i = 1, ..., m)$ s.t. $p_m(q_A, x_A, q_B, x_B) = x_A x_B$ are admissible polynomials, the proposed ID-AKE protocol $\Pi_{p_1,...,p_m}$ constructed from p_i $(i = 1, ..., m)$ is secure in the id-eCK model.*

The proof of Theorem 1 is provided in Appendix A. We provide a intuitive discussion here.

Proof (Sketch). From the first condition of admissible polynomials, both users can compute the shared values as follows. User U_A, who knows secret keys D_A, x_A, can compute shared values

$$Z_i = e(D_A^{c_i,0,0} Z^{c_i,1,0 x_A}, Q_B) \cdot e(D_A^{c_i,0,1} Z^{c_i,1,1 x_A}, X_B).$$

User U_B, who knows secret keys D_B, x_B, can compute shared values

$$Z_i = e(Q_A, D_B^{c_i,0,0} Z^{c_i,0,1 x_B}) \cdot e(X_A, D_B^{c_i,1,0} Z^{c_i,1,1 x_B}).$$

The gap BDH solver $\mathcal{S}$ extracts the answer g_T^{uvw} of an instance $(U = g^u, V = g^v, W = g^w)$ of the gap BDH problem using adversary $\mathcal{A}$. For instance, we assume the case that test session $\mathtt{sid}^*$ has no matching session $\overline{\mathtt{sid}}^*$, adversary $\mathcal{A}$ is given D_A, and adversary $\mathcal{A}$ does not obtain x_A and D_B from the condition of freshness. In this case, solver $\mathcal{S}$ embeds the instance as $Z = U (= g^u)$, $X_A = V$ $(= g^v)$ and $Q_B = W (= g^w)$, and extracts g_T^{uvw} from the shared values $Z_i = g_T^{zp_i}$ $(i = 1, ..., m - 1)$, $g_T^{zp_m} = e(Z, Z_m)$.

Solver $\mathcal{S}$ can perfectly simulate StaticKeyReveal query by selecting random q_i and setting $Q_i = H_1(ID_i) = g^{q_i}$ and $D_i = Z^{q_i}$. Solver $\mathcal{S}$ can perfectly simulate EphemeralKeyReveal query by selecting random x_i and setting $X_i = g^{x_i}$.

From the second condition of admissible polynomials, solver $\mathcal{S}$ can extract the answer of the gap BDH instance as follows. From the second condition, there exist i, j $(1 \le i, j \le m)$, s.t. $(c_{i,1,0}, c_{i,1,1})$ and $(c_{j,1,0}, c_{j,1,1})$ are linearly independent. Using q_A, solver $\mathcal{S}$ can compute

$$Z_i' = g_T^{z(c_{i,1,0}x_Aq_B + c_{i,1,1}x_Ax_B)} = Z_i / (e(Z, Q_B)^{c_{i,0,0}q_A} e(Z, X_B)^{c_{i,0,1}q_A}),$$

$$Z_j' = g_T^{z(c_{j,1,0}x_Aq_B + c_{j,1,1}x_Ax_B)} = Z_j / (e(Z, Q_B)^{c_{j,0,0}q_A} e(Z, X_B)^{c_{j,0,1}q_A}).$$

Solver $\mathcal{S}$ can compute $g_T^{zx_Aq_B}$ from Z_i', Z_j' as

$$(Z'_i{}^{c_{j,1,1}} / Z'_j{}^{c_{i,1,1}})^{1/(c_{i,1,0}c_{j,1,1} - c_{j,1,0}c_{i,1,1})} = g_T^{zx_Aq_B}$$

since $(c_{i,1,0}, c_{i,1,1})$ and $(c_{j,1,0}, c_{j,1,1})$ are linearly independent, and successfully outputs the answer $g_T^{zx_Aq_B} = g_T^{uvw}$ of the gap BDH problem.

From the third condition of admissible polynomials, solver $\mathcal{S}$ can check if the shared values are correctly formed w.r.t. IDs and ephemeral public keys, and can simulate H and SessionKeyReveal queries consistently, i.e., in the simulation of the $H(Z_1, ..., Z_m, \Pi, ID_A, ID_B, X_A, X_B)$ query, solver $\mathcal{S}$ needs to check that the shared values Z_i $(i = 1, ..., m)$ are correctly formed, and if so return session key K being consistent with the previously answered SessionKeyReveal($\Pi, \mathcal{I}, ID_A, ID_B, X_A, X_B$) and SessionKeyReveal($\Pi, \mathcal{R}, ID_B, ID_A, X_A, X_B$) queries.

For all i $(= 1, ..., m)$, solver $\mathcal{S}$ performs the following procedure. If condition a) of the third condition holds, $p_i(u_0, u_1, v_0, v_1) = \ell_i(u_0, u_1)\ell_i'(v_0, v_1)$, where $\ell_i(u_0, u_1)$ and $\ell_i'(v_0, v_1)$ are linear combinations of u_0, u_1 and v_0, v_1, respectively. Then, solver $\mathcal{S}$ can check if shared value Z_i is correctly formed w.r.t. the IDs and ephemeral public keys by asking BDDH oracle

$$\text{BDDH}(Z, g^{\ell_i(q_A, x_A)}, g^{\ell_i'(q_B, x_B)}, Z_i) = 1.$$

Here solver $\mathcal{S}$ can compute $g^{\ell_i(q_A, x_A)} = Q_A^{d_1} X_A^{d_2}$, $g^{\ell_i'(q_B, x_B)} = Q_B^{d_3} X_B^{d_4}$ since $\ell_i(q_A, x_A)$, $\ell_i'(q_B, x_B)$ are linear, that is, they are expressed as $\ell_i(q_A, x_A) = d_1 q_A + d_2 x_A$, $\ell_i'(q_B, x_B) = d_3 q_B + d_4 x_B$.

Otherwise, from condition b) of the third condition,

$$c_{i,f,0}u_fv_0 + c_{i,f,1}u_fv_1 = \ell_{i,f,*}(u_0, u_1)\ell_{i,f,*}'(v_0, v_1),$$

where $\ell_{i,f,*}(u_0, u_1)$ and $\ell_{i,f,*}'(v_0, v_1)$ are linear combinations of u_0, u_1 and v_0, v_1, respectively. Using q_A, solver $\mathcal{S}$ can compute

$$Z_i' = g_T^{z(c_{i,1,0}x_Aq_B + c_{i,1,1}x_Ax_B)} = Z_i / (e(Z, Q_B)^{c_{i,0,0}q_A} e(Z, X_B)^{c_{i,0,1}q_A}).$$

Then, solver $\mathcal{S}$ can check if shared value Z_i' is correctly formed w.r.t. the IDs and ephemeral public keys, by asking BDDH oracle

$$\text{BDDH}(Z, g^{\ell_{i,1,*}(q_A, x_A)}, g^{\ell_{i,1,*}'(q_B, x_B)}, Z_i') = 1,$$

and this implies Z_i is correctly formed. Here solver $\mathcal{S}$ can compute $g^{\ell_{i,1,*}(q_A,x_A)} = Q_A^{d_1} X_A^{d_2}$, $g^{\ell'_{i,1,*}(q_B,x_B)} = Q_B^{d_3} X_B^{d_4}$ since $\ell_{i,1,*}(q_A,x_A)$, $\ell'_{i,1,*}(q_B,x_B)$ are linear, that is, they are expressed as $\ell_{i,1,*}(q_A,x_A) = d_1 q_A + d_2 x_A$, $\ell'_{i,1,*}(q_B,x_B) = d_3 q_B + d_4 x_B$.

Solver $\mathcal{S}$ can check if shared value $Z_m = g^{x_A x_B}$ is correctly formed w.r.t. the ephemeral public keys, by checking $e(X_A, X_B) = e(g, Z_m)$.

Finally, we assume the case that adversary $\mathcal{A}$ issues MasterKeyReveal query. In this case, solver $\mathcal{S}$ embeds the instance as $X_A = V\ (= g^v)$, $X_B = W\ (= g^w)$, and extracts answer g^{vw} of the computational Diffie-Hellman problem from the shared values $Z_m = g^{x_A x_B}$.   □

3.4 Remarks

The resulting ID-AKE protocols based on admissible polynomials are id-eCK secure, however their security requires a gap assumption, i.e., the gap BDH assumption, not a conventional computational assumption, such like the BDH assumption. To remove the gap assumption, the twin DH technique [10] is applicable to the constructed protocols. Although this modification requires twice as many public keys as the original protocols, the modified protocol are id-eCK secure under the BDH assumption.

In the consideration of the sufficient condition, we assume that the coefficients of the admissive polynomials, $c_{i,j,k}$, are constant. However, it is possible that these coefficients are generated from a publicly available value, such like the session id. For instance, when the following admissive polynomials

$$p_1(a,x,b,y) = (c_a a + x)(b + y), \quad p_2(a,x,b,y) = (a+x)(c_b b + y), \quad p_3(a,x,b,y) = xy$$

are used to construct an ID-AKE protocol and the coefficients, c_a, c_b, are generated with a additional hash function H' such that $c_a = H'(X_A)$, $c_b = H'(X_B)$, the resulting protocol is the same with protocol Π_1 in [20].

In Table 1, we show the comparison with the existing id-eCK secure id-AKE protocols, [21] and [20]. In the table, "model" means security model, "#SK, #EK, #SV, #exp, #pairing" mean the number of static key, ephemeral key, shared value, exponential operation (including computation of ephemeral public key), and pairing operation, and "assumption" means required assumption. The proposed protocol requires m shared values, $4(m-1)+2$ exponential operations, and $2(m-1)$ pairing operations in general. However, it contains efficient one, e.g., "Example 3" that is same as Π_2 of [20].

Table 1. Comparison with the existing id-eCK secure id-AKE protocols

protocol	model	#SK	#EK	#SV	#exp	#pairing	assumption
HC [21]	id-eCK	2	1	3	3	2	BDH+ROM
Π_1 of FSU [20]	id-eCK	1	1	3	5	2	GBDH+ROM
Π_2 of FSU [20]	id-eCK	1	1	3	3	2	GBDH+ROM
Proposed protocol	id-eCK	1	1	m	$4(m-1)+2$	$2(m-1)$	GBDH+ROM
Example 3 of proposed	id-eCK	1	1	3	3	2	GBDH+ROM

4   Conclusion

We presented a sufficient condition for constructing id-eCK secure two-pass AKE protocols with a single static (secret) key and a single ephemeral key using a single hash function. The constructed protocols consist of several two-dimensional versions of the DH key exchange protocol, and their security proofs do not depend on the Forking Lemma. As a result, our protocols provide strong security assurances without compromising too much on efficiency.

References

1. Baek, J., Safavi-Naini, R., Susilo, W.: Universal Designated Verifier Signature Proof (or How to Efficiently Prove Knowledge of a Signature). In: Roy, B. (ed.) ASIACRYPT 2005. LNCS, vol. 3788, pp. 644–661. Springer, Heidelberg (2005)
2. Bao, F., Deng, R.H., Zhu, H.: Variations of Diffie-Hellman Problem. In: Qing, S., Gollmann, D., Zhou, J. (eds.) ICICS 2003. LNCS, vol. 2836, pp. 301–312. Springer, Heidelberg (2003)
3. Bellare, M., Rogaway, P.: Random oracles are practical: A paradigm for designing efficient protocols. In: CCS 1993: Proceedings of the 1st ACM Conference on Computer and Communications Security, pp. 62–73 (1993)
4. Bellare, M., Rogaway, P.: Entity Authentication and Key Distribution. In: Stinson, D.R. (ed.) CRYPTO 1993. LNCS, vol. 773, pp. 232–249. Springer, Heidelberg (1994)
5. Blake-Wilson, S., Johnson, D., Menezes, A.: Key Agreement Protocols and their Security Analysis. In: Darnell, M. (ed.) Cryptography and Coding 1997. LNCS, vol. 1355, pp. 30–45. Springer, Heidelberg (1997)
6. Boyd, C., Choo, K.-K.R.: Security of Two-Party Identity-Based Key Agreement. In: Dawson, E., Vaudenay, S. (eds.) Mycrypt 2005. LNCS, vol. 3715, pp. 229–243. Springer, Heidelberg (2005)
7. Boyd, C., Cliff, Y., Gonzalez Nieto, J.M., Paterson, K.G.: Efficient One-Round Key Exchange in the Standard Model. In: Mu, Y., Susilo, W., Seberry, J. (eds.) ACISP 2008. LNCS, vol. 5107, pp. 69–83. Springer, Heidelberg (2008), Full version available at http://eprint.iacr.org/2008/007/
8. Bresson, E., Lakhnech, Y., Mazaré, L., Warinschi, B.: A Generalization of DDH with Applications to Protocol Analysis and Computational Soundness. In: Menezes, A. (ed.) CRYPTO 2007. LNCS, vol. 4622, pp. 482–499. Springer, Heidelberg (2007)
9. Canetti, R., Krawczyk, H.: Analysis of Key-Exchange Protocols and Their Use for Building Secure Channels. In: Pfitzmann, B. (ed.) EUROCRYPT 2001. LNCS, vol. 2045, pp. 453–474. Springer, Heidelberg (2001)
10. Cash, D., Kiltz, E., Shoup, V.: The Twin Diffie-Hellman Problem and Applications. In: Smart, N.P. (ed.) EUROCRYPT 2008. LNCS, vol. 4965, pp. 127–145. Springer, Heidelberg (2008)
11. Chen, L., Cheng, Z., Smart, N.P.: Identity-based key agreement protocols from pairings. International Journal of Information Security 6(4), 213–241 (2007)
12. Chow, S.S.M., Choo, K.-K.R.: Strongly-Secure Identity-Based Key Agreement and Anonymous Extension. In: Garay, J.A., Lenstra, A.K., Mambo, M., Peralta, R. (eds.) ISC 2007. LNCS, vol. 4779, pp. 203–220. Springer, Heidelberg (2007)

13. Choo, K.-K.R., Boyd, C., Hitchcock, Y.: Examining Indistinguishability-Based Proof Models for Key Establishment Protocols. In: Roy, B. (ed.) ASIACRYPT 2005. LNCS, vol. 3788, pp. 585–604. Springer, Heidelberg (2005)
14. Cremers, C.J.F.: Session-state Reveal Is Stronger Than Ephemeral Key Reveal: Attacking the NAXOS Authenticated Key Exchange Protocol. In: Abdalla, M., Pointcheval, D., Fouque, P.-A., Vergnaud, D. (eds.) ACNS 2009. LNCS, vol. 5536, pp. 20–33. Springer, Heidelberg (2009)
15. Cremers, C.J.F.: Examining indistinguishability-based security models for key exchange protocols: The case of CK, CK-HMQV, and eCK. In: 6th ACM Symposium on Information, Computer and Communications Security, pp. 80–91. ACM, New York (2011)
16. Diffie, W., Hellman, H.: New directions in cryptography. IEEE Transactions of Information Theory 22(6), 644–654 (1976)
17. Fiore, D., Gennaro, R.: Making the Diffie-Hellman Protocol Identity-Based. In: Pieprzyk, J. (ed.) CT-RSA 2010. LNCS, vol. 5985, pp. 165–178. Springer, Heidelberg (2010)
18. Fiore, D., Gennaro, R.: Identity-Based Key Exchange Protocols without Pairings. In: Gavrilova, M.L., Tan, C.J.K., Moreno, E.D. (eds.) Transactions on Computational Science X. LNCS, vol. 6340, pp. 42–77. Springer, Heidelberg (2010)
19. Fujioka, A., Suzuki, K.: Designing Efficient Authenticated Key Exchange Resilient to Leakage of Ephemeral Secret Keys. In: Kiayias, A. (ed.) CT-RSA 2011. LNCS, vol. 6558, pp. 121–141. Springer, Heidelberg (2011)
20. Fujioka, A., Suzuki, K., Ustaoğlu, B.: Ephemeral Key Leakage Resilient and Efficient ID-AKEs That Can Share Identities, Private and Master Keys. In: Joye, M., Miyaji, A., Otsuka, A. (eds.) Pairing 2010. LNCS, vol. 6487, pp. 187–205. Springer, Heidelberg (2010)
21. Huang, H., Cao, Z.: An ID-based authenticated key exchange protocol based on bilinear Diffie-Hellman problem. In: Safavi-Naini, R., Varadharajan, V. (eds.) ASIACCS 2009: Proceedings of the 2009 ACM Symposium on Information, Computer and Communications Security, New York, NY, USA, pp. 333–342 (2009)
22. LaMacchia, B., Lauter, K., Mityagin, A.: Stronger Security of Authenticated Key Exchange. In: Susilo, W., Liu, J.K., Mu, Y. (eds.) ProvSec 2007. LNCS, vol. 4784, pp. 1–16. Springer, Heidelberg (2007)
23. Pointcheval, D., Stern, J.: Security arguments for digital signatures and blind signatures. J. of Cryptology 13(3), 361–396 (2000)

A   Proof of Theorem 1

We need the gap Bilinear Diffie-Hellman (gap BDH) assumption in pairing groups G, G_T of order q with generator g, g_T, where one tries to compute $BCDH(U, V, W)$ accessing the BDDH oracle. Here, we denote $BCDH(g^u, g^v, g^w) = g_T^{uvw}$, and the BDDH oracle on input (g^u, g^v, g^w, g_T^x) returns bit 1 if $uvw = x$, or bit 0 otherwise. We also need a variant of the gap BDH assumption, where one tries to compute $BCDH(U, V, V)$ instead of $BCDH(U, V, W)$. We call the variant as the *square gap BDH assumption*, which is equivalent to the gap BDH assumption if groups G, G_T have prime order q [2] as follows. Given a challenge U, V of the square gap BDH assumption, one sets $W = V^s$ for random integers $s \in_R [1, q-1]$ and can compute $BCDH(U, V, V) = BCDH(U, V, W)^{1/s}$. Given a challenge U, V, W of the gap BDH assumption, one sets $V_1 = VW, V_2 = VW^{-1}$ and can compute $\text{BCDH}(U, V, W) = (\text{BCDH}(U, V_1, V_1)/\text{BCDH}(U, V_2, V_2))^{1/4}$.

We show that if polynomially bounded adversary $\mathcal{A}$ can distinguish the session key of a fresh session from a randomly chosen session key, we can solve the gap BDH problem. Let κ denote the security parameter, and let $\mathcal{A}$ be a polynomial-time bounded adversary w.r.t. security parameter κ. We use adversary $\mathcal{A}$ to construct the gap BDH solver $\mathcal{S}$ that succeeds with non-negligible probability. Adversary $\mathcal{A}$ is said to be successful with non-negligible probability if adversary $\mathcal{A}$ wins the distinguishing game with probability $\frac{1}{2} + f(\kappa)$, where $f(\kappa)$ is non-negligible, and the event M denotes that an adversary $\mathcal{A}$ is successful.

Let the test session be $\texttt{sid}^* = (\Pi, \mathcal{I}, ID_A, ID_B, X_A, X_B)$ or $(\Pi, \mathcal{R}, ID_B, ID_A, X_A, X_B)$, which is a completed session between honest users U_A and U_B, where user U_A is the initiator and user U_B is the responder of the test session $\texttt{sid}^*$. Let H^* be the event that adversary $\mathcal{A}$ queries $(Z_1, ..., Z_m, \Pi, ID_A, ID_B, X_A, X_B)$ to H. Let $\overline{H^*}$ be the complement of event H^*. Let $\texttt{sid}$ be any completed session owned by an honest user such that $\texttt{sid} \neq \texttt{sid}^*$ and $\texttt{sid}$ is non-matching to $\texttt{sid}^*$. Since $\texttt{sid}$ and $\texttt{sid}^*$ are distinct and non-matching, the inputs to the key derivation function H are different for $\texttt{sid}$ and $\texttt{sid}^*$. Since H is a random oracle, adversary $\mathcal{A}$ cannot obtain any information about the test session key from the session keys of non-matching sessions. Hence, $\Pr(M \wedge \overline{H^*}) \leq \frac{1}{2}$ and $\Pr(M) = \Pr(M \wedge H^*) + \Pr(M \wedge \overline{H^*}) \leq \Pr(M \wedge H^*) + \frac{1}{2}$, whence $f(\kappa) \leq \Pr(M \wedge H^*)$. Henceforth, the event $M \wedge H^*$ is denoted by M^*.

We denote a user as U_i, and user U_i and other parties are modeled as probabilistic polynomial-time Turing machines w.r.t. security parameter κ. We denote master secret (public) key as z (Z). For user U_i, we denote static secret keys as D_i and ephemeral secret (public) keys as x_i (X_i, respectively). We also denote the session key as K. Assume that adversary $\mathcal{A}$ succeeds in an environment with n users and activates at most s sessions within a user.

We consider the non-exclusive classification of all possible events in Tables 2 and 3. Here, users U_A and U_B are initiator and responder of the test session $\texttt{sid}^*$, respectively. Table 2 classifies events when identities ID_A, ID_B are distinct, and Table 3 classifies events when identities $ID_A = ID_B$ are the same, i.e., reflection attacks. In these tables, "ok" means the secret key is not revealed, or

Table 2. Classification of attacks when IDs ID_A, ID_B are distinct. "ok" means the secret key is not revealed. "r" means the secret key may be revealed. "n" means no matching session exists. The "instance embedding" row shows how the simulator embeds an instance of the gap BDH problem.

	z	D_A	x_A	D_B	x_B	instance embedding
E_1	ok	r	ok	ok	n	$Z = U, X_A = V, Q_B = W$
E_2	ok	ok	r	ok	n	$Z = U, Q_A = V, Q_B = W$
E_3	ok	r	ok	ok	r	$Z = U, X_A = V, Q_B = W$
E_4	ok	ok	r	ok	r	$Z = U, Q_A = V, Q_B = W$
E_5	r	r	ok	r	ok	$X_A = V, X_B = W$
E_6	ok	ok	r	r	ok	$Z = U, Q_A = V, X_B = W$

Table 3. Classification of attacks when IDs $ID_A = ID_B$ are the same.

	z	D_A	x_A	D_B	x_B	instance embedding
E_2'	ok	ok	r	ok	n	$Z = U, Q_A = V, Q_B = V$
E_4'	ok	ok	r	ok	r	$Z = U, Q_A = V, Q_B = V$
E_5'	r	r	ok	r	ok	$X_A = V, X_B = V$

the matching session exists and the ephemeral key is not revealed. "r" means the secret key may be revealed. "n" means no matching session exists. The "instance embedding" row shows how the simulator embeds an instance of the gap BDH problem.

Since the classification covers all possible events, at least one event $E_i \wedge M^*$ in the tables occurs with non-negligible probability if event M^* occurs with non-negligible probability. Thus, the gap BDH problem can be solved with non-negligible probability, which means that the proposed protocol is secure under the gap BDH assumption. We investigate each of these events in the following subsections.

A.1   Event $E_1 \wedge M^*$

In event E_1, test session $\texttt{sid}^*$ has no matching session $\overline{\texttt{sid}^*}$, adversary $\mathcal{A}$ obtains D_A, and adversary $\mathcal{A}$ does not obtain x_A and D_B from the condition of freshness. In this case, solver $\mathcal{S}$ embeds the instance as $Z = U \ (= g^u)$, $X_A = V \ (= g^v)$ and $Q_B = W \ (= g^w)$, and extracts g_T^{uvw} from the shared values $Z_i = g_T^{zp_i}$ $(i = 1, ..., m-1)$, $g_T^{zp_m} = e(Z, Z_m)$ using the knowledge of q_A. In event $E_1 \wedge M^*$, solver $\mathcal{S}$ performs the following steps.

Setup. The gap BDH solver $\mathcal{S}$ embeds instance $(U = g^u, V = g^v, W = g^w)$ of the gap BDH problem as follows. $\mathcal{S}$ set the master public key $Z = U$, establishes n honest users $U_1, ..., U_n$. $\mathcal{S}$ randomly selects two users U_A and U_B and integer $t \in_R [1, s]$, which is a guess of the test session with probability $1/n^2 s$. $\mathcal{S}$ sets the ephemeral public key of t-th session of user U_A as $X_A = V$, and sets hash value $Q_B = W$ of ID_B of of user U_B. $\mathcal{S}$ selects random q_i, sets $Q_i = H_1(ID_i) = g^{q_i}$ and $D_i = Z^{q_i}$, and assigns the static secret key D_i to user U_i.

$\mathcal{S}$ activates adversary $\mathcal{A}$ on this set of users and awaits the actions of $\mathcal{A}$. We next describe the actions of $\mathcal{S}$ in response to user activations and oracle queries.

Simulation. $\mathcal{S}$ maintains a list L_H that contains queries and answers of H oracle, and a list L_S that contains queries and answers of SessionKeyReveal. $\mathcal{S}$ simulates oracle queries as follows.

1. Send($\Pi, \mathcal{I}, ID_i, ID_j$): $\mathcal{S}$ selects ephemeral secret key $x_i \in_U \mathbb{Z}_q$, computes ephemeral public key X_i honestly, records (Π, ID_i, ID_j, X_i), and returns it.
2. Send($\Pi, \mathcal{R}, ID_j, ID_i, X_i$): $\mathcal{S}$ selects ephemeral secret key $x_j \in_U \mathbb{Z}_q$, computes ephemeral public key X_j honestly, records $(\Pi, ID_i, ID_j, X_i, X_j)$, and returns it.

3. Send$(\Pi, \mathcal{I}, ID_i, ID_j, X_i, X_j)$: If (Π, ID_i, ID_j, X_i) is not recorded, $\mathcal{S}$ records the session $(\Pi, \mathcal{I}, ID_i, ID_j, X_i, X_j)$ as not completed. Otherwise, $\mathcal{S}$ records the session as completed.

4. $H(Z_1, ..., Z_m, \Pi, ID_i, ID_j, X_i, X_j)$:
   (a) If $(Z_1, ..., Z_m, \Pi, ID_i, ID_j, X_i, X_j)$ is recorded in list L_H, then return recorded value K.
   (b) Else if the session $(\Pi, \mathcal{I}, ID_i, ID_j, X_i, X_j)$ or $(\Pi, \mathcal{R}, ID_j, ID_i, X_i, X_j)$ is recorded in list L_S, then $\mathcal{S}$ checks that the shared values Z_i $(i = 1, ..., m)$ are correctly formed w.r.t. IDs and ephemeral public keys ID_i, ID_j, X_i, X_j using knowledge of secret keys q_i or x_i by the procedure Check described below.
   If the shared values are correctly formed, then return recorded value K and record it in list L_H.
   (c) Else if $i = A, j = B$, and the session is t-th session of user U_A, then $\mathcal{S}$ checks that the shared values Z_i $(i = 1, ..., m)$ are correctly formed w.r.t. IDs and ephemeral public keys ID_A, ID_B, X_A, Y_B using knowledge of secret key q_A by the procedure Check described below.
   If the shared values are correctly formed, then $\mathcal{S}$ computes the answer of the gap BDH instance from the shared values and public keys using knowledge of secret key q_A by the procedure Extract described below, and is successful by outputting the answer.
   (d) Otherwise, $\mathcal{S}$ returns random value K and records it in list L_H.

5. SessionKeyReveal$((\Pi, \mathcal{I}, ID_i, ID_j, X_i, X_j)$ or $(\Pi, \mathcal{R}, ID_j, ID_i, X_i, X_j))$:
   (a) If the session $(\Pi, \mathcal{I}, ID_i, ID_j, X_i, X_j)$ or $(\Pi, \mathcal{R}, ID_j, ID_i, X_i, X_j)$ $(=$ sid$)$ is not completed, return error.
   (b) Else if sid is recorded in list L_S, then return recorded value K.
   (c) Else if $(Z_1, ..., Z_m, \Pi, ID_i, ID_j, X_i, X_j)$ is recorded in list L_H, then $\mathcal{S}$ checks that the shared values Z_i $(i = 1, ..., m)$ are correctly formed w.r.t. IDs and ephemeral public keys ID_i, ID_j, X_i, X_j using knowledge of secret keys q_i or x_i by the procedure Check described below.
   If the shared values are correctly formed, then return recorded value K and record it in list L_S.
   (d) Otherwise, $\mathcal{S}$ returns random value K and records it in list L_S.

6. SessionStateReveal(sid): If sid is t-th session of U_A, then $\mathcal{S}$ aborts with failure. Otherwise, $\mathcal{S}$ computes all session states as the protocol with s_i and returns it.

7. Corrupt(U_i): If $i = A$ or $i = B$, then $\mathcal{S}$ aborts with failure. Otherwise, $\mathcal{S}$ returns D_i and all session states of sessions owned by U_i.

8. MasterKeyReveal(): $\mathcal{S}$ aborts with failure.

9. Test(sid): If sid is not t-th session of U_A, then $\mathcal{S}$ aborts with failure. Otherwise, $\mathcal{S}$ responds to the query faithfully.

10. If adversary $\mathcal{A}$ outputs a guess γ, $\mathcal{S}$ aborts with failure.

Extract : The procedure Extract computes $g_T^{zu_f v_0}$ from the shared values $Z_i = g_T^{zp_i}$ $(i = 1, ..., m-1)$, $g_T^{zp_m} = e(Z, Z_m)$ and public values $Q_A = U_0 = g^{u_0}, X_A = U_1 = g^{u_1}, Q_B = V_0 = g^{v_0}, X_B = V_1 = g^{v_1}$ using knowledge of secret key u_f as follows.

From the second condition of admissible polynomials, there exist i, j $(1 \leq i, j \leq m)$, s.t. $(c_{i,\overline{f},0}, c_{i,\overline{f},1})$ and $(c_{j,\overline{f},0}, c_{j,\overline{f},1})$ are linearly independent. Using u_f, the procedure Extract computes

$$Z_i' = g_T^{z(c_{i,\overline{f},0}u_{\overline{f}}v_0 + c_{i,\overline{f},1}u_{\overline{f}}v_1)} = Z_i / (e(Z, V_0)^{c_{i,f,0}u_f} e(Z, V_1)^{c_{i,f,1}u_f}),$$

$$Z_j' = g_T^{z(c_{j,\overline{f},0}u_{\overline{f}}v_0 + c_{j,\overline{f},1}u_{\overline{f}}v_1)} = Z_j / (e(Z, V_0)^{c_{j,f,0}u_f} e(Z, V_1)^{c_{j,f,1}u_f}).$$

The procedure Extract computes $g^{zu_{\overline{f}}v_0}$ from Z_i', Z_j' as

$$(Z_i'^{c_{j,\overline{f},1}} / Z_j'^{c_{i,\overline{f},1}})^{1/(c_{i,\overline{f},0}c_{j,\overline{f},1} - c_{j,\overline{f},0}c_{i,\overline{f},1})} = g_T^{zu_{\overline{f}}v_0}$$

since $(c_{i,\overline{f},0}, c_{i,\overline{f},1})$ and $(c_{j,\overline{f},0}, c_{j,\overline{f},1})$ are linearly independent.

The procedure Extract can compute $g_T^{zu_{\overline{f}}v_1}$ using knowledge of secret key u_f same as above. The procedure Extract can compute $g_T^{zu_0v_{\overline{f'}}}$ and $g_T^{zu_1v_{\overline{f'}}}$ using knowledge of secret key $v_{f'}$ same as above. Thus, we can compute the answer of the gap BDH problem.

Check : The procedure Check checks that the shared values $Z_i = g_T^{zp_i}$ $(i = 1, ..., m-1)$, $g_T^{zp_m} = e(Z, Z_m)$ are correctly formed w.r.t. public values $Q_A = U_0 = g^{u_0}, X_A = U_1 = g^{u_1}, Q_B = V_0 = g^{v_0}, X_B = V_1 = g^{v_1}$ using knowledge of secret key u_f as follows.

For all $i = 1, ..., m$, the procedure Check performs the following. If condition a) of the second condition of admissible polynomials holds, there exist linear combination $\ell_i(u_0, u_1)$ of u_0, u_1 and linear combination $\ell_i'(v_0, v_1)$ of v_0, v_1, s.t. $p_i(u_0, u_1, v_0, v_1) = \ell_i(u_0, u_1)\ell_i'(v_0, v_1)$. Then, the procedure Check checks if shared value Z_i is correctly formed w.r.t. public values by asking BDDH oracle

$$\text{BDDH}(Z, g^{\ell_i(u_0, u_1)}, g^{\ell_i'(v_0, v_1)}, Z_i) = 1.$$

Here, we can compute $g^{\ell_i(u_0, u_1)} = U_0^{d_{u_0}}U_1^{d_{u_1}}$ and $g^{\ell_i'(v_0, v_1)} = V_0^{d_{v_0}}V_1^{d_{v_1}}$, since $\ell_i(u_0, u_1)$ and $\ell_i'(v_0, v_1)$ are expressed as $\ell_i(u_0, u_1) = d_{u_0}u_0 + d_{u_1}u_1$ and $\ell_i'(v_0, v_1) = d_{v_0}v_0 + d_{v_1}v_1$.

Otherwise, from condition b) of the second condition of admissible polynomials, there exist linear combination $\ell_{i,\overline{f},*}(u_0, u_1)$ of u_0, u_1 and linear combination $\ell_{i,\overline{f},*}'(v_0, v_1)$ of v_0, v_1, s.t. $c_{i,\overline{f},0}u_{\overline{f}}v_0 + c_{i,\overline{f},1}u_{\overline{f}}v_1 = \ell_{i,\overline{f},*}(u_0, u_1)\ell_{i,\overline{f},*}'(v_0, v_1)$. Using knowledge of secret key u_f, the procedure Check computes

$$Z_i' = g_T^{z(c_{i,\overline{f},0}u_{\overline{f}}v_0 + c_{i,\overline{f},1}u_{\overline{f}}v_1)} = Z_i / (e(Z, V_0)^{c_{i,f,0}u_f} e(Z, V_1)^{c_{i,f,1}u_f}).$$

Then, the procedure Check checks that shared value Z_i' is correctly formed w.r.t. public values by asking BDDH oracle

$$\text{BDDH}(Z, g^{\ell_{i,\overline{f},*}(u_0, u_1)}, g^{\ell_{i,\overline{f},*}'(v_0, v_1)}, Z_i') = 1,$$

and this implies that shared value Z_i is correctly formed w.r.t. public values. Here, we can compute $g^{\ell_{i,\overline{f},*}(u_0, u_1)} = U_0^{d_{u_0}}U_1^{d_{u_1}}$ and $g^{\ell_{i,\overline{f},*}'(v_0, v_1)} = V_0^{d_{v_0}}V_1^{d_{v_1}}$, since $\ell_{i,\overline{f},*}(u_0, u_1)$ and $\ell'_{i,\overline{f},*}(v_0, v_1)$ are expressed as $\ell_{i,\overline{f},*}(u_0, u_1) = d_{u_0} u_0 + d_{u_1} u_1$ and $\ell'_{i,\overline{f},*}(v_0, v_1) = d_{v_0} v_0 + d_{v_1} v_1$.

The procedure Check can check that the shared values are correctly formed w.r.t. the public values using knowledge of secret key v_f same as above.

The procedure Check can check also that the shared value $Z_m = g^{x_A x_B}$ is correctly formed w.r.t. the ephemeral public keys, by checking $e(X_A, X_B) = e(g, Z_m)$.

Analysis. The simulation of the environment for adversary $\mathcal{A}$ is perfect except with negligible probability. The probability that adversary $\mathcal{A}$ selects the session, where U_A is initiator, U_B is responder, and ephemeral public key X_A is V, as the test session sid^* is at least $\frac{1}{n^2 s}$. Suppose this is indeed the case, solver $\mathcal{S}$ does not abort in Step 9.

Suppose event E_1 occurs, solver $\mathcal{S}$ does not abort in Steps 6, 7 and 8.

Suppose event M^* occurs, adversary $\mathcal{A}$ queries correctly formed $Z_1, ..., Z_m$ to H. Therefore, solver $\mathcal{S}$ is successful as described in Step 4c, and does not abort as in Step 10.

Hence, solver $\mathcal{S}$ is successful with probability $Pr(S) \geq \frac{p_1}{n^2 s}$, where p_1 is probability that $E_1 \wedge M^*$ occurs.

A.2   Event $E_2 \wedge M^*$

In event E_2, test session sid^* has no matching session $\overline{\mathrm{sid}^*}$, $\mathcal{A}$ obtains x_A, and $\mathcal{A}$ does not obtain either D_A or D_B. The reduction to the gap BDH assumption is similar to event $E_1 \wedge M^*$ in Subsection A.1, except the following points.

In Setup and Simulation, $\mathcal{S}$ embeds gap BDH instance U, V, W as $Z = U$, $Q_A = V$, and $Q_B = W$.

In Simulation, using knowledge of x_A, $\mathcal{S}$ extracts answer $g_T^{z q_A q_B}$ of the gap BDH problem.

A.3   Event $E_3 \wedge M^*$

In event E_3, test session sid^* has matching session $\overline{\mathrm{sid}^*}$, $\mathcal{A}$ obtains D_A and x_B, and $\mathcal{A}$ does not obtain either x_A or D_B. The reduction to the gap BDH assumption is similar to event $E_1 \wedge M^*$ in Subsection A.1, except the following points.

In Setup and Simulation, $\mathcal{S}$ embeds gap BDH instance U, V, W as $Z = U$, $X_A = V$, and $Q_B = W$.

In Simulation, using knowledge of q_A or x_B, $\mathcal{S}$ extracts answer $g_T^{z x_A q_B}$ of the gap BDH problem.

A.4   Event $E_4 \wedge M^*$

In event E_4, test session sid^* has matching session $\overline{\mathrm{sid}^*}$, $\mathcal{A}$ obtains x_A and x_B, and $\mathcal{A}$ does not obtain either D_A or D_B. The reduction to the gap BDH assumption is similar to event $E_1 \wedge M^*$ in Subsection A.1, except the following points.

In Setup and Simulation, $\mathcal{S}$ embeds gap BDH instance U, V, W as $Z = U$, $Q_A = V$, and $Q_B = W$.

In Simulation, using knowledge of x_A or x_B, $\mathcal{S}$ extracts answer $g_T^{z q_A q_B}$ of the gap BDH problem.

A.5   Event $E_5 \wedge M^*$

In event E_5, test session $\mathtt{sid}^*$ has matching session $\overline{\mathtt{sid}}^*$, $\mathcal{A}$ obtains z, D_A, and D_B, and $\mathcal{A}$ does not obtain either x_A or x_B. The reduction to the gap BDH assumption is similar to event $E_1 \wedge M^*$ in Subsection A.1, except the following points.

In Setup and Simulation, $\mathcal{S}$ embeds gap BDH instance U, V, W as $X_A = V$ and $X_B = W$.

In Simulation, using knowledge of q_A or q_B, $\mathcal{S}$ extracts $g^{x_A x_B}$ from shared value Z_m and can compute answer $e(U, g^{x_A x_B})$ of the gap BDH problem.

A.6   Event $E_6 \wedge M^*$

In event E_6, test session $\mathtt{sid}^*$ has matching session $\overline{\mathtt{sid}}^*$, $\mathcal{A}$ obtains x_A and D_B, and $\mathcal{A}$ does not obtain either D_A or x_B. The reduction to the gap BDH assumption is similar to event $E_1 \wedge M^*$ in Subsection A.1, except the following points.

In Setup and Simulation, $\mathcal{S}$ embeds gap BDH instance U, V, W as $Z = U$, $Q_A = V$, and $X_B = W$.

In Simulation, using knowledge of x_A or q_B, $\mathcal{S}$ extracts answer $g_T^{z q_A x_B}$ of the gap BDH problem.

A.7 $ID_A = ID_B$ Cases

In the case of $ID_A = ID_B$, i.e., $Q_A = Q_B$, we make the reduction to the square gap BDH assumption.

In event E_2', the reduction to the square gap BDH assumption is similar to event E_2. In Setup and Simulation, $\mathcal{S}$ embeds square gap BDH instance U, V as $Z = U$, $Q_A = V$, and $Q_B = V$. In Simulation, using knowledge of x_A, $\mathcal{S}$ extracts answer $g_T^{z q_A q_B}$ of the square gap BDH problem.

In event E_4', the reduction to the square gap BDH assumption is similar to event E_4. In Setup and Simulation, $\mathcal{S}$ embeds square gap BDH instance U, V as $Z = U$, $Q_A = V$, and $Q_B = V$. In Simulation, using knowledge of x_A or x_B, $\mathcal{S}$ extracts answer $g_T^{z q_A q_B}$ of the square gap BDH problem.

In event E_5', the reduction to the square gap BDH assumption is similar to event E_5. In Setup and Simulation, $\mathcal{S}$ embeds square gap BDH instance U, V as $X_A = V$ and $X_B = V$. In Simulation, using knowledge of q_A or q_B, $\mathcal{S}$ extracts $g^{x_A x_B}$ from shared value Z_m and can compute answer $e(U, g^{x_A x_B})$ of the square gap BDH problem. $\qquad\square$

Author Index